Membranes and Transport

Volume 2

Membranes and Transport

Volume 2

Edited by

Anthony N. Martonosi

State University of New York
Syracuse, New York

Plenum Press • New York and London

Library of Congress Cataloging in Publication Data

Main entry under title:

Membranes and transport.

Bibliography: p.
   Includes index.
   1. Membranes (Biology). 2. Biological transport. I. Martonosi, Anthony, 1928–
      . [DNLM: 1. Membranes—Metabolism. 2. Biological transport.   QH 509
M5335]
QH601.M4817   574.87′5   82-3690
ISBN 0-306-40854-6 (v. 2)   AACR2

© 1982 Plenum Press, New York
A Division of Plenum Publishing Corporation
233 Spring Street, New York, N.Y. 10013

Printed in the United States of America

Contributors

Najmutin G. Abdulaev
Shemyakin Institute of Bioorganic Chemistry
USSR Academy of Sciences
117988 Moscow V-334, USSR

Hans-Erik Åkerlund
Department of Biochemistry
University of Lund
S-220 07 Lund, Sweden

Per-Åke Albertsson
Department of Biochemistry
University of Lund
S-220 07 Lund, Sweden

Edson X. Albuquerque
Department of Pharmacology and Experimental
    Therapeutics
University of Maryland School of Medicine
Baltimore, Maryland 21201

Giovanna Ferro-Luzzi Ames
Department of Biochemistry
University of California
Berkeley, California 94720

J. W. Anderson
Botany Department
La Trobe University
Bundoora, Victoria 3083, Australia

Bertil Andersson
Department of Biochemistry
University of Lund
S-220 07 Lund, Sweden

Thomas E. Andreoli
Department of Internal Medicine
University of Texas Medical School
Houston, Texas 77025

Yasuhiro Anraku
Department of Biology
Faculty of Science
University of Tokyo
Hongo, Tokyo 113, Japan

David J. Anstee
South Western Regional Blood Transfusion Centre
Southmead, Bristol BS10 5ND, England

Feroza Ardeshir
Department of Biochemistry
University of California
Berkeley, California 94720

Peter S. Aronson
Departments of Medicine and Physiology
Yale University School of Medicine
New Haven, Connecticut 06510

Gilbert Ashwell
Laboratory of Biochemistry
National Institute of Arthritis, Metabolism and
    Digestive Diseases
National Institutes of Health
Bethesda, Maryland 20205

P. F. Baker
Department of Physiology
King's College
Strand, London WC2R 2LS, England

Ted Begenisich
Department of Physiology
University of Rochester School of Medicine and
    Dentistry
Rochester, New York 14642

Winfried Boos

Department of Biology
University of Konstanz
D-7750 Konstanz, West Germany

I. R. Booth

Department of Microbiology
University of Aberdeen
Marischal College
Aberdeen AB9 1AS, Scotland

Roscoe O. Brady

Developmental and Metabolic Neurology Branch
National Institute of Neurological and Communicative
    Disorders and Stroke
National Institutes of Health
Bethesda, Maryland 20205

Daniel Branton

The Biological Laboratories
Harvard University
Cambridge, Massachusetts 02138

Volkmar Braun

Mikrobiologie II
Universität Tübingen
D-7400 Tübingen, West Germany

Richard P. Bunge

Departments of Biological Chemistry, and Anatomy
    and Neurobiology
Division of Biology and Biomedical Sciences
Washington University School of Medicine
St. Louis, Missouri 63110

Sally E. Carty

Department of Biochemistry and Biophysics
University of Pennsylvania School of Medicine
Philadelphia, Pennsylvania 19104

Dan Cassel

Department of Biological Chemistry
The Hebrew University of Jerusalem
Jerusalem, Israel

H. N. Christensen

Department of Biological Chemistry
The University of Michigan Medical School
Ann Arbor, Michigan 48109

Michael G. Clark

Division of Human Nutrition
CSIRO
Adelaide, South Australia

Carl M. Cohen

Department of Research
Division of Hematology/Oncology
St. Elizabeth's Hospital
Boston, Massachusetts 02146

Frederick L. Crane

Department of Biological Sciences
Purdue University
West Lafayette, Indiana 47906

Robert K. Crane

Department of Physiology and Biophysics
College of Medicine and Dentistry of New Jersey–
    Rutgers Medical School
Piscataway, New Jersey 08854

Frans J. M. Daemen

Department of Biochemistry
University of Nijmegen
Nijmegen, The Netherlands

John W. Daly

Department of Pharmacology and Experimental
    Therapeutics
University of Maryland School of Medicine
Baltimore, Maryland 21201

Willem J. De Grip

Department of Biochemistry
University of Nijmegan
Nijmegen, The Netherlands

Richard A. Dilley

Department of Biological Sciences
Purdue University
West Lafayette, Indiana 47907

Frederick C. Dorando

Department of Physiology and Biophysics
College of Medicine and Dentistry of New Jersey–
    Rutgers Medical School
Piscataway, New Jersey 08854

Thomas G. Ebrey

Department of Physiology and Biophysics
University of Illinois
Urbana, Illinois 61801

A. Alan Eddy

Department of Biochemistry
University of Manchester Institute of Science and
    Technology
Manchester M60 1QD, England

Amira T. Eldefrawi

Department of Pharmacology and Experimental
    Therapeutics
University of Maryland School of Medicine
Baltimore, Maryland 21201

Mohyee E. Eldefrawi

Department of Pharmacology and Experimental
    Therapeutics
University of Maryland School of Medicine
Baltimore, Maryland 21201

Martin Engelhard
Max-Planck-Institut für Ernährungsphysiologie
4600 Dortmund, West Germany

Wolfgang Epstein
Departments of Biochemistry and of Biophysics and
  Theoretical Biology
The University of Chicago
Chicago, Illinois 60637

J. Eveloff
Department of Physiology and Biophysics
Albert Einstein College of Medicine
Bronx, New York 10461

Robert Fraley
Cancer Research Institute and Department of
  Pharmacology
University of California School of Medicine
San Francisco, California 94143
Present address:
Monsanto Co.
St. Louis, Missouri 63166

F. Scott Furbish
Developmental and Metabolic Neurology Branch
National Institute of Neurological and Communicative
  Disorders and Stroke
National Institutes of Health
Bethesda, Maryland 20205

Gerhard Giebisch
Department of Physiology
Yale University School of Medicine
New Haven, Connecticut 06510

Luis Glaser
Departments of Biological Chemistry, and Anatomy
  and Neurobiology
Division of Biology and Biomedical Sciences
Washington University School of Medicine
St. Louis, Missouri 63110

Mary Catherine Glick
Department of Pediatrics
University of Pennsylvania School of Medicine and
  Children's Hospital of Philadelphia
Philadelphia, Pennsylvania 19104

E. B. Goldberg
Department of Molecular Biology and Microbiology
Tufts University School of Medicine
Boston, Massachusetts 02111

L. Grinius
Department of Biochemistry and Biophysics
Vilnius University
Vilnius 232031
Lithuanian SSR, USSR

Eugene V. Grishin
Shemyakin Institute of Bioorganic Chemistry
USSR Academy of Sciences
117988 Moscow V-334, USSR

W. A. Hamilton
Department of Microbiology
University of Aberdeen
Marischal College
Aberdeen AB9 1AS, Scotland

Donald J. Hanahan
Department of Biochemistry
The University of Texas Health Science Center
San Antonio, Texas 78284

Ulf-Peter Hansen
Institut für Angewandte Physik
Neue Universität
23 Kiel, West Germany

Klaus Hantke
Mikrobiologie II
Universität Tübingen
D-7400 Tübingen, West Germany

Franklin M. Harold
Department of Molecular and Cellular Biology
National Jewish Hospital and Research Center
Denver, Colorado 80206

Steven C. Hebert
Department of Internal Medicine
University of Texas Medical School
Houston, Texas 77025

Georgia L. Helmer
Department of Biochemistry
The University of Chicago
Chicago, Illinois 60637

Benno Hess
Max-Planck-Institut für Ernährungsphysiologie
4600 Dortmund, West Germany

Christopher F. Higgins
Department of Biochemistry
University of California
Berkeley, California 94720

Peter C. Hinkle
Section of Biochemistry, Molecular and Cell Biology
Cornell University
Ithaca, New York 14853

S. B. Hladky
Department of Pharmacology
University of Cambridge
Cambridge CB2 2QD, England

Jen-shiang Hong
Department of Cell Physiology
Boston Biomedical Research Institute
Boston, Massachusetts 02114.

Arthur G. Hunt
Department of Cell Physiology
Boston Biomedical Research Institute
Boston, Massachusetts 02114

Vadim T. Ivanov
Shemyakin Institute of Bioorganic Chemistry
USSR Academy of Sciences
117988 Moscow V-334, USSR

K. Iwasa
Laboratory of Neurobiology
National Institute of Mental Health
Bethesda, Maryland 20205

Aya Jakobovits
Department of Biophysics
The Weizmann Institute of Science
Rehovoth, Israel

Robert G. Johnson
Department of Biochemistry and Biophysics
University of Pennsylvania School of Medicine
Philadelphia, Pennsylvania 19104

H. R. Kaback
The Laboratory of Membrane Biochemistry
Roche Institute of Molecular Biology
Nutley, New Jersey 07110

Frederick C. Kauffman
Department of Pharmacology and Experimental
   Therapeutics
University of Maryland School of Medicine
Baltimore, Maryland 21201

A. Kepes
Centre National de la Recherche Scientifique
Institut de Recherche en Biologie Moléculaire
75251 Paris Cedex 05, France

George A. Kimmich
Department of Radiation Biology and Biophysics
University of Rochester Medical Center
Rochester, New York 14642

R. Kinne
Department of Physiology and Biophysics
Albert Einstein College of Medicine
Bronx, New York 10461

Philip A. Knauf
Department of Radiation Biology and Biophysics
University of Rochester School of Medicine and
   Dentistry
Rochester, New York 14642

D. E. Knight
Department of Physiology
King's College
Strand, London WC2R 2LS, England

William J. Knowles
Department of Pathology
Yale University School of Medicine
New Haven, Connecticut 06510

Tetsuro Kono
Department of Physiology
Vanderbilt Medical School
Nashville, Tennessee 37232

A. Kotyk
Department of Cell Physiology
Institute of Microbiology
Czechoslovak Academy of Sciences
Prague, Czechoslovakia

Terry Ann Krulwich
Department of Biochemistry
Mount Sinai School of Medicine of the City University
   of New York
New York, New York 10029

Michael Krupp
Department of Physiological Chemistry
The Johns Hopkins University School of Medicine
Baltimore, Maryland 21205

Dietrich Kuschmitz
Max-Planck-Institut für Ernährungsphysiologie
4600 Dortmund, West Germany

B. Labedan
Institut de Microbiologie
Universite Paris XI
91405 Orsay, France

Laimonis A. Laimins
Department of Biophysics and Theoretical Biology
The University of Chicago
Chicago, Illinois 60637

Abel Lajtha
Center for Neurochemistry
Rockland Research Institute
Ward's Island, New York 10035

√ **Robert Landick**
Department of Biological Chemistry
University of Michigan
Ann Arbor, Michigan 48109

M. Daniel Lane
Department of Physiological Chemistry
The Johns Hopkins University School of Medicine
Baltimore, Maryland 21205

Robert T. Leonard

Department of Botany and Plant Sciences
University of California
Riverside, California 92521

Julia E. Lever

Department of Biochemistry and Molecular Biology
University of Texas Medical School
Houston, Texas 77025

J. T. Lin

Department of Physiology and Biophysics
Albert Einstein College of Medicine
Bronx, New York 10461

J. Lindstrom

Receptor Biology Laboratory
The Salk Institute for Biological Studies
San Diego, California 92158

Laszlo Lorand

Department of Biochemistry, Molecular and Cell
  Biology
Northwestern University
Evanston, Illinois 60201

Hans Löw

Karolinska Hospital
Stockholm, Sweden

S. E. Luria

Department of Biology
Massachusetts Institute of Technology
Cambridge, Massachusetts 02139

Richard E. McCarty

Section of Biochemistry, Molecular and Cell Biology
Division of Biological Sciences
Cornell University
Ithaca, New York 14853

Robert I. Macey

Department of Physiology–Anatomy
University of California
Berkeley, California 94720

Vincent T. Marchesi

Department of Pathology
Yale University School of Medicine
New Haven, Connecticut 06510

Toshio Matsumoto

Departments of Medicine and Biology
Yale University School of Medicine
New Haven, Connecticut 06510

William J. Mawby

Department of Biochemistry
University of Bristol
Bristol BS8 1TD, England

M. Montal

Departments of Biology and Physics
University of California at San Diego
La Jolla, California 92093

Jon S. Morrow

Department of Pathology
Yale University School of Medicine
New Haven, Connecticut 06510

Nathan Nelson

Department of Biology
Technion-Israel Institute of Technology
Haifa, Israel

Garth L. Nicolson

Department of Tumor Biology
University of Texas System Cancer Center
M. D. Anderson Hospital and Tumor Institute
Houston, Texas 77030

Hiroshi Nikaido

Department of Microbiology and Immunology
University of California
Berkeley, California 94720

Franklin F. Offner

Biomedical Engineering Center
Northwestern University
Evanston, Illinois 60201

Kathleen Ottina

Department of Physiology
Harvard Medical School
Boston, Massachusetts 02115

Yuri A. Ovchinnikov

Shemyakin Institute of Bioorganic Chemistry
USSR Academy of Sciences
117988 Moscow V-334, USSR

Dale L. Oxender

Department of Biological Chemistry
University of Michigan
Ann Arbor, Michigan 48109

Etana Padan

Department of Microbial and Molecular Ecology
Institute of Life Sciences
Hebrew University
Jerusalem 91904, Israel

Demetrios Papahadjopoulos

Cancer Research Institute and Department of
  Pharmacology
University of California School of Medicine
San Francisco, California 94143

Arthur B. Pardee
Sidney Farber Cancer Institute and Department of
   Pharmacology
Harvard Medical School
Boston, Massachusetts 02115

Hermann Passow
Max-Planck-Institut für Biophysik
Frankfurt am Main, West Germany

Camillo Peracchia
Department of Physiology
University of Rochester Medical Center
Rochester, New York 14642

Robert D. Perry
Department of Biology
Washington University
St. Louis, Missouri 63130
Present address:
Department of Microbiology and Public Health
Michigan State University
East Lansing, Michigan 48823

R. Neal Pinckard
Department of Pathology
The University of Texas Health Science Center
San Antonio, Texas 78284

M. G. Pitman
School of Biological Sciences
University of Sydney
Sydney, New South Wales 2006, Australia

Peter G. W. Plagemann
Department of Microbiology
University of Minnesota
Minneapolis, Minnesota 55455

Ronald J. Poole
Biology Department
McGill University
Montreal H3A 1B1, Canada

G. B. Ralston
Department of Biochemistry
University of Sydney
Sydney, New South Wales 2006, Australia

Howard Rasmussen
Departments of Medicine and Biology
Yale University School of Medicine
New Haven, Connecticut 06510

Barry P. Rosen
Department of Biological Chemistry
University of Maryland School of Medicine
Baltimore, Maryland 21201

Terrone L. Rosenberry
Department of Pharmacology
Case Western Reserve University
Cleveland, Ohio 44106

Aser Rothstein
Research Institute
The Hospital for Sick Children
Toronto, Canada

Bertram Sacktor
Laboratory of Molecular Aging
Gerontology Research Center
National Institute on Aging
National Institutes of Health
Baltimore City Hospitals
Baltimore, Maryland 21224

Milton H. Saier, Jr.
Department of Biology
The John Muir College
University of California at San Diego
La Jolla, California 92093

Joshua R. Sanes
Department of Physiology and Biophysics
Washington University Medical Center
St. Louis, Missouri 63110

Antonio Scarpa
Department of Biochemistry and Biophysics
University of Pennsylvania School of Medicine
Philadelphia, Pennsylvania 19104

E. Schoffeniels
Laboratory of General and Comparative Biochemistry
University of Liège
B-4020 Liège, Belgium

Michael Schramm
Department of Biological Chemistry
The Hebrew University of Jerusalem
Jerusalem, Israel

Shimon Schuldiner
Department of Molecular Biology
Hadassah Medical School
Jerusalem 91000, Israel

Zvi Selinger
Department of Biological Chemistry
The Hebrew University of Jerusalem
Jerusalem, Israel

Giorgio Semenza
Laboratorium für Biochemie der E.T.H.
ETH-Zentrum
CH-8092 Zurich, Switzerland

Henry Sershen

Center for Neurochemistry
Rockland Research Institute
Ward's Island, New York 10035

Nathan Sharon

Department of Biophysics
The Weizmann Institute of Science
Rehovoth, Israel

Simon Silver

Department of Biology
Washington University
St. Louis, Missouri 63130

M. Silverman

Department of Medicine
University of Toronto
Toronto, Ontario M5S 1A8, Canada

William S. Sly

Departments of Pediatrics, Genetics, and Medicine
Washington University School of Medicine
   and St. Louis Children's Hospital,
St. Louis, Missouri 63110

Erik N. Sorensen

Department of Biological Chemistry
University of Maryland School of Medicine
Baltimore, Maryland 21201

David W. Speicher

Department of Pathology
Yale University School of Medicine
New Haven, Connecticut 06510

Clifford J. Steer

Laboratory of Biochemistry
National Institute of Arthritis, Metabolism and
   Digestive Diseases
National Institutes of Health
Bethesda, Maryland 20205

Paul K. Stumpf

Department of Biochemistry and Biophysics
University of California
Davis, California 95616

Joan L. Suit

Department of Biology
Massachusetts Institute of Technology
Cambridge, Massachusetts 02139

Sergei V. Sychev

Shemyakin Institute of Bioorganic Chemistry
USSR Academy of Sciences
117988 Moscow V-334, USSR

Michael J. Tanner

Department of Biochemistry
University of Bristol
Bristol BS8 1TD, England

I. Tasaki

Laboratory of Neurobiology
National Institute of Mental Health
Bethesda, Maryland 20205

Jonathan M. Tyler

The Biological Laboratories
Harvard University
Cambridge, Massachusetts 02138
Present address:
Department of Genetics
University of Alberta
Edmonton, Alberta
Canada T6G-2E9

Dan W. Urry

Laboratory of Molecular Biophysics
University of Alabama
Birmingham, Alabama 35294

Hans H. Ussing

Institute of Biological Chemistry A
University of Copenhagen
Copenhagen, Denmark

N. A. Walker

Biophysics Laboratory
School of Biological Sciences
University of Sydney
Sydney, New South Wales 2006, Australia

Jordan E. Warnick

Department of Pharmacology and Experimental
   Therapeutics
University of Maryland School of Medicine
Baltimore, Maryland 21201

Thomas J. Wheeler

Section of Biochemistry, Molecular and Cell Biology
Cornell University
Ithaca, New York 14853

Dorothy M. Wilson

Department of Physiology
Harvard Medical School
Boston, Massachusetts 02115

T. Hastings Wilson

Department of Physiology
Harvard Medical School
Boston, Massachusetts 02115

Robert M. Wohlhueter

Department of Microbiology
University of Minnesota
Minneapolis, Minnesota 55455

C. R. Worthington

Department of Biological Sciences and Physics
Carnegie-Mellon University
Pittsburgh, Pennsylvania 15213

Preface

This work is a collection of short reviews on membranes and transport. It portrays the field as a mosaic of bright little pieces, which are interesting in themselves but gain full significance when viewed as a whole. Traditional boundaries are set aside and biochemists, biophysicists, physiologists, and cell biologists enter into a natural discourse.

The principal motivation of this work was to ease the problems of communication that arose from the explosive growth and interdisciplinary character of membrane research. In these volumes we hope to provide a readily available comprehensive source of critical information covering many of the exciting, recent developments on the structure, biosynthesis, and function of biological membranes in microorganisms, animal cells, and plants. The 182 reviews contributed by leading authorities should enable experts to check up on recent developments in neighboring areas of research, allow teachers to organize material for membrane and transport courses, and give advanced students the opportunity to gain a broad view of the topic. Special attention was given to developments that are expected to open new areas of investigation. The result is a kaleidoscope of facts, viewpoints, theories, and techniques, which radiates the excitement of this important field. Publication of these status reports every few years should enable us to follow progress in an interesting and easygoing format.

I am grateful to the authors, to Plenum Publishing Corporation, and to several of my colleagues for their thoughtful suggestions and enthusiastic cooperation, which made this work possible.

Anthony N. Martonosi

Lovell, Maine

Contents

VIII · The Transport of Metabolites and Ions in Microorganisms

IX · *Membrane-Linked Metabolite and Ion Transport Systems in Animal Cells*

X · Channels, Pores, Intercellular Communication

XI · Excitable Membranes

XII · The Structure and Permeability of Blood Cell Membranes

XIII · *Properties of Cell Surfaces. Recognition and Interaction of Cells*

XIV · The Control of Cell Surfaces. Growth Regulation, Hormones, Hormone Receptors, and Transformation

XV · Membrane-Linked Metabolic and Transport Processes in Plants

Contents of Volume I

II · The Physical Properties of Biological and Artificial Membranes

III · *Biosynthesis of Cell Membranes: Selected Membrane-Bound Metabolic Systems*

IV · The Structure, Composition, and Biosynthesis of Membranes in Microorganisms

V · Bioenergetics of Electron and Proton Transport in Mitochondria

VI · *Energy-Transducing ATPases and Electron Transport in Microorganisms*

VII · Ion Transport Systems in Animal Cells

Membranes
and Transport

Volume 2

VIII

The Transport of Metabolites and Ions in Microorganisms

ATP-Linked Transport of Metabolites and Ions in Bacteria

Franklin M. Harold

The liveliest issue in bacterial transport was, until very recently, the role of the proton-motive force in energy coupling. The period of conflict and confusion ended about 1975, when the majority of investigators became convinced that many bacterial transport systems are energized by the proton circulation as envisaged in Mitchell's chemiosmotic theory (recent reviews: Eddy, 1978; Harold, 1977, 1978; Hinkle and McCarty, 1978; Mitchell, 1979; Rosen and Kashket, 1978; West, 1980). Ironically, at about the same time it became clear that bacteria also produce a variety of transport systems that perform work at the expense of either ATP or a related phosphoryl donor. This important insight was first attained by Berger (1973; Berger and Heppel, 1974), who found that *Escherichia coli* contains two classes of transport systems that differ with respect to both structure and energy coupling. The first class comprises the systems that are unaffected when the cells are subjected to osmotic shock and are retained in membrane vesicles prepared by Kaback's well-known procedure; we now recognize that these transport systems are energized by the protonmotive force. The second class consists of transport systems that are inactivated by osmotic shock; Berger and Heppel (1974) presented convincing evidence that these transport systems operate only in cells that generate ATP.

Subsequently, transport systems dependent upon ATP production were discovered in gram-positive bacteria as well (Harold and Spitz, 1975; Henderson *et al.*, 1979; Guffanti *et al.*, 1979). These are not particularly sensitive to osmotic shock and may well have quite a different structural basis. In fact, it seems likely that the rubric "ATP-linked transport" designates a heterogeneous collection of systems that will be sorted out as the underlying molecular mechanisms are discovered. As matters stand, ATP itself has been implicated in a few cases, but in general neither the identity of the energy donor nor the mechanism of energy coupling is clearly established. For just this reason "ATP-linked"

Franklin M. Harold • Department of Molecular and Cellular Biology, National Jewish Hospital and Research Center, Denver, Colorado 80206, and Department of Biochemistry, Biophysics and Genetics, University of Colorado Medical School, Denver, Colorado 80262.

transport of ions and metabolites deserves a place in this volume: it is one aspect of bacterial transport in which much fundamental work remains to be done.

1. ENERGY COUPLING TO "SHOCK-SENSITIVE" TRANSPORT SYSTEMS

This class of transport systems, apparently confined to gram-negative bacteria, was originally defined by the inactivating effects of osmotic shock; substrates include sugars, amino acids, and inorganic ions (reviewed in several articles in Rosen, 1978). As data accumulate, an underlying structural unity is becoming apparent. Shock-sensitive transport systems are composed of a small periplasmic protein together with two (?) additional proteins buried in the membrane. The periplasmic proteins, many of which have been purified, bind one or two molecules of the transport substrate; their loss when cells are subjected to osmotic shock accounts for the inactivation of transport. The consensus is that the periplasmic binding proteins recognize the substrate and are responsible for the high affinity and specificity of transport, but are not carriers in the traditional sense. Just how translocation occurs is not known but a transmembrane channel of some kind appears likely (Wilson, 1978). What concerns us here is the nature of the energy source.

Berger and Heppel (1974) based their conclusions on a comparison of representative transport systems with respect to the effects of inhibitors and genetic lesions. In *E. coli*, uptake of proline, serine, and cysteine is mediated by systems resistant to osmotic shock. Accumulation of these metabolites was inhibited by proton conductors whether respiration or glycolysis provided metabolic energy; in a mutant devoid of ATPase, oxidizable substrates such as lactate or ascorbate supported accumulation but glycolysis did not; finally, accumulation driven by respiration was resistant to arsenate. These characteristics all point to energy coupling via the proton circulation. By contrast, the uptake of glutamine, arginine, and histidine, which require periplasmic binders, was relatively resistant to proton conductors; in a mutant lacking the ATPase, transport was supported by glycolysis but not by respiration; and transport was blocked by arsenate. It was therefore concluded that "shockable" transport systems depend either upon ATP or upon an unidentified metabolite derived from ATP (Berger, 1973; Berger and Heppel, 1974).

Subsequent research has reinforced the distinction between proton-linked and "ATP"-linked systems, but has yet to establish the chemical identity of the energy donor. There is no particular correlation between transport and the cellular ATP level: According to Plate (1979; Plate *et al.*, 1974), for example, cells of *E. coli* treated under certain conditions with either colicin K or valinomycin retain high ATP levels, but both the membrane potential and glutamine uptake are depressed; Plate (1979) suggests that accumulation may require both ATP and a sufficient electrical potential across the membrane. Lieberman and Hong (1976) likewise obtained evidence that ATP generation, while necessary, is not sufficient to support transport. In a recent paper, Hong *et al.* (1979) reported that synthesis of acetyl phosphate is required for the accumulation of glutamine and of histidine. *E. coli* normally has two pathways for acetyl phosphate synthesis; transport is normal in mutants lacking either the acetate kinase or the phosphotransacetylase, but is severely impaired in the double mutant. The authors cautiously suggest that acetyl phosphate may be the energy donor for shock-sensitive transport systems but recognize that alternative explanations must be kept in mind. It seems likely that this problem will remain refractory until the shock-sensitive systems can either be studied in everted vesicles, or reconstituted into liposomes with the appropriate orientation.

2. ATP-DEPENDENT TRANSPORT OF INORGANIC IONS

Bacteria, like other free-living cells and organisms, tend to expel sodium and protons while accumulating high concentrations of potassium in the cytoplasm (0.2 M to 1 M K^+ is common, halobacteria attain 5 M). Animal cells exchange sodium for potassium with the aid of a vectorial Na^+/K^+-ATPase. This enzyme is not found in bacteria, fungi, protists, or plant cells, all of which nevertheless accumulate K^+ avidly and establish steep concentration gradients (10^5 or better in some cases). Two physiological functions of major importance are known to depend on K^+ uptake: the activation of protein synthesis and the control of cell turgor; regulation of the cytoplasmic pH is probably another (see Smith and Raven, 1979; Harold, 1982). For none of the microorganisms do we fully understand the mechanism of K^+ transport, but recent findings with bacteria indicate that it is a prime example of ATP-linked transport. Nor is it the only one: ATP has been implicated as the energy donor for sodium and calcium transport as well, not in all bacteria but in some. We seem to be looking at a whole series of ion-translocating ATPases, distinct from the familiar proton-translocating F_1F_0 complex. It is a promising but confusing field, whose implications for cellular physiology and evolution have still to emerge.

Let us begin with K^+ transport in *E. coli*, which is comparatively well worked out thanks to the meticulous researches of Epstein and his associates (Epstein and Laimins, 1980). Genetic studies identified no fewer than 10 cistrons concerned with K^+ uptake, which fall into two groups. Four cistrons specify a K^+ transport system of high affinity designated Kdp that is normally produced only by cells growing in low-potassium media (Rhoads *et al.*, 1978). The other six were first thought to represent at least several additional systems, but it now appears more likely that each specifies one element of the major constitutive K^+ transport system designated TrkA (Epstein and Laimins, 1980).

The Kdp system has recently been found to consist of three polypeptides (molecular weights 90,000, 47,000, and 22,000), all integral proteins of the cytoplasmic membrane; there is no periplasmic binder (Laimins *et al.*, 1978). Studies with intact cells (Rhoads and Epstein, 1977) placed the Kdp system into the ATP-linked category by the criteria of Berger and Heppel (1974). A large step forward was then taken with the discovery that the Kdp system can exhibit K^+-activated ATPase activity (Epstein *et al.*, 1978; Wieczorek and Altendorf, 1979). In order to demonstrate this, it was necessary to remove the F_1F_0-ATPase, either by mutation or by extraction with urea. Identity of the K^+-ATPase with the Kdp system was assured by several criteria: both are repressed in cells grown in K^+-rich medium, both are specific for K^+ over Rb^+, and a mutant with lowered K^+ affinity for transport also has an altered ATPase. The only active nucleotide is ATP itself (Epstein *et al.*, 1978; Wieczorek and Altendorf, 1979). Very recent observations summarized by Epstein and Laimins (1980) indicate that the mechanism involves a phosphorylated intermediate. Thanks to this splendid work the Kdp system has suddenly become one of the best characterized of all bacterial transport systems.

The TrkA system appears to be both genetically and physiologically much more complex and exhibits a feature that had never been clearly recognized before: in intact cells, K^+ accumulation by the TrkA system requires the cells to generate both ATP and the protonmotive force (Rhoads and Epstein, 1977); autologous exchange of external for internal K^+ requires only ATP. We do not know what this means, but one may, to begin with, envisage either an ATP-driven transport system that is regulated by the transmembrane electrical potential, or else a system energized by the protonmotive force but functional only if ATP is present.

For neither Kdp nor TrkA is it known whether or not K^+ transport is obligatorily linked to the movement of any other ion, nor whether the overall process is electrogenic. Considerable work has, however, shown that sodium expulsion by *E. coli* is independent of K^+, and mediated by secondary antiport for protons (reviewed by Lanyi, 1979).

For the past decade, one of the chief concerns of my own laboratory has been ion transport in the gram-positive bacterium *Streptococcus faecalis*. These organisms lack oxidative phosphorylation and generate metabolic energy by substrate-level phosphorylation only, chiefly by glycolysis. Their simple metabolic pattern, coupled with the absence of internal energy reserves and sensitivity to a wide range of ionophores, makes the streptococci convenient for the dissection of energy-coupling mechanisms in both intact cells and everted membrane vesicles. Early work led to the proposal (Harold and Papineau, 1972), squarely based on the chemiosmotic theory, that the primary process is the active and electrogenic extrusion of protons by the proton-translocating F_1F_0-ATPase. Potassium would then accumulate in response to the electrical potential (interior negative), while sodium exits by antiport for protons in response to the pH gradient (interior alkaline). However, there were some difficulties with this hypothesis from the start, chiefly the fact that movements of both Na^+ and K^+ were seen only in glycolyzing cells, suggesting some dependence on ATP (Harold and Papineau, 1972); we lacked the wit to see the implications of this curious finding.

The picture resulting from more recent research is considerably more elaborate. Bakker and Harold (1980) found that K^+ accumulation is an electrogenic process that can attain a concentration gradient of some 50,000, a gradient far too steep to be in equilibrium with the electrical potential across the membrane (about -150 mV). Like the TrkA system of *E. coli*, the streptococcal K^+ transport system depends on the generation of both ATP and the protonmotive force, but I would emphasize that in neither case has ATP itself been implicated; the phosphorylated compound remains to be identified. The results would be consistent either with a primary, "ATP"-driven pump, or with an appropriate secondary porter regulated by "ATP": a K^+/H^+ symporter was suggested (Bakker and Harold, 1980).

Extrusion of sodium by *S. faecalis* has also proven to be ATP-linked, and in this case there is now good evidence that ATP itself is the energy source for a vectorial system that exchanges Na^+ for another cation, probably protons (Heefner and Harold, 1980; Heefner *et al.*, 1980). Data come from concurrent studies with both intact cells and everted membrane vesicles, and suggest that the sodium transport system is far more sophisticated than was originally anticipated. Briefly, intact cells can expel sodium only if they are supplied with a source of metabolic energy; even efflux of sodium down the concentration gradient requires the cells to generate ATP. Under proper conditions, metabolizing cells can expel Na^+ against a steep concentration gradient even if the proton circulation has been completely short-circuited and both pH gradient and membrane potential are effectively zero (Heefner and Harold, 1980). One can also find conditions such that sodium extrusion requires both ATP and the protonmotive force, the latter probably serving to elevate the cytoplasmic pH. A complementary study with everted vesicles (Heefner *et al.*, 1980) showed that these structures accumulate $^{22}Na^+$ when incubated with ATP. Here again, under some conditions ATP appears to be the sole energy donor for $^{22}Na^+$ transport, while under others both ATP and the protonmotive force are required. It is not clear what physical meaning should be assigned to this elusive interplay of ATP and the proton circulation. Perhaps the simplest is the notion that sodium is expelled by antiport for protons, by an ATP-driven system of modular construction such that the components can come apart during vesicle preparation (Heefner *et al.*, 1980). One virtue of this suggestion is that it minimizes the gap between the clearly secondary Na^+/H^+ antiporter of *E. coli* and other bacteria, and the ATP-driven system that appears (so far) to be unique to streptococci.

In this context, mention must also be made of the remarkable sodium-dependent ATPase of *Acholeplasma* (Jinks *et al.*, 1978). The authors suggest that this enzyme is a cation pump whose energy-coupling function would be akin to that of the Na^+/K^+-ATPase of animal cells (there appears to be no F_1F_0-ATPase in this organism). A serious study of K^+ and Na^+ movements in *Acholeplasma* would be very welcome, the more so as ion movements in *Mycoplasma mycoides* were interpreted (Leblanc and Le Grimellec, 1979) pretty much in conventional chemiosmotic terms.

Calcium transport is another odd case to ponder: There is good evidence that, in several aerobic bacteria, calcium extrusion is energized by the protonmotive force (review: Silver, 1978). In streptococci, however, calcium extrusion is apparently mediated by an ATP-driven primary pump (Kobayashi *et al.*, 1978).

3. AFTERTHOUGHTS

If the reader feels somewhat bewildered by the variety of bacterial transport mechanisms, the fault is neither his nor mine. There is no escaping the conclusion that bacteria have devised a multiplicity of ways to link energy metabolism to membrane transport: secondary coupling to currents of H^+ and Na^+, group translocation pathways, and primary pumps energized by ATP and acetyl phosphate (?), with other mechanisms likely to be discovered. Not only have they diverse mechanisms, they seem at liberty to pick and choose. Several organisms are known to possess both secondary and primary transport systems for a given substrate (e.g., phosphate in *E. coli*); a particular substance may be carried by an H^+-linked system in one organism and by an ATP-linked one in another (Na^+, Ca^{2+}); and it is likely that phylogenetically remote organisms such as halobacteria and *Acholeplasma* do things in ways substantially different from those of conventional bacteria.

These developments raise questions at several levels. Current interest centers chiefly on molecular mechanisms, with the answers still elusive. But one must also wonder about the reasons for the diversity of transport systems and their particular distribution. Sometimes variety can be rationalized on grounds of adaptation. Transport systems that involve periplasmic binding proteins establish steeper gradients than secondary porters do, presumably because of the direct input of energy from ATP. Primary systems are likely to be advantageous when nutrient traces must be scavenged, and there is evidence (see West, 1980) that secondary systems are superior when substrates are present in excess; a prudent organism may find it profitable to produce both. Differences between *S. faecalis* and *E. coli* presumably reflect the divergent life-styles of an organism limited to fermentative metabolism and one capable of respiration, but the nature of the relationship has yet to be analyzed. Finally, one should mention, at least, the prospect that ATP-linked transport systems will be found in eukaryotic microorganisms and plants as well. Students of transport who prefer blazing trails to paving the roads may find this an intriguing and challenging bit of wilderness.

4. RECENT DEVELOPMENTS

Studies in this laboratory (D. L. Heffner and F. M. Harold, in preparation) go far toward clarifying the mechanism of sodium extrusion by *S. faecalis*. Everted membrane vesicles prepared in the presence of protease inhibitors accumulate $^{22}Na^+$ by a process that depends on ATP, but does not require the protonmotive force. Such vesicles also contain a sodium-stimulated ATPase activity that is distinct from the F_1F_0-ATPase and genetically correlated with ATP-dependent sodium transport. The data provide quite persuasive evidence for an ATP-driven sodium pump that mediates electroneutral exchange of Na^+ for H^+.

REFERENCES

Bakker, E. P., and Harold, F. M. (1980). *J. Biol. Chem.* **255**, 433–440.
Berger, E. A. (1973). *Proc. Natl. Acad. Sci.* USA **70**, 1514–1518.
Berger, E. A., and Heppel, L. A. (1974). *J. Biol. Chem.* **249**, 7747–7755.
Eddy, A. A. (1978). *Curr. Top. Membr. Transp.* **10**, 279–360.
Epstein, W., and Laimins, L. (1980). *Trends Biochem. Sci.* **5**, 21–23.
Epstein, W., Whitelaw, V., and Hesse, J. (1978). *J. Biol. Chem.* **253**, 6666–6668.
Guffanti, A. A., Blanco, R., and Krulwich, T. A. (1979). *J. Biol. Chem.* **254**, 1033–1037.
Harold, F. M. (1977). *Curr. Top. Bioenerg.* **6**, 84–149.
Harold, F. M. (1978). In *The Bacteria* (L. N. Ornston and J. R. Sokatch, eds.), Vol. VI, pp. 463–521, Academic Press, New York.
Harold, F. M. (1982). *Curr. Top. Membr. Transp.*, **16**, 485–516.
Harold, F. M., and Papineau, D. (1972). *J. Membr. Biol.* **8**, 45–62.
Harold, F. M., and Spitz, E. (1975). *J. Bacteriol.* **122**, 266–277.
Heefner, D. A., and Harold, F. M. (1980). *J. Biol. Chem.* **255**, 11396–11402.
Heefner, D. L., Kobayashi, H. K., and Harold, F. M. (1980). *J. Biol. Chem.* **255**, 11403–11407.
Henderson, G. B., Zevely, E. M., and Huenekens, F. M. (1979). *J. Bacteriol.* **139**, 552–559.
Hinkle, P. C., and McCarty, R. E. (1978). *Sci. Am.* **238**, 104–123.
Hong, J.-s., Hunt, A. G., Masters, P. S., and Lieberman, M. A. (1979). *Proc. Natl. Acad. Sci.* USA **76**, 1213–1217.
Jinks, D. C., Silvius, J. R., and McElhaney, R. N. (1978). *J. Bacteriol.* **136**, 1027–1036.
Kobayashi, H., Van Brunt, J., and Harold, F. M. (1978). *J. Biol. Chem.* **253**, 2085–2092.
Laimins, L., Rhoads, D. B., Altendorf, K., and Epstein, W. (1978). *Proc. Natl. Acad. Sci.* USA **75**, 3216–3219.
Lanyi, J. (1979). *Biochim. Biophys. Acta* **559**, 377–398.
Leblanc, A., and LeGrimellec, C. (1979). *Biochim. Biophys. Acta* **559**, 168–179.
Lieberman, M. A., and Hong, J.-s. (1976). *Arch. Biochem. Biophys.* **172**, 312–315.
Mitchell, P. (1979). *Science* **206**, 1148–1159.
Plate, C. A. (1979). *J. Bacteriol.* **137**, 221–225.
Plate, C. A., Suit, J. L., Jetten, A. M., and Luria, S. E. (1974). *J. Biol. Chem.* **249**, 6138–6143.
Rhoads, D. B., and Epstein, W. (1977). *J. Biol. Chem.* **252**, 1394–1401.
Rhoads, D. B., Laimins, L., and Epstein, W. (1978). *J. Bacteriol.* **135**, 445–452.
Rosen, B. P. (ed.) (1978). *Bacterial Transport,* Dekker, New York.
Rosen, B. P., and Kashket, E. R. (1978). In *Bacterial Transport* (B. P. Rosen, ed.), pp. 559–620, Dekker, New York.
Silver, S. (1978). In *Bacterial Transport* (B. P. Rosen, ed.), pp. 221–324, Dekker, New York.
Smith, F. A., and Raven, J. A. (1979). *Annu. Rev. Plant Physiol.* **30**, 289–311.
West, I. C. (1980). *Biochim. Biophys. Acta* **604**, 91–126.
Wieczorek, L., and Altendorf, K. (1979). *FEBS Lett.* **98**, 233–236.
Wilson, D. B. (1978). *Annu. Rev. Biochem.* **47**, 933–965.

The Energetics of Osmotic-Shock-Sensitive Active Transport in Escherichia coli

Arthur G. Hunt and Jen-shiang Hong

Active transport systems in *Escherichia coli* can, in general, be divided into two classes: those whose activity is greatly diminished by a cold osmotic shock of cells, and those whose activity is relatively unaffected by it. The sensitivity of the former class to osmotic shock can be explained by the apparent involvement of a periplasmic, shock-releasable substrate binding protein in transport. The latter class is unaffected by osmotic shock because the systems in this class are composed solely of membrane-bound proteins. [For reviews discussing this distinction, examples of systems from each class, and properties of the binding proteins involved in transport, see Simoni and Postma (1975), Wilson (1978), Harold (1977), Silhavy *et al.* (1978), and Wilson and Smith (1978).]

Besides this difference in the requirement for a periplasmic binding protein, osmotic-shock-sensitive and -resistant transport systems can be distinguished by their energetic requirements (Berger, 1973; Berger and Heppel, 1974). Shock-resistant transport systems are able to utilize an electrochemical proton gradient to drive transport. Shock-sensitive systems, on the other hand, require a form of chemical energy in order to function. This review will discuss research centering on the energetics of osmotic-shock-sensitive active transport in *E. coli,* and will deal with two questions: what compound (or compounds) supplies the chemical energy required for transport?, and is an electrochemical potential of protons ($\Delta\bar{\mu}_{H^+}$) required for shock-sensitive transport?

The arsenate sensitivity of osmotic-shock-sensitive transport, along with the requirement for a source of ATP, be it oxidative or glycolytic, has been generally interpreted to mean that shock-sensitive transport is driven by ATP (Berger, 1973; Berger and Heppel, 1974). This interpretation has, however, been questioned by results from other laboratories. Reports by Lieberman and Hong (1976) and Plate *et al.* (1974) demonstrated a lack

Arthur G. Hunt and Jen-shiang Hong • Department of Cell Physiology, Boston Biomedical Research Institute, Boston, Massachusetts 02114.

of correlation between ATP levels and shock-sensitive transport under certain conditions. Specifically, Plate and colleagues observed that, in a strain of *E. coli* lacking the Ca^{2+}/Mg^{2+}-ATPase, colicin K abolished shock-sensitive transport but had no such effect on ATP levels. Similarly, Lieberman and Hong saw that, in a temperature-sensitive *ecf* mutant, shock-sensitive transport was inhibited by incubation of cells at the nonpermissive temperature, while ATP levels actually increased under identical conditions. Plate and colleagues suggested that, besides ATP, an "energized membrane state" was required for shock-sensitive transport (see below). Hong's group, on the other hand, proposed the involvement of the *ecf* gene product in shock-sensitive transport. Subsequent observations made in this laboratory showed that this was not the case (Hong *et al.*, 1979). Instead, these observations suggested the involvement of a small molecule other than ATP in shock-sensitive transport. Specifically, these authors saw that, when treated with cyanide and fluoride, strains of *E. coli* lacking a functional phosphotransacetylase (*pta*) were unable to use pyruvate as an energy source for shock-sensitive transport. Strains lacking acetate kinase (*ack*), on the other hand, were able to use pyruvate to stimulate transport just as well as the wild-type. In both *pta* and *ack* mutants, ATP levels remained unchanged while acetyl CoaA levels rose dramatically upon the addition of pyruvate. Furthermore, acetyl phosphate levels increased severalfold in the *ack* mutants but remained unchanged in the *pta* mutants (Hong and Hunt, 1980). Because, in the presence of cyanide and fluoride, *E. coli* can derive energy from pyruvate only by the *pta–ack* pathway, these authors concluded that acetyl phosphate was required for transport. It was noted, however, that a role for ATP in transport could not be ruled out by the results obtained. To date, the possible roles of acetyl phosphate and ATP in shock-sensitive transport in *E. coli* await clarification.

The original observations of Berger (1973) and Berger and Heppel (1974) that shock-sensitive transport was relatively insensitive to uncouplers of oxidative phosphorylation suggested that an electrochemical gradient of protons was not required by shock-sensitive systems. Subsequently, this suggestion has been contradicted by several reports. Plate *et al.* (1974) observed that colicin K was equally effective in abolishing both shock-sensitive and shock-resistant transport in an *uncA* strain of *E. coli,* even though ATP levels actually increased under the same conditions. These authors proposed that this inhibition was a consequence of the effect of colicin K on the membrane potential in *E. coli,* and that shock-sensitive as well as shock-resistant transport requires a membrane potential. Singh and Bragg (1977), using cytochrome-deficient cells grown anaerobically, arrived at a similar conclusion. In this system, these authors saw that pyruvate-stimulated shock-sensitive transport was sensitive to N,N'-dicyclohexylcarbodiimide, an inhibitor of the Ca^{2+}/Mg^{2+}-ATPase, and to both carbonyl cyanide *m*-chlorophenylhydrazone and 2,4-dinitrophenol, uncouplers of oxidative phosphorylation. These results prompted the suggestion that an "energized membrane state" is required for shock-sensitive transport. Plate (1979) has elaborated still further on these possibilities. Here, he showed that glutamine transport, a characteristic shock-sensitive system, was inhibited by treatment of cells with valinomycin plus potassium. This inhibition was apparent even under conditions where no appreciable effect on ATP levels was seen. Because valinomycin plus potassium affects just membrane potential, and has no effect on ΔpH, Plate concluded that glutamine transport, besides requiring ATP, was also dependent on a membrane potential. Singh and Bragg (1979), using tributyltin chloride (TBTC) as a tool for altering the energetic state of the cell, observed effects exactly analogous to those seen by Plate, the only difference being that, under the conditions described by Singh and Bragg, ΔpH and not $\Delta \Psi$ was affected by TBTC. Noting that under Plate's conditions $\Delta \bar{\mu}_{H^+}$ is essentially identical to $\Delta \Psi$, whereas in Singh and Bragg's report ΔpH makes a large contribution to $\Delta \bar{\mu}_{H^+}$, the results obtained by Plate and Singh and Bragg indicate a requirement for $\Delta\bar{\mu}_{H^+}$ by shock-sensitive transport systems.

Other results, however, are at variance with the conclusions reached by Plate and Singh and Bragg. Hong and colleagues have observed conditions under which proline transport, a characteristic shock-resistant system, is nonfunctional, but glutamine transport is normal. Two systems have been studied in their laboratory. The first utilized an *ecf* mutant and showed that, under conditions where the glucose-stimulated transport of proline is completely inhibited, glutamine transport remains unaffected (Hong *et al.*, 1979). In this strain, proline transport, and presumably $\Delta\bar{\mu}_{H^+}$, are abolished by treatment of cells with KCN. Such treatment had no effect on glutamine transport. In the other system, the effect of cyanide on transport in *unc* mutants yielded a similar result (Masters and Hong, unpublished results). Here, cyanide was seen to stimulate shock-sensitive transport in *unc* mutants given low concentrations of glucose. Under identical conditions, shock-resistant transport is inactive, and $\Delta\bar{\mu}_{H^+}$ presumably absent. In both cases, the observation of functional shock-sensitive transport under conditions where shock-resistant transport was abolished led to the conclusion that $\Delta\bar{\mu}_{H^+}$ is not necessary for shock-sensitive transport.

Henderson *et al.* (1977) also suggested that $\Delta\bar{\mu}_{H^+}$ is not involved in shock-sensitive transport. These authors showed that cells able to transport galactose solely by a shock-resistant system extruded protons into the medium in the presence of sugar, whereas cells possessing only a shock-sensitive system for galactose uptake did not. Because no proton movement was seen under conditions where shock-sensitive transport was presumably active, these authors concluded that shock-sensitive transport is not driven by $\Delta\bar{\mu}_{H^+}$. These results, however, do not address the possible involvement of $\Delta\bar{\mu}_{H^+}$ in the maintenance of the structural integrity of shock-sensitive transport systems.

There has been, to date, no resolution of the conflicting conclusions presented above. However, criticisms of each experimental approach make a resolution possible, if not inevitable. The experiments described by Plate and Singh and Bragg, on the one hand, neglected to take into account possible secondary effects of the treatment of cells in question. It is possible that the treatments described (colicin K, valinomycin plus potassium, tributyltin chloride, removal of cytochromes by mutation) alter the level of a compound or compounds required for shock-sensitive transport. Such an effect could arise from a direct alteration of an enzyme or enzymes involved in the formation of such a compound or compounds, or from an efflux of substrates required for the formation of this (these) compound. Indeed, an efflux of glycolytic intermediates has been demonstrated in colicin K-treated *E. coli* (Fields and Luria, 1969).

The results discussed above that argue against the involvement of $\Delta\bar{\mu}_{H^+}$ in shock-sensitive transport, on the other hand, suffer from the fact that, in all cases, ΔpH or $\Delta\Psi$ was not measured directly, but was assumed to be reflected by shock-resistant transport activity. A recent report from Schweiger's laboratory indicates that this assumption may not be accurate (Wagner *et al.*, 1980). This report suggests that $\Delta\bar{\mu}_{H^+}$ need only be decreased by 40–50% in order to abolish shock-resistant transport. The possible involvement of some sort of inhibitor in shock-resistant transport, both in Schweiger's report and in the results described by Berger, Berger and Heppel, and *Hong* and colleagues, also cannot be ruled out. In either case, the basic assumption behind these arguments may be questionable.

The experimental approach for the resolution of these questions may be the study of shock-sensitive transport in membrane vesicles. This approach, however, would require the addition of binding protein, as well as ATP, acetyl phosphate, or whatever small molecules are required for transport to the vesicles. To date, conditions for the restoration of transport in binding-protein-dependent systems have been worked out with spheroplasts, for phosphate (Gerdes *et al.*, 1977), ribose (Galloway and Furlong, 1979), and glutamine transport (Masters and Hong, 1981). However, reconstitution of shock-sensitive transport in isolated membrane vesicles has yet to be accomplished, even with the incorporation of diverse mixtures of small molecules into vesicles (Masters and Hong, unpublished results).

There are three possible explanations for the failure to see shock-sensitive transport in isolated vesicles in the presence of binding protein. It is possible that a required small molecule is being omitted in such experiments. Indeed, Hong and Hunt (1980) have suggested the involvement of some unidentified small molecule in shock-sensitive transport. Also possible is a requirement for a cytoplasmic macromolecule in shock-sensitive transport. Recent results dealing with histidine transport in *Salmonella typhimurium* (Ames and Nikaido, 1978) and maltose transport in *E. coli* (Bavoil *et al.*, 1980; Silhavy *et al.*, 1979) indicate the involvement of unidentified gene products in these transport systems. It is possible that these unidentified products might in fact be cytoplasmic components of these systems. Finally, it may be that subjecting spheroplasts to the osmotic shocks involved in vesicle formation may disrupt membrane-bound components in such a way as to render them insensitive to binding protein or otherwise nonfunctional.

The energetics of osmotic shock-sensitive active transport in *E. coli* have yet to be clearly established. Most investigators agree that there is a requirement for a high-energy compound, be it ATP, acetyl phosphate, or some other molecule. However, the identity of this high-energy compound as well as the role of $\Delta\tilde{\mu}_{H^+}$ in shock-sensitive transport remain to be determined.

ACKNOWLEDGMENTS. Work performed in this laboratory was supported by Grant GM22576 from the National Institute of General Medical Sciences. A.G.H. is a predoctoral trainee supported by U.S. Public Health Service Training Grant GM212.

REFERENCES

Ames, G. F.-L., and Nikaido, K. (1978). *Proc. Natl. Acad. Sci. USA* **75**, 5447–5451.
Bavoil, P., Hofnung, M., and Nikaido, H. (1980). *J. Biol. Chem.* **255**, 8366–8369.
Berger, E. A. (1973). *Proc. Natl. Acad. Sci. USA* **70**, 1514–1518.
Berger, E. A., and Heppel, L. A. (1974). *J. Biol. Chem.* **249**, 7747–7755.
Fields, K. L., and Luria, S. E. (1969). *J. Bacteriol.* **97**, 64–77.
Galloway, D. R., and Furlong, C. E. (1979). *Arch. Biochem. Biophys.* **197**, 158–162.
Gerdes, R. G., Strickland, K. P., and Rosenberg, H. (1977). *J. Bacteriol.* **131**, 512–518.
Harold, F. M. (1977). *Curr. Top. Bioenerg.* **6**, 83–149.
Henderson, P. F., Giddnes, R. A., and Jones-Mortimer, M. C. (1977). *Biochem. J.* **162**, 309–320.
Hong, J.-s., and Hunt, A. G. (1980). *J. Supramol. Struct.* Suppl. 4, p. 77, Abstr. No. 189.
Hong, J.-s., Hunt, A. G., Master, P. S., and Liberman, M. A. (1979). *Proc. Natl. Acad. Sci. USA* **76**, 1213–1217.
Lieberman, M. A., and Hong, J.-s. (1976). *Arch. Biochem. Biophys.* **172**, 312–315.
Masters, P. S., and Hong, J.-s. (1981). *Biochemistry* **20**, 4900–4904.
Plate, C. A. (1979). *J. Bacteriol.* **137**, 221–225.
Plate, C. A., Suit, J. L., Jetten, A. M., and Luria, S. E. (1974). *J. Biol. Chem.* **249**, 6138–6143.
Silhavy, T. J., Ferenci, T., and Boos, W. (1978). In *Bacterial Transport* (B. P. Rosen, ed.), pp. 127–169, Dekker, New York.
Silhavy, T. J., Brickman, E., Bassford, P. J., Jr., Casadaban, M. J., Shuman, H. A., Schwartz, V., Guarente, L., Schwatz, M., and Beckwith, J. R. (1979). *Mol. Gen. Genet.* **174**, 249–259.
Simoni, R. D., and Postma, P. W. (1975). *Annu. Rev. Biochem.* **44**, 523–544.
Singh, A. P., and Bragg, P. D. (1977). *J. Supramol. Struct.* **6**, 389–398.

Singh, A. P., and Bragg, P. D. (1979). *Can. J. Biochem.* **57,** 1376–1383.

Wagner, E. F., Ponta, H., and Schweiger, M. (1980). *J. Biol. Chem.* **255,** 534–539.

Wilson, D. B. (1978). *Annu. Rev. Biochem.* **47,** 933–965.

Wilson, D. B., and Smith, J. B. (1978). In *Bacterial Transport* (B. P. Rosen, ed.), pp. 495–557, Dekker, New York.

89

Tribulations of Bacterial Transport

A. Kepes

During the last 10 years the field of energy coupling to uphill transport of solutes by bacteria was dominated by the progress toward general acceptance of the proton cotransport model of Mitchell. Its adoption did not occur without some reluctance not only as stated by Mitchell (1972) "because this simple unifying concept was not investigated experimentally in the established schools of bacterial membrane transport," but also because in the growing literature there appeared a deleterious mixture of accurate observations accompanied by unlikely interpretations, and of approximate observations serving as the foundation for rigorous reasoning. In many instances, mechanistic or thermodynamic considerations superseded the requirements of physiological function. Fervent supporters, designing experiments to help victory, did not always reach the intended effect.

In the following I would like to recall some of the topics that occupied the foreground and diverted the discussion from the real problems, and to attempt a critical evaluation of some reports in the literature aspects of which need to be put in harmony either with theory or with reality.

Reality is a physiological nutrient uptake. Let us consider a sugar capable of supporting the growth of *Escherichia coli* as the sole carbon source. In a minimal medium, this carbon source will allow a doubling of the cell mass every T minutes and the utilization of 1 g sugar will yield y gram dry weight of cells. The rate of uptake of the carbon source (Kepes, 1964) should be at least

$$\text{Rate} = \ln 2/T \cdot y \qquad (\text{min} \cdot \text{g dry wt})^{-1}$$

In a typical case of one doubling per hour and a yield of 0.4, the rate should be 28.9 mg/min per g or about 160 μmoles/min per g for a hexose. Assuming that 1 g dry cells contains 0.5 g total protein, or 0.05 g membrane protein, the physiological rate of uptake for a hexose should be in the vicinity of 320 μmoles/min per g protein or 3200 nmoles/min per mg membrane protein (or roughly half this figure for a disaccharide like lactose). These

---

A. Kepes • Centre National de la Recherche Scientifique, Institut de Recherche en Biologie Moléculaire, 75251 Paris Cedex 05, France.

predictions were found fulfilled in uptake experiments with lactose, melibiose, glucose, galactose, gluconate, and glucuronate, showing that the conditions of incubation for uptake measurements were not dramatically different from growth conditions.

Very different results can be reached if there is an unaccounted loss of intracellular solute in the form of CO_2 (Kay and Kornberg, 1971), when inadequate washing of the filter causes a loss (Anraku, 1968) due, for example, to a sudden hypoosmotic shock (Tsapis and Kepes, 1977), or when very unphysiological conditions of incubation are used (West and Mitchell, 1972, 1973).

In the case of membrane vesicles, the discrepancy with the expected physiological figures can be two orders of magnitude (Kaback and Milner, 1970). While this discrepancy may not endanger the conclusions, in some cases it can lead to surprising deductions, especially when combined with assumptions concerning the quality of the vesicle preparation. As an example let us consider the group translocation by the glucose phosphotransferase system of *E. coli* (Kundig and Ghosh, 1964). This process, which is dependent on cytoplasmic generation of phospho-Hpr, was observed upon addition of phosphoenolpyruvate and glucose to vesicles, but its apparent K_m as well as its rate was two orders of magnitude lower than *in vivo* (Kaback, 1968). The activity was attributed to hypothetical multienzyme assemblies of EI, Hpr, and EII in the membrane, instead of assuming a contamination of the vesicle content by 1% of cytoplasm, which would explain the low rate and also the low apparent K_m, the activity being limited by the P donor phospho-Hpr instead of the acceptor substrate, glucose. It seemed logical that the group translocation by vesicles did not respond to the addition of purified cytoplasmic components to the medium, but the production of extravesicular glucose-6-P attributed by the authors to a fluoride-dependent reorientation of the system could more simply be explained by a small proportion of inverted vesicles or of unsealed membrane fragments, with their cytoplasmic face accessible to phospho-Hpr.

The undeniable breakthrough of reproducing lactose transport, notwithstanding its low rate in membrane vesicles, was the origin of a wealth of new, significant experimental results. It was possible for the first time to rule out that a phosphorylated metabolite was indispensable for energization of uphill lactose transport. Clearly a source of respiratory energy such as D-lactate together with molecular oxygen was sufficient. This observation led the authors to the concept of D-lactate dehydrogenase-dependent transport. Other respiratory substrates worked to a lesser extent, except for reduced phenazine methosulfate, which gave the highest results. The concept was later relaxed to include a hypothetical respiratory carrier close to lactate dehydrogenase in the respiratory chain as the transport carrier (Barnes and Kaback, 1971; Kaback and Barnes, 1971). As D-lactate stimulation was subsequently extended to a variety of transport systems (Kaback, 1972), this concept was carried over to probably more than 100 reports in the literature and was quoted until 1977 (Therisod *et al.*, 1977), in spite of overwhelming evidence accumulating against essentiality of D-lactate dehydrogenase, not to mention the long recognized fact that uncouplers inhibited uphill lactose uptake without interfering with lactate oxidation. From the very beginning it was "difficult to see why D-lactate, or the part of the respiratory chain associated with D-lactate oxidation, would be singled out as having a special relationship with the system catalyzing galactoside translocation" (Mitchell, 1972).

During a subsequent period starting in 1973, the use of membrane vesicles raised hopes of dissecting individual steps within the transport process, thanks to the use of fluorescent analogs of lactose, the dansyl galactosides (Reeves *et al.*, 1973). These compounds, when put into contact with D-lactate-energized membrane vesicles, gave rise to an increase of fluorescence, together with a blue shift of the emission spectrum as if transferred from water to a less polar environment. This fluorescence increase was attributed to a binding of the substrate to the specific sites of the lactose carrier, which therefore had a binding site available only when the membrane was energized; this binding was supposed to represent the first step of transport (Schuldiner *et al.*, 1976).

The amount of dansyl galactoside bound to the vesicles was small, only about twice the number of active sites, as evaluated by Jones and Kennedy (1969), which in turn was some 7 times higher than the thiodigalactoside binding sites measured by Kennedy *et al.* (1974). More disturbing, the number of binding sites varied slightly with the probe utilized, and also with pH and temperature (Therisod *et al.*, 1978). Moreover, the time course of binding, 10–30 sec, more resembled that of an uptake than that of a protein–ligand reaction. The nonpenetration of dansyl galactosides was reinforced by the finding that they were behaving as inducers of the *lac* operon *in vitro* but not in intact cells. This latter negative finding was, however, of dubious value, for the competition of dansyl galactosides against lactose uptake in whole cells was also very poor compared to vesicles (Schuldiner *et al.*, 1975), so that the outer membrane was the probable barrier. This controversy was settled in 1979, when the *lacY* gene was cloned and its gene product amplified 10-fold so that more reliable data about its natural abundance became available. As stated by Kaczorowski and Kaback (1979), "notwithstanding recent experiments (Overath *et al.*, 1979) indicating that *N*-dansyl-aminoalkyl-D-galactopyranosides may be translocated across the membrane, these molecules are extremely useful probes" . . . except that the isolation of a first step of energy-dependent binding turned out to be utopian.

Turning to the phase of testing experimentally the proton symport hypothesis of Mitchell (Mitchell, 1963), three classes of experiments must be distinguished. The first includes those experiments made with intact cells, where physiological active transport of lactose (or lactose analog) is the reference. These produce qualitative and quantitative fits with the model. Maximal lactose gradients are up to 2000-fold, i.e., close to 200 mV, and approach the gradient postulated for ATP synthesis. The lack of correlation with intracellular level of ATP and the collapse of lactose gradients by H^+-conducting uncouplers (Pavlasova and Harold, 1969), the essentiality of one of the two primary proton pumps, membrane ATPase and respiratory chain (Schairer and Haddock, 1972; Haddock and Schairer, 1973; Kepes, 1974), converge to show the key role of protonmotive force.

The second class includes experiments with whole cells, when parameters of H^+ gradient and/or H^+ flux are measured or when starting from a deenergized state an artificially imposed protonmotive force is employed. In these experiments, there is a discrepancy of 1 to 2 orders of magnitude from the physiological facts (Flagg and Wilson, 1977). Measured $\Delta\mu_{H^+}$ values without inhibitors often are about 100–130 mV instead of the 180 mV expected from the maximal lactose gradients or instead of the 210–240 mV postulated for ATP synthesis (with $2H^+/ATP$). In this respect it is worth mentioning that substrate gradients were generally not extrapolated to maximum by decreasing substrate concentrations toward zero to minimize discrepancies, although it is obvious that at concentrations that approach saturation, substrate gradients become inversely proportional to substrate concentration in the medium. Another generalized omission is the lack of simultaneous scrutiny of gradients and fluxes. High gradients with inadequate fluxes are as unphysiological as insufficient gradients. An example is the Na^+–H^+ antiport described by West and Mitchell (1974), where fluxes are difficult to calculate, but a rate of some 0.7 μmole/min per g can be estimated from a figure at 20 mM Na^+ in the medium; for supporting melibiose symport, an Na^+ flux of some 100 μmoles/min per g would be necessary at an Na^+ concentration of 0.5–1 mM. Transient active transport in deenergized whole cells by an artificial protonmotive force, such as pH shift, gives fluxes and gradients about $1/10$ of the active transport, while lactose–lactose countertransport can reach a stoichiometry larger than 0.5 (Bentaboulet and Kepes, 1977). This is in sharp contrast with the heterologous lactose–proline countertransport (Bentaboulet *et al.*, 1979; Flagg and Wilson, 1977), which hardly gives $^1/_{100}$ of the maximal active proline transport or $^1/_{1000}$ of the maximal lactose transport.

The third class of experiments, which gives a qualitative fit with the model but is a factor of 10^2–10^3 away from physiological phenomena, includes the generation of artificial protonmotive forces in membrane vesicles.

Thus, Hirata *et al.* (1973, 1974) observed a transient thiomethylgalactoside uptake of 1.5 nmoles/min per mg membrane protein as a result of the efflux of some 250 mmoles K^+. This is about $^1/_{50}$ of the lactose uptake with phenazinemethosulfate, which is itself about $^1/_{50}$ of the lactose uptake in intact cells. Amino acid uptake ranged between 0.2 and 0.8 nmole/mg membrane protein as a result of the efflux of 0.5 mole K^+ ions, as compared to 15 μmoles/g dry weight or 300 nmoles/mg membrane protein *in vivo*.

While there is no doubt that thermodynamics compels the utilization of ΔpH and of $\Delta\Psi$ to accomplish the work of lactose accumulation, the specific role played by the H^+ ion implies a cycle of protonation and deprotonation of the carrier. This aspect has hardly been touched upon until now. Disturbing statements have been made about variable H^+/lactose stoichiometry (Ramos and Kaback, 1977), and about the net charge of the carrier, which are not even tentatively integrated into the background. A protein of the size of lactose permease has dozens of residues that could contribute to net electric charge, and the protonation of these residues changes according to the pH of the environment and at any given pH, statistically on a nanosecond time scale. If protonation is obligatory for symport, it should involve a strategically placed unique residue in the carrier, and it should depend on the local proton potential. Therefore, $\Delta\Psi$, which contributes to the protonmotive force, must do it via the local proton potential; the immediate neighborhood of the protonable site should be exposed to the electric field across the dielectric of the membrane, but shielded against other species of cations which could compete against protons. In addition to these general postulates, the site should have a local pH, which varies in parallel with the pH of the bulk solution. This statement can be formulated by saying that transport should be pH dependent and that beyond the apparent pK of the strategic protonable group, transport efficiency (K_m or V_{max} or a combination of the two) should change 10-fold per pH unit. This correlation has now been found in the lactose transport system (Bentaboulet and Kepes, 1981). Above a pK of 8.4 K_m for external substrate increases proportionally to pH whether measured during uphill accumulation of thiomethyl-β-D-galactoside or during downhill flux of o-nitrophenyl-β-D-galactoside in energized or in energy-inhibited cells.

Considering the mass and variety of experimental evidence that supports the proton symport model, the quantitative discrepancies can be judged unimportant, but it would be presumptuous to say that no place is left for doubt. Given the ubiquitous presence of H^+ ions both in the medium and in the cytoplasm and a more than millionfold excess of potential H^+ ions in the form of water, a tracer for this symported substrate is unthinkable. The identification of the pK of a strategic protonable group is a first step toward a molecular identification of the symport mechanism.

REFERENCES

Anraku, Y. (1968). *J. Biol. Chem.* **243**, 3128–3135.
Barnes, E. M., and Kaback, H. R. (1971). *J. Biol. Chem.* **246**, 5518–5528.

Bentaboulet, M., and Kepes, A. (1977). *Biochem. Biophys. Acta* **471**, 125–134.
Bentaboulet, M., and Kepes, A. (1981). *Eur. J. Biochem.* **117**, 233–238.
Bentaboulet, M., Robin, A., and Kepes, A. (1979). *Biochem. J.* **178**, 103–107.
Flagg, J. L., and Wilson, T. H. (1977). *Membr. Biochem.* **1**, 61–72.
Flagg, J. L., and Wilson, T. H. (1977). *J. Membr. Biol.* **31**, 233–255.
Haddock, B. A., and Schairer, H. U. (1973). *Eur. J. Biochem.* **35**, 34.
Hirata, H., Altendorf, K., and Harold, F. M. (1973). *Proc. Natl. Acad. Sci. USA* **70**, 1804–1808.
Hirata, H., Altendorf, K., and Harold, F. M. (1974). *J. Biol. Chem.* **249**, 2939–2945.
Jones, T. D. H., and Kennedy, E. P. (1969). *J. Biol. Chem.* **244**, 5981–5987.
Kaback, H. R. (l968). *J. Biol. Chem.* **243**, 3711.
Kaback, H. R. (1972). *Biochim. Biophys. Acta* **265**, 367–416.
Kaback, H. R., and Barnes, E. M. (1971). *J. Biol. Chem.* **246**, 5523–5531.
Kaback, H. R., and Milner, L. S. (1970). *Proc. Natl. Acad. Sci. USA* **66**, 1008.
Kaczorowski, G. J., and Kaback, H. R. (1979). *Biochemistry* **18**, 3691–3704.
Kay, W. K., and Kornberg, H. K. L. (1971). *Eur. J. Biochem.* **18**, 274–281.
Kennedy, E. P., Rumley, M. K., and Armstrong, J. B. (1974). *J. Biol. chem.* **249**, 33–37.
Kepes, A. (1964). In *The Cellular Functions of Membrane Transport* (J. F. Hoffman, ed.), pp. 155–169, Prentice–Hall, Englewood Cliffs, N.J.
Kepes, A. (1974). In *Membrane Proteins in Transport and Phosphorylation* (G. F. Azzone, M. E. Klingenberg, E. Quagliariello, N. Siliprandi, eds.) North-Holland, Amsterdam.
Kundig, W., and Ghosh, S. (1964). *Proc. Natl. Acad. Sci. USA* **52**, 1067.
Mitchell, P. (1963). *Biochem. Soc. Symp.* **22**, 142.
Mitchell, P. (1972). *Bioenergetics* **4**, 265–293.
Overath, P., Teather, R. M., Simoni, R. D., Aichele, G., and Wilhelm, U. (1979). *Biochemistry* **18**, 1.
Pavlasova, E., and Harold, F. M. (1969). *J. Bacteriol.* **98**, 198.
Ramos, s., and Kaback, H. R. (1977). *Biochemistry* **16**, 4271–4274.
Reeves, J. P., Schechter, E., Weil, R., and Kaback, H. R. (1973). *Proc. Natl. Acad. Sci. USA* **70**, 2722–2726.
Schairer, H. V., and Haddock, B. A. (1972). *Biochem. Biophys. Res. Commun.* **48**, 544–551.
Schuldiner, S., Kung, H. F., and Kaback, H. R. (1975). *J. Biol. Chem.* **250**, 3679–3682.
Schuldiner, S., Rudnick, G., Weil, R., and Kaback, H. R. (1976). *Trends Biochem. Sci.* **1**, 41–45.
Therisod, H. L., Letellier, L., Weil, R., and Shechter, E. (1977). *Biochemistry* **16**, 3772–3776.
Therisod, H. L., Weil, R., and Shechter, E. (1978). *Proc. Natl. Acad. Sci. USA* **75**, 4214–4218.
Tsapis, A., and Kepes, A. (1977). *Biochim. Biophys. Acta* **469**, 1–12.
West, I., and Mitchell, P. (1972). *J. Bioenerg.* **3**, 445.
West, I., and Mitchell, P. (1973). *Proc. Natl. Acad. Sci. USA* **70**, 1804–1808.
West, I., and Mitchell, P. (1974). *Biochem. J.* **144**, 87–90.

Mechanisms of Active Transport in Isolated Bacterial Membrane Vesicles

H. R. Kaback

According to Mitchell's chemiosmotic hypothesis (Mitchell, 1961, 1968, 1973), the immediate driving force for many bacterial transport systems, as well as a number of other processes, is an electrochemical gradient of protons ($\Delta\bar{\mu}_{H^+}$, interior negative and alkaline), and over the past few years, virtually unequivocal support for the hypothesis has been presented (Harold, 1976; Kaback, 1976; Konings and Boonstra, 1977; Lanyi, 1979). As the energetics of these transport systems are now resolved to a great extent, focus is shifting to a more mechanistic level, and studies with isolated cytoplasmic membrane vesicles (Kaback, 1971, 1974; Owen and Kaback, 1978, 1979a,b) are beginning to provide certain insights in this regard. It is the purpose of this contribution to review some of these recent observations.

1. ENERGETICS

Although chemiosmotic mechanisms play a central, obligatory role in active transport, certain aspects of the general chemiosmotic hypothesis remain unresolved in the vesicle system. Thus, it has been demonstrated that ATP hydrolysis leads to the generation of $\Delta\bar{\mu}_{H^+}$ in inverted vesicles (Hertzberg and Hinkle, 1974; Rosen and McClees, 1974; Singh and Bragg, 1976, 1979; Schuldiner and Fishkes, 1978; Reenstra et al., 1980), but this phenomenon has not been elucidated in right-side-out vesicles, despite numerous and varied attempts to make ATP accessible to the inner surface of the vesicle membrane.

Saier et al. (1975) have described an inducible P-glycerate transport system in *Salmonella typhimurium* that catalyzes the transport of 3-P-glycerate, 2-P-glycerate, and P-enolpyruvate. Using *S. typhimurium* grown on 3-P-glycerate, membrane vesicles were loaded with pyruvate kinase and ADP by lysing spheroplasts under appropriate conditions (Hugen-

H. R. Kaback • The Laboratory of Membrane Biochemistry, Roche Institute of Molecular Biology, Nutley, New Jersey 07110.

holtz *et al.*, 1981). Vesicles prepared in this manner catalyze proline transport in the presence of *P*-enolpyruvate, and transport is abolished by treatment of the vesicles with *N,N'*-dicyclohexylcarbodiimide, but not by anoxia or cyanide. Importantly, moreover, *P*-enolpyruvate does not drive transport in vesicles prepared under similar conditions from cells that are not induced for *P*-glycerate transport. The results are consistent with an overall mechanism in which *P*-enolpyruvate gains rapid access to the interior of the vesicles by means of the *P*-glycerate transporter where it is acted on by pyruvate kinase to phosphorylate ADP. ATP formed inside of the vesicles is then hydrolyzed by the Ca^{2+}/Mg^{2+}-stimulated ATPase, leading to the generation of $\Delta\bar{\mu}_{H^+}$ and H^+/proline symport.

Utilizing pBr322 as the vector and *E. coli* as host, a fragment of *S. typhimurium* DNA containing the gene for the *P*-glycerate transport system has been cloned (Hugenholtz *et al.*, 1981). Expression and amplification of the transport system should prove invaluable for studies of ATP-driven transport, as well as the mechanism of *P*-glycerate transport.

2. MECHANISTIC STUDIES

2.1 H⁺/Substrate Symport

Under completely deenergized conditions (i.e., facilitated diffusion), the β-galactoside transport system exhibits a high apparent K_m for either lactose or β-D-galactopyranosyl 1-thio-β-D-galactopyranoside (TDG), and generation of $\Delta\bar{\mu}_{H^+}$ via the respiratory chain results in at least a 100-fold decrease in apparent K_m (Kaczorowski *et al.*, 1979; Robertson *et al.*, 1980). Furthermore, a low apparent K_m is observed when the membrane potential ($\Delta\Psi$) or the pH gradient (ΔpH) is dissipated selectively with an appropriate ionophore and when either $\Delta\Psi$ or ΔpH is imposed artificially across the membrane (Robertson *et al.*, 1980). Thus, either component of $\Delta\bar{\mu}_{H^+}$ is able to elicit the low apparent K_m characteristic of the energized system. A detailed series of kinetic experiments were carried out in which initial rates of lactose transport were studied as a function of lactose concentration under conditions where ΔpH and/or $\Delta\Psi$ were varied systematically at pH 5.5 and pH 7.5 (Roberston *et al.*, 1980). Surprisingly, the results demonstrate that the apparent K_m remains constant from about -180 mV to -30 mV, while the maximum velocity of transport varies to the second power with either component of $\Delta\bar{\mu}_{H^+}$ at both pHs, even though the maximum velocity is about 10-fold higher at pH 7.5 over a comparable range of $\Delta\bar{\mu}_{H^+}$ values. Because a high apparent K_m is observed under completely deenergized conditions, the findings appear to be paradoxical; however, studies carried out over an extended range of lactose concentrations demonstrate that when $\Delta\bar{\mu}_{H^+}$ is dissipated partially, the system exhibits biphasic kinetics (Robertson *et al.*, 1980). One component of the overall process exhibits the kinetic parameters typical of $\Delta\bar{\mu}_{H^+}$-driven active transport and the other has the characteristics of facilitated diffusion. It is apparent, therefore, that in addition to acting thermodynamically as the driving force for active transport, $\Delta\bar{\mu}_{H^+}$ alters the distribution of the *lac* carrier between two different kinetic states. Although studied in less detail, similar observations have been reported with intact cells (Ghazi and Schechter, 1981). In addition, a number of other transport systems respond kinetically to $\Delta\bar{\mu}_{H^+}$ and its components in a manner similar to that described for the β-galactoside transport system (Robertson *et al.*, 1980).

In view of the paucity of knowledge regarding the molecular basis of $\Delta\bar{\mu}_{H^+}$-driven active transport, it is impossible to provide a completely satisfactory explanation for these phenomena. However, the uniqueness of some of the observations, in particular the second-power relationship between $\Delta\tilde{\mu}_{H^+}$ and V_{max}, calls for some attempt at conceptualization. Thus, it is suggested very tentatively and with no firm support that the *lac* carrier may exist in two forms, monomer and dimer, that the monomer catalyzes facilitated diffusion and the dimer active transport, and finally, that $\Delta\tilde{\mu}_{H^+}$ promotes aggregation of monomers to dimers. Although much of the present data can be explained by this simplistic notion, it is readily apparent that other possibilities exist. For instance, $\Delta\tilde{\mu}_{H^+}$ might cause a change in the rate-limiting step for transport without a change in the structural state of the carrier. Under conditions of facilitated diffusion, transport might be limited by deprotonation of the carrier on the inner surface of the membrane, while in the presence of $\Delta\tilde{\mu}_{H^+}$, the rate-limiting step might involve a reaction corresponding kinetically to the return of an unloaded, negatively charged carrier to the external surface of the membrane (Kaczorowski and Kaback, 1979; Kaczorowski *et al.*, 1979). As the *lacY* gene has been cloned and its product amplified (Teather *et al.*, 1978), the *lac* carrier protein is readily visualized by SDS-polyacrylamide gel electrophoresis, particularly after photoaffinity labeling (see below); therefore, cross-linking experiments carried out under energized and deenergized conditions might allow a test of the former suggestion. On the other hand, it would be difficult to test the latter suggestion even if the carrier were reconstituted in a purified form. In other words, the monomer–dimer idea is not intended as a formal hypothesis, but as a suggestion that is potentially testable.

2.2 Na⁺/Substrate Symport

Although many bacterial transport systems catalyze H^+/substrate symport, certain other bacterial systems utilize Na^+ as the symported ion (Stock and Roseman, 1971; Halpern *et al.*, 1973; Lanyi *et al.*, 1976; Tokuda and Kaback, 1977; Guffanti *et al.*, 1978; Lopilato *et al.*, 1978). For these systems, it has been proposed that the driving force is an electrochemical gradient of Na^+ ($\Delta\tilde{\mu}_{Na^+}$), created from $\Delta\tilde{\mu}_{H^+}$ by exchange of H^+ for Na^+ by an antiport mechanism (Mitchell, 1973).

In a novel approach toward understanding H^+/β-galactoside symport, passive lactose movements were used to drive turnover of the *lac* carrier under a variety of conditions in isolated membrane vesicles (Kaczorowski and Kaback, 1979; Kaczorowski *et al.*, 1979). From studies on efflux and exchange and the effects of $\Delta\tilde{\mu}_{H^+}$ on these processes and on the kinetics of lactose uptake, the following results and conclusions were presented: (1) Efflux of lactose is much slower than exchange, indicating that the rate-limiting step for efflux is associated with the return of the unloaded carrier to the inner surface of the membrane. (2) Rates of efflux, *but not exchange,* are altered by imposition of $\Delta\Psi$, ΔpH, or $\Delta\tilde{\mu}_{H^+}$, implying that the unloaded carrier may catalyze a reaction involving the movement of a negative charge and that the loaded carrier is neutral. (3) Energy, in the form of $\Delta\tilde{\mu}_{H^+}$ or a lactose concentration gradient ($\Delta\mu_{lac}$), decreases the apparent K_m for lactose uptake by about two orders of magnitude with little effect on the maximum velocity of transport.

The melibiose transport system of *Escherichia coli* catalyzes Na^+ (or Li^+)/methyl-l-thio-β-D-galactopyranoside (TMG) symport, and the cation is required not only for respiration-driven transport, but also for binding of substrate to the carrier in the absence of energy and for carrier-mediated TMG efflux (Cohn and Kaback, 1980). As opposed to the H^+/β-galactoside symport system, efflux and exchange of TMG occur at the same rate, implying that the rates of the two processes are limited by a common step, most likely the translocation of substrate across the membrane. Furthermore, *the rate of exchange, as well as efflux,* is influenced by imposition of a membrane potential ($\Delta\Psi$; interior negative), suggesting that the ternary complex between Na$^+$, TMG, and the porter may bear a net positive charge. Consistently, energization of the vesicles leads to a large increase in the V_{max} for TMG influx, with little or no change in the apparent K_m of the process. It is proposed that the Na$^+$ gradient (Na$^+_{out}$ < Na$^+_{in}$) and the $\Delta\Psi$ (interior negative) may affect different steps in the overall mechanism of active TMG accumulation in the following manner: The Na$^+$ gradient causes an increased affinity for TMG on the outer surface of the membrane relative to the inside, and the $\Delta\Psi$ facilitates a reaction involved with the translocation of the positively charged ternary complex to the inner surface of the membrane.

The Na$^+$/H$^+$ antiporter of the obligate alkalophile *Bacillus alcalophilus* facilitates growth at alkaline pH and precludes growth below 8.5. Thus, nonalkalophilic mutant strains do not exhibit Na$^+$/H$^+$ antiport activity, and interestingly, such strains concomitantly lose the ability to catalyze Na$^+$-dependent accumulation of α-aminoisobutyrate (Krulwich *et al.*, 1979). Very recently, Guffanti *et al.* (1981) have documented several other Na$^+$-dependent transport systems in vesicles from wild-type *B. alcalophilus* and demonstrated that all of these systems are defective in vesicles from the nonalkalophilic point mutant KM23. Surprisingly, the defect does not seem to result from the loss of Na$^+$/H$^+$ antiport activity *per se*, but from a pleiotropic defect in the Na$^+$/substrate symporters themselves. Monensin, an ionophore that catalyzes Na$^+$–H$^+$ exchange, does not restore respiration-driven Na$^+$/substrate symport in KM23 vesicles. Moreover, with KM23 vesicles, efflux of various substrates down their respective concentration gradients is not stimulated by Na$^+$, as observed with wild-type vesicles. Because monensin should ameliorate a simple defect in Na$^+$/H$^+$ antiport activity and the antiporter should not be required for Na$^+$/substrate symport down a concentration gradient, the results suggest that there may be a direct relationship between the antiporter and various Na$^+$/substrate symporters. One possibility is that the systems share a common Na$^+$-translocating subunit.

3. CHEMICAL STUDIES

3.1. Effect of $\Delta\bar{\mu}_{H^+}$ on Diethylpyrocarbonate (DEPC) and Maleimide Inactivation of Active Transport

Lactose transport in *E. coli* membrane vesicles is inhibited by DEPC, a reagent that is relatively specific for histidinyl residues at pH 6.0, and by exposure of the vesicles to light in the presence of rose bengal, an operation that leads to photooxidation of histidinyl residues (Padan *et al.*, 1979). In both instances, moreover, an increase in the apparent K_m for $\Delta\bar{\mu}_{H^+}$- and/or $\Delta\mu_{lac}$-driven transport is observed with no effect on V_{max} or on facilitated diffusion. Thus, it appears that a histidinyl residue(s) may be involved either in the binding and translocation of H$^+$ or in a conformational change that may occur upon protonation. In addition, inhibition of transport by DEPC is accelerated by $\Delta\bar{\mu}_{H^+}$, implying that the reactivity or accessibility of a functionally required histidinyl group(s) is altered.

Many transport systems are also inhibited by maleimides and other sulfhydryl reagents under conditions where there is no interference with the generation of $\Delta\bar{\mu}_{H^+}$ (Kaback and Patel, 1978). Furthermore, with certain systems, substrate protects against inactivation, demonstrating further that inhibition occurs at the carrier level (Cohn *et al.*, 1981). The rate of inactivation of the H$^+$/β-galactoside transport system is increased in the presence of $\Delta\bar{\mu}_{H^+}$, and similar effects are observed with the proline and melibiose transport systems (Cohn *et al.*, 1981). Thus, it appears that either the reactivity or the accessibility of a sulfhydryl group(s) in each of these carriers is altered by the presence of a transmembrane $\Delta\bar{\mu}_{H^+}$.

The findings are consistent with the notion that $\Delta\bar{\mu}_{H^+}$, in addition to acting as the immediate driving force for active transport, may bring about structural or conformational changes in certain membrane proteins that catalyze the phenomenon. In other words, these effects may represent the chemical basis for some of the kinetic results discussed above.

3.2. Photoaffinity Labeling of the lac Carrier Protein

4-Nitrophenyl-α-D-galactopyranoside (NPG) is a photoaffinity reagent that specifically inactivates the β-galactoside transport system in *E. coli* membrane vesicles (Kaczorowski *et al.*, 1980). Photolysis of NPG produces time-dependent, irreversible loss of transport activity with corresponding incorporation of [^{3}H]-NPG into the membrane. Both processes are blocked by TDG, and inactivation of lactose transport is specific, for NPG photolysis does not affect proline uptake or the ability of the vesicles to generate $\Delta\bar{\mu}_{H^+}$. Arylation of the *lac* carrier protein is stoichiometric and results in the formation of 0.25 nmole NPG adduct/mg membrane protein. All attempts to regenerate transport activity by reillumination in the presence of externally added nucleophiles failed, indicating that arylation is functionally irreversible. When vesicles labeled with [^{3}H]-NPG under appropriate experimental conditions are solubilized and analyzed by gel electrophoresis, only one radioactive peak with an apparent molecular weight of 30,000 is observed, confirming that the reaction is highly specific. The availability of this probe should allow structural studies of the *lac* carrier protein.

REFERENCES

Cohn, D., and Kaback, H. R. (1980). *Biochemistry* **19**, 4237–4243.
Cohn, D., Kaczorowski, G. J., and Kaback, H. R. (1981). *Biochemistry* **20**, 3308–3313.
Ghazi, A., and Shechter, E. (1981). *Biochim. Biophys. Acta* **645**, 305–315.
Guffanti, A. A., Susman, P., Blanco, R., and Krulwich, T. A. (1978). *J. Biol. Chem.* **253**, 708–715.
Guffanti, A. A., Cohn, D. E., Kaback, H. R., and Krulwich, T. A. (1981). *Proc. Natl. Acad. Sci. USA*, **78**, 1481–1484.
Halpern, Y. S., Barash, H., Dover, S., and Druck, K. (1973). *J. Bacteriol.* **114**, 53–58.
Harold, F. M. (1976). *Curr. Top. Bioenerg.* **6**, 83–149.
Hertzberg, E., and Hinkle, P. (1974). *Biochem. Biophys. Res. Commun.* **58**, 178–184.
Hugenholtz, J., Hong, J.-s. and Kaback, H. R. (1981). *Proc. Natl. Acad. Sci. USA* **78**, 3446–3449.
Kaback, H. R. (1971). *Methods Enzymol.* **22**, 99–120.
Kaback, H. R. (1974). *Science* **186**, 882–892.
Kaback, H. R. (1976). *J. Cell. Physiol.* **89**, 575–594.
Kaback, H. R., and Patel, L. (1978). *Biochemistry,* **17**, 1640–1646.
Kaczorowski, G. J., and Kaback, H. R. (1979). *Biochemistry* **18**, 3691–3697.
Kaczorowski, G. J., Robertson, D. E., and Kaback, H. R. (1979). *Biochemistry* **18**, 3697–3704.
Kaczorowski, G. J., LeBlanc, G., and Kaback, H. R. (1980). *Proc. Natl. Acad. Sci. USA* **77**, 6319–6323.
Konings, W. N., and Boonstra, J. (1977). *Curr. Top. Membr. Transp.* **9**, 177–231.
Krulwich, T. A., Mandel, K. G., Bornstein, R. F., and Guffanti, A. A. (1979). *Biochem. Biophys. Res. Commun.* **91**, 58–62.
Lanyi, J. K. (1979). *Biochim. Biophys. Acta* **559**, 377–397.
Lanyi, J. K., Renthal, R., and MacDonald, R. I. (1976). *Biochemistry* **15**, 1603–1610.
Lopilato, J., Tsuchiya, T., and Wilson, T. H. (1978). *J. Bacteriol.* **134**, 147–156.
Mitchell, P. (1961). *Nature (London)* **191**, 144–148.
Mitchell, P. (1968). *Chemiosmotic Coupling and Energy Transduction,* Glynn Research, Bodmin, England.
Mitchell, P. (1973). *J. Bioenerg.* **4**, 63.

Owen, P., and Kaback, H. R. (1978). *Proc. Natl. Acad. Sci. USA* **75,** 3148–3152.

Owen, P., and Kaback, H. R. (1979a). *Biochemistry* **18,** 1413–1422.

Owen, P., and Kaback, H. R. (1979b). *Biochemistry* **18,** 1423–1426.

Padan, E., Patel, L., and Kaback, H. R. (1979). *Proc. Natl. Acad. Sci. USA,* **76,** 6221–6225.

Reenstra, W., Patel, L., Rottenberg, H., and Kaback, H. R. (1980). *Biochemistry* **19,** 1–9.

Robertson, D. E., Kaczorowski, G. J., Garcia, M. L., and Kaback, H. R. (1980). *Biochemistry* **19,** 5692–5702.

Rosen, B. P., and McClees, J. S. (1974). *Proc. Natl. Acad. Sci. USA* **71,** 5042–5046.

Saier, M. H., Wentzel, D. L., Feucht, B. U, and Judice, J. J. (1975). *J. Biol. Chem.* **250,** 5089–5096.

Schuldiner, S., and Fishkes, H. (1978). *Biochemistry* **17,** 706–711.

Singh, A. P., and Bragg, P. D. (1976). *Eur. J. Biochem.* **67,** 177–186.

Singh, A. P., and Bragg, P. D. (1979). *Arch. Biochem. Biophys.* **195,** 74–80.

Stock, J., and Roseman, S. (1971). *Biochem. Biophys. Res. Commun.* **44,** 132–138.

Teather, R. M., Muller-Hill, B., Abrutsch, V., Aichele, G., and Overath, P. (1978). *Mol. Gen. Genet.* **159,** 239.

Tokuda, H., and Kaback, H. R. (1977). *Biochemistry* **16,** 2130–2136.

91

The Bacterial Phosphotransferase System in Regulation of Carbohydrate Permease Synthesis and Activity

Milton H. Saier, Jr.

*Willst du dich am Ganzen erquicken, So musst du das
Ganze im Kleinsten erblicken.*
Goethe

1. CARBOHYDRATE TRANSPORT REGULATION—AN OVERVIEW

Exogenous carbohydrates must permeate the cytoplasmic membrane of a cell before metabolic enzymes can yield utilizable sources of carbon and energy (Atkinson, 1977). Studies carried out in numerous laboratories over the past two decades have revealed that in gram-negative bacteria, five distinct mechanisms are operative that couple carbohydrate uptake to some form of metabolic energy (Dills *et al.*, 1980; Rosen, 1978). Further, corresponding studies on the regulation of carbohydrate uptake have led us to recognize five distinct mechanisms by which the transmembrane translocation process can be controlled (Dills *et al.*, 1980; Rosen, 1978). In all five mechanisms, regulation is effected by substances or quantities that the bacterium recognizes as available sources of energy. Thus, carbohydrate transport may be subject to feedback control by the end products of sugar metabolism.

In addition to regulation of permease activities, the biosyntheses of these proteins are subject to regulation. In the wild-type *E. coli* cell, synthesis of virtually every carbohydrate permease is reliant upon the presence of an intracellular or exogenous inducer as well as cytoplasmic cyclic AMP (Pastan and Adhya, 1976; Saier, 1977). This fact and the routes of metabolism of cyclic AMP and the inducers are depicted schematically in Fig. 1. While the mechanisms of transcriptional regulation by intracellular inducer and cyclic AMP are

Milton H. Saier, Jr. • Department of Biology, The John Muir College, University of California at San Diego, La Jolla, California 92093.

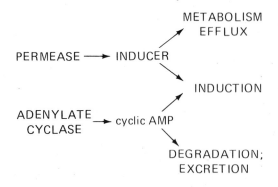

Figure 1. Schematic diagram of processes controlling the intracellular concentrations of the two molecules (cyclic AMP and inducer) that, in general, control the rates of synthesis of carbohydrate permeases and catabolic enzymes in bacteria. The scheme illustrates that intracellular inducer enters the cell via its specific permease and leaves the cytoplasmic pool either upon metabolism to inactive catabolites, or upon efflux via a transmembrane transport system. Cyclic AMP is synthesized as a result of the catalytic activity of adenylate cyclase, is degraded by cytoplasmic cyclic AMP phosphodiesterase, and can exit from the cell via the cyclic nucleotide transport system.

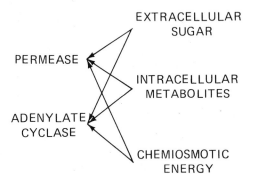

Figure 2. Schematic depiction of regulatory agents that control the activities of adenylate cyclase and carbohydrate permeases in bacteria. Extracellular sugar substrates of the PTS control these activities by the PTS-mediated mechanisms discussed in this review and also by simple competition at the extracellular sugar binding site of the permease. Intracellular metabolites and chemiosmotic energy function in regulatory capacities by quite distinct mechanisms. For detailed discussion of these processes, see Dills *et al.* (1980).

fairly well understood (Pastan and Adhya, 1976), and concrete hypotheses are available to explain induction from without (Dietz, 1976; Saier, 1979), there is still considerable confusion concerning the mechanisms controlling the cytoplasmic concentrations of cyclic AMP (Leonard *et al.,* 1981; Pastan and Adhya, 1976). Concrete evidence for two such mechanisms, operative in *E. coli,* has been published (Peterkofsky and Gazdar, 1978, 1979; Postma and Roseman, 1976; Saier and Feucht, 1975), and the existence of a third such mechanism has been proposed (Saier, 1979). These diverse regulatory interactions, controlling the activities of the permeases, responsible for inducer uptake as well as the cyclic AMP biosynthetic enzyme, are illustrated in Fig. 2. In this synopsis, one regulatory mechanism, involving the protein constituents of the bacterial phosphotransferase system, will be considered. Detailed reviews of the subject may be consulted for more extensive literature surveys and critical evaluations of the available experimental data (Dills *et al.*, 1980; Leonard *et al.*, 1981; Peterkofsky and Gazdar, 1978; Postma and Roseman, 1976; Saier, 1977).

2. THE BACTERIAL PHOSPHOTRANSFERASE SYSTEM

The phosphoenolpyruvate : sugar phosphotransferase system (PTS) was first described in *E. coli* (Kundig *et al.*, 1964) and subsequently demonstrated in bacterial genera of divergent evolutionary origins (Dills *et al.*, 1980). The system catalyzes the phosphorylation of a number of hexoses, hexitols, pentitols, aminohexoses, acetylated aminohexoses, and disaccharides in a process that is coupled to the transmembrane translocation of the sugar—from the extracellular milieu into the cytoplasm (Dills *et al.*, 1980). In the *E. coli* cell, several distinct proteins are required: Enzyme I and HPr are two cytoplasmic energy-coupling proteins that exhibit no sugar specificity and are themselves phosphorylated during the overall reaction. Both proteins have been purified to near homogeneity (Anderson *et al.*, 1979; Robillard *et al.*, 1979; Waygood *et al.*, 1979; Waygood and Steeves, 1980). In addition to these general proteins of the PTS, the phosphorylation and transport of some sugars (such as mannitol) depend on the presence of a single, integral membrane protein [i.e., the Enzyme II^{Mtl} (Jacobson *et al.*, 1979)]. Other sugars, such as glucose and fructose, are phosphorylated if and only if a sugar-specific Enzyme III as well as an Enzyme II are present (Waygood *et al.*, 1979). The preparation of homogeneous Enzyme II^{Mtl} has been reported (Jacobson *et al.*, 1979), and the partial purification of several Enzymes II and Enzymes III has been attained (Waygood *et al.*, 1979; Kundig and Roseman, 1971; Kundig, 1974). The genes that code for the general energy-coupling proteins of the PTS as well as that for the mannitol Enzyme II have been cloned (C. A. Lee, G. R. Jacobson, S. S. Dills, and M. H. Saier, unpublished results). Evidence that the mannitol Enzyme II is the only sugar-specific membrane protein required for sugar phosphorylation and transport resulted from protein biosynthetic studies with cloned DNA (C. A. Lee, G. R. Jacobson, and M. H. Saier, unpublished results) and from experiments in which the purified protein was reconstituted in an artificial lipid bilayer (J. E. Leonard and M. H. Saier, unpublished results). These investigations serve to define the functions of the individual proteins of the bacterial phosphotransferase systems and provide experimental systems for detailed mechanistic studies on PTS-mediated sugar transport.

3. PTS-MEDIATED REGULATION OF SUGAR PERMEASE ACTIVITIES

As discussed in greater detail elsewhere (Dills *et al.*, 1980; Peterkofsky and Gazdar, 1978; Postma and Roseman, 1976; Rosen, 1978; Saier, 1977), the PTS is thought to function catalytically to regulate the activities of several permeases that can function normally in the total absence of the protein constituents of the PTS. In *E. coli* and *Salmonella typhimurium*, these permeases include those specific for glycerol, lactose, melibiose, maltose, and galactose. The PTS also regulates the activity of adenylate cyclase, the cyclic AMP biosynthetic enzyme, by an apparently coordinate mechanism (Saier and Feucht, 1975). One possible mechanism first proposed in 1975 (Saier and Feucht, 1975) and subsequently suggested by others (Postma and Roseman, 1976; Peterkofsky and Gazdar, 1978) is illustrated in Fig. 3 and discussed below. Although some evidence has been interpreted to suggest that the regulatory role of the PTS is indirect (Yang *et al.*, 1979), other results indicate a direct, catalytic function (Castro *et al.*, 1976; Peterkofsky and Gazdar, 1978; Postma and Roseman, 1976; Saier and Feucht, 1975). Genetic studies, together with biochemical and physiological analyses (Saier, 1977), have led to the following results:

1. Reduced cellular activities of either Enzyme I or HPr render target permeases and adenylate cyclase *hypersensitive* to inhibition by any extracellular sugar substrate of the PTS.
2. Inhibition by a particular sugar substrate of the PTS requires that the Enzyme II, which recognizes, transports, and phosphorylates that sugar, be catalytically active.
3. Mutation of a gene (designated *crrA*, which maps adjacent to the *pts* operon) renders all sensitive permeases resistant to PTS-mediated regulation by all sugar substrates of the PTS, although other regulatory mechanisms may still be operative. While *crrA* mutants exhibit permease activities that are comparable to maximal

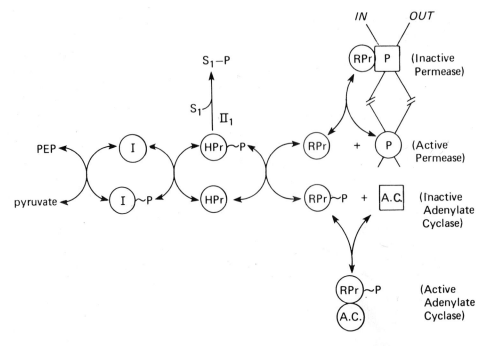

Figure 3. Proposed mechanism of PTS-mediated regulation of carbohydrate permeases and adenylate cyclase in gram-negative bacteria. PEP, phosphoenolpyruvate; I, Enzyme I; HPr, the heat-stable phosphoryl carrier protein of the PTS; RPr, postulated allosteric regulatory protein of the PTS; II, an Enzyme II of the PTS that exhibits specificity toward one of its substrates, S.

activities observed in the wild type, adenylate cylase activity is usually greatly reduced. *crrA* mutants exhibit reduced activity of a particular PTS protein, the glucose Enzyme III, thus implicating this protein in the regulatory process.

4. Regulatory mutations, which map within or very near the genes that code for the individual target permease proteins, abolish the PTS-mediated control of that permease without altering the regulatory constraints imposed on other target permeases (Saier *et al.*, 1978).

Until recently, this last class of mutations was poorly characterized. However, fine-structure mapping of mutations that specifically abolish PTS-mediated regulation of the lactose permease in *E. coli* showed that two such mutations mapped within and to one side of the *lacY* gene (which codes for the lactose permease) (R. C. Tuttle and M. Pfahl, unpublished results). The *lac*-specific regulatory mutations have thus been termed *lacY*[r]. Additionally, four independently isolated mutations that specifically abolish PTS-mediated regulation of the maltose permease in *E. coli* were found to map within the *malK* gene (M. Schwartz, unpublished results). Evidence for codominance of the mutant allele with the wild-type allele was obtained. These mutations have thus been termed *malK*[r]. Employing a procedure developed by Bavoil *et al.* (1980), one mutant *malK* gene product was preliminarily found to exhibit an altered electrophoretic mobility, possessing the same size as the wild-type protein but an increased acidity (T. Osumi and P. Bavoil, unpublished results). As the *malK* gene product is an essential constituent of the maltose permease, these results suggest that the target permeases are subject to allosteric regulation by a regulatory protein that is presumably coded for by the *crrA* gene. Missense mutations that appropriately alter the allosteric binding site of the permease so that the regulatory protein is not recognized, would be expected to abolish permease regulation without affecting its transport function. This explanation provides insight into the presumed consequences of the permease-specific regulatory mutations discussed above.

4. MODEL FOR PTS-MEDIATED REGULATION

The results described in the preceding section are consistent with a mechanism in which the product of the *crrA* gene, termed RPr for regulatory protein, interacts with the allosteric binding site of the target permease, thereby converting the permease to a conformation with low activity (Saier, 1977; Saier and Feucht, 1975). Phosphorylation of RPr (at the expense of phosphoenolpyruvate in a process catalyzed by Enzyme I and HPr) is thought to give rise to RPr~P, which either is not effective in binding to the allosteric site or else cannot induce the conformational change that reduces permease activity. This model, which is illustrated in Fig. 3, appears to account for most of the data that bear on the PTS-mediated regulatory process.

Recently, direct biochemical evidence for a regulatory involvement of the glucose Enzyme III (presumed but not proven to be RPr) has been obtained (Dills *et al.*, 1982). Partially purified Enzyme III^{glc} was osmotically shocked into *E. coli* membrane vesicles that possessed high activity of the lactose permease. Intravesicular (but not extravesicular) Enzyme III^{glc} inhibited lactose uptake, and in the presence of Enzyme I and HPr, inhibition was abolished by intravesicular phosphoenolpyruvate. These results provide preliminary biochemical evidence for the involvement of Enzyme III^{glc} as RPr in the proposed regulatory process.

Evidence bearing on adenylate cyclase regulation by the PTS led to the proposal that a mechanism was operative that was analogous to that proposed for permease regulation (Peterkofsky and Gazdar, 1978; Postma and Roseman, 1976; Saier, 1977; Saier and Feucht, 1975). The only fundamental difference was that phospho-RPr was thought to *activate* adenylate cyclase, while free RPr was believed to *inhibit* permease function. Thus, while adenylate cyclase may be under *positive* control by the phosphorylated form of RPr, the permeases appear to be subject to *negative* control by RPr itself.

5. DESENSITIZATION TO PTS-MEDIATED REGULATION

We have shown that, under appropriate physiological conditions (when cellular inducer and cyclic AMP levels are high), target permeases and adenylate cyclase may become desensitized (insensitive or less sensitive) to PTS-mediated regulation (Keeler *et al.*, 1977), and these results have recently been confirmed employing quite different experimental approaches (Postma and Scholte, 1979). A detailed discussion of this phenomenon can be found in Dills *et al.* (1980). While the desensitization mechanism(s) is (are) not yet understood, demonstration of this phenomenon reveals that in bacteria, as in animal cells (Catt *et al.*, 1979), regulatory systems may themselves be subject to regulation. Even in the simplest of living organisms, evolution tends toward complexity when complexity favors the simple.

6. BIOCHEMICAL DEMONSTRATION OF THE REGULATORY MECHANISM

Substantial progress has been made toward establishing biochemical assays for PTS-mediated regulation as follows: (1) Direct binding of Enzyme III^{glc} to the lactose permease has been demonstrated in a strain that has elevated levels of the lactose permease. Binding of Enzyme III^{glc} was cooperative with lactose binding (Osumi and Saier, 1982). (2) Cooperative interactions between lactose and Enxyme III^{glc} binding to the lactose permease has been demonstrated *in vivo* by showing that lactose binding relieves inhibition of glycerol and maltose uptake by methyl α-glucoside (Comeau-Fuhrman *et al.*, 1982). (3) The complete regulatory system has been reconstituted in a defined proteolipose preparation containing homogeneous proteins. Inhibition of lactose permease activity in proteoliposomes is effected by Enzyme III^{glc} and is demonstrable in the absence of any other protein. Complete relief from inhibition results when phosphoenolpyruvate, Enzyme I, and HPr are included in the incubation buffer. All three components were required for restoration of activity (M. J. Newman *et al.*, manuscript in preparation). These results establish the mechanism depicted in Figure 3.

REFERENCES

Anderson, B., Weigel, N., Kundig, W., and Roseman, S. (1971). *J. Biol. Chem.* **246,** 7023–7033.
Atkinson, D. E. (1977). *Cellular Energy Metabolism and Its Regulation*, Academic Press, New York.
Bavoil, P., Hofnung, M., and Nikaido, H. (1980). *J. Biol. Chem.* **255,** 8366–8369.
Castro, L., Feucht, B. U., Morse, M. L., and Saier, M. H., Jr. (1976). *J. Biol. Chem.* **251,** 5522–5527.
Catt, K. J., Harwood, J. P., Aguilera, G., and Dufau, M. L. (1979). *Nature (London)* **280,** 109–116.
Comeau-Fuhrman, D. E., Osumi, T., and Saier, M. H. (1982). *J. Bacteriol.* Submitted for publication.
Dietz, G. W., Jr. (1976). *Adv. Enzymol.* **44,** 237–259.
Dills, S. S., Apperson, A., Schmidt, M. R., and Saier, M. H., Jr. (1980). *Microbiol. Rev.* **44,** 385–418.
Dills, S. S., Schmidt, M. R., and Saier, M. H. (1982). *J. Cellular Biochem* **18,** 239–244.
Jacobson, G. R., Lee, C. A., and Saier, M. H., Jr. (1979). *J. Biol. Chem.* **254,** 249–252.
Keeler, D. K., Feucht, B. U., and Saier, M. H., Jr. (1977). *Fed Proc.* **36,** 685.
Kundig, W. (1974). *J. Supramol. Struct.* **2,** 695–714.
Kundig, W., and Roseman, S. (1971). *J. Biol. Chem.* **246,** 1407–1418.
Kundig, W., Ghosh, S., and Roseman, S. (1964). *Proc. Natl. Acad. Sci. USA* **52,** 1067–1074.
Leonard, J. E., Lee, C. A., Apperson, A. J., Dills, S. S., and Saier, M. H., Jr. (1981). In *Organixation of Prokaryotic Cell Membranes*, Vol. 1 (B. K. Ghosh, ed.), pp. 1–52, CRC Press, West Palm Beach, Fla.
Osumi, T., and Saier, M. H. (1982). *Proc. Natl. Acad. Sci. USA* **79,** 1457–1461.
Pastan, I., and Adhya, S. (1976). *Bacteriol. Rev.* **40,** 527–551.
Peterkofsky, A., and Gazdar, C. (1978). *J. Supramol. Struct.* **9,** 219–230.
Peterkofsky, A., and Gazdar, C. (1979). *Proc. Natl. Acad. Sci. USA* **76,** 1099–1103.
Postma, P., and Roseman, S. (1976). *Biochim. Biophys. Acta* **457,** 213–257.
Postma, P. W., and Scholte, B. J. (1979). In *Function and Molecular Aspects of Biomembrane Transport* (E. Quagliariello, F. Palmieri, S. Papa, and M. Klingenberg, eds.), pp. 249–257, Elsevier/North-Holland, Amsterdam.
Robillard, G. T., Dooijewaard, G., and Lolkema, J. (1979). *Biochemistry* **18,** 2984–2989.
Rosen, B. P. (ed.) (1978). *Bacterial Transport,* Dekker, New York.
Saier, M. H., Jr. (1977). *Bacteriol. Rev.* **41,** 856–871.
Saier, M. H., Jr. (1979). In *The Bacteria* (S. R. Sokatch and L. N. Ornstein, eds.), Vol. 7, pp. 168–227, Academic Press, New York.
Saier, M. H., Jr., and Feucht, B. U. (1975). *J. Biol. Chem.* **250,** 7078–7080.
Saier, M. H., Jr., Stroud, H., Massman, L. S., Judice, J. J., Newman, M. J., and Feucht, B. U. (1978). *J. Bacteriol.* **133,** 1358–1367.
Scholte, B. J., and Postma, P. W. (1980). *J. Bacteriol,* **141,** 751–757.
Waygood, E. B.,, and Steeves, T. (1980). *Can. J. Biochem.* **58,** 40–48.
Waygood, E. B., Meadow, N. D., and Roseman, S. (1979). *Anal. Biochem.* **95,** 293–304.
Yang, J. K., Bloom, R. W., and Epstein, W. (1979). *J. Bacteriol.* **138,** 275–279.

Melibiose Transport in Bacteria

T. Hastings Wilson, Kathleen Ottina, and Dorothy M. Wilson

1. INTRODUCTION

Although it was known for many years that strains of *Escherichia*, *Salmonella*, *Klebsiella*, and other bacteria were able to grow on melibiose, it was not clear whether this disaccharide entered the cell via diffusion or by means of a carrier such as the one for lactose transport. It was not until 1965 that Prestidge and Pardee discovered that there was indeed a distinct melibiose transport system that could be induced in *E. coli* strains B and K_{12}. This carrier was produced under conditions that did not induce the lactose operon, and furthermore could be found in a lactose-deleted K_{12} strain when grown in the presence of melibiose or galactinol. Studies of this carrier revealed several extremely interesting and unusual features that will be discussed in this review.

2. TEMPERATURE SENSITIVITY OF CARRIER

A most unexpected characteristic of the melibiose transport system of *E. coli* K_{12} is the instability of the carrier at 37°C. Prestidge and Pardee (1965) found that cells grown at 37°C did not possess a melibiose transport system, in contrast to cells grown at 25°C in which active transport was present. Later work by Tsuchiya *et al.* (1978) confirmed the heat lability of the transport protein. Cells previously grown on melibiose at 30°C show a 50% loss of their transport activity when incubated for 30 min at 37°C and a 90% loss after 2 hr, with either thiomethyl-β-galactoside or melibiose as the substrate. However, in spite of the instability of the melibiose transport protein, K_{12} strains grow well on this disaccharide at 37°C (Beckwith, 1963). This is due to the induction of the lactose transport system, which can accumulate melibiose. It is interesting that *E. coli* K_{12} should have retained the ability to produce a melibiose carrier that would be inactive in the normal habitat of this strain (at 37°C in the mammalian small intestine). In other strains, carriers

T. Hastings Wilson, Kathleen Ottina, and Dorothy M. Wilson • Department of Physiology, Harvard Medical School, Boston, Massachusetts 02115.

for melibiose have been found that are not temperature sensitive. The melibiose carrier proteins of *E. coli* B, *Salmonella typhimurium*, and *Klebsiella aerogenes* are active at 37°C.

3. α-GALACTOSIDASE

When *E. coli* K$_{12}$ is grown at 30°C in the presence of melibiose, there is a coordinate induction of α-galactosidase and melibiose carrier. α-Galactosidase of *E. coli* has not been studied in great detail, partly because of difficulty in retaining activity during the process of purification. Although the enzyme from *Salmonella* is more stable than that from *E. coli* (Levinthal, 1971), no major purification has been reported. Burstein and Kepes (1971) found that nicotinamide adenine dinucleotide was a very potent activator of the enzyme in *E. coli*. Genetic analysis by Schmitt and Rotman (1966; Schmitt, 1968), and by Levinthal (1971) indicate that specific genes code for α-galactosidase and for melibiose transport and that the position on the *E. coli* chromosome is at about 92 minutes (Schmitt, 1968).

4. SUBSTRATE AND INDUCER SPECIFICITIES

Thiomethyl-β-galactoside (TMG), a nonmetabolizable compound, is accumulated by melibiose-induced cells of *E. coli* to an internal concentration as high as 200 times that in the incubation medium. A summary of the compounds tested as substrates is given in Table I. Both α- and β-galactosides are transported by the melibiose carrier in *E. coli*, lactose being the prominent exception.

A striking difference between the melibiose permease of *K. aerogenes* V9A and that of *E. coli* is that lactose is a substrate for the *Klebsiella* melibiose carrier (Table 1). This was first demonstrated by the growth experiments of Reeve and Braithwaite in 1973. A more direct approach was the measurement of uptake of [^{14}C]lactose in *lacY⁻* V9A (grown

Table I. Substrate and Inducer Specificity for the Melibiose Carrier in *E. coli* and *Klebsiella* [a]

Sugar	Substrate		Inducer	
	E. coli	*Klebsiella*	*E. coli*	*Klebsiella*
Melibiose	+	+	+	+
Lactose	−	+	−	−
Melibiitol	+		+	−
Thiomethyl-β-galactoside	+	+	−	−
Thioisopropyl-β-galactoside			+	−
Galactose	+		+	+
D-Fucose	−		−	−
Raffinose	+	+	+	+
p-Nitrophenyl-α-galactoside	+	+		

[a] Data for *E. coli*: Pardee (1957), Rotman *et al.* (1968), Sheinin and Crocker (1961), Leder and Perry (1967). Data for *Klebsiella*: Reeve and Braithwaite (1973), Wilson and Wilson (unpublished).

on melibiose), which has confirmed this finding (Wilson and Wilson, unpublished results). It has been shown in addition that a melibiose-negative mutant failed to give comparable uptake of lactose.

The induction of the melibiose operon of *E. coli* has been studied in some detail (Pardee, 1957; Rotman *et al.*, 1968; Sheinin and Crocker, 1961; Leder and Perry, 1967). Several α-galactosides induce the melibiose operon without inducing the lactose system. Among these are galactinol, melibiitol, and α-phenylgalactoside. On the other hand, many β-galactosides that do induce the *lac* operon are also capable of inducing the melibiose operon. Table I lists some of the compounds tested as inducers of the melibiose system in *E. coli* and *K. aerogenes*.

5. CATION REQUIREMENTS

In 1971 Stock and Roseman reported the first study of cation requirements for melibiose transport. They showed that Na^+ or Li^+ stimulated TMG uptake and that TMG stimulated Na^+ uptake in *S. typhimurium*. This provided strong evidence for Na^+/TMG cotransport in these cells. These observations were confirmed and extended by further studies in both *S. typhimurium* (Tokuda and Kaback, 1977, 1978; Silva and Dobrogosz, 1978) and *E. coli* (Tsuchiya *et al.*, 1978; Lopilato *et al.*, 1978; Tsuchiya and Wilson, 1980; Wilson *et al.*, 1980; Tsuchiya *et al.*, 1980; Ottina *et al.*, 1980; Tanaka *et al.*, 1980; Cohn and Kaback, 1980).

Tokuda and Kaback (1977) showed that membrane vesicles isolated from *S. typhimurium* grown in the presence of melibiose, transport TMG in the presence of Na^+ or Li^+. Similar studies with intact cells were reported by Van Thienen *et al.* (1978). Tokuda and Kaback (1978) also demonstrated Na^+-dependent binding of *p*-nitrophenyl-α-galactoside to the melibiose carrier. Experiments were carried out in *E. coli* by Tsuchiya *et al.* (1977) who showed that when an inwardly directed electrochemical potential difference of Na^+ was imposed across the membrane of energy-depleted cells of *E. coli*, transient uptake of TMG was observed. Conversely, addition of TMG induced Na^+ uptake by these cells (Tsuchiya and Wilson, 1980). Li^+ was found to substitute for Na^+ in TMG/cation cotrans-

Table II. Effects of Cations on Sugar Transport in the Melibiose System of *E. coli* K_{12} [a]

Sugar	Na^+	Li^+	H^+
Melibiose	stimulate	inhibit	stimulate
p-Nitrophenyl-α-galactoside	inhibit	inhibit	
Methyl-α-galactoside	stimulate	stimulate	stimulate
Thiomethyl-β-galactoside	stimulate	stimulate	no effect
Methyl-β-galactoside	stimulate		no effect
Thiodigalactoside	stimulate		no effect
p-Nitrophenyl-β-galactoside	stimulate	no effect	
Raffinose	stimulate	no effect	no effect
Galactose	stimulate		

[a] Taken from Tsuchiya and Wilson (1978).

port. On the other hand, Li$^+$ is a potent inhibitor of melibiose entry and blocks growth of cells on this disaccharide (Tsuchiya *et al.*, 1978).

One of the interesting features of the melibiose transport carrier is the remarkable difference between the cation requirements for the transport of different substrates. While most of the substrates can use Na$^+$ for cotransport, some can also use Li$^+$ or H$^+$ (Table II). Measurement of pH changes associated with transport of several sugars indicates that in the absence of Na$^+$, melibiose and α-methylgalactoside uptake is associated with proton entry into *E. coli*, while no such H$^+$ cotransport was observed with TMG, thiodigalactoside (TDG), or β-methylgalactoside (Tsuchiya and Wilson, 1978). There is evidence that this melibiose/H$^+$ cotransport is not via the galactose (*galP*) or L-arabinose (*araE*) transport systems.

5.1. Melibiose/H$^+$ Cotransport Is via the Melibiose Carrier and Not by Way of Other Sugar/H$^+$ Systems

There are many transport systems in bacteria for galactose and galactosides. In *E. coli*, for example, galactose is recognized by the following five carriers: lactose (*lacY*), melibiose (*melB*), galactose (*galP*), methylgalactoside (*mglP*), and L-arabinose (*araE*). Melibiose is recognized by both the lactose and the melibiose carriers, but has not been adequately tested in the other three. Four of the carriers (*lacY*, *melB*, *galP*, and *araE*) carry out sugar/proton cotransport. Thus, it was important to demonstrate clearly that the melibiose carrier was indeed capable of melibiose/H$^+$ cotransport.

The first experiment (Fig. 1A) demonstrates the temperature sensitivity of melibiose/H$^+$ cotransport, a property unique to the melibiose carrier (*melB*). Strain W3133 (*lacY$^-$*), possessing temperature sensitivity of the melibiose carrier, was grown on tryptone plus 10mM melibiose at 30 or 42°C. When melibiose was added to an anaerobic suspension of cells grown at 30°C, proton uptake was observed. In a similar experiment with cells grown at 42°C, no proton movement was noted.

The second experiment (Fig. 1B) demonstrates that melibiose/H$^+$ cotransport can occur in the absence of the galactose transport system (*galP*). A *galP$^-$ lacY$^-$* mutant was grown in tryptone plus 10 mM melibiose. There was no proton uptake with D-fucose, a substrate for the *galP* and *araE* systems (Henderson, 1974), but proton/melibiose cotransport was present.

In the final experiment (Fig. 1C), melibiose/H$^+$ cotransport was observed under conditions where the galactose and arabinose systems were not induced. Protons were not taken up on the addition of D-fucose, a substrate for both the galactose and the L-arabinose transport systems. It is concluded from these experiments that proton/melibiose cotransport can occur by the melibiose transport system (*melB* gene product). In an additional experiment (not shown), strain RA11, which is partially constitutive for the melibiose system, grown on tryptone with no inducer, showed melibiose/H$^+$ cotransport. All of the other sugar/H$^+$ cotransport systems are inducible and would therefore not be present under these conditions.

5.2. Klebsiella aerogenes

The melibiose carrier of *Klebsiella* is quite different from that found in *E. coli* and *Salmonella*. Na$^+$ and Li$^+$ have no effect on the transport of either TMG or *p*-nitrophenyl-α-galactoside in *Klebsiella*, and Li$^+$ has little effect on the growth of this organism in melibiose, at concentrations of Li$^+$ that completely inhibit the growth of *E. coli* in this

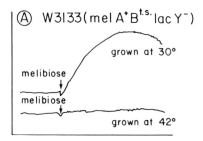

Figure 1. The melibiose transport system carries out H$^+$/melibiose cotransport. Washed cells were suspended in 100 mM KC1, 30 mM KSCN, 3 mM Tris–HC1 pH 7, and 0.5 mM dithiothreitol. Measurements of extracellular pH were made in a narrow glass vessel (2-ml capacity) with a plastic lid through which was passed a combined glass and reference electrode. The cell suspension was mixed with a small magnetic stirring bar, and O$_2$-free nitrogen was continually passed over the surface to maintain an anaerobic environment. After a 30-min incubation anaerobically at 23°C, 20-μl samples of anaerobic 0.5 M solutions of melibiose or D-fucose were added. (A) W3133 ($melA^+B^{t.s.}lacY^-$) was grown on tryptone plus 10 mM melibiose at 30 or 42°C. (B) JM1100-1 ($galP^-\ lacY^-$) was grown in tryptone plus 10 mM melibiose at 30°C. (C) RA11 ($melA^-B^+\ lacY^-$) was grown in tryptone plus 10 mM melibiose at 37°C. In this strain the melibiose carrier ($melB^+$) is fully active at 37°C. In (A) and (B) the cells contain α-galactosidase and following the proton uptake with melibiose, acidification occurred as a result of subsequent metabolism.

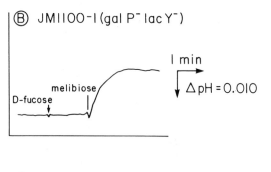

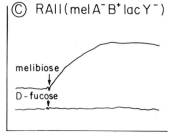

disaccharide. These cells were tested for sugar/H$^+$ cotransport by measuring the change in pH of anaerobic cell suspensions on the addition of sugar. A $lacY^-$ mutant of *Klebsiella* grown in melibiose showed TMG/H$^+$ cotransport (Wilson and Wilson, unpublished). Thus, the *Klebsiella* carrier is presumed to be a sugar/H$^+$ cotransport system, but further work is necessary to clarify the details of the cation requirements in this organism.

6. MUTANTS

The melibiose system is also of interest from an evolutionary point of view. It has been postulated (Wilson and Maloney, 1976) that the earliest cells in evolution used proton/substrate cotransport as is the case among many present-day microorganisms. On the other hand, animal cells utilize Na$^+$/substrate cotransport. Presumably in the evolutionary transition between cotransport with H$^+$ to Na$^+$ there would have been systems that could utilize both cations. The melibiose transport system of *E. coli* appears to be an example of such a system [as is alanine transport of thermophilic bacteria (Hirata, 1979)]. From an evolutionary point of view it would be interesting to find mutants in which cation specificity is altered. One technique for the isolation of mutants with altered cation recognition is

Table III. Effect of Na⁺ and Li⁺ on Galactoside Transport by Parent and an Li⁺-Resistant Mutant

			Transport activity (%)	
Substrate	Cation	Plasmid ($melA^+B^-$)	Parent[a] RA11	Mutant[a] RL24
α-pNPG[b] (0.5 mM)	Control	+	100	200
	Na⁺ (10 mM)	+	20	200
	Li⁺ (10 mM)	+	6	50
TMG (35 μM)	Control	+	100	30
	Na⁺ (10 mM)	+	500	28
	Li⁺ (10 mM)	+	700	19

[a] The parent is RA11/plc 16-37 ($melA^-B^+$/pl $melA^+B^-$) and the mutant RL24/plc 16-37. The plasmid plc 16-37 is from the collection of Clarke and Carbon (1976). The parent control is taken as 100% for each sugar (Ottina and Wilson, unpublished experiments).
[b] α-pNPG, p-nitrophenyl-α-galactoside.

to grow cells in melibiose in the presence of 10 mM Li⁺. Normal cells fail to grow under these conditions and Li⁺-resistant mutants appear (Tsuchiya *et al.*, 1978). Table III illustrates some of the altered cation effects in an Li⁺-resistant mutant. Na⁺ has no effect on transport of these two substrates, and presumably the carrier functions as a sugar/H⁺ co-transport system. The opposite type of mutant recognizing Na⁺ only has not yet been found. It is clear, however, that mutations can occur in the cation recognition of the melibiose carrier.

7. PERSPECTIVES FOR FUTURE RESEARCH

Current experiments in several laboratories are aimed at reconstituting the melibiose carrier into artificial phospholipid vesicles or planar bilayers. Such systems will facilitate the purification of the carrier and study of the mechanism of transport. Preliminary studies (Newman and Wilson, 1981; Tsuchiya *et al.*, unpublished observations) indicate that reconstitution of the carrier into phospholipid vesicles is feasible. These methods will permit a variety of interesting studies of the biochemistry and physiology of the carrier.

REFERENCES

Beckwith, J. (1963). *Biochim. Biophys. Acta* **76**, 162–164.
Burstein, C., and Kepes, A. (1971). *Biochim. Biophys. Acta* **230**, 52–63.
Clarke, L., and Carbon, J. (1976). *Cell* **9**, 91–99.
Cohn, D. E., and Kaback, H. R. (1980). *Biochemistry* **19**, 4237–4243.
Henderson, P. J. F. (1974). In *Comparative Biochemistry and Physiology of Transport* (L. Bolis, K. Bloch, S. E. Luria, and F. Lynen, eds.), pp. 409–424, North-Holland, Amsterdam.
Hirata, H. (1979). In *Function and Molecular Aspects of Biomembrane Transport* (E. Quagliariello, F. Palmieri, S. Papa, and M. Klingenberg, eds.), pp. 505–512, Elsevier/North-Holland, Amsterdam.
Leder, I. G., and Perry, J. W. (1967). *J. Biol. Chem.* **242**, 457–462.
Levinthal, M. (1971). *J. Bacteriol.* **105**, 1047–1052.
Lopilato, J., Tsuchiya, T., and Wilson, T. H. (1978). *J. Bacteriol.* **134**, 147–156.

Newman, M., and Wilson, T. H. (1981). *Abstracts, Annual Meeting of the American Society of Microbiology*, pp. 158.

Ottina, K., Lopilato, J., and Wilson, T. H. (1980). *J. Membr. Biol.* **56**, 169–175.

Pardee, A. B. (1957). *J. Bacteriol.* **73**, 376–385.

Prestidge, L. S., and Pardee, A. B. (1965). *Biochim. Biophys. Acta* **100**, 591–593.

Reeve, E. C. R., and Braithwaite, J. A. (1973). *Genet. Res.* **21**, 273–285.

Rotman, B., Ganesan, A. K., and Guzman, R. (1968). *J. Mol. Biol.* **36**, 247–260.

Schmitt, R. (1968). *J. Bacteriol.* **96**, 462–471.

Schmitt, R., and Rotman, B. (1966). *Biochem. Biophys. Res. Commun.* **22**, 473–479.

Sheinin, R., and Crocker, B. F. (1961). *Can. J. Biochem. Physiol.* **39**, 63–72.

Silva, D., and Dobrogosz, W. J. (1978). *Biochem. Biophys. Res. Commun.* **81**, 750–755.

Stock, J., and Roseman, S. (1971). *Biochem. Biophys. Res. Commun.* **44**, 132–138.

Tanaka, K., Niiya, S., and Tsuchiya, T. (1980). *J. Bacteriol.* **141**, 1031–1036.

Tokuda, H., and Kaback, H. R. (1977). *Biochemistry* **16**, 2130–2136.

Tokuda, H., and Kaback, H. R. (1978). *Biochemistry* **17**, 698–705.

Tsuchiya, T., and Wilson, T. H. (1978). *Membr. Biochem.* **2**, 63–79.

Tsuchiya, T., Raven, J., and Wilson, T. H. (1977). *Biochem. Biophys. Res. Commun.* **76**, 26–31.

Tsuchiya, T., Lopilato, J., and Wilson, T. H. (1978). *J. Membr. Biol.* **42**, 45–59.

Tsuchiya, T., Takeda, K., and Wilson, T. H. (1980). *Membr. Biochem.* **3**, 131–146.

Van Thienen, G. M., Postma, P. W., and Van Dam, K. (1978). *Biochim. Biophys. Acta* **513**, 395–400.

Wilson, T. H., and Maloney, P. C. (1976). *Fed. Proc.* **35**, 2174–2179.

Wilson, T. H., Tsuchiya, T., Lopilato, J., and Ottina, K. (1980). *Ann. N.Y. Acad. Sci.* **341**, 465–472.

A Consideration of the Role of Thermodynamic and Kinetic Factors in Regulating Lactose Accumulation in Escherichia coli

W. A. Hamilton and I. R. Booth

The cytoplasmic membrane of the prokaryotic cell, and the mitochondrial and chloroplast membranes of eukaryotes, play a cardinal role in biological energy transduction. These membranes form the phase boundary that delimits the cytoplasm of the organism, or the matrix of the organelle. Transport systems constitute a means of communication between the compartments separated by the membrane. Changes in the composition of either the external or the internal environments may be sensed by the transport proteins and an appropriate response elicited. In prokaryotes this capacity of the transport systems is expressed in the relatively constant internal pH (pH_i), cation balance, and amino acid pools of the cytoplasm (Harold 1976). Such a condition of negative entropy can only be maintained, however, by the input of energy. It is evident, therefore, that the interaction of energy with transport proteins must be an important facet of its transduction within the cell.

The most common energy supply for transport is the force composed of elements of both chemical and electrical potential gradients across the membrane. In bacteria the predominant gradients are those of charge ($\Delta\Psi$, interior negative), of protons (ΔpH, interior alkaline), and of sodium ions ($\Delta\mu_{Na^+}$, low Na^+ in interior). The chemiosmotic concept of bacterial active transport was developed by Mitchell (1963), and principally, though not exclusively, envisages transport occurring via proton symport. The energy used for transport of neutral solutes is thus the flow of protons down their electrochemical gradient. The role of the carrier protein in transport is usually seen from the viewpoint of the ionophore model as a vehicle for the equilibration of the gradients of charge, pH, and solute (Rotten-

W. A. Hamilton and I. R. Booth • Department of Microbiology, University of Aberdeen, Marischal College, Aberdeen AB9 1AS, Scotland.

berg, 1976). This article seeks to elaborate on this theme, with particular reference to the proton-linked transport of β-galactosides in *Escherichia coli*.

The protonmotive force, Δp, which is the driving force for β-galactoside transport, is described by the equation

$$\Delta p = \Delta \Psi - Z \Delta pH \tag{1}$$

where Z equals $2.303RT/F$ and converts $\log_{10}$ concentration ratios to mV. In *E. coli* the effect of metabolic activity is to generate $\Delta \Psi$ and maintain pH_i relatively constant. Consequently, ΔpH is strongly dependent upon external pH (pH_o), and therefore the magnitude and composition of Δp changes with pH_o. The magnitude of the change in the protonmotive force is lessened by a compensating increase in $\Delta \Psi$ over the pH range in which ΔpH declines (Booth *et al.*, 1979). These changes are of very great significance in consideration of the mechanism of active transport.

The uptake of solutes, such as β-galactosides, neutral amino acids, etc., by a proton symport should be responsive to both components of the protonmotive force. It is possible, therefore, to compare the work done in terms of the concentration gradient of a solute, S, attained at the steady state, with the magnitude of the driving force. Thus:

$$Z \log_{10} \frac{[S]_i}{[S]_o} = (m + n)\Delta \Psi - nZ\Delta ph \tag{2}$$

where $[S]_i$ and $[S]_o$ are the internal and external concentrations of solute, m is the charge on the solute, and n is the number of protons cotransported (Mitchell, 1973). This equation is a statement of thermodynamic equilibrium between solute gradient and driving force. It has been used as the experimental and theoretical basis of a transport model (Ramos and Kaback, 1977a,b; Rottenberg, 1976). However, its use as an experimental model rests on three assumptions:

1. That the solute does in fact achieve thermodynamic equilibrium with the driving force.
2. That the solute–carrier interaction plays no role in determining the level of accumulation of the solute.
3. That $\Delta \Psi$ and ΔpH are equivalent in terms of the energy-coupling mechanism.

We have been investigating the relationship in *E. coli* between Δp and the accumulation of β-galactosides in terms of equation (2) as a transport model. The details of our work have been published elsewhere and will not be exhaustively reviewed here (Booth *et al.*, 1979; Booth and Hamilton, 1980; Booth, 1980). Our results showed that in endogenously respiring cells, at no value of pH_o (pH range 5.9–8.7) did the lactose accumulation appear to be in equilibrium with the protonmotive force. These results have led us to question the validity of the assumptions inherent in the use of equation (2).

1. Thermodynamic Equilibrium

A precondition that must be met before this assumption can be valid is that the solute must enter and leave the cell by a single transport system (Christensen, 1976). In the case of β-galactosides this condition is fulfilled by lactose (Rickenberg *et al.*, 1956) but not by the analog, thiomethylgalactoside (TMG), which shows an appreciable passive permeability through the membrane (Maloney and Wilson, 1974; Booth and Hamilton, 1980). Such a leak will reduce $[S]_i$ to a level below that which would be catalyzed by the carrier alone. Additionally a leak process results in a net proton influx at the steady state equal to the rate of TMG leakage (Eddy *et al.,* 1979). The accumulation level of TMG can thus only come into equilibrium with Δp if the proton leak is sufficiently great to cause an equivalent partial collapse of the protonmotive force. This point has not been examined experimentally.

2. Interaction of Solute and Carrier

Inherent in the use of equation (2) is the assumption that the carrier is energetically passive, in the sense that it only facilitates the coupled flux of solute and protons toward a predetermined equilibrium set by the size and direction of $\Delta\Psi$ and ΔpH. However, it is now considered to be a general property of ion-linked solute transport systems that their behavior shows nonpassive characteristics with the solute–carrier interaction directly influencing the apparent efficiency of energy coupling (Eddy, 1980; Booth, 1980; Ahmed and Booth, 1981).

a. Exit

This is manifested in the marked decrease in the solute accumulation ratio as the external concentration of solute is increased. Eddy (1980) has attributed this to uncoupling of the flow of solute from the movement of coupling ions when $[S]_o$ is large. In the absence of a leak pathway, the uncoupled flux of solute must occur via the carrier and has been termed *slip* (Eddy, 1980). Clearly the presence of slip reactions points to a direct role for the carrier–solute interaction in determining the level of solute accumulation.

Evidence for this characteristic of the β-galactoside carrier in *E. coli* has been available for many years (Rickenberg *et al.,* 1956). These authors observed that the level of accumulation of β-galactosides was a function both of the external sugar concentration and of the identity of the sugar. At the steady state the following equation was found to be satisfied:

$$\frac{V_m^i[S]_o}{K_t^i+[S]_o}=k[S]_i \tag{3}$$

where V_m^i and K_t^i are respectively the maximal rate of β-galactoside influx and the apparent affinity constant of the uptake process; k is the first-order rate constant for exit. Under fully energized conditions, exit appears to be first order at all vales of $[S]_i$ (Rickenberg *et al.,* 1956). However, the equation that would be expected to apply if thermodynamic equilibrium were attained is as follows:

$$\frac{V_m^i[S]_o}{K_t^i+[S]_o}=\frac{V_m^e[S]_i}{K_t^e+[S]_i} \tag{4}$$

where V_m^e and K_t^e are the equivalent kinetic parameters for carrier-mediated efflux. That the data fit equation (3) and not equation (4) indicates that efflux is primarily responsive to $[S]_i$. Recently we have further investigated this point. At values of $[S]_o < K_t^i$, equation (3) can be simplified to

$$\frac{V_m^i}{K_t^i\cdot k}=\frac{[S]_i}{[S]_o} \tag{5}$$

This gives the maximal value of the accumulation ratio, and allows one to substitute v_i, the initial rate of influx, for V_m^i/K_t^i. Using equation (2) it is possible to predict the maximum theoretical value of $[S]_i$ from a knowledge of $[S]_o$ over a range of experimentally determined values of Δp. Substitution of this value of $[S]_i$, $[S]_o$, and measured values of v_i into equation (5) then gives the theoretical value of k, the rate constant for exit. This can be compared with the value of k measured experimentally by diluting cells preloaded with radioactive lactose into buffer containing the same β-galactoside at $[S]_o$ and observing the initial first-order loss of counts. We were able to show by this method that exit was approximately sixfold faster than the rate predicted from the assumption of thermodynamic equilibrium, as expressed in equations (2) and (4) (Booth and Hamilton, 1980; Booth, 1980). This confirms the relative independence of exit from energy coupling. However, the system is not completely independent because exit can be stimulated by uncouplers. A model of this system must thus envisage only a portion of the lactose efflux being via a slip reaction.

b. Entry

The early studies on the β-galactoside transport system demonstrated that each sugar tested gave unique values of V_m^i, K_t^i, and k, resulting in different maximal accumulation levels (Rickenberg et al., 1956). If the identity of the solute dictates the ratio $[S]_i/[S]_o$, then the carrier cannot be wholly unresponsive to the solute. Rather, the coupling of energy to solute accumulation via the carrier must be directly affected by the nature of the interaction of the solute with the carrier.

Recent studies with membrane vesicles of E. coli have shown that K_t^i may be an energy-dependent parameter in the lactose transport system (Wright et al., 1979; Kaczorowski et al., 1979; Wright and Overath, 1980; Toci et al., 1980). Thus, upon energization of the vesicles the K_t^i for lactose decreased 200-fold (Kaczorowski et al., 1979; Wright et al., 1979; Wright and Overath, 1980). However, the K_t^i values for other solutes were not significantly affected by the initiation of respiration in vesicles (Wright et al., 1979; Toci et al., 1980). These solutes, thiodigalactoside, p-nitrophenylgalactoside, and dansylgalactosides, are all poorly accumulated as predicted by equation (5).

From the foregoing it is clear that the nature of the solute binding to the carrier modifies the carrier such that its response to the gradients of charge and pH is less than maximal as predicted from the passive or ionophore model.

3. The Equivalence of $\Delta\Psi$ and ΔpH

The previous section dealt with the interaction of the solute with the carrier. In the following we are solely concerned with the effects of $\Delta\Psi$ and ΔpH on the carrier, i.e., the right-hand side of equation (2). Energy coupling is generally viewed as the component gradients of the protonmotive force directly affecting the carrier such that apparent changes in the kinetic parameters for influx and efflux occur. Energization of the cell has been shown to increase V_m^i in both whole cells (Winkler and Wilson, 1966; Lancaster et al., 1975) and membrane vesicles (Kaczorowski et al., 1979). Effects on K_t^i, discussed above, have not been demonstrated in whole cells to date. When cells are treated with uncouplers, efflux is accelerated (Winkler and Wilson, 1966) probably as a result of a change from first-order to saturable kinetics (Lancaster et al., 1975). In the presence of a protonmotive force, V_m^i and K_t^e are apparently increased and K_t^i and V_m^e appear to decrease leading to an acceleration of influx and a slowing of efflux resulting in solute accumulation. A question that arises from this is whether $\Delta\Psi$ and ΔpH are equivalent in driving transport. This point has been discussed elsewhere (Booth, 1980); however, new evidence has since been presented that suggests that $\Delta\Psi$ and ΔpH do not have equivalent effects on the carrier. In membrane vesicles of *E. coli*, Wright and Overath (1980) have demonstrated that the apparent change in K_t^i for lactose is almost wholly due to the interaction of $\Delta\Psi$ with the carrier. Thus, in vesicles respiring lactate, gradual reduction of $\Delta\Psi$ by titration with valinomycin in the presence of high potassium concentrations resulted in a 200-fold increase in K_t^i (Wright and Overath, 1980). As ΔpH was essentially constant throughout the experiment, it was concluded that the affinity constant was solely responsive to the membrane potential. On the other hand, other workers have reported that collapsing $\Delta\Psi$ in vesicles did not affect K_t^i and that the observed enhancement of affinity for lactose could be caused by either $\Delta\Psi$ or ΔpH (Kaczorowski *et al.*, 1979). Thus, the issue of equivalence of $\Delta\Psi$ and ΔpH must remain an open one at this time (Ahmed and Booth, 1981).

In summary, we have examined critically the assumptions inherent in accepting the concept of thermodynamic equilibrium as a working model for transport. It is important to recognize that the level of accumulation of a solute is the result of interactions between the solute and the carrier, and the carrier and the components of the protonmotive force. Both pairs of interactions may influence each other and thus contribute to the level of accumulation of the solute. Clearly the thermodynamic model is too simple, and more meaningful models that take into consideration these interactions are essential.

Recent advances in the primary structure of the lactose carrier protein (Ehring *et al.*, 1980; Beyreuther, 1980; Buchel *et al.*, 1980) and in techniques for obtaining crystals of membrane proteins (Garavito and Rosenbusch, 1980; Michel and Oesterhelt, 1980) make one optimistic that in quite a short time the molecular architecture of the carrier will be known. It is our hope that kinetic studies can proceed apace to yield a complementary model of energy coupling.

REFERENCES

Ahmed, S., and Booth, I. R. (1981). *Biochem. J.* **200**, 583–589.

Beyreuther, K. (1980). *Biochem. Soc. Trans.*, **8, 675–676.**

Booth, I. R. (1980). *Biochem. Soc. Trans.* **8**, 276–278.

Booth, I. R., and Hamilton, W. A. (1980). *Biochem. J.* **188**, 467–473.

Booth, I. R., Mitchell, W. J., and Hamilton, W. A. (1979). *Biochem. J.* **182**, 687–696.

Buchel, D. E., Gronenborn, B., and Müller-Hill, B. (1980). *Nature (London)* **283**, 541–545.

Christensen, H. N. (1976). *J. Theor. Biol.* **57**, 419–432.

Eddy, A. A. (1980). *Biochem. Soc. Trans.* **8**, 271–273.

Eddy, A. A., Seaston, A., Gardner, D., and Hacking, C. (1979). In *Function and Molecular Aspects of Biomembrane Transport* (E. Quagliariello, F. Palmieri, S. Papa, and M. Klingenberg, eds.), pp. 219–228, Elsevier/North-Holland, Amsterdam.

Ehring, R., Beyreuther, K., Wright, J. K., and Overath, P. (1980). *Nature (London)* **283**, 537–540.

Garavito, R. M., and Rosenbusch, J. P. (1980). *J. Cell Biol.* **86**, 327–329.

Harold, F. M. (1976). *Curr. Top. Bioenerg.* **6**, 83–149.

Kaczorowski, G. J., Robertson, D. E., and Kaback, H. R. (1979). *Biochemistry* **18**, 3697–3704.

Lancaster, J. R., Hill, R. J., and Struve, N. G. (1975). *Biochim. Biophys. Acta* **401**, 285–298.

Maloney, P. C., and Wilson, T. H. (1974). *Biochim. Biophys. Acta* **330**, 196–205.

Michel, H. and Oesterhelt, D. (1980). *Proc. Natl. Acad. Sci. USA* **77**, 1283–1286.

Mitchell, P. (1963). In *Organisation and Control in Prokaryotic and Eukaryotic Cells* (P. Charles and B. C. J. G. Knight, eds.), pp. 122–166, Cambridge University Press, London.

Mitchell, P. (1973). *J. Bioenerg.* **4**, 53–91.

Ramos, S., and Kaback, H. R. (1977a). *Biochemistry* **16**, 854–859.

Ramos, S., and Kaback, H. R. (1977b). *Biochemistry* **16**, 4271–4275.

Rickenberg, H. V., Cohen, G. N., Buttin, G., and Monod, J. (1956). *Ann. Inst. Pasteur* **91,** 829–857.

Rottenberg, H. (1976). *FEBS Let.* **66,** 159–163.

Toci, R., Belaich, A., and Belaich, J. P. (1980). *J. Biol. Chem.* **255,** 4603–4606.

Winkler, H. H., and Wilson, T. H. (1966). *J. Biol. Chem.* **241,** 2200–2211.

Wright, J. K., and Overath, P. (1980). *Biochem. Soc. Trans.* **8,** 279–281.

Wright, J. K., Teather, R. M., and Overath, P. (1979). In *Function and Molecular Aspects of Biomembrane Transport* (E. Quagliariello, F. Palmieri, S. Papa, and M. Klingenberg, eds.), pp. 239–248, Elsevier/North-Holland, Amsterdam.

94

Symport Mechanisms as the Basis of Solute Transport in Yeast and Mouse Ascites-Tumor Cells

A. Alan Eddy

1. THE Na$^+$ SYMPORTS OF MOUSE ASCITES-TUMOR CELLS

The notion that amino acid transport in mouse ascites-tumor cells might be driven by the ionic gradients of Na$^+$ or K$^+$ acting across the plasmalemma was first proposed more than 20 years ago and has since been surrounded by controversy (recent reviews: Christensen, 1979; Heinz *et al.*, 1977; Johnstone, 1979; West, 1980). The matter is complicated by the presence in the tumor cells of at least three distinct systems for absorbing neutral amino acids (Christensen, 1979). Two of these systems, A and ASC, are markedly stimulated by the presence of Na$^+$ whereas the third, system L, is much less affected by changes in the extracellular concentrations of Na$^+$ or K$^+$. The L system nevertheless concentrates its amino acid substrates extensively (Christensen, 1979; Hacking and Eddy, 1981). As many of the common amino acids are substrates of more than one of these transport systems, the problem of defining the role of Na$^+$ in their uptake and efflux is complex. Most work has been done with glycine, 2-aminoisobutyrate, or L-methionine. All three amino acids are substrates of the Na$^+$-dependent A system whereas L-methionine is also transported readily by the L system. A further nonsaturable kinetic component is evident in the uptake and efflux of these compounds from solutions containing relatively large amino acid concentrations (Christensen and Handlogten, 1977). This pathway contributes significantly to the efflux of amino acids concentrated through the other systems (Hacking and Eddy, 1981).

A. Alan Eddy • Department of Biochemistry, University of Manchester Institute of Science and Technology, Manchester M60 1QD, England.

1.1. Coupling to the Na$^+$ Gradient

Early work on the uptake of glycine by preparation of the mouse ascites-tumor cells depleted of ATP showed that this was accompanied by the absorption of about one equivalent of Na$^+$ and the efflux of a smaller amount of K$^+$. Substantial gradients of glycine formed across the plasma membrane under these conditions, apparently at the expense of the two ionic gradients. Ouabain was without effect on this process. Studies with valinomycin and with other ionophores, both with energy-depleted and with respiring cellular preparations, showed that the Na$^+$-dependent mode of amino acid uptake is electrogenic and is not obligatorily coupled to the efflux of K$^+$ through the amino acid carrier (Heinz *et al.*, 1977; Heinz and Geck, 1978; Johnstone and Laris, 1980; Reid *et al.*, 1974). The mouse ascites-tumor cells depleted of ATP can concentrate L-methionine transiently in the presence of valinomycin to about the same extent as cell preparations that maintain their ionic gradients by means of energy metabolism (Reid *et al.*, 1974). These observations have led to the view that the electrochemical gradient of Na$^+$ is indeed competent to drive amino acid accumulation in the mouse ascites-tumor cells up to the normal physiological levels. Qualitative support for this conclusion has come from several studies with plasma membrane vesicles in which the response of the system both to Na$^+$ and to changes in the membrane potential was demonstrated (see Im and Spector, 1980).

Work with carbocyanine dyes that serve as a probe of the membrane potential has shown that the uptake of various amino acids by the intact tumor cells tends to depolarize the plasma membrane whereas their efflux tends to hyperpolarize it (Johnstone and Laris, 1980; Philo and Eddy, 1978a). One interesting situation occurs when cellular [Na$^+$] is large and the cellular and extracellular concentrations of K$^+$ are also both large. The action of the electrogenic sodium pump then hyperpolarizes the tumor cells and amino acids are concentrated appreciably despite the virtual absence of a concentration gradient of either Na$^+$ or K$^+$ across the plasma membrane (Gibb and Eddy, 1972; Heinz *et al.*, 1977; Laris *et al.*, 1978; Philo and Eddy, 1978a). Under these conditions amino acid uptake is instantly inhibited by ouabain and appears to be coupled electrically to the functioning of the sodium pump and thus to the hydrolysis of ATP.

The above evidence favors the view that the A system of amino acid transport of the mouse ascites-tumor cells is an Na$^+$ symport. It is possible, though not certain, that all the effects of extracellular K$^+$ on the functioning of the symport can be explained in terms of changes in the membrane potential (see West, 1980). Potassium ions evidently do not serve the role of Na$^+$ in the carrier mechanism. They probably do not traverse the plasma membrane through the amino acid carrier, a process that may, however, occur in the kidney (Burckhardt *et al.*, 1980). Anions such as Cl$^-$ have not been implicated in the mouse ascites carrier mechanism, in contrast with the behavior of some other Na$^+$-dependent vertebrate transport systems (West, 1980).

1.2. Quantitative Relations between the Amino Acid Accumulation Ratio ([A]$_i$/[A]$_o$) and the Electrochemical Gradient of Na$^+$

The application of a method for assaying the membrane potential fluorimetrically in the presence of carbocyanine dyes* (Henius and Laris, 1979; Philo and Eddy, 1978a,b) together with the use of specialized microelectrode techniques (Hoffman *et al.*, 1979) has shown that the membrane potential of the mouse ascites-tumor cells probably lies in the

---

*A recent objection by Smith *et al.* (1981) to the use of these dyes as quantitative probes does not apply to the circumstances in which Philo and Eddy (1978a,b) worked.

range 25–55 mV in the different cell lines involved. Values at the lower end of that range are consistent with those deduced from the distribution of Cl^- across the plasa membrane if this is assumed to be governed by a Nernst relationship. However, Geck *et al.* (1980) have suggested that the cellular content of Cl^- can become larger than would be expected on that basis. They propose that the tumor cells also use an electroneutral mechanism to move K^+, Na^+, and 2 Cl^- simultaneously across the plasma membrane. This mechanism is insensitive to ouabain and is recognized by the fact that it is inhibited by diuretics such as furosemide. That the permeability of the plasma membrane to Cl^- as such is relatively small seems likely in view of the evidence that the sodium pump can hyperpolarize the preparations. However, the exact relationship between the chloride distribution and the membrane potential is uncertain (see Hoffman *et al.*, 1979).

Comparison of the magnitude of the prevailing sodium electrochemical gradient and the corresponding steady-state amino acid gradient formed from dilute solutions of methionine or 2-aminoisobutyrate has revealed situations in which the former is slightly larger than the latter (Hacking and Eddy, 1981; Henius and Laris, 1979; Philo and Eddy, 1978b). The discrepancies involved, when account is taken of leak pathways for the amino acid, correspond to a factor of 2 or less and can plausibly be explained by supposing that the activity of cellular Na^+ is about half its apparent value. Of greater significance is the close correlation found between the respective magnitudes of the amino acid gradient and its putative driving force (Fig. 1). This behavior is strong evidence in favor of the ion gradient hypothesis. The supposition underlying the above comparison is that the amino acid gradient formed from the dilute solutions of amino acids and the sodium electrochemical gradient are close to thermodynamic equilibrium. However, this is evidently not the case

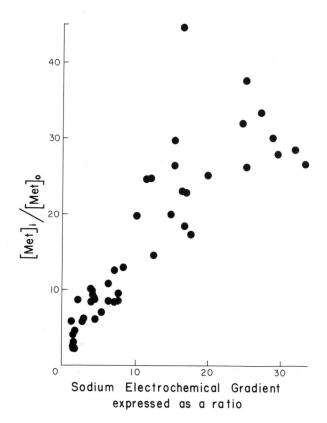

Figure 1. The steady-state ratio of the cellular to extracellular methionine concentrations formed during energy metabolism when the electrochemical gradient of Na^+ acting across the plasma membrane of the mouse ascites-tumor cells was systematically varied (Philo and Eddy, 1978b).

at larger amino acid concentrations (Eddy *et al.*, 1980; Hacking and Eddy, 1981). The sodium gradient is then considerably larger than the amino acid gradient although both parameters have achieved steady values and no significant metabolism of the relevant amino acid takes place (see Kotyk and Struzinsky, 1977). A related problem arises with amino acids such as leucine or isoleucine, which are concentrated to a smaller extent, even from dilute solutions, than are glycine, methionine, and 2-aminoisobutyrate. Nevertheless, the magnitude of the sodium electrochemical gradient maintained by the tumor cells is similar, though not identical, in all these cases (Hacking and Eddy, 1981). Furthermore, leucine like glycine is probably absorbed with 1 equivalent of Na^+.

A clue to the nature of the factors controlling amino acid uptake through the Na^+ symport is that the presence of large amino acid concentrations depolarizes the cells both initially and, to a significant extent, when a steady state of amino acid distribution is reached (Eddy *et al.*, 1980). This behavior serves to refute the suggestion that amino acid uptake may become uncoupled from the net uptake of Na^+ as the process proceeds (Christensen, 1979). One interpretation of the steady-state behavior (Eddy, 1980) is that the electrical depolarization is due to a net flow both of Na^+ and of the amino acid itself through the symport, the steady state of amino acid distribution being maintained by an outflow of amino acid through a separate carrier-mediated leak pathway situated elsewhere in the plasma membrane. It is amino acids like leucine and isoleucine which, on the one hand, exhibit a relatively small affinity for the Na^+ symport (pump) and, on the other hand, exhibit a relatively fast uptake in the absence of Na^+ (leak), that the tumor cells accumulate poorly from concentrated solutions (1–10 mM) of the respective amino acids (Hacking and Eddy, 1981). It seems possible that one component of the above "leak" is related to the L system of amino acid transport (Christensen, 1979). Paradoxically the rapid absorption and concentration of isoleucine from relatively dilute solutions, in the physiological range below 1 mM, appears to depend mainly on the L system of amino acid transport, the mechanism of which has not been established (Christensen, 1979; Hacking and Eddy, 1981; Johnstone, 1979; Ohsawa *et al.*, 1980). Further work on the complex relations between these different transport systems is evidently needed. One unresolved issue concerns the extent to which the uptake of amino acids through the L system depends on their exchange with endogenous amino acids.

2. THE PROTON SYMPORTS OF YEAST

Whereas the mouse ascites-tumor cells concentrate amino acids no more than about 30- to 50-fold with respect to their extracellular concentration, much larger gradients of up to 5×10^4-fold were produced when a certain strain of *Saccharomyces* was given 0.1 μM glycine. Preparations of the same yeast depleted of ATP formed similar amino acid gradients when the cells were suspended at pH 4.5 under conditions where (1) a substantial inwardly directed pH gradient persisted across the plasma membrane and (2) cellular $[K^+]$ was about 0.2 M with extracellular $[K^+]$ very small (Seaston *et al.*, 1976). The uptake of glycine was accompanied by the simultaneous absorption of about 2 equivalents of H^+ and the efflux of about 2 equivalents of K^+. There are grounds for thinking that the mechanism of glycine uptake in this yeast is based on an electrogenic symport with 2 H^+ and that the observed efflux of 2 K^+ is due to electrical depolarization of the plasma membrane. The outflow of K^+ represents the major route for balancing the flow of protons accompanying the uptake of the amino acid when the proton pump of the plasma membrane is not functioning owing to the lack of ATP (Eddy, 1978). A number of other amino acid symports

Table I. Proton Stoichiometry of Various Symport Mechanisms Found in Saccharomyces [a]

System	Equivalents of H^+/mole of substrate
General amino acid permease	
Glycine	2
L-Methionine	2
L-Lysine	2
L-Leucine	2
Specific amino acid permeases	
L-Lysine	1
L-Methionine	1
L-Proline	1
L-Glutamate	2 or 3
Other permeases	
Maltose	1
α-Methylglucoside	1
Hypoxanthine	1
Phosphate	2 or 3
Sulfate	3

[a] From Eddy (1978) and Eddy *et al.* (1980).

have been demonstrated in various yeasts (Table I). The fast uptake of glycine occurs through the so-called general amino acid permease, which is known to absorb various other amino acids with two proton equivalents. These yeasts also carry more specific amino acid permeases exhibiting a proton stoichiometry of one equivalent.

The amount of amino acid absorbed by yeast is regulated, not by its efflux, which is a very slow process, but by the phenomenon of transinhibition whereby the activity of the carrier (permease) is progressively inhibited and probably inactivated as the cellular amino acid content rises (Crabeel and Grenson, 1970; Indge *et al.*, 1977; Morrison and Lichstein, 1976). While the mechanism involved is obscure, the need for such a process is dramatically demonstrated by the fact that yeast cells fed with large amounts of glycine tend to burst when the normal feedback control is interrupted by treating the yeast in certain ways (Indge *et al.*, 1977).

Reference to Table I shows that the proton stoichiometries of the various symport mechanisms for amino acids, carbohydrates, sulfate, and phosphate are small integers. The anions glutamate, phosphate, and sulfate are absorbed with a larger number of protons than would merely neutralize their negative charges (Cockburn *et al.*, 1975; Eddy *et al.*, 1980; Roomans and Borst-Pauwels, 1979; Roomans *et al.*, 1979). The evidence for this is not merely the number of proton equivalents that are absorbed with the anion in question but also the fact that K^+ ions are simultaneously displaced from the yeast. In effect the anions are absorbed as positively charged complexes. It may be inferred that the driving forces acting on the proton symport with, for instance, $H_2PO_4^-$ are relatively large. They appear to be sufficient to sustain the largest gradients of phosphate concentration (10^6-fold) that yeast has been reported to form (Cockburn *et al.*, 1975). There is some evidence for the presence of a distinct sodium ion symport with phosphate in *Saccharomyces* (Roomans and Borst-Pauwels, 1979). A similar dichotomy is better known in bacteria (Eddy, 1978; Hamilton, 1977; West, 1980). Little progress has been made with attempts to demonstrate symport mechanism in vesicles derived from the yeast plasma membrane. The related fungus *Neurospora* carries an electrogenic, proton-translocating ATPase in the plasma membrane (Scarborough, 1980) that resembles the superficial ATPase of yeast (Dufour *et al.*, 1980) for which a similar proton-translocating function has been assumed.

The factors governing the concentration by yeast of α-thioethylglucoside, an unfermented analog of the carbohydrate α-methylglucoside, are quite different from those controlling the uptake of amino acids. The methylglucoside proton symport concentrates the analog by no more than a factor of about 200-fold. When the specific activity of the symport was varied by physiological means, the extent to which the yeast concentrated α-thioethylglucoside changed roughly in proportion (Eddy *et al.*, 1979). This phenomenon appears to be due to the circumstance that the symport functions in parallel with a leak mechanism, a situation analogous to that obtaining during amino acid transport in the mouse ascites-tumor cells (Eddy *et al.*, 1980). The steady state of α-thioethylglucoside absorption is associated with a net uptake of protons, presumably through the symport mechanism itself, which is revealed in the presence of metabolic inhibitors that stop the recycling of the protons, out of the yeast, through the proton pump (Eddy *et al.*, 1979). This phenomenon may arise from the same causes as the steady-state electrical depolarization of the mouse ascites-tumor cells in the presence of glycine or other amino acids. In both experimental systems the steady state of substrate distribution was probably not in thermodynamic equilibrium with the distribution of H^+ or Na^+ across the plasma membrane (Eddy *et al.*, 1980).

REFERENCES

Burckhardt, G., Kinne, R., Stange, G., and Murer, H. (1980). *Biochim. Biophys. Acta* **599,** 191–201.

Christensen, H. N. (1979). *Adv. Enzymol. Relat. Areas Mol. Biol.* **49,** 41–101.

Christensen, H. N., and Handlogten, M. E. (1977). *Biochim. Biophys. Acta* **469,** 216–220.

Cockburn, M., Earnshaw, P., and Eddy, A. A. (1975). *Biochem. J.* **146,** 705–712.

Crabeel, M., and Grenson, M. (1970). *Eur. J. Biochem.* **14,** 197–204.

Dufour, J.-P., Boutry, M., and Goffeau, A. (1980). *J. Biol. Chem.* **255,** 5735–5741.

Eddy, A. A. (1978). *Curr. Top. Membr. Transp.* **10,** 279–360.

Eddy, A. A. (1980). *Biochem. Soc. Trans.* **8,** 271–273.

Eddy, A. A., Seaston, A., Gardner, D., and Hacking, C. (1979). In *Function and Molecular Aspects of Biomembrane Transport* (E. Quagliariello, F. Palmieri, S. Papa, and M. Klingenberg, eds.), pp. 219–228, Elsevier/North-Holland, Amsterdam.

Eddy, A. A., Seaston, A., Gardner, D., and Hacking, C. (1980). *Ann. N.Y. Acad. Sci.* **341,** 494–509.

Geck, P., Pietrzyk, C., Burckhardt, B.-C., Pfeiffer, E., and Heinz, E. (1980). *Biochim. Biophys. Acta* **600,** 432–447.

Gibb, L. E., and Eddy, A. A. (1972). *Biochem. J.* **129,** 979–981.

Hacking, C., and Eddy, A. A. (1981). *Biochem. J.,* **194,** 415–426.

Hamilton, W. A. (1977). *Symp. Soc. Gen. Microbiol.* **27,** 185–216.

Heinz, E., and Geck, P. (1978). In *Membrane Transport Processes* (J. F. Hoffman, ed.), Vol. 1, pp. 13–30, Raven Press, New York.

Heinz, E., Geck, P., Pietrzyk, C., Burckhardt, G., and Pfeiffer, B. (1977). *J. Supramol. Struct.* **6,** 125–133.

Henius, G. V., and Laris, P. C. (1979). *Biochem. Biophys. Res. Commun.* **91,** 1430–1436.

Hoffman, E. K., Simonsen, L. O., and Sjohølm, C. (1979). *J. Physiol.* **296,** 61–84.

Im, W. B., and Spector, A. A. (1980). *J. Biol. Chem.* **255,** 764–770.

Indge, K., Seaston, A., and Eddy, A. A. (1977). *J. Gen. Microbiol.* **99,** 243–255.

Johnstone, R. M. (1979). *Can. J. Physiol. Pharmacol.* **57,** 1–15.

Johnstone, R. M., and Laris, P. C. (1980). *Biochim. Biophys. Acta* **599,** 715–730.

Kotyk, A., and Struzinsky, R. (1977). *Biochim. Biophys. Acta* **470,** 484–491.

Laris, P. C., Bootman, M., Pershadsingh, H. A., and Johnstone, R. M. (1978). *Biochim. Biophys. Acta* **512,** 397–414.

Morrison, C. E., and Lichstein, H. C. (1976). *J. Bacteriol.* **125,** 864–871.

Ohsawa, M., Kilberg, M. S., Kimmel, G., and Christensen, H. N. (1980). *Biochim. Biophys. Acta* **599,** 175–190.

Philo, R., and Eddy, A. A. (1978a). *Biochem. J.* **174,** 801–810.

Philo, R., and Eddy, A. A. (1978b). *Biochem. J.* **174,** 811–817.

Reid, M., Gibb, L. E., and Eddy, A. A. (1974). *Biochem. J.* **140,** 383–393.

Roomans, G. M., and Borst-Pauwels, G. W. F. H. (1979). *Biochem. J.* **178,** 521–527.

Roomans, G. M., Kuypers, G. A. J., Theuvenet, A. P. R., and Borst-Pauwels, G. W. F. H. (1979). *Biochim. Biophys. Acta* **551,** 197–206.

Scarborough, G. A. (1980). *Biochemistry* **19,** 2925–2931.

Seaston, A., Carr, G., and Eddy, A. A. (1976). *Biochem. J.* **154,** 669–676.

Smith, T. C., Herlihy, J. T., and Robinson, S. C. (1981). *J. Biol. Chem.* **256,** 1108–1110.

West, I. C. (1980). *Biochim. Biophys. Acta* **604,** 91–126.

95

Sugar Transport in Yeast

A. Kotyk

Although different species of yeasts (in particular *Saccharomyces cerevisiae, S. carlsbergensis, Candida* sp.) have been studied for many decades and in spite of the fact that utilization of sugars is a major aspect of their metabolism, it is only quite recently that quantitative and sufficiently reliable data have begun to emerge concerning the various uptake mechanisms and their role in the biochemistry of the yeast cell.

I will attempt to give here an overall picture of the state of our knowledge in the field, emphasizing the work of the last 3–5 years.

1. CARRIERS AND THEIR MULTIPLICITY

Virtually all sugars and related substances are transported by one or another type of carrier mechanism, specificity, saturability, and countertransport being the distinctive features of such mechanisms. Only D-ribose, occurring to as much as 10% in the linear form even in solution, and acyclic polyols apparently can penetrate by nonmediated diffusion (Horák and Kotyk, 1969; Canh *et al.*, 1975).

It is now clear that in every yeast species so far investigated there is a multiplicity of transport systems for practically every class of substances. In this respect yeasts lie halfway between bacteria where this is a general phenomenon and higher organisms where the "unnecessary" systems seem to have been weeded out during evolution.

The well-known baker's yeast contains the following systems for monosaccharides: one that takes up all pyranose-type monosaccharides, another for pyranose-type sygars with an equatorial hydroxyl group at C-4, and another, inducible, system preferring sugars with an axial hydroxyl at C-4 (Kotyk *et al.*, 1975). It is now suspected that there may be parallel systems present even within the above categories (Serrano and DelaFuente, 1974). The same holds for *Torulopsis candida* (Haškovec and Kotyk, 1973), *Candida utilis* (Barnett and Sims, 1976), and *Rhodotorula glutinis* (= *gracilis* = *Rhodosporidium toruloides*) (Janda

A. Kotyk • Department of Cell Physiology, Institute of Microbiology, Czechoslovak Academy of Sciences, Prague, Czechoslovakia.

et al., 1976; Alcorn and Griffin, 1978). A fine demonstration of this fact was achieved by Bhandari and Hayashibe (1977) who isolated two glucose-binding proteins and one mannose-binding protein from the plasma membrane of *Schizosaccharomyces pombe*.

It is now assumed that most, if not all, yeast species may contain a general system for monosaccharide transport—something in the nature of general amino acid permease (Grenson *et al.*, 1970)—plus various specific systems that may be more energy-dependent but of low capacity (Ehwald and Mavrina, 1981).

Indications of multiple parallel carriers are now found in disaccharide transport (Kotyk and Michaljaničová, 1979).

2. KINETICS AND ENERGETICS

Practically all sugar transports in yeasts proceed at the expense of metabolically stored energy, the only major exceptions being the mediated diffusion systems for monosaccharides in baker's yeast, *S. cerevisiae,* and brewer's bottom yeast, *S. carlsbergensis* ($= uvarum$).

The energy dependence may involve either the carrier (changes of its effective affinity) or the substrate (phosphorylation).

Many instances of the first type are known. Free monosaccharides are transported uphill in *R. glutinis* (Kotyk and Höfer, 1965), *S. fragilis* (Jaspers and van Steveninck, 1977), *T. candida* (Haškovec and Kotyk, 1973), *C. parapsilosis* (Kotyk and Michaljaničová, 1978), and a number of others (Kotyk, 1978). Free disaccharides are transported uphill in *S. cerevisiae* (Okada and Halvorson, 1964; Kotyk and Michaljaničová, 1979) and in *R. glutinis* (Janda and von Hedenström, 1974). Acyclic polyols were described to use active transport systems in *T. candida* (Haškovec and Kotyk, 1973), *R. gracilis* (Klöppel and Höfer, 1976a), and *C. guillermondii* (Miersch, 1977).

Substrate phosphorylation (a type of group translocation conceptually similar to the phosphotransferase systems of bacteria) was described in the case of 2-deoxy-D-glucose, and perhaps glucose itself, in *S. fragilis* (Jaspers and van Steveninck, 1975).

While in this last case the involvement of polyphosphate as the phosphate donor is assumed, the other listed active transport processes appear to derive their energy from an electrochemical gradient of protons. The first to observe transient alkalification of the suspension medium (suggesting proton symport of sugars) were Misra and Höfer (1975), working with D-xylose uptake in *R. gracilis*. Further support for the involvement of both components of the protonmotive force, i.e., the membrane potential,* $\Delta\Psi$, and the concentration term, $(RT/F)\ln([H^+]_i/[H^+]_o)$, was provided by the same group (Hauer and Höfer, 1978; Höfer and Misra, 1978).

Proton-driven uptake of L-sorbose and 2-deoxy-D-galactose was reported to occur in *S. fragilis* by Jaspers and van Steveninck (1977). The stoichiometry (H^+ : sugar) ranged from 0.5 or 1.0 at low pH values to practically zero at higher pH values. A surprisingly constant stoichiometry of 1.0 for the case of maltose uptake by baker's yeast was found by Serrano (1977). The uptake of protons with sugars is probably compensated by an efflux of potassium ions from cells.

---

*Membrane potentials of 50–80 mV were found in *R. gracilis* using indirect methods (Hauer and Höfer, 1978), of 50–120 mV in *S. cerevisiae,* and of 40–65 mV, using direct microelectrode measurement, in *Endomyces magnusii* (Vacata *et al.*, 1981).

In other cases, particularly in nonstarved cells, no major proton uptake concomitant with the uptake of sugars could be observed (Kotyk and Michaljaničová, 1978, 1979).

It remains to be elucidated what the mechanism of generation of the proton gradient in yeast might be (an H^+-translocating ATPase, polyphosphatase, pyrophosphatase?).

While most of the transports probably involve single-site carriers, cooperativity phenomena were observed with the transport of trehalose in baker's yeast (Kotyk and Michaljaničová, 1979) and of D-xylose and D-galactose in *R. glutinis* (Janda *et al.*, 1976).

3. FACTORS INFLUENCING TRANSPORT OF SUGARS

Like most enzyme activity, transport of sugars in yeasts is temperature-dependent, with optima at 28–40°C, but so far no breaks on Arrhenius plots that might correspond to phase transitions of membrane lipids have been reported.

The pH dependence is either featureless (flat maximum of monosaccharide transport in baker's yeast) or shows a clear maximum, usually at pH 4.5 to 6.0. This shift toward more acid values may reflect the involvement of the proton gradient in secondary active transport. However, it should be pointed out that the accumulation ratio of sugars in many species (Kotyk, 1980) reaches its maximum at higher pH values (about 6.0–6.5) where the free energy available from the electrochemical gradient of protons may be near its minimum (Kotyk, 1979). This may indicate that neither the overall membrane potential nor the ΔpH measured in bulk solutions constitutes the actual driving force for the active transports.

All sugar transport systems appear to be sensitive to heavy metal ions, in particular Th^{4+}, La^{3+}, and UO_2^{2+}.

The activity of existing transport proteins can apparently be altered by intracellular factors. Aerobic incubation will increase the half-saturation constant of glucose uptake by baker's yeast (Kotyk and Kleinzeller, 1967), possibly along with a decrease of maximum rate of efflux which is related to glucose-6-phosphate concentration in cells (Azam and Kotyk, 1969). A similar phenomenon (decrease of half-saturation constant in the presence of metabolism) was described by Serrano and DelaFuente (1974).

4. SYNTHESIS AND DEGRADATION OF TRANSPORT SYSTEMS

A surprising number of transport systems for sugars are inducible or derepressible. This is particularly true for the disaccharide transports, e.g., maltose, trehalose, α-methyl-D-glucoside, and melibiose in *Saccharomyces* (Okada and Halvorson, 1964; Kew and Douglas, 1976; Kotyk and Michaljaničová, 1979).

A great deal of effort has been expended on the galactose transport and metabolism in baker's yeast. The transport protein, coded by a gene on chromosome XII, is induced before the metabolic enzymes (Kotyk and Haškovec, 1968) and it does not require a mitochondrial corepressor (unlike the metabolic enzymes; Puglisi and Algeri, 1971). The structural genes for the galactose metabolism enzymes are clustered on chromosome II.

A case of inducible pentitol uptake is known from *R. gracilis* (Klöppel and Höfer, 1976b).

It appears that the synthesis of a number of transport systems can be derepressed in the absence of superior nutrients (much like in bacteria or filamentous fungi), but facultatively anaerobic yeasts exemplified by the genus *Saccharomyces* show a peculiar positive effect of preincubation with sources of energy, most typically with glucose. Such a preincubation results in *de novo* synthesis leading to the appearance of transport systems for monosaccharides (Spoerl *et al.*, 1973), disaccharides (Kotyk and Michaljaničová, 1979), as well as amino acids, purines, sulfate and phosphate anions (cf. Kotyk, 1979). This synthesis requires the presence of active mitochondria.

The last thing to be said in this chapter concerns the final fate of carrier molecules, their proteolytic degradation. While constitutive carrier proteins show very little turnover (the monosaccharide carriers in baker's yeast have a half-life of more than 2 days), the inducible ones may be degraded at rapid rates (half-lives of the order of tens of minutes) (Alonso and Kotyk, 1978; Kotyk and Michaljaničová, 1979). It is a common finding that the degradation of these systems is speeded up by the presence of glucose (apparently through mitochondrial ATP), which is a form of catabolite inactivation described by Holzer (e.g., 1976). This is not the only factor playing a role in the degradation rates. Some systems are degraded faster when glucose plus cycloheximide is present (those induced by galactose and by α-methyl-D-glucoside or trehalose) while others are degraded faster with glucose alone (that induced by maltose). Apparently, the synthesis of a carrier-hydrolyzing proteinase is blocked more efficiently than the resynthesis of the transport protein.

REFERENCES

Alcorn, M. E., and Griffin, C. C. (1978). *Biochim, Biophys. Acta* **510**, 361–371.
Alonso, A., and Kotyk, A. (1978). *Folia Microbiol. (Prague)* **23**, 118–125.
Azam, F., and Kotyk, A. (1969). *FEBS Lett.* **2**, 333–335.
Barnett, J. A., and Sims, A. P. (1976). *Arch. Microbiol.* **111**, 193–194.
Bhandari, H. C., and Hayashibe, M. (1977). *J. Biochem.* **82**, 1197–1205.
Canh, D. S., Horák, J., Kotyk, A., and Říhová, L. (1975). *Folia Microbiol. (Prague)* **20**, 320–325.
Ehwald, R., and Mavrina L. (1981). *Folia Microbiol. (Prague)* **26**, 95–102.
Grenson, M., Hou, C., and Crabeel, M. (1970). *J. Bacteriol.* **103**, 770–777.
Haškovec, C., and Kotyk, A. (1973). *Folia Microbiol. (Prague)* **18**, 118–124.
Hauer, R., and Höfer, M. (1978). *J. Membr. Biol.* **43**, 335–349.
Höfer, M., and Misra, P. C. (1978). *Biochem. J.* **172**, 15–22.
Holzer, H. (1976). *Trends Biochem. Sci.* **1**, 178–181.
Horák, J., and Kotyk, A. (1969). *Folia Microbiol. (Prague)* **14**, 291–296.
Janda, S., and von Hedenström, M. (1974). *Arch. Microbiol.* **101**, 273–280.
Janda, S., Kotyk, A., and Tauchová, R. (1976). *Arch. Microbiol.* **111**, 151–154.
Jaspers, H. T. A., and van Steveninck, J. (1975). *Biochim. Biophys. Acta* **406**, 370–385.
Jaspers, H. T. A., and van Steveninck, J. (1977). *Biochim. Biophys. Acta* **469**, 292–300.
Kew, O. M., and Douglas, H. C. (1976). *J. Bacteriol.* **125**, 33–41.
Klöppel, R., and Höfer, M. (1976a). *Arch. Microbiol.* **107**, 329–334.
Klöppel, R., and Höfer, M. (1976b). *Arch. Microbiol.* **107**, 335–342.
Kotyk, A. (1978). *VIth International Biophysics Congress*, Abstr. VII-7-(H), Kyoto.
Kotyk, A. (1979). In *Cell Compartmentation and Metabolic Channeling* (L. Nover, F. Lynen, and K. Mothes, eds.), pp. 63–74, Gustav Fischer Verlag, Jena.
Kotyk, A. (1980). In *Frontiers of Bio-organic Chemistry and Molecular Biology* (S. N. Ananchenko, ed.), pp. 417–425. Pergamon Press, Elmsford, N.Y.
Kotyk, A., and Haškovec, C. (1968). *Folia Microbiol. (Prague)* **13**, 12–19.
Kotyk, A., and Höfer, M. (1965). *Biochim. Biophys. Acta* **102**, 410–422.
Kotyk, A., and Kleinzeller, A. (1967). *Biochim. Biophys. Acta* **135**, 106–111.
Kotyk, A., and Michaljaničová, D. (1978). *Folia Microbiol. (Prague)* **23**, 18–26.
Kotyk, A., and Michaljaničová, D. (1979). *J. Gen. Microbiol.* **110**, 323–332.
Kotyk, A., Michaljaničová, D., Vereš, K., and Soukupová, V. (1975). *Folia Microbiol. (Prague)* **20**, 496–503.
Miersch, J. (1977). *Folia Microbiol. (Prague)* **22**, 363–372.
Misra, P. C., and Höfer, M. (1975). *FEBS Lett.* **52**, 95–97.
Okada, H., and Halvorson, H. O. (1964). *Biochim. Biophys. Acta* **82**, 547–555.

Puglisi, P. P., and Algeri, A. (1971). *Mol. Gen. Genet.* **110,** 110–117.
Serrano, R. (1977). *Eur. J. Biochem.* **80,** 97–102.
Serrano, R., and DelaFuente, G. (1974). *Mol. Cell. Biochem.* **5,** 161–171.
Spoerl, E., Williams, J. P., and Benedict, S. H. (1973). *Biochim. Biophys. Acta* **298,** 956–966.
Vacata, V., Kotyk, A., and Sigler, K. (1981). *Biochim. Biophys. Acta,* **643,** 265–268.

96

Cation/Proton Antiport Systems in Escherichia coli

Erik N. Sorensen and Barry P. Rosen

In *Escherichia coli*, primary proton pumps convert chemical energy into an electrochemical gradient of protons, composed of an electrical gradient or membrane potential ($\Delta\Psi$) and a chemical gradient (ΔpH). Energy stored in the electrochemical proton gradient can be used to drive secondary transport systems (Mitchell, 1973). In *E. coli*, three secondary antiport systems that catalyze the exchange of protons for inorganic cations have been described (Brey *et al.*, 1978). These three systems are the Ca^{2+}/H^+, Na^+/H^+, and K^+/H^+ antiporters, and each is responsible for the extrusion of the cationic substrates from intact cells. The properties of each system will be discussed below.

1. Ca^{2+}/H^+ ANTIPORTER

Silver and Kralovic (1969) first suggested the existence of a Ca^{2+} extrusion system in *E. coli*. They found that more Ca^{2+} is associated with cells at 4°C than at higher temperatures. Rosen and McClees (1974) reasoned that extrusion from intact cells should be equivalent to uptake into everted membrane vesicles, i.e., vesicles in which the membrane has an orientation opposite to that of the intact cell. Vesicles made by lysis of intact cells with a French press have been shown to be everted (Futai, 1974). Such vesicles actively accumulate Ca^{2+} (Rosen and Tsuchiya, 1979). Accumulation can be driven by oxidation of reduced compounds by the electron transport chain or by hydrolysis of ATP via the BF_0F_1, implicating the protonmotive force as the actual driving force. Phosphate or oxalate in the reaction media stimulates accumulation, probably by intravesicular formation of insoluble calcium salts. Ca^{2+} uptake by everted membrane vesicles can also be energized by an artificially imposed ΔpH, acid interior, directly demonstrating that the extrusion system is a Ca^{2+}/H^+ exchanger (Tsuchiya and Rosen, 1975, 1976).

Erik N. Sorensen and Barry P. Rosen • Department of Biological Chemistry, University of Maryland School of Medicine, Baltimore, Maryland 21201.

Further characterization of the Ca^{2+}/H^+ antiporter was carried out by studying the perturbation of the ΔpH established by respiration caused by addition of divalent cations (Brey and Rosen, 1979). ΔpH was monitored by measurement of the distribution of weak lipophilic bases. In these assays the weak lipophilic bases were the aminoacridines quinacrine and 9-aminoacridine. The fluorescence of these bases is quenched during the formation of a pH gradient that is acid inside. Using this assay, Brey and Rosen (1979) found that the divalent cations Ca^{2+}, Mn^{2+}, Sr^{2+}, and Ba^{2+} are all substrates of antiport systems. Competition experiments showed that they are all substrates of the same system, i.e., a divalent cation/proton antiporter. Sigmoidal kinetics are observed when the initial rate of fluorescence change is measured as a function of cation concentration. For each substrate the value of the Hill coefficient approaches 2. The $s_{0.5}$ values indicate that the order of affinity of the antiporter for substrates is $Ca^{2+} \simeq Mn^{2+} > Sr^{2+} > Ba^{2+}$. Mg^{2+} is not a substrate, but acts as an inhibitor, decreasing the V_{max}. At lower concentrations of Mg^{2+}, the Hill coefficients for Ca^{2+} and Sr^{2+} decrease to approximately 1. This suggests an allosteric mechanism in which both homotropic and heterotropic effects occur. Mg^{2+}, the major intracellular divalent cation, may thus be a regulator of the Ca^{2+}/H^+ antiporter, increasing the affinity but decreasing the V_{max}. Under physiological conditions the ranges of intracellular Mg^{2+} and Ca^{2+} are such that slight differences in the internal Mg^{2+} concentration can turn on or off the antiporter.

Brey and Rosen (1979) also found that the Ca^{2+}/H^+ antiporter has a pH optimum of about 8. Dissipation of the membrane potential with permeant anions inhibits the activity, suggesting either that the antiporter operates by an electrogenic mechanism in which $H^+/Ca^{2+} > 2$ or that the presence of a potential may be required for activation of the antiporter, or both.

Aside from the fact that high internal Ca^{2+} concentrations would result in calcium phosphate precipitation, the need for a calcium extrusion system is not understood, although it has been postulated that Ca^{2+} fluxes are in some way involved in chemotaxis (Rosen and McClees, 1974; Brey and Rosen, 1979; Rosen, 1981).

2. Na⁺/H⁺ ANTIPORTER

The Na^+/H^+ antiporter was first described by West and Mitchell (1974). They observed that intracellular Na^+ stimulates the equilibration of protons displaced from the cell by a respiratory pulse. Also, addition of Na^+ to an anaerobic cell suspension causes the expulsion of protons against their electrochemical gradient.

Beck and Rosen (1979) were able to observe $^{22}Na^+$ uptake in response to the formation of a protonmotive force in everted vesicles from *E. coli* by using flow dialysis. Tsuchiya and Takeda (1979b) observed Na^+ extrusion from whole cells using an Na^+ electrode. Schuldiner and Fishkes (1978) observed energy-dependent extrusion of $^{22}Na^+$ from right-side-out membrane vesicles.

The Na^+/H^+ antiporter has been further characterized by observing the effects of monovalent cations on ΔpH using fluorescent probes (Schuldiner and Fishkes, 1978; Tsuchiya and Takeda, 1979a; Beck and Rosen, 1979). The substrates of the antiporter include both Li^+ and Na^+, but not other monovalent cations. The system appears to be electrophoretic, in that dissipation of the membrane potential with permeant anions inhibits the effect of Na^+ on the ΔpH (Beck and Rosen, 1979).

The function of the Na^+/H^+ antiporter is clear. There are several known Na^+–solute symporters in *E. coli*, where accumulation of solute is coupled to the electrochemical Na^+

gradient (Hasan and Tsuchiya, 1977; Tsuchiya *et al.*, 1977). *E. coli* lacks a primary Na^+ pump. The Na^+/H^+ antiporter functions as a transformer to convert the energy of the electrochemical proton gradient into an electrochemical Na^+ gradient, which is then available to energize uptake of solutes utilizing Na^+–solute symporters. Zilberstein *et al.* (1980) have recently suggested that another function of the Na^+/H^+ antiporter is regulation of intracellular pH when cells are grown under alkaline conditions.

3. K^+/H^+ ANTIPORTER

The K^+/H^+ antiporter of *E. coli* was first observed by Brey *et al.* (1978) by observing the effects of monovalent cations on ΔpH in everted membrane vesicles, where ΔpH was monitored by observing changes in the fluorescence of quinacrine. Substrates of the K^+/H^+ antiporter include K^+, Na^+, Li^+, Rb^+, and Tl^+ (Brey *et al.*, 1980). Although the antiporter appears to transport most monovalent cations, it seems likely that *in vivo* it acts as a K^+/H^+ antiporter. Normally the intracellular K^+ concentration exceeds 0.1 M, whereas the intracellular concentration of Na^+, the only other monovalent cation normally present inside *E. coli*, is in the millimolar range.

The K^+/H^+ antiporter is rapidly inactivated by trypsin treatment of everted membrane vesicles, whereas other membrane activities such as the Na^+/H^+ and Ca^{2+}/H^+ antiporters, the respiratory chain, and the BF_0F_1 are not affected by the same treatment (Brey *et al.*, 1980). Prior energization of the membrane protects the antiporter from trypsin inactivation. These results suggest that a lysyl or arginyl residue of the antiporter is exposed on the cytoplasmic side of the membrane (i.e., the outside of an everted membrane vesicle), but energization causes a conformational change in the antiporter such that the residue is no longer accessible to trypsin. Butanedione, an arginine-modifying reagent, inactivates the antiporter, while pyridoxal phosphate, a lysine-modifying reagent, neither affects the activity of the antiporter nor prevents inactivation by trypsin (Young and Rosen, unpublished). Both energization and K^+ prevent inactivation by butanedione. Thus, it appears that an arginyl residue essential for transport activity is accessible to both trypsin and butanedione in unenergized vesicles. Formation of a protonmotive force alters the conformation of the antiporter so that the arginyl residue is no longer exposed to large molecules, such as trypsin, and small molecules, such as butanedione.

Because dissipation of $\Delta\Psi$ with permeant anions does not change the effects of K^+ on ΔpH, it is likely that the K^+/H^+ antiporter is electroneutral, with a stoichiometry of $K^+/H^+ = 1$. The pH optimum for the antiporter is about 8. In addition to the quinacrine assay, the activity of the K^+/H^+ antiporter can be measured by the accumulation of $^{204}Tl^+$ in everted vesicles (Brey *et al.*, 1980).

As there are a number of K^+ uptake systems in *E. coli* (Rhoads *et al.*, 1976) that maintain high intracellular K^+ levels, it is unlikely that the function of the K^+/H^+ antiporter is to form an electrochemical K^+ gradient. A likely function is regulation of the intracellular pH. The intracellular pH in *E. coli* remains relatively constant (Padan *et al.*, 1976), and a mechanism is needed to return protons, pumped out to establish the electrochemical proton gradient, to the cytosol when the external pH rises above about 7.5. This could be accomplished by the electroneutral K^+/H^+ antiporter. In support of this hypothesis an alkaline-sensitive strain of *E. coli* has been isolated that lacks the K^+/H^+ antiporter activity (Plack and Rosen, 1980).

The basic properties of the three cation/proton antiporters in *E. coli* have been determined. Further work will be required to elucidate the molecular mechanisms of these antiporters and in some cases to establish their physiological role. This work will be facilitated by the genetic manipulations that are possible in *E. coli*.

REFERENCES

Beck, J. C., and Rosen, B. P. (1979). *Arch. Biochem. Biophys.* **194,** 208–214.

Brey, R. N., and Rosen, B. P. (1979). *J. Biol. Chem.* **254,** 1957–1963.

Brey, R. N., Beck, J. C., and Rosen, B. P. (1978). *Biochem. Biophys. Res. Commun.* **83,** 1588–1594.

Brey, R. N., Rosen, B. P., and Sorensen, E. N. (1980). *J. Biol. Chem.* **255,** 39–44.

Futai, M. (1974). *J. Membr. Biol.* **15,** 15–28.

Hasan, S. M., and Tsuchiya, T. (1977). *Biochem. Biophys. Res. Commun.* **78,** 122–128.

Mitchell, P. (1973). *J. Bioenerg.* **4,** 63–91.

Padan, E., Zilberstein, D., and Rottenberg, H. (1976). *Eur. J. Biochem.* **63,** 533–541.

Plack, R. H., Jr., and Rosen, B. P. (1980). *J. Biol. Chem.* **255,** 3824–3825.

Rhoads, D. B., Waters, F. B., and Epstein, W. (1976). *J. Gen. Physiol.* **67,** 325–341.

Rosen, B. P. (1982). In *Membrane Transport of Calcium* (E. Carafoli, ed.), pp. 187–216, Academic Press, New York.

Rosen, B. P., and McClees, J. S. (1974). *Proc. Natl. Acad. Sci. USA* **71,** 5042–5046.

Rosen, B. P., and Tsuchiya, T. (1979). In *Methods in Enzymology* (S. Fleischer and L. Packer, eds.), Vol. 56, pp. 233–241, Academic Press, New York.

Schuldiner, S., and Fishkes, H. (1978). *Biochemistry* **17,** 706–711.

Silver, S., and Kralovic, M. L. (1969). *Biochem. Biophys. Res. Commun.* **34,** 640–645.

Tsuchiya, T., and Rosen, B. P. (1975). *J. Biol. Chem.* **250,** 7687–7692.

Tsuchiya, T., and Rosen, B. P. (1976). *J. Biol. Chem.* **251,** 962–967.

Tsuchiya, T., and Takeda, K. (1979a). *J. Biochem. (Tokyo)* **85,** 943–951.

Tsuchiya, T., and Takeda, K. (1979b). *J. Biochem. (Tokyo)* **86,** 225–230.

Tsuchiya, T., Raven, J., and Wilson, T. H. (1977). *Biochem. Biophys. Res. Commun.* **76,** 26–31.

West, I. C., and Mitchell, P. (1974). *Biochem. J.* **144,** 87–90.

Zilberstein, D., Padan, E., and Schuldiner, S. (1980). *FEBS Lett.* **116,** 177–180.

97

How Does *Escherichia coli* Regulate Internal pH?

Shimon Schuldiner and Etana Padan

1. INTRODUCTION

The concentration of protons plays a fundamental role in essential physicochemical reactions such as ionization, oxidoreduction, and solvolysis. Remarkably, the aquatic habitats in which bacteria proliferate show a very wide difference in proton concentration, which can reach nine orders of magnitude. In the acidic range, there are the sulfur springs with a pH of between 1 and 2, while in the alkaline extreme, the soda lakes have a pH of up to 11 (Brock, 1969; Talling *et al.*, 1973; Hargreaves *et al.*, 1975; Langworthy, 1978; Imhoff *et al.*, 1979; Tindall *et al.*, 1980). Microorganisms exist all over this range, yet it is clear that the more extreme the pH the less abundant the species, and most microorganisms flourish at around neutrality (Hargreaves *et al.*, 1975; Langworthy, 1978). On the other hand, because proton concentration is altered by biological activity, drastic and rapid fluctuations in pH, which can reach 3–4 units within a diurnal cycle, are often encountered in many habitats (Abeliovich and Azov, 1976); yet many bacteria survive. Hence, it must be concluded that bacteria are equipped with mechanisms that allow them to cope with extreme pH ranges and/or fluctuations in the medium pH.

Because most proteins, even those isolated from alkalophiles and acidophiles, have a narrow pH range of optimum activity and/or stability that falls around neutrality (Langworthy, 1978), it is not surprising that we have adopted the notion (Padan *et al.*, 1976) that adaptation to medium pH in bacteria involves a mechanism that maintains the cytoplasmic pH constant at around neutrality.

The extensively studied genetics, physiology, and biochemistry of *Escherichia coli* led us to choose this bacterium for pursuing the mechanism of adaptation to pH in a neutrophile. It grows optimally between pH 6 and 8 and, remarkably, over this range its

Shimon Schuldiner • Department of Molecular Biology, Hadassah Medical School, Jerusalem 91000, Israel.   *Etana Padan* • Department of Microbial and Molecular Ecology, Institute of Life Sciences, Hebrew University, Jerusalem 91904, Israel.

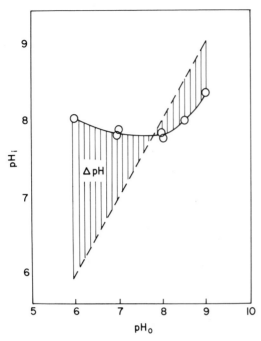

Figure 1. Internal pH of *E. coli* at various external pH values. Internal pH was measured in energized cells as described by Padan *et al.* (1976). The dashed line represents the theoretical equilibrium line. The difference between the equilibrium level and the actual experimental value is the ΔpH across the membrane.

internal pH is kept constant at pH 7.6 (Padan *et al.*, 1976; Navon *et al.*, 1977). The fact that the internal pH of *E. coli* remains constant as the external pH changes over a very wide range implies that a proton gradient, ΔpH, exists across the membrane that changes drastically with external pH. Thus, in *E. coli,* ΔpH is 1.6, basic inside, at an external pH of 6. When the external pH increases to 7.6, the ΔpH is reduced to 0 and the polarity is then reversed to being acidic inside and reaches 0.7 unit at an external pH of 9 (Fig. 1) (Padan *et al.*, 1976; Navon *et al.*, 1977). Understanding how this ΔpH across the cytoplasmic membrane is maintained and varies with external pH is therefore the goal of all those studying the mechanism of adaptation to pH.

2. PRIMARY PROTON PUMPS IN THE REGULATION OF CYTOPLASMIC pH

Primary proton pumps linked to electron transport, ATP hydrolysis, and photochemical reactions have been shown to exist in the cytoplasmic membrane of bacteria, and to play a major role in the energy conversion reactions that occur by means of the chemiosmotic mechanism suggested by Mitchell (Mitchell, 1968, 1979; Harold, 1978). Thus, both in intact *E. coli* cells (Padan *et al.*, 1976; Zilberstein *et al.*, 1979; Felle *et al.*, 1980) and in isolated membrane vesicles (Ramos *et al.*, 1976; Ramos and Kaback, 1977), these pumps have been shown to maintain an electrochemical gradient of protons, $\Delta\bar{\mu}_{H^+}$, comprised of both ΔpH and $\Delta\Psi$. The ΔpH in both experimental systems shows a similar marked sensitivity to external pH, and the systems exhibit similar propensity to maintain a constant internal pH at pH 7.5–7.6. When the proton pumps are inactivated, or in the presence of uncouplers, membranous energy conversions stop concomitant to rapid equilibration of protons across the membrane. Hence it has been deduced that in bacteria in addition to energy conversion, the proton pumps are also active in the maintenance of a constant internal pH (Fig. 1).

The proton pumps are outwardly directed throughout the external pH range. Accordingly the potential is always negative inside (Padan *et al.*, 1976; Zilberstein *et al.*, 1979; Felle *et al.*, 1980). This outward directionality of the pumps is consonant with the inward alkaline orientation of the ΔpHs observed up to an external pH of 7.6. The rate of respiration is unaltered through this external pH range, suggesting a constant rate of proton pumping. However, the increase in $\Delta\Psi$ with increasing pH (Zilberstein *et al.*, 1979; Felle *et al.*, 1980) is in accordance with the decrease in the ΔpH, for, as in many other systems, here also $\Delta\Psi$ limits the amounts of protons that can be pumped out. Indeed, when this $\Delta\Psi$ is collapsed by valinomycin in the presence of K^+, the ΔpH increases (Padan *et al.*, 1976). However, it is clear that this increase in $\Delta\Psi$ does not account for all the changes observed in the ΔpH, for (1) the increase in $\Delta\Psi$ with pH is less than the decrease in ΔpH and (2) most importantly, a $\Delta\Psi$ effect cannot explain what occurs beyond pH 7.6: $\Delta\Psi$ hardly changes and is still negative inside even when the ΔpH reverses and becomes acidic inside. In any event, the increase in $\Delta\Psi$ with pH is interesting and implies that an electrogenic movement of an ion (or ions) changes with external pH.

Another mechanism that may account for both the decrease in the ΔpH with external pH and the reversion of the ΔpH at alkaline pHs is the operation of a cation/proton antiporter or anion/proton symporter (Mitchell, 1968; Padan *et al.*, 1976; Skulachev, 1978).

Whatever mechanism is active, it appears that the signal for the internal pH control mechanism is the pH at the periplasmic side of the membrane. Thus, the ΔpH profile at different external pHs at the *periplasmic side* of the membrane is identical in both right-side-out and inverted membrane vesicles (Reenstra *et al.*, 1980).

3. A Na$^+$/H$^+$ ANTIPORTER AND REGULATION OF CYTOPLASMIC pH

An electroneutral Na^+/H^+ antiporter for example will be driven by the ΔpH initially formed by the primary proton pumps: Na^+ will be extruded and H^+ will be returned to the cell. If the rate of such an antiporter increases with pH, it could account for the decrease in the ΔpH observed. At an external pH of 7.6, it can explain the zero ΔpH, if the back leak of protons via the antiporter equates the primary proton extrusion rate (Padan *et al.*, 1976). However, the more alkaline the pH, the smaller the ΔpH becomes, and above pH 7.6 only $\Delta\Psi$ remains across the membrane. Under such conditions an electrogenic antiport activity must be postulated (Schuldiner and Fishkes, 1978).

Electrogenicity of the antiporter can be readily explained by assuming a certain stoichiometry between the Na^+ and the antiported protons. Furthermore, the relative rates of such an electrogenic antiporter and the primary electrogenic proton pumps may yield an overall net influx of protons whereas the net positive charge will still be directed outward. The exact stoichiometry of the antiporter is not yet known, and a hypothetical example is depicted in Fig. 2.

Evidence supporting the presence of the Na^+/H^+ antiporter has been presented in many energy-transducing membranes (Lanyi, 1979) including that of *E. coli* (West and Mitchell, 1974; Schuldiner and Fishkes, 1978; Beck and Rosen, 1979). Energy-dependent extrusion of Na^+ from cells has long been observed (Schultz and Solomon, 1961) and has been shown both in intact cells (West and Mitchell, 1974; Zilberstein *et al.*, 1980) and in right-side-out vesicles (Schuldiner and Fishkes, 1978) to be dependent on the proton gradient maintained across the membrane. In an anaerobic cell suspension of *E. coli*, after H^+

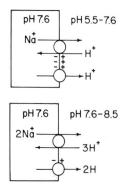

Figure 2. Changes in the stoichiometry of the Na$^+$/H$^+$ antiporter and in its rate relative to the proton pump play a role in regulating internal pH.

ions have been translocated outwards across the cytoplasmic membrane by a respiratory pulse (West and Mitchell, 1974), or inwards by a lactose pulse (Zilberstein *et al.*, 1979), reequilibration is catalyzed by the presence of Na$^+$. Similarly, Na$^+$ flux reduces the ΔpH (acidic inside) across right-side-out membrane vesicles (Schuldiner and Fishkes, 1978). The H$^+$/Na$^+$ antiporter is specific to Na$^+$ and Li$^+$ with apparent equal affinities of 3–7 mM (Schuldiner and Fishkes, 1978; Beck and Rosen, 1979). It should be emphasized that proton fluxes from the intact cell have been followed directly by a pH-sensitive electrode (West and Mitchell, 1974; Zilberstein *et al.*, 1979). However, in the membrane vesicles, ΔpH has been monitored by following the changes in fluorescence of the weak bases (Schuldiner and Fishkes, 1978; Beck and Rosen, 1979), which were shown in other systems to distribute across the membrane in response to the ΔpH (Rottenberg, 1979). These measurements in *E. coli* membrane vesicles are only qualitative in view of the more recent data (Reenstra *et al.*, 1980).

Dependency of the Na$^+$/H$^+$ antiporter activity on external pH has been observed with intact energy-depleted *E. coli* cells. After protons were translocated inwards, by a lactose pulse, their reequilibration was much faster at basic pHs and Na$^+$ markedly enhanced this movement (Zilberstein *et al.*, 1979). Furthermore, working with right-side-out isolated membrane vesicles, Schuldiner and Fishkes (1978) showed that the sodium efflux at low pH is driven by ΔpH, whereas at high pH it is driven by ΔΨ alone. These results may imply that the antiporter is electroneutral at low pH, whereas at alkaline pHs it changes stoichiometry with protons and becomes electrogenic. In studies on inverted isolated membrane vesicles, electrogenicity of the antiporter has been noted (Reenstra *et al.*, 1980).

4. MUTANTS IMPAIRED IN THE Na$^+$/H$^+$ ANTIPORTER CANNOT GROW AT ALKALINE pH

A more direct way to assess the participation of the Na$^+$/H$^+$ antiporter in the regulation of internal pH in *E coli* is to either inhibit or eliminate the antiporter and then study the consequences of such a situation on the internal pH. Unfortunately, no inhibitor of the antiporter has yet been found. Nevertheless, Zilberstein *et al.* (1980) developed a powerful selection method for isolation of mutants defective in the antiporter.

In addition to its suggested role in controlling internal pH, the antiporter must fulfill two other functions concerned with Na$^+$ metabolism, as no primary Na$^+$ pumps have been documented in *E. coli*. The antiporter driven by the ΔpH lowers the internal sodium concentration while at the same time forming a sodium gradient directed inwards across the cytoplasmic membrane. The Na^+ gradient thus generated may be used as an energy source to drive other reactions. Indeed, several symport systems have been described in bacteria in which sodium is the symported ion (Lanyi, 1979). In *E. coli*, two such systems have been documented: the melibiose (Tsuchiya *et al.*, 1977) and glutamate (Halpern *et al.*, 1973; MacDonald *et al.*, 1977) uptake systems.

This last function of the antiporter in allowing utilization of both glutamate and melibiose affords very powerful enrichment and selection procedures. After mutagenesis, penicillin enrichment can be performed in the presence of both substrates. Mutated organisms that are simultaneously defective in growth on both carbon sources can then be isolated. This phenotype is most likely impaired in a common function, indispensable for utilization of both substrates, which is the Na^+/H^+ antiporter activity. Several such mutants have been isolated, one of which, DZ3, has been further characterized (Zilberstein *et al.*, 1980).

The mutant DZ3 of *E. coli* is impaired in growth on melibiose and glutamate. Conjugation studies showed that the defect is a result of a mutation in a single locus designated *phs*, which is linked to the *metB* locus and is halfway between the far apart genes *gltC* and *melA,B*, which are the structural genes of the respective carrier proteins of glutamate and melibiose transport (Zilberstein *et al.*, 1980). Indeed, DZ3 conjugates with *gltC* to give wild-type recombinants (unpublished). A preliminary analysis of the product of the *phs* locus on SDS-polyacrylamide gels indicates that a polypeptide with an apparent molecular weight of 24,000 is missing in strain DZ3.

The mutant is impaired in its ability to extrude sodium actively and is therefore most likely impaired in the antiporter. Strikingly, the adaptability to alkaline pHs has also been lost: The wild type grows almost optimally up to pH 8.6, but the mutant is drastically and progressively inhibited in its growth with increasing pH above 8.2 (Zilberstein *et al.*, 1980). Hence, the wild-type allele of the *phs* locus in *E. coli* is responsible for maintaining the sodium gradient across the cell membrane needed for the operation of the Na^+ symporters and at the same time for growth at alkaline pH. Both capacities do, however, reappear in recombinants that are capable of growth on glutamate and melibiose.

Although the mutant DZ3 is defective only in the *phs* locus, interestingly, it is impaired in the transport of both melibiose and glutamate (Zilberstein, Schuldiner, and Padan, unpublished results). The initial rate and the steady state of transport of both substrates are low without addition of Na^+ and are similar to those observed in the wild type. However, whereas Na^+ increases the initial rates four- to sixfold in the wild type, it only increases the transport rates of the mutant twofold at most. Furthermore, whereas Na^+ reduces the K_m for glutamate from 100 μM to 13 μM, the K_m for glutamate of the mutant remains 118 μM even in the presence of Na^+. It should also be noted that the K_m for Na^+ in the mutant is 200 μM, tenfold higher than that of the wild type. In all cases, the V_{max} values are hardly affected.

The diminished steady-state levels of the Na^+-dependent accumulation are expected to be secondary to the loss of Na^+/H^+ antiport activity. Such defects caused by impaired Na^+ extrusion should be repaired by addition of monensin, which, like the antiporter, catalyzes Na^+/H^+ exchange. However, this ionophore did not improve the impaired transport activities of the mutant. Furthermore, passive, carrier-mediated efflux of glutamate down the glutamate chemical gradient was also impaired in the mutant as compared to the wild type even in the absence of imposed ion gradients. Hence, it is suggested that there may be a more direct relationship between the Na^+-translocating antiport and symport systems. A similar pleiotropic effect of a mutation in the Na^+/H^+ antiport has been observed with the alkalophile *Bacillus alcalophilus*, and the model suggested by Guffanti *et al.* (1981) for the alkalophile may thus also hold for *E. coli*.

The Na^+/H^+ antiporter and the individual symporters Na^+/glutamate and Na^+/melibiose are depicted as oligomers consisting of at least two dissimilar subunits. One subunit, common to each "translocation complex" and coded presumably by the locus *phs*, is responsible for Na^+ translocation. The other subunits, encoded by a series of unrelated genes, are unique to each complex and are involved in the translocation of specific substrates. It is obvious from this scheme that a mutation in the gene *phs* coding for the Na^+ subunit would result in a pleiotropic defect in all Na^+-translocating systems. On the other hand, a mutation in the gene coding for the substrate-specific subunits would lead to unique defects in individual symporters. This model is somewhat analogous to the suggestion of Hong (1977) that the H^+/substrate symport systems in *E. coli* share a common H^+-translocating subunit. A mutation located outside the structural genes coding for the carrier proteins pleiotropically affects many transport systems coupled to $\Delta\bar{\mu}_{H^+}$. Finally, chemical modification of histidine in *E. coli*, which affects the coupling of lactose transport to $\Delta\bar{\mu}_{H^+}$, similarly "uncouples" many transport systems that are driven by the electrochemical proton gradient (Padan *et al.*, 1979).

5. A K^+/H^+ ANTIPORTER IN REGULATION OF CYTOPLASMIC pH

A potassium/proton antiporter has also been suggested to play a role in regulation of internal pH in *E. coli* (Brey *et al.*, 1978, 1980; Plack and Rosen, 1980). This antiporter appears electroneutral over pH 7–9 and is less specific than the Na^+/H^+ antiporter for it antiports K^+, Na^+, Li^+, Rb^+, and Tl^+. The activity of this antiporter is optimal at about 8 (Brey *et al.*, 1980). It has been suggested that the pronounced alkaline pH optimum of this antiporter will regulate its function in accordance with the external pH. However, it should be pointed out that this pH optimum has been obtained with inverted membrane vesicles of which the membrane side facing the medium is the cytoplasmic side, and as shown by Reenstra *et al.* (1980), it is the periplasmic side of the membrane that is sensitive to external pH.

Alkaline-sensitive mutants have been selected (Plack and Rosen, 1980) by mutagenesis followed by penicillin killing at pH 8.3. One such strain was found to lack K^+–H^+ exchange. At neutral pH, the doubling time of the mutant was similar to the wild type, but at pH 8.3 it increased over fivefold. These results are consistent with the suggestion that the K^+/H^+ antiporter contributes to the regulation of cytosolic pH in *E. coli*.

6. BOTH THE K^+/H^+ AND Na^+/H^+ ANTIPORTERS REGULATE INTERNAL pH

Several reasons led Brey *et al.* (1980) to believe that the K^+/H^+ antiporter is the main antiporter controlling the internal pH of *E. coli*. Potassium is an absolute requirement for *E. coli* growth. The well-known (Rhoads *et al.*, 1976) primary pump for potassium, K^+-ATPase, will use ATP to replenish the internal K^+ concentration, depleted by the antiporter. Being electroneutral, the K^+/H^+ antiporter will affect only the ΔpH component of the $\Delta\bar{\mu}_{H^+}$ across the membrane and may therefore even increase $\Delta\Psi$ while decreasing ΔpH. However, at alkaline external pH when only $\Delta\Psi$ exists across the membrane (Zilberstein *et al.*, 1979), the immediate driving force for reimporting protons via such an electroneutral antiporter must be only the outwardly directed K^+ gradient. Hence, this antiporter will act efficiently only when the external K^+ concentration is below a certain limit.

On the other hand, the operation of the Na^+/H^+ antiporter appears to be directly linked to the proton gradient as no primary Na^+ pump is known. Furthermore, this antiporter, while governing internal pH, maintains both a low internal Na^+ concentration as well as the inwardly directed Na^+ gradient that is needed for the uptake of certain substrates (Lanyi, 1979). Although a Na^+ requirement for growth has not been shown in *E. coli*, it should be emphasized that almost every commercially available compound as well as the glassware are contaminated with Na^+ salts that can yield up to 0.1 mM Na^+ in the medium (Mandel *et al.*, 1980). Therefore, the possibility that bacteria require a low Na^+ concentration has not yet been ruled out.

In any event, the participation of the K^+/Na^+ and Na^+/H^+ antiporters in regulating internal pH is not mutually exclusive, and their relative contributions might change under varying environmental conditions. For instance, at a high external K^+ concentration, the Na^+/H^+ antiporter might function exclusively, whereas the K^+/H^+ antiporter would dominate when external Na^+ is low.

7. CONTROL OF INTERNAL pH IN BACTERIA LIVING IN DIFFERENT pH RANGES

The capacity of maintaining a constant cytoplasmic pH over a wide range of external pH is not unique to *E. coli*. Data were obtained for other bacteria that prefer the neutral pH range for growth. These include *Streptococcus faecalis* (Harold *et al.*, 1970; Harold and Papineau, 1972), *Micrococcus lysodeikticus* (Friedberg and Kaback, 1980), *Bacillus subtilis* (Khan and Macnab, 1980), and *Halobacterium halobium* (Bakker *et al.*, 1976). Clearly the $\Delta\bar{\mu}_{H^+}$ generated by the proton pumps of many bacteria react to external pH in a pattern identical to that observed in *E. coli* (Padan *et al.*, 1976; Zilberstein *et al.*, 1979): $\Delta\Psi$ increases with pH while ΔpH drastically decreases. The resulting internal pH remains constant at around 7.6 over a wide range of external pH.

If the mechanism described in *E. coli* is the general mechanism chosen by nature for the control of internal pH, it is tempting to suggest that acidophiles and alkalophiles possess essentially the same mechanism, when drawn to the extreme at either the acidic or the alkaline range. This does appear to be the case. In the acidophiles, the membrane potential decreases with lowering external pH, as for *E. coli*. It is even reversed to being positive inside at pH 2. The ΔpH increases to the maximal values, of 4.5 units, observed in biological systems. The resulting internal pH remains constant at 6.5 throughout the extreme acidic range (Krulwich *et al.*, 1978; Cox *et al.*, 1979).

As with *E. coli* in the alkaline range, in the alkalophiles, the membrane potential increases with pH to a maximum of about 130 mV and the ΔpH reverses (acidic inside), reaching a maximal reversed ΔpH of about 2 pH units (Guffanti *et al.*, 1978, 1980). The internal pH is again kept constant throughout the alkaline pH range but at a higher level, pH 9.5. An electrogenic Na^+/H^+ antiporter has been demonstrated in these bacteria (Guffanti *et al.*, 1978; Mandel *et al.*, 1980). Furthermore, mutants have been isolated by Krulwich and collaborators (Krulwich *et al.*, 1979; Guffanti *et al.*, 1980) that lack the Na^+/H^+ antiporter activity and like the *E. coli* mutant have lost the capacity to grow at the extreme alkaline pH range.

Two very interesting bioenergetic problems must be pointed out: in the acidophiles living only with ΔpH, transport cannot be conducted by the $\Delta\Psi$-dependent mechanism, and in the alkalophiles at pH 10–11 life proceeds essentially without $\Delta\bar{\mu}_{H^+}$. Finally, with respect to internal pH, three groups of microorganisms may be discerned: the acidophiles. neutrophiles, and alkalophiles with internal pHs of 6.5, 7.6, and 9.5, respectively.

ACKNOWLEDGMENT. Supported by a grant from the United States–Israel Binational Foundation (BSF). The authors thank Dan Zilberstein for critically reading the review and for many stimulating discussions.

REFERENCES

Abeliovich, A., and Azov, J. (1976). *Appl. Env. Microbiol.* **31**, 801–806.

Bakker, E. P., Rottenberg, H., and Caplan, S. R. (1976). *Biochim. Biophys. Acta* **440**, 557–572.

Beck, J. C., and Rosen, B. P. (1979). *Arch. Biochem.* **194**, 208–214.

Brey, R. N., Beck, J. C., and Rosen, B. P. (1978). *Biochem. Biophys. Res. Commun.* **83**, 1588–1594.

Brey, R. N., Rosen, B. P., and Sorensen, E. N. (1980). *J. Biol. Chem.* **255**, 39–44.

Brock, T. D. (1969). *Sym. Soc. Gen. Microbiol.* **19**, 15–41.

Cox, J. C., Nicholls, D. G., and Ingledew, W. J. (1979). *Biochem. J.* **178**, 195–200.

Felle, H., Porter, J. S., Slayman, C. L., and Kaback, H. R. (1980). *Biochemistry* **19**, 3585–3590.

Friedberg, I., and Kaback, I. (1980). *J. Bacteriol.* **142**, 651–658.

Guffanti, A. A., Susman, P., Blanco, R., and Krulwich, T. A. (1978). *J. Biol. Chem.* **253**, 708–715.

Guffanti, A. A., Blanco, R., Benenson, R. A., and Krulwich, T. A. (1980). *J. Gen. Microbiol.* **119**, 79–86.

Guffanti, A. A., Cohn, D. E., Kaback, H. R., and Krulwich, T. A. (1981). *Proc. Natl. Acad. Sci. USA*, **78**, 1481–1484.

Halpern, Y. S. H., Barash, H., Dover, S., and Druck, K. (1973). *J. Bacteriol.* **114**, 53–58.

Hargreaves, J. W., Lloyd, E. J. H., and Whitton, B. A. (1975). *Freshwater Biol.* **5**, 563–576.

Harold, F. M. (1978). In *The Bacteria* (I. C. Gunsalus, ed.), Vol. 6, 463–521, Academic Press, New York.

Harold, F. M., and Papineau, D. (1972). *J. Membr. Biol.* **8**, 27–44.

Harold, F. M., Pavlasova, E., and Baarda, J. R. (1970). *Biochim. Biophys. Acta* **196**, 235–244.

Hong, J. S. (1977). *J. Biol. Chem.* **252**, 8582–8588.

Imhoff, J. F. Sahl, H. G., Soliman, G. S. H., and Truper, H. G. (1979). *Geomicrobiol. J.* **1**, 219–234.

Khan, S., and Macnab, R. M. (1980). *J. Mol. Biol.* **138**, 599–614.

Krulwich, T. A., Davidson, L. F., Filip, S. J., Jr., Zuckerman, R. S., and Guffanti, A. A. (1978). *J. Biol. Chem.* **253**, 4599–4603.

Krulwich, T. A., Mandel, K. G., Bornstein, R. F., and Guffanti, A. A. (1979). *Biochem. Biophys. Res. Commun.* **91**, 58–62.

Langworthy, T. A. (1978). In *Microbial Life in Extreme pH Values* (D. J. Kushner, ed.), pp. 279–315, Academic Press, New York.

Lanyi, J. K. (1979). *Biochim. Biophys. Acta* **559**, 377–397.

MacDonald, R. E., Lanyi, J. K., and Greene, R. V. (1977). *Proc. Natl. Acad. Sci. USA* **74**, 3167–3170.

Mandel, K. G., Guffanti, A. A., and Krulwich, T. A. (1980). *J. Biol. Chem.* **255**, 7391–7396.

Mitchell, P. (1968). In *Chemiosmotic Coupling and Energy Transduction,* Glynn Research, Bodmin.

Mitchell, P. (1979). *Eur. J. Biochem.* **95**, 1–20.

Navon, G., Ogawa, S., Shulman, R. G., and Yamane, T. (1977). *Proc. Natl. Acad. Sci. USA* **74**, 888–891.

Padan, E., Zilberstein, D., and Rottenberg, H. (1976). *Eur. J. Biochem.* **63**, 533–541.

Padan, E., Patel, L., and Kaback, H. R. (1979). *Proc. Natl. Acad. Sci. USA* **76**, 6221–6225.

Plack, R. H., Jr., and Rosen, B. P. (1980). *J. Biol. Chem.* **255**, 3824–3825.

Ramos, S., and Kaback, H. R. (1977). *Biochemistry* **16**, 848–854.

Ramos, S., Schuldiner, S., and Kaback, H. R. (1976). *Proc. Natl. Acad. Sci. USA* **73**, 1892–1896.

Reenstra, W. W., Patel, L., Rottenberg, H., and Kaback, H. R. (1980). *Biochemistry* **19**, 1–9.

Rhoads, D. B., Waters, F. B., and Epstein, W. (1976). *J. Gen. Physiol.* **67**, 325–341.

Rottenberg, H. (1979). *Methods Enzymol.* **55**, 547–569.

Schuldiner, S., and Fishkes, H. (1978). *Biochemistry* **17**, 706–711.

Schultz, S. G., and Solomon, A. K. (1961). *J. Gen. Physiol.* **45**, 355–369.

Skulachev, V. P. (1978). *FEBS Lett.* **87**, 171–179.

Talling, J. F., Wood, R. B., Prosser, M. V., and Baxter, R. M. (1973). *Freshwater Biol.* **3**, 53–76.

Tindall, B. J., Mills, A. A., and Grant, W. D. (1980). *J. Gen. Microbiol.* **116,** 257–260.
Tsuchiya, T., Raven, J., and Wilson, T. H. (1977). *Biochem. Biophys. Res. Commun.* **76,** 26–31.
West, I. C., and Mitchell, P. (1974). *Biochem. J.* **144,** 87–90.
Zilberstein, D., Schuldiner, S., and Padan, E. (1979). *Biochemistry* **18,** 669–673.
Zilberstein, D., Padan, E., and Schuldiner, S. (1980). *FEBS Lett.* **116,** 177–180.

98

Bioenergetic Problems of Alkalophilic Bacteria

Terry Ann Krulwich

1. INTRODUCTION

The problems of obligately alkalophilic bacteria pose a challenge with respect to chemiosmotic principles. There is considerable support for Mitchell's proposal that an electrochemical gradient of protons ($\Delta\bar{\mu}_{H^+}$, outside acid and positive for a bacterial cell) is involved in the energization of many solute transport systems, ATP synthesis, and several other bioenergetic processes (e.g., Mitchell, 1968; Harold, 1977; Hinkle and McCarty, 1978). Yet, as outlined prospectively by Garland (1977), obligate alkalophiles would not be able to maintain the usual $\Delta\bar{\mu}_{H^+}$. Such organisms grow optimally at pH 10 to 11, often exhibit growth up to pH 12, and cannot grow below pH 8.5. Clearly their cytoplasmic pH would have to be lower than the external pH under optimal growth conditions. Thus, if primary proton extrusion occurs, concomitantly with respiration, as it does in other bacteria and in mitochondria, the alkalophiles would require some secondary ion movements to achieve a relatively acidified cytoplasm. Moreover, the "reversed" chemical gradient of protons (ΔpH) thus produced would lower the net total $\Delta\bar{\mu}_{H^+}$ that is generated by alkalophiles. Perhaps they could compensate with an elevation of the second component of the $\Delta\bar{\mu}_{H^+}$, the transmembrane electrical potential ($\Delta\Psi$). If so, an extraordinarily high $\Delta\Psi$ would probably be required, and its generation would be quite interesting. On the other hand, if such a compensatory relationship between a reversed ΔpH and a correspondingly high $\Delta\Psi$ were not feasible, then the problem of a low $\Delta\bar{\mu}_{H^+}$ would exist vis-à-vis transport processes, ATP synthesis, etc. Several groups of investigators (Horikoshi, 1971; Ohta *et al.*, 1975; Guffanti *et al.*, 1980) have isolated and characterized alkalophilic species of *Bacillus* with a view toward using these organisms as a system for studying bioenergetics in *extremis*. The results thus far suggest that alkalophiles possess specific adaptations of

Terry Ann Krulwich • Department of Biochemistry, Mount Sinai School of Medicine of the City University of New York, New York, New York 10029.

interest in the respiratory chain, in cation/proton antiporters, and perhaps in the ATP synthetic enzyme.

2. THE ELECTROCHEMICAL PROTON GRADIENT IN ALKALOPHILES

The variation in the $\Delta\bar{\mu}_{H^+}$ as a function of the external pH has been studied in two obligately alkalophilic bacilli (Guffanti *et al.*, 1978, 1980) and in two alkaline-tolerant bacilli (Guffanti *et al.*, 1979a, 1980). The alkaline-tolerant strains differ from the obligate alkalophiles by their ability to grow in the neutral range of pH and their inability to grow above pH 9.5. In all four strains, the cytoplasmic pH was maintained at pH 9.5 or below. The alkalophilic species, *Bacillus alcalophilus* and *B. firmus* RAB, could maintain this internal pH at external pHs as high as pH 11.5. Alkaline-tolerant species, by contrast, were apparently unable to generate a "reversed" ΔpH. Their upper pH limit for growth, pH 9.5, probably corresponds to the highest cytoplasmic pH that is compatible with viability.

The $\Delta\Psi$ exhibited by L-malate-grown whole cells of *B. alcalophilus* increased from -84 mV at pH 9.0 to -152 mV at pH 11.5 (Guffanti *et al.*, 1978); a generally similar pattern was observed in *B. firmus* RAB (Guffanti *et al.*, 1980). In neither alkalophile did the increases in the $\Delta\Psi$ offset the progressively more reversed ΔpH, as the external pH was raised. Therefore, the total $\Delta\bar{\mu}_{H^+}$ decreased at the more alkaline pHs, and was between -15 and -80 mV at pHs at which the bacteria grow.

Isolated membrane vesicles from L-malate-grown cells of *B. alcalophilus* were used to study the ions that are involved in producing the observed ΔpH and $\Delta\Psi$ patterns (Mandel *et al.*, 1980). Vesicles that were prepared in the absence of K^+ and Na^+ generated a conventional $\Delta\bar{\mu}_{H^+}$, outside *acid* and positive, upon energization with ascorbate/N,N,N',N'-tetramethyl-p-phenylenediamine (TMPD). The presence of K^+ in the preparations resulted in a $\Delta\bar{\mu}_{H^+}$ that was entirely composed of the $\Delta\Psi$, with ΔpH $= 0$. Addition of Na^+ resulted in generation of a $\Delta\Psi$ together with a ΔpH, inside acid; i.e., in the presence of Na^+, the $\Delta\bar{\mu}_{H^+}$ pattern of vesicles resembled that of whole cells. Interestingly, vesicles prepared from lactose-grown cells exhibited $\Delta\bar{\mu}_{H^+}$ patterns, upon energization, that were completely comparable to those from L-malate-grown cells (Mandel, 1980). Whole lactose-grown cells, by contrast, exhibited a generally lower cytoplasmic pH than L-malate-grown cells (Guffanti *et al.*, 1979b). Growth on lactose was optimal at pH 8.5 to 9.5, and the magnitude of $\Delta\Psi$ increased with increasing external pH only up to pH 10.0. The differences between the two growth substrates, manifest in whole cells but not in vesicles, were interpreted as reflecting a somewhat fermentative utilization of lactose.

3. CATION/PROTON ANTIPORTERS

The $\Delta\bar{\mu}_{H^+}$ patterns observed in vesicles from *B. alcalophilus* suggested that these membranes possess a K^+/H^+ antiporter and an Na^+/H^+ antiporter, the latter being involved in acidifying the intravesicular space or cytoplasm (Mandel *et al.*, 1980). Everted membrane vesicles from L-malate-grown *B. alcalophilus* catalyzed $\Delta\Psi$-dependent uptake of $^{22}Na^+$ concomitant with H^+ extrusion. A K_m of about 0.7 mM for Na^+ was observed. A nonalkalophilic mutant straing lacked these activities (Mandel *et al.*, 1980). The nonalkalophilic strain had also lost the ability to grow above pH 8.5 while gaining the ability to grow in the neutral range of pH (Krulwich *et al.*, 1979). A dozen separate nonalkalophilic isolates were found to have lost apparent Na^+/H^+ antiport activity. Reversion to alkalophily was invariably accompanied by return of the antiporter activity. Parallel nonalkalophilic strains have been isolated from *B. firmus* RAB (Guffanti *et al.*, 1980). Thus, the Na^+/H^+ antiporter has been strongly implicated in the regulation of cytoplasmic pH in alkalophiles. The role of the K^+/H^+ antiporter is not yet clear. Nor is the mechanism for the increase in the $\Delta\Psi$ with increasing external pH understood. Any explanation of the latter observation will have to take into account the consumption of energy involved in the activity of the Na^+/H^+ antiporter.

4. SOLUTE TRANSPORT

The problem of a low net $\Delta\bar{\mu}_{H^+}$ vis-à-vis solute transport has been examined in several alkalophiles. Transport of β-galactosides by lactose-grown *B. alcalophilus* was found to depend upon ATP per se rather than upon the $\Delta\bar{\mu}_{H^+}$ (Guffanti *et al.*, 1979b). In view of the potential for fermentative production of ATP from lactose, this transport system thus bypasses the problem of a low $\Delta\bar{\mu}_{H^+}$.

For a variety of nonfermentable solutes, Na^+/solute symport systems have been found in alkalophilic bacteria. Such a symport system has implicated in the uptake of α-aminoisobutyric acid by several alkalophiles (Koyama *et al.*, 1976; Kitada and Horikoshi, 1977; Guffanti *et al.*, 1978, 1980). The relevant energetic parameter for $\Delta\Psi$-dependent Na^+/solute symport is the sodium-motive force rather than the $\Delta\bar{\mu}_{H^+}$; the reversed ΔpH would therefore be energetically inconsequential for such systems.

Recently, Na^+/solute symport of α-aminoisobutyric acid, L-malate, L-aspartate, L-serine, and L-leucine has been demonstrated in membrane vesicles from L-malate-grown *B. alcalophilus* (Guffanti *et al.*, 1981b). Interestingly, the nonalkalophilic mutant strain pleiotropically loses Na^+ coupling for all of these transport systems; some solute transport remains, at neutral pH, but it is no longer coupled to Na^+. The Na^+ coupling is restored for all the transport systems in revertants of the mutant that have also regained the Na^+/H^+ antiporter and the alkalophilic growth characteristics. In the nonalkalophilic mutant, no Na^+ coupling occurs even at every initial time points or in the presence of monensin, which should restore the capacity for Na^+ efflux (Krulwich *et al.*, 1979; Guffanti *et al.*, 1981b). Passive solute efflux from unenergized vesicles has recently been employed to probe various properties of solute transport, including cation coupling (Kaczorowski and Kaback, 1979; Kaczorowski *et al.*, 1979; Cohn and Kaback, 1980). In *B. alcalophilus,* efflux of α-aminoisobutyric acid, L-malate, and L-aspartate from wild-type vesicles was Na^+ dependent, whereas efflux of these solutes from vesicles from the nonalkalophilic strain was Na^+ independent (Guffanti *et al.*, 1981b). Taken together, the data on the nonalkalophilic mutant suggest a relationship between Na^+-translocating antiporters and symporters in *B. alcalophilus*. One possible model for this relationship is a common subunit, as depicted in Fig. 1. If such a subunit exists, it may well be a major membrane protein in *B. alcalophilus* and be amenable to detailed characterization. It will also be of interest to examine the energy coupling of transport in the nonalkalophilic strain. Preliminary data suggest that H^+/solute symport occurs (Krulwich *et al.*, 1979), thus representing a mutational change in coupling ion. Tsuchiya and Wilson (1978) have shown that either Na^+ or H^+ can normally be cotransported with melibiose in *Escherichia coli*, but the mechanism underlying this flexibility is unknown.

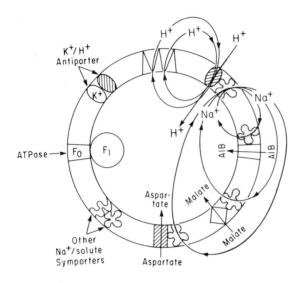

Figure 1. A model indicating the cation-translocating processes in the membranes from *Bacillus alcalophilus*. Primary proton pumping is shown to occur by the respiratory chain. An Na^+/H^+ antiporter catalyzes net proton uptake in exchange for Na^+. Na^+/solute symport systems are suggested to possess Na^+-translocating subunit in common with that of the antiporter.

5. SPECIAL PROBLEMS RELATED TO ATP SYNTHESIS

It was possible that the problem of a low $\Delta\bar{\mu}_{H^+}$ in connection with ATP synthesis would also be solved by energetic coupling to some cation other than protons. However, evidence to date indicates that alkalophiles utilize a BF_1F_0 that translocates protons concomitantly with ATP synthesis, albeit at low $\Delta\bar{\mu}_{H^+}$ values. Koyama *et al.* (1980) have purified the BF_1 from an alkalophilic *Bacillus* and found it to possess a conventional subunit composition. In ADP + P_i-loaded vesicles from *B. alcalophilus*, synthesis of ATP has been demonstrated upon energization by ascorbate/TMPD (Guffanti *et al.*, 1981a). ATP synthesis was not dependent upon added Na^+, was DCCD sensitive, and occurred concomitantly with inward proton movements. The phosphate potentials (ΔGp) observed were much higher than could be accounted for by energization of ATP synthesis by the observed $\Delta\bar{\mu}_{H^+}$ using $2H^+$/ATP. In this respect, the *B. alcalophilus* poses an energetic dilemma that has been raised by others (e.g., Pick *et al.*, 1974; Azzone *et al.*, 1978; Decker and Lang, 1978; Deutsch and Kula, 1978; Kell *et al.*, 1978; Michel and Oesterhelt, 1980). The resolution of this dilemma will be of general importance with respect to oxidative phosphorylation.

6. PROPERTIES OF THE RESPIRATORY CHAIN

The alkalophiles clearly have unusually high energy demands. It was therefore notable that isolated membranes from two obligate alkalophiles, *B. alcalophilus* and *B. firmus* RAB, have extraordinarily high concentrations of respiratory chain components (e.g., 5.5 nmoles cytochrome heme/mg membrane protein) whereas their nonalkalophilic mutant derivatives have much lower levels (Lewis *et al.*, 1980). The *b*- and *c*-type cytochromes showed a greater reduction, relative to cytochrome *a*, in the nonalkalophilic strains. A detailed study of the respiratory chain of *B. alcalophilus* and its nonalkalophilic derivative has been conducted (Lewis *et al.*, 1981). The results indicate that the wild-type strain possesses cytochromes with the following midpoint potentials at pH 9.0: cytochrome aa_3, +240 mV; cytochrome *b*, +20, −120 mV, −240 mV, and −320 mV; and cytochrome

c, $+70$ mV. Membranes of this strain also contained a Rieske iron–sulfur protein with a midpoint potential of $+75$ mV at pH 9.0 and several other iron–sulfur proteins as well as ubiquinone, The nonalkalophilic strain appeared to have qualitatively fewer b-type cytochromes as well as generally lower levels of respiratory chain components. Yet, both whole cells and vesicles of the two strains exhibit comparable respiratory rates at their respective pH optima (Lewis *et al.*, 1980, 1981). It is an intriguing possibility that the obligate alkalophiles possess a respiratory chain that is adapted to transduce energy with unusual efficiency (e.g., pump more protons?). Jones *et al.* (1975) have proposed that the H^+/O ratios bacterial respiratory chains depend upon the array of cytochromes and may vary greatly between species. The properties of the respiratory chain will relate importantly to the problem of ATP synthesis in the alkalophiles. In view of the greatly reduced level of respiratory chain components in the nonalkalophilic mutants, the alkalophiles may also offer a productive system for study of the regulation of the electron transport chain.

ACKNOWLEDGMENT. The work from the author's laboratory was supported by NSF Grant PCM7810213.

REFERENCES

Azzone, G. F., Pozzan, T., and Massari, S. (1978). *Biochim. Biophys. Acta* **501**, 307–316.

Cohn, D. E., and Kaback, H. R. (1980). *Biochemistry* **19**, 4237–4243.

Decker, S. J., and Lang, D. R. (1978). *J. Biol. Chem.* **253**, 6738–6743.

Deutsch, C., and Kula, T. (1978). *FEBS Lett.* **87**, 145–151.

Garland, P. B. (1977). *Symp. Soc. Gen. Microbiol.* **27**, 1–21.

Guffanti, A. A., Susman, P., Blanco, R., and Krulwich, T. A. (1978). *J. Biol. Chem.* **253**, 708–715.

Guffanti, A. A., Monti, L. G., Blanco, R., Ozick, D., and Krulwich, T. A. (1979a). *J. Gen. Microbiol.* **112**, 161–169.

Guffanti, A. A., Blanco, R., and Krulwich, T. A. (1979b). *J. Biol. Chem.* **254**, 1033–1037.

Guffanti, A. A., Blanco, R., Benenson, R. A., and Krulwich, T. A. (1980). *J. Gen. Microbiol.* **119**, 79–86.

Guffanti, A. A., Bornstein, R. F., and Krulwich, T. A. (1981a). *Biochim. Biophys. Acta,* **635**, 619–630.

Guffanti, A. A., Cohn, D. E., Kaback, H. R., and Krulwich, T. A. (1981b). *Proc. Natl. Acad. Sci. USA,* **78**, 1481–1484.

Harold, F. M. (1977). *Curr. Top. Bioenerg.* **6**, 84–149.

Hinkle, P. C., and McCarty, R. E. (1978). *Sci. Am.* **238**, 104–123.

Horikoshi, K. (1971). *Agric. Biol. Chem,* **35**, 1407–1414.

Jones, C. W., Brice, J. M., Downs, A. J., and Drozd, J. W. (1975). *Eur. J. Biochem.* **52**, 265–271.

Kaczorowski, G. J., and Kaback, H. R. (1979). *Biochemistry* **18**, 3691–3697.

Kaczorowski, G. J., Robertson, D. E., and Kaback, H. R. (1979). *Biochemistry* **18**, 3697–3704.

Kell, D. B., John, P., and Ferguson, S. U. (1978). *Biochem. J.* **174**, 257–266.

Kitada, M., and Horikoshi, K. (1977). *J. Bacteriol.* **131**, 784–788.

Koyama, N., Kiyomiya, A., and Nosoh, Y. (1976). *FEBS Lett.* **72**, 77–78.

Koyama, N., Koshiya, K., and Nosoh, Y. (1980). *Arch. Biochem. Biophys.* **199**, 103–109.

Krulwich, T. A. Mandel, K. G., Bornstein, R. F., and Guffanti, A. A. (1979). *Biochem. Biophys. Res. Commun.* **91**, 58–62.

Lewis, R. J., Belkina, S., and Krulwich, T. A. (1980). *Biochem. Biophys. Res. Commun.* **95**, 857–863.

Lewis, R. J., Prince, R., Dutton, B. L., Knaff, D., and Krulwich, T. A. (1981). *J. Biol. Chem.* **256**, 10543–10549.

Mandel, K. G. (1980). Ph.D. dissertation, Mount Sinai School of Medicine of the City University of New York.

Mandel, K. G., Guffanti, A. A., and Krulwich, T. A. (1980). *J. Biol. Chem.* **255**, 7391–7396.

Michel, H., and Oesterhelt, D. (1980). *Biochemistry* **19**, 4615–4619.

Mitchell, P. (1968). *Chemiosmotic Coupling and Energy Transduction,* Glynn Research, Bodmin.

Ohta, K., Kiyomiya, A., Koyama, N., and Nosoh, Y. (1975). *J. Gen. Microbiol.* **86**, 259–266.

Pick, U., Rottenberg, H., and Avron, M. (1974). *FEBS Lett.* **48**, 32–36.

Tsuchiya, T., and Wilson, T. H. (1978). *Membr. Biochem.* **2**, 63–79.

99

Bacterial Periplasmic Binding Proteins

Robert Landick and Dale L. Oxender

The cell envelope of gram-negative bacteria includes inner and outer membranes with the periplasmic space in between. The transport systems for many nutrients in these organisms include components in both the periplasm and the inner membrane. In some cases, e.g., maltose, transport components in the outer membrane are also involved. The scope of this review will be limited to a progress report on our understanding of the nature and the role of the periplasmic components that serve as receptors or binding proteins (BPs)* for the various nutrients. Bacterial periplasmic BPs for a large variety of carbohydrates, amino acids, inorganic ions, and enzyme cofactors have been isolated and characterized. More complete reviews on these BPs have appeared recently (Oxender and Quay, 1976; Wilson and Smith, 1978; Furlong and Schellenberg, 1980).

1. PERIPLASMIC BINDING PROTEIN FUNCTION

1.1. Substrate Binding

Both carbohydrate and amino acid BPs have extremely high affinities for their ligands with K_ds of 10^{-7} to 10^{-6} M. It is now clear that nondenaturing equilibrium dialysis is insufficient for removal of bound ligand, which is usually retained during purification and even crystallization. Guanidinium HCl (1–5 M) can partially denature BPs and allow ligand removal. Binding activity is usually regained upon renaturation. This procedure has been used to prepare Gal-BP and Ara-BP for stopped-flow binding studies (Miller *et al.*, 1979, 1980). These experiments have demonstrated unequivocally that Gal-BP binds both glucose and galactose with K_ds of 4×10^{-8} and 1.4×10^{-7} M, respectively. Thus, the Gal-

---

*Abbreviations: BP, binding protein; Ara-BP, arabinose binding protein; Gal-BP, galactose binding protein; His-BP, histidine binding protein; LAO-BP, lysine-arginine-ornithine binding protein; LIV-BP, leucine-isoleucine-valine binding protein; LS-BP, leucine-specific binding protein; Mal-BP, maltose binding protein.

Robert Landick and Dale L. Oxender • Department of Biological Chemistry, University of Michigan, Ann Arbor, Michigan 48109.

BP actually has a much higher affinity for glucose than galactose. When the same technique was applied to the Ara-BP, the pseudo-first-order rate constant for arabinose binding was found to decrease as the fraction of Ara-BP containing bound arabinose increased. Two explanations of this variable binding, which has also been observed with the LIV-BP (Amanuma *et al.*, 1976), are possible. BP molecules may be loosely associated and thus capable of allosteric interactions, which could account for changes in K_ds. Alternatively, some as yet undetected interaction between the BP and its ligand may cause the observed changes in binding affinity. Further characterization of this phenomenon may shed light on the possible physiological role of such behavior.

1.2. Interaction of Binding Proteins with the Inner Membrane

For many BP transport systems studied, the involvement of two different inner-membrane protein components in the solute translocation step has been implicated by genetic studies. The HisP protein for high-affinity histidine transport in *Salmonella* described by Ames was the first such inner-membrane component discovered (Ames and Nikaido, 1978). Both inner-membrane components of the maltose transport system have been identified (Schuman *et al.*, 1980; Bavoil *et al.*, 1980). We have recently cloned the *livH* gene, coding for an inner-membrane component of the LIV-I transport system, on a 1.5-kilobase insert in the plasmid pBR322. Relative to the number of BP molecules present in the periplasm, small amounts of the inner-membrane proteins are present (for example, 500 copies of the inner-membrane component to 10,000 BPs per cell have been estimated).

The interaction of BPs with the inner-membrane components is assumed to be an important step in the transport of substrates from the periplasm to the cytoplasm. Ames has isolated a His-BP mutant defective for transport but still capable of normal histidine binding. One pseudorevertant of this mutation was mapped in *hisP*, the gene for a histidine-transport inner-membrane component (Ames and Spudich, 1976). This is currently the strongest evidence for the interaction of BPs with inner-membrane proteins during solute translocation. The recent cloning of the genes for many of the inner-membrane components should aid in their identification and purification (Higgins *et al.*, this volume).

1.3. Outer-Membrane Interactions

The outer membrane is a permeability barrier for large molecules but not for amino acids and simple sugars. The presence of outer-membrane proteins such as porin apparently provides pores for the entry of many low-molecular-weight nutrients. Schwartz and colleagues have isolated a mutant in the Mal-BP with a phenotype similar to mutants in *lamB*, the gene for the outer-membrane maltodextran pore (Wandersman *et al.*, 1979). This altered Mal-BP exhibits near-normal maltose binding, leading these workers to suggest that the defect is in a region of the Mal-BP that interacts with the LamB protein in a cooperative fashion to facilitate the entry of maltodextrans through the LamB pore (Ferenci and Boos, 1980). A similar interaction of the dicarboxylate-BP with the outer membrane has been suggested (Bewick and Lo, 1979). The possibility remains that such an interaction will prove important for other BP high-affinity transport systems.

1.4. Binding Proteins in Chemotaxis

It is now well established that carbohydrate BPs serve a dual role as receptors for both chemotaxis and high-affinity transport. Amino acid BPs, however, do not appear to function in chemotactic response systems. The Mal-BP, ribose-BP, and Gal-BP, when combined with substrate, interact with the membrane protein gene products of the *tar* gene (Mal-BP) or the *tsr* gene (ribose-BP and Gal-BP), which are signal transducers for chemotaxis (Koshland, 1979; Springer *et al.*, 1979). W. Mahoney and M. Hermodson (personal communication) have recently determined a partial amino acid sequence for a mutant of the Gal-BP that shows normal transport but is defective in chemotaxis. They discovered amino acid changes of *lys* and *glu* at residue 156 and *glu* to *ala* at residue 157. This region of the Gal-BP is not thought to be involved in transport.

Attempts have been made in our laboratory to determine if the leucine-BPs are involved the chemotactic response of *E. coli* away from leucine. Mutants that lack either the LIV-BP or the LS-BP show normal chemotactic avoidance of leucine (unpublished results). Hence, the role of BPs in chemotaxis may be limited to the carbohydrate BPs. This is in agreement with the finding that aspartate interacts directly with the *tar* gene product to stimulate the avoidance response (Clark and Koshland, 1979).

1.5. Proposed Role of Periplasmic Binding Proteins in Transport

The dual role of carbohydrate BPs in chemotaxis and high-affinity accumulation of carbon sources is now well established. In contrast to these findings, the amino acid BPs may serve a somewhat different role. A good example is the high-affinity leucine transport system of *E. coli*. Wild-type *E. coli* is capable of synthesizing all the amino acids and thus has a greater need to retain the internal amino acid pool than to accumulate them from external sources. In fact, extracellular leucine represses the synthesis of the LIV-BP and the LS-BP. The low-affinity LIV-II system is present regardless of the external leucine concentration. Under conditions where greater external concentrations of isoleucine or valine are present, this system may actually facilitate the loss of cytoplasmic leucine by exchange. Thus, the finding that the LIV-BP and LS-BP are present at about 70 to 80% of their derepressed levels under conditions where all of the cell's leucine is supplied biosynthetically is reasonable. We suggest that the main role of the high-affinity LIV-I transport system under these conditions is to prevent the loss of internal branched-chain amino acids. This is accomplished by recapture and transport from the periplasm back into the cell through the inner membrane. Such a model suggests that regulation of the high-affinity LIV-I system is paramount in the precise control of the internal level of leucine, known to be involved in a unique regulatory relationship with numerous other cellular functions (Quay and Oxender, 1980). This may account for our recent discovery that the LIV-BP and LS-BP genes are preceded by regulatory regions suggestive of a transcription attenuation mechanism distinct from that found for biosynthetic operons. A different situation may well be found for other amino acids, especially for those that serve as nitrogen sources (His and Trp).

The periplasmic BPs serve as receptors for high-affinity transport and must interact with membrane components to carry out solute translocation across the inner membrane. We have described the genetic evidence for the interaction of the His-BP with an inner-membrane protein. The finding that BPs are composed of two domains with a cleft in between containing the substrate (see Section 2) has led to the suggestion that the BP interacts with two subunits in the inner membrane causing the BP to hinge open and deliver the solute to the inner-membrane components. These components may then form a pore through which the solute passes. One of these steps is apparently coupled to the hydrolysis of high-energy phosphate bonds.

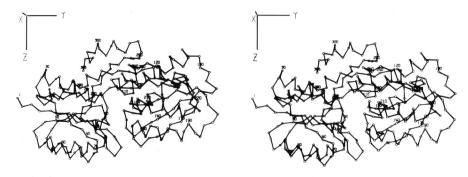

Figure 1. A stereo drawing of the carbon backbone of the Ara-BP kindly supplied by Dr. F. Quiocho of Rice University. The three-dimensional structure is best viewed with the aid of stereo glasses, but can also be observed by placing an opaque divider between the images and viewing from approximately 6 inches above the page.

2. STRUCTURE OF PERIPLASMIC BINDING PROTEINS

The application of X-ray crystallography and NMR spectroscopy as structural probes promises to increase our knowledge of BP structure and conformation dramatically in the next few years. Quiocho and co-workers have determined the 2.8-Å-resolution structure of crystalline L-Ara-BP from *E. coli* (Quiocho *et al.*, 1977). A stereo drawing of the carbon backbone of this molecule is shown in Fig. 1. It is composed of two well-defined domains that are connected by three short peptide chains. It is interesting to note that the NH_2 terminus of the mature protein projects outward. Hence, in this conformation, the precursor peptide (signal sequence, leader peptide) would be exposed on the surface of the molecule.

Recently Quiocho and co-workers have used small angle X-ray scattering to show that Ara-BP rotates approximately 18° about a hinge deep in the base of the cleft to bind L-arabinose by side chain hydrogen bonding to all four substrate hydroxyls (Newcomer *et al.*, 1979, 1981a,b). Low-resolution diffraction patterns of the Gal-BP crystals suggest it is similar in structure to the Ara-BP (Quiocho and Pflugrath, 1980). Mal-BP and LIV-BP of *E. coli* and the sulfate-BP of *Salmonella* have been subjected to X-ray crystallography, and preliminary results suggest they may also be bilobate.(Quiocho *et al.*, 1979; Gilliland and Quiocho, 1981).

NMR studies of the periplasmic BPs may prove an even richer source of information on the role of protein structure and conformation in bacterial transport in the near future. Recent work by Ho and co-workers has shown that conformational changes of the His-BP induced by the binding of histidine can be monitored by significant shifts in its proton magnetic resonance spectrum (Manuck and Ho, 1979; Ho *et al.*, 1981). That such a significant conformational shift occurs with a glutamine-BP was suggested in 1971 by Weiner and Heppel based on the quenching of its internal tryptophan fluorescence when glutamine was added. Further support for the notion that BPs undergo a significant conformational change on binding ligand has been found by Quiocho during attempts to prepare the LIV-BP X-ray diffraction analysis. Two distinct crystal types have been observed depending on whether leucine is present or absent during crystallization.

Table I. Precursor Peptides of Bacterial Periplasmic Binding Proteins

	-25				-20					-15					-10					-5				-1	Binding protein
Met	Lys	Ile	Lys	Thr	Gly	Ala	Arg	Ile	Leu	Ala	Leu	Ser	Ala	Leu	Thr	Thr	Met	Met	Phe	Ser	Ala	Ser	Ala	Leu	Ala	Maltose[a]
				Met	Lys	Lys	Leu	Ala	Leu	Ser	Leu	Ser	Leu	Val	Leu	Ala	Phe	Ser	Ser	Ala	Thr	Ala	Ala	Phe	Ala	Histidine[b]
					Met	Lys	Lys	Thr	Val	Leu	Ala	Leu	Ser	Leu	Leu	Ile	Gly	Leu	Ser	Thr	Ala	Ala	Ser	Thr	Ala	LAO[b]
			Met	Asn	Ile	Lys	Gly	Lys	Ala	Leu	Leu	Ala	Gly	Cys	Ile	Ala	Leu	Ala	Phe	Ser	Asn	Met	Ala	Leu	Ala	LIV[c]
			Met	Lys	Ala	Asn	Ala	Lys	Thr	Ile	Ile	Ala	Gly	Met	Ile	Ala	Leu	Ala	Ile	Ser	His	Thr	Ala	Met	Ala	LS[d]
			Met	Lys	X	Thr	Lys	Leu	Val	Leu	Gly	Ala	Val	Ile	Leu	Gly	Ala	Thr	Leu	Ser	X	Gly	Ala	X	Ala	Arabinose[e]

The sequence region is headed **Precursor peptide**.

[a] *E. coli*, Bedouelle *et al.* (1980).
[b] *S. typhimurium*, Higgins and Ames, 1981.
[c] *E. coli*, R. Landick (unpublished results).
[d] *E. coli*, Oxender *et al.* (1980b).
[e] *E. coli*, Wilson and Hogg (1980).

3. PERIPLASMIC BINDING PROTEIN SECRETION

3.1. Precursor Peptide Sequences

As of this writing, five complete BP precursor peptide sequences have been determined. A partial NH_2-terminal sequence for the precursor of the Ara-BP of *E. coli* was recently derived from amino acid analysis by Wilson and Hogg (1980). Table I presents these precursor peptide sequences. Three sequences from *E. coli* contain 23 amino acids while the two from *Salmonella* contain only 22; only the COOH-terminal alanine is conserved in all six sequences. These sequences resemble closely those found for other secreted proteins in both prokaryotes and eukaryotes. Thus, no evidence exists to suggest a unique secretory mechanism for the periplasmic BPs. To the contrary, both the LS-BP and the LIV-BP are processed by a peptidase active on M13 procoat protein isolated by Zwizinski and Wickner (1980) and Daniels *et al.* (1981a).

3.2. Role of the Precursor Peptide in Secretion

Genetic approaches have been used by Hofnung, Beckwith, and colleagues (Bedouelle *et al.*, 1980) to isolate amino acid substitutions in the precursor peptide region of the Mal-BP that led to defective secretion. Four of these substitutions were changes of hydrophobic residues near the center of the sequence to charged residues, and the fifth was the introduction of proline at residue 10. These experiments provide genetic proof that the hydrophobic nature of the precursor peptide and perhaps its α-helical structure play a role in BP secretion. Another approach to this problem is the construction of chimeric proteins by gene fusion. Beckwith and colleagues have obtained such gene fusions between the Mal-BP gene and *lacZ*, the gene for β-galactosidase (Bassford *et al.*, 1979). Chimeric proteins containing only a small portion of the NH_2 terminus of the Mal-BP are retained in the cytoplasm while chimeras containing more than half of the Mal-BP appear to begin the secretory process but are unable to complete it, leaving the chimeric protein lodged in the membrane. Induction by maltose is toxic to cells containing the latter type of gene fusions. We have obtained a consistent result when fusions of the gene BP and β-galactosidase were constructed. Attempts to fuse the LIV-BP at amino acid 194 have led only to frameshifts at the site of fusion (R. Landick, unpublished results). This is most likely because in-frame fusion produces a toxic protein chimera. The results of these and other studies suggest that the signal sequence alone is not sufficient to direct the protein across the inner membrane. The possibility that more than the primary/secondary structure of the precursor peptide is also necessary for transport is supported by our finding that the precursor and mature forms of the LIV-BP differ significantly in conformation (Oxender *et al.*, 1980b). We have recently found that the LS-BP can be shown to be completely synthesized prior to processing by *in vivo* pulse–chase labeling experiments.

3.3. Current Theories of Secretion of Binding Protein

The advantages of genetic approaches have made this study of the secretion and processing of BPs a useful model system for determining the structural requirements for protein secretion in prokaryotes. Early theories of protein secretion emphasized the importance of the primary and secondary structure of the precursor peptide in directing secretion (Blobel, 1980). An alternative view that emphasizes the role of protein conformation has been suggested by Wickner (1979). If the emphasis is placed on the definition of structural requirements for secretion, then the question of whether the initiation of secretion occurs cotranslationally or posttranslationally becomes a separate and less important issue. One may envisage that the conformation that defines these structural requirements, whether achieved as an incomplete nascent chain on the ribosome or as a freely soluble protein, directs the protein to associate with the membrane or some component in it. It has recently been determined that disruption of electrochemical potentials by carbonyl cyanide *m*-chlorophenylhydrazone can be correlated with the interruption of BP secretion at this stage and the subsequent accumulation of its precursor form (Daniels *et al.*, 1981b; Randall, 1981). It thus appears that electrochemical potentials are important for proper secretion and may serve as the energy source for the transport of the newly synthesized BPs to the periplasm. Removal of the precursor peptide allows the protein to assume its mature conformation as a soluble periplasmic BP.

4. CLONED BINDING PROTEIN GENES

The *E. coli* LIV-, LS-BP (Oxender *et al.*, 1980a; Landick, 1981) and the *S. typhimurium* LAO-, His-BP (Ardeshir and Amos, 1980; Higgins *et al.*, this volume) gene loci have been cloned and the complete nucleotide sequences of these BP genes have been determined (Higgins and Ames, 1981; Landick and Oxender, manuscript in preparation). The availability of these cloned BP genes should greatly facilitate advances in our understanding of BP structure and function.

REFERENCES

Amanuma, H., Itoh, J., and Anraku, Y. (1976). *J. Biochem (Tokyo)* **79**, 1167–1182.
Ames, G. F.-L., and Nikaido, H. (1978). *Proc. Natl. Acad. Sci. USA* **75**, 5447–5451.
Ames, G. F.-L., and Spudich, E. N. (1976). *Proc. Natl. Acad. Sci. USA* **73**, 1877–1881.
Ardeshir, F., and Ames, G. F.-L. (1980). *J. Supramol. Struct.* **13**, 117–130.
Bassford, P., Jr., Silhavy, T. J., and Beckwith, J. (1979). *J. Bacteriol.* **139**, 19–39.
Bavoil, P., Hofnung, M., and Nikaido, H. (1980). *J. Biol. Chem.* **255**, 8366–8369.
Bedouelle, H., Bassford, P., Jr., Fowler, A. V., Zabin, I., Beckwith, J., and Hofnung, M. (1980). *Nature (London)* **285**, 78–81.
Bewick, M. A., and Lo. T. C. Y. (1979). *Can. J. Biochem.* **57**, 653–661.
Blobel, G. (1980). *Proc. Natl. Acad. Sci. USA* **77**, 1496–1500.
Clark, S., and Koshland, D. E., Jr. (1979). *J. Biol. Chem.* **254**, 9695–9702.
Daniels, C. J., Anderson, J. J., Landick, R., and Oxender, D. L. (1981a). *J. Supramol. Struct.*, **14**, 305–311.
Daniels, C. J., Bole, D. G., Quay, S. C., and Oxender, D. L. (1981b). *Proc. Natl. Acad. Sci. USA* **78**, 5396–5400.
Enequist, H. G., Hirst, T. R., Harayama, S., Hardy, S. J. S., and Randall, L. L. (1981). *Eur. J. Biochem.* **116**, 227–233.
Ferenci, T., and Boos, W. (1980). *J. Supramol. Struct.* **13**, 101–116.
Furlong, C. E., and Schellenberg, G. D. (1980). In *Microorganisms and Nitrogen Sources* (J. W. Payne, ed.), pp. 89–123, Wiley, New York.
Gilliland, G. L., and Quiocho, F. A. (1981). *J. Mol. Biol.* **146**, 341–362.
Higgins, C. F., and Ames, G. F.-L. (1981). *Proc. Natl. Acad. Sci. USA* **78**, 6038–6042.
Ho, C., Giza, Y., Takahashi, S., Ugen, K. E., Cottam, P. F., and Dowd, S. R. (1980). *J. Supramol. Struct.* **13**, 131–145.
Koshland, D. E., Jr. (1979). *Physiol. Rev.* **59**, 812–862.
Landick, R., Anderson, J. J., Mayo, M. M., Gunsalus, R. P., Mavromara, P., Daniels, C. J., and Oxender, D. L. (1981). *J. Supramol. Struct.*, **14**, 527–537.

Manuck, B. A., and Ho, C. (1979). *Biochemistry* **18,** 566–573.

Miller, D. M., Newcomer, M. E., and Quiocho, F. A. (1979). *J. Biol. Chem.* **254,** 7521–7528.

Miller, D. M., Olson, J. S., and Quiocho, F. A. (1980). *J. Biol. Chem.* **255,** 2465–2471.

Newcomer, M. E., Miller, D. M., and Quiocho, F. A. (1979). *J. Biol. Chem.* **254,** 7529–7533.

Newcomer, M. E., Gilliland, G. L., and Quiocho, F. A. (1981a). *J. Biol. Chem.* **256,** 13213–13217.

Newcomer, M. E., Lewis, B. A., and Quiocho, F. A. (1981b). *J. Biol. Chem.* **256,** 13218–13222.

Oxender, D. L., and Quay, S. C. (1976). In *Methods of Membrane Biology* (E. D. Korn, ed.), Vol. 6, pp. 183–242, Plenum Press, New York.

Oxender, D. L., Anderson, J. J., Daniels, C. J., Landick, R., Gunsalus, R. P., Zurawski, G., Selker, E., and Yanofsky, C. (1980a). *Proc. Natl. Acad. Sci. USA* **77,** 1412–1416.

Oxender, D. L., Anderson, J. J., Daniels, C. J., Landick, R., Gunsalus, R. P., Zurawski, G., and Yanofsky, C. (1980b). *Proc. Natl. Acad. Sci. USA* **77,** 2005–2009.

Quay, S. C., and Oxender, D. L. (1980). In *Biological Regulation and Development* (R. F. Goldberger, ed.), Vol. 2, pp. 413–436, Plenum Press, New York.

Quiocho, F. A., and Pflugrath, J. W. (1980). *J. Biol. Chem.* **255,** 6559–6561.

Quiocho, F. A., Gilliland, G. L., and Phillips, G. N. (1977). *J. Biol. Chem.* **252,** 5142–5149.

Quiocho, F. A., Meador, W. E., and Pflugrath, J. W. (1979). *J. Mol. Biol.* **133,** 181–184.

Shuman, H. A., Silhavy, T. J., and Beckwith, J. R. (1980). *J. Biol. Chem.* **255,** 168–174.

Springer, M. S., Gor, M. F., and Adler, J. (1979). *Nature (London)* **280,** 279–284.

Wandersman, C., Schwartz, M., and Ferenci, T. (1979). *J. Bacteriol.* **140,** 1–13.

Weiner, J. H., and Heppel, L. A. (1971). *J. Biol. Chem.* **246,** 6933–6941.

Wickner, W. (1979). *Annu. Rev. Biochem.* **48,** 23–46.

Wilson, D. B., and Smith, J. B. (1978). In *Bacterial Transport* (B. P. Rosen, ed.), pp. 495–557, Dekker, New York.

Wilson, V. G., and Hogg, R. W. (1980). *J. Biol. Chem.* **255,** 6745–6750.

Zwizinski, C., and Wickner, W. (1980). *J. Biol. Chem.* **255,** 7973–7977.

Active Transport of Amino Acids in Bacteria

Advances Resulting from Genetic and Recombinant DNA Techniques

Christopher F. Higgins, Feroza Ardeshir, and Giovanna Ferro-Luzzi Ames

1. INTRODUCTION

The elucidation of the molecular mechanisms of active transport has been, and still is, a difficult problem. The main difficulty in studying active transport from a biochemical point of view is the lack of an *in vitro* assay for transport activity. Thus, indirect approaches have generally proven most fruitful, particularly those aimed at identifying and characterizing the genes coding for transport components. Concomitant with such studies has been the identification and characterization of the proteins coded by these genes, together with studies on the behavior of specifically mutated proteins and on the regulation of gene expression.

Salmonella typhimurium and *Escherichia coli* have generally been the organisms of choice for such studies, both because of the extensive body of background information available on their biology and genetics and because many of the recent advances in recombinant DNA technology have been developed using these species. The two organisms are similar enough that most conclusions drawn from studies on one of them are usually directly applicable to the other.

In both species, any given amino acid is transported by a multiplicity of transport

Christopher F. Higgins, Feroza Ardeshir, and Giovanna Ferro-Luzzi Ames • Department of Biochemistry, University of California, Berkeley, California 94720.

systems of different specificities and affinities. This pattern of multiplicity, which was initially established for histidine and aromatic amino acids in *S. typhimurium* (Ames, 1964; Ames and Roth, 1968), has been subsequently shown to be a general phenomenon, and also to apply to substrates other than amino acids (several reviews in Rosen, 1978; Payne, 1980).

The isolation and characterization of transport mutants can be achieved using a variety of techniques: those that have been used in the study of histidine transport (the system that has received the most extensive genetic characterization) are generally applicable to the analysis of any other transport system. Thus, mutants can be selected by (1) their resistance to inhibitory analogs transported by the system (Ames, 1964; Ames and Roth, 1968), (2) their improved ability to utilize an essential metabolite that is a poor substrate of the system (Krajewska-Grynkiewicz *et al.*, 1971), or (3) loss of the ability to use an essential metabolite that is transported by the system (Ames and Lever, 1970). In addition, transposons (e.g., Tn*10*) can be used to create insertions and deletions in transport genes (Noel and Ames, 1978).

The amino acid transport systems most extensively characterized using such approaches are the histidine and the branched-chain amino acid transport systems of *S. typhimurium* and *E. coli*. In this review an attempt is made to cover only the most recent advances in the study of these transport systems that have resulted from genetic and recombinant DNA techniques and to indicate the potential of such approaches in contributing to our understanding of the mechanisms of active transport.

2. HIGH-AFFINITY HISTIDINE PERMEASE

The high-affinity histidine permease of *S. typhimurium* is the best characterized system of several distinct histidine permeases that function in this organism. An analogous system has been identified in *E. coli* (Ardeshir and Ames, 1980). A detailed genetic map of the histidine transport operon has been constructed (Ames *et al.*, 1976). The operon, located at 48.5 minutes on the *Salmonella* linkage map, consists of four genes, *hisJ*, *hisQ*, *hisM*, and *hisP*, which are under the control of a promoter/operator locus, *dhuA*. Expression of this transport operon, and that of several other transport systems, is regulated by nitrogen availability. The *hisJ* gene codes for a periplasmic histidine-binding protein (molecular weight 25,000), which has recently been sequenced (Hogg, 1981). The *hisP* gene product is a basic cytoplasmic-membrane protein (molecular weight 24,000) and has been identified by two-dimensional gel electrophoretic analysis of the proteins from *hisP* mutant strains (Ames and Nikaido, 1978). The *hisQ* gene product, which has been identified with the help of recombinant DNA techniques, is a basic membrane protein of molecular weight 21,000. The *hisM* gene product has not yet been identified: its existence is inferred from nucleotide sequences (Higgins and Ames, submitted for publication).

Detailed characterization of mutations in this operon has been important in developing models for transport. Thus, it has been shown genetically that the periplasmic J protein possesses a site, in addition to its histidine-binding site, that is involved in an interaction with the membrane-bound P protein during transport (Ames and Spudich, 1976). Mutant J proteins have also been important in demonstrating that conformational changes occur during transport (Manuch and Ho, 1979). Several possible models for the mechanisms of translocation of histidine across the membrane are consistent with all available data (Ames and Spudich, 1976). The P protein may form a pore that opens to allow the passage of histidine only upon interaction with molecules of the J protein that carry bound histidine.

Alternatively, the P protein may itself bind histidine after receiving it from the binding site on the J protein. No function has been assigned to the Q and M proteins as yet.

The Q, M, and P proteins appear to be multifunctional as they are also essential components for the transport of arginine as a nitrogen source. This process requires, in addition to the Q, M, and P proteins, a periplasmic binding protein, the lysine-arginine-ornithine-binding protein (LAO protein), which is immunologically related to the J protein (Kustu and Ames, 1973; Kustu *et al.*, 1979). The LAO protein is encoded by the *argT* gene, which is immediately upstream from the histidine transport operon and which is also under nitrogen control. The LAO protein is believed to interact with the P protein in the course of arginine transport in an analogous manner to the J–P interaction during histidine transport (see scheme below).

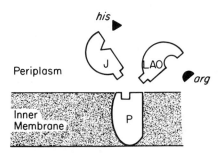

The entire histidine transport operon has been cloned as a 12.4-kb DNA fragment in phage and plasmid vectors (Ardeshir and Ames, 1980). The exact location of the transport genes on this fragment was determined making use of the many genetically characterized deletions of the transport operon (Ardeshir *et al.*, 1981). This information has been used in designing strategies for sequencing the DNA from specific regions of the transport operon. The complete DNA sequence of the *hisJ* gene has now been obtained, confirming the protein sequence data (Higgins and Ames, 1981). The complete DNA sequence of the *argT* gene shows that the LAO protein is 70% homologous with J. In addition, the DNA sequences reveal the existence of typical signal peptides, 22 amino acid residues long, at the amino termini of both the J and the LAO proteins. Such signal peptides are presumably necessary for the secretion of periplasmic proteins (Landick and Oxender, this volume). The close proximity of the *argT* and *hisJ* genes, and their extensive homology, indicates that they originated from a single ancestral gene by tandem duplication and subsequently diverged such that their protein products displayed different substrate-binding specificities. Three lines of evidence have now allowed us to assign both the amino acid-binding site and the site involved in interacting with the membrane-bound P protein to specific regions of each binding protein: (1) the distribution of amino acid differences between the two proteins; (2) the properties of a functional chimeric protein, produced by a deletion mutant in which the first half of the *argT* gene is fused to the second half of the *hisJ* gene; (3) the sequence change in a mutant J protein unable to interact with P (Higgins and Ames, 1981).

An interesting feature of the transport operon is that the J protein is produced in much greater quantities than the products of the *hisQ* and *hisP* genes, which map downstream from *hisJ*. The nucleotide sequence of the 182 bases in the intergenic region between *hisJ* and *hisQ* has been determined (Higgins and Ames, submitted for publication). An extensive region of dyad symmetry is found, capable of forming a very large and stable stem-loop structure ($\Delta G = -54$ kcal/mole) and which is presumed to be responsible for the observed decrease in expression of the downstream genes. In addition, analysis of DNA sequences from the regulatory regions 5' to the histidine transport operon (the *dhuA* region) and 5' to *argT* shows that (1) there is no similarity with the regulatory regions (attenuators) of the amino acid biosynthetic operons, and (2) the two regions contain regions of homology that are believed to be the sites where nitrogen regulation is effected (Higgins and Ames, submitted for publication, 1982).

3. LEUCINE, ISOLEUCINE, AND VALINE

Extensive studies by Oxender and his co-workers have provided the following picture of the systems and components involved in the transport of leucine, isoleucine, and valine in *E. coli*. Three distinct transport systems have been identified, two high-affinity periplasmic systems (LIV-I and LS) and a low-affinity membrane-bound system (LIV-II) (reviewed in Payne, 1980; Anderson and Oxender, 1977; Oxender *et al.*, 1977).

The structural genes for the two high-affinity systems are closely linked at 74 minutes on the *E. coli* linkage map. The LIV-I system requires the *livJ* product, a periplasmic binding protein with affinity for L-leucine, L-isoleucine, L-valine, L-alanine, and L-threonine. The LS system also requires a binding protein, the *livK* gene product, which specifically binds L-leucine. These two binding proteins have been fully and partially sequenced, respectively, and show 80% homology (Ovchinnikov *et al.*, 1977; Landick *et al.*, 1981). Both the LIV-I and the LS systems also require the function of two additional genes, *livH* and *livG*, which code for two as yet unidentified proteins, possibly membrane-bound. Preliminary DNA sequence data indicate that, like the *hisP* protein, the *livH* protein is strongly basic (Oxender, personal communication): these two proteins may therefore perform analogous functions.

The cluster of genes, *livJ*, *livK*, *livH*, and *livG*, has recently been cloned in the plasmid vector pACYC184 (Oxender *et al.*, 1980a). Although the gene order has not been established genetically, *in vitro* experiments with the cloned genes indicate the order *livJ–livK–livH*. Both the DNA sequence of the beginning of *livK* and the NH_2-terminal sequence of the precursor form of its product, the LS-binding protein (synthesized *in vitro*), show that this protein possess a 23-amino-acid signal peptide similar to those of other periplasmic binding proteins (Oxender *et al.*, 1980b). Genes *livJ* and *livK*, like the *hisJ* and *argT* genes of the histidine transport system, seem to be under the control of separate promoters (Landick *et al.*, 1981).

These two high-affinity transport systems are regulated by two *trans*-acting genes that map at 22 minutes on the *E. coli* linkage map: *livR*, which regulates the level of LIV-binding protein, and *1stR*, which regulates synthesis of the LS-binding protein (Anderson *et al.*, 1976). In addition, regulation of both the LIV-I and the LS systems involves $tRNA^{leu}$, the rho factor, and the cellular levels of guanosine tetraphosphate, implying that an attenuation mechanism may be involved, possibly similar to that for amino acid biosynthetic operons (Quay and Oxender, 1977, 1979; Quay *et al.*, 1975, 1978).

The third system identified by Oxender's group is a low-affinity, membrane-bound system (LIV-II). While this system has not been characterized in as great detail as the two high-affinity systems discussed above, it can be distinguished from them both kinetically and genetically. The system can be identified by mutations in the *livP* gene, which maps at 76 minutes on the *E. coli* chromosome. Amino acid uptake via this system does not appear to require any periplasmic components and can be observed in membrane vesicles.

Genetic data obtained independently by two other groups seem to confirm the results described above, despite the use of a different nomenclature (Guardiola *et al.*, 1974a,b; Yamato *et al.*, 1979; Yamato and Anraku, 1980). In addition, a regulatory gene, *hrbA*, located at 9 minutes, has been identified by Anraku's group (Yamato *et al.*, 1979) and may be identical to gene *brnR* (Guardiola *et al.*, 1974b).

Leucine-isoleucine-valine transport in *S. typhimurium* has not been as extensively characterized as in *E. coli*. However, mutants have been isolated and kinetic data indicate the existence of multiple systems (Kiritani and Ohnishi, 1977). It seems likely that branched-chain amino acid transport will be found to be similar in these two organisms.

4. CONCLUSIONS

From the above, it can be seen that genetic studies have contributed enormously toward an understanding of the mechanisms of active transport of amino acids in bacteria. Although only the two most extensively characterized systems have been discussed, several transport systems for other amino acids have also been identified genetically and should be amenable to similar characterization.

A detailed background of genetic analysis is invaluable in exploiting recent developments in recombinant DNA techniques. In the near future the amino acid sequences of several membrane-bound transport proteins, normally present in amounts too small for conventional protein analysis, will be obtained from the nucleic acid sequences of their genes. The amino acid alterations (via nucleic acid sequences) of transport-defective mutant proteins will be important in devising models for the mechanism of transport. Studies on cloned DNA fragments should also lead to a clearer understanding of the mechanisms by which the cellular levels of transport components are regulated. In addition, the cloning of genes for transport components provides a promising approach toward the overproduction of these components for purification purposes. Until now the purification of membrane-bound transport proteins has proved extremely difficult due to the very low levels of these proteins in the cell and to the lack of a suitable assay for activity. With the purification of these proteins and a concomitant structural and functional analysis, the possibility of understanding the mechanisms of membrane transport at a molecular level now becomes a much more realistic goal.

REFERENCES

Ames, G. F.-L. (1964). *Arch. Biochem. Biophys.* **104,** 1–18.
Ames, G. F.-L., and Lever, J. (1970). *Proc. Natl. Acad. Sci. USA* **66,** 1096–1103.
Ames, G. F.-L., and Nikaido, K. (1978). *Proc. Natl. Acad. Sci. USA* **75,** 5447–5451.
Ames, G. F.-L., and Roth, J. R. (1968). *J. Bacteriol.* **96,** 1742–1749.
Ames, G. F.-L., and Spudich, E. N. (1976). *Proc. Natl. Acad. Sci. USA* **73,** 1877–1881.
Ames, G. F.-L., Noel, K. D., Taber, H., Spudich, E. N., Nikaido, K., Afong, J., and Ardeshir, F. (1976). *J. Bacteriol.* **129,** 1289–1297.
Anderson, J. J., and Oxender, D. L. (1977). *J. Bacteriol.* **130,** 384–392.
Anderson, J. J., Quay, S. C., and Oxender, D. L. (1976). *J. Bacteriol.* **126,** 80–90.
Ardeshir, F., and Ames, G. F.-L. (1980). *J. Supramol. Struct.* **13,** 117–130.
Ardeshir, F., Higgins, C. F., and Ames, G. F.-L. (1981). *J. Bacteriol.* **147,** 401–409.

Guardiola, J., DeFelice, M., Klopotowski, T., and Iaccarino, M. (1974a). *J. Bacteriol.* **117**, 382–392.

Guardiola, J., DeFelice, M., Klopotowski, T., and Iaccarino, M. (1974b). *J. Bacteriol.* **117**, 393–405.

Higgins, C. F., and Ames, G. F.-L. (1981). *Proc. Natl. Acad. Sci. USA,* **78**, 6038–6042.

Higgins, C. F., and Ames, G. F.-L. (1982). *Proc. Natl. Acad. Sci. USA* **79**, in press.

Hogg, R. W. (1981). *J. Biol. Chem.* **256**, 1935–1939.

Kiritani, K., and Ohnishi, K. (1977). *J. Bacteriol.* **129**, 589–598.

Krajewska-Grynkiewicz, K., Walczak, W., and Klopotowski, T. (1971). *J. Bacteriol.* **105**, 28–37.

Kustu, S. G., and Ames, G. F.-L. (1973). *J. Bacteriol.* **116**, 107–113.

Kustu, S. G., McFarland, N. C., Hui, S. P., Esmon, B., and Ames, G. F.-L. (1979). *J. Bacteriol.* **138**, 218–234.

Landick, R., Anderson, J. J., Mayo, M. M., Gunsalus, R. P., Mavcomara, P., Daniels, C. J., and Oxender, D. L. (1981). *J. Supramol. Struct.* **14**, 527–537.

Manuch, B. A., and Ho, C. (1979). *Biochemistry* **18**, 566–573.

Noel, K. D., and Ames, G. F.-L. (1978). *Mol. Gen. Genet.* **166**, 217–223.

Ovchinnikov, Y. A., Aldanova, N. A., Grinkevich, V. A., Arzamazova, N. M., Moroz, I. N., and Nazimov, I. V. (1977). *Bioorg. Chem. USSR* **3**, 564–570.

Oxender, D. L., Anderson, J. J., Mayo, M. M., and Quay, S. C. (1977). *J. Supramol. Struct.* **6**, 419–431.

Oxender, D. L., Anderson, J. J., Daniels, C. J., Landick, R., Gunsalus, R. P., Zurawski, G., Selker, E., and Yanofsky, C. (1980a). *Proc. Natl. Acad. Sci. USA* **77**, 1412–1416.

Oxender, D. L., Anderson, J. J., Daniels, C. J., Landick, R., Gunsalus, R. P., Zurawski, G., and Yanofsky, C. (1980b). *Proc. Natl. Acad. Sci. USA* **77**, 2005–2009.

Payne, J. (ed.) (1980). *Microorganisms and Nitrogen Sources*, Wiley, New York.

Quay, S. C., and Oxender, D. L. (1977). *J. Bacteriol.* **130**, 1024–1029.

Quay, S. C., and Oxender, D. L. (1979). *J. Bacteriol.* **137**, 1059–1062.

Quay, S. C., Kline, E. C., and Oxender, D. L. (1975). *Proc. Natl. Acad. Sci. USA* **72**, 3921–3924.

Quay, S. C., Lawther, R. P., Hatfield, G. W., and Oxender, D. L. (1978). *J. Bacteriol.* **134**, 683–686.

Rosen, B. P. (ed.) (1978). *Bacterial Transport*, Dekker, New York.

Yamato, I., and Anraku, Y. (1980). *J. Bacteriol.* **144**, 36–44.

Yamato, I., Ohki, M., and Anraku, Y. (1979). *J. Bacteriol.* **138**, 24–32.

101

Recent Trends in Genetic Studies of Bacterial Amino Acid Transport Mutants

Yasuhiro Anraku

1. INTRODUCTION

Much of the current interest in bacterial amino acid transport is directed at the mechanism of energy coupling through transport carrier proteins in the cytoplasmic membrane. Through the functions of specific transport systems, bacteria maintain amino acid pools intracellularly against a concentration gradient of three to four orders of magnitude (Anraku, 1980). A carrier protein is the primary component of the active transport system and performs this osmotic work as a secondary chemiosmotic solute pump via the symporter hypothesis of Peter Mitchell (Anraku, 1979; Rosen and Kashket, 1978). Thus, the major problem to be argued is how the carrier molecule is coupled with the protonmotive force and induces a downhill proton flow simultaneously with an uphill influx of coupled solute across the cytoplasmic membrane. The proline transport system of *Escherichia coli* K_{12} has been the most favored system for testing this hypothesis because the proline carrier protein in the cytoplasmic membrane (1) couples with a protonmotive force (Hirata *et al.*, 1974; Kasahara and Anraku, 1974); (2) can be solubilized in a reconstitutively active form in liposomes: active transport occurs when energy is supplied as membrane potential generated by K^+ diffusion via valinomycin (Amanuma *et al.*, 1977b); and (3) binds H^+ and proline in the presence and absence of sodium ions in the nonenergized state (Amanuma *et al.*, 1977a; Anraku *et al.*, 1980). Biochemical studies of the proline carrier protein have also been advanced greatly by isolation of mutants defective in the gene proT for proline carrier (Motojima *et al.*, 1978) and by utilization of strains containing a ColEl–*proT*+ hybrid plasmid, in which synthesis of a *proT* product is increased by about sixfold (Motojima *et al.*, 1979).

I will describe briefly the recent trends in genetic isolation of amino acid transport mutants from *E. coli* and *Salmonella typhimurium*. Emphasis will be placed upon the strategy

Yasuhiro Anraku • Department of Biology, Faculty of Science, University of Tokyo, Hongo, Tokyo 113, Japan.

for isolating mutants with a phenotype of high concentration requirement of L-amino acid for growth (Motojima *et al.*, 1978) together with newer, D-amino acid-hypersensitive mutants in strains that lacked "nitrogen control" (Kustu *et al.*, 1979). Readers should consult comprehensive reviews (Anraku, 1978, 1980) for the terminology of bacterial transport systems and collected data on transport mutants.

2. EARLIER WORK

Since 1968, Ames and co-workers (Ames, 1974) have developed the genetics of histidine transport of *S. typhimurium* systematically. They made use of phenotypes with elevated utilization of D-histidine for growth, followed by second screening for resistance to the histidine analog 2-hydrazino-3-(4-imidazolyl)propionic acid (HIPA) (Ames and Lever, 1970). The phenotypes mentioned above do not necessarily correlate directly with the defect of the primary part of carrier function, and rather reflect some changes in sensitivity or resistance to metabolic interference caused by the amino acid analog and D-amino acid used. In fact, they first isolated the *dhuA* mutants as strains that are capable of growing with D-histidine and are supersensitive to HIPA, and identified the gene *dhuA* as the regulatory gene for the histidine transport operon. Using this type of strain as a parent and selecting HIPA-resistant mutations, they further identified the genes *hisJ* and *hisK* for histidine-specific binding proteins and the gene *hisP* for membrane-bound transport component.

3. CARRIER PROTEIN MUTANTS

Lubin *et al.* (1960) first attempted to isolate proline transport mutants from a proline auxotroph strain of *E. coli* as strains having a phenotype of high concentration requirement of L-proline (250 μg/ml medium) for growth. Kaback and Stadtman (1966) demonstrated that the membrane vesicles prepared from one such mutant, strain W157, could not concentrate proline against a concentration gradient with glucose as energy source. Motojima *et al.* (1978) revived this idea under careful experimental protocol with the help of presently available, highly sensitive analytical methods. *E. coli* K_{12} strain JE2133 (*proA* Put$^+$) was mutagenized with *N*-methyl-*N*-nitro-*N'*-nitrosoguanidine, and mutants requiring a high concentration of L-proline for growth were selected. Three mutants, which were isolated independently, were unable to grow in inorganic salts–glucose medium supplemented with 5 μg proline/ml, but grew as well as the parental strain with 400 μg proline/ml medium. Genetic studies indicated that all the mutations (*proT*) were point mutations, and these were mapped at 82 min of the *E. coli* genetic map. The biochemical defect due to the mutation of gene *proT* was studied, using right-side out cytoplasmic membrane vesicles of high purity (Yamato *et al.*, 1975, 1978). The results indicated that (1) strain PT21 (*proA proT* Put$^-$ ProT$^-$) and two strains mentioned above lack D-lactate-driven proline uptake activity completely, (2) a pH-dependent binding of proline to the membranes is also defective, and (3) both proline uptake and proline binding activities of the cytoplasmic membranes from strain MinS/pLC4-45 (ColEl-*proT*$^+$), which specifically contain a 24,000-molecular-weight polypeptide as a *proT*$^+$ product, are amplified simultaneously by about sixfold more than those from strain MinS (*proT*$^+$) (Motojima *et al.*, 1979). Thus, it appears that gene *proT* is the structure gene for the proline carrier protein of 24,000 molecular weight that is located in the cytoplasmic membrane and catalyzes stereospecific binding of substrate and translocation of bound substrate driven by a protonmotive force. More recently, Anraku *et al.* (1980) isolated a novel type of mutant defective in the function of proline carrier, which was also caused by the mutation of *proT*. This mutant strain PT32 (*proA proT*) lacks proline uptake activity completely as strain PT21, but retains proline binding activity with altered dependence for pH and sodium ions. This finding strongly suggests that in this mutant the binding of proline to carrier is uncoupled from the subsequent translocation of bound substrate, therefore providing a new source of genetic material for studies of the mechanism of H^+/proline symport.

The same strategy for isolating transport mutants with use of the phenotype mentioned above has been applied successfully to select branched-chain amino acid transport carrier mutants of *E. coli* K-$_{12}$. After mutagenesis of a leucine auxotroph, strain Wl-1 (*leu-6*), with *N*-methyl-*N*-nitro-*N'*-nitrosoguanidine and selection of Ile$^-$ Val$^-$ mutants, Yamato *et al.* (1979) screened them further for a phenotype of Hrb$^-$; they isolated mutants that did not grow with 20 μg each of isoleucine and valine in the presence of 90 μg leucine/ml medium, but grew in the presence of 100 μg each of isoleucine and valine per milliliter of medium. Strain 176 (*leu-6 hrbA* Hrb$^-$ Ile$^-$ Val$^-$), one of mutants selected as above, was found to have a defect of a leucine-nonrepressible transport carrier activity which is driven by a protonmotive force and detectable in the cytoplasmic membrane vesicles. The gene *hrbA* is the structure gene for branched-chain amino acid carrier protein and was mapped at 8.9 min of the *E. coli* genetic map. The activity of this leucine-nonrepressible *hrbA* system is not inhibited by threonine, for the transport carrier encoded by gene *hrbA* does not require the leucine-isoleucine-valine-threonine (LIVT)-binding protein for substrate recognition. They also found that strain B762 (*hrbA* Hrb$^-$ Ile$^-$ Val$^-$), which was constructed independently by transduction with P*lkc* phage grown on strain 176, has a normal, leucine-repressible transport activity detectable in the cytoplasmic membrane vesicles when cells are grown in the absence of added leucine (Yamato *et al.*, 1979).

To develop further the genetic studies of leucine-repressible branched-chain amino acid transport carrier systems, Yamato and Anraku (1980) constructed strain B763 (*ileA hrbA*), which is able to grow well with isoleucine at a concentration of only 2 μg/ml medium. Adopting this strain as a parental strain and treating it with ethylmethane sulfonate, they isolated mutants that did not grow well with 10 μg isoleucine/ml, but grew with 500 μg isoleucine/ml medium. Strain B7634 (*ileA hrbA hrbBC hrbD* Hrb$^-$), one of mutants obtained in this manner, was found to have additional defects of two independent genetic loci other than *hrbA*, *hrbBC*, and *hrbD*, all of which encode the leucine-repressible transport carrier proteins for branched-chain amino acids. The genes *hrbBC* were mapped at 76 min near *malT*, and the gene *hrbD* mapped at 77 min near *xyl* of the *E. coli* genetic map. The *hrbB* and *hrbC* systems catalyze ATP-driven, LIVT-binding-protein-requiring active transport with a high and a low K_m value, respectively, while the *hrbD* system is responsible for the uptake activity which is driven by a protonmotive force and is stimulated by the LIVT-binding protein (Yamato and Anraku, 1980).

4. BINDING PROTEIN MUTANTS

Independent of this work, Oxender and co-workers (Rahmanian *et al.*, 1973; Anderson *et al.*, 1976) studied the D-leucine-utilizing strains (Dlu$^+$) of *E. coli* K$_{12}$ genetically and found that this phenotype is due to mutations of regulatory genes *livR* and *lstR* for synthesis of LIVT-binding protein (Amanuma *et al.*, 1976) and leucine-specific binding protein (Furlong and Weiner, 1970), respectively. Both genes were mapped near *aroA* of the *E. coli* genetic map (Anderson *et al.*, 1976). Furthermore, Anderson and Oxender (1977) found that a mutant strain (*livR* or *lstR,* Dlu⁺), which has elevated level of a high-affinity leucine transport activity and relating binding protein, is hypersensitive to valine. Taking advantage of this strain's background and using mutator phage Mu to induce mutation, they isolated a new class of transport mutants for branched-chain amino acids with lesions in the binding proteins. The genes *livK* and *livJ* were thus identified genetically, which encode leucine-specific binding protein and LIVT-binding protein, respectively. They also isolated a mutant lacking a high-affinity leucine transport system requiring functional binding protein. The gene *livH*, which codes for an additional component of the transport system, is closely linked to *livK* and *livJ*, and all were mapped in the region near *malT* of the *E. coli* genetic map. It should be noted here that the procedure for screening transport mutations with a phenotype of Dlu⁺ enriches mutants with elevated level of the binding proteins, while with a phenotype of Hrb⁻ mainly affords mutants defective in the carrier protein activities. Therefore, both procedures are of independent genetic value and complementary to each other.

5. NITROGEN-REGULATORY MUTANTS

The regulatory mechanisms involved in the assimilation of ammonia into glutamate and in the formation of glutamate and ammonia from various amino acids via degradative enzyme pathways are of principal importance in living organisms and are, in general, called nitrogen control (Magasanik, 1976; Tyler, 1978). Nitrogen control affects pleiotropically the expression of several unlinked genes for glutamine synthetase, amino acid degradative enzymes, and other proteins required for nitrogen fixation. Recently, Kustu *et al.* (1979) presented clear evidence that synthesis of some transport components including binding proteins for amino acids are also under nitrogen control. They found that nitrogen-regulatory mutants, which have elevated level of glutamine synthetase activity and uptake activities for glutamate, arginine, histidine, and glutamine under all growth conditions, express several characteristic phenotypes such as high efficiency in D-histidine utilization and increased sensitivity to the glutamate analog γ-glutamylhydrazide. Taking advantage of these phenotypes, they isolated transport-defective mutants that specifically lack the binding proteins for histidine (*hisJ*), glutamine (*glnH*), and arginine (*argT*). Considering the fact that the structure genes encoding binding proteins make a cluster with the genes encoding other transport components (presumably carrier proteins) (Ames and Lever, 1970; Ames *et al.*, 1977; Anderson and Oxender, 1977; Oxender *et al.*, 1980), this novel screening strategy is effective in identifying transport operons for binding-protein-dependent transport systems (Anraku, 1968, 1971, 1978; Yamato and Anraku, 1977, 1980). Unfortunately, however, most binding-protein-dependent transport systems so far studied are driven by ATP as coupling energy (Berger and Heppel, 1974; Yamato and Anraku, 1980), and a proper *in vitro* analytical method for ensuring measurement of this ATP-driven carrier activity has not yet been available.

In summary, the current trends in genetic studies of bacterial amino acid transport systems provide new strategies for the isolation of useful mutants to study the mechanisms of molecular assembly and energy coupling of the secondary chemiosmotic solute pumps, and the role of transport carrier proteins in these processes.

ACKNOWLEDGMENTS. Work from the author's laboratory discussed herein was supported in part by a grant-in-aid for a Special Research Project from the Ministry of Education, Science, and Culture of Japan and a grant from the Toray Science Foundation of Japan.

Note Added in Proof. Most recently, strains PT21 and PT32 were found to have additional defect of independent genetic locus, *putP,* near *pyrC* of the *E. coli* genetic map (H. Yamamoto and Y. Anraku, manuscript in preparation).

REFERENCES

Amanuma, H., Itoh, J., and Anraku, Y. (1976). *J. Biochem.* **79,** 1167–1182.

Amanuma, H., Itoh, J., and Anraku, Y. (1977a). *FEBS Lett.* **78,** 173–176.

Amanuma, H., Motojima, K., Yamaguchi, A., and Anraku, Y. (1977b). *Biochem. Biophys. Res. Commun.* **74,** 366–373.

Ames, G. F.-L. (1974). In *Methods in Enzymology* (S. Fleischer and L. Packer, eds.), Vol. 32, pp. 849–856, Academic Press, New York.

Ames, G. F.-L., and Lever, J. E. (1970). *Proc. Natl. Acad. Sci. USA* **66,** 1096–1103.

Ames, G. F.-L., Noel, K. D., Taker, H., Spudich, E. N., Nikaido, K., Afong, J., and Adeshir, F. (1977). *J. Bacteriol.* **129,** 1289–1297.

Anderson, J. J., and Oxender, D. L. (1977). *J. Bacteriol.* **130,** 384–390.

Anderson, J. J., Quay, S. C., and Oxender, D. L. (1976). *J. Bacteriol.* **126,** 80–90.

Anraku, Y. (1968). *J. Biol. Chem.* **243,** 3128–3135.

Anraku, Y. (1971). *J. Biochem.* **70,** 855–866.

Anraku, Y. (1978). In *Bacterial Transport* (B. P. Rosen, ed.), pp. 171–219, Dekker, New York.

Anraku, Y. (1979). *Taisha* **16,** 1697–1709.

Anraku, Y. (1980). In *Microorganisms and Nitrogen Sources* (J. W. Payne, ed.), pp. 9–33, Wiley, New York.

Anraku, Y., Mogi, T., Yamamoto, H., and Fujimura, T. (1980). *Abstract Japan Bioenergetics Group* **6,** 70–72.

Berger, E. A., and Heppel, L. A. (1974). *J. Biol. Chem.* **249,** 7747–7755.

Furlong, C. E., and Weiner, J. H. (1970). *Biochem. Biophys. Res. Commun.* **38,** 1076–1083.

Hirata, H., Altendorf, K., and Harold, F. M. (1974). *J. Biol. Chem.* **249,** 2939–2945.

Kaback, H. R., and Stadtman, E. R. (1966). *Proc. Natl. Acad. Sci. USA* **55,** 920–927.

Kasahara, M., and Anraku, Y. (1974). *J. Biochem.* **76,** 977–983.

Kustu, S. G., McFarland, N. C., Hui, S. P., Esmon, B., and Ames, G. F.-L. (1979). *J. Bacteriol.* **138,** 218–234.

Lubin, M., Kessel, D. H., Budreau, A., and Gross, J. D. (1960). *Biochim. Biophys. Acta* **42,** 535–538.

Magasanik, B. (1976). *Proc. Nucleic Acid Res. Mol. Biol.* **17,** 99–115.

Motojima, K., Yamato, I., and Anraku, Y. (1978). *J. Bacteriol.* **136,** 5–9.

Motojima, K., Yamato, I., Anraku, Y., Nishimura, A., and Hirota, Y. (1979). *Proc. Natl. Acad. Sci. USA* **76,** 6255–6259.

Oxender, D. L., Anderson, J. J., Daniels, C. J., Landick, R., Gunsalus, R. P., Zurawski, G., Selker, E., and Yanofsky, C. (1980). *Proc. Natl. Acad. Sci. USA* **77,** 1412–1416.

Rahmanian, M., Claus, D. R., and Oxender, D. L. (1973). *J. Bacteriol.* **116,** 1258–1266.

Rosen, B. P., and Kashket, E. R. (1978). In *Bacterial Transport* (B. P. Rosen, ed.), pp. 559–620, Dekker, New York.

Tyler, B. (1978). *Annu. Rev. Biochem.* **47,** 1127–1162.

Yamato, I., and Anraku, Y. (1977). *J. Biochem.* **81,** 1517–1523.

Yamato, I., and Anraku, Y. (1980). *J. Bacteriol.* **144,** 36–44.

Yamato, I., Anraku, Y., and Hirosawa, K. (1975). *J. Biochem.* **77,** 705–718.

Yamato, I., Futai, M., Anraku, Y., and Nonomura, Y. (1978). *J. Biochem.* **83,** 117–128.

Yamato, I., Ohki, M., and Anraku, Y. (1979). *J. Bacteriol.* **138,** 24–32.

sn-Glycerol-3-Phosphate Transport in Escherichia coli

Winfried Boos

1. INTRODUCTION

sn-Glycerol-3-phosphate (G3P) is an essential intermediate of phospholipid biosynthesis (Raetz, 1978). This requires a finely tuned regulatory mechanism of anabolic and catabolic pathways to maintain a suitable pool of this compound for continuing phospholipid synthesis. In the absence of glycerol or G3P in the growth medium, G3P is synthesized via an NADH-dependent enzyme that reduces dihydroxyacetone phosphate (Kito and Pizer, 1969). However, in the presence of exogenous glycerol or G3P, the enzymes of the *glp* regulon are induced. This leads to the uptake of glycerol and its phosphorylation to G3P and further degradation via an aerobic or alternatively an anaerobic and flavin-dependent enzyme to form dihydroxyacetone phosphate that enters glycolysis. As part of the *glp* regulon, an active transport system for G3P coded for by the *glpT* gene is also induced. Figure 1 summarizes the activities and regulation of the *glp*-controlled enzymes and transport systems that have been studied in great detail by Lin and his collaborators (1976). Obviously, the function of the *glp* regulon is geared toward an effective degradation of glycerol or G3P as a carbon source. In the case of G3P, more phosphate may be taken up than is needed by the cell. This excess phosphate can be removed via the formation of methylglyoxal and inorganic phosphate from dihydroxyacetone phosphate (Cooper and Anderson, 1970). Exit of phosphate may then be catalyzed by the *glpT* transport system (Hayashi *et al.*, 1964). In contrast to the function of the *glpT* transport system as a provider of carbon source, there exists a second transport system for G3P, coded for by the *ugp* gene(s) (Argast *et al.*, 1978), that is clearly geared toward a utilization of phosphate in the G3P molecule. This system has recently been found to belong to the *pho* regulon (Argast and Boos, 1980). Thus, it is derepressed concomitantly with alkaline phosphatase under conditions of phosphate limitation or by mutations leading to constitutive synthesis of this

Winfried Boos • Department of Biology, University of Konstanz, D-7750 Konstanz, West Germany.

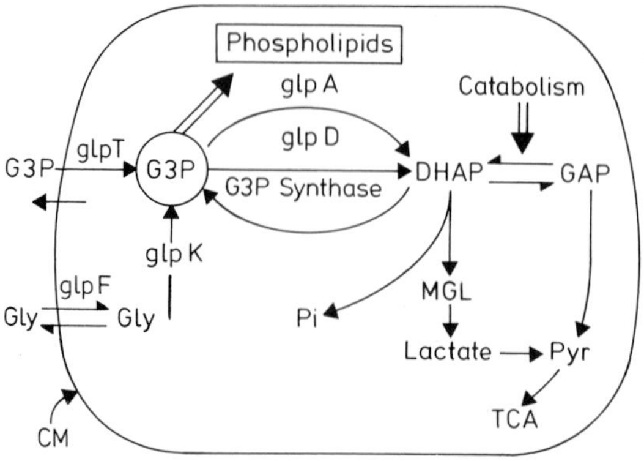

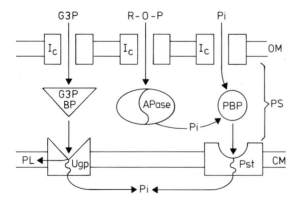

Figure 1. *sn*-Glycerol-3-phosphate (G3P) as carbon source. G3P is actively taken up by the *glpT* transport system. Its conversion to dihydroxyacetone phosphate (DHAP) is aerobically mediated by the *glpD*-encoded G3P dehydrogenase, and anaerobically by the *glpA*-encoded G3P dehydrogenase. Excess inorganic phosphate (P_i) may be excreted by the methylglyoxal (MGL) pathway (Cooper and Anderson, 1970). P_i may then exit through the *glpT*-encoded transport system. With glycerol (Gly) as carbon source, G3P is formed by the *glpK*-encoded glycerolkinase. Gly enters through the *glpF*-encoded glycerol facilitator (Heller *et al.*, 1980). All *glp* genes are under the control of the common regulator *glpR*. For details see Lin (1976). CM, cytoplasmic membrane; GAP, glyceraldehyde-3-phosphate; Pyr, pyruvate; TCA, tricarbonic acid cycle.

enzyme. Figure 2 shows the different activities and transport systems that are under the control of the *pho* regulon.

Besides the *glpT*- and *ugp*-dependent transport systems, there is a third system to recognize G3P that is coded for by the *uhp* gene, responsible for the uptake of glucose-6-phosphate and other sugar phosphates (Guth *et al.*, 1980). This transport system is only turned on in the presence of exogenous glucose-6-phosphate (Winkler, 1971; Dietz and Heppel, 1971), and the specificity for G3P might be fortuitous.

The following discussion deals with the biochemical, genetic, and physiological properties of the two G3P transport systems that belong to the *glp* and *pho* regulon, respectively.

Figure 2. *sn*-Glycerol-3-phosphate (G3P) as phosphate source. During phosphate starvation or by mutation to constitutivity, several proteins appear in the cell envelope: alkaline phosphatase (APase); the phosphate-binding protein (PBP); the G3P-binding protein (G3PBP); and a pore protein (I_c) in the outer membrane (OM). These proteins are geared to an optimal scavenge of phosphate in free or bound form. A pore facilitates the entry into the periplasm. There, G3P can be picked up by a binding protein of high affinity. Further transport into the cell occurs via membrane-bound components of a transport system coded for by the *ugp* gene(s). G3P entering in this way appears immediately in phospholipids. The phosphate moiety of G3P can be used as the sole source of phosphate for growth, while the carbon in G3P does not enter the cells' general pool of catabolism. Other larger organic phosphates (R–O–P), after entering the periplasm, will be split by alkaline phosphatase to liberate P_i. P_i, after entering the periplasm or liberated in the periplasm, is recognized by the phosphate-binding protein and further transported through the *pst*-encoded transport system (Gerdes *et al.*, 1977; Willsky and Malamy, 1980). CM, cytoplasmic membrane; PS, periplasmic space; PL, phospholipids.

2. THE glpT-DEPENDENT G3P TRANSPORT SYSTEM

Lin and his co-workers discovered that G3P is taken up by *E. coli* cells against the concentration gradient, in an energy-dependent fashion and without chemical alteration (Hayashi *et al.*, 1964). This was particularly surprising because the space between inner and outer membrane, the periplasm, is known to contain phosphatases that can hydrolyze G3P to glycerol and inorganic phosphate prior to its uptake into the cell (Torriani and Rothman, 1961). Mutants that lack the transport system can easily be isolated by selecting for resistance against the cell wall antibiotic fosfomycin, which is an analog of G3P that is recognized by the transport system (Kahan *et al.*, 1974). When analyzed by genetic crosses, these mutants have been found to map in a region called *glpT* that is about 50% cotransducible with the marker *gyrA* (Kistler and Lin, 1971). Other G3P⁻ mutants have been isolated as fosfomycin-resistant clones that do not map in *glpT*. However, these mutants exhibit pleiotropic carbohydrate-negative phenotypes and carry mutations in the nonspecific genes for the phosphotransferase system (Cordaro *et al.*, 1976), the adenylate cyclase or the catabolite gene activator protein (CAP). This demonstrates that the transcription of the *glpT* operon is subjected to catabolite repression and to positive regulation by the cAMP–CAP complex. The substrates that can be transported by the *glpT* permease are *sn*-glycerol-3-phosphate (L-α-glycerolphosphate) (K_m 12 μM) but not the corresponding D isomer (Lin *et al.*, 1962), glyceraldehyde-3-phosphate, inorganic phosphate (P_i) (Hayashi *et al.*, 1964), and arsenate (Bennet, 1973). In addition, toxic analogs can be recognized, for instance fosfomycin (Kahan *et al.*, 1974) and 3,4-dihydroxybutylphosphonate (Leifer *et al.*, 1977).

The transport system mediates entry as well as exit of G3P (Hayashi *et al.*, 1964). The latter may serve as an overflow system when the level of endogenous G3P becomes too high, for instance during growth on glycerol. This can be seen particularly under anaerobic conditions, where growth on glycerol in the presence of fumarate is severely reduced in mutants that lack the *glpT*-dependent transport system. The energy coupling of the system is provided by the protonmotive force (PMF). This can be seen by the system's sensitivity toward uncouplers of the proton conductive type as well as by the fact that membrane vesicles exhibit high transport activity that is driven by the oxidation of respiratory substrates, D-lactate, and also by G3P itself (Boos *et al.*, 1977; Heppel *et al.*, 1972). Experiments designed to identify and isolate proteins that are part of the transport machinery have lead to the discovery of a periplasmic protein (GLPT protein) the role of which in the transport mechanism remains a mystery. The GLPT protein is released from the cell envelope by the cold osmotin shock procedure (Silhavy *et al.*, 1976) in analogy to many substrate-binding proteins that establish the recognition site of many active transport systems (Wilson and Smith, 1978). However, the GLPT protein is clearly not a binding protein: it does not bind its substrate, and in contrast to monomeric substrate-binding proteins, it is a dimer of apparently identical subunits of 40,000 molecular weight (Ludtke *et al.*, 1981). [Due to the unusual behavior of the undissociated form of GLPT protein on SDS gels, the complex's subunit structure was previously (Boos *et al.*, 1977) reported to be a tetramer.] The complex is remarkably stable against SDS at room temperature and can "renature" to the dimer after urea treatment or heating in SDS (Boos *et al.*, 1977). The gene for the GLPT protein has been cloned and used as matrix in an *in vitro* synthesizing system (Schumacher and Bussmann, 1978) as well as in minicells (Larson, 1981). Under these conditions, the protein is formed as mature protein as well as in a form of slightly higher molecular weight. The latter may represent the biosynthetic precursor in analogy to numerous periplasmic and outer-membrane proteins that are synthesized and secreted as precursors with a signal sequence of 15–30 amino acids (Silhavy *et al.*, 1979). The correlation for a possible participation of the GLPT protein in the transport process is indirect: (1) most G3P⁻ mutants lack the protein and all G3P⁺ revertants produce it again; (2) the synthesis of the GLPT protein is under the same control as G3P transport and all mutants missing the GLPT protein map in the *glpT* region.

The smallest cloned piece of DNA that allows complementation for growth on G3P is 2000 bases long. Strains carrying this plasmid do not synthesize the GLPT protein (Larson, 1982). In addition, from complementation analysis with a large number of mutants that cannot grow on G3P, it is clear that only one gene is essential for G3P transport (Ludtke *et al.*, 1982). Obviously, this gene cannot be the structural gene for the GLPT protein. It is located proximal to the structural gene for GLPT on a polycystronic operon.

Therefore, if at all, the GLPT protein can only play a subtle role in the transport of G3P through the cell envelope of *E. coli*. In analogy to the function of the λ receptor in the transport of maltose and maltodextrins (Ferenci and Boos, 1980), the GLPT protein may function by increasing the diffusion rate of G3P through the outer membrane. In this respect it is significant that *Salmonella typhimurium* also exhibits a *glp*-regulated G3P system even though it does not produce periplasmic GLPT protein. It maps at the corresponding position of the *Salmonella* chromosome and exhibits the same specificity as the *E. coli* system. Concomitantly, the K_m of G3P transport is elevated by a factor of about five in comparison to the (GLPT-containing) *E. coli* system (Hengge and Boos, unpublished).

3. THE ugp-DEPENDENT TRANSPORT SYSTEM

Recently, a second transport system for G3P has been discovered (Argast *et al.*, 1978). It has a higher affinity for G3P (5 μM) but a different specificity range than the *glpT*-dependent system. Thus, inorganic phosphate, fosfomycin, and glyceraldehyde-3-phosphate are not substrates. On the other hand, 3,4-dihydroxybutylphosphonate can be recognized by the system (Guth *et al.*, 1980). Due to the toxic nature of this substance, mutants lacking the *ugp*-dependent transport system can easily be isolated (Schweizer and Boos, 1982). Wild-type bacteria that grow in the presence of P_i at concentrations greater than 1 mM do not synthesize the *ugp* transport system. Only starvation for P_i or the introduction of mutations that lead to constitutive synthesis of alkaline phosphatase allow the expression of the *ugp*-dependent system. Unlike most catabolite operons, *ugp* is not under the control of catabolite repression and is highly active in cells grown on glucose. Mechanistically, the system is operating via a classical periplasmic binding protein for G3P (Argast and Boos, 1979) and is, in contrast to the *glpT*-dependent system, not active in membrane vesicles. Surprisingly, mutants that lack alkaline phosphatase as well as glycerolkinase but that are still derepressed for *ugp*-dependent G3P transport are unable to grow on G3P in the absence of an additional carbon source. Despite the fact that G3P is taken up as the intact molecule and at a rate that is even higher than measured for a fully induced *glpT* system, G3P taken up by the *ugp* system does not allow growth. Even though these cells are unable to grow on G3P, ¹⁴C label in G3P can easily be recovered in phospholipids as well as in protein. On the other hand, the phosphate moiety of G3P can serve as the only source of phosphate in the presence of a derepressed *ugp* system and in the presence of an alternate carbon source (Schweizer and Boos, 1982). This is true in the absence of alkaline phosphatase and any other G3P-transporting system. It is clear that phosphate is released from G3P only after uptake from the periplasm. Glycerol or phosphate in large excess does not inhibit the uptake of G3P via the *ugp*-dependent system (Argast *et al.*, 1978).

The *ugp* transport system is only one of several cell envelope-localized proteins that are synthesized in response to phosphate starvation or by mutation to constitutivity (Argast and Boos, 1980). So far, identified are the periplasmic alkaline phosphatase, the *ugp*-dependent binding protein for G3P, a transport-related phosphate-binding protein, and another periplasmic protein of as yet unknown function. Most interesting, however, is the simultaneous induction (Tommassen and Lugtenberg, 1980) of a novel outer-membrane protein, previously called Ic, E, or e (Henning *et al.*, 1977; van Alphen *et al.*, 1978; Foulds and Chai, 1978), that has properties of a pore for small hydrophilic substances, but with preferences for those that are negatively charged (Argast and Boos, 1980; van Alphen *et al.*, 1978). This demonstrates that the bacterial cell is very well equipped to face times of limited phosphate supply.

The uptake of G3P by *E. coli* under different nutritional situations, be it as carbon or phosphate source, represents an interesting model of intricate regulatory circles.

REFERENCES

Argast, M., and Boos, W. (1979). *J. Biol. Chem.* **254,** 10931–10935.
Argast, M., and Boos, W. (1980). *J. Bacteriol.* **143,** 142–150.
Argast, M., Ludtke, D., Silhavy, T. J., and Boos, W. (1978). *J. Bacteriol.* **136,** 1070–1083.
Bennet, R. L. (1973). Ph.D. thesis, Tufts University, Boston, Mass.
Boos, W., Hartig-Beecken, I., and Altendorf, K. (1977). *Eur. J. Biochem.* **72,** 571–581.
Cooper, R. A., and Anderson, A. (1970). *FEBS Lett.* **11,** 273–276.
Cordaro, J. C., Melton, T., Stratis, J. P., Atagün, M., Gladding, C., Hartman, P. E., and Roseman, S. (1976). *J. Bacteriol.* **128,** 785–793.
Dietz, G. W., and Heppel, L. A. (1971). *J. Biol. Chem.* **246,** 2885–2890.
Ferenci, T., and Boos, W. (1980). *J. Supramol. Struct.* **13,** 101–116.
Foulds, J., and Chai, T. J. (1978). *J. Bacteriol.* **133,** 1478–1483.
Gerdes, R. G., Strickland, K. P., and Rosenberg, H. (1977). *J. Bacteriol.* **131,** 512–518.
Guth, A., Engel, R., and Tropp, B. E. (1980). *J. Bacteriol.* **143,** 538–539.
Hayashi, S., Koch, J. P., and Lin, E. C. C. (1964). *J. Biol. Chem.* **239,** 3098–3105.
Heller, K. B., Lin, E. C. C., and Wilson, T. H. (1980). *J. Bacteriol.* **144,** 274–278.
Henning, U., Schmidmayr, W., and Hindenach, J. (1977). *Mol. Gen. Genet.* **154,** 293–298.
Heppel, L. A., Rosen, B. P., Friedberg, I., Berger, E. A., and Weiner, J. H. (1972). In *The Molecular Basis of Biological Transport* (J. F. Woessner and F. Huijing, eds.), Academic Press, New York.
Kahan, F. M., Kahan, J. S., Cassidy, P. J., and Kropp, H. (1974). *Anal. N.Y. Acad. Sci.* **235,** 364–386.
Kistler, W. S., and Lin, E. C. C. (1971). *J. Bacteriol.* **108,** 1224–1234.
Kito, M., and Pizer, L. I. (1969). *J. Biol. Chem.* **244,** 3316–3323.
Larson, T. (1982). In preparation.
Leifer, Z., Engel, R., and Tropp, B. E. (1977). *J. Bacteriol.* **130,** 968–971.
Lin, E. C. C. (1976). *Annu. Rev. Microbiol.* **30,** 535–578.
Lin, E. C. C., Koch, J. P., Chused, T. M., and Jorgensen, S. E. (1962). *Proc. Natl. Acad. Sci. USA* **48,** 2145–2150.
Ludtke, D., Beck, C. F., and Boos, W. (1982). In preparation.
Raetz, C. R. H. (1978). *Microb. Rev.* **42,** 614–659.
Schumacher, G., and Bussmann, K. (1978). *J. Bacteriol.* **135,** 239–250.
Schweizer, H., and Boos, W. (1982). *J. Bacteriol.*, in press.
Silhavy, T. J., Hartig-Beecken, I., and Boos, W. (1976). *J. Bacteriol.* **126,** 951–958.
Silhavy, T. J., Bassford, P. J., and Beckwith, J. R. (1979). In *Bacterial Outer Membranes: Biogenesis and Functions* (M. Inouye, ed.), Wiley, New York.
Tommassen, A., and Lugtenberg, B. (1980). *J. Bacteriol.* **143,** 151–157.

Torriani, A., and Rothman, F. (1961). *J. Bacteriol.* **81,** 835–836.
van Alphen, W., van Selm, N., and Lugtenberg, B. (1978). *Mol. Gen. Genet.* **159,** 75–83.
Willsky, G. R., and Malamy, M. H. (1980). *J. Bacteriol.* **144,** 356–365.
Wilson, D. B., and Smith, J. B. (1978). In *Bacterial Transport* (B. P. Rosen, ed.), Dekker, New York.
Winkler, H. H. (1971). *J. Bacteriol.* **107,** 74–78.

103

Receptor-Dependent Transport Systems in Escherichia coli for Iron Complexes and Vitamin B_{12}

Volkmar Braun and Klaus Hantke

1. INTRODUCTION

The outer membrane of *Escherichia coli* contains phospholipids, glycolipids (lipopolysaccharide), and proteins in amounts and arrangement similar to other biological membranes. It has long been considered merely as a protective skin. Recent studies, however, have revealed that it is a functionally versatile membrane that forms a permeability barrier. Certain metabolites and antagonists can overcome the barrier by utilizing proteins in the membrane. Each molecule of nucleic acid when supplied in a phage particle can enter the cell in an intact form despite the presence of membrane-bound nucleases. Few molecules of toxic proteins, the so-called colicins, suffice to kill a cell, and they penetrate into cells through two membranes that are otherwise impermeable to macromolecules. Proteins in the outer membrane serve as receptors for phages and colicins. The mechanisms by which the macromolecules enter cells are largely unknown (Braun and Hantke, 1977; Konisky, 1979; Schwartz, 1980; Grinius, 1980). In 1973, a beneficial function for the cell was ascribed to one protein, the receptor for the phage BF23 and the E colicins, which facilitates the uptake of vitamin B_{12} (DiMasi *et al.*, 1973). In the following years, functions have been identified for most of the outer-membrane proteins. They form pores for the permeation of small hydrophilic molecules (Nikaido and Nakae, 1979), they participate in the active transport of scarce but essential metals and molecules, and they are important for the exchange of genetic information between cells in that they serve as binding sites during conjugation and transduction (Reeves, 1979; Stocker and Mäkelä, 1978).

When one compares the known translocation systems across the outer membrane (Fig. 1), one recognizes a gradation in specificity of the proteins. The most specific interactions

Volkmar Braun and Klaus Hantke • Mikrobiologie II, Universität Tübingen, D-7400 Tübingen, West Germany.

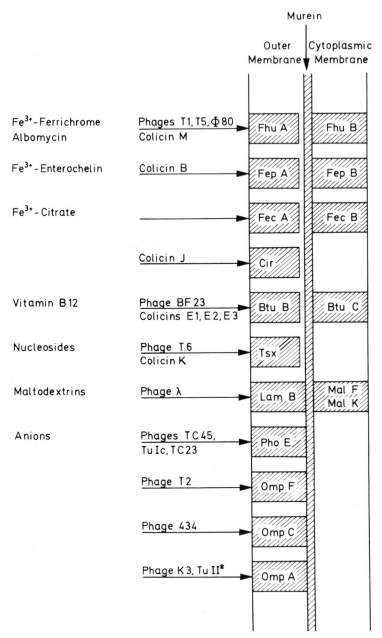

Figure 1. The functions related to outer-membrane proteins. The proteins are designated as the structural genes (Reeves, 1979). The terms FhuA and FhuB, formerly tonA and Sid, have been introduced in a recent publication (Kadner *et al.*, 1980) to describe the functions of these proteins in ferric hydroxamate uptake (ferrichrome and its structural analogs ferricrocin, ferrichrysin, and albomycin). The designations FecA and FecB have recently been proposed (Wagegg and Braun, 1981). An additional protein, 83K, synthesized under iron-stress conditions, has not been included because no compound has been identified that uses the 83K protein as receptor. The term PhoE was proposed by B. Lugtenberg (private communication) because it is under the same control as the *pho* genes (Tommassen and Lugtenberg, 1980; Argast and Boos, 1980). No proteins (permeases) in the cytoplasmic membrane have been drawn in the complex uptake systems for nucleosides (Munch-Petersen *et al.*, 1979) and the many transport systems of the substrates that pass through the outer membrane via the porins (OmpC, OmpF). Also, not all known phages are included in the scheme for the sake of clarity.

occur between the iron complexes of enterochelin, ferrichrome, and citrate, and the cobalt complex of vitamin B_{12}, and their receptor proteins. Mutants that lack these proteins are unable to take up these metal complexes. The transport of nucleosides (Krieger-Brauer and Braun, 1980) and of maltodextrins (Szmelcman *et al.*, 1976) is only facilitated by the outer-membrane proteins. However, these proteins speed up the rate of permeation of these substrates through the outer membrane. In favor of the conclusion that they function as part of these transport systems is the finding that their synthesis is under the same genetic control as the respective transport systems in the cytoplasmic membrane. The third functional class of proteins are the porins, which form channels through which pass hydrophilic substrates, ions, amino acids, peptides, and sugars. Their selectivity is mainly determined by the size of the substrates. Substances with molecular weights greater than 600 are largely excluded.

In the following, we discuss the receptor-dependent uptake of the iron complexes and of vitamin B_{12}. These transport systems are distinguished by the strict dependence on the outer-membrane protein, the binding of the substrates to the proteins, the competition between the substrates and the phages, the colicins for the adsorption to the receptors, and the requirement for the mysterious function(s) coded by the *tonB* gene.

2. THE IRON TRANSPORT SYSTEMS

Iron is an abundant metal in nature but occurs under aerobic conditions at pH 7 as a ferric hydroxy aquo complex with a very low solubility. The equilibrium concentration of iron is 10^{-12} μM. Cells of *E. coli* require for growth a concentration of 0.1 μM. To gain sufficient iron, they produce an iron chelator, enterochelin (enterobactin), which is extremely effective in binding iron (formation constant $K_f = 52$). Enterochelin, concomitant with the transport system, is only synthesized under iron-limiting growth conditions (Rosenberg and Young, 1974). Furthermore, two other transport systems have been identified, and studied in some detail. *E. coli* actively transports ferrichrome, an iron chelator produced by certain molds, and ferric citrate. Three *outer-membrane proteins* (Fig. 1) have been related to these transport systems, which are independent of each other. In addition, two other proteins appear in the outer membrane when cells are grown at low iron concentrations, the Cir protein (Fig. 1) and a protein with an apparent molecular weight of 83,000 (83K). It is very likely that the latter two proteins belong to two additional, hitherto unidentified iron transport systems. Recently, a new plasmid-coded iron transport system has been found (Williams, 1979; Stuart *et al.*, 1980) that increases the *pathogenicity* of the plasmid-containing *E. coli* strains. Iron seems to be the major nutritional factor for microorganisms that invade mammals (Weinberg, 1978). For example, in the human serum, iron is so tightly bound to transferrin that the concentration of free iron is only about 10^{-8} μM. The availability of iron is low in most environments. No wonder that cells of *E. coli* probably contain four to six iron transport systems!

The evidence for the participation of outer-membrane proteins in the transport of the iron complexes of enterochelin, ferrichrome, and citrate is convincing. They are synthesized in response to iron limitation (Braun *et al.*, 1976; Hancock *et al.*, 1976; Pugsley and Reeves, 1976; McIntosh and Earhart, 1977); their absence in mutants inactivates the transport systems entirely; and binding of ferric enterochelin (Ichihara and Mizushima, 1978), of ferrichrome (Braun and Hantke, 1977; Neilands, 1977), and of ferric citrate (Wagegg and Braun, 1981) to the receptor proteins have been demonstrated.

The localization of the transport functions in the *cytoplasmic membrane* (Fig. 1) is indirect. Ferric enterochelin has been found in the cytoplasm of mutants devoid of the esterase that hydrolyzes enterochelin (Rosenberg and Young, 1974). The esterase is probably a cytoplasmic enzyme because it cannot be released by osmotic-shock like periplasmic enzymes.

Uptake into spheroplasts of *fepB*$^+$ cells has been demonstrated, whereas spheroplasts of *fepB* mutants are inactive (Wookey and Rosenberg, 1978). In the case of ferrichrome transport, the requirement for the FhuA protein could be bypassed by treating cells with Pronase but they still required the *fhuB* gene product (Wookey, Hussein, and Braun, unpublished). The action of Pronase destroyed the outer-membrane proteins related to iron transport except the FecA protein (Wagegg and Braun, 1981). Consequently, Pronase treatment rendered the outer membrane impermeable to the ferric citrate complex. Mutants lacking the FecA protein remained impermeable to the ferric citrate upon Pronase treatment. There is only, still incomplete, evidence for a second gene beside the *fecA* gene, and the product of the *fecB* gene has tentatively been ascribed to the cytoplasmic membrane (Wagegg and Braun, 1981).

Concerning the *control of the iron transport systems*, it is of interest that they are regulated by the concentration of the metal ion in the cell. For the quantitative study, we inserted the phage Mu carrying a gene conferring ampicillin resistance and the lactose structural genes without the lactose promoter [Mu(Ap, *lac*) of Casadaban and Cohen (1979)] into the *fhuA, fepA,* and *cir* genes. We then studied the expression of β-galactosidase in response to the iron concentration supplied in the growth medium. The expression of the β-galactosidase gene in the *fepA* and the *cir* gene increased by a factor of 25 at iron-limiting growth conditions. In the *fhuA* :: Mu(Ap, *lac*) configuration, the enzyme activity increased only by a factor of 3. These results agree with the amounts of these proteins observed on gels. Evidence for a common regulator that senses the intracellular iron concentration was gained by selecting mutants that produced more β-galactosidase irrespective of the iron concentration in the medium. Further studies showed that beside the common regulation, there exists an additional control of the individual systems.

A similar mutant was fortuitously isolated from *Salmonella typhimurium* that showed derepressed activities for enterochelin synthesis, for ferric enterochelin, and for ferrichrome uptake. Three outer-membrane proteins, analogous to those of *E. coli*, were constitutively synthesized (Ernst *et al.*, 1978).

The ferric citrate transport system is peculiarly regulated. It is induced by growing cells in the presence of 0.1 mM citrate (Frost and Rosenberg, 1974; Woodrow *et al.*, 1978) although citrate does not enter the cells in measurable quantities. We collected the following additional data (Hussein and Braun, unpublished). Fluorocitrate induced the ferric citrate transport system but did not inhibit cells because it was not taken up. Fluorocitrate was also inefficient in ferric ion transport. To maintain the transport system in the induced state, the same high concentration of citrate was required as for the primary induction. Mutants devoid of ferric citrate transport (*fecB*) still expressed the FecA protein in the outer membrane in response to citrate in the growth medium. The intracellular concentration of citrate (1 mM) did not induce the ferric citrate transport system. It is unknown which compound is the true inducer and how it acts.

Modification of the iron-chelating agents has been demonstrated for enterochelin and ferrichrome. Enterochelin is hydrolyzed to its basic subunit 2,3-dihydroxy-*N*-benzoyl-L-serine, which is excreted (Rosenberg and Young, 1974). The destruction of the chelator and the reduction of iron from the ferric to the ferrous form releases the iron intracellularly from the strong ligand. Ferrichrome is acetylated at one of the six iron-binding sites and excreted (Hartmann and Braun, 1980). Reduction is a prerequisite for acetylation, and both reactions occur *in vitro* in the membrane fraction (Schneider and Braun, unpublished). The acetylation reduces strongly the affinity for iron. This could be a mechanism by which cells rid themselves of a potentially deleterious ligand for iron in the cytoplasm.

3. THE TRANSPORT SYSTEM FOR VITAMIN B$_{12}$

The products of the known genes controlling vitamin B$_{12}$ transport are depicted in Fig. 1. Mutants in the *btuB* gene were impaired in initial binding of vitamin B$_{12}$ and in transport (Kadner, 1978). *btuC* mutants showed normal binding but little secondary uptake. They accumulated vitamin B$_{12}$ in the periplasmic space between the outer and the cytoplasmic membrane (Reynolds *et al.*, 1980). The addition of arsenate stimulated the uptake into the periplasmic space 3- to 12-fold and impaired the transport across the cytoplasmic membrane. It was concluded that *btuC*-dependent transport across the cytoplasmic membrane requires phosphate bond energy. The release of vitamin B$_{12}$ from the BtuB protein into the periplasm was dependent on a different functional domain of the receptor protein than the initial binding (revealed by *btuA* mutants).

4. THE *tonB* FUNCTION

The function controlled by the *tonB* gene has only been found in relation to uptake processes across the outer membrane which require receptor proteins. The designation of the gene was derived from the phage T1, which binds irreversibly only to *tonB*$^+$ cells. Irreversible binding also required energy which could be derived either from respiration or from ATP hydrolysis. These two observations led to the hypothesis that the *tonB*-controlled function might act like a coupling factor which mediates the energy state of the cytoplasmic membrane to some outer-membrane proteins and thus regulates their function (Hancock and Braun, 1976). In support of this conclusion were the following findings. Host-range mutants of phage T1, which adsorbed already irreversibly to the isolated outer membrane, infected cells independent of the *tonB* function. Ferrichrome only inhibited adsorption of phage T5 to the FhuA protein when the receptor was uncoupled from the *tonB* function (*tonB* mutants) or from the energization of the cell (unenergized cells or isolated outer membrane) (Hantke and Braun, 1978). The requirement of the FhuA protein and also of the *tonB* function for the killing of cells by colicin M could be bypassed by osmotic shock treatment. Furthermore, colicin M remained bound to the FhuA protein and prevented adsorption of phage T5 in unenergized cells and in *tonB* mutants (Braun *et al.*, 1980). With regard to the transport of ferric enterochelin, the outer-membrane FepA protein (Fig. 1) could also be bypassed by adjusting the synthesis of enterochelin to a low level so that it remained presumably in the periplasm. Under these conditions no *tonB* function was required (Hancock *et al.*, 1977). The release of vitamin B$_{12}$ from the BtuB outer-membrane protein also required the *tonB* gene product, and the authors suggested that the mechanisms may involve a diffusible messenger that is generated in the periplasm by the action of the *tonB* protein and the protonmotive force in the cytoplasmic membrane (Reynolds *et al.*, 1980). Spheroplasts obtained by treating *E. coli tonB* mutants with EDTA and lysozyme in Tris–HCl buffer transported ferrichrome. The authors suggest that in spheroplasts, ferrichrome has direct access to the cytoplasmic membrane and that this is the reason why the transport became *tonB* independent (Weaver and Konisky, 1980).

The following observations cannot be reconciled with the above-mentioned hypothesis of *tonB* function. Spheroplasts prepared in glycylglycine buffer with half the concentration of lysozyme and one tenth of the EDTA concentration described before were still dependent on the *tonB* function in the uptake of ferric enterochelin. It was concluded that the *tonB* function is involved in the transport of ferric enterochelin through the cytoplasmic membrane (Wookey and Rosenberg, 1978). In line with this interpretation is the finding that after treatment of *fhuA* mutant cells with Pronase, they remained fully viable and transported ferrichrome with similar rates as the *fhuA*$^+$ parent cells. The outer membrane of the Pronase-treated cells probably became permeable to ferrichrome but the uptake still required the *tonB* function (Wookey, Hussein, and Braun, unpublished). It is questionable whether the *tonB* product functions in all the translocation processes in which it is involved in the same manner. This may explain the different results. In fact, *tonB* mutants were isolated that were impaired in some *tonB*-dependent functions but not in others (Hantke and Braun, 1978). Considerations of its mode of action have to account for the observation that the *tonB* gene product is unstable (Bassford *et al.*, 1977) and apparently consumed by the processes in which it participates (Kadner and McElhaney, 1978).

The fascination of the studies on the receptors of the outer membrane of *E. coli* is derived from the finding that by their mediation many compounds enter the cell very specifically through a membrane that is normally impermeable to such substances. The compounds include nucleic acids, colicins, and metal chelates. The mechanisms by which they pass the outer membrane are especially interesting, for according to current thoughts the outer membrane is separated from the metabolism of the cell by the murein layer and the periplasm. However, the work described in this paper indicates a functional interaction of both membranes.

ACKNOWLEDGMENTS. The authors' work was supported by the Deutsche Forschungsgemeinschaft (SFB 76) and by the Fonds der Chemischen Industrie. We thank P. Wookey for his comments on the manuscript.

REFERENCES

Argast, M., and Boos, W. (1980). *J. Bacteriol.* **143,** 142–150.
Bassford, P. J., Schnaitman, C. A., and Kadner, R. J. (1977). *J. Bacteriol.* **130,** 750–758.
Braun, V., and Hantke, K. (1977). In *Microbial Interactions* (J. L. Reissig, ed.), Series B, Vol. 3, pp. 99–137, Chapman & Hall, London.
Braun, V., and Hantke, K. (1981). In *Organization of the Prokaryotic Cell Membrane* (B. K. Ghosh, ed.), Vol. II, pp. 1–73, CRC Press, Boca Raton, Fla.
Braun, V., Hancock, R. E. W., Hantke, K., and Hartmann, A. (1976). *J. Supramol. Struct.* **5,** 37–58.
Braun, V., Frenz, J., Hantke, K., and Schaller, K. (1980). *J. Bacteriol.* **142,** 162–168.
Casadaban, M. J., and Cohen, S. N. (1979). *Proc. Natl. Acad. Sci. USA* **76,** 4530–4533.
DiMasi, D. R., White, J. C., Schnaitman, C. A., and Bradbeer, C. (1973). *J. Bacteriol.* **115,** 506–513.
Ernst, J. F., Bennett, R. L., and Rothfield, L. I. (1978). *J. Bacteriol.* **135,** 928–934.
Frost, F. E., and Rosenberg, H. (1974). *Biochim. Biophys. Acta* **330,** 90–101.
Grinius, L. (1980). *FEBS Lett.* **113,** 1–10.
Hancock, R. E. W., and Braun, V. (1976). *J. Bacteriol.* **125,** 409–415.
Hancock, R. E. W., Hantke, K., and Braun, V. (1976). *J. Bacteriol.* **127,** 1370–1375.
Hancock, R. E. W., Hantke, K., and Braun, V. (1977). *Arch. Microbiol.* **114,** 231–239.
Hantke, K., and Braun, V. (1978). *J. Bacteriol.* **135,** 190–197.
Hartmann, A., and Braun, V. (1980). *J. Bacteriol.* **143,** 246–255.
Ichihara, S., and Mizushima, S. (1978). *J. Biochem.* **83,** 137–140.
Kadner, R. J. (1978). In *Bacterial Transport* (B. P. Rosen, ed.), pp. 463–493, Dekker, New York.
Kadner, R. J., and McElhaney, G. (1978). *J. Bacteriol.* **134,** 1020–1029.

Kadner, R. J., Heller, K., Coulton, J. W., and Braun, V. (1980). *J. Bacteriol.* **143,** 256–264.

Konisky, J. (1977). In *Bacterial Outer Membranes* (M. Inouye, ed.), pp. 319–359, Wiley–Interscience, New York.

Krieger-Brauer, H. J., and Braun, V. (1980). *Arch. Microbiol.* **124,** 233–242.

McIntosh, M. A., and Earhart, C. F. (1977). *J. Bacteriol.* **131,** 331–339.

Munch-Petersen, A., Mygind, B., Nicolasen, A., and Pihl, N. J. (1979). *J. Biol. Chem.* **254,** 3730–3737.

Neilands, J. B. (1977). In *Bioorganic Chemistry-II* (K. N. Raymond, ed.), pp. 1–32, American Chemical Society, Washington, D.C.

Nikaido, H., and Nakae, T. (1979). *Adv. Microb. Physiol.* **20,** 163–250.

Pugsley, A. P., and Reeves, P. (1976). *Biochem. Biophys. Res. Commun.* **70,** 846–853.

Reeves, P. (1979). In *Bacterial Outer Membranes* (M. Inouye, ed.), pp. 256–291, Wiley–Interscience, New York.

Reynolds, P. R., Mottur, G. P., and Bradbeer, C. (1980). *J. Biol. Chem.* **255,** 4314–4319.

Rosenberg, H., and Young, I. G. (1974). In *Microbial Iron Metabolism* (J. B. Neilands, ed.), pp. 67–82, Academic Press, New York.

Schwartz, M. (1980). In *Virus Receptors* (L. L. Randall and L. Philipson, eds.), Part 1, pp. 61–94, Chapman & Hall, London.

Stocker, B. A. D., and Mäkelä, P. H. (1978). *Proc. R. Soc. London* **202,** 5–30.

Stuart, S. J., Greenwood, K. T., and Luke, R. K. J. (1980). *J. Bacteriol.* **143,** 35–42.

Szmelcman, S., Schwartz, M., Silhavy, T. J., and Boos, W. (1976). *Eur. J. Biochem.* **65,** 13–19.

Tommassen, J., and Lugtenberg, B. (1980). *J. Bacteriol.* **143,** 151–157.

Wagegg, W., and Braun, V. (1981). *J. Bacteriol.* **145,** 156–163.

Weaver, C. H., and Konisky, J. (1980). *J. Bacteriol.* **143,** 1513–1518.

Weinberg, E. D. (1978). *Microb. Rev.* **42,** 45–66.

Williams, P. (1979). *Infect. Immun.* **26,** 925–932.

Woodrow, G. C., Langman, L., Young, I. G., and Gibson, F. (1978). *J. Bacteriol.* **133,** 1524–1526.

Wookey, P., and Rosenberg, H. (1978). *J. Bacteriol.* **133,** 661–666.

104

Bacterial Inorganic Cation and Anion Transport Systems

A Bug's Eye View of the Periodic Table

Simon Silver and Robert D. Perry

1. INTRODUCTION

Free-living cells, whether bacterial or eukaryotic, use energy-dependent transport processes to accumulate required inorganic nutrients from their environment and, simultaneously, to exclude toxic or nonessential ions.

There appear to be quite separate transport systems for all required inorganic nutrients. The specificities of these ion transport systems are analogous to specificities shown toward organic nutrients. Transport systems for normally abundant ions sometimes have minor specificity overlaps that may function in accumulation of less abundant trace ions. The specificities and other kinetic parameters of these systems appear tailored by evolution for the roles that the transport systems play in cellular metabolism. This is the thesis of this "bug's eye" view of the periodic table and several more detailed recent reviews (Jasper and Silver, 1977; Silver, 1977, 1978; Silver and Jasper, 1977). We will limit our discussion to developments that have been published during the last few years.

Many of the transport systems that will be considered are diagrammed in Fig. 1. Taken as a whole, the separate uptake and efflux transport systems carefully regulate the cellular ionic environment to meet the needs of cellular metabolism, osmotic pressure, and pH. Divalent cation transport systems do not accept monovalent cations and vice versa. The transport of Ca^{2+}, Fe^{3+}, K^+, Mg^{2+}, Mn^{2+}, Na^+, PO_4^{3-}, and SO_4^{2-} will be described in separate sections below. Note also the several closely related chapters in this volume

Simon Silver and Robert D. Perry • Department of Biology, Washington University, St. Louis, Missouri 63130. *Present address of R. D. P.:* Department of Microbiology and Public Health, Michigan State University, East Lansing, Michigan 48823.

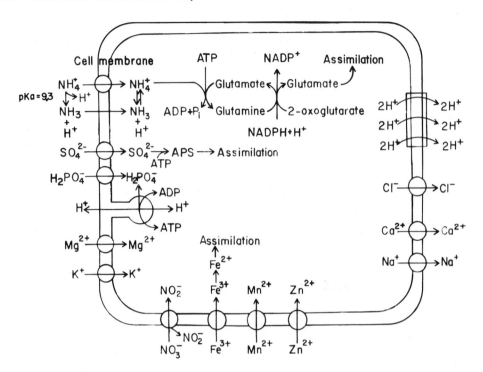

Figure 1. Some known membrane transport systems for cations and anions.

(Chapters 87, 96, 105, 117) that report in greater detail recent understanding of several ion transport systems.

2. AMMONIUM, CHLORIDE, HYDROGEN, AND ZINC TRANSPORT

The active transport of NH_4^+ and methylamine, $CH_3NH_3^+$ (an analog of ammonia), has been reported for several eukaryotic microbes. While ammonia is thought to enter as charged NH_4^+, this may vary from organism to organism (see Silver and Perry, 1981, for a short review).

Chloride is the major inorganic anion found in microbial environments for which there is no known intracellular role. Direct studies on outwardly oriented Cl^- transport systems have been performed only in eukaryotic microbes (Miller and Budd, 1975, 1976). The hypothesized bacterial Cl^- efflux system (summarized by Silver, 1978) is based on indirect evidence that intracellular Cl^- levels are below those predicted from the known membrane potential (intracellular negative by 125 to 180 mV), and the distribution prediction from the Nernst equilibrium equation.

Protons are a special case because they are used as energy-coupling "currency" in Mitchell's chemiosmotic hypothesis (Harold, 1978; Mitchell, 1968, 1979). In Fig. 1, H^+ is shown leaving or entering the cell via the reversible ATPase ATP synthetase. H^+ efflux is also shown occurring at three points in a box representing the respiratory chain. Each "coupling site" is now pictured as a site for the outward pumping of two protons per electron transported down the chain (Mitchell, 1979). Protons are also used in symport and antiport mechanisms for the accumulation or exclusion of other inorganic or organic compounds (Rosen and Kashket, 1978).

Although a specific Zn^{2+} transport system has been demonstrated in eukaryotic microbes (Failla, 1977), no one has followed up the few published reports of bacterial Zn^{2+} transport (Bucheder and Broda, 1974; Weiss *et al.*, 1978). The putative bacterial Zn^{2+} transport system is difficult to study because contaminating Zn^{2+} in growth media and uptake buffers may be sufficient to saturate or repress Zn^{2+} accumulation. Very little information is available concerning other possibly required trace cations including those of chromium, cobalt, copper, and nickel (Silver, 1978).

3. POTASSIUM TRANSPORT

All bacterial cells actively accumulate K^+; many bacterial cells including both gram negative (Helmer *et al.*, this volume) and gram positive (see Silver, 1978) have two separate K^+ transport systems. The systems in *Escherichia coli* are called Kdp (for K^+-dependent growth) and TrkA (for transport of K^+). The Kdp system is a very-high-affinity transport system (K_m about 2 μM K^+) that is subject to derepression by K^+ starvation. It can concentrate K^+ to nearly $10^6 : 1$ inside to outside the cells, or greatly above the few-hundred-to-one ratio that would be obtained by Nernst equation equilibrium with the existing membrane potential. A major achievement in understanding this system was the discovery that the Kdp system functions as an ATPase (Epstein *et al.*, 1978), and the identification of one of the polypeptide components of this system, the product of the *kdpB* gene, as a phosphorylated membrane protein (Epstein and Laimins, 1980). Two other membrane polypeptides, the products of the *kdpA* and *kdpC* genes, have also been identified, and a fourth gene, *kdpD*, plays a regulatory role by governing the synthesis of the proteins of this system. The Kdp multicomponent ATPase transport system is the first such found in bacterial cells and may be the paradigm for many others. As Harold describes in Chapter 87 of this monograph, a similar K^+ transport system exists in *Streptococcus faecalis* and depends upon both the membrane potential and directly on ATP. The second K^+ transport system in *E. coli*, the TrkA system, appears to be a conventional, relatively low-affinity system dependent upon the membrane potential as its driving force (Epstein and Laimins, 1980). Its K_m of 1.5 mM K^+ is adequate for many growth situations, but not for competition with other species under conditions of potassium starvation. The TrkA system is affected by mutations in many genes, *trkA* through *trkF*, but it is not clear how the products of these genes interact. Mutations in two of the genes, *trkB* and *trkC*, affect only the retention of intracellular K^+ and not uptake rates (Epstein and Laimins, 1980). Recently, a K^+ efflux system has been identified that probably regulates intracellular pH by K^+–H^+ exchange (Brey *et al.*, 1980).

4. MAGNESIUM TRANSPORT

Unfortunately, there has been little progress in understanding bacterial Mg^{2+} transport systems in the last couple of years. The current understanding was summarized by Jasper and Silver (1977) and by Silver (1978). The key report was that of Park *et al.* (1976), which defined the two Mg^{2+} transport systems of *E. coli*. The magnesium transport system, Mgt, is specific for Mg^{2+}, repressible by growth in high Mg^{2+}, and missing in *mgt* mutants. Its K_m, however, is no different (30 μM Mg^{2+}) than that for the less specific Cor Mg^{2+} transport system, which accepts Mg^{2+} and also Co^{2+}, Mn^{2+}, and Ni^{2+} as alternative substrates (Nelson and Kennedy, 1972; Park *et al.*, 1976). The only recent Mg^{2+} transport paper was that of Jasper and Silver (1978) reporting that *Rhodopseudomonas capsulata* has energy-dependent Mg^{2+} transport (inhibited by Mn^{2+} and Co^{2+}) basically similar to that of *E. coli*. Every microorganism that has been studied has a specific Mg^{2+} transport system. However, relatively little research has been done on these systems in recent years.

5. MANGANESE TRANSPORT

Manganese transport systems have also been found in all bacteria that have been tested (Silver and Jasper, 1977). These systems differ from the Mg^{2+} transport systems in that they have higher affinities (K_m's about 1/100 those for Mg^{2+} transport) and that extracellular Mg^{2+} is totally without effect on Mn^{2+} uptake. Recently it has been found that Cd^{2+} enters *S. aureus* cells through the Mn^{2+} transport system (Weiss *et al.*, 1978). Cd^{2+} is not accumulated by *E. coli* cells at levels comparable to those accumulated by *S. aureus* (unpublished experiments). This observation may account for the relative Cd^{2+}-resistance of gram-negative bacteria. In studies of right-side-out membrane vesicles of *S. aureus*, a careful kinetic analysis suggests that Cd^{2+} and Mn^{2+} are alternative substrates for a common, chemiosmotic transport system with K_m's of 0.25 μM Cd^{2+} and 1.1 μM Mn^{2+} (Perry and Silver, 1982). Mn^{2+} transport is, then, the first example of a trace element, high-affinity transport system functioning as a membrane-potential-dependent system.

6. IRON TRANSPORT

With the solubility of $Fe(OH)_3$ near 10^{-39} M, essentially no free ionic Fe^{3+} exists in solution; Fe^{2+} is generally oxidized in normal growth media. Because animal transferrins and other host iron-binding compounds avidly chelate ionic iron, microbial pathogens must compete to survive in an iron-deficient environment (Weinberg, 1978). Microbes have devised highly specific and complex transport systems that extract iron from bound and precipitated forms, making it available to the cells. *E. coli* synthesizes at least four different Fe^{3+} transport systems.

1. The enterochelin system uses an iron chelate synthesized by *E. coli* and related bacteria. Enterochelin is a cyclic triester of 2,3-dihydroxy-*N*-benzoyl-L-serine and is synthesized by a series of enzymes coded for by the *entA* through *entG* genes (Rosenberg and Young, 1974). This siderophore (iron chelate) is quantitatively excreted by the cells. It chelates extracellular Fe^{3+}, which is then translocated (with a K_m of 0.1 μM) back into the cytoplasm by an array of outer- and inner-membrane proteins (Rosenberg and Young, 1974; Konisky, 1979). Iron is released from the chelate by hydrolytic cleavage of enterochelin. In this cycle only a single Fe^{3+} enters the cell at the expense of synthesis and destruction of a molecule of enterochelin (Rosenberg and Young, 1974; Byers and Arceneaux, 1977; Neilands, 1981).

2. *E. coli* and other bacteria have additional systems for the transport of extracellular hydroxamate siderophores. (Enterochelin is a phenolate siderophore.) In *E. coli*, several hydroxamate siderophores are used despite the fact that they are not synthesized by this bacterium. *Bacillus megaterium* is an example of a bacterium that synthesizes and uses its own hydroxamate siderophore. The iron chelate is generally translocated into the cytoplasm where iron is reductively released from the hydroxamate (Byers and Arceneaux, 1977).

3. Some strains of *E. coli* (K-12 and B/r but not W) possess a transport system

induced by extracellular citrate during iron starvation. This system is quite specific for citrate and has an apparent K_m for Fe^{3+} of 0.2 μM (Rosenberg and Young, 1974).

4. Several microorganisms can accumulate ionic iron by a very-low-affinity process, whose energy dependence has not been clearly established. This "system" is clearly useful only during growth in an iron-sufficient environment (Rosenberg and Young, 1974).

5. Recently a new plasmid-coded, high-affinity transport system has been identified. During iron starvation, *E. coli* cells harboring colV plasmids synthesize a cell-associated hydroxamate siderophore [recently identified as aerobactin (Braun, 1981), the same siderophore originally found in *Aerobacter aerogenes* (Gibson and Magrath, 1969)] and show enhanced iron uptake (Williams, 1979; Stuart *et al.*, 1980). The increased pathogenicity of bacteria with plasmid colV has reinitiated an interest in the role of iron availability in infectious processes.

7. PHOSPHATE TRANSPORT

E. coli has two primary inorganic phosphate transport systems: one highly specific, and the other less stringent in its substrate specificity (Silver, 1978). The basic properties of the Phosphate-specific transport system, Pst, are a dependence upon ATP for energy, a very low K_m of about 0.2 μM PO_4^{3-}, and a much higher K_i for the analog AsO_4^{3-} of about 25 μM (Rosenberg *et al.*, 1977). The Pst system is derepressed by phosphate starvation (Rosenberg *et al.*, 1977) or by arsenate toxicity (Willsky and Malamy, 1980). The less specific phosphate transport system, Pit, has a similar K_i for arsenate but a much lower affinity (K_m 25 μM PO_4^{3-}) for phosphate. Thus, the basis for the specificity is not that the affinity toward the toxic analog is reduced, but rather that the affinity toward the natural substance is increased. Although only a single gene is known that affects the membrane-potential-dependent Pit system, two or possibly three genes affect the Pst system (Rosenberg *et al.*, 1977; Willsky and Malamy, 1980; Cox et al., 1981). The *phoS* gene determines a periplasmic phosphate-binding protein analogous to those found for other ATP-dependent systems. In fact, the Kdp K^+ transport system is currently the only ATP-dependent system that apparently lacks such an extracellular binding protein (Epstein and Laimins, 1980). *phoT* and *pst* are the other two genes that are thought to govern inner-membrane proteins that constitute the Pst transport system itself.

8. SULFATE TRANSPORT

Salmonella typhimurium has a high-affinity sulfate transport system ($K_m = 36$ μM) with a relatively nonstringent specificity; CrO_4^{2-}, SeO_4^{2-}, SO_3^{2-}, and $S_2O_3^{2-}$ inhibit SO_4^{2-} transport (Dreyfuss, 1964; Pardee *et al.*, 1966). A number of genes that affect the regulation and transport activity of this system as well as SO_4^{2-} assimilation have been identified (Ohta *et al.*, 1971). *E. coli* also has a relatively high-affinity sulfate transport system ($K_m = 50$ μM); SeO_4^{2-} inhibits SO_4^{2-} transport (Springer and Huber, 1972).

9. CALCIUM AND SODIUM EFFLUX SYSTEMS

Ca^{2+} and Na^+ are major cations found in natural environments and tissues of higher animals. Both Ca^{2+} and Na^+ are separately excreted from bacterial cells by efflux transport systems (Beck and Rosen, 1979; Brey and Rosen, 1979; Lanyi, 1979; Lindley and MacDonald, 1979). As these systems are described in great detail in Chapters 96 and 117, we need not describe the significant progress that has been made since the quite recent suggestion that such systems should exist (see Silver, 1977, 1978).

10. CONCLUSION

For every inorganic cation and anion that microbial cells need for nutritional purposes or that exist in common growth milieus, highly specific membrane transport systems exist with directionality, energization, specificity, and other properties tailored to cellular needs. One need only find the conditions for defining and studying these systems.

REFERENCES

Beck, J. C., and Rosen, B. P. (1979). *Arch. Biochem. Biophys.* **194,** 208–214.
Braun, V. (1981). *FEBS Lett.* **11,** 225–228.
Brey, R. N., and Rosen, B. P. (1979). *J Biol. Chem.* **254,** 1957–1963.
Brey, R. N., Rosen, B. P., and Sorensen, E. N. (1980). *J. Biol. Chem.* **255,** 39–44.
Bucheder, F., and Broda, E. (1974). *Eur. J. Biochem.* **45,** 555–559.
Byers, B. R., and Arceneaux, J. E. L. (1977). In *Microorganisms and Minerals* (E. D. Weinberg, ed.), pp. 215–249, Dekker, New York.
Cox, G. B., Rosenberg, H., Downie, J. A., and Silver, S. (1981). *J. Bacteriol.* **148,** 1–9.
Dreyfuss, J. (1964). *J. Biol. Chem.* **239,** 2292–2297.
Epstein, W., and Laimins, L. (1980). *Trends Biochem. Sci.* **5,** 21–23.
Epstein, W., Whitelaw, V., and Hesse, J. (1978). *J. Biol. Chem.* **253,** 6666–6668.
Failla, M. L. (1977). In *Microorganisms and Minerals* (E. D. Weinberg, ed.), pp. 151–214, Dekker, New York.
Gibson, F., and Magrath, D. I. (1969). *Biochim. Biophys. Acta* **192,** 175–184.
Harold, F. M. (1978). In *The Bacteria* (L. N. Ornston and J. R. Sokatch, eds.), Vol. 6, pp. 463–521, Academic Press, New York.
Jasper, P., and Silver, S. (1977). In *Microorganisms and Minerals* (E. D. Weinberg, ed.), pp. 4–47, Dekker, New York.
Jasper, P., and Silver, S. (1978). *J. Bacteriol.* **133,** 1323–1328.
Konisky, J. (1979). In *Bacterial Outer Membranes* (M. Inouye, ed.), pp. 319–359, Wiley–Interscience, New York.
Lanyi, J. K. (1979). *Biochim. Biophys. Acta* **559,** 377–397.
Lindley, E. V., and MacDonald, R. E. (1979). *Biochem. Biophys. Res. Commun.* **88,** 491–499.
Miller, A. G., and Budd, K. (1975). *J. Bacteriol.* **121,** 91–98.
Miller, A. G., and Budd, K. (1976). *J. Bacteriol.* **128,** 741–748.
Mitchell, P. (1968). *Chemiosmotic Coupling and Energy Coupling,* Glynn Research, Bodmin.
Mitchell, P. (1979). *Science* **206,** 1148–1159.
Neilands, J. B. (1981). *Ann. Rev. Biochem.* **50,** 715–731.
Nelson, D. L., and Kennedy, E. P. (1972). *Proc. Natl. Acad. Sci. USA* **69,** 1091–1093.
Ohta, N., Galsworthy, P. R., and Pardee, A. B. (1971). *J. Bacteriol.* **105,** 1053–1062.
Pardee, A. B., Prestridge, L. S., Whipple, M. B., and Dreyfuss, J. (1966). *J. Biol. Chem.* **241,** 3962–3969.
Park, M. H., Wong, B. B., and Lusk, J. E. (1976). *J. Bacteriol.* **126,** 1096–1103.
Perry, R. D., and Silver, S. (1982). *J. Bacteriol.* **150(2),** in press.
Rosen, B. P., and Kashket, E. R. (1978). In *Bacterial Transport* (B. P. Rosen, ed.), pp. 559–620, Dekker, New York.
Rosenberg, H., and Young, I. G. (1974). In *Microbial Iron Metabolism* (J. B. Neilands, ed.), pp. 67–82, Academic Press, New York.
Rosenberg, H., Gerdes, R. G., and Chegwidden, K. (1977). *J. Bacteriol.* **131,** 505–511.
Silver, S. (1977). In *Microorganisms and Minerals* (E. D. Weinberg, ed.), pp. 49–103, Dekker, New York.
Silver, S. (1978). In *Bacterial Transport* (B. P. Rosen, ed.), pp. 221–324, Dekker, New York.

Silver, S., and Jasper, P. (1977). In *Microorganisms and Minerals* (E. D. Weinberg, ed.), pp. 105–149, Dekker, New York.

Silver, S., and Perry, R. D. (1981). In *Recent Developments in Biological and Chemical Research with Short-Lived Radioisotopes* (J. W. Root and K. A. Krohn, eds.), American Chemical Society, Washington, D.C., in press.

Springer, S. E., and Huber, R. E. (1972). *FEBS Lett.* **27,** 13–15.

Stuart, S. J., Greenwood, K. T., and Luke, R. K. J. (1980). *J. Bacteriol.* **143,** 35–42.

Weinberg, E. D. (1978). *Microbiol. Rev.* **42,** 45–66.

Weiss, A. A., Silver, S., and Kinscherf, T. G. (1978). *Antimicrob. Agents Chemother.* **14,** 856–865.

Williams, P. H. (1979). *Infect. Immun.* **26,** 925–932.

Willsky, G. R., and Malamy, M. H. (1980). *J. Bacteriol.* **144,** 356–365.

Mechanisms of Potassium Transport in Bacteria

Georgia L. Helmer, Laimonis A. Laimins, and Wolfgang Epstein

1. INTRODUCTION

Bacteria accumulate large amounts of K^+ to achieve intracellular concentrations of about
0.25 M in *Escherichia coli* and *Salmonella,* 0.7 M in *Staphylococcus aureus,* and 4 M in
the extreme halophiles (Christian and Waltho, 1962; Schultz and Solomon, 1961). Intracellular K^+ has several functions including the activation of intracellular enzymes (Steinbach, 1962), control of intracellular osmolarity (Epstein and Schultz, 1968), and the regulation of internal pH (Plack and Rosen, 1980). Osmotic needs determine the amount of
K^+ accumulated by many bacteria. Bacteria maintain an internal osmolarity greater than
that outside resulting in an osmotic pressure difference. This difference, called the turgor
pressure, must be maintained for cells to grow (Knaysi, 1951). To maintain turgor pressure, *E. coli* regulates internal osmolarity by adjusting the internal concentration of K^+.
This role for K^+ is suggested by the dependence of intracellular K^+ concentration on medium osmolarity and by the immediate stimulation of K^+ uptake when external osmolarity
is increased (Epstein and Schultz, 1965). There is suggestive evidence that K^+ has an
osmotic role in other species. In a number of species the intracellular K^+ concentration is
positively correlated with the highest osmolarity at which the species can grow (Christian
and Waltho, 1961). Some bacteria, however, probably do not use K^+ to regulate osmolarity. These bacteria can adapt to high external osmolarity although they have only low
internal concentrations of K^+, suggesting some other solute has the major role in osmore-
gulation (Christian and Waltho, 1962). Unpublished experiments from our laboratory on
Bacillus subtilis showed that normal growth rates and therefore presumably normal turgor
pressure occur over a wide range of intracellular K^+ concentrations, suggesting that K^+ is

Georgia L. Helmer • Department of Biochemistry, The University of Chicago, Chicago, Illinois 60637.
Laimonis A. Laimins • Department of Biophysics and Theoretical Biology, The University of Chicago, Chi-
cago, Illinois 60637.   ***Wolfgang Epstein*** • Departments of Biochemistry and of Biophysics and Theoretical
Biology, The University of Chicago, Chicago, Illinois 60637.

not the primary solute controlling internal osmolarity in this species. The wall-less organism *Mycoplasma mycoides,* which cannot tolerate a significant turgor pressure, also accumulates K^+ (LeBlanc and LeGrimellec, 1979); in this case the ion may serve to maintain internal osmolarity close to that outside.

2. MODEST-AFFINITY K⁺ TRANSPORT SYSTEMS

Bacteria transport K^+ by uptake systems whose K_m for K^+ is near 1 mM (Jasper, 1978; Erecińska *et al.*, 1981; for reviews of earlier work see Harold and Altendorf, 1974; Silver, 1978). These systems generally transport Rb^+ and Tl^+ as well, but with a higher K_m and/or a lower V_{max} than for K^+ (Rhoads *et al.*, 1977; Jasper, 1978; Kashket, 1979; Damper *et al.*, 1979). Detailed information is available for two species, *Streptococcus faecalis* and *E. coli.* In both species modest-affinity uptake has unusual energy requirements: both ATP and the protonmotive force are required for net K^+ uptake. K^+ exchange in the steady state, however, requires only ATP (Rhoads and Epstein, 1977, 1978; Bakker and Harold, 1980). Mechanistic interpretations of this dual energy requirement are speculative. We now favor the idea that the protonmotive force provides the driving energy while ATP regulates to turn off K^+ transport when cell energy levels are low (Silver, 1978; Bakker and Harold, 1980). Without such control, deenergized cells would lose K^+ rapidly. Other models have been suggested, such as energy coupling to ATP with regulation by the protonmotive force (Rhoads and Epstein, 1977; Bakker and Harold, 1980) or that both ATP and the protonmotive force are energy coupled and that either can provide the energy for transport [suggested by Heefner *et al.* (1980) for Na^+ transport but equally applicable to K^+ transport].

The large membrane potential across the bacterial membrane, magnitude $\sim$ 150 mV and interior negative, could drive K^+ accumulation by a K^+-specific channel allowing electrogenic movement of the cation. Such a K^+-uniport mechanism, attractive in its simplicity (Harold and Altendorf, 1974), is an oversimplification for *S. faecalis, Streptococcus lactis,* and *E. coli.* In these organisms the potential is too low to account for the maximum K^+ concentration ratios the cells can generate (Kashket and Barker, 1977; Bakker and Harold, 1980; Bakker, 1980). One mode of energy coupling that can explain K^+ accumulation is symport with another cation. This cation could be a proton (Silver, 1978; Bakker, 1980). Recent findings on the effect of Na^+ and Li^+ on K^+ transport have been interpreted in terms of a Na^+-symport mechanism (Sorensen and Rosen, 1980).

The modest-affinity Trk system (formerly TrkA system) of *E. coli* has been analyzed by genetic means (Epstein and Kim, 1971; Rhoads *et al.*, 1976). Mutations at four unlinked loci, *trkA, trkD, trkE,* and *trkG,* reduce the rate of K^+ uptake. Mutations at two other unlinked loci, *trkB* and *trkC,* do not have any apparent effect on uptake kinetics but produce an abnormal rate of K^+ efflux. It is not known whether the *trkB* and *trkC* loci alter the Trk system, or whether they cause K^+ leakage by paths unrelated to the Trk system. Mutations at several of the *trk* loci alter the substrate specificity and energetics of Trk transport. The wild-type Trk system accepts Rb^+ and Tl^+ as substrates but significantly prefers K^+; the residual transport activity in *trkA trkD* double mutants somewhat prefers Rb^+ to K^+ (Rhoads *et al.*, 1977). This same double mutant has also lost the ATP requirement; in such mutants net uptake requires only the protonmotive force, and exchange requires no energy source (Rhoads and Epstein, 1977).

Kinetic analysis of *trk* mutants allows the description of three components of Trk transport: a modest-affinity, high-rate system seen in wild-type strains; a modest-affinity, modest-rate system seen in *trkA* mutants and formerly called the TrkD system; and a low-rate, nonsaturable system seen in *trkA trkD* double mutants and formerly called the TrkF system (Rhoads *et al.*, 1976). Recent work in our laboratory, showing that mutations at the *trkE* and *trkG* loci alter both saturable components, suggests that these "systems" represent different levels of function of a common structure. Only when mutation abolishes one level of function is the next level apparent.

The *trkA, trkE*, and *trkG* loci are of primary importance for Trk function; mutations at any one of these loci significantly reduce Trk system uptake. Mutations at *trkD* appear to have an effect only when a *trkA* mutation is also present. Increasing gene dosage at all three genes (*trkA, trkE*, and *trkG*) significantly increases the V_{max} of uptake, whereas increasing the dosage of only two has no effect (Helmer and Epstein, 1981). We conclude that these three genes code for rate-limiting components of the Trk system. The product of one of these genes, *trkA*, has been identified as an integral membrane protein.

3. HIGH-AFFINITY K⁺ TRANSPORT

E. coli possess a high-affinity system called Kdp ($K_m = 2$ μM) that enables the bacteria to scavenge K⁺ and grow in media of low K⁺ concentration. This type of K⁺ transport system has not been described in other bacteria. The Kdp system is not expressed when K⁺ requirements are met by the Trk system, such as at relatively high external K⁺ concentrations. At low K⁺ concentrations, however, where Trk function is inadequate, Kdp is expressed. The cells detect K⁺ need and hence determine the need to express Kdp by sensing changes in the turgor pressure (Laimins *et al.*, 1981). Neither internal nor external K⁺ concentrations alone determine Kdp expression, though each has an indirect role through effects on turgor pressure. The product of the *kdpD* gene, which codes for a positive regulator of Kdp (Rhoads *et al.*, 1978), has been tentatively identified as a membrane protein. The *kdpD* protein may sense changes in turgor pressure and, through a conformational change, translate these changes into a genetic signal to turn on expression of the *kdp* structural genes.

The three structural genes, *kdpA*, *kdpB*, and *kdpC*, code for inner-membrane proteins of molecular weight 47,000, 90,000, and 22,000, respectively (Laimins *et al.*, 1978). Transport of K⁺ requires only ATP *in vivo* (Rhoads and Epstein, 1977). *In vitro*, the Kdp system exhibits K⁺-dependent ATPase activity (Epstein *et al.*, 1978; Wieczorek and Altendorf, 1979). An acyl phosphate derivative of the 90,000-dalton *kdpB* protein is a presumed intermediate of the transport cycle (Epstein *et al.*, 1979). The Kdp system can create very large K⁺ gradients; concentration ratios as high as $4 \times 10^6 : 1$ have been measured (Rhoads *et al.*, 1976).

A three-dimensional picture of the structure of the Kdp system is emerging (L. A. Laimins and W. Epstein, unpublished). Genetic analysis demonstrating intracistronic complementation between *kdpA* mutations (Epstein and Davies, 1970) indicates that at least two *kdpA* protein monomers are in physical contact. Recent work, indicating that Kdp has an equal number of each of the three subunits, requires that the minimum stoichiometry of the functional complex be $A_2B_2C_2$. Proteolytic accessibility studies suggest that two of the subunits, the *kdpA* and *kdpB* proteins, are exposed to the cytoplasm. Part of the *kdpB* protein must be exposed to the cytoplasm to explain phosphorylation by ATP. Mutations that reduce affinity for K⁺ are in the *kdpA* gene, suggesting that the *kdpA* protein forms the site of initial binding of K⁺ in the periplasmic space. This function of the *kdpA* protein requires that it also be exposed to the periplasmic space, thus spanning the membrane. No structural information is available on the location of the *kdpC* protein.

4. REGULATORY ASPECTS OF K⁺ TRANSPORT

The osmotic role of K^+ is inferred in part from the striking effect of osmotic changes on K^+ transport. Ørskov (1948) first reported that increasing external osmolarity leads to a rapid uptake of K^+. This is an immediate response of the Trk and Kdp systems (Rhoads and Epstein, 1978). The finding of a similar effect in plant cells (Gutknecht, 1968) suggests that this mechanism is characteristic of walled cells which use K^+ as an osmoregulatory solute. Kinetic studies show a stimulation of influx while efflux continues unchanged (Kepes *et al.*, 1976; Rhoads and Epstein, 1978). One model for this regulation proposes a periplasmic protein that blocks K^+-pump sites when turgor pressure is high (Martirosov, 1979). Rather, osmotic control of influx may be a property of the membrane proteins of these transport systems. Changes in either transmembrane pressure or stretch in the plane of the membrane could produce conformational changes in the transport proteins to regulate influx. The mechanism producing these conformational changes could be analogous to that proposed to account for turgor pressure regulation of the genetic expression of the Kdp system (Laimins *et al.*, 1981).

Monovalent cations have various effects on the activity of K^+ transport systems. Na^+-loaded, K^+-depleted *E. coli* have a fivefold higher V_{max} of Trk activity than do cells with the normal high-K^+, low-Na^+ contents. This effect suggests stimulation by internal Na^+ or inhibition by internal K^+. The Kdp system is inhibited by external K^+ at concentrations far above the K_m for transport, but only in cells that have normal cell cation levels; K^+-depleted, Na^+-loaded cells are not inhibited. This result argues for allosteric sites for Na^+ inside and for K^+ outside (Rhoads and Epstein, 1978). We doubt that Na^+ is a substrate of Kdp, because the ATPase associated with the system is indifferent to Na^+ (Epstein *et al.*, 1978). Na^+ and Li^+ stimulate the uptake of Tl^+, a K^+ analog, in *S. lactis* (Kashket, 1979) and *E. coli* (Sorensen and Rosen, 1980). In *E. coli*, Na^+ greatly stimulates uptake of Tl^+ and of K^+ by the Trk system as compared to uptake in the presence of choline; for the Kdp system, Li^+ was more effective in stimulating Tl^+ uptake, while Na^+ was better in stimulating K^+ uptake. These results suggest the stimulatory ions are cosubstrates for transport (Sorensen and Rosen, 1980), but allosteric effects can also explain them.

A process of equal importance to transporting K^+ *in* is that of pumping this ion *out*. All cells have leakage paths for K^+ movement across their membranes. When cells that maintain a large membrane potential are grown in high-K^+ media, overaccumulation of K^+ will occur unless a way exists to pump K^+ out against its concentration gradient. Systems that perform this function must exist, but none have been identified yet. Looking to Ca^{2+} export for models, we must consider antiporters (Tsuchiya and Rosen, 1976) and ATP-driven systems (Kobayashi *et al.*, 1978). A K^+/H^+ antiporter in *E. coli*, and a mutant defective in this activity have been described (Brey *et al.*, 1980; Plack and Rosen, 1980). The mutant cannot grow at alkaline pH, but can grow at neutral or acid pH in high- or low-K^+ media. This suggests the antiporter has a role in regulating internal pH, but probably not in K^+ export. Interestingly, an Na^+/H^+ antiporter seems to have a pH-regulating function in *Bacillus alkalophilus* (Krulwich *et al.*, 1979).

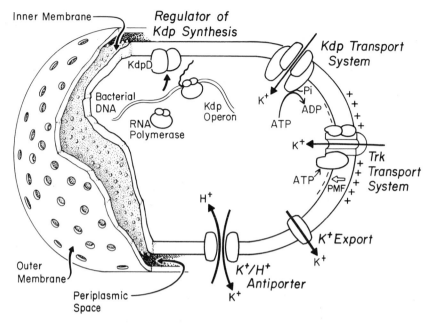

Figure 1. A schematic view of K$^+$ transport in *E. coli*. The outer membrane, drawn with exaggeratedly large pores, probably does not restrict K$^+$ movement into the periplasmic space. Four types of transport systems are shown spanning the inner membrane. The K$^+$/H$^+$ antiporter can use the K$^+$ gradient to pump H$^+$ in to prevent excessively high internal pH. The K$^+$ export system is hypothetical; no such activity has been described as yet. At the right is shown the constitutive, high-rate Trk system, which requires both ATP and the protonmotive force for net K$^+$ uptake and consists of at least three protein components. The Kdp system is ATP-driven and consists of three membrane proteins. An acyl phosphate derivative of one of these proteins is a presumed intermediate in transport. The synthesis of the Kdp proteins is turned on when turgor pressure is low. This regulation is mediated by the KdpD protein, drawn as a dimer. This protein is presumed to interact with the promoter for the structural genes to turn on transcription when turgor pressure is low.

5. SUMMARY

Our current view of bacterial K$^+$ transport as seen in *E. coli* is illustrated in Fig. 1. Two systems are shown for inward, energy-coupled uptake, Trk and Kdp. Included are a K$^+$/H$^+$ antiporter, and a postulated K$^+$ export pump that may be more than a single system. The paths for energetically passive leakage of K$^+$ probably occur at many types of sites and are not shown. We also illustrate in Fig. 1 the unusual genetic regulation of the Kdp system whose synthesis is controlled by turgor pressure and is mediated by the *kdpD* gene product. The large number of systems for the transport of K$^+$ found in bacteria show a remarkable complexity in a supposedly simple cell.

REFERENCES

Bakker, E. P. (1980). *J. Supramol. Struct.* Suppl. 4, p. 80.
Bakker, E. P., and Harold, F. M. (1980). *J. Biol. Chem.* **255,** 433–440.
Brey, R. N., Rosen, B. P., and Sorensen, E. N. (1980). *J. Biol. Chem.* **255,** 39–44.
Christian, J. H. B., and Waltho, J. A. (1961). *J. Gen. Microbiol.* **25,** 97–102.
Christian, J. H. B., and Waltho, J. A. (1962). *Biochim. Biophys. Acta* **65,** 506–508.

Damper, P. D., Epstein, W., Rosen, B. P., and Sorensen, E. K. (1979). *Biochemistry* **18**, 4165–4169.

Epstein, W., and Davies, M. (1970). *J. Bacteriol.* **101**, 836–843.

Epstein, W., and Kim, B. S. (1971). *J. Bacteriol.* **108**, 639–644.

Epstein, W., and Schultz, S. G. (1965). *J. Gen. Physiol.* **49**, 221–234.

Epstein, W., and Schultz, S. G. (1968). In *Microbial Protoplasts, Spheroplasts and L-forms* (L. B. Guze, ed.), pp. 186–193, Williams & Wilkins, Baltimore.

Epstein, W., Whitelaw, V., and Hesse, J. (1978). *J. Biol. Chem.* **253**, 6666–6668.

Epstein, W., Laimins, L. A., and Hesse, J. E. (1979). *Abstracts, XIth International Congress of Biochemistry*, p. 449.

Erecińska, M., Deutsch, C. J., and Davis, J. S. (1981). *J. Biol. Chem.* **256**, 278–284.

Gutknecht, J. (1968). *Science* **160**, 68–70.

Harold, F. M., and Altendorf, K. (1974). *Curr. Top. Membr. Transp.* **5**, 1–50.

Heefner, D. L., Kobayashi, H., and Harold, F. M. (1980). *J. Biol. Chem.* **255**, 11403–11407.

Helmer, G. L., and Epstein, W. (1981). *Biophys. J.* **33**, 61a.

Jasper, P. (1978). *J. Bacteriol.* **133**, 1314–1322.

Kashket, E. R. (1979). *J. Biol. Chem.* **254**, 8129–8131.

Kashket, E. R., and Barker, S. L. (1977). *J. Bacteriol.* **130**, 1017–1023.

Kepes, A., Meury, A., Robin, A., and Jimeno, J. (1976). In *FEBS Symposium on the Biochemistry of Membrane Transport* (G. Semenza and E. Carafoli, eds.), pp. 633–647, Springer-Verlag, New York.

Knaysi, G. (1951). *Elements of Bacterial Cytology*, 2nd ed., Cornell University Press (Comstock), Ithaca, N.Y.

Kobayashi, H., Van Brunt, J., and Harold, F. M. (1978). *J. Biol. Chem.* **253**, 2085–2092.

Krulwich, T. A., Mandel, K. G., Bornstein, R. F., and Guffanti, A. A. (1979). *Biochem. Biophys. Res. Commun.* **95**, 857–863.

Laimins, L., Rhoads D. B., Altendorf, K., and Epstein, W. (1978). *Proc. Natl. Acad. Sci. USA* **75**, 3216–3219.

Laimins, L. A., Rhoads, D. B., and Epstein, W. (1981). *Proc. Natl. Acad. Sci. USA* **78**, 464–468.

LeBlanc, G., and LeGrimellec, C. (1979). *Biochim. Biophys. Acta* **554**, 156–167.

Martirosov, S. M. (1979). *J. Electroanal. Chem.* **104**, 315–321.

Ørskov, S. L. (1948). *Acta Pathol. Microbiol. Scand.* **25**, 277–283.

Plack, R. H., Jr., and Rosen, B. P. (1980). *J. Biol. Chem.* **255**, 3824–3825.

Rhoads, D. B., and Epstein, W. (1977). *J. Biol. Chem.* **252**, 1394–1401.

Rhoads, D. B., and Epstein, W. (1978). *J. Gen. Physiol.* **72**, 283–295.

Rhoads, D. B., Waters, F. W., and Epstein, W. (1976). *J. Gen. Physiol.* **67**, 325–341.

Rhoads, D. B., Woo, A., and Epstein, W. (1977. *Biochim. Biphys. Acta* **469**, 45–51.

Rhoads, D. B., Laimins, L. A., and Epstein, W. (1978). *J. Bacteriol.* **135**, 445–452.

Schultz, S. G., and Solomon, A. K. (1961). *J. Gen. Physiol.* **45**, 355–369.

Silver, S. (1978). In *Bacterial Transport* (B. P. Rosen, ed.), pp. 221–234, Dekker, New York.

Sorensen, E. N., and Rosen, B. P. (1980). *Biochemistry* **19**, 1458–1462.

Steinbach, H. B. (1962). In *Comparative Biochemistry* (M. Florkin and H. Mason, eds.), Vol. 5, pp. 677–720, Academic Press, New York.

Tsuchiya, T., and Rosen, B. P. (1976). *J. Biol. Chem.* **251**, 962–967.

Wieczorek, L., and Altendorf, K. (1979). *FEBS Lett.* **98**, 233–236.

106

Protonmotive-Force-Dependent DNA Transport across Bacterial Membranes

L. Grinius

1. DNA TRANSPORT PROCESSES

DNA transport across the bacterial cell membrane occurs at the early stages of several genetic processes: transformation, transfection, phage invasion, and sexual conjugation. For genetic transformation and transfection, DNA is isolated either from a bacterium or from a phage, respectively. During phage invasion, the recipient bacterium gains the genetic material of the viral particle, whereas the DNA donor involved in conjugation is an intact bacterial cell. The uptake of exogenous DNA starts with the adsorption of the donor to the recipient cell surface. Therefore, the donor DNA reaches the cytoplastmic membrane of the recipient and is linearly transported into the cytoplasm (see Goldberg, 1980; Notani and Setlow, 1974). Osmotic shocking of gram-positive bacteria, which are able to take up DNA and thereafter to transform genetically, releases water-soluble polypeptides with molecular weights of about 10,000 (see Notani and Setlow, 1974). When isolated, these polypeptides exhibit DNA-uptake-stimulating activity after being added to suspensions of bacterial cells devoid of the ability to take up DNA.

In the early stages of *S. pneumoniae* and *S. sanguis* transformation, the donor DNA is complexed with recipient cell material that appears to be a protein of molecular weight about 19,500 and 15,500, respectively (Morrison and Baker, 1979; Raina and Ravin, 1980). Initially, the complex formed upon uptake is located outside the cytoplasmic membrane of *S. sanguis* and can be released from the cell under conditions promoting spheroplast formation (Raina *et al.*, 1979). The donor DNA complexed with the protein has also been found inside *S. sanguis* cells. This enabled the authors to suggest that to enter the cell the complex must move through the inner membrane. Despite the isolation of similar complexes from *B. subtilis* (Pieniążek *et al.*, 1977), additional experiments are necessary to prove the ability of the complex to penetrate the membrane.

L. Grinius • Department of Biochemistry and Biophysics, Vilnius University, Vilnius 232031, Lithuanian SSR, USSR.

A protein that binds to either single-stranded or double-stranded DNA but not to ribonucleic acid has been isolated by osmotic shock treatment of growing *H. influenzae* cells (Sutrina and Scocca, 1979). Certain mutant strains of *H. influenzae* defective in DNA uptake were found to be deficient in DNA-binding protein, suggesting that the protein participates in the transport of DNA.

During some phage invasion, both nucleic acid and some coat proteins are transported into the host cell in approximately equimolar amounts and their kinetics of penetration are similar (see Goldberg, 1980); this suggests that the penetration of phage nucleic acid into the cell involves the transport of a protein—nucleic acid complex, rather than of free nucleic acid and protein.

Studies on *E. coli* conjugation have shown (Silver, 1963) that the transferred material consisted of 9% of total cell DNA, as well as 1.5% of total cell RNA and 1% of total cell protein. Taking into account that the number of protein molecules greatly exceeds the number of DNA molecules in the cell, these findings could be interpreted as evidence that some proteins of the donor cell, presumably the proteins of sex-pili, form the complex with the donor DNA and thereafter this complex is transferred from the donor to the recipient.

The initiation of DNA transport requires divalent ions, among which Mg^{2+} and Ca^{2+} are often involved. According to Morrison (1971), binding of divalent ions by EDTA blocks only the initiation of entry of the DNA molecule into *B. subtilis,* the succeeding stages of the entry being resistant to EDTA. On the other hand, Tomasz (1978) points out the requirement for Ca^{2+} during DNA transport across the membrane of pneumococci.

DNA transport processes do not depend specifically on K^+ and Na^+. The only role of these ions is to maintain the appropriate ionic strength.

The tremendous variety of DNA donors and the existence of many different pathways through the cell wall to the cytoplasmic membrane prompted Tomasz (1978) to conclude "whether or not there might be . . . a basic similarity between conjugation, phage-mediated processes, and transformation is not clear at the present time." It was often taken for granted that each DNA transport process proceeded via its own mechanism. Recently, a general mechanism of DNA transport applicable to all of the above-mentioned genetic processes has been formulated (Grinius, 1976, 1980). According to this concept, the transmembrane movement of DNA is driven by the protonmotive force.

2. ROLE OF PROTONMOTIVE FORCE

The postulate of the chemiosmotic theory of energy conservation (Mitchell, 1966) that respiration as well as ATP hydrolysis are coupled with H^+-ion extrusion from the bacterial cell has now been proven experimentally (see Harold, 1977). The transport of H^+ occurs across the cytoplasmic membrane and leads to the establishment of a transmembrane electrostatic potential ($\Delta\Psi$), which reaches 180 mV (negative inside), and a transmembrane pH gradient (ΔpH). Both $\Delta\Psi$ and ΔpH create a protonmotive force pushing H^+ ions back into the cell. The protonophorous uncouplers of oxidative phosphorylation as well as the mixture of ionophore antibiotics valinomycin and nigericin dissipate the protonmotive force.

It is well known that uncouplers have an inhibitory effect on the early stages of genetic transformation (see Notani and Setlow, 1974). The inhibitory effect of uncouplers on the transformation of *B. subtilis* correlates with their effect on $\Delta\Psi$ of the cell (Chaustova *et al.*, 1980). The inhibition of *B. subtilis* transformation by uncouplers occurs despite the fact that the phosphorylation potential of the glycolyzing host cell remains at a constant level (Chaustova *et al.*, 1980). On the other hand, the transformation of *B. subtilis* in arsenate-containing medium proceeds despite the decrease of the intracellular ATP level.

The ability of valinomycin to carry K$^+$ ions across the membrane permits gradual changes of $\Delta\Psi$ by adding different amounts of either this ionophore or K$^+$ ions. Nigericin as an ionophore exchanging K$^+$ for H$^+$ reduces only the ΔpH of the bacterial cell. In order to determine the role of both $\Delta\Psi$ and ΔpH in genetic transformation of *B. subtilis*, the effects of valinomycin and nigericin were studied (Chaustova *et al.*, 1980). Valinomycin caused a strong inhibitory effect on DNA uptake during *B. subtilis* transformation, nigericin being less effective. One can interpret these findings as an indication of a requirement for $\Delta\Psi$, while the role of ΔpH is not clear.

The inhibitory effect of valinomycin on the transformation of *B. subtilis* (Chaustova *et al.*, 1980) is potentiated by nigericin and vice versa, indicating a requirement for both and ΔpH in the uptake of DNA.

The growing body of evidence also indicates the involvement of the protonmotive force in the entrance of phage DNA into the cell. Uncouplers inhibited the irreversible absorption of phages T1 and ϕ80 to *E. coli*, while phage absorption proceeded efficiently after lowering the intracellular ATP by arsenate (Hancock and Braun, 1976).

The entrance of phage T4 DNA into *E. coli* was also shown (Kalasauskaite *et al.*, 1979, 1980; Kalasauskaite and Grinius, 1979) to be dependent on the protonmotive force but not the intracellular ATP. Measurements of DNA penetration from T4 phage absorbed to *E. coli* at different $\Delta\Psi$ values have shown at precipitous reduction of the process between 110 and 60 mV (Labedan *et al.*, 1980). This threshold of $\Delta\Psi$ for DNA penetration was rather independent of ΔpH. The effective inhibition of DNA penetration by colicin K and valinomycin plus K$^+$ (not by nigericin) has been taken (Labedan and Goldberg, 1979) as evidence that $\Delta\Psi$ is solely required in the penetration of phage T4 DNA. Despite these findings, the possible involvement of ΔpH in the phage invasion is not excluded at the present time.

DNA transport during bacterial conjugation is sensitive to uncouplers and depends on the protonmotive force in the recipient cell (Fisher, 1957; Grinius and Beržinskiene, 1976). Recent experimental data (Beržinskiene *et al.*, 1980) indicate, however, that conjugation depends also on the intracellular ATP. Additional experiments are needed to reveal the stage(s) of the conjugation that is (are) dependent on the intracellular ATP and protonmotive force, respectively.

In summary, recent experimental evidence discussed above indicates that different processes of DNA transport manifest a dependence on the protonmotive force across the host cell membrane. It is too early, however, to conclude that the protonmotive force drives DNA transport. More experimental data are needed to reveal the molecular mechanism of protonmotive force involvement in DNA transport.

REFERENCES

Beržinskienė, J. A., Zizaitė, L. J., Baronaitė, Z. A., and Grinius, L. (1980). *Biokhimiya* **45**, 1103–1112.

Chaustova, L. P., Grinius, L. L., Griniuvienė, B. B., Jasaitis, A. A., Kadziauskas, J. P., and Kiaušinytė, R. V. (1980). *Eur. J. Biochem.* **103**, 349–357.

Fisher, K. W. (1957). *J. Gen. Microbiol.* **16**, 136–145.

Goldberg, E. B. (1980). In *Receptors and Recognition: Virus Receptors* (L. Randall and L. Philipson, eds.), Series B, Vol. 7, pp. 115–141, Chapman & Hall, London.

Grinius, L. (1976). *Biokhimiya* **41**, 1539–1547.

Grinius, L. (1980). *FEBS Lett.* **113**, 1–10.

Grinius, L., and Beržinskienė, J. (1976). *FEBS Lett.* **72,** 151–154.

Hancock, R. E. W., and Braun, V. (1976). *J. Bacteriol.* **125,** 409–415.

Harold, F. M. (1977). *Curr. Top. Bioenerg.* **4,** 83–149.

Kalasauskaitė, E., and Grinius, L. (1979). *FEBS Lett.* **99,** 287–291.

Kalasauskaitė, E. V., Armalytė, V. K., and Grinius, L. L. (1979). *Biokhimiya* **44,** 221–232.

Kalasauskaitė, E., Grinius, L., Kadišaitė, D., and Jasaitis, A. (1980). *FEBS Lett.* **117,** 232–236.

Labedan, B., and Goldberg, E. B. (1979). *Proc. Natl. Acad. Sci. USA* **76,** 4669–4674.

Labedan, B., Heller, K. B., Jasaitis, A. A., Wilson, T. H., and Goldberg, E. B. (1980). *Biochem. Biophys. Res. Commun.* **93,** 625–630.

Mitchell, P. (1966). *Biol. Rev. Cambridge Philos. Soc.* **41,** 445–502.

Morrison, D. A. (1971). *J. Bacteriol.* **108,** 38–44.

Morrison, D. A., and Baker, M. F. (1979). *Nature (London)* **282,** 215–217.

Notani, N. K., and Setlow, J. (1974). *Prog. Nucleic Acid. Res.* **14,** 39–100.

Pieniążek, D., Piechowska, M., and Venema, G. (1977). *Mol. Gen. Genet.* **156,** 251–261.

Raina, J. L., and Ravin, A. W. (1980). *Biochem. Biophys. Res. Commun.* **93,** 228–234.

Raina, J. L., Metzer, E., and Ravin, A. W. (1979). *Mol. Gen. Genet.* **170,** 249–259.

Silver, S. (1963). *J. Mol. Biol.* **6,** 349–360.

Sutrina, S. L., and Scocca, J. A. (1979). *J. Bacteriol.* **139,** 1021–1027.

Tomasz, A. (1978). In *Transport of Macromolecules in Cellular Systems* (S. C. Silverstein, ed.), pp. 21–34, Verlag Chemie, Berlin, Dahlem Konferenzen 1978.

107

DNA Transport across Bacterial Membranes

B. Labedan and E. B. Goldberg

In this review we will outline the most recent experimental approaches toward elucidation of the mechanism(s) of DNA transport across the bacterial envelope into the cytoplasm. Transformation (transport of free DNA from the medium into the cell), conjugation (transfer of DNA directly between a donor cell and a recipient cell), and phage infection (transfer of DNA from a phage adsorbed to a host cell) are three natural types of DNA transport that have been studied extensively.

In order to be brief and yet facilitate the readers' entry into the literature, we will usually cite the most recent articles and reviews dealing with a subject and leaving it to the reader to work his way back. We apologize to those whose more original or more convincing work we did not cite.

1. PILOT PROTEIN LEADS THE DNA TERMINUS TO AND THROUGH THE MEMBRANE

The steps toward recognition and stabilization of DNA donor vehicles in conjugation and phage infection have most recently been discussed in an article by Willets and Skurray (1980) and in a short book edited by Randall and Philipson (1980), respectively. We shall deal mainly with the subsequent processes that are shared by all DNA transport phenomena: how does DNA find its entry port, initiate entry, and traverse the cytoplasmic membrane, nucleotide by nucleotide? A general theme running through the more recent literature has been that a protein attached to the DNA tip that is involved in: recognition of the entry port, transport through it, and aid in various DNA-related functions once in the cell cytoplasm. This type of protein has been denoted *pilot protein* by Kornberg, who together

B. Labedan • Institut de Microbiologie, Université Paris XI, 91405 Orsay, France.
E. B. Goldberg • Department of Molecular Biology and Microbiology, Tufts University School of Medicine, Boston, Massachusetts 02111.

with Jazwinski and Marco first applied this concept to phages ϕX174 and M13 (Kornberg, 1980). In brief, a phage coat protein (gpH for ϕX174 and gp3 for M13) involved in recognition and attachment to the host cell is altered in configuration upon attachment leading to uncoating of the particle and penetration of the DNA, led by the terminal DNA–protein complex into the cell. (In the case of M13, gp3 is cleaved in the process.) It is not clear if the protein–DNA terminus complex stays attached to the membrane while the DNA is threaded in, which would form a loop; nor is it clear if the protein becomes a component of the pore or if it is passed through a preexisting host pore. (One exception to this type of entry process seems to be the transport of covalent closed circular DNA into Ca^{2+}-treated cells, and we will not deal with this phenomenon here.)

2. BACTERIAL TRANSFORMATION

Proteins have often been implicated in the process of DNA uptake during bacterial transformation. It has recently been shown that covalent closed circular plasmid DNA is transformed very well in *S. pneumoniae* as is restriction cleaved linear plasmid DNA, though at somewhat reduced efficiency. Transformation of these plasmids is sensitive to trypsin and dependent upon the presence of cellular endonuclease (Barany and Tomasz, 1980). Lacks (1977) has presented a model for transformation in gram-positive cells that shows naked DNA first being adsorbed horizontally in a rather nonspecific manner to the cell surface and then nicked by a nuclease located at the site of the uptake pore to create a terminus that combines with a DNA-binding protein for transport of a single strand into the cell powered by phosphodiester cleavage. As a result, nonspecific DNA can compete with homologous DNA and prevent genetic transformation.

The situation in the gram-negative *H. influenzae* is somewhat different. Here, too, a protein that binds to DNA (but not RNA) was isolated during the growth phase by osmotic shock from competent *H. influenzae* but not from some incompetent strains (Sutrina and Scocca, 1979). Sisco and Smith (1979) have recently offered an explanation of why non-specific DNA will not compete with homologous DNA in transformation of *H. influenzae*. They add a specificity restriction type cleavage to Lacks' model. Thus, when a specific 11-nucleotide sequence (Danner *et al.*, 1980) of the loosely bound *H. influenzae* DNA contacts the specific nuclease located at the site of the entry port, both strands are cleaved, the terminus is bound to the receptor and enters the cell as a double-stranded exonuclease V-resistant DNA.

3. BACTERIAL CONJUGATION

For conjugal F transfer in *E. coli*, the the DNA of the F plasmid in the donor must be nicked in a specific DNA sequence called *oriT* (origin of transfer). This nicking involves four transfer genes: *traY* and *traM*, whose products are located in the cell membrane, and *traZ* and *traI*, whose products are in the cytoplasm. It has been proposed that the DNA termini created by phosphodiester cleavage during nicking are linked to proteins (possibly located in the donor–recipient membrane junction), but there is no direct positive evidence for such a complex. On the other hand, Col El (a nonconjugative plasmid, mobilizable by F plasmids) has its own *oriT*, which needs all the F *tra* genes for its transfer except *tra M*, *I*, *Z*, and possibly *Y*. It has, however, three genes on its own DNA that are required for F-mobilized transfer. When Col El plasmid DNA is nicked, a 60,000-molecular-weight protein is covalently attached to the 5' terminus. (The work on conjugal transfer of plasmid DNA is summarized in Willets and Skurray, 1980.) For conjugal transfer of the bacterial chromosome, a specific *oriT* sequence must be nicked probably by the *traI* gene product of the F plasmid. Whether the TraI protein binds to the 5' terminus like the 60,000-molecular-weight protein does to Col El *oriT*, to pilot the protein into the recipient cell, remains to be seen (Achtman and Skurray, 1977).

4. BACTERIOPHAGE INFECTION

Many phages carry, in their heads, proteins that seem to be involved in DNA transport. These are, of course, *cis* acting in the genetic sense. The *Salmonella* phage P22, however, has a *trans*-acting protein gp16 that seems to act (in a hexamer) as a pore for the P22 DNA gp20–gp7 protein complex. Thus, it is possible that some phages may carry with them their own DNA pore or a specific modifier of a preexisting host DNA pore. Another important indication of the existence of DNA transport pores is superinfection exclusion. Both T4 and P22 phage genome expression lead to an alteration in the infected host membrane that prevents entry of new (superinfecting) phages, which, however, can attach normally and even eject their DNA abortively into the periplasm. Elucidation of the mechanism of this phenomenon preventing DNA entry should help us to understand the mechanism of DNA transport (see Goldberg, 1980).

Labedan *et al.* (1973) showed an electron micrograph of a centrifuged infected cell with T5 second-step transfer DNA attached at one end, stretched out on the grid with the phage capsid at the other end. The same observation was obtained for total T5 DNA after adsorption of the phage at 0°C and centrifugation (Labedan, 1976). Subsequent incubation of such complexes at 37°C produced infected cells and phage progeny if DNase was not added first. This showed not only that initial phage DNA attachment was from one end, and subsequent DNA traversal was therefore probably nucleotide by nucleotide, but that the energy for DNA penetration (at least for T5) must come from the recipient cell, even though there is enough energy for ejection *in vitro* stored in the phage.

To sum up, there have been many solutions to the overall mechanism of delivery of DNA to the site of its uptake. It is probable that the specificity and sophistication of the delivery mechanism determines the efficiency of genetic transfer. This may vary from 1 for phages like T4, to 0.2 for phages like λ, to orders of magnitude less for transformation systems that depend on the degree of preparation (competence) of the cells. Thus, the efficiency of transfer is determined by the state of the recipient surface as well as by the positioning of the appropriately prepared DNA terminus against a pore properly primed to receive it, transport it, and protect it once it is inside the cytoplasm.

5. THE ROLE OF MEMBRANE POTENTIAL ($\Delta\Psi$) IN PHAGE T4 DNA TRANSPORT

How is cellular energy transduced to bring DNA into the cell? This problem has been studied recently for T4 phage. There are two main methods in use for measuring T4 DNA injection. In the classical technique, injected DNA is separated from noninjected DNA by violent agitation of the infected bacteria in a blender. The virion (with or without DNA in its head) is rent from the host cell, which is then sedimented at low speed to separate it from the phage heads. This method actually measures ejection from the phage head and association with the host cell. More recently, a faster and simpler method was devised by Labedan and Goldberg (1979) that measures DNA actually injected into the cytoplasm. DNA is injected from a mutant phage, defective in a protein that normally protects the terminus of the entering DNA from exonucleolytic restriction in the host by the cytoplasmic exonuclease V (recBC). The amount of acid-soluble DNA reflects the quantity of DNA that actually penetrated the cytoplasm. To determine the entry requirements for phage DNA injection, these two techniques were used in the presence of different metabolic inhibitors.

In normal cells, high concentrations (10 mM or more) of cyanide (a well-known respiratory inhibitor) were needed to inhibit DNA injection. However, in *uncA* cells (deficient for membrane ATPase), DNA injection was sensitive to 1 mM cyanide (Kalasauskaitė and Grinius, 1979; Labedan and Goldberg, 1979). This suggested that the protonmotive force is involved in DNA injection. It was also shown (Labedan and Goldberg, 1979) that proton ionophore uncouplers like CCCP, azide, and 2,4-DNP, which are known to directly collapse the protonmotive force, also abolish T4 DNA penetration, in unc^+ and also in unc^- cells where internal ATP was maintained at a high concentration (Plate *et al.*, 1974). More recently, it has been shown directly that phosphate potential is not a source of energy for phage T4 DNA transport (Kalasauskaitė *et al.*, 1980).

The protonmotive force (Δp) is composed of a chemical gradient (ΔpH) and a membrane potential ($\Delta \Psi$). Which of these components is required for phage T4 DNA penetration? Addition of high concentrations (1 mM or more) of the lipophilic cation TPMP$^+$, colicin K (20 molecules per cell), or high concentrations of external potassium in the presence of its ionophore, valinomycin, abolishes DNA injection in each case (Labedan and Goldberg, 1979). Thus, the collapse of $\Delta \Psi$ without significantly changing ΔpH is sufficient to prevent DNA transport. On the contrary, dissipation of ΔpH (with a corresponding increase of $\Delta \Psi$), either by addition of nigericin or by changing the external pH from 6 to 8, has no significant influence on the efficiency of the T4 DNA injection.

Direct measurement of both $\Delta \Psi$ and ΔpH when correlated with DNA injection at different levels of protonmotive force demonstrated a precipitous reduction in DNA penetration when $\Delta \Psi$ was reduced from 110 mV to about 60 mV. The threshold value of $\Delta \Psi$ for DNA penetration was almost independent of ΔpH (Labedan *et al.*, 1980). The requirement for $\Delta \Psi$ at the initiation of the DNA transport process was postulated because adsorption of phage in the absence of membrane energy does not seem to expose the DNA to the periplasm (as measured by susceptibility to DNase before or after spheroplasting), whereas the DNA or its pilot protein does seem to have exited the head and possibly to have unplugged the tail tip as well. On the basis of these experiments, Labedan *et al.* (1980) speculate that the cytoplasmic membrane has specific DNA channels (probably made of proteins) that are "gated," that is, the configuration that determines their interaction with the tip of the DNA or the physical opening is voltage dependent. In the case of phage T4 DNA, the postulated pilot proteins (which precede the DNA during the injection process) will specifically interact with the DNA channels thus initiating DNA traversal of the channel. This traversal may also require additional energy input by the host.

6. THE ROLE OF PROTONMOTIVE FORCE IN BACTERIAL TRANSFORMATION AND CONJUGATION

Protonmotive force seems to be required in some bacterial transformations. Transformation of *B. subtilis* is unaffected by arsenate (which reduces the intracellular ATP levels twofold), but it is inhibited by uncouplers that reduce $\Delta\Psi$ while maintaining high intracellular ATP levels (Chaustova *et al.*, 1980). It is not yet possible to conclude from these experiments whether $\Delta\Psi$ alone or both $\Delta\Psi$ and ΔpH components are required for transformation of *B. subtilis* or whether some ATP is required as well. However, it would seem from the data that $\Delta\Psi$ plays a major role in the early stages of transformation. Nigericin by itself has no effect in transformation but it potentiates the ability of valinomycin to inhibit transformation. Though this may imply a role for ΔpH in DNA uptake it may also be interpreted as an effect of nigericin on potentiating the ability of valinomycin to reduce $\Delta\Psi$ under these conditions.

DNA transport during bacterial conjugation is sensitive to uncouplers and depends on the protonmotive force in the recipient cell (Grinius and Beržinskienė, 1976). Recent experiments indicate, however, that conjugation depends on intracellular ATP as well (Beržinskienė *et al.*, 1980).

7. CONCLUSION

The process of DNA transport can be subdivided into different stages. The early stages depend on the particular donor vehicle while the later stages may correspond to a more universal process. The fact that a reduction in membrane energy can prevent DNA transfer by conjugation, transformation, and phage infection does not mean that the same stage is affected in each case. For example, phages T1 and ϕ80 cannot attach irreversibly to *E. coli* if the protonmotive force has been dissipated (Hancock and Braun, 1976). We must await more precise independent experimental definitions of the stages of DNA transport to interpret the stages and molecular interactions for which ATP, $\Delta\Psi$, or ΔpH is required in each system. Thus, the general hypothesis of Grinius (1976) that DNA traversal is a proton-symport-mediated transport system still remains to be tested.

REFERENCES

Achtman, M., and Skurray, R. (1977). In *Microbial Interactions* (J. L. Reissig, ed.), pp. 235–279, Chapman & Hall, London.
Barany, F., and Tomasz, A. (1980). *J. Bacteriol.* **144,** 698–709.
Beržinskienė, J. A., Zizaite, L. J., Baronaitė, Z. A., and Grinius, L. (1980). *Biokhimiya* **45,** 1103–1112.
Chaustova, L. P., Grinius, L. L., Griniuvienė, B. B., Jasaitis, A. A., Kadziauskas, J. P., and Kiaušinytė, R. V. (1980). *Eur. J. Biochem.* **103,** 349–357.
Danner, D. B., Deich, R. A., Sisco, K. L., and Smith, H. O. (1980). *Gene* **11,** 311–318.
Goldberg, E. B. (1980). In *Receptors and Recognition: Virus Receptors* (L. Randall and L. Philipson, eds.), Series B, Vol. 7, pp. 115–141, Chapman & Hall, London.
Grinius, L. (1976). *Biokhimiya* **41,** 1539–1547.
Grinius, L., and Beržinskienė, J. (1976). *FEBS Lett.* **72,** 151–154.
Hancock, R. E. W., and Braun, V. (1976). *J. Bacteriol.* **125,** 409–413.
Kalasauskaitė, E., and Grinius, L. (1979). *FEBS Lett.* **99,** 287–291.
Kalasauskaitė, E., Grinius, L., Kadišaitė, D. and Jasaitis, A. (1980). *FEBS Lett.* **177,** 232–236.
Kornberg, A. (1980). *DNA Replication,* Freeman, San Francisco.
Labedan, B. (1976). *Virology* **75,** 368–375.
Labedan, B., and Goldberg, E. B. (1979). *Proc. Natl. Acad. Sci. USA* **76,** 4669–4674.
Labedan, B., Crochet, M., and Legault-Demare, J. (1973). *J. Mol. Biol.* **75,** 213–234.
Labedan, B., Heller, K. B., Jasaitis, A. A., Wilson, T. H., and Goldberg, E. B. (1980). *Biochem. Biophys. Res. Commun.* **93,** 625–630.
Lacks, S. A. (1977). In *Microbial Interactions* (J. L. Reissig, ed.), Series B, Vol. 3, pp. 177–232, Chapman & Hall, London.

Plate, C. A., Suit, J. L., Jetten, A. M., and Luria, S. E. (1974). *J. Biol. Chem.* **249,** 6138–6143.

Randall, L. L., and Philipson, L. (eds.) (1980). *Virus Receptors, Part 1, Bacterial Viruses,* Chapman & Hall, London.

Sisco, K. L., and Smith, H. O. (1979). *Proc. Natl. Acad. Sci. USA* **76,** 972–976.

Sutrina, S. L., and Scocca, J. A. (1979). *J. Bacteriol.* **139,** 1021–1027.

Willets, N., and Skurray, R. (1980). *Annu. Rev. Genet.* **14,** 41–76.

IX

Membrane-Linked Metabolite and Ion Transport Systems in Animal Cells

108

Nature of the Frog Skin Potential

Hans H. Ussing

1. INTRODUCTION

Early work by our group (Ussing, 1948, 1949a,b) showed that the inward transport of sodium through the isolated frog skin was due to active transport, whereas the chloride transport might be passive. This view was substantiated when it was shown (Ussing and Zerahn, 1951) that the entire current output by the short-circuited frog skin was due to active inward transport of sodium. The flux ratio for chloride, on the other hand, fitted the flux ratio equation under all the conditions tested, indicating passive behavior of that ion (Koefoed-Johnsen *et al.*, 1952).

Thus, the conclusion was that the entire electric asymmetry of the frog skin—and thus the maintenance of an electric potential across the skin—was due to active sodium transport.

A more detailed analysis of the phenomenon was made possible by the two-membrane model of the epithelium (Ussing and Koefoed-Johnsen, 1956; Koefoed-Johnsen and Ussing, 1958). The model assumes the transport to be performed by a continuous layer of cells, whose inward- and outward-facing membranes have different transport properties: The outward-facing membrane is passively, but selectively, permeable to sodium, and impermeable to potassium. The inward-facing membrane, on the other hand, is permeable to potassium and impermeable to sodium. It is the seat of a sodium pump that from the outset was assumed to be of the same nature as the sodium–potassium exchange pump of erythrocytes. Both membranes were assumed to be permeable to chloride.

Originally the two-membrane model was based on the potential responses to changes in the ionic composition of the outside and inside bathing solutions. To this came the finding (Koefoed-Johnsen, 1957) that the ouabain sensitivity and thus the sodium pump was located at the inward-facing boundary of the epithelium. Originally we thought that the stratum germinativum of the epithelium was the site of active ion transport and potential development. Careful microelectrode studies (Ussing and Windhager, 1964) showed, however, that the sodium-selective membrane was the outer barrier of the outermost living

Hans H. Ussing • Institute of Biological Chemistry A, University of Copenhagen, Copenhagen, Denmark.

cell layer and that all the living cell layers were coupled via permeable cell contacts. Thus, the epithelium seemed to act like a sort of syncytium. A similar model was advanced independently and nearly simultaneously by Farquhar and Palade (1964). The Ussing–Windhager model possessed, however, additional new features, namely that there must be ionic leak paths through the epithelium, both via the cells and via the "tight" seals between the outermost living cells.

Over the years there has been considerable dispute with respect to the correctness of the assumptions underlying the two-membrane model. The problem has been reviewed recently (Ussing *et al.*, 1974; Ussing and Leaf, 1978; Erlij and Ussing, 1978). These articles discuss most of the pertinent literature and conclude that the evidence is in favor of the theory.

In the present review we shall only consider some papers that have appeared after the publication of the reviews mentioned, and which have put the membrane model on an even firmer footing.

2. MICROPROBE STUDIES

The microprobe group at the Department of Physiology, University of Munich (Rick *et al.*, 1978; Dörge *et al.*, 1981), has studied the sodium and potassium concentrations in freeze-dried sections of frog skins that had been exposed to different experimental conditions. In every case the distribution was exactly as predicted from the Ussing–Windhager version of the two-membrane model. The cornified layer simply followed the composition of the outside bath, indicating that this layer is dead and does not present any significant diffusion resistance. All the other layers behaved as expected for a syncytium. Under normal conditions the sodium concentration in all layers was low (about 10 mM) and the potassium concentration high (about 110 mM). All concentrations are given as recalculated on a wet weight basis. After inhibition of the sodium pump with ouabain, the cellular concentration of sodium rose in all layers to 110 mM whereas the potassium concentration dropped to about 20. However, if ouabain inhibition was performed in the presence of the sodium-channel inhibitor amiloride, the concentration of sodium in all cell layers remained low and the potassium concentration high, indicating that sodium reaches all cell layers exclusively via the sodium-selective, amiloride-sensitive membrane of the outermost living cell layer. During ouabain poisoning, sodium-free choline Ringer in the outside bath will also keep the cellular potassium high and the cellular sodium low, even if the inside bath contains sodium Ringer. These results are in good agreement with the finding (Ussing, 1978) that excessive swelling of all cell layers in ouabain-inhibited frog skins can be elicited by a moderate dilution of the inside medium if the outside medium is sodium Ringer, but not if it is potassium Ringer. Again the implication is that sodium, but not potassium, can pass the outward-facing cell membrane of the outermost living cells and from there can pass on from one cell layer to the next via conducting cell junctions.

3. COMPUTER SIMULATION OF THE MODEL

A computer simulation model of the two-membrane theory was worked out by Larsen and Kristensen (1978) and Larsen (1978). It predicts the steady-state electrochemical properties of the skin quite satisfactorily. More recently, Lew *et al.* (1979) have given a comprehensive treatment of computer-simulated non-steady-state situations. They have shown that many potential and current transients that had been thought to be in disagreement with the theory are actually consequences of it. The model can be used not only for frog skin, but also for several other tight epithelia.

4. FLUCTUATION ANALYSIS OF EPITHELIAL ION TRANSPORT

In the model the entry of sodium via the outward-facing membrane is assumed to be passive but highly selective. All recent studies agree on this point. The fact that the process seems to show saturation kinetics has induced some investigators to propose a carrier mechanism. The careful study by Fuchs *et al.* (1977) showed, however, that the apparent saturation is due to the effect of sodium on an inhibitory site on the sodium channel. This site seems to be identical to the site acted upon by the channel inhibitor amiloride (Cuthbert and Shum, 1976). The view that sodium actually enters through a channel has been further strengthened by a study of the spontaneous fluctuations of membrane voltage and membrane current (van Driessche and Lindemann, 1979; Lindemann, 1980). The statistical analysis of the sodium transport fluctuations at the outward-facing membrane was made possible by reducing the number of open channels through the addition of amiloride. The channels seemed to open and close randomly, and the number of open channels seemed to decrease with increasing outside sodium concentration. Incidentally, it has turned out that the sodium-selective membrane (the apical membrane) of skins of *Rana temporaria* (but not *Rana esculenta*) possesses a small number of spontaneously switching K^+ channels (Zeiske and van Dreissche, 1979; Hirschmann and Nagel, 1978). These channels are insensitive to amiloride but can be inhibited by Ba^{2+}, H^+, Rb^+, and Cs^+. The fact that such channels are virtually absent in the apical membrane of *R. esculenta* and totally absent in the toad urinary bladder makes it likely that in *R. temporaria* they are just leftovers from the premolt period when the cells in question belonged to the inward-facing K^+-selective part of the syncytium.

5. Na⁺–K⁺ COUPLING RATIO OF THE PUMP

One of the consequences of the two-membrane theory is that the potassium that the pump transfers to the cell in return for sodium must normally diffuse back to the inside bathing medium via the specific K^+ channels; this is because the outward-facing membrane is virtually tight to potassium. If the apical membrane was made leaky to potassium, a fraction of it would escape that way, and one would observe active outward pumping of potassium pari passu with the inward pumping of sodium. This in fact has been shown to be correct. Robert Nielsen (1971, 1979a), working in our institute, has shown that if the outward-facing membrane of the frog skin is made leaky to potassium with the polyene antibiotics filipin or amphotericin B, an outward-directed active transport of potassium is "created." This potassium transport can be inhibited by lack of sodium in the outside bath or by ouabain added to the inside bath. These experiments indicate that the two-membrane model is essentially correct, but it cannot yield a precise value for the coupling ratio between Na^+ and K^+ in the pump. Quite recently, however, Nielsen (1979b) has been able to measure the coupling ratio in a very simple way. Let us assume that the pump is of the same nature as that of red cells, where three sodium ions are pumped out and two potassium ions pumped in for each ATP broken down. During short-circuiting, the net electric current through the basolateral membrane must then be carried by one sodium ion (which represents the electrogenic contribution of the pump) and two potassium ions returning from the cell to the inside bath via a potassium-selective channel. The latter can be selectively inhibited by Ba^{2+} added to the inside bath. Thus, barium addition should reduce the short-circuited current to one third. This is in fact what happens. Preferably the experiment should be performed on isolated epithelium rather than on whole skins. If so, barium addition leads within seconds to the predicted drop in current. The barium effect is maximal at 3 mM, and further addition of Ba^{2+} does not alter the degree of inhibition of the current. Barium obviously does not inhibit the sodium pump per se.

REFERENCES

Cuthbert, A. W., and Shum, W. K. (1976). *J. Physiol.* (*London*) **255**, 587–604.

Dörge, A., Rick, R., Katz, U., and Thurau, K. (1981). In *Water Transport across Epithelia*, Alfred Benzon Symposium 14 (H. H. Ussing, N. Bindslev, N. A. Lassen, and O. Sten-Knudsen, eds.), Munksgaard, Copenhagen, in press.

Erlij, D., and Ussing, H. H. (1978). In *Membrane Transport in Biology* (G. Giebisch, D. C. Tosteson, and H. H. Ussing, eds.), Vol. III, pp. 175–208, Springer-Verlag, Berlin.

Farquhar, M. G., and Palade, G. E. (1964). *Proc. Natl. Acad. Sci. USA* **51**, 569–577.

Fuchs, W., Larsen, E. H., and Lindemann, B. (1977). *J. Physiol.* (*London*) **267**, 137–166.

Hirschmann, W., and Nagel, W. (1978). *Pfluegers Arch.* **373**, R48.

Koefoed-Johnsen, V. (1957). *Acta Physiol. Scand.* **42**(Suppl. 145), p. 87.

Koefoed-Johnsen, V., and Ussing, H. H. (1958). *Acta Physiol. Scand.* **42**, 298–308.

Koefoed-Johnsen, V., Levi, H., and Ussing, H. H. (1952). *Acta Physiol. Scand.* **25**, 150–163.

Larsen, E. H. (1978). *11th Alfred Benzon Symposium*, pp. 438–456, Munksgaard, Copenhagen.

Larsen, E. H., and Kristensen, P. (1978). *Acta Physiol. Scand.* **102**, 1–21.

Lew, V. L., Ferreira, H. G., and Moura, T. (1979). *Proc. R. Soc. London Ser. B* **206**, 53–83.

Lindemann, B. (1980). *J. Membr. Biol.* **54**, 1–11.

Nielsen, R. (1971). *Acta Physiol. Scand.* **83**, 106–114.

Nielsen, R. (1979a). *J. Membr. Biol.* **51**, 161–184.

Nielsen, R. (1979b). *Acta Physiol. Scand.* **107**, 189–191.

Rick, R., Dörge, A. V., Arnim, E., and Thurau, K. (1978). *J. Membr. Biol.* **39**, 313–331.

Ussing, H. H. (1948). *Cold Spring Harbor Symp. Quant. Biol.* **13**, 193–200.

Ussing, H. H. (1949a). *Acta Physiol. Scand.* **17**, 1–37.

Ussing, H. H. (1949b). *Acta Physiol. Scand.* **19**, 43–56.

Ussing, H. H. (1978). *J. Membr. Biol.* **40S**, 5–14.

Ussing, H. H., and Koefoed-Johnsen, V. (1956). Abstracts Communications 20th International Physiology Congress, Brussels.

Ussing, H. H., and Leaf, A. (1978). In *Membrane Transport in Biology* (G. Giebisch, D. C. Tosteson, and H. H. Ussing, eds.), Vol. III, pp. 1–26, Springer-Verlag, Berlin.

Ussing, H. H., and Windhager, E. E. (1964). *Acta Physiol. Scand.* **61**, 484–504.

Ussing, H. H., and Zerahn, K (1951). *Acta Physiol. Scand.* **23**, 110–127.

Ussing, H. H., Erlij, D., and Lassen, U. (1974). *Annu. Rev. Physiol.* **36**, 17–49.

van Driessche, W., and Lindemann, B. (1979). *Nature* (*London*) **282**, 519–520.

Zeiske, W., and van Driessche, W. (1979). *J. Membr. Biol.* **47**, 77–96.

109

The Analog Inhibition Strategy for Discriminating Similar Transport Systems: Is It Succeeding?

H. N. Christensen

To examine the question posed above, I will use as examples the several Na$^+$-dependent transport systems for neutral amino acids detected in the plasma membrane of the cells of higher animals. The first five pages of my earlier review (Christensen, 1975a) can serve for immediate background. The strategy used here for discrimination was outlined in my 1966 essay, and in more detail in 1969, and in pages 176 to 196 of my monograph (1975b). The title question refers to the situation in the cells of the higher animal where mutants restricted to each of the transport systems usually are not readily available.

We will assume for our example that the investigator has measured the initial rate of uptake of a neutral amino acid, and furthermore that he had deducted the uptake observed in parallel tests in which Na$^+$ had been omitted from the suspending medium. Transport measured without deduction of this component must be interpreted with exceptional care. We will assume in line with usual experience that the Na$^+$-dependent uptake applies to several amino acids. Does this Na$^+$-dependent uptake occur by a single agency? Most readers will appreciate that we cannot prove that a single agency is responsible merely by showing that substrate influx bears a hyperbolic relation to its concentration, whether we are assisted by a computer or not. The risk is too high that the sets of kinetic parameters are not as different as they need to be to distort the hyperbolic relation unequivocally. The strategy usually employed is instead to explore a fairly wide range of analogs for their uptake and their inhibition of uptake. Some of these may need to be synthesized ad hoc. By including the full range of effective analogs, one will probably encounter a pair each of which inhibits, exclusively one hopes, a different part of the Na$^+$-dependent uptake. We will assume that in this way two distinct transport components or mediations are provisionally identified, and ideally also a model substrate for each of them, i.e., an analog with its

H. N. Christensen • Department of Biological Chemistry, The University of Michigan Medical School, Ann Arbor, Michigan 48109.

transport and inhibition of transport restricted to that system. In completing this strategy one may at some stage try also to inhibit a single transport system, or all but one system, by a nonsubstrate inhibitor, or to change transport specifically by a given system through lowering the pH, adding a hormone, or starving the cell under test with regard to amino acids.

Having by differential inhibition resolved the Na^+-dependent transport flux for a group of amino acids into two components characterized as mutually exclusive, we should ideally resume the process for each component until we consider we have exhausted the possibility of heterogeneity in each. At any point, analytical imprecision in exploring for further heterogeneity, or statistical deficiencies in detecting it, may set limits for a given comparison, so that one may need to turn to another substrate–inhibitor combination, or to another cell to extend the challenge. I suppose the conclusion that a single, homogeneous transport system has been identified will always remain provisional. Ultimately, some unexpected heterogeneity, for example separate regulation for two similar substrates, may yet be encountered.

In the instances of diversity of routes first encountered, optimism rose rapidly because satisfactory model substrates were soon found, for example 2-(methylamino)isobutyric acid (MeAIB) or N-methylalanine for System A (Christensen et al., 1965, 1967), the $(-)$ or $(\pm)$ amino-endo isomer of 2-aminobicyclo-(2,2,1)-heptane-2-carboxylic acid (BCH) for Na^+-independent System L, and taurine for the *beta* amino acid system. When the discrimination of a system serving for glycine from one serving for alanine was encountered in the pigeon red blood cell, it was appreciated early that these two amino acids made a useful although imperfect discrimination between the two components in that cell and also in the rabbit reticulocyte, first as inhibitors (Vidaver et al., 1964) and then as substrates (Eavenson and Christensen, 1967; Wheeler et al., 1956; Wheeler and Christensen, 1967a,b). Discrimination of these two components, soon designated as Systems Gly and ASC, can, however, present difficulties to be considered in this essay. Subsequently, imperfections in model substrates have been encountered all too often, so that one even begins to wonder whether a model substrate as good as MeAIB is for A will ever be attained for Gly or ASC.

Our original characterization of System ASC in the Ehrlich cell (Christensen et al., 1967) recognized a discrimination problem more serious than the one presented by the ASC–Gly dichotomy in the two red blood cell types, or by the A–L dichotomy in any cell. Although MeAIB inhibits System A with little if any effect on ASC, each of the substrates we encountered for ASC proved also a substrate for A. Therefore, we had to rely on a negative criterion for ASC, i.e., to assign provisionally to it the Na^+-dependent uptake retained in the presence of a substantial excess (25 mM) of MeAIB. We recognized here not only a quantitative but also a qualitative risk, as follows:

Is System ASC a real entity, or instead a possible artifact of the presence of all this MeAIB? Our caveat on this score opened as follows:

"The reactive sites of System A could occur in sets . . . . , so that filling part of them [with an α-methylamino acid] might change the characteristics of those remaining unfilled" (Christensen et al., 1967).

The presence of excess MeAIB or N-methylalanine in theory could certainly change the maximal rates, and perhaps even the relative transport affinities of System A for various amino acids. It seems to us unlikely it would do so in the internally consistent pattern we and others have observed. Dr. Rose Johnstone of McGill University presented me personally with a new and extended version of this argument. She proposed that in the Ehrlich cell, depolarization of the plasma membrane by MeAIB could artifactually produce one or another of several features of the residual Na^+-dependent uptake by System A (for example, differences in relative pH sensitivity or stereospecificity) which we ascribed to the

ASC component, and which supported our conclusion for two distinct mediating systems. Dr. Johnstone chose to restrict her argument to the Ehrlich cell, where *ASC* awkwardly is a minor component; but our continuing concern for the discrimination problem understandably is not restricted to that cell. Various other cells have shown evidence for the *A–ASC* dichotomy. Although it would be pleasant to return for the Ehrlich cell and other cells to the simpler picture of a single Na^+-dependent component for α-amino acids, we believe we have gone beyond the point at which this view is generally tenable. While we are waiting for the possible challenge to the reality of System *ASC* in the Ehrlich cell to evolve, I will in this essay consider the question, how well can we demonstrate the *A–ASC* dichotomy without recourse to MeAIB or N-methylalanine? I will not cite here studies from elsewhere that have confirmed the dichotomy through the use of a methylamino acid.

The basic difficulty I have already described, our lack of a universal model substrate restricted in its action to System *ASC*, has often forced us to depend on the negative criterion of resistance to MeAIB inhibition. Elsewhere I have summarized some previous efforts to repair that defect (Christensen, 1979). We originally limited the action of several ordinary amino acids to transport by System *ASC* by using pigeon red blood cells and rabbit reticulocytes. In these cells System *A* appears to be entirely absent, so no analog needs to be added to limit Na^+-dependent uptake of alanine, serine, or cysteine largely to *ASC*. The confidence we needed to decide originally for the intrinsic character of the *A–ASC* dichotomy flowed to a major extent from our concurrent direct observations on these red blood cells. Most of our early results describing System *ASC* were in fact obtained using these cells. Because the system showed features quite different from System *A* but quite similar to those we found for the *ASC* component of the Ehrlich cell, we came to accept the validity of the other evidence we had accumulated for the reality of System *ASC* in that cell.

Our search for the needed model *ASC* substrate has now turned to other cells, especially hepatocytes. In the latter cell, the *ASC* components tend to dominate Na^+-dependent uptake, and the *A* component is subject to repression. Hence, the analytical situation can be made much more favorable than that presented by the Ehrlich cell. The search was also intensified because other investigators have sometimes concluded in favor of the *A–ASC* division of uptake on what seemed to us insufficient grounds. An example is the to us unexpected division of Na^+-dependent 2-aminoisobutyrate (AIB) uptake between Systems *A* and *ASC* in the isolated rat hepatocyte (LeCam and Freychet, 1977). We have confirmed their inference on this point, and also incidentally confirmed a substantial uptake of AIB by Na^+-independent System *L* in the hepatocyte (Kelley and Potter, 1978; Kilberg *et al.*, 1981). These findings place the burden of proof on anyone choosing AIB as a system-specific transport substrate in any new context. An additional factor accentuating the need for a model substrate allowing direct measure of System *ASC* activity was the finding in a number of instances of unexpectedly wide scope for the system among neutral amino acids, results that logically place even leucine, phenylalanine, and its analog melphalan among its proposed substrates (for summaries see Christensen, 1979, pp. 83–88; Kilberg *et al.*, 1981). Shotwell *et al.*, (1981) also find a broad range for the *ASC* component of cultured Chinese hamster ovary cells. This scope, wider in general than we found for the cells studied earlier, does not involve any changes, however, in the identity of the amino acids characteristically transported most rapidly by *ASC* (Christensen *et al.*, 1969; Thomas and Christensen, 1970, 1971). Perhaps the most important proposal of a widened range for *ASC* is the suggestion that Na^+-dependent intestinal absorption of neutral amino acids occurs principally by this route (Christensen and Handlogten, 1977; Sepúlveda and Smith, 1978; Kilberg *et al.*, 1980c; Mircheff *et al.*, 1980). Note that Fig. 5 in the first of these references shows how much more closely the absorptive activity for methionine in the rabbit ileum (Preston *et al.*, 1974) corresponds to System *ASC* than *A* in its structural selectivity. Our original term for System *ASC* for the pigeon red cell was *ASCP* (Eavenson and Christensen, 1967), illustrating one way to recognize differences in *ASC* from one tissue to another. Instead of proliferating such code terms, we now suggest that nonidentical versions of what appears to be a single transport system be recognized by qualifications in words, e.g., *System ASC of the rat hepatocyte*. Where variants of a system occur together, such terms as Systems A_1 and A_2, or L_1 and L_2, may serve.

Our recent characterization of a unique Na^+-dependent route ("System *N*") for the transport of glutamine, asparagine, and histidine in the isolated rat hepatocyte (Kilberg *et al.*, 1980b) warns further of the danger of assigning to *ASC* all the MeAIB-insensitive uptake of neutral amino acids. The same warning comes from the demonstration that in the presence of MeAIB, glycine still shows a relatively large uptake by System *Gly* in the rat hepatocyte and the hepatoma cell line HTC (Christensen and Handlogten, 1981). These findings add further strength to the call for a model substrate allowing direct measures of uptake by System *ASC*. Both of these components might have been included erroneously in the *ASC* component by the older criterion. Other such cases may yet be encountered. To us the risk that another system might hide in the *ASC* component, measured by its MeAIB insensitivity, seems much greater than the risk that *ASC* might not exist at all.

We were pleased therefore to find the Na^+-dependent cysteine uptake largely specific to System *ASC* in the rat hepatocyte (Kilberg *et al.*, 1980a, 1981), and apparently in the rat ileum (Kilberg *et al.*, 1980c). We were able to show with cysteine used alone (no MeAIB added) that System *ASC* has a different pH sensitivity than either Systems *A* or *N*, and that System *ASC* fails to be stimulated, as Systems *A* and *N* are, by amino acid starvation. Furthermore, cysteine used alone fails to respond in its uptake to insulin or glucagon as System *A* does (Kilberg *et al.*, 1981). These properties for *ASC* closely resemble those previously observed using inhibition by an *N*-methylalanine derivative to isolate the *ASC* component. It was Gazzola *et al.* (1973) who first observed the failure of the *ASC* component to respond to "adaptive regulation."

The astonishing aspect is that two different amino acids serve best as substrates for observing System *ASC* in liver cells: cysteine in the surviving hepatocyte from Sprague–Dawley rats, and threonine in the hepatoma cell line HTC, but neither amino acid is a good discriminator in the other cell. The difference apparently arises from transformation, because the amino acid transport properties of the inbred Buffalo rat, from which the HTC cell is derived, appear much like those of the Sprague–Dawley rat. For service in HTC, cysteine is much too strong an inhibitor of MeAIB uptake, $K_i = 0.6$ mM, hence by definition, of System *A*. Even in the normal rat hepatocyte, cysteine inhibits System *A*, $K_i = 12$ mM, compared with its K_m of 2 mM for Na^+-dependent uptake. No significant amount of MeAIB-inhibitable uptake of cysteine could, however, be observed. Threonine also shows the latter imperfection as a test substrate in HTC, but here the K_i in inhibiting MeAIB uptake, also 12 mM, has a value of 60 times as high as the K_m for its own uptake, 0.1 mM. This relatively stronger selectivity of threonine for *ASC* in HTC is enough to make it a good model substrate, and to make the direct demonstration of *ASC* even more impressive in HTC than in ordinary rat hepatocytes (Handlogten *et al.*, 1981).

Conversely, MeAIB showed no appreciable inhibition of threonine uptake by HTC. If a System *A* uptake of threonine with a K_m of 12 mM actually occurs, its relative contribution might be rather too small to yield a distinct sensitivity to inhibition by MeAIB. Hence, we cannot tell whether threonine inhibits System *A* without being transported at all by it. In the case of cysteine and the ordinary hepatocyte, presenting a ratio of six between the K_i for System A and the K_m of 2 mM for ASC, the inhibition appears dissociated from any substrate action, unless it were one with a very low V_{max}. The phenomenon of analog inhibition of an apparently unshared transport system has been observed repeatedly and especially often in the relation between Systems ASC and A. Therefore, a model substrate must not only have its transport mainly restricted to one system; its inhibition of transport must also be restricted. Furthermore, the ratio of six shown by cysteine in its selectivity for ASC over A for inhibition is small enough to cause analytical difficulties in its use in discriminating these transport routes.

By using threonine as test substrate, we have also been able to show a difference of ASC from System A in pH sensitivity in HTC, and as usual in its lack of response to amino acid starvation and to insulin. The good agreement in the list of amino acids that show uptake assignable to ASC on this basis, and the list showing uptake assignable to ASC on the basis of resistance to MeAIB, and also the good agreement between the properties of the ASC component under the two sets of conditions, support not only the reality of the A–ASC dichotomy, but also the approximate validity of the use made heretofore of the N-methylalanine derivatives for the discrimination.

Let us now return to the discrimination of transport routes for glycine. In the Ehrlich cell, in contrast to the pigeon red blood cell and the rabbit reticulocyte, glycine is a substrate for transport System A (Oxender and Christensen, 1963), and this cell has failed to show clear evidence for System Gly. Its K_m for Na$^+$-dependent glycine uptake attributed to System A is rather high, 3 to 4 mM. By analogy with AIB and alanine, one would not expect the N-methylation of glycine to block its access to System A. In agreement, sarcosine shows a normal V_{max} and a K_m of about 5 mM for uptake by the Ehrlich cell (Christensen *et al.*, 1965). Vidaver *et al.* (1964) pointed out that in the pigeon red blood cell, glycine produces little inhibition of alanine uptake, hence of what later came to be called System ASC. In agreement, Eavenson and I (1967) found the K_i of sarcosine in inhibiting the uptake of serine, proline, and alanine to be about 3 mM in the pigeon red blood cell, about 18 times the K_m this amino acid shows for uptake. Although as expected sarcosine readily inhibits the System A uptake of serine in the Ehrlich cell (i.e., the same portion inhibited by MeAIB), its action on the ASC (i.e., MeAIB-insensitive) component was relatively weak, $K_i = 21$ mM (Christensen *et al.*, 1967, documented in text, p. 5242).

When Reichberg and Gelehrter (1980) recently showed that two parallel but independent routes participate in glycine uptake by the hepatoma cell line HTC, we were stimulated to show, as seemed predictable from our prior observations, that the two routes are Gly and A and only to a minor extent ASC (Christensen and Handlogten, 1981). Reichberg and Gelehrter considered the possibility that the high-affinity system was Gly, but obtained unfavorable evidence on the point and therefore considered it to be ASC. In our experiments with HTC (Handlogten *et al.*, 1981), glycine uptake at 0.05 mM showed little slowing on lowering the pH to 6 or even 5, quite in contrast to the pronounced falls shown by MeAIB (System A, the fall occurring between pH 7.4 and 6) and by threonine (System ASC, the fall between pH 6.5 and 4.5). Furthermore, we found sarcosine more effective by 15-fold in inhibiting the System Gly component than the uptake of MeAIB by HTC. Conversely and as expected, MeAIB proved in turn very much weaker in inhibiting the Gly than the A component. Threonine inhibited only a small portion of glycine uptake to the high intensity expected for its effect on System ASC. Its action was for the most part no stronger than could be expected for the A contribution. Glycine in turn showed only a weak inhibition of threonine uptake. The system observed by Reichberg and Gelehrter to have the higher affinity for glycine accordingly was System Gly, known to be a high-affinity system in other cells, and not ASC as they concluded. They found the system we identify as *Gly* more resistant than the other (*A*) component to inhibition by dexamethasone, and to inhibition by lowering the pH to 6. So far we find the glycine and sarcosine uptake assignable to *Gly* (i.e., at low test concentrations) in HTC unresponsive to amino acid starvation, and only weakly if at all responsive to insulin in comparison with the System *A* uptake predominating at high levels. System *Gly* could be demonstrated also in the cultured rat hepatocyte.

Such pairs of selective inhibitors, in this case sarcosine and MeAIB, need to be used at intermediate and not excessive concentrations, except where their selectivity between systems has proved very strong. We raised the level of MeAIB to 25 mM in the Ehrlich cell only because System *A* uptake dominated the *ASC* component to the point where even 98% inhibition of *A* seemed barely enough.

The partitioning of glycine uptake between these two systems in liver cells (with probably at most a minor third component assignable to *ASC*) presents a degree of analogy to the partitioning for example of serine uptake between Systems *A* and *ASC*. Historically viewed, however, the case of glycine transport presented an advantage to investigators, namely that the harder problem of a cell showing a glycine uptake divided rather unevenly between two systems was encountered later, whereas for serine and alanine uptake we faced first the more difficult problem of an admixture of two systems in one cell, indeed an admixture in unfavorable proportions. Only later did we develop the opportunities for seeing, first by using other cells and then by using other test amino acids, each of these systems operating alone. The two cases differ also in that a symmetrical pair of inhibitor–substrates, sarcosine and MeAIB, was already available when the case of dual glycine transport arose, but is even now not uniformly available for the *A–ASC* dichotomy, each amino acid analog to be specific to one of the two transport components, according to my 1966 model. The circumstance that both Systems *Gly* and *A* tolerate *N*-methylation of their substrates could have complicated their discrimination; but nevertheless the selectivity of sarcosine and MeAIB proved adequate. It is perhaps also worth emphasizing here that this strategy has by no means recommended reliance on curved Lineweaver–Burk plots for demonstrating or excluding heterogeneity in the mediators producing membrane transport, even with computer assistance.

I trust that this discussion has indicated that the strategy of inhibition analysis is succeeding in the discrimination of the Na^+-dependent systems for the uptake of neutral amino acids. For MeAIB and BCH in many observations by a series of different observers in this laboratory we have encountered only recently clearcut cases of significant inhibition of initial rates of uptake other than by the systems for which these model substrates were defined. For the problems of discriminating System *ASC*, a provisional reliance on a negative trait, toleration of MeAIB in excess, seems so far not to have seriously misled us, although direct measures must be urged. For Systems *ASC* and *Gly* in some important cell types, we are still forced to use test substrates (e.g., cysteine, threonine, sarcosine) whose reactivity with amino acid transport agencies is not entirely limited to one of these systems. In line with this experience, we recommend that the selectivity of a transport substrate serving to measure the contribution of a given system to transport be established for each new occurrence.

ACKNOWLEDGMENT. The results here that derive from my laboratory are from research supported by Grant HD01233 from the Institute of Child Health and Human Development, National Institutes of Health.

REFERENCES

Christensen, H. N. (1966). *Fed. Proc.* **25,** 850–853.

Christensen, H. N. (1969). *Adv. Enzymol.* **32,** 1–20.

Christensen, H. N. (1975a). *Curr. Top. Membr. Transp.* **6,** 227–258.

Christensen, H. N. (1975b). *Biological Transport*, 2nd ed., pp. 176–196, Reading, Mass.

Christensen, H. N. (1979). *Adv. Enzymol.* **49,** 41–101 (see pp. 75–77, 83–88).

Christensen, H. N., and Handlogten, M. E. (1977). *J. Membr. Biol.* **37,** 193–211.

Christensen, H. N., and Handlogten, M. E. (1981). *Biochem. Biophys. Res. Commun.* **98,** 102–107.

Christensen, H. N., Oxender, D. L., Liang, M., and Vatz, K. A. (1965). *J. Biol. Chem.* **240,** 3609–3616.

Christensen, H. N., Liang, M., and Archer, E. G. (1967). *J. Biol. Chem.* **242,** 5237–5246.

Christensen, H. N., Thomas, E. L., and Handlogten, M. E. (1969). *Biochim. Biophys. Acta* **193,** 228–230.

Eavenson, E., and Christensen, H. N. (1967). *J. Biol. Chem.* **242,** 5386–5396.

Gazzola, G. C., Franchi-Gazzola, R., Ronchi, R., and Guidotti, G. G. (1973). *Biochim. Biophys. Acta* **311,** 292–301.

Handlogten, M. E., Garcia-Cañero, R., Lancaster, K. T., and Christensen, H. N. (1981). *J. Biol. Chem.* **256,** 7905–7909.

Kelley, D. S., and Potter, V. R. (1978). *J. Biol. Chem.* **253,** 9009–9017.

Kilberg, M. S., Christensen, H. N., and Handlogten, M. E. (1980a). *Biochem. Biophys. Res. Commun.* **88,** 744–751.

Kilberg, M. S., Handlogten, M. E., and Christensen, H. N. (1980b). *J. Biol. Chem.* **255,** 4011–4019.

Kilberg, M. S., Lancaster, K. T., and Christensen, H. N. (1980c). *Fed. Proc.* **39,** 1712a.

Kilberg, M. S., Handlogten, M. E., and Christensen, H. N. (1981). *J. Biol. Chem.* **256,** 3304–3312.

LeCam, A., and Freychet, P. (1977). *J. Biol. Chem.* **252,** 148–156.

Mircheff, A. K., van Os, C. H., and Wright, E. M. (1980). *J. Membr. Biol.* **52,** 83–92.

Oxender, D. L., and Christensen, H. N. (1963). *J. Biol. Chem.* **238,** 3686–3699.

Preston, R. L., Schaeffer, J. F., and Curran, P. F. (1974). *J. Gen. Physiol.* **64,** 443–467.

Reichberg, S. B., and Gelehrter, T. D. (1980). *J. Biol. Chem.* **255,** 5708–5714.

Sepúlveda, F. V., and Smith, M. W. (1978). *J. Physiol. (London)* **282,** 73–90.

Shotwell, M. A., Jayme, D. W., Kilberg, M. S., and Oxender, D. L. (1981). *J. Biol. Chem.* **256,** 5422–5427.

Thomas, E. L., and Christensen, H. N. (1970). *Biochem. Biophys. Res. Commun.* **40,** 277–283.

Thomas, E. L., and Christensen, H. N. (1971). *J. Biol. Chem.* **246,** 1682–1688.

Vidaver, G. A., Romain, L. F., and Haurowitz, F. (1964). 'Arch. Biochem. Biophys. **107,** 82–87.

Wheeler, K. P., and Christensen, H. N. (1967a). *J. Biol. Chem.* **242,** 1450–1457.

Wheeler, K. P., and Christensen, H. N. (1967b). *J. Biol. Chem.* **242,** 3782–3788.

Wheeler, K. P., Inui, Y., Hollenberg, P. F., Eavenson, E., and Christensen, H. N. (1965). *Biochim. Biophys. Acta* **109,** 620–622.

110

The Kinetics and Mechanism of Na$^+$ Gradient-Coupled Glucose Transport

Robert K. Crane and Frederick C. Dorando

Studies of the kinetics of Na$^+$ gradient-coupled glucose transport in membrane vesicles derived from jejunal and proximal kidney tubular brush borders have begun to show great promise of deepening our understanding of the transport system. Not only do these studies seem finally to resolve the long-standing question of whether high- and low-affinity glucose transports reflect the presence of multiple, kinetically simple systems or a single, kinetically complex system but also they seem to provide a basis for revision in concepts of plausible molecular mechanisms of glucose transport.

High- and low-affinity glucose transport was first described by Crane *et al.* (1965) in studies of glucose uptake by rings of everted hamster small intestine incubated *in vitro*. The first interpretation of the data was in terms of a single transport system in which high-affinity glucose transport occurred via a ternary complex formed with the carrier and Na$^+$, CNaG, and low-affinity glucose transport via a binary complex formed without Na$^+$, CG. Lyon and Crane (1966) corrected this interpretation by showing, in rat intestine, that there were reciprocal effects of the concentrations of each substrate, glucose and Na$^+$, on the apparent affinity of the other. The data now fitted a formulation of Na$^+$ gradient-coupled sugar transport as a random bi bi rapid equilibrium (Stein, 1967) in which there are two pathways to the common ternary complex, CNaG, through which all transport takes place. For each of the substrates, glucose and Na$^+$, the low-affinity pathway (LAP) is that in which it binds first, the high-affinity pathway (HAP) is that in which it binds second. Thus, LAP$_G$ and HAP$_{Na}$ are one and the same pathway as HAP$_G$ and LAP$_{Na}$ are the other (Fig. 1).

Although the possibility of a random bi bi mechanism was clearly seen by Schultz and Curran (1970), a failure to appreciate that this mechanism uniquely allows for reciprocal affinity effects (Segel, 1975) together with a failure to collect sufficient data led Schultz and Curran (1970) as well as others (Heinz *et al.*, 1972, Geck and Heinz, 1976)

Robert K. Crane and Frederick C. Dorando • Department of Physiology and Biophysics, College of Medicine and Dentistry of New Jersey–Rutgers Medical School, Piscataway, New Jersey 08854.

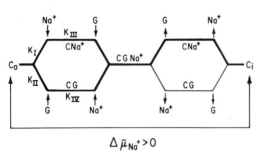

$$\Delta \bar{\mu}_{Na^+} > 0$$

Figure 1. Diagrammatic representation of Na$^+$ gradient-coupled glucose transport as a random bi bi kinetic mechanism. In the general case, all steps are assumed to be freely reversible. The arrows indicate the expected sequence for initial rate measurement where there is an inward electrochemical gradient of Na$^+$; i.e., $\Delta\mu_{Na+} > 0$. Preliminary studies of the effect of internal Na$^+$ and glucose, separately and together, on the rate of initial uptake strongly suggest that the transport system is not symmetrical, as is intended to be indicated by the difference in line width for the lower leg of the inner process. The evidence so far suggests that this leg is inoperable. If this is confirmed the kinetic mechanism would then be random bi ordered bi. For a discussion of the general case for enzymes see Segel (1975, p. 299 et seq.).

to formulate a concept of "V_{max}," "K_m," and "mixed" systems depending upon whether, with the limited data available, Na$^+$ appeared to influence the transport rate, the affinity of the substrate for the carrier, or both. The question was further complicated by Honegger and Semenza (1973) who reinterpreted the two kinetic events found by Crane *et al.* (1965) in hamster intestine as two separate systems on the basis of an apparent difference in sugar specificity between the high- and low-affinity pathways without realizing that the limited range of sugar concentrations used might not reveal the low-affinity pathway for sugars with a substantially lower affinity than glucose. More recently, Hopfer and Groseclose (1980) have studied glucose transport with rabbit jejunal brush border membrane vesicles by isotope exchange under zero gradient conditions. Equal concentrations of glucose and Na$^+$ were placed inside and out and any residual membrane potential was collapsed by the addition of monensin. Under these conditions they obtained results that they interpreted as demonstrating an ordered bi bi mechanism with glide symmetry. This single pathway had, for glucose, the affinity characteristics of LAP$_G$ (see below).

More definitive work in this laboratory and others now returns us to the original concept as revised by Lyon and Crane (1966). Turner and Silverman (1978) in studies with dog kidney brush border membrane vesicles have demonstrated the existence of two Na$^+$-dependent glucose pathways with high- and low-affinity characteristics like those found by Lyon and Crane (1966). They did not, however, draw a conclusion as to kinetic mechanism owing to an absence of information about the reciprocal effects of glucose on Na$^+$ affinity. Dorando and Crane (1980) have similarly found two Na$^+$-dependent kinetic events for glucose in jejunal brush border membranes that represent, respectively, a high- and a low-affinity pathway for glucose (see Fig. 2). In other studies not yet reported (however, see below), they have demonstrated the reciprocal effects of glucose concentration on the affinity of Na$^+$. From these and other considerations they conclude that one glucose transport system with a random bi bi mechanism as in Fig. 1 satisfies the data. They differ in one unexpected respect from Lyon and Crane (1966). The data are consistent with a steady-state rather than a rapid-equilibrium system (Crane and Dorando, 1979, 1980). In a rapid-equilibrium system, translocation is rate limiting for transport, in a steady system it is not.

The long lifetime of the assumption that translocation is rate limiting (Geck and Heinz, 1976; Crane, 1977) is surely the result of the pervasive influence of the sense of motion inherent in the word "carrier," in spite of the understanding that the distance involved in translocation may be very short (Crane, 1977). Data in the literature on the turnover num-

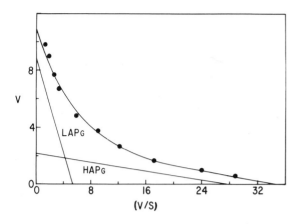

Figure 2. Representative V vs. V/S plot of the corrected 15-sec rate of glucose transport over the concentration range from 20 μM to 8 mM. The curve is resolved into two linear components by a computer program using the assumption that the curvilinear plot is the sum of two Michaelis–Menten kinetic events. The actual rates and the proportionation between LAP_G and HAP_G depend upon the conditions used. In this case, rabbit jejunal brush border membrane vesicles containing 300 mM mannitol were incubated in 100 mM Na_2SO_4 plus glucose. V is in nanomoles/15 sec per mg protein. $n = 18$.

ber of the carrier precede by a substantial period of time and indeed predict the current finding that translocation is not rate limiting. Using phlorizin binding to estimate the number of glucose carriers in dog kidney, Diedrich (1966) calculated a turnover number for the carrier of 1390 μmoles glucose reabsorbed/μmole carrier per min. Bode *et al.* (1972) revised the turnover number upward, in the rat, to a value (4320 μmoles glucose/μmole carrier per min) that is in the range of many enzymes. The basic assumption in these studies that phlorizin binding measures the number of glucose "carriers" has been amply confirmed (Frasch *et al.*, 1970; Bode *et al.*, 1972; Glossman and Neville, 1972; Chesney *et al.*, 1974; Silverman and Black, 1975; Toggenburger *et al.*, 1978).

Being non-rate limiting, translocation may have the characteristics of a fast, enzymatic reaction (Crane and Dorando, 1979, 1980). On the point of rate, they and Hopfer and Groseclose (1980) are in agreement. The latter have determined in a direct way that translocation is not rate limiting. However, on the kinetic mechanism they are not in agreement. Dorando and Crane (1980) see two pathways, Hopfer and Groseclose (1980) appear to see only one. The probable answer to this fundamental disagreement lies in the real nature of a random bi bi steady-state mechanism and the particular characteristics of Na⁺ gradient-coupled transport. Na⁺ gradient-coupled transport is subject to the forces in the transmembrane electrochemical gradient of Na⁺; that is, $\Delta\bar{\mu}_{Na^+} = \Delta\Psi + \Delta\bar{\mu}_{Na^+}$. Depending upon the distribution of forces in the gradient and the actual Na⁺ concentration, the transport of glucose should, indeed must, proportionate between the two pathways (Segel, 1975). At low Na⁺ concentrations, the LAP_G (HAP_{Na})* should predominate. At high Na⁺ concentrations, both LAP_G and HAP_G (LAP_{Na}) should be expressed. A demonstration of this expected proportionation has been made under a variety of conditions of $\Delta\Psi$ and $\Delta[Na^+]$ including $\Delta\bar{\mu}_{Na^+} = 0$ (Dorando and Crane, 1980, and unpublished; also see below). It turns out that under the conditions chosen by Hopfer and Groseclose (1980), i.e., $\Delta\bar{\mu}_{Na^+} = 0$ and $[Na^+] \simeq 100$ mM, glucose transport as measured by uptake occurs so predominantly by way of LAP_G that HAP_G is virtually undetectable (Dorando and Crane, 1980). Thus, the apparent ordered mechanism of Hopfer and Groseclose (1980) is probably only a reflection of

---

*The values of K_{app}(glucose) and K_{app}(Na⁺) in the two pathways vary with the rate of transport as is expected for steady-state systems (Segel, 1975). The value of K_{app}(Na⁺) also varies with the membrane potential. The range of values for K_{app}(glucose) is 76–115 μM in HAP_G and LAP_G, respectively (Dorando and Crane, 1980). The range of values for K_{app}(Na⁺) in HAP_{Na} is from about 8 mM in the presence of a membrane potential to about 35 mM in its absence; the values in LAP_{Na} are 10–20 times higher (Dorando and Crane, unpublished).

the proportionation effect in a random bi bi steady-state mechanism. This view is consistent with an analysis of Hopfer and Groseclose given by Turner and Silverman (1981) in which it is shown that Hopfer and Groseclose have not validly excluded a random mechanism.

A number of possible problems in studies of Na^+ gradient-coupled transport by uptake measurements have been noted by Hopfer (1977a), which led him to choose isotope exchange under zero gradient conditions to study the system (Hopfer, 1977b). These include, to choose from the extremes, the possible nonuniformity of the vesicle population and the certain collapse of at least a portion of the driving membrane potential, inside negative, by the entry of Na^+ cotransported with glucose (Rose and Schultz, 1971). These and other possible problems have been put to experimental test in this laboratory and found either not to exist or to be controllable and not to affect significantly the results and conclusions. The tests are arduous and time-consuming and have not been submitted for final editorial review. However, the problems themselves are circumvented when the interactions of the transporter with a nonpenetrating inhibitor such as phlorizin are studied (Turner and Silverman, 1980). The random bi bi mechanism based on uptake studies (Crane and Dorando, 1979, 1980; Dorando and Crane, 1980) is substantiated by studies with phlorizin.

The essence of the test with phlorizin is that at low or zero Na^+ concentrations, low-affinity binding sites for phlorizin, representing LAP_G, should be seen. As the NA^+ concentration is raised, high-affinity binding sites, representing HAP_G, should appear. However, there should be no change in the total number of binding sites. In 1970, Frasch and colleagues reported this result in studies with isolated brush border membrane of rat kidney. The same result is now reported in studies with brush border membrane vesicles from dog kidney (Turner and Silverman, 1981). In the latter studies, the experimental results are consistent with a random binding scheme in which the ratio of phlorizin to sodium is unity. The apparent binding constant for phlorizin is approximately 16 μm at $[Na^+] = 0$, 0.1 μm at $[Na^+] = 100$ mM, and approaches 0.05 μm as $[Na^+] \rightarrow \infty$.

Toggenburger *et al.* (1978) and Aronson (1978) have shown that under zero *trans* sodium conditions, phlorizin binding to brush border membrane vesicles is enhanced by increases in the membrane potential, inside negative. Within the confines of conventional carrier theory (e.g., Geck and Heinz, 1976; Heinz, 1978), these results indicate to some that the free carrier has a negative charge (Aronson, 1978; Turner and Silverman, 1980) and orients its binding site externally under the influence of the membrane potential. Others (Toggenburger *et al.*, 1978) are more cautious. As pointed out earlier (Crane, 1977), conventional carrier theory makes no demand that the carrier, in its operation, expose its binding sites directly to the bulk-phase solutions bathing the two membrane surfaces—the only demand is alternation of exposure; the carrier may be discrete in space within the membrane. In an elaboration of this point of view, Crane and Dorando (1979, 1980) have extended it to include the possibility that energy transduction from the flux of Na^+ to the flux of glucose may be accomplished entirely by chemical bonds formed in the ternary complex and not at all by translational motion of a carrier. In the present context, what is important to recognize is that if the ternary complex is formed within, rather than at the surfaces of the membrane, then the bulk of the transporter molecule may provide channels leading to and away from the ternary complex. If there are channels, then the effect of the membrane potential may be substantially, if not entirely, upon the concentration of Na^+ within the channel. The membrane potential need not affect at all the transporter per se.

The decision turns on the difference between recruitment or reproportionation as being the principal effect of the membrane potential on the absolute maximal rate of transport or its equivalent, the total number of phlorizin-binding sites. If the absolute maximal rate of transport and the total number of phlorizin-binding sites increase in response to increases in the membrane potential, that is recruitment. A negatively charged carrier responsive in its orientation to the membrane potential may be assumed. On the other hand, if the absolute maximal rate of transport and the total number of phlorizin-binding sites remain the same but the proportion of transport by HAP$_G$ and the number of high-affinity phlorizin-binding sites increase with increase in the membrane potential, that is reproportionation. It would then be clear that carrier orientation is not affected by the membrane potential whatever its charge state may be. A neutral carrier is not proved, though it may be assumed. However, reproportionation would indicate an effect of the membrane potential on the local Na$^+$ concentration at the binding site of the carrier.

We appear to have achieved measurement of the absolute maximal rate of glucose transport for rabbit jejunal brush border membrane vesicles at a value of 17.5 ± 0.9 nmoles/15 sec per mg by extrapolation of V vs. V/S plots both in the presence and in the absence of a membrane potential. For the zero membrane potential situation, vesicles containing 500 mM KCl were incubated in 500 mM KCl with the addition of valinomycin to provide a high-rate leak pathway for K$^+$. Glucose transport was then measured over the range of 40 μM to 5 mM D-glucose in the presence of graded concentrations of NaCl from 100 to 500 mM with the results shown in Fig. 3. At 100 mM Na$^+$, HAP$_G$ would not be detected. However, at 200 mM Na$^+$, HAP$_G$ began to be expressed (2.7 out of a total of 16.7 nmoles/15 sec per mg) and reached a value of 6.8 out of a total of 18.5 nmoles/15

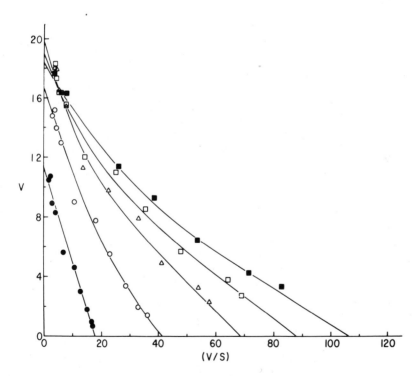

Figure 3. The effect of Na$^+$ concentration on the rate of transport and its proportionation between LAP$_G$ and HAP$_G$ with the membrane potential shorted out. Na$^+$ concentrations were as follows: ●, 100 mM; ○, 200 mM; $\triangle$, 300 mM; $\square$, 300 mM; ■, 500 mM. The vesicles contained 1 M mannitol. External mannitol (1 M) was adjusted for addition of NaCl. Note that the apparent absolute maximal rate is achieved at 300 mM Na$^+$, which is about 8–9 times K_{app}(Na$^+$) in the LAP$_G$ (HAP$_{Na}$). Further increases in Na$^+$ concentration increase HAP$_G$ at the expense of LAP$_G$ as theory would require. Further details are in the text. V is in nanomoles/15 sec per mg protein. $n = 7$.

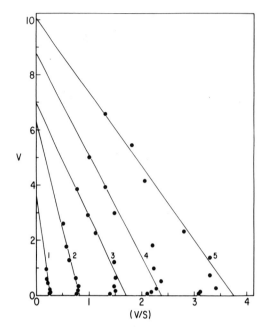

Figure 4. When $[Na^+]$ is restricted to concentrations below $K_{app}(Na^+)$ in the pathway, LAP_G (HAP_{Na}), the other pathway, HAP_G (LAP_{Na}), is not expressed even in the presence of a membrane potential. The numbers in the figure represent the Na^+ concentration in mM. V is in nanomoles/15 sec per mg protein. $n = 10$. Other details are in the text.

sec per mg at 500 mM Na^+. In another experiment (Fig. 4) with 150 mM K_2SO_4, inside only, and valinomycin added to provide a K^+ diffusion membrane potential, glucose transport was measured in the presence outside of graded concentrations of Na_2SO_4 (plus mannitol for osmotic equilibrium) from 1 to 5 mM. Primary intercept replot of the data (Fig. 5) gave a V_{max} of 17.2 nmoles/15 sec per mg entirely through LAP_G. HAP_G could not be detected. In still another experiment with 250 mM K_2SO_4 inside and valinomycin added or not, glucose transport was measured in the presence outside of 250 mM Na_2SO_4 with the results shown in Fig. 6. The extrapolated V_{max} in the presence of valinomycin was 16.7 nmoles/15 sec per mg, distributing 10.7 through HAP_G and 6.0 through LAP_G. Thus, it is seen (1) that HAP_G (LAP_{Na}) requires high concentrations of Na^+ to be expressed when

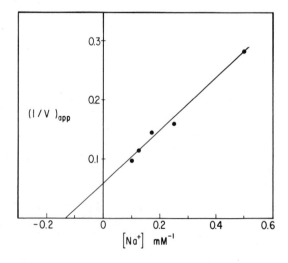

Figure 5. Determination of V_{max} under conditions of restricted Na^+ concentration by primary intercept replot. The extrapolated intercepts in Fig. 4 are replotted. See Segel (1975, p. 278) for validation of this method.

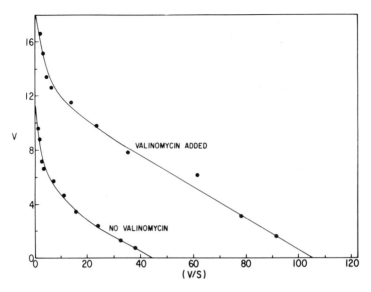

Figure 6. The effect of the membrane potential on rate and proportionation of glucose transport at high concentrations of Na⁺. In the presence of valinomycin, the total rate is increased to the apparent absolute maximum by an effect almost entirely upon HAP_G. HAP_G in the upper curve is clearly more dominant than in the curve in Fig. 3 at the same external Na⁺ concentration, 500 mM, where the membrane potential was shorted out. The curve in the absence of valinomycin serves only as a control for the effect of valinomycin. It is not strictly comparable to the 500 mM Na⁺ curve in Fig. 3 because under the ionic conditions imposed in the above experiment in the absence of valinomycin (see the text) a membrane potential, inside positive, may be produced by Na⁺ entering with glucose, act to reduce the apparent inward electrochemical gradient of Na⁺, and thus limit the forward rate as is seen. V is in nanomoles/15 sec per mg protein. $n = 12$. Other details are in the text.

there is no membrane potential, (2) that it cannot be expressed even in the presence of a membrane potential if the Na⁺ concentration is limited, and (3) that it dominates when both adequate Na⁺ and a membrane potential are provided, all at the same approximate total maximal transport rate of 17.5 ± 0.9 nmoles/15 sec per min. These results are consistent with reproportionation of transport through LAP_G (HAP_{Na}) and HAP_G (LAP_{Na}) as a result of changes in the Na⁺ concentration to which the carrier is exposed either by addition of Na⁺ to the outside or to the increase of [Na⁺] in a channel due to the expected action of the membrane potential. The results are not consistent with recruitment. What this conclusion leads to is the view that there is no special or direct effect of the membrane potential at the active center or elsewhere on the carrier. In this view, the total force in the transmembrane electrochemical potential gradient of Na⁺, $\Delta\bar{\mu}_{Na^+} + \Delta\Psi$, is converted to and interacts with the carrier as $\Delta\bar{\mu}_{Na^+}$.

Studies of phlorizin binding under the influence of the membrane potential comparable to the above are not available. Toggenburger *et al.* (1978) have come closest to providing the needed data but in their focus on high-affinity phlorizin-binding sites, the total number of sites was left in doubt. Nonetheless, it may be pointed out that their extrapolated data show a greater number of glucose-protectable phlorizin-binding sites under the influence of a Cl⁻ potential than of an SCN⁻ potential. This result would suggest reproportionation rather than recruitment. Recruitment would have given a higher value with the higher membrane potential provided by SCN⁻. It may be hoped that this weakness in the overall data base will soon be corrected.

ACKNOWLEDGMENTS. The work in our laboratory is supported by a grant from NIAMDD. We thank Dr. James Turner for making available to us a prepublication copy of Turner and Silverman (1981).

REFERENCES

Aronson, P. S. (1978). *J. Membr. Biol.* **41**, 81–98.
Bode, F., Baumann, K., and Diedrich, D. F. (1972). *Biochim. Biophys. Acta* **290**, 134–149.
Chesney, R., Sacktor, B., and Kleinzeller, A. (1974). *Biochim. Biophys. Acta* **332**, 263–277.
Crane, R. K. (1977). *Rev. Physiol. Biochem. Pharmacol.* **78**, 99–159.
Crane, R. K., and Dorando, F. C. (1979). In *Function and Molecular Aspects of Biomembrane Transport* (E. Quagliariello, F. Palmieri, S. Papa, and M. Klingenberg, eds.), pp. 271–277, Elsevier/North-Holland, Amsterdam.
Crane, R. K., and Dorando, F. C. (1980). *Ann. N.Y. Acad. Sci.* **339**, 46–52.
Crane, R. K., Forstner, G., and Eichholz, A. (1965). *Biochim. Biophys. Acta* **109**, 467–477.
Diedrich, D. F. (1966). *Am. J. Physiol.* **211**, 581–587.
Dorando, F. C., and Crane, R. K. (1980). *Fed. Proc.* **39**, 2159.
Frasch, W., Frohnert, P. P., Bode, F., Baumann, K., and Kinne, R. (1970). *Pfleugers Arch.* **320**, 265–284.
Geck, P., and Heinz, E. (1976). *Biochim. Biophys. Acta* **443**, 49–53.
Glossman, H., and Neville, D. M., Jr. (1972). *J. Biol. Chem.* **247**, 7779–7789.
Heinz, E. (1978). *Mol. Biol. Biochem. Biophys.* **29**, 1–159.
Heinz, E., Geck, P., and Wilbrandt, W. (1972). *Biochim. Biophys. Acta* **255**, 442–461.
Honegger, P., and Semenza, G. (1973). *Biochim. Biophys. Acta* **318**, 390–410.
Hopfer, U. (1977a). *J. Supramol. Struct.* **7**, 1–13.
Hopfer, U. (1977b). *Am. J. Physiol.* **233**, 445–449.
Hopfer, U., and Groseclose, R. (1980). *J. Biol. Chem.* **255**, 4453–4462.
Lyon, I., and Crane, R. K. (1966). *Biochim. Biophys. Acta* **112**, 278–291.
Rose, R. C., and Schultz, S. G. (1971). *J. Gen. Physiol.* **57**, 639–663.
Schultz, S. G., and Curran, P. (1970). *Physiol. Rev.* **50**, 637–717.
Segel, I. H. (1975). *Enzyme Kinetics,* Wiley, New York.
Silverman, M., and Black, J. (1975). *Biochim. Biophys. Acta* **394**, 10–30.
Stein, W. D. (1967). *The Movement of Molecules across Cell Membranes,* Academic Press, New York.
Toggenburger, G., Kessler, M., Rothstein, A., Semenza, G., and Tannenbaum, C. (1978). *J. Membr. Biol.* **40**, 269–290.
Turner, R. J., and Silverman, M. (1978). *Biochim. Biophys. Acta* **511**, 470–486.
Turner, R. J., and Silverman, M. (1980). *Biochim. Biophys. Acta* **596**, 272–291.
Turner, R. J., and Silverman, M. (1981). *J. Membr. Biol.* **58**, 43–55.

111

The Glucose Transporter of Mammalian Cells

Thomas J. Wheeler and Peter C. Hinkle

The plasma membrane of most mammalian cells contains a protein that catalyzes the transport of D-glucose and some other hexoses across the membrane. The kinetics and specificity of this "facilitated diffusion" transporter have been studied extensively, particularly in human red cells, which have unusually high rates of glucose transport (for reviews see Jung, 1975; LeFevre, 1975; Naftalin and Holman, 1977; Widdas, 1980). In recent years the emphasis has shifted to identification of the transporter, determination of its structure, and study of its regulation. We will review recent results concerned with these topics and with the characterization of the isolated transporter.

1. COVALENT LABELING

Identification of the glucose transporter was first attempted by affinity labeling methods. Taverna and Langdon (1973b) used radioactive glucosylisothiocyanate as an affinity label, although the transporter does not have high affinity for sugar substrates. A variety of proteins were labeled. The highest activity per slice of the SDS-polyacrylamide gel was associated with a protein of 100,000 molecular weight (band 3), but the bulk of the label was in a broad band of lower molecular weight. More recent experiments by this group have employed maltosylisothiocyanate as an affinity label, which is not transported but irreversibly inhibits the transporter (Mullins and Langdon, 1980a,b). It was found to label almost exclusively a protein of 100,000 molecular weight. This reaction was partially inhibited by glucose or cytochalasin B, an inhibitor that acts both on the cytoskeleton and on the glucose transporter (Taverna and Langdon, 1973a) and has been used extensively for identification of the transporter.

In another type of labeling experiment, substrates or inhibitors have been used to

Thomas J. Wheeler and Peter C. Hinkle • Section of Biochemistry, Molecular and Cell Biology, Cornell University, Ithaca, New York 14853.

modify the reactivity of the transporter with covalent labels, analogous to the studies of the lactose transporter by Fox and Kennedy (1965). The first studies with 1-fluoro-2,4-dinitrobenzene (FDNB) showed that glucose enhanced the labeling of a 180,000-molecular-weight polypeptide (Jung and Carlson, 1975). Subsequently, a polypeptide of 50,000 molecular weight was identified when cytochalasin B was used as the modulating agent (Lienhard *et al.*, 1977). Shanahan and Jacquez (1978) identified three proteins with FDNB, two of 200,000 and 90,000 molecular weight when glucose is used to enhance reactivity and one of 60,000 molecular weight when cytochalasin B or maltose is used to inhibit reactivity.

Impermeant maleimides were used by Batt *et al.* (1976) in similar experiments. These reagents inhibit glucose transport; D-glucose and cytochalasin B protected against this inhibition and decreased the labeling of a broad band at 60,000 molecular weight by maleimides. This group also found a 250,000-molecular-weight polypeptide labeled by maleimides (Batt *et al.*, 1978), possibly similar to the high-molecular-weight polypeptide found with FDNB. The labeling studies have thus implicated several polypeptides in glucose transport.

2. RECONSTITUTION

An approach for isolation of the glucose transporter is to incorporate it in active form into liposomes. The reconstituted transport activity can then be used as an assay during purification. This approach was developed by Kasahara and Hinkle (1976, 1977a,b), who first used a sonication procedure and then developed a freeze–thaw technique for reconstitution. The technique consists of adding protein fractions with detergent removed (which are usually vesicular because of endogenous lipids) to sonicated soybean phospholipid liposomes and then freezing the mixture in the absence of any cryoprotectants such as glycerol or sucrose. This causes fusion of the liposomes with the vesicles of transporter protein, giving a turbid suspension of large vesicles several micrometers in diameter. These are then broken up by a very brief sonication to form 500- to 3000-Å liposomes, which have stereospecific and cytochalasin B-sensitive D-glucose permeability [see also Carter-Su *et al.* (1980) for discussion of the freeze–thaw procedure]. Glucose permeability was assayed by mixing liposomes with radioactive D- or L-glucose, stopping with a cold $HgCl_2$ solution, filtering and washing the vesicles as in a conventional bacterial transport assay.

Using these methods, Kasahara and Hinkle (1977a,b) solubilized the red cell membrane with Triton X-100 and chromatographed the extract on DEAE-cellulose. The fraction that was not retarded by the column had all of the reconstitutable glucose transport activity and consisted of one broad band at about 55,000 molecular weight on SDS-polyacrylamide gel electrophoresis, a component of zone 4.5. An improved purification procedure (Sogin and Hinkle, 1978) resulted in a preparation that had an equilibrium exchange activity of 50 μmoles/mg protein per min at 20 mM glucose (Wheeler and Hinkle, 1980). This is about five times the specific activity of the red cell membrane, and is 10–40% of that expected based on kinetic studies of red cells.

Three other groups have also isolated the red cell glucose transporter using a variety of methods. All of them found that the most active fractions contained predominantly a 55,000-molecular-weight polypeptide. Kahlenberg and Zala (1977) prepared Triton extracts of red cell membranes and reconstituted fractions from DEAE-cellulose chromatography into liposomes of red cell lipids by simple sonication. They assayed transport activity by the efflux of D- versus L-glucose during gel filtration. This assay measured extent of transport (i.e., the fraction of liposomes with at least one transporter), rather than rates of transport. Baldwin *et al.* (1980, 1981) have used Triton, Nikkol, or octaethyleneglycol-*n*-dodecyl ether to solubilize the transporter from red cells, followed by DEAE-cellulose chromatography and cholate gel filtration-dialysis for reconstitution into dioleoylphosphatidylcholine liposomes. They assayed uptake of D- and L-glucose, stopping with cytochalasin B and centrifuging through small Sephadex columns. Goldin and Rhoden (1978) solubilized the red cell membrane with cholate and used hollow-fiber dialysis for formation of liposomes. They assayed uptake of D- and L-glucose using an ice bath stop and gel filtration to separate liposomes from the medium. They purified the transporter by a novel "transport specificity fractionation" procedure. Edwards (1977) used deoxycholate to solubilize the red cell proteins and dialysis to reconstitute liposomes that transported D-glucose, but did not purify the transporter.

Red cell membrane proteins solubilized with Triton X-100 have also been reconstituted into planar lipid membranes to give stereospecific D-glucose permeability (Jones and Nickson, 1978). Crude extracts gave the best activity, although fractions containing predominantly zone 4.5 polypeptides (50,000 molecular weight) were also active (Phutrakul and Jones, 1979).

Czech and co-workers have used reconstituted transport activity as an assay in the partial purification of the adipocyte glucose transporter (Shanahan and Czech, 1977; Carter-Su *et al.*, 1980). Liposomes were prepared by mixing cholate-solubilized membrane proteins with cholate-dispersed phospholipids, passing through a Sephadex column, freezing, thawing, and sonicating briefly. Membrane extracts were fractionated by gel filtration, hydroxylapatite chromatography, and lectin adsorption. These studies indicated that the transporter is not one of the major adipocyte membrane proteins (Carter-Su *et al.*, 1980).

3. CYTOCHALASIN B BINDING

The glucose transporter can also be identified and quantitated by measuring the binding of radioactive cytochalasin B to protein fractions after removal of detergents. In this way, Sogin and Hinkle (1978, 1980b) showed that the 55,000-molecular-weight protein purified by Kasahara bound 9 nmoles cytochalasin B/mg protein in a glucose-reversible reaction. This amount of binding corresponds to about 1 mole cytochalasin B/2 moles polypeptide, but that could be fortuitous. The binding was also reversed by the transport inhibitors phloretin, diethylstilbestrol, maltose, 6-*o*-propyl-D-galactose, and propyl-β-D-glucopyranoside. These studies indicated that of the various ligands studied, only one can bind to the transporter at a time. Baldwin *et al.* (1979) measured cytochalasin B binding to a similar DEAE-cellulose-purified glucose transporter.

4. POSSIBLE PROTEOLYSIS

Thus, the reconstitution and cytochalasin B binding studies with red cell proteins have indicated that a single polypeptide of approximately 55,000 molecular weight is the glucose transporter, and labeling studies have indicated the same polypeptide plus others of higher molecular weights. It is possible that the higher-molecular-weight polypeptides have sugar binding activity unrelated to transport. It was also suggested recently that the 55,000-molecular-weight polypeptide is a proteolytic fragment of the native transporter (Phutrakul and Jones, 1979; Mullins and Langdon, 1980b). Any experiments using detergents to isolate the transporter are potentially open to this possibility, so two groups independently made antibodies against the isolated transporter and used the antibodies to stain SDS-polyacrylamide gels of fresh red cells, which should not have any proteolysis (Baldwin and Lienhard, 1980; Sogin and Hinkle, 1980a). Both studies showed that only the 55,000-molecular-weight broad band, the same as the isolated transporter, was stained. Because the antibody binding takes place after denaturation of the proteins with SDS, antigenic sites of a possible fragment should be exposed in the native transporter. A possible explanation for the results of Mullins and Langdon (1980a,b) is that maltosylisothiocyanate may react with the anion transporter as the main product at 100,000 molecular weight and the labeled glucose transporter may have been such a broad band that it was missed.

5. COMPOSITION AND STRUCTURE

Characterization of the isolated transporter has proven difficult. The amino acid and carbohydrate composition (Sogin and Hinkle, 1978), lectin binding, and sensitivity to endo-β-galactosidase (Gorga *et al.*, 1979) have been reported. There may be two different polypeptide chains of similar molecular weight, for the glycosylation prevents high-resolution SDS-polyacrylamide gel electrophoresis. Electron microscopy using freeze-fracture of reconstituted vesicles has indicated that the native protein is about 60 Å in diameter in the membrane (Hinkle *et al.*, 1979), which would be consistent with a dimer of the 55,000-molecular-weight polypeptide. Alternatively, electron radiation inactivation studies of glucose transport and cytochalasin B binding in red cells have indicated a molecular weight of 185,000 for the native transporter, consistent with a tetramer of the 55,000-molecular-weight polypeptide (Cuppoletti *et al.*, 1981).

6. KINETICS

Some progress has been made in the characterization of the kinetic properties of the reconstituted transporter. One concern is whether the asymmetric kinetics seen in erythrocytes (reviewed by Widdas, 1980) are retained. To examine this question it is necessary to determine the orientation of the reconstituted transporter. Using trypsin, which inactivates the transport activity in red cells only from the cytoplasmic side of the membrane (Masiak and LeFevre, 1977), Baldwin *et al.* (1980) observed either 50 or 75% of the molecules oriented with the cytoplasmic face outward, depending on the method of reconstitution, and Wheeler and Hinkle (1981) concluded that their reconstituted transporter had a 50–50 scrambled orientation. Trypsin treatment generates a population of active transporters oriented as in the red cell. By comparing the kinetic parameters for zero-*trans* uptake and efflux before and after trypsin treatment, Wheeler and Hinkle (1981) determined a fourfold asymmetry in both K_m and V_{max} for the two faces of the transporter. This is similar to the asymmetry in erythrocyte V_{max} values for net flux. Asymmetry of transport kinetics is a natural consequence of the asymmetric nature of the transporter in the membrane. The inward-facing and outward-facing conformations are presumably rather different and are not expected to have equal probabilities or equal affinities for glucose. Because the transporter does not concentrate glucose on one side of the membrane, the kinetic constants of a simple carrier model are restricted by the Haldane relationship, i.e., the ratio of K_m to V_{max} must be equal in each direction. This relationship was observed in the reconstituted system (Wheeler and Hinkle, 1981) but not in studies of red cells (Karlish *et al.*, 1972; Lacko *et al.*, 1972).

Equilibrium exchange in the reconstituted liposomes had a high K_m (30 mM) and a

V_{max} greater than the V_{max} values for net flux (Wheeler and Hinkle, 1981); these are also properties of the transporter in red cells (Miller, 1968). The kinetics of net flux and equilibrium exchange in the reconstituted system could be fitted by the asymmetric carrier model described by Eilam and Stein (1974). Baldwin *et al.* (1981) also observed higher rates for exchange than for net flux.

7. INHIBITORS

The reconstituted glucose transporter has been shown to be sensitive to several of the inhibitors of red cell glucose transport: mercuric ions and cytochalasin B (Kasahara and Hinkle, 1976), FDNB, *N*-ethylmaleimide, *p*-chloromercuribenzene sulfonate, and phloretin (Kahlenberg and Zala, 1977). The apparent lack of inhibition by phloretin and diethylstilbestrol reported by Kasahara and Hinkle (1977b) was due to the binding of these inhibitors to liposomes (Wheeler and Hinkle, 1981). Asymmetric phloretin inhibition (Benes *et al.*, 1972; Kahlenberg and Zala, 1977; Goldin and Rhoden, 1978) can probably also be explained by the binding of phloretin to lipids (Wheeler and Hinkle, 1981). Goldin and Rhoden (1978) reported that Hg^{2+} inhibited to a maximum of about 50% in a presumably scrambled reconstituted system. Wheeler and Hinkle (1981) observed complete inhibition by $HgCl_2$; however, the external face of the transporter appeared to be more sensitive.

8. REGULATION

In adipocytes it has been demonstrated that insulin increases the V_{max} of glucose transport with little effect on K_m, suggesting either activation of transporters already in the plasma membrane or recruitment of additional transporter molecules. Melchior and Czech (1979) reported a correlation between membrane fluidity and transport activity in a reconstituted preparation, suggesting that insulin might increase transport activity by increasing fluidity. However, the studies are complicated by the large amount of nonspecific uptake and the fact that initial rates were not measured. In support of the fluidity hypothesis, Pilch *et al.* (1980) compared transport activity in membrane vesicles from control and insulin-treated cells. Addition of unsaturated fatty acids increased only the basal activity, while addition of saturated fatty acids decreased only the insulin-stimulated activity. Insulin-treated adipocytes also show a different dependence of transport on temperature than untreated cells (Amatruda and Finch, 1979; Whitesell and Gliemann, 1979).

Two groups have presented evidence that insulin may stimulate glucose uptake by recruitment of transporters from an intracellular membrane pool. Suzuki and Kono (1980) found that insulin treatment of adipocytes increased the transport activity that could be reconstituted from plasma membranes, while decreasing a second and much larger activity fractionating with Golgi membranes. Cushman and Wardzala (1980) found an increase in cytochalasin B binding sites in plasma membranes upon insulin treatment which was equaled by a decrease in sites in microsomal membranes; again, most of the transporters appeared to be initially intracellular. A third study (Carter-Su and Czech, 1980) also found a large amount of reconstitutable transport activity in microsomal membranes, but it was not decreased by insulin. Both the plasma membrane activity and the activity of the combined fractions were increased by insulin, in support of the activation hypothesis. The disagreement between these results and those of Suzuki and Kono (1980) may involve the different reconstitution and assay procedures used.

Another study (I. A. Simpson, S. W. Cushman, T. J. Wheeler, and P. C. Hinkle, unpublished results) used the antibody against the erythrocyte transporter (Sogin and Hinkle, 1980a) to identify cross-reacting polypeptides in adipocyte membranes after SDS-polyacrylamide gel electrophoresis (Towbin *et al.*, 1979). The major stained band at about 50,000 molecular weight increased in the plasma membrane and decreased in the microsomal fraction upon insulin treatment, although the apparent relative amounts of antigenic activity in the various fractions are not in agreement with the cytochalasin B binding data (Cushman and Wardzala, 1980).

Glucose transport increases in various cells upon transformation (Hatanaka, 1974) and glucose starvation (Martineau *et al.*, 1972); antibody studies may also prove useful in investigating these areas of regulation. Salter and Weber (1979) observed more glucose-inhibitable cytochalasin B binding sites in Rous sarcoma virus-transformed chick embryo fibroblasts than in normal cells, suggesting an increase in the number of transporters. Inui *et al.* (1979) found that membrane vesicles from simian virus 40-transformed mouse fibroblasts (3T3) had higher glucose transport and cytochalasin B binding activities than vesicles from normal cells. Lever (1979) found no difference in transport in similar vesicles, but obtained much lower rates than Inui *et al.* (1979).

9. OTHER STUDIES

The antibody against the red cell transporter, which both precipitated all of the cytochalasin B binding activity in membrane extracts (Sogin and Hinkle, 1980a) and inhibited reconstituted glucose transport (Wheeler and Hinkle, 1981), has proven useful in identifying the transporter in other cells. Both HeLa (Sogin and Hinkle, 1980a) and rat adipocytes (see above) showed cross-reacting proteins at about 50,000 molecular weight. In addition, such polypeptides have been observed in mouse Ehrlich ascites tumor cells, brain, and normal rat kidney (NRK) cells (M. Spector, T. J. Wheeler, and P. C. Hinkle, unpublished results). Thus, a similar transporter protein occurs in a wide variety of cell types.

10. SUMMARY

The reconstitution of the red cell glucose transporter has led to its purification and identification. Further studies with the purified protein have revealed some of its structural and kinetic characteristics. The antibody against this protein has identified very similar proteins in several other cells and is a powerful tool for studies of the synthesis, processing, and regulation of the glucose transporter in those cells.

REFERENCES

Amatruda, J. M., and Finch, E. D. (1979). *J. Biol. Chem.* **254**, 2619–2625.
Baldwin, J. M., Lienhard, G. E., and Baldwin, S. A. (1980). *Biochim. Biophys. Acta* **599**, 699–714.
Baldwin, J. M., Gorga, J. C., and Lienhard, G. E. (1981). *J. Biol. Chem.* **256**, 3685–3689.
Baldwin, S. A., and Lienhard, G. E. (1980). *Biochem. Biophys. Res. Commun.* **94**, 1401–1408.
Baldwin, S. A., Baldwin, J. A., Gorga, F. R., and Lienhard, G. E. (1979). *Biochim. Biophys. Acta* **552**, 193–199.
Batt, E. R., Abbott, R. E., and Schachter, D. (1976). *J. Biol. Chem.* **251**, 7184–7190.
Batt, E. R., Abbott, R. E., and Schachter, D. (1978). *Fed. Proc.* **37**, 312.
Benes, J., Kolínská, J., and Kotyk, A. (1972). *J. Membr. Biol.* **8**, 303–309.
Carter-Su, C., and Czech, M. P. (1980). *J. Biol. Chem.* **255**, 10382–10386.

Carter-Su, C., Pillion, D. J., and Czech, M. P. (1980). *Biochemistry* **19,** 2374–2385.
Cuppoletti, J., Jung, C. Y., and Green, F. A. (1981). *J. Biol. Chem.* **256,** 1305–1306.
Cushman, S. W., and Wardzala, L. J. (1980). *J. Biol. Chem.* **255,** 4758–4762.
Edwards, P. A. (1977). *Biochem. J.* **164,** 125–129.
Eilam, Y., and Stein, W. D. (1974). *Methods Membr. Biol.* **2,** 283–354.
Fox, C. F., and Kennedy, E. P. (1965). *Proc. Natl. Acad. Sci. USA* **54,** 891–899.
Goldin, S. M., and Rhoden, V. (1978). *J. Biol. Chem.* **253,** 2575–2583.
Gorga, F. R., Baldwin, S. A., and Lienhard, G. E. (1979). *Biochem. Biophys. Res. Commun.* **91,** 955–961.
Hatanaka, M. (1974). *Biochim. Biophys. Acta* **355,** 77–104.
Hinkle, P. C., Sogin, D. C., Wheeler, T. J., and Telford, J. N. (1979). In *Function and Molecular Aspects of Biomembrane Transport* (E. Quagliariello, F. Palmieri, S. Papa, and M. Klingenberg, eds.), pp. 487–494, Elsevier/North-Holland, Amsterdam.
Inui, K.-I., Moller, D. E., Tillotson, L. G., and Isselbacher, K. J. (1979). *Proc. Natl. Acad. Sci. USA* **76,** 4350–4354.
Jones, M. N., and Nickson, J. K. (1978). *Biochim. Biophys. Acta* **550,** 188–200.
Jung, C. Y. (1975). In *The Red Blood Cell* (D. M. Surgenor, ed.), 2nd ed., pp. 705–749, Academic Press, New York.
Jung, C. Y., and Carlson, L. M. (1975). *J. Biol. Chem.* **250,** 3217–3220.
Kahlenberg, A., and Zala, C. A. (1977). *J. Supramol. Struct.* **7,** 287–300.
Karlish, S. J. D., Lieb, W. R., Ram, D., and Stein, W. D. (1972). *Biochim. Biophys. Acta* **255,** 126–132.
Kasahara, M., and Hinkle, P. C. (1976). *Proc. Natl. Acad. Sci. USA* **73,** 396–400.
Kasahara, M., and Hinkle, P. C. (1977a). In *Biochemistry of Membrane Transport* (G. Semenza and E. Carafoli, eds.), pp. 346–350, Springer-Verlag, Berlin.
Kasahara, M., and Hinkle, P. C. (1977b). *J. Biol. Chem.* **252,** 7384–7390.
Lacko, L., Wittke, B., and Kromphardt, H. (1972). *Eur. J. Biochem.* **25,** 447–454.
LeFevre, P. G. (1975). *Curr. Top. Membr. Transp.* **7,** 109–215.
Lever, J. E. (1979). *J. Biol. Chem.* **254,** 2961–2967.
Lienhard, G. E., Gorga, F. R., Oransky, J. E., Jr., and Zoccoli, M. A. (1977). *Biochemistry* **16,** 4921–4926.
Martineau, R., Kohlbacher, M., Shaw, S. N., and Amos, H. (1972). *Proc. Natl. Acad. Sci. USA* **69,** 3407–3411.
Masiak, S. J., and LeFevre, P. G. (1977). *Biochim. Biophys. Acta* **465,** 371–377.
Melchior, D. L., and Czech, M. P. (1979). *J. Biol. Chem.* **254,** 8744–8747.
Miller, D. M. (1968). *Biophys. J.* **8,** 1339–1352.
Mullins, R. E., and Langdon, R. G. (1980a). *Biochemistry* **19,** 1199–1205.
Mullins, R. E., and Langdon, R. G. (1980b). *Biochemistry* **19,** 1205–1212.
Naftalin, R. J., and Holman, G. D. (1977). In *Membrane Transport in Red Cells* (J. C. Ellory and V. L. Lew, eds.), pp. 257–300, Academic Press, New York.
Phutrakul, S., and Jones, M. N. (1979). *Biochim. Biophys. Acta* **550,** 188–200.
Pilch, P. F., Thompson, P. A., and Czech, M. P. (1980) *Proc. Natl. Acad. Sci. USA* **77,** 915–918.
Salter, D. W., and Weber, M. J. (1979). *J. Biol. Chem.* **254,** 3554–3561.
Shanahan, M. F., and Czech, M. P. (1977). *J. Biol. Chem.* **252,** 8341–8343.
Shanahan, M. F., and Jacquez, J. A. (1978). *Membr. Biochem.* **1,** 239–267.
Sogin, D. C., and Hinkle, P. C. (1978). *J. Supramol. Struct.* **8,** 447–453.
Sogin, D. C., and Hinkle, P. C. (1980a). *Proc. Natl. Acad. Sci. USA* **77,** 5725–5729.
Sogin, D. C., and Hinkle, P. C. (1980b). *Biochemistry* **19,** 5417–5420.
Suzuki, K., and Kono, T. (1980). *Proc. Natl. Acad. Sci. USA* **77,** 2542–2545.
Taverna, R. D., and Langdon, R. G. (1973a). *Biochim. Biophys. Acta* **323,** 207–219.
Taverna, R. D., and Langdon, R. G. (1973b). *Biochem. Biophys. Res. Commun.* **54,** 593–599.
Towbin, H., Staehelin, T., and Gordon, J. (1979). *Proc. Natl. Acad. Sci. USA* **76,** 4350–4354.
Wheeler, T. J., and Hinkle, P. C. (1980). *Fed. Proc.* **39,** 1813.
Wheeler, T. J., and Hinkle, P. C. (1981) **256,** 8907–8914.
Whitesell, R. R., and Gliemann, J. (1979). *J. Biol. Chem.* **254,** 5276–5283.
Widdas, W. F. (1980). *Curr. Top. Membr. Transp.* **14,** 165–223.

112

Intestinal Transport of Sugar

The Energetics of Epithelial "Pump–Leak" Systems

George A. Kimmich

1. INTRODUCTION

Absorption of monosaccharides from dietary constituents to the circulatory system involves the sequential transfer of sugar molecules across two plasma membranes that function in series at opposite poles of the intestinal epithelium (for a recent general review see Kimmich, 1981). Sugar is first accumulated in the epithelial cell by a concentrative Na^+-dependent "pump" system localized in the mucosal boundary. The accumulated sugar can then be transferred to the circulatory system via a facilitated diffusion "leak" system located in the serosal cell boundary. The characteristics of *net* sugar absorption reflect to some degree the characteristics of each separate transport event. Thus, net absorption is Na^+-dependent and inhibited by phlorizin, which are characteristic of mucosal sugar accumulation (Goldner *et al.*, 1969). By the same token, net absorption can also be inhibited by theophylline, which selectively inhibits sugar transfer at the serosal cell boundary (Holman and Naftalin, 1975; Randles and Kimmich, 1978).

In contrast, the energetics of epithelial sugar transport has usually been considered solely in terms of understanding the function of the mucosal Na^+-dependent transport system. There has been a tendency to evaluate the energetic requirements for this system as an independent entity without regard to the function of the serosal system. This tendency has led to a number of experimental enigmas during the past decade with consequent confusion regarding the energetics of intestinal sugar transport in particular, but of gradient-coupled transport systems in general. This short review is aimed at describing recent developments that identify the limitation of conventional approaches to epithelial transport

George A. Kimmich • Department of Radiation Biology and Biophysics, University of Rochester Medical Center, Rochester, New York 14642.

energetics, and the necessity for full consideration of both "pump" and "leak" flux pathways.

2. THE Na+-COUPLED SUGAR PUMP

Intestinal absorption of sugar against a concentration gradient exhibits an absolute dependence on Na^+. The first explicit model proposed to explain this dependence was described by Crane *et al.* (1961) with subsequent revisions (1965, 1968). His model suggested that active sugar accumulation was driven by flow of Na^+ into the cell down a chemical concentration gradient. Energy transduction was postulated to be a result of coupled Na^+ and sugar flows via a membrane component (carrier) with binding sites for both solutes. Kinetic evidence by Crane *et al.* (1965) and Goldner *et al.* (1969) indicated that Na^+ binding changed either the affinity of the carrier for sugar, or the velocity of sugar entry on the carrier. Because K^+ was known to alter the properties of the sugar carrier (Bosackova and Crane, 1965), some versions of the transport model (Crane *et al.*, 1965) included the possibility that K^+ might compete for the Na^+ site such that outward flow of K^+ on the carrier could provide a contributory energy input. In either case, ion flow down a gradient of chemical potential was envisioned as providing the energy for movement of sugar against a gradient of chemical potential.

Later, work with isolated intestinal epithelial cells was important in demonstrating that cellular K^+ gradients cannot energize sugar transport in a manner consistent with the notion of a carrier with K^+-binding sites (Kimmich and Randles, 1973). On the other hand, abundant evidence accumulated that indicated that Na^+ gradients can energize sugar accumulation in systems lacking ATP or capability for generating ATP (Murer *et al.*, 1974; Lücke *et al.*, 1978; Sacktor, 1977; Hopfer *et al.*, 1973; Carter-Su and Kimmich, 1979). Experiments with membrane vesicles prepared from the brush border boundary of the intestinal epithelium were particularly compelling in this regard. Indeed, membrane preparations from each pole of the epithelial cell have been instrumental in documenting the fact that the brush border is the only locus for an Na^+-dependent, phlorizin-sensitive, concentrative sugar transport system (Hopfer *et al.*, 1975, 1976).

Recently, the Na^+-dependent sugar transport system has been shown to be rheogenic (i.e., current-generating) in character and to be "driven" by the cellular membrane potential as well as by chemical gradients for Na^+. This conclusion was drawn primarily from numerous investigations that demonstrated that diffusion potentials induced by ionophores or gradients of lipophilic anions will induce transient sugar accumulation in ATP-depleted intact intestinal cells (Carter-Su and Kimmich, 1979, 1980; Kimmich *et al.*, 1977) or in the membrane vesicle preparations (Lücke *et al.*, 1978; Sacktor, 1977; Hopfer *et al.*, 1975, 1976; Murer and Hopfer, 1974). Function of the sugar carrier under these conditions still requires Na^+. An Na^+ gradient is not mandatory but enhances the magnitude of the sugar gradient established by an electrical potential alone. Thus, the "modern" view of energy transduction for sugar accumulation is via a carrier that delivers Na^+ down a gradient of *electrochemical* potential with concomitant flow of sugar *up* a smaller gradient of chemical potential (for a short recent review see Kimmich and Carter-Su, 1978). The sugar carrier has specificity for the pyranose ring form of monosaccharides and has an absolute requirement for a hydroxyl group at carbon 2 in the same configuration as that in D-glucose. Sugars with small substituents in the α or β configuration at position 1 (e.g., α- or β-Ch_3 glycosides) can be transported (Kimmich and Randles, 1981; Holman and Naftalin, 1976).

The system is also tolerant to small substituents or epimeric inversions of configuration at carbons 3, 4, and 6 of the sugar molecule (Crane, 1960; Kimmich, 1981).

3. Na$^+$-INDEPENDENT SUGAR "LEAK" PATHWAYS

Transepithelial sugar transport obviously requires the transfer of hydrophilic molecules across the hydrophobic serosal cell membrane. The first clue to the nature of this process was derived from studies by Bihler and Cybulsky (1973) who utilized isolated epithelial cells that had been treated with HgCl$_2$ at their mucosal boundary prior to isolation in order to inhibit mucosal Na$^+$-dependent sugar transport. They were able to show that these cells have a saturable influx mechanism for glucose. The pathway is nonconcentrative, does not require Na$^+$, and can be competed for by several other monosaccharides. One of the sugars that exhibits competitive entry is 2-deoxyglucose, which is not a substrate for the mucosal Na$^+$-dependent sugar pump. The authors concluded that the sugar entry they observed under these conditions reflected the function of a facilitated diffusion system for sugars that is operative at the serosal cell boundary and that has a different specificity than the mucosal transport system. Subsequent work from other laboratories with isolated cells (Kimmich and Randles, 1975, 1976) and with lateral-serosal membrane vesicle preparations (Hopfer *et al.*, 1975, 1976) confirmed these basic conclusions. Work with isolated cells was instrumental in defining a number of agents that selectively inhibit the serosal sugar carrier with little or no effect on the Na$^+$-dependent system. These agents include phloretin, theophylline, cytochalasin B, and a number of flavones and flavanones (Kimmich and Randles, 1975, 1978, 1979; Randles and Kimmich, 1978). The most potent and selective of these is cytochalasin B.

The serosal transport inhibitors and isolated cell preparation have been extremely useful for exploring the relative magnitude of sugar fluxes via the mucosal and serosal transport systems under different circumstances. Using these tools we have shown that at modest extracellular sugar concentrations (0.1–1 mM), approximately 90% of the total unidirectional sugar influx is via the Na$^+$-dependent carrier, 7% via the serosal carrier, and 3% by a third route that has the characteristics of a diffusional pathway. At the steady state, however, about two-thirds of the total sugar *efflux* is via the serosal carrier and diffusional routes. The Na$^+$-dependent pump therefore ordinarily operates against a very substantial facilitated diffusion and diffusional "leak." Physiologically, the leak pathways are important for delivering sugar to the circulatory system, but in terms of understanding transport energetics they have proven troublesome, as described below.

Typically, the isolated cells establish 7- to 10-fold sugar gradients. When serosal transport inhibitors are employed to partially block a gradient-dissipating leak pathway, the Na$^+$-dependent pump can establish much higher sugar gradients. With cytochalasin B, gradients as high as 60- to 70-fold are established (Kimmich and Randles, 1979).

Recently, we have found that methylglucosides are good substrates for the Na$^+$-dependent carrier, but they do not satisfy the serosal carrier. Using these sugars, 70-fold sugar gradients can be established even in the absence of serosal inhibitors (Kimmich and Randles, 1981). It is important to recognize that even under these conditions, about 70% of the total steady-state *efflux* is occurring via a diffusional "leak" rather than by backflux on the Na$^+$-dependent carrier. If the diffusional pathway could be blocked, much higher sugar gradients would be expected.

4. THE ENERGETIC ADEQUACY QUESTION

4.1. Na$^+$ Electrochemical Potentials and Sugar Gradients

If flow of Na$^+$ into the cell down a gradient of electrochemical potential is the *only* driving force for sugar accumulation (and there is no compelling evidence to the contrary), then the following thermodynamic relationship must be obeyed:

$$RT \ln \frac{[S]_i}{[S]_o} \leq (RT \ln \frac{[Na^+]_o}{[Na^+]_i} + FV)n \tag{1}$$

where R, T, and F have their usual meaning, V is the membrane potential, and n is the number of Na$^+$ ions coupled to the flow of each sugar molecule. The subscripts i and o refer to intracellular and extracellular compartments, respectively. Typically, when the energetics of intestinal sugar transport have been analyzed, steady-state values for $[S]_i/[S]_o$, $[Na]_o/[Na]_i$, and V have been determined and used to examine whether the relationship shown in equation (1) is upheld. A value of 1.0 has been used for n, based on results obtained with intact tissue by Goldner *et al.* (1969) and more recently for brush border vesicles by Hopfer and Groseclose (1979). In the most complete reported study of this type (Armstrong *et al.*, 1973), it was concluded that sufficient energy for sugar transport could be provided by the electrochemical potential for Na$^+$, although in this case a sugar gradient of only fourfold was established at the steady state. A transport efficiency of 60% was calculated, which was considered reasonable.

4.2. Transport Efficiency—Use and Misuse

For a given membrane potential, Na$^+$ gradient, and coupling ratio (n), equation (1) can be used to calculate a theoretical maximum sugar gradient. Intuitively, one might predict that a transport efficiency could be calculated by taking the ratio between observed sugar gradients and calculated theoretical sugar gradients, and it is this procedure that has often been employed (Armstrong *et al.*, 1973; Kessler and Semenza, 1979). This approach does not take into account the fact that a high percentage of steady-state sugar fluxes are occurring by routes other than via the Na$^+$-dependent carrier. In fact, as we have already mentioned, inhibition of the serosal sugar carrier converts isolated cells from a capability for maintaining steady-state sugar gradients near 10-fold to a capability for maintaining gradients near 70-fold. *Calculated* efficiencies for the Na$^+$-dependent carrier therefore improve dramatically, even though the serosal inhibitors do not alter cellular Na$^+$ gradients or membrane potentials. Obviously, the intuitive and conventional method for transport efficiency is a measure for efficiency of the system rather than for only the Na$^+$-dependent sugar carrier. In fact, if all sugar flux routes except that catalyzed by the Na$^+$-dependent pump could be totally inhibited, then one would expect the sugar pump to establish an equilibrium between the transmembrane difference in chemical potential for sugar and the difference in electrochemical potential for Na$^+$ (Kimmich, 1980). An equilibrium between these two potentials does not imply that the sugar carrier is 100% efficient any more than an enzymatic reaction at equilibrium implies that the enzyme is 100% efficient.

4.3 Na$^+$: Sugar Coupling Stoichiometry

Measurements of the Na$^+$ gradient (5- to 10-fold) (Gupta *et al.*, 1968; Gupta and Hall, 1979; O'Doherty *et al.*, 1978) and plasma membrane potential (-35 mV) (Rose and Schultz,

1971; Rose *et al.*, 1977; White and Armstrong, 1971; Okada *et al.*, 1977) maintained by the intestinal epithelium indicate that theoretical sugar gradients no more than 20- to 40-fold could be established if the Na^+ : sugar coupling stoichiometry is 1.0. Obviously, the 70-fold gradients observed for α-methylglucosides are incompatible with these measurements unless the value of n is greater than 1.0. This has led us to focus attention on the procedure by which the coupling stoichiometry has been determined. The usual procedure does not take account of the fact that sugar entry is a rheogenic process that can interfere with the measurement of Na^+ fluxes by other routes (Kimmich, 1981). This error results in an incorrect determination of Na^+ flux via the sugar carrier and a calculated Na^+ : sugar coupling ratio that is too low. When appropriate precautions are taken to avoid concomitant changes in membrane potential during the measurement, the coupling stoichiometry proves to be 2.0 (Kimmich and Randles, 1980).

4.4. Theoretical Sugar Gradients—The Implications

Because of the exponential relationship between theoretical sugar gradients and Na^+ : sugar coupling stoichiometry, a value of 2.0 for n results in an energy input that is much greater than that expected for 1 : 1 coupling. Indeed, for a 5-fold Na^+ gradient and a membrane potential of -35 mV, sugar gradients as great as 400-fold could be expected. Considering that the thermodynamic activity of intracellular Na^+ is less than its concentration (O'Doherty *et al.*, 1979), then even larger theoretical limits are possible. Such limits could only be realized experimentally if methods for limiting the diffusional leak of sugar can be found so that the Na^+-dependent sugar pump can realize its full capability. It is possible but unlikely that such selective control can be achieved. On the other hand, we have recently described a procedure by which the theoretical sugar gradient can be calculated for isolated intestinal cells even under conditions in which sugar leak pathways are not fully controlled (Kimmich, 1981). Values calculated by this procedure will reflect the exact magnitude of the Na^+ electrochemical potential that the cells maintain. By utilizing equation (1), one can use the theoretical sugar gradient in conjunction with a value for the membrane potential to calculate the Na^+ activity gradient. Conversely, a value for the Na^+ gradient would allow calculation of the membrane potential. The gradient-coupled transport of sugar therefore offers a potentially useful approach for the determination of cellular ion activities or membrane potentials by noninvasive techniques. To the extent that similar ideas can be applied to the more widely occurring Na^+-dependent amino acid transport systems, the concept will have application to a variety of cell types.

5. CONCLUSIONS

Facilitated diffusion and diffusion of sugars across epithelial cell membranes represent significant "leak" pathways against which the Na^+-dependent mucosal sugar "pump" ordinarily must operate. The innate gradient-forming capacity and therefore the analysis of transport energetics can *only* be accomplished under circumstances in which the flow of sugar by *all* routes is taken into account. As we have gained better experimental control over serosal sugar flux routes, we have at the same time gained better insight into the energetics of the Na^+-dependent sugar carrier. Notably, Na^+ : sugar coupling stoichiometries have proved to be 2 : 1 rather than 1 : 1 as commonly assumed and the thermodynamic power developed by flow of Na^+ on the sugar carrier is much greater than believed previously. Work in the eighties will focus on this and related "pump–leak" systems as tools for exploring cellular electrochemical Na^+ potentials.

ACKNOWLEDGMENTS. This work was supported in part by NIH Grant AM-15365 and U.S. Department of Energy Contract DE/-AC02-76-EV03490 and has been assigned Report No. UR-3490-1946.

REFERENCES

Armstrong, W. M., Byrd, B. J., and Hamang, P. M. (1973). *Biochim. Biophys. Acta* **330**, 237–241.
Basackova, J., and Crane, R. K. (1965). *Biochim. Biophys. Acta* **102**, 423–435.
Bihler, J., and Cybulsky, R. (1973). *Biochim. Biophys. Acta* **298**, 429–433.
Carter-Su, C., and Kimmich, G. A. (1979). *Am. J. Physiol.* **237**, C67–C74.
Carter-Su, C., and Kimmich, G. A. (1980). *Am. J. Physiol.* **238**, C73–C80.
Crane, R. K. (1960). *Physiol. Rev.* **40**, 789–825.
Crane, R. K. (1965). *Fed. Proc.* **24**, 1000–1005.
Crane, R. K. (1968). In *Handbook of Physiology* (C. F. Code, ed.), Sect. 6, Vol. III, pp. 1323–1351, American Physiology Society, Washington, D.C.
Crane, R. K., Miller, D., and Bihler, I. (1961). In *Membrane Transport and Metabolism* (A. Kleinzeller and A. Kotyk, eds.), pp. 439–449, Academic Press, New York.
Crane, R. K., Forstner, G., and Eicholz, A. (1965). *Biochim. Biophys. Acta* **109**, 467–477.
Goldner, A. M., Schultz, S. H., and Curran, P. F. (1969). *J. Gen. Physiol.* **53**, 362–383.
Gupta, B. J., and Hall, T. A. (1978). *Fed. Proc.* **38**, 144–153.
Gupta, B. J., Hall, T. A., and Naftalin, R. J. (1968). *Nature (London)* **272**, 70–73.
Holman, G. D., and Naftalin, R. J. (1975). *Biochim. Biophys. Acta* **406**, 386–401.
Holman, G. D., and Naftalin, R. J. (1976). *Biochim. Biophys. Acta* **433**, 597–614.
Hopfer, U., and Groseclose, R. (1979). *J. Biol. Chem.* **255**, 4453–4462.
Hopfer, U., Nelson, K., Perotta, J., and Isselbacher, K. J. (1973). *J. Biol. Chem.* **248**, 25–32.
Hopfer, U., Sigrist-Nelson, K., and Murer, H. (1975). *Ann. N.Y. Acad. Sci,* **264**, 414–427.
Hopfer, U., Sigrist-Nelson, K., Amman, E., and Murer, H. (1976). *J. Cell Physiol.* **89**, 805–810.
Kessler, M., and Semenza, G. (1979). *FEBS Lett.* **108**, 205–208.
Kimmich, G. A. (1981a). *Fed. Proc.* **40**, 2474–2479.
Kimmich, G. A. (1981b). In *Physiology of the Digestive Tract* (L. Johnson, M. Grossman, E.. Jacobson, J. Christensen, and S. Schultz, eds.), pp, 1035–1061, Raven Press, New York.
Kimmich, G. A., and Carter-Su, C. (1978). *Am. J. Physiol.* **235**, C73–C81.
Kimmich, G. A., and Randles, J. (1973). *J. Membr. Biol.* **12**, 23–46.
Kimmich, G. A., and Randles, J. (1975). *J. Membr. Biol.* **23**, 57–76.
Kimmich, G. A., and Randles, J. (1976). *J. Membr. Biol.* **27**, 363–379.
Kimmich, G. A., and Randles, J. (1978). *Membr. Biochem.* **1**, 221–237.
Kimmich, G. A., and Randles, J. (1979). *Am. J. Physiol.* **237**, C56–C63.
Kimmich, G. A., and Randles, J. (1980). *Biochim. Biophys. Acta* **596**, 439–444.
Kimmich, G. A., and Randles, J. (1981). *J. Am. Physiol.* **241**, C222–C237.
Kimmich, G. A., Carter-Su, C., and Randles, J. (1977). *Am. J. Physiol.* **238**, E357–E362.
Lücke, H., Beiner, W., Menge, H., and Murer, H. (1978). *Pfluegers Arch.* **373**, 243–248.
Murer, H., and Hopfer, U. (1974). *Proc. Natl. Acad. Sci. USA* **71**, 484–488.
Murer, H., Hopfer, U., Kinne-Saffran, E., and Kinne, R. (1974). *Biochim. Biophys. Acta* **345**, 170–179.
O'Doherty, J., Garcia-Diaz, J. F., and Armstrong, W. M. (1979). *Science* **263**, 1349–1351.
Okada, Y. W., Tsuchiza, W., Irimajiri, A., and Inouye, A. (1977). *J. Membr. Biol.* **31**, 205–219.
Randles, J., and Kimmich, G. A. (1978). *Am. J. Physiol.* **234**, C64–C72.
Rose, R. C., and Schultz, S. G. (1971). *J. Gen. Physiol.* **57**, 639–663.
Rose, R. C., Nahrwald, D. L., and Koch, M. J. (1977). *Am. J. Physiol.* **232**, E5–E12.
Sacktor, B. (1977). *Curr. Top. Bioenerg.* **6**, 39–81.
White, J. F., and Armstrong, W. M. (1971). *Am. J. Physiol.* **221**, 194–201.

113

The Small-Intestinal Na$^+$/D-Glucose Carrier Is Inserted Asymmetrically with Respect to the Plane of the Brush Border Membrane

A Model

Giorgio Semenza

"Carrier" is a membrane transport agency with a binding site that is exposed alternately to one or to the other side of the membrane (but not to both sides simultaneously) (Läuger, 1980), and which displays some particular kinetic properties (e.g., Kotyk and Janáček, 1975; Läuger, 1980), the most characteristic and the oldest known of which is the counterflow phenomenon (see Wilbrandt and Rosenberg, 1961).

As their functional and structural properties are being unraveled, it is becoming increasingly evident that carriers [even those not leading to accumulation of substrate(s)] have different characteristics at the two sides of the membrane, i.e., the structures exposed to the two sides of the membrane are not identical. The best characterized examples of carriers are probably the anion transporter of the erythrocyte membrane, the functional arrangement of which is now known in some detail (see the reviews by Cabantchik *et al.*, 1978; Rothstein and Ramjeesingh, 1980) and the ADP/ATP translocator of mitochondria (see Klingenberg, 1976; Klingenberg *et al.*, 1980).

In the following I will summarize the present state of knowledge on the asymmetry of the Na$^+$/D-glucose carrier of the small-intestinal brush border with respect to the plane of the membrane. As is well known, the cotransport mechanism, originally proposed by Crane (Crane *et al.*, 1961; Crane, 1962, 1965; see also Murer and Hopfer, 1974; for a review see Crane, 1977), is now universally accepted.

Giorgio Semenza • Laboratorium für Biochemie der E.T.H., ETH-Zentrum, CH-8092 Zurich, Switzerland.

1. ASYMMETRIC IRREVERSIBLE INACTIVATION OF THE Na⁺/D-GLUCOSE CARRIER

In addition to catalyzing the translocation of the two substrates across the membrane, the carrier binds phlorizin, a fully competitive inhibitor (Alvarado and Crane, 1962). Phlorizin binding to membrane vesicles or to membrane fragments requires the presence of Na^+ (Tannenbaum *et al.*, 1977; Toggenburger *et al.*, 1978; Klip *et al.*, 1979a), is inhibited by D-glucose (Toggenburger *et al.*, 1978; Klip *et al.*, 1979a), and shows a K_d corresponding to the K_i for phlorizin inhibition of D-glucose uptake (Toggenburger *et al.*, 1978; Klip *et al.*, 1979a). It is thus possible to monitor the function of this carrier by either of two properties: the Na^+-dependent D-glucose translocation or the Na^+-dependent phlorizin binding. Essentially, two approaches were used:

1. Comparison of the effect of impermeant (or slowly permeant) reagents on the Na^+-dependent phlorizin binding in right-side-out sealed vesicles (Klip *et al.*, 1979b) vs. deoxycholate-disrupted membranes (Klip *et al.*, 1979a). Three kinds of reagents were tested:

a. Proteases: trypsin, chymotrypsin, and papain were without effect on phlorizin binding to intact vesicles, but each inactivated it in deoxycholate-disrupted membranes (Table I).

b. *p*-Chloromercuribenzyl sulfonate, a slowly permeant reagent (Rothstein, 1970), at low concentrations has little or no effect on phlorizin binding in intact vesicles (Fig. 1D), but inactivates it in deoxycholate-disrupted membranes (Fig. 1B).

c. $HgCl_2$ inactivates phlorizin binding both in intact vesicles and in deoxycholate-disrupted membranes (Klip *et al.*, 1979a). Thiols reverse this effect of $HgCl_2$. Of the thiols tested, glutathione, which is slowly permeant, is relatively ineffective in restoring phlorizin binding activity in $HgCl_2$-treated vesicles, but is very effective in deoxycholate-disrupted membranes (Table II).

It is conceivable that deoxycholate, in addition to disrupting the brush border membranes, may also partially dislodge the carrier within the membrane, thereby making it accessible to the action of particular reagents. This "dislodging," if it occurs under the

Table I. Effect of Proteases on Phlorizin Binding in Intact and Deoxycholate-Treated Membranes [a,b]

	Protein released to the super-natant (%)	Specific binding activity in the pellet (%)	Binding recovery in the pellet (%)
Intact vesicles			
Control	2.3 ± 1.6	100.0	97.7
Trypsin	3.8 ± 1.9	118.7 ± 8.4	114.2
Chymotrypsin	8.2 ± 3.6	113.7 ± 4.5	104.4
Papain	27.8 ± 3.9	165.8 ± 7.6	121.4
Deoxycholate-treated membranes			
Control	2.8 ± 2.7	100.0	97.2
Trypsin	10.9 ± 2.0	19.6 ± 7.3	17.5
Chymotrypsin	11.0 ± 2.3	21.6 ± 4.8	19.2
Papain	46.0 ± 5.9	26.4 ± 1.4	14.3

[a] From Klip *et al.* (1979c).
[b] Brush border vesicles or deoxycholate-disrupted membranes were exposed to the protease indicated. After centrifugation, the pellets were washed and used for determination of phlorizin binding. "Specific" phlorizin binding is defined as the phlorizin bound in the presence of 100 mM NaCl minus that bound in the presence of 100 mM KCl. Means ± S.E. (4 determinations).

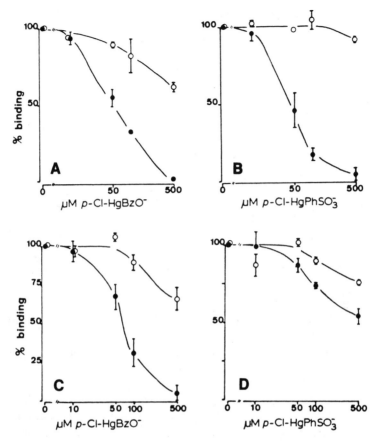

Figure 1. Concentration dependence of the inhibition of Na⁺-dependent phlorizin binding to deoxycholate-disrupted membranes (A, B) or to intact, sealed, right-side-out vesicles (C, D) by two organomercurials (solid symbols). In some experiments (open symbols) the membranes were washed with 5 mM dithioerythritol prior to phlorizin binding. The points are the mean ± S.E. of three experiments performed in duplicate. [From Klip *et al.* (1979a) with permission of *Biochimica et Biophysica Acta.*]

conditions of our experiments, must be minimal, because the phlorizin binding capacity and the K_d value are unaffected by deoxycholate treatment (Klip *et al.*, 1979a,b) and because *lysis* of the vesicles correlates closely with the inactivation of phlorizin binding by trypsin (Klip *et al.*, 1979b). Furthermore, a second approach also led to the conclusion that the Na⁺/ᴅ-glucose carrier is indeed differently accessible from the two sides of the membrane to the action of inactivating reagents, as follows.

2. Comparison of the effect of highly permeant vs. slowly permeant reagents on right-side-out sealed vesicles. This approach (Klip *et al.*, 1979a,b) allows us to test with some reagents, ᴅ-glucose transport, along with phlorizin binding, when the treatments do not severely change the permeability of the membrane.

a. *p*-Chloromercuribenzoate and *p*-chloromercuribenzyl sulfonate are equally effective in inhibiting phlorizin binding in deoxycholate-disrupted membranes (Figs. 1A and B), 50% inactivation being observed at approximately 50 μM concentration. Thus, whatever differences may be found in their effect on intact vesicles cannot be attributed to a difference in their chemical reactivity. In intact vesicles at nearly neutral pHs, the slowly permeant sulfonate is less effective in inhibiting phlorizin binding (Figs. 1C and D) or ᴅ-glucose equilibrium exchange (Klip *et al.*, 1979a) than the slowly permeant (Rothstein,

Table II. Reversibility of Mercurial-Inhibited Phlorizin Binding by Two Different Thiols [a,b]

	n	% Na[+]-dependent binding (mean ± S.E.)
Sealed vesicles		
Control	5	100
HgCl$_2$	5	0 ± 0
HgCl$_2$, then dithioerythritol	5	79.2 ± 5.0
HgCl$_2$, then glutathione	4	10.7 ± 6.4
Deoxycholate-treated membranes		
Control	4	100
HgCl$_2$	4	3.2 ± 2.2
HgCl$_2$, then dithioerythritol	4	88.2 ± 9.5
HgCl$_2$, then glutathione	4	80.7 ± 13.8

[a] From Klip *et al.* (1979a).
[b] Membranes (1.2 mg/ml) were incubated with 0.05 mM HgCl$_2$ for 5 min at ice temperature in mannitol–Tris–HEPES buffer, pH 7.0, then diluted with 5–10 volumes of buffer with or without the indicated thiol (1 mM); spun down, and tested for phlorizin binding. *n*, number of experiments.

1970) benzoate derivative. The benzoate derivative inhibits phlorizin binding in intact vesicles and in deoxycholate-disrupted membranes equally well (Figs. 1A and C).

b. Glutathione and dithioerythritol are equally effective in restoring phlorizin binding to deoxycholate-disrupted membranes in which prior HgCl$_2$ treatment had inactivated this binding. In vesicles, the more permeant dithioerythritol is more effective than the slowly permeant glutathione in restoring HgCl$_2$-inactivated phlorizin binding (Table II).

It seems certain, therefore, that the Na/D-glucose cotransporter has one or more SH groups, and probably Arg or Lys and Tyr or Phe residues at the cytoplasmic and/or hydrophobic surface, which are essential for phlorizin binding and for transport activity. Neither phlorizin nor D-glucose affords any protection from inactivation by mercurials. Whatever the role of the thiol group(s), these observations show that the small-intestinal Na[+]/D-glucose carrier exposes different structures on the two sides of the D-membrane (Klip *et al.*, 1980a).

The exact location of the phlorizin binding site has not been established with certainty. On the basis of indirect evidence, it is generally believed that the binding occurs at the outside surface of both intestinal and renal brush borders (see Kinne, 1976). If the protease-sensitive part of the carrier is indeed located at the cytosolic face of the membrane, this implies that the carrier, in addition to being asymmetric with respect to the plane of the membrane, also spans it.

Also, the small-intestinal Na[+]/D-glucose carrier is an intrinsic membrane protein, for it is not extracted or inactivated by buffers of low or high ionic strengths or by the moderate concentrations of deoxycholate that are used in the preparation of the vesicles and disrupted membranes.

2. FUNCTIONAL ASYMMETRY

It is to be expected that a structurally asymmetric carrier should show at least some degree of functional asymmetry, in addition, of course, to that which may be imposed onto the carrier by asymmetric distributions of the substrates or by a $\Delta\Psi \neq 0$. This is indeed the case also for the carrier discussed here. Both symmetric and asymmetric models have been proposed for the small-intestinal Na⁺/D-glucose carrier. Naturally, whatever functional asymmetries exist in a two-substrate transport system, they must be mutually related. It is an easy matter to show, on the basis of the principle of microreversibility, that at equilibrium (G′ = G″; N′ = N″; $\Delta\Psi = 0$), the following relation must hold (for example, in the case that the translocation probabilities of the binary complexes are negligibly small):

$$\frac{p''_C}{p'_C} \cdot \frac{p'_{CGN}}{p''_{CGN}} = \frac{K^{N'}_{CGN} K'_{CN}}{K^{N''}_{CGN} K''_{CN}} = \frac{K^{G'}_{CGN} K'_{CG}}{K^{G''}_{CGN} K''_{CG}}$$

where the p's are the translocation probabilities from either the ′ or the ″ side of the carrier form indicated in the subscript; the K's are the dissociation constants (at either the ′ or the ″ side) of the carrier form indicated in the subscript (the superscript N or G, if present, indicates the substrate binding last); and C, G, and N are carrier, D-glucose, and Na⁺, respectively.

Naturally, the equation above, short of setting the frame for possible functional, i.e., kinetic, asymmetries, does not preclude any of the models suggested. At least one of them (iso-ordered bi bi with glide symmetry, Hopfer and Groseclose, 1980) implies that the carrier is asymmetric with respect to the plane of the membrane. Hopfer's model is physically plausible, is in keeping with current ideas on carrier mechanisms (Singer, 1974, 1977; Läuger, 1980), and is presented with beautiful logic. It rests, however, on rather weak experimental observations. In fact, the discrimination between "random" and "ordered" sequences of binding rests solely on the biphasic activation curves of the one substrate for the equilibrium exchange rates of the other, i.e., on their mutual inhibitions at very high concentrations—clearly, a phenomenon that can be given alternative explanations.* Furthermore, the very high "leak" rates required sizable corrections to be introduced in the total equilibrium exchange rates. This was done either by subtracting the exchange rates in the presence of phlorizin, or by manipulating the data so as to obtain "the best fits" on the assumption that the carrier-catalyzed rates are a single strictly Michaelian function of the substrate concentrations. [The random bi bi mechanism suggested by Crane and Dorando (1979, 1980) can lead to a non-Michaelian behavior.]

How far these corrections are responsible for the different K_m values for D-glucose reported by the same author working under identical experimental conditions, it cannot be established at this moment (14.2 mM, Hopfer, 1977; 2.4 mM, Hopfer and Groseclose, 1980). Be as it may, it is unfortunate that the experimental uncertainties and the related circular reasoning prevent one from accepting Hopfer's attractive model on the basis of the evidence at hand at this moment.

Although the random bi bi model of Crane and Dorando (1979, 1980) is now available in a preliminary form only, it can be safely predicted that its kinetic parameters will not be identical at both sides of the membrane. Indeed, whatever model will eventually turn out to be correct, it will have to accommodate some sort of asymmetry.

The basis for this statement is the high degree of asymmetry in the *trans*-effects by the substrates (Kessler and Semenza, 1982): (1) within reasonable concentration ranges, *trans*-inhibitions and *trans*-stimulations were observed in influx experiments only, i.e., by D-glucose and /or Na⁺ acting on the *cytosolic* side of the carrier; the efflux of D-glucose proved singularly little responsive to *external* D-glucose or Na⁺; (2) the influx of D-glucose is accelerated by $\Delta\Psi$ (negative at the *trans* side) much more than its efflux; (3) the estimated K_m for efflux is approximately one order of magnitude larger than the K_m for influx; (4) uptake and efflux rates may differ by a factor 10 when measured at equivalent, but

---

* See, for example, Fig. 2,e in Hopfer and Liedtke (1981).

mirrored conditions. These and additional observations will be discussed in detail else-where.

3. A PROBABLY "GATED CHANNEL"

The occurrence of a *stable* asymmetry of the carrier with respect to the plane of the membrane implies that it does not tumble or flip over freely across this plane. Indeed, no such free movement would be expected in an intrinsic protein, if it is inserted vectorially into the membrane during or after its biosynthesis (see e.g. Sabatini and Kreibich, 1976; Blobel and Dobberstein, 1975; Wickner, 1979, 1980; DiRienzo *et al.*, 1978; Inouye and Halegoua, 1980; Engelman and Steitz, 1981), and if this protein has both hydrophobic and hydrophilic domains on its surface (as one can expect in the case of a carrier-translocating hydrophilic substrate). Also, the very size of the small-intestinal Na^+/D-glucose carrier, which may contain at least one polypeptide chain of approximately 72,000 molecular weight (Klip *et al.*, 1979c, 1980b; recently identified by photoaffinity labeling by Hosang *et al.*, 1981), makes "diffusive" or "rotatory" models less likely.

The physical model that is most compatible with current ideas on the biosynthesis and insertion of membrane intrinsic proteins and with the properties of the Na^+/D-glucose carrier is that of a "gated channel" [Crane and Dorando's model (1979, 1980) is also that of a "gated channel"], i.e., of a "channel with multiple conformational states", which is a limiting case of the general channel mechanisms (Läuger, 1980).

Figure 2, which is taken from Läuger's review (1980) with minor modifications, il-lustrates the simple type of change in the potential profile (from the "s" to the "r" state, and vice versa) due, for example, to protein structure fluctuations, leading to the alternate exposure of the substrate binding site(s) to one or to the other side of the membrane. The resulting effect we could call an "equivalent gate."

There are kinetic indications (Kessler and Semenza, 1982) that the "equivalent gate" in the unloaded form is negatively charged, presumably a carboxylate group. Neutralization of this charge by Na^+ would trigger an alteration in the transport protein such that changes

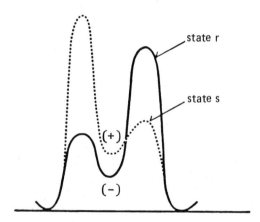

Figure 2. Energy profile of a channel with two conformational states (r, s). The drawing, which is taken from Läuger (1980), has been modified in order to indicate that in the case of the small-intestinal Na^+ D-glucose carrier, the "gated channel" has a negatively charged site (−) that binds Na^+ (+). The D-glucose binding site is sug-gested to be located near the Na^+ binding site.

(fluctuations) in the potential profile become possible, with resulting exposure of the binding sites on the other side of the membrane and release of the substrates there. The translocation probability of the Na$^+$-free, ᴅ-glucose-loaded carrier would thus be much smaller than that of the Na$^+$/ᴅ-glucose-loaded carrier. Finally, the effect of $\Delta\Psi$ on Na$^+$/ᴅ-glucose transport (Murer and Hopfer, 1974; Kessler and Semenza, 1979) and on Na$^+$-dependent phlorizin binding (Tannenbaum *et al.*, 1977; Toggenburger *et al.*, 1978) can also be rationalized on the basis of the model of Fig. 2: if the shapes of the potential profiles are influenced by the presence or absence of a charge at the "equivalent gate," it is to be expected that these shapes (i.e., the distribution of conformational fluctuations) would be different, depending on the $\Delta\Psi$ present across the membrane. A part of the mechanistic model for Na$^+$/ᴅ-glucose cotransport is now in press (Toggenburger *et al.* 1981).

Acknowledgments. The reviewer's work was partially supported by the SNSF, Bern. Thanks are due to the reviewer's co-workers, in particular to Drs. A. Klip, S. Grinstein, M. Kessler, M. Hosang, G. Toggenburger, and Mrs. O'Neill.

REFERENCES

Alvarado, F., and Crane, R. K. (1962). *Biochim. Biophys. Acta* **56,** 170–172.
Blobel, G., and Dobberstein, B. (1975). *J. Cell Biol.* **67,** 835–851.
Cabantchik, Z. I., Knauf, P. A., and Rothstein, A. (1978). *Biochim. Biophys. Acta* **515,** 239–302.
Crane, R. K. (1962). *Fed. Proc.* **21,** 891–895.
Crane, R. K. (1965). *Fed. Proc.* **24,** 1000–1006.
Crane, R. K. (1977). *Rev. Physiol. Biochem. Pharmacol.* **78,** 101–159.
Crane, R. K., and Dorando, F. (1979). In *Functional and Molecular Aspects of Biomembrane Transport* (E. Quagliarielo, F. Palmieri, S. Papa, and M. Klingenberg, eds.), pp. 271–278, Elsevier/North-Holland, Amsterdam.
Crane, R. K., and Dorando, F. C. (1980). *Ann. N.Y. Acad. Sci.* **339,** 46–52.
Crane, R. K., Miller, D., and Bihler, I. (1961). In *Symposium on Membrane Transport and Metabolism* (A. Kleinzeller and A. Kotyk, eds.), pp. 439–449, Academic Press, New York.
DiRienzo, J. M., Nakamura, K., and Inouye, M. (1978). *Annu. Rev. Biochem.* **47,** 411–422.
Engelman, D. M., and Steitz, T. A. (1981). *Cell* **23,** 411–422.
Hopfer, U. (1977). *J. Supramol. Struct.* **7,** 1–13.
Hopfer, U., and Groseclose, R. (1980). *J. Biol. Chem.* **255,** 4453–4462.
Hopfer, U., and Liedtke, C. M. (1981). *Membrane Biochemistry* **4,** 11–29.
Hosang, M., Gibbs, E. M., Diedrich, D. F., and Semenza, G. (1981). *FEBS Lett.* **130,** 244–248.
Inouye, M., and Halegoua, S. (1980). *CRC Crit. Rev. Biochem.* **10,** 339–371.
Kessler, M., and Semenza, G. (1979). *FEBS Lett.* **108,** 205–208.
Kessler, M., and Semenza, G. (1981). Submitted for publication.
Kinne, R. (1976). *Curr. Top. Membr. Transp.* **8,** 209–267.
Klingenberg, M. (1976). *The Enzymes of Biological Membranes* (A. Martonosi, ed.), Vol. 3, pp. 383–438, Plenum Press, New York.
Klingenberg, M., Hackenberg, H., Kramer, R., Lin, C. S., and Aquila, H. (1980). *Ann. N.Y. Acad. Sci.* **358,** 83–95.
Klip, A., Grinstein, S., and Semenza, G. (1979a). *Biochim. Biophys. Acta* **558,** 233–245.
Klip, A., Grinstein, S., and Semenza, G. (1979b). *FEBS Lett.* **99,** 91–96.
Klip, A., Grinstein, S., and Semenza, G. (1979c). *J. Membr. Biol.* **51,** 47–73.
Klip, A., Grinstein, S., and Semenza, G. (1980a). *Ann. N.Y. Acad. Sci.* **358,** 374–377.
Klip, A., Grinstein, S., Biber, J., and Semenza, G. (1980b). *Biochim. Biophys. Acta* **589,** 100–114.
Kotyk, A., and Janáček, K. (1975). *Cell Membrane Transport,* 2nd ed., Plenum Press, New York.
Läuger, P. (1980). *J. Membr. Biol.* **57,** 163–178.
Murer, H., and Hopfer, U. (1974). *Proc. Natl. Acad. Sci. USA* **71,** 484–488.
Rothstein, A. (1970). *Curr. Top. Membr. Transp.* **1,** 135–176.
Rothstein, A., and Ramjeesingh, M. (1980). *Ann. N.Y. Acad. Sci.* **358,** 1–12.

Sabatini, D. D., and Kreibich, G. (1976). In *The Enzymes of Biological Membranes* (A. Martonosi, ed.), Vol. 2, pp. 531–579, Plenum Press, New York.

Singer, S. J. (1974). *Annu. Rev. Biochem.* **43,** 805–833

Singer, S. J. (1977). *J. Supramol. Struct.* **6,** 313–323.

Tannenbaum, C., Toggenburger, G., Kessler, M., Rothstein, A., and Semenza, G. (1977). *J. Supramol. Struct.* **6,** 519–533.

Toggenburger, G., Kessler, M., Rothstein, A., Semenza, G., and Tannenbaum, C. (1978). *J. Membr. Biol.* **40,** 269–290.

Toggenburger, G., Kessler, M. and Semenza, G. (1982). *Biochim. Biophys. Acta,* in press.

Wickner, W. (1979). *Annu. Rev. Biochem.* **48,** 23–45.

Wickner, W. (1980). *Science* **210,** 861–867.

Wilbrandt, W., and Rosenberg, T. (1961). *Pharmacol. Rev.* **13,** 109–183.

114

Sodium Cotransport Systems in the Kidney

Mapping of the Nephron and First Steps to a Molecular Understanding

R. Kinne, J. T. Lin, and J. Eveloff

1. INTRODUCTION

During the last decade it has become obvious that Na^+-cotransport systems play an essential role in tubular reabsorption. Electrophysiological studies and studies with membrane vesicles isolated from the luminal and contraluminal cell pole demonstrated the presence of Na^+-cotransport systems for a large variety of organic and inorganic solutes in the proximal tubule (for recent reviews see Ullrich, 1979; Murer and Kinne, 1980). The most extensively studied system is the Na^+/D-glucose cotransport system, and as a consequence, it is also this system where attempts to isolate the carrier molecule are most advanced. However, the membrane-molecular investigation of the transport mechanisms in other segments of the nephron remained undefined because of the difficulties associated with obtaining purified renal tubular segments from the heterogeneous renal tissue. This obstacle has been overcome recently with studies on plasma membranes from the thick ascending limb of Henle's loop. The presence of an Na^+/Cl^- cotransport system in these cells constitutes the second part of the contribution.

2. PARTIAL PURIFICATION OF THE Na^+/D-GLUCOSE COTRANSPORT SYSTEM FROM PROXIMAL TUBULE

The main obstacle in the attempt to isolate transport systems from the proximal tubule is the low concentration of the carrier protein in the brush border membrane. Assuming a molecular weight of 100,000, the contribution of the transport protein to the total protein of the membrane has been estimated to be 0.1% or one part per million. Methods to

R. Kinne, J. T. Lin, and J. Eveloff • Department of Physiology and Biophysics, Albert Einstein College of Medicine, Bronx, New York 10461.

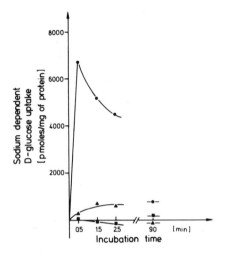

Figure 1. Na⁺-dependent D-glucose uptake into proteolipo-somes formed out of asolectin and the membrane extract (▲), the protein fraction not retained by the phlorizin polymer (■), or the protein fraction eluted by D-glucose (●). Glucose uptake was determined by the rapid filtration method in the presence of an NaSCN gradient and is corrected for the uptake observed in the presence of a KSCN gradient. For further details see Lin *et al.* (1981).

achieve isolation therefore include preparation of large quantities of high-purity membranes and the development of isolation methods that concentrate the transport system, such as affinity chromatography. The former step was realized by scaling up the well-known methods for the preparation of brush border membranes, and now gram quantities of the membranes are prepared and used for the starting material (Lin *et al.*, 1979). The chromatographic approach has been made possible by synthesizing a matrix in which the phlorizin molecule is cross-linked with formaldehyde and urea (Lin and Kinne, 1980). Phlorizin is a competitive inhibitor of glucose transport with a binding affinity of 10^{-6} M (Lin and Kinne, 1980). In the polymer the local concentration of the ligand, D-glucose, is about 10^{-3} M. The gel retains about 2% of a brush border membrane extract. When 0.5 M D-glucose is added to the elution buffer, 0.4% of the original protein is eluted. If the Na⁺-dependent D-glucose transport activity is measured in liposomes that have incorporated the original membrane extract, the nonadsorbed protein fraction, and the fraction eluted with 0.5 M D-glucose, the results shown in Fig. 1 are obtained. The fraction eluted with D-glucose shows an uptake activity of 7700 pmoles/30 sec per mg protein and the original membrane extract a transport activity of 300 pmoles/30 sec per mg protein. This indicates an enrichment of approximately 26-fold in transport activity. In an SDS-gradient gel, several polypeptide bands are observed in the purified fraction. One protein band corresponding to a molecular weight of about 60,000–70,000 seems to be particularly enriched. The positive identification of this protein as the Na⁺/D-glucose cotransporter requires further investigation. The actual enrichment of the transport molecule remains also to be determined. Studies with different lipid-to-protein ratios and different lipids gave similar results, indicating that the increased transport activity is not simply due to an activation of the system. Furthermore, other Na⁺-cotransport systems, such as L-alanine and phosphate, are enriched to a much smaller extent. This fact provides evidence that a relative enrichment of the D-glucose transport system above other brush border membrane transport systems occurs. The nature of the interaction of the transport molecule with the phlorizin matrix in the presence of the detergent is also not clear. Interactions with the glucose ligand as well as with the hydrophobic matrix provided by the phenolic aglucone part of phlorizin are possible. Thus, a total purification of the Na⁺/D-glucose cotransport system has still to be achieved. Nevertheless, the studies can be regarded as a sign that in the not too distant future it will be possible to study the Na^+/D-glucose cotransport system in its isolated state. Such studies will provide important information about the events involved in the binding, transport, and debinding of Na^+ and the cotransported solute.

3. Na^+/Cl^--COTRANSPORT IN THE MEDULLARY THICK ASCENDING LIMB OF HENLE'S LOOP

The most recently discovered Na^+ cotransport system in the renal tubule is the co-transport of Na^+/Cl^- in the medullary thick ascending limb of Henle's loop (TALH) (Eveloff *et al.*, 1980, 1981).

The Na^+/Cl^- cotransporter in the TALH has been investigated using isolated cells prepared from the TALH and plasma membrane vesicles derived from the TALH cells. Using oxygen consumption as a measure of the energy expenditure of the TALH cells due to the Na^+ and Cl^- transport processes, it was found that the maximal oxygen consumption in the cells was dependent on the presence of *both* Na^+ and Cl^- in the incubation medium. The replacement of either ion or both ions produced an equal decrease in oxygen consumption of 50%. Further, furosemide, a potent loop diuretic, inhibited oxygen consumption by 50% and required both Na^+ and Cl^- for its inhibitory effect. The same was observed for ouabain. Thus, it appears that half of the oxygen consumption of the cell maintains the movement of Na^+ and Cl^- across the cell membrane and that the movement of Na^+ and Cl^- is coupled.

The Cl^- dependency of Na^+ movement can also be observed in plasma membrane vesicles prepared from the TALH cells. Na^+ uptake into the vesicles is greater in the presence of Cl^- than when nitrate or gluconate is the accompanying anion (Fig. 2). Furosemide inhibits Na^+ uptake only in the presence of Cl^- (Fig. 2). In addition, a decrease of tracer $^{22}Na^+$ uptake by increasing concentrations of unlabeled NaCl was found. Tracer replacement was not observed with increasing concentrations of $NaNO_3$ (Fig. 3), indicating that a limited number of transport sites for Na^+ exist within the membrane and require Cl^- to interact with Na^+.

The Na^+/Cl^- cotransporter is unaffected by amiloride or 4-acetamido-4'-isothiocy-anostilbene-2,2'-disulfonate (SITS) at 10^{-4} M. This result suggests that the coupled Na^+/Cl^- cotransporter is not a combination of two antiport systems, i.e., Na^+/H^+ and Cl^-/OH^- antiport, as suggested for NaCl transport in the rat intestine (Liedtke and Hopfer, 1980), but represents a true cotransport system.

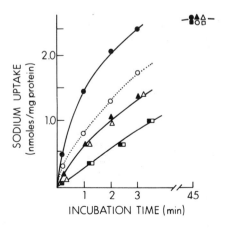

Figure 2. Na^+ uptake into plasma membrane vesicles prepared from cells of the TALH of the rabbit renal medulla. Membranes were prepared in 100 mM sucrose, 20 mM Tris–Hepes, pH 7.4, and 1 mM $Mg(NO_3)_2$. The incubation medium contained in addition: 2 mM ^{22}Na salt (12 μCi/150 μl) and 98 mM K salt. The uptake of Na^+ into the vesicles was studied in the presence of an inwardly directed gradient of NaCl–KCl (circles), $NaNO_3$–KNO_3 (triangles), or Na gluconate–K gluconate (squares) in the absence (solid symbols) or presence (empty symbols) of 1 mM furosemide. Values are the means of two experiments.

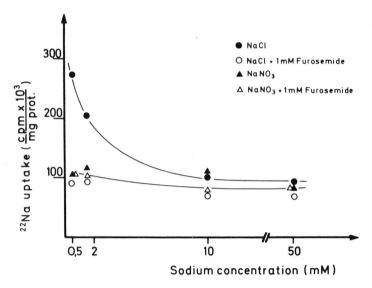

Figure 3. Tracer replacement of $^{22}Na^+$ uptake into plasma membrane vesicles prepared from cells of the TALH. The membranes were prepared as in Fig. 2. The incubation medium contained 100 mM sucrose, 1 mM $Mg(NO_3)_2$, 20 mM Tris–Hepes, pH 7.4, and increasing concentrations of NaCl (•) or $NaNO_3$ (▲) (the total salt concentration was 100 mM with KCl or KNO_3) plus or minus the presence of 1 mM furosemide [NaCl (○) or $NaNO_3$ (△)]. The reaction was terminated at 10 sec. Values are the means of three experiments.

4. CONCLUSIONS

Thus, for the reabsorption of solutes along the nephron, Na^+-cotransport systems are not only present in the proximal tubule but also in the TALH. The analysis of further segments is dependent on the isolation of homogeneous cell populations and the purification of plasma membranes from these cell populations. It is also anticipated that the experience gained in the attempts to isolate the Na^+-D-glucose cotransport system will be applicable to the isolation of the Na^+-Cl^- cotransport system. Similarities and/or dissimilarities in the molecular properties of the cotransport systems will shed light on their molecular arrangement and on the basic mechanisms involved in membrane transport.

ACKNOWLEDGMENT. This work was supported in part by NIH Grants AM 27441 and GM 27859

REFERENCES

Eveloff, J., Bayerdorffer, E., Haase, W., and Kinne, R. (1980). *Int. J. Biochem.* **12,** 55–59.
Eveloff, J., Bayerdorffer, E., Silva, P., and Kinne, R. (1981). *Pfluegers Arch.* **389,** 263–270
Liedtke, C. M., and Hopfer, U. (1980). *Fed. Proc.* **39,** 734.
Lin, J. T., and Kinne, R. (1980). *Angew Chem. Int. Ed. Engl.* **19,** 540–541.
Lin, J. T., Riedel, S., and Kinne, R. (1979). *Biochim. Biophys. Acta* **557,** 179–187.
Lin, J. T., DaCruz, M. E. M., Riedel, S., and Kinne, R. (1981). *Biochim. Biophys. Acta.* **640,** 39–54.
Murer, H., and Kinne, R. (1980). *J. Membr. Biol.* **55,** 81–95.
Ullrich, K. J. (1979). *Annu. Rev. Physiol.* **41,** 181–195.

115

Salt and Water Transport in the Mouse Medullary Thick Ascending Limb

Role of Antidiuretic Hormone

Steven C. Hebert and Thomas E. Andreoli

1. INTRODUCTION

The purpose of this chapter is to summarize some recent observations from our laboratory, as well as those from other laboratories, on the action of antidiuretic hormone (ADH) on salt and water absorption in single medullary thick ascending limbs of Henle (mTALH) isolated from mouse kidney. These results are consistent with the view that, in this nephron segment: ADH enhances the rate of conservative NaCl absorption; that NaCl transport from lumen to cells involves a coupled NaCl entry step across luminal plasma membranes; that the Na^+ gradient between luminal fluid and cytoplasm provides the energy source for Cl^- transport from luminal fluid into cells; and that a negative feedback system, determined by peritubular osmolality, modulates the rate of ADH-stimulated NaCl absorption in the mTALH.

Several lines of evidence have suggested that ADH might affect NaCl absorption in the ascending limb of Henle's loop, and therefore might act as a regulator of the countercurrent multiplication process. Wirz (1957), and subsequently others (Bray, 1960; Kessler, 1960; Perlmutt, 1962; Ruiz-Guiñazú et al., 1964), observed that outer medullary and papillary NaCl concentrations increased during antidiuresis, as compared to water diuresis, and that this effect could be demonstrated within 30 min after inducing the antidiuretic state. Second, the recent observations of an ADH-induced increase in adenylate cyclase activity (Chabardés et al., 1977; Morel, 1981) and protein kinase activity (Edwards et al., 1980) in medullary but not cortical thick ascending limbs of Henle, at least in some mammalian

Steven C. Hebert and Thomas E. Andreoli • Department of Internal Medicine, University of Texas Medical School, Houston, Texas 77025.

species, not only supported this earlier contention but also implied that an ADH effect on salt absorption might be restricted to the mTALH. Finally, the preliminary observations of Hall (1979) and Hall and Varney (1979, 1980) first established that, in mouse mTALH, vasopressin (ADH) increased the spontaneous, lumen-positive transepithelial voltage and concomitantly increased net rates of chloride absorption without affecting the low water permeability of that segment. Subsequently, Sasaki and Imai (1980) reported similar effects of ADH and cyclic adenosine monophosphate (cAMP) on the transepithelial voltage and unidirectional lumen to bath Cl^- fluxes in mouse mTALH.

Recently, we (Hebert *et al.*, 1980) have confirmed and extended these latter observations for the mouse mTALH. Based on these results, we have proposed a model for both basal and ADH-stimulated NaCl absorption in the mouse mTALH in which NaCl absorption depends on a furosemide-sensitive, coupled NaCl apical membrane entry process of indeterminate stoichiometry, which is driven by the Na^+/K^+-ATPase-generated transmembrane electrochemical gradient for sodium.

2. PASSIVE PERMEABILITY PROPERTIES OF THE MOUSE mTALH

2.1. Permselectivity of the Shunt Pathway

The passive permeability characteristics of the *in vitro* microperfused mouse mTALH are shown in Table I. For a given transepithelial salt gradient, salt dilution voltages in the mTALH are symmetrical, that is, are equal in magnitude but opposite in sign for lumen to bath versus bath to lumen dilutions (Hebert *et al.*, 1980). This finding, consistent with the Ussing–Windhager formulation about the route for passive ion flows in epithelia (Ussing and Windhager, 1964), therefore indicates that the route for passive ion permeation in the mouse mTALH is paracellular. The permselective characteristics of this paracellular (shunt) route, whether calculated from salt dilution voltages or from bath to lumen tracer Na^+ and Cl^- fluxes, are similar and indicate that the junctional complexes in the mouse mTALH are sodium rather than chloride-permselective (Table II). These findings are qualitatively

Table I. Passive Permeability Properties of the Mouse mTALH [a]

	Tracer	Electrical
P_{Na} (μm/sec)	0.24[b]	1.12[c]
P_{Cl} (μm/sec)	0.11[b]	0.56[c]
P_{Na}/P_{Cl}	2.2[b]	2.0[c]
	(tracer)	(salt dilution)
R_T (ohm cm²)	50[d]	11[c]
G_T (mS)	19.6[d]	90.9[e]

[a] Data are from Hebert *et al.* (1980).
[b] The isotopic ionic permeability coefficients for sodium (P_{Na}) and chloride (P_{Cl}) were determined from bath to lumen fluxes of $^{22}Na^+$ and $^{36}Cl^-$, respectively, in isolated segments of mouse mTALH at 37°C.
[c] The electrical P_{Na} and P_{Cl} values were calculated from the measured electrical resistance, R_T (calculated by cable analysis), and the electrical P_{Na}/P_{Cl} ratio (determined from salt dilution voltages).
[d] The tracer total ionic conductance was calculated from the sum of the partial ionic conductances (gNa + gCl). The tracer transepithelial resistance was calculated as $1(g$Na + gCl).
[e] The electrical total ionic conductance was calculated as $1/R_T$.

Table II. Transport Characteristics of the Mouse mTALH[a]

	−ADH	+ADH
V_e (mV)	4.9 ± 0.4	10.7 ± 0.5
Net Cl⁻ absorption	2630	10,790
(J_{Cl}^{net}, pEq/sec/cm²)		
P_f (μm/sec)	10–15	10–15

[a]Isolated mouse mTALH segments were perfused and bathed in symmetrical KRB solutions at 37°C. ADH, when added, was at a concentration of 250 μU/ml. Data are from Hebert et al. (1980).

similar to those demonstrated for the rabbit cTALH and/or mTALH (Burg and Green, 1973; Rocha and Kokko, 1973; Gregor and Frömter, 1980).

2.2. Nature of the Shunt Pathway in the Mouse mTALH

It is generally accepted that epithelia may be classified as "tight" or "leaky" based on their ionic permeability characteristics (Ussing and Windhager, 1964; Barry et al., 1971; Schultz, 1977). By such criteria, the mTALH, having a transepithelial resistance in the range of 11–50 ohm cm² (Table I), might be classified as leaky. But leaky epithelia, such as the proximal tubule, generally have remarkably high transepithelial hydraulic conductances (Andreoli et al., 1978; Schafer et al., 1978). In contrast, the hydraulic conductances of both the rabbit mTALH (Rocha and Kokko, 1973) and the mouse mTALH (Table II; Hall, 1979; Hebert et al., 1980), with or without ADH, are on the order of 10–15 μm/sec, i.e., 100-fold smaller than that of proximal renal tubules and similar to the hydraulic conductances of unmodified lipid bilayer membranes (Hanai et al., 1966; Cass and Finkelstein, 1967). The low hydraulic conductance of the mTALH permits the mTALH to act as a diluting segment, and confers on the mTALH shunt pathway a hybrid nature: the paracellular pathway has a high ionic conductance (G_T, Table I) but a low water conductance (P_f, Table II).

As the route for passive ion permeation in the mTALH is paracellular in location, it is clear that the junctional complexes in this segment must have a remarkably high ion/water selectivity. This degree of ion/water selectivity might be achieved if ion and water permeation through these junctional complexes involved transport through aqueous channels sufficiently narrow to require single-file movement of both salt and water. Using 2-Å-radius gramicidin A channels, which exhibit single-file ion and water transport kinetics (Rosenberg and Finkelstein, 1978a,b), as a model for these putative junctional complex channels, one can calculate (Hebert et al., 1980) that 2.4×10^{10} gramicidin A-like channels/cm² luminal surface area would be required to account for the measured electrical resistance of 11 ohm cm² (cf. Table I). This channel number would result in a transepithelial hydraulic conductance of approximately 2.3 μm/sec for the mTALH, that is, a value similar to the hydraulic conductance of this segment. In other words, these enabling calculations are consistent with the view that the hybrid permeability characteristics of junctional complexes in mTALH (i.e., leaky to NaCl but tight to water) may be referable to ion and water permeation through rather narrow (e.g., 2-Å radius) channels.

The nature of the passive ionic permeation pathways in mouse mTALH junctional complexes may also account for another unusual passive ionic permeability property of this nephron segment. The value for the ionic conductance of the paracellular pathway computed from the tracer ionic permeability coefficients for Na^+ and Cl^- is about one-fifth that computed from the measured electrical resistance (cf. Table I), i.e., 19.6 mS versus 90.9 mS, respectively. If, as indicated above, junctional complexes in mouse mTALH contain narrow 2-Å-radius channels, it is possible that passive ion permeation through these junctional complexes involves single-file kinetics.

Under these circumstances, one can show (Hebert *et al.*, 1980), using a quantitative argument formally identical to one developed previously (Hebert and Andreoli, 1980) for the relation between osmotic water permeability coefficients (P_f) and THO diffusion coefficients (P_{Dw}), that: just as the P_f/P_{Dw} ratio in narrow 2-Å-radius channels exceeds unity by n_w, the number of water molecules in the channel, the ratio of the partial ionic conductance of the *i*th ion computed electrically, g_i^e, to that measured by tracer flux data, g_i^t, will be given by n_i, the number of current-carrying *i*th ions in the channel. For the more complex case where the single-file channels are not uniquely permselective for a given ion, the ratio of the total electrical conductance measured electrically to that measured using tracer flux data is given by $\Sigma n_i t_i$, where t_i is the transference number for the *i*th ion (Hebert *et al.*, 1980).

This latter disparity, when taken in conjunction with the combination of electrical leakiness and water-impermeability of the junctional complexes in this nephron segment, is at least internally consistent with the possibility that passive ion permeation through junctional complexes may occur through narrow channels, having effective radii in the range of 2 Å, which preclude side-by-side passage of water, Na^+, or Cl^- ions.

3. ADH-STIMULATED NaCl ABSORPTION IN MOUSE mTALH

Table II summarizes the effects of ADH on salt transport in the *in vitro* microperfused mouse mTALH. In accord with the differing effects of ADH on adenylate cyclase activation in mTALH and cTALH (cf. above), the addition of peritubular ADH, at concentrations seen during ordinary antidiuresis in mammalian species (10 μU/ml, 2.6×10^{-11} M, or 28 pg/ml), increased the spontaneous lumen-positive transepithelial voltage (V_e) twofold in the mTALH (Table II), but had no effect on V_e in the cTALH. Similar increases in V_e could be obtained in the mTALH by the addition of either 8-bromo-cAMP or dibutyryl-AMP to peritubular media (Hebert *et al.*, 1980). In the mTALH, this ADH-stimulated increase in V_e was accompanied by a four- to five-fold increase in the net rate of NaCl absorption (J_{Cl}^{net}, Table II).

These data indicate that the mTALH and cTALH are functionally heterogeneous with respect to the effect of ADH on net NaCl absorption; and that the stimulation of net NaCl absorption in the mouse mTALH by ADH is mediated via cAMP, in accord with the second-messenger hypothesis. Because the stimulating effect of ADH on NaCl absorption in the mTALH occurred at hormone concentrations found in mammalian species during ordinary antidiuresis, this hormone may play an important regulating role in the countercurrent multiplication process as originally suggested by Wirz (1957).

Because the ADH-mediated increase in V_e was paralleled by a comparable increase in J_{Cl}^{net}, the effects of ion substitutions and inhibitors on V_e were used to provide an index to the effects of these agents on net salt absorption. These data are shown in Figs. 1 and 2.

Figure 1 illustrates the effects of ion substitutions on the spontaneous transepithelial voltage: V_e was completely or nearly completely abolished by the symmetrical removal of either Na^+ or Cl^-, and by the removal of K^+ from the peritubular medium. However, removal of CO_2/HCO_3^- from the external media was without effect on V_e. Likewise, re-

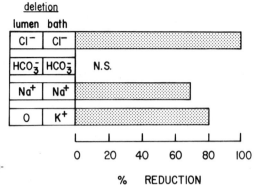

Figure 1. Effect of ion replacement on the ADH-dependent transepithelial voltage in mouse mTALH.

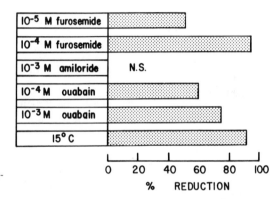

Figure 2. Effect of inhibitors on the ADH-dependent transepithelial voltage in mouse mTALH.

moval of CO_2/HCO_3^- had no effect on the measured net rate of NaCl absorption (Hebert *et al.*, 1980). These data therefore indicate a requirement for the presence of both Na^+ and Cl^- for the generation of V_e and for net salt absorption, and hence contrast sharply with classical electrogenic systems involving primary active transport processes (Ussing and Zerahn, 1951; Koefoed-Johnson and Ussing, 1958).

The effects of inhibitors on V_e, and by inference the net rates of NaCl absorption, in the mTALH are shown in Fig. 2. The addition of luminal furosemide inhibited V_e in a dose-dependent fashion. In contrast, luminal amiloride, at a doze 100-fold greater than that needed to inhibit either net Na^+ absorption in rabbit cortical collecting tubules (Stoner *et al.*, 1974) or basal and ADH-stimulated increases in net Na^+ transport in amphibian epithelia (Bentley, 1968; Cuthbert and Shum, 1974, 1975), was without effect. These inhibitor data indicate that Na^+ entry across apical plasma membranes of the mTALH does not involve amiloride-sensitive channels, but may be coupled to a furosemide-sensitive anion transport process. The fact that cooling reduces V_e to near zero implies that net NaCl absorption ultimately depends on an active transport process. And because either peritubular ouabain (Fig. 2) or peritubular K^+ removal (Fig. 1) abolished V_e, one may argue that the major energetic source for salt transport involves basolateral Na^+/K^+-ATPase, an enzyme unit that is present in remarkably high activity in this segment (Schmidt and Dubach, 1969; Katz *et al.*, 1979).

Finally, the observations (Hebert *et al.*, 1980) that ADH and furosemide had no effect on either the electrical (i.e., from salt dilution voltages) P_{Na}/P_{Cl} ratio, tracer P_{Na} and P_{Cl} values, or the tracer P_{Na}/P_{Cl} ratio suggest that ADH and/or furosemide affect conservative transcellular NaCl transport rather than the passive permeability properties of the mTALH.

3.1. A NaCl Cotransport Model for Apical Plasma Membranes

When taken together, the data cited above are consistent with the model for NaCl absorption in the mTALH illustrated in Fig. 3. In this model, NaCl absorption involves a furosemide-sensitive coupled NaCl entry process across apical plasma membranes in which the energy for conservative Cl^- entry from lumen to cells is provided by the electrochemical Na^+ gradient between extracellular fluid and cytoplasm; in other words, Cl^- entry into mTALH cells is a secondary active transport process. These data, however, do not provide explicit information regarding the stoichiometry for the putative Na^+/Cl^- cotransport process, and therefore whether this step is electrogenic or electrically silent. Basolateral Na^+/K^+-ATPase may provide both the route for Na^+ exit from cells to peritubular media and for maintaining the electrochemical Na^+ gradient across plasma membranes. The route for Cl^- exit from mTALH cells is unknown, but presumably passive; and given the ion substitution data (cf. Fig. 1), Cl^- exit from cells appears to be independent of HCO_3^-/CO_2.

The terms aNa^+ and bCl^- in Fig. 3 refer to the stoichiometry for NaCl entry across apical plasma membranes. Evidently, if the aNa^+/bCl^- ratio is unity, the apical salt entry step is electroneutral. In this case, the lumen-positive transepithelial voltage observed in these mTALH might depend on NaCl accumulation in paracellular fluid, provided that the postjunctional complex diffusion resistance of the paracellular pathway in mouse mTALH is sufficiently great to permit NaCl hypertonicity in paracellular fluid. Under these circumstances, the observed V_e in the mTALH might be the sum of two diffusion voltages in series: one between paracellular solutions and luminal fluid involving Na^+-permselective (Table I) junctional complexes; and one between paracellular solution (assuming that the Na^+/Cl^- permselectivity ratio in paracellular fluids is less than unity, as it is in free solution) and peritubular media.

Alternatively, Gregor and Frömter (1980) have argued that, in the isolated rabbit cTALH, apical NaCl entry involves a cotransport step in which the ratio aNa^+/bCl^- (cf. Fig. 3) is less than unity, and that this latter stoichiometry accounts entirely for the lumen-positive spontaneous transepithelial voltage observed in the rabbit cTALH. Thus, according to these workers, secondary active transport of Cl^- across apical plasma membranes results

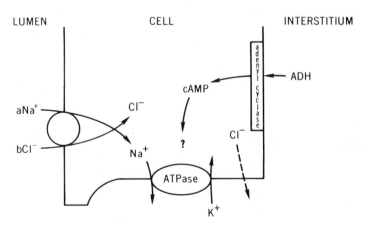

Figure 3. Model for Na^+Cl^- transport in mTALH and modulation of the process by ADH and/or cAMP.

in an entirely electrogenic driving force for paracellular Na^+ absorption. It should be recognized that an implicit requirement of the Gregor–Frömter model is that the diffusion resistance of the paracellular pathway in the rabbit cTALH is sufficiently low to permit the NaCl concentrations in paracellular solutions to be near diffusion equilibruim with peritubular media during NaCl absorption in that nephron segment.

Thus, a number of uncertainties exist with respect to the origin of V_e in the mouse mTALH. In the case of a model that invokes paracellular solution hypertonicity as the origin of the transepithelial V_e, accumulation of salts other than NaCl in paracellular solutions could, in principle, result in salt dilution voltages across junctional complexes and between paracellular solutions and peritubular media. Conversely, an entirely electrogenic voltage might arise if the apical NaCl entry step were entirely electroneutral and the parcellular solutions were near diffusion equilibrium with peritubular media, but apical membranes contained a conductive pathway for secretion of another cation besides Na^+; or a conductive pathway for absorption of another anion besides Cl^-. We consider these alternatives unlikely, given the virtually complete dependence of V_e on NaCl (Fig. 1); the virtually complete independence of V_e and NaCl absorption on HCO_3^- (cf. above, Fig. 1); and entirely comparable effects of furosemide on net NaCl absorption and on V_e (Fig. 2.) Thus, when taken together, these data indicate that V_e was determined primarily by the system modulating NaCl transport.

The present data do not permit us to conclude whether, for the mouse mTALH, the NaCl-dependent V_e depends primarily on electrogenic events, i.e., on an aNa^+/bCl^- apical membrane entry ratio less than unity in combination with near-diffusion equilibrium between NaCl concentrations in paracellular solutions and peritubular media; primarily on salt dilution voltages, i.e., on NaCl accumulation in paracellular solutions associated with an aNa^+/bCl^- unity ratio of the NaCl apical step shown in Fig. 3; or on a hydrid model in which an aNa^+/bCl^- apical entry ratio less than unity, in combination with NaCl hypertonicity in paracellular fluids, results in the observed spontaneous V_e.

Finally, the present results do not allow us to specify the site, or sites, where ADH and cAMP act to enhance conservative transcellular NaCl absorption. Because either ADH or cAMP enhances V_e within less than 2 min (Hebert *et al.*, 1980), one may conclude that the hormone, acting via the second messenger cAMP, enhances NaCl absorption by activating transport units already present in renal tubular cells, rather than by stimulating *de novo* protein synthesis. However, whether ADH affects primarily apical membrane NaCl entry, basolateral Cl^- exit, basolateral Na^+/K^+-ATPase, or some other process, is unknown at present.

3.2. Modulation of the ADH Effect by Peritubular Osmolality

In symmetrical, isotonic external solutions, ADH and cAMP analogs increase the rate of net NaCl absorption by increasing the rate of conservative NaCl transport from lumen through cells to intercellular spaces (cf. above). In other words, when the mTALH is exposed to symmetrical external solutions, ADH and/or cAMP increase the diluting power of that nephron segment. However, *in vivo* micropuncture studies in a number of different mammalian species have shown that the osmolalities of early distal tubular fluids, which obviously depend on the diluting power of the mTALH, are comparable during water diuresis and antidiuresis (Clapp and Robinson, 1966; Gottschalk, 1961; Wirz, 1957). Furthermore, it is generally believed that steady-state external Na^+ balance is not altered significantly by the presence or absence of ADH (Peters and Roch-Ramel, 1970). How can these two findings seemingly contradictory to the fact that ADH enhances conservative NaCl absorption in the mTALH, be rationalized?

One mechanism for maintaining a constant diluting power for the *in vivo* mTALH might depend on the fact that, during antidiuresis, interstitial NaCl concentrations in the mammalian renal medulla rise as a consequence of the ADH-induced increase in conservative NaCl transport in the mTALH. Evidently, an increase in peritubular NaCl concentration would tend to increase the passive backleak (J_{NaCl}^{leak} of NaCl from peritubular media to lumen. Thus, in the *in vivo* circumstance, the maintenance of a relatively constant mTALH diluting power during water diuretic and antidiuretic states might depend in part on NaCl accumulation in the medullary interstitium during antidiuresis, and consequently on a greater J_{NaCl}^{leak} during antidiuresis with respect to water diuresis.

There may also exist another mechanism for modulating the effect of ADH on net NaCl absorption by the mTALH. As noted by Dousa (1972), increases in peritubular osmolality inhibit the activation of adenylate cyclase by ADH in slices of bovine medulla. Consequently, we (Hebert *et al.*, 1981) studied the effects of increasing bath osmolality on the passive and active transport characteristics of these tubules. The results, shown in Table III, indicate that increases in bath osmolality with urea have no effect on either absolute or relative ionic permeability characteristics of the mTALH, but reduce strikingly both ADH-stimulated and cAMP-stimulated J_{NaCl}^{net} and V_e in these tubules.

In other studies not shown in Table III, we (Hebert *et al.*, 1981) found that: (1) The effect of increasing peritubular osmolality on net salt absorption was not referable to transepithelial urea gradients, but rather to the increase in peritubular osmolality per se, for symmetrical increases in the osmolality of luminal and peritubular media, using urea, also reduced the spontaneous V_e in mTALH. (2) The inhibition of V_e produced by increasing peritubular osmolality occurred when either urea, NaCl, or mannitol was used to raise osmolality; however, cell volume remained the same with hypertonic bath urea but was reduced with hypertonic bath NaCl or mannitol. (3) Supramaximal concentrations of ADH or urea did not reverse the inhibitory effect of peritubular urea on V_e.

When taken together, these observations are consistent with a negative feedback system in which ADH-dependent increases in conservative transcellular NaCl absorption are modulated downward by increases in peritubular (i.e., interstitial) osmolality. That is, increases in peritubular osmolality with urea modulate NaCl transport downward in a manner

Table III. Effects of Hypertonic Bath Urea on Net Cl^- Absorption in mTALH[a]

	Bath urea		
	0		600 mM
Tracer P_{Na} (μm/sec)	0.24 ± 0.02	(NS)	0.22 ± 0.02
Electrical P_{Na}/P_{Cl}	1.9 ± 0.2	(NS)	1.7 ± 0.2
J_{Cl}^{net} (pEq/sec/cm^2)	$12,132 \pm 1438$	($p < 0.02$)	2027 ± 571
V_e (mV)	8.7 ± 0.7	($p < 0.05$)	3.0 ± 0.5

[a] Data are from isolated segments of mouse mTALH perfused and bathed with symmetrical phosphate-buffered (pH 7.40) 145 mM NaCl solution gassed with 100% O_2 at 37°C. Urea, where indicated, was added to peritubular media at a concentration of 600 mM. All bathing solutions contained ADH at 10 μU/ml. Data are from Hebert *et al.* (1981).

that is independent of cell volume and noncompetitive with respect to ADH and to cAMP. Moreover, the reduction in J_{NaCl}^{net} produced by increasing peritubular NaCl osmolality may be referable both to a reduction of conservative NaCl transport and to an increase in the backleak of NaCl into luminal fluids. This construct may thus explain partially the fact that, in general, *in vivo* the mTALH is able to maintain a constant diluting power during either water diuresis or antidiuresis; and the fact that ADH has no significant effect on steady-state external salt balance.

4. GENERAL CONCLUSIONS

In summary, these data are compatible with three general sets of conclusions. First, ADH, operating via cAMP, increases strikingly the NaCl pump rate by the medullary thick ascending limb. This phenomenon, operating *in vivo*, can effect an increase in outer medullary NaCl concentration, and thus would enhance the concentrating power of the nephron. An ADH-mediated increase in the steady-state outer medullary osmolality might result in both an increase in NaCl backleak from interstitium to lumen and a noncompetitive down-modulation of the ADH-stimulated NaCl pump rate in the medullary thick ascending limb. This kind of a negative feedback system may account for the maintenance of medullary hypertonicity and the diluting power of the medullary thick ascending limb without affecting external salt balance.

Second, the mode of Na^+/Cl^- absorption in the mTALH involves a secondary active transport process that is dependent on the transcellular Na^+ gradient maintained by basolateral Na^+/K^+-ATPase. Finally, the unusual combination of electrical leakiness and water impermeability in the mTALH, as well as the disparity in electrical vs. tracer resistance values, may be due to the nature of the passive permeation pathways in mTALH junctional complexes: the latter may contain approximately 2-Å-radius channels through which water, Na^+, and Cl^- pass by single-file diffusion.

REFERENCES

Andreoli, T. E., Schafer, J. A., and Troutman, S. L. (1978). *Kidney Int.* **14,** 263–269.

Barry, P. H., Diamond, J. M., and Wright, E. M. (1971). *J. Membr. Biol.* **4,** 358–394.

Bentley, P. J. (1968). *J. Physiol. (London)* **195,** 317–330.

Bray, G. A. (1960). *Am. J. Physiol.* **199,** 915–918.

Burg, M. B., and Green, N. (1973). *Am. J. Physiol.* **224,** 659–668.

Cass, A., and Finkelstein, A. (1967). *J. Gen. Physiol.* **50,** 1765–1784.

Chabardés, D., Imbert-Teboul, M., Gagnan-Brunette, M., and Morel, F. (1977). *Curr. Probl. Clin. Biochem.* **8,** 447–454.

Clapp, J. R., and Robinson, R. R. (1966). *J. Clin. Invest.* **45,** 1847–1853.

Cuthbert, A. W., and Shum, W. K. (1974). *Mol. Pharmacol.* **10,** 880–891.

Cuthbert, A. W., and Shum, W. K. (1975). *Proc. R. Soc. London Ser. B* **189,** 543–575.

Dousa, T. P. (1972). *Am. J. Physiol.* **222,** 657–662.

Edwards, R. M., Jackson, B. A., and Dousa, T. P. (1980). *Am. J. Physiol.: Renal Fluid Electrolyte Physiol.* **7,** F269–F278.

Gottschalk, C. W. (1961). *Physiologist* **4,** 35–55.

Gregor, R., and Frömter, G. (1980). *Abstracts, 28th International Congress of Psychology, Budapest*, p. 445.

Hall, D. A. (1979). *Clin. Res.* **27,** 416A.

Hall, D. A., and Varney, D. (1979). *Kidney Int.* **16,** 818.

Hall, D. A., and Varney, D. M. (1980). *J. Clin. Invest.* **66,** 792–802.

Hanai, T., Haydon, D. A., and Redwood, W. R. (1966). *Ann. N.Y. Acad. Sci.* **137,** 731–739.

Hebert, S. C., and Andreoli, T. E. (1980). *Am. J. Physiol.: Renal Fluid Electrolyte Physiol.* **7,** F470–F480.
Hebert, S. C., Culpepper, R. M., and Andreoli, T. E. (1981). *Kidney Int.* **19,** 244.
Hebert, S. C., Culpepper, R. M., Misanko, B. S., and Andreoli, T. E. (1980). *Clin. Res.* **28,** 533A.
Katz, A. I., Doucet, A., and Morel, F. (1979). *Am. J. Physiol.* **6,** F114–F120.
Kessler, R. H. (1960). *Am. J. Physiol.* **199,** 1215–1218.
Koefoed-Johnson, V., and Ussing, H. H. (1958). *Acta Physiol. Scand.* **42,** 298–308.
Morel, F. (1981). *Am. J. Physiol.: Renal Fluid Electrolyte Physiol.* **240,** F159.
Perlmutt, J. H. (1962). *Am. J. Physiol.* **202,** 1098–1104.
Peters, G., and Roch-Ramel, F. (1970). In *Pharmacology of the Endocrine System and Related Drugs: The Neurophypophysis* (H. Heller and B. T. Pickering, eds.), pp. 229–278, Pergamon Press, New York.
Rocha, A. S., and Kokko, J. P. (1973). *J. Clin. Invest.* **52,** 612–623.
Rosenberg, P. A., and Finkelstein, A. (1978a). *J. Gen. Physiol.* **72,** 327–340.
Rosenberg, P. A., and Finkelstein, A. (1978b). *J. Gen. Physiol.* **72,** 341–350.
Ruiz-Guiñazú, A., Arrizurieta, E. E., and Yelinek, L. (1964). *Am. J. Physiol.* **206,** 725–730.
Sasaki, S., and Imai, M. (1980). *Pfluegers Arch.* **383,** 215–221.
Schafer, J. A., Patlak, C. S., Troutman, S. L., and Andreoli, T. E. (1978). *Am. J. Physiol.: Renal Fluid Electrolyte Physiol.* **3,** F340–F348.
Schmidt, U., and Dubach, U. C. (1969). *Pfluegers Arch.* **306,** 219–226.
Schultz, S. G. (1977). *Yale J. Biol. Med.* **50,** 99–113.
Stoner, L. C., Burg, M. B., and Orloff, J. (1974). *Am. J. Physiol.* **227,** 453–459.
Ussing, H. H., and Windhager, E. E. (1964). *Acta Physiol. Scand.* **61,** 484–504.
Ussing, H. H., and Zerahn, K. (1951). *Acta Physiol. Scand.* **23,** 110–127.
Wirz, H. (1957). In *The Neurophypophysis. Proceedings of the 8th Symposium of the Colston Research Society* (H. Heller, ed.), pp. 157–182, Academic Press, New York.

116

Na⁺ Gradient-Dependent Transport Systems in Renal Proximal Tubule Brush Border Membrane Vesicles

Bertram Sacktor

Recent studies of the uptake of solutes by renal proximal tubule brush border (microvillus, luminal) membrane vesicles have contributed significantly to our understanding of the mechanisms by which ions and metabolites are removed from the glomerular filtrate and translocated across the epithelium to the interstitial fluid. The isolated membrane vesicle represents an ideal experimental model to examine transport dissociated from the complications of cell metabolism. Additionally, because the composition of the intravesicular and extravesicular media can be manipulated at will and membrane potentials can be induced predictably with specific ionophore-generated diffusion potentials, the ionic and electrical forces that drive the transport can be better defined. Lastly, uptake of a solute by the isolated luminal segment of the tubular cell plasma membrane can be studied independently of movement of the same chemical species across the isolated basolateral segment of the plasma membrane; thus, the mechanism and energetics of the vectorial transepithelial transport process in the polar epithelial cell can be described.

Classical physiological observations of renal reabsorption systems indicate that the transport of many solutes is highly dependent on the presence of Na^+ in the tubular fluid (Ullrich, 1979). Early studies of intestinal sugar uptake led to the formulation of the Na^+ gradient transport hypothesis, which postulates that the nonelectrolyte is symported with Na^+, and the Na^+ electrochemical gradient across the luminal membrane provides the energy for the uphill transport of the sugar (Crane, 1977). The Na^+ gradient theory has since been found to be applicable in explaining the concentrative uptakes by the kidney of many substances, including sugars, amino acids, ions, and various organic compounds. In the intact tissue, the ouabain-sensitive active extrusion of Na^+ across the basolateral membrane of the tubule presumably contributes to the Na^+ chemical concentration difference and to

Bertram Sacktor • Laboratory of Molecular Aging, Gerontology Research Center, National Institute on Aging, National Institutes of Health, Baltimore City Hospitals, Baltimore, Maryland 21224.

the electrical potential required for uphill transport. In the membrane vesicle, the imposed Na^+ electrochemical gradient (extravesicular $>$ intravesicular) provides the driving force to effect the movement of the cotransported compound into the membrane vesicle against its concentration gradient (Kinne, 1976; Sacktor, 1977).

To date, the various solutes that have been reported to be cotransported with Na^+ are listed in Table I. It is important to emphasize that the Na^+ gradient-dependent uptake systems in the isolated renal brush border membrane vesicle possess the specificities characteristic of the transport systems in more physiologically intact preparations. For example, in the vesicle the D-glucose analogs, D-galactose and α-methyl-D-glucoside, inhibit the Na^+ electrochemical gradient-dependent uptake of D-glucose and support accelerated exchange diffusion. These findings indicate that these sugars share the same carrier. Different sugars, on the other hand, including L-glucose, D-fructose, D-mannose, and 2-deoxy-D-glucose, do

Table I. Na^+ Gradient-Dependent Transport Systems in Renal Proximal Tubule Brush Border Membrane Vesicles

Transported solute	Reference
Sugars	
D-Glucose	Aronson and Sacktor (1975)
D-Galactose	Sacktor et al. (1974)
Myoinositol	Hammerman et al. (1980)
Amino acids	
Neutral	
L-Alanine	Fass et al. (1977)
L-Phenylalanine	Evers et al. (1976)
L-Glutamine	Weiss et al. (1978)
Acidic	
L-Glutamate	Schneider et al. (1980)
L-Aspartate	Schneider et al. (1980)
Basic	
L-Arginine	Hilden and Sacktor (1981)
L-Ornithine	Leopolder et al. (1980)
Imino acids	
L-Proline	Hammerman and Sacktor (1977)
β-Amino acids	
β-Alanine	Hammerman and Sacktor (1978)
Taurine	Rozen et al. (1979)
Cystine	Segal et al. (1977)
Glycine (L-alanylglycine)	McNamara et al. (1976); Welch and Campbell (1980)
Ions	
Phosphate	Hoffmann et al. (1976)
Sulfate	Lücke et al. (1979)
Hydroxyl (Na^+/H^+ antiport)	Murer et al. (1976)
Organic metabolites	
L-Lactate	Barac-Nieto et al. (1980)
Ketone bodies	
Acetoacetate	Garcia et al. (1980)
3-Hydroxybutyrate	Garcia et al. (1980)
Tricarboxylate cycle intermediates	
Succinate	Wright et al. (1980)
α-Ketoglutarate	Kippen et al. (1979)
Citrate	Kippen et al. (1979)

not inhibit and, therefore, presumably do not share the D-glucose carrier (Aronson and Sacktor, 1975). These findings are in accord with results from studies with renal cortical slices (Kleinzeller, 1970) and with intact animals, *in vivo* (Silverman *et al.*, 1970). D-Glucose is also an inhibitor of the Na^+ electrochemical gradient-dependent uptake of myoinositol, but myoinositol does not significantly inhibit the uptake of D-glucose. The D-isomer, but not L-glucose, elicits the countertransport of myoinositol. This suggests that D-glucose inhibits myoinositol uptake by sharing the myoinositol carrier and by dissipating the membrane potential (Hammerman *et al.*, 1980). The inhibition of myoinositol transport across the brush border membrane by D-glucose is of additional significance for it explains the clinical observation that glycosuria produces inosituria in patient with diabetes mellitus.

Multiple systems for the transport of amino acids by brush border membrane vesicles have been identified (Sacktor, 1978). The Na^+ gradient-dependent systems are concentrative, stereospecific, saturable, and highly specific for the different classes of amino acids. Uptake of the neutral amino acid L-alanine is competitively inhibited by other neutral amino acids, including L-valine, L-leucine, L-serine, L-phenylalanine, L-histidine, and glycine, which presumably share the same carrier (Fass *et al.*, 1977). The Na^+ gradient-dependent acidic amino acid transporter for L-glutamate is shared by L- and D-aspartate and L-cysteate, but not by D-glutamate nor by neutral, basic, and imino acids (Schneider *et al.*, 1980). L-Proline uptake is strongly inhibited only by L-hydroxyproline (Hammerman and Sacktor, 1977). β-Alanine and taurine share a common carrier (Hammerman and Sacktor, 1978). The Na^+ gradient-dependent L-arginine transport system is shared by specific basic amino acids and, in part, by L-cystine, but not by other classes of amino acids (Hilden and Sacktor, 1981). L-Arginine also inhibits the uptake of L-cystine (Segal *et al.*, 1977). The observation that L-cystine is not as effective an inhibitor of L-arginine uptake as is L-lysine and L-ornithine may be due to the presence of multiple (high and low affinity) transport systems for both L-cystine and L-arginine and that these are not equally shared (Foreman *et al.*, 1980). The uptake of glycine is strongly inhibited by both neutral amino acids and imino acids, but is insensitive to the copresence of acidic and basic amino acids (Sacktor, 1977). Glycine, presented to the membrane in the form of the dipeptide, L-alanylglycine, is also taken up by the vesicle following the action of a membrane-bound dipeptidase (Welch and Campbell, 1980). In summary, these findings suggest that the ability of the proximal tubule to transport amino acids selectively may be ascribed to an intrinsic property of the brush border membrane. It may be hypothesized additionally that clinically significant metabolic disorders such as specific aminoacidurias, Hartnup's disease, and the Franconi syndrome may have their etiology in a defect in the Na^+ gradient-dependent amino acid transport systems in the brush border membrane of the renal proximal tubule.

The brush border membrane vesicle has been used to study the transport of various ions. The uptakes of phosphate and sulfate are Na^+ gradient-dependent, and independent of each other. In the animal, phosphate homeostasis and reabsorption in the kidney are regulated by multiple physiological–pathological factors, including parathyroid hormone, diet, metabolites of vitamin D_3, acid–base status, and genetic mutants. The actions of these effectors are manifest, at least in part, by alterations in the Na^+ gradient-dependent transport process in the microvillus membrane (Stoll *et al.*, 1979; Kempson and Dousa, 1979; Cheng and Sacktor, 1981; Tenenhouse and Scriver, 1978). The transport of several organic anions, e.g., urate, *p*-aminohippurate, across the brush border membrane, however, is not dependent on an Na^+ gradient (Blomstedt and Aronson, 1980). To be noted, the well-characterized Na^+–H^+ exchange reaction in the brush border membrane is listed in Table I, for it cannot be distinguished from an Na^+/OH^- cotransport mechanism.

Intermediates of the tricarboxylate cycle are taken up into the microvillus membrane

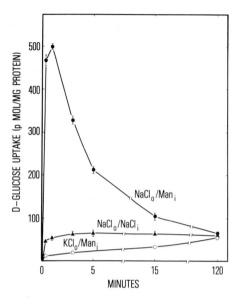

Figure 1. General properties of the Na^+ electrochemical gradient-dependent uptake of D-glucose by renal brush border membrane vesicles. The intravesicular medium is 300 mM mannitol containing 5 mM Hepes–Tris, pH 7.5 (Man$_i$), or 100 mM NaCl in 100 mM buffered mannitol (NaCl$_i$). The extravesicular medium is 100 mM NaCl (NaCl$_o$) or 100 mM KCl (KCl$_o$) in 100 mM buffered mannitol. The D-glucose concentration is 50 μM.

vesicle by an Na^+ gradient-dependent process (Wright et al., 1980). The dicarboxylate, succinate, and the tricarboxylate, citrate, are transported by a common mechanism. Specificity studies suggest that the system is relatively specific for four-carbon, terminal dicarboxylates in the *trans* configuration. The system is comparatively insensitive to monocarboxylates. It is claimed that pyruvate does not compete with the Na^+ gradient-dependent uptake of L-lactate, but that D-lactate inhibits the transport of the L isomer (Barac-Nieto *et al.*, 1980).

Figure 1 illustrates with the uptake of D-glucose several properties of the Na^+ electrochemical gradient-dependent transport of a solute by renal brush border membrane vesicles. When the intravesicular medium is isotonic buffered mannitol and the extravesicular medium contains 100 mM KCl replacing some buffered mannitol, the initial (approximated) rate of uptake is extremely low and this slow rate of uptake continues for 1 to 2 hr. In the presence of a 100 mM NaCl gradient (extravesicular > intracellular), the uptake is markedly stimulated. The initial rate is increased about 50-fold. Accumulation of the sugar is maximal in about 1 min. Thereafter, the amount of D-glucose in the vesicles decreases, indicating efflux. The final level of uptake of D-glucose in the presence and absence of the Na^+ electrochemical gradient is the same, approximately 65 pmoles/mg membrane protein, suggesting that equilibrium has been established and the membrane vesicles in this preparation have an average intravesicular volume of 1.3 μl/mg protein. At the peak of the "overshoot," the accumulation of D-glucose reaches 8 to 10 times the final equilibrium value. This finding demonstrates that the imposition of a large extravesicular > intravesicular Na^+ electrochemical gradient provides the driving force to effect the transient movement of the sugar into the membrane vesicle against its concentration gradient. This effect is specific for Na^+, although Li^+ is a poor substitute (Aronson and Sacktor, 1975). When the membrane vesicles are preloaded with NaCl so that the extravesicular and intravesicular concentrations of Na^+ are equal, the overshoot is abolished. This observation indicates that it is not the concentration of Na^+ per se, but the Na^+ electrochemical gradient that is crucial in energizing uphill transport into the membrane vesicle. This view is supported by findings that gramicidin or other effectors that increase the permeability of the membrane to Na^+ and dissipate the Na^+ electrochemical gradient, inhibit uphill transport (Beck and Sacktor, 1975).

In the Na^+/D-glucose cotransport system, not only should the Na^+ electrochemical gradient enhance the uptake of the sugar, but D-glucose should stimulate the flux of Na^+. Evidence for this sugar-dependent transport of Na^+ is shown in Fig. 2. In this experiment, extravesicular and intravesicular media are equilibrated with $^{22}Na^+$; then, either 50 mM D- or L-glucose is added to the reaction medium, and the intravesicular content of $^{22}Na^+$ is determined at various times. The addition of D-glucose, but not L-glucose, elicits the accumulation of $^{22}Na^+$ in the membrane vesicles, reaching almost a 100% increase in $^{22}Na^+$ uptake in 3 min of incubation. Thereafter, the $^{22}Na^+$ content gradually decreases until the intravesicular $^{22}Na^+$ concentration, in the presence of D- or L-glucose, is essentially the same. The properties of this D-glucose gradient-dependent transport of Na^+ in membrane vesicles are the same as those for the Na^+ electrochemical gradient-dependent uptake of D-glucose (Hilden and Sacktor, 1979).

Because in the intact animal the reabsorption of the filtered sugar (amino acids, etc.) is obligatorily coupled to the reabsorption of Na^+, it follows that these cotransport systems may be responsible for a significant fraction of the net volume uptake in the proximal tubule. Assuming that 1 (or 2) Eq of Na^+ is cotransported with each mole of D-glucose (and amino acid, etc.) reabsorbed and the cation is pumped out of the cell via the Na^+/K^+-ATPase in the basolateral membrane, then the reabsorption of each mole of metabolite results in the uptake of 1 or more equivalent of Na^+ and 1 or more equivalent of accompanying anion. Considering that transport in the proximal tubule is essentially isoosmotic, it is estimated that the cotransport mechanism may account for 10 to 20% of the fluid reabsorbed (Hilden and Sacktor, 1979).

An important aspect of the coupling mechanism is whether the cotransport system is electrogenic or electroneutral. If the process is electrogenic, then the Na^+ gradient-dependent uptake results in the net transfer of charge across the membrane and, therefore, it is affected by the membrane potential. In contrast, if the system is electroneutral, the cotransport across the microvillus membrane is not associated with a net transfer of electrical charge and, hence, it is largely independent of the membrane potential. Both types of Na^+ cotransport mechanisms are found in the renal brush border membrane (Sacktor, 1980).

The effects of the membrane potential on the Na^+ gradient-dependent uptake of phosphate and D-glucose are compared in Fig. 3. In this experiment the intravesicular pH is 5.5 and the extravesicular pH is 7.5, and H^+ diffusion potentials are induced by the mitochon-

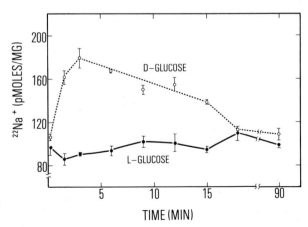

Figure 2. The D-glucose gradient-dependent uptake of $^{22}Na^+$ in renal brush border membrane vesicles. [From Hilden and Sacktor, 1979.]

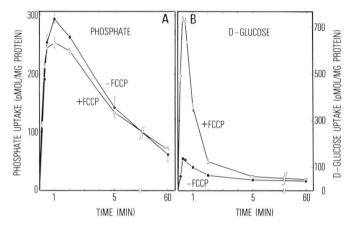

Figure 3. A comparison of the effects of an FCCP-generated H^+ diffusion potential (interior negative) on the Na^+ gradient-dependent uptakes of phosphate and D-glucose. [From Cheng and Sacktor, 1981.]

drial uncoupler, carbonyl cyanide p-trifluoromethoxyphenylhydrazone (FCCP). The uncoupler-induced H^+ diffusion potential (inside negative) has no effect on phosphate uptake (Cheng and Sacktor, 1981). In contrast, the H^+ ionophore greatly enhances the overshoot when D-glucose is the transported solute. FCCP presumably induces the efflux of H^+ down its electrochemical gradient with concomitant generation of a membrane potential (interior negative), and in the case of D-glucose uptake, the development of this potential accelerates the influx into the vesicles of the positively charged Na^+ coupled to the transport of the uncharged sugar. Identical results are obtained when a membrane potential (interior negative) is generated by a K^+ diffusion potential, in the presence of valinomycin, or by using Na^+ salts comprised of anions of different modes of permeability (Beck and Sacktor, 1975; Cheng and Sacktor, 1981). Moreover, when a K^+ gradient is established (extravesicular $>$ intravesicular) and valinomycin induces the inward movement of the cation resulting in the generation of a membrane potential (interior positive), the uptake of phosphate is again unaffected (Cheng and Sacktor, 1981). In contrast, the uptake of D-glucose is markedly inhibited (Schneider et al., 1980). These findings show that the Na^+ gradient-dependent uptake of D-glucose is affected by a membrane potential in a predictable fashion and demonstrate that the $Na^+/$D-glucose cotransport system is electrogenic. On the other hand, $Na^+/$phosphate cotransport is unaffected by the membrane potential and thus is electroneutral, i.e., the cotransport across the microvillus membrane is not associated with the net transfer of electrical charge.

In the case of the $Na^+/$D-glucose cotransport system, we have shown that the Na^+ electrochemical gradient across the brush border membrane provides the energy for the concentrative transport of the sugar. Furthermore, it is known that the Na^+ electrochemical gradient, $\Delta\bar{\mu}_{Na^+}$, is comprised of two components: an electrical component, $\Delta\Psi$, and a chemical component, $\Delta\mu_{[Na^+]}$, a function of the transmembrane activity ratio of Na^+. Each component of the electrochemical potential, $\Delta\Psi$ (inside negative) or the $\Delta\mu_{[Na^+]}$, when assayed independently, supports the concentrative uptake of the sugar (Beck and Sacktor, 1978a). That the same is true for other Na^+ gradient-dependent cotransport systems is illustrated in Fig. 4 for the uptake of L-proline (Sacktor, Thomas, and Beck, unpublished observations). Curve A shows the time course of L-proline uptake driven by both an Na^+ concentration gradient, $\Delta\mu_{[Na^+]}>0$ (10/1, medium $>$ vesicle), and a K^+ diffusion potential,

$\Delta\Psi > 0$ (10/1, vesicle > medium), in the presence of valinomycin. Relative to the mannitol control (curve E), in which Na$^+$ and K$^+$ were deleted from the intravesicular and extravesicular media, the initial rate of uptake is increased markedly and at the peak of the overshoot (1 min), L-proline accumulates to a value about 7 times the equilibrium value (60 min). In contrast, the uptake of the imino acid never exceeds the equilibrium value when there is no Na$^+$ electrochemical potential ($\Delta\bar{\mu}_{Na^+} = 0$) (curve D) although Na$^+$ and K$^+$ are present but $[Na^+]_i = [Na^+]_o$ and $[K^+]_i = [K^+]_o$ and the ionophores valinomycin and nigericin are added to ensure the absence of ionic gradients. In the presence of only an Na$^+$ chemical gradient, $\Delta\mu_{[Na^+]} > 0$ (10/1, medium > vesicle), but in the absence of a membrane potential, $\Delta\Psi = 0$ ($[K^+]_i = [K^+]_o$) (curve B), the initial rate of uptake is much enhanced relative to rate in the absence of an Na$^+$ gradient (curve D) and a significant overshoot, about fourfold the equilibrium value, is found. This demonstrates that an Na$^+$ chemical gradient independently can support the uphill transport of the imino acid. In the presence of only a membrane potential, $\Delta\Psi > 0$ (10/1, vesicle > medium), but in the absence of an Na$^+$ chemical gradient, $\Delta\mu_{[Na^+]} = 0$ ($[Na^+]_i = [Na^+]_o$) (curve C), both a stimulation in initial rate and an overshoot are seen. This demonstrates that the membrane potential independently can provide the energy to accumulate L-proline against a concentration gradient. It is important to point out, however, that there is no effect of a membrane potential on L-proline uptake in the absence of Na$^+$. For both L-proline and D-glucose, uptakes are a direct function of the $\Delta\mu_{[Na^+]}$ and $\Delta\Psi$ driving forces, expressed as $\log[Na^+]_o/[Na^+]_i$ and $\log[K^+]_i/[K^+]_o$, respectively (Beck and Sacktor, 1978a; Sacktor, Thomas, and Beck, unpublished observations).

The findings that Na$^+$ is symported with nonelectrolytes and the brush border membrane potential is negative on the inside (Beck and Sacktor, 1975) have raised the question whether the electrogenic cotransport of Na$^+$ and D-glucose will depolarize the membrane.

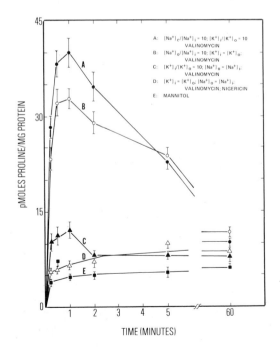

Figure 4. The effects of the Na$^+$ electrochemical gradient and its component driving forces, the Na$^+$ chemical gradient and the membrane potential, on the uptake of L-proline.

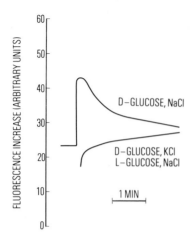

Figure 5. Demonstration using the fluorescent probe $DiS-C_3-(5)$ that Na^+/D-glucose cotransport depolarizes the membrane. [Redrawn from Beck and Sacktor, 1978b.]

Because electrophysiological techniques to monitor potentials in isolated microvillar vesicles are precluded, optical probes, such as 3,3'-dipropylthiodicarbocyanine iodide [Di5-C_3-(5)], have been employed to assess changes in potential (Beck and Sacktor, 1978b). As shown in Fig. 5, when D-glucose is added to a suspension of membrane vesicles equilibrated in an Na^+-containing medium and with dye, there is a rapid transient increase of fluorescence indicative of the vesicle interior becoming more positive. This is explainable by the entrance into the vesicles of Na^+ being cotransported with the sugar, depolarizing the membrane. The increase in fluorescence is transient, presumably reflecting the subsequent movement of Cl^- into the membrane vesicle to compensate for charge or the efflux of the accumulated Na^+, or both. This response to D-glucose is stereospecific, for addition of L-glucose, whose uptake is Na^+-independent, does not alter the fluorometric trace. The enhancement in fluorescence is absolutely dependent on the presence of Na^+, and is inhibited by phlorizin and by ionophores that dissipate ionic gradients (Beck and Sacktor, 1978b). In contrast, when L-glutamate, an amino acid that is cotransported with Na^+ by an electroneutral mechanism, is added to a suspension of membrane vesicles, there is no specific increase in fluorescence. This indicates that the Na^+ cotransport of L-glutamate does not depolarize the membrane potential (Schneider et al., 1980).

The various Na^+ gradient cotransport systems also differ with respect to the effect of other cation gradients on the uptake processes. For example, a K^+ gradient (intravesicular > extravesicular) specifically enhances the Na^+-dependent uptake of L-glutamate (Sacktor and Schneider, 1980; Schneider and Sacktor, 1980). Figure 6 illustrates this effect by comparing the Na^+ gradient-dependent transport of L-glutamate with those of L-proline and D-glucose when the membrane vesicles are preloaded with either KC1 or mannitol. L-Glutamate transport is increased severalfold when the vesicles contain intravesicular K^+. In contrast, the uptakes of L-proline and D-glucose are relatively decreased in the KC1-preloaded vesicles, presumably because a Cl^- electrochemical gradient, which contributes to the driving force in the electrogenic Na^+/D-glucose and Na^+/L-proline cotransports, is lacking. The effect of the K^+ gradient in enhancing the uptake of L-glutamate has an absolute requirement for Na^+. Importantly, in the presence of Na^+ but in the absence of an Na^+ gradient, i.e., $[Na^+]_o = [Na^+]_i$, the K^+ gradient can drive the concentrative uptake of the acidic amino acid. The internal K^+ gradient increases the maximal flux of the carrier without affecting the affinity of the carrier for L-glutamate. Because the K^+ gradient energizes the uphill uptake of L-glutamate in the absence of membrane potential changes, and

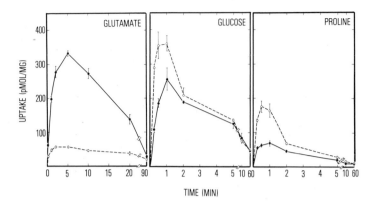

Figure 6. The effect of a K$^+$ gradient (intravesicular > extravesicular) on the uptakes of L-glutamate, L-proline, and D-glucose. [From Schneider and Sacktor, 1980.]

in the absence of Na$^+$ and anion gradients, it is suggested that the cotransport of Na$^+$/L-glutamate is coupled to the transmembrane flux of K$^+$.

In addition to the specific effect of K$^+$ on L-glutamate transport, the presence of an H$^+$ gradient (intravesicular > extravesicular) specifically stimulates the Na$^+$-dependent uptake of phosphate (Sacktor and Cheng, 1981). Again, in the presence of Na$^+$ but in the absence of an Na$^+$ gradient, the pH gradient can provide the driving force for the transient uphill transport of phosphate. In this system, however, a transmembrane flux of H$^+$ need not be necessary. Rather, the effect of intravesicular H$^+$ may be attributed to a difference in the transport of two of the anionic forms of phosphate, HPO$_4^{2-}$ and H$_2$PO$_4^-$. It is suggested that HPO$_4^{2-}$ is rapidly taken up and is converted intravesicularly to H$_2$PO$_4^-$, which effluxes slowly.

In conclusion, this report on the current status of Na$^+$ coupled transport systems in renal brush border membrane vesicles documents that the Na$^+$ electrochemical gradient is nearly the universal driving force to effect the secondary active uptakes of various metabolites. At the same time, although Na$^+$ coupled cotransport systems represent a common biological modus operandi, the precise mechanism by which a specific solute is transported may differ from the mechanisms by which other compounds are taken up, and, in fact, each mechanism may be unique. Furthermore, specific physiological effectors regulate the individual transport systems and this is reflected in changes in the membrane. It is in this important regulatory area that investigations with the isolated membrane can be most productive, especially when integrated into studies of the entire phenomenon.

REFERENCES

Aronson, P. S., and Sacktor, B. (1975). *J. Biol. Chem.* **250**, 6032–6039.
Barac-Nieto, M., Murer, H., and Kinne, R. (1980). *Am. J. Physiol.* **239**, F496–F506.
Beck, J. C., and Sacktor, B. (1975). *J. Biol. Chem.* **250**, 8674–8680.
Beck, J. C., and Sacktor, B. (1978a). *J. Biol. Chem.* **253**, 5531–5535.
Beck, J. C., and Sacktor, B. (1978b). *J. Biol. Chem.* **253**, 7158–7162.
Blomstedt, J. W., and Aronson, P. S. (1980). *J. Clin. Invest.* **65**, 931–934.
Cheng, L., and Sacktor, B. (1981). *J. Biol. Chem.* **256**, 1156–1564.
Crane, R. K. (1977). *Rev. Physiol. Biochem. Pharmacol.* **78**, 99–159.
Evers, J., Murer, H., and Kinne, R. (1976). *Biochim. Biophys. Acta* **426**, 598–615.

Fass, S. J., Hammerman, M. R., and Sacktor, B. (1977). *J. Biol. Chem.* **252,** 583–590.

Foreman, J. W., Hwang, S., and Segal, S. (1980). *Metabolism* **29,** 53–61.

Garcia, M. L., Benavides, J., and Valdivieso, F. (1980). *Biochim. Biophys. Acta* **600,** 922–930.

Hammerman, M. R., and Sacktor, B. (1977). *J. Biol. Chem.* **252,** 541–545.

Hammerman, M. R., and Sacktor, B. (1978). *Biochim. Biophys. Acta* **509,** 338–347.

Hammerman, M. R., Sacktor, B., and Daughaday, W. H. (1980). *Am. J. Physiol.* **239,** F113–F120.

Hilden, S. A., and Sacktor, B. (1979). *J. Biol. Chem.* **254,** 7090–7096.

Hilden, S. A., and Sacktor, B. (1981). *Arch. Biochem. Biophys.* **210,** 289–297.

Hoffman, N., Thees, M., and Kinne, R. (1976). *Pfluegers Arch.* **362,** 147–156.

Kempson, S. A., and Dousa, T. P. (1979). *Life Sci.* **24,** 881–888.

Kinne, R. (1976). *Curr. Top. Membr. Transp.* **8,** 209–267.

Kippen, I., Hirayama, B., Klineberg, J. R., and Wright, E. M. (1979). *Proc. Natl. Acad. Sci. USA* **76,** 3397–3400.

Kleinzeller, A. (1970). *Biochim. Biophys. Acta* **211,** 264–276.

Leopolder, A., Burckhardt, G., and Murer, H. (1980). *Renal Physiol.* **2,** 157–158.

Lücke, H., Stange, G., and Murer, H. (1979). *Biochem. J.* **182,** 223–229.

McNamara, P. D., Ozegovic, B., Pepe, L. M., and Segal, S. (1976). *Proc. Natl. Acad. Sci. USA* **73,** 4521–4525.

Murer, H., Hopfer, U., and Kinne, R. (1976). *Biochem. J.* **154,** 597–604.

Rozen, R., Tenenhouse, H. S., and Scriver, C. R. (1979). *Biochem. J.* **180,** 245–248.

Sacktor, B. (1977). *Curr. Top. Bioenerg.* **6,** 39–81.

Sacktor, B. (1978). In *Renal Function* (G. H. Giebisch and E. F. Purcell, eds.), pp. 221–229, Josiah Macy, Jr. Foundation, New York.

Sacktor, B. (1980). *Curr. Top. Membr. Transp.* **13,** 291–299.

Sacktor, B., and Cheng, L. (1981). *J. Biol. Chem.* **256,** 8080–8084.

Sacktor, B., and Schneider, E. G. (1980). *Int. J. Biochem.* **12,** 229–234.

Sacktor, B., Chesney, R. W., Mitchell, M. E., and Aronson, P. S. (1974). In *Recent Advances in Renal Physiology and Pharmacology* (L. G. Wesson and G. M. Fanalli, eds.), pp. 13–26, University Park Press, Baltimore.

Schneider, E. G., and Sacktor, B. (1980). *J. Biol. Chem.* **255,** 7645–7649.

Schneider, E. G., Hammerman, M. R., and Sacktor, B. (1980). *J. Biol. Chem.* **255,** 7650–7656.

Segal, S., McNamara, P. D., and Pepe, L. M. (1977). *Science* **197,** 169–171.

Silverman, M., Aganon, M. A., and Chinard, F. P. (1970). *Am. J. Physiol.* **218,** 735–750.

Stoll, R., Kinne, R., Murer, H., Fleisch, H., and Bonjour, J.-P. (1979). *Pfluegers Arch.* **380,** 47–52.

Tenenhouse, H. S., and Scriver, C. R. (1978). *Can. J. Biochem.* **56,** 640–646.

Ullrich, K. J. (1979). In *Membrane Transport in Biology* (G. H. Giebisch, D. C. Tosteson, and H. H. Ussing, eds.), Vol. IVA, pp. 413–448, Springer-Verlag, Berlin.

Weiss, S. D., McNamara, P. D., Pepe, L. M., and Segal, S. (1978). *J. Membr. Biol.* **43,** 91–105.

Welch, C. L., and Campbell, B. J. (1980). *J. Membr. Biol.* **54,** 39–50.

Wright, S. H., Kippen, I., Klineberg, J. R., and Wright, E. M. (1980). *J. Membr. Biol.* **57,** 73–82.

117

Sugar Transport in the Mammalian Nephron

M. Silverman

1. INTRODUCTION

Active reabsorption of sugars by the renal tubule is effected by two different carrier-mediated steps operating in series. At the brush border membrane of the proximal tubule there is a Na^+-dependent secondary active transport system, while at the basolateral surface there is a facilitated carrier mechanism.

Specificity characteristics of these different sugar transporters were reviewed in detail some years ago (Silverman, 1976; Ullrich, 1976; Kinne, 1976b) and in abbreviated form more recently (Ullrich, 1979; Silverman, 1980). Figure 1 provides a schematic, somewhat speculative representation of the individual carrier systems at the luminal and contraluminal surfaces. This diagram is based on a synthesis of new and old data and is intended to summarize current knowledge.

2. LUMINAL MEMBRANE

2.1. Glucose Carrier Systems

We place a low-affinity (K_m range 1–4 mM, pH 7.4, 25°C), Na^+-dependent, phlorizin-sensitive glucose carrier at the level of the superficial proximal convoluted tubule and designate it G_1^{Na}. The Na^+-dependent, high-affinity "phlorizin receptor" (Frasch et al., 1970; Glossmann and Neville, 1972; Vick et al., 1973; Silverman and Black, 1975; Turner and Silverman, 1978a) is probably identical to that part of the G_1^{Na} transporter exposed at the external (urine) face of the luminal membrane. It is further postulated that the G_1^{Na} carrier is shared principally by α (or β)-methyl-D-glucose (MG), D-galactose, and 5-thio-D-glucose.

M. Silverman • Department of Medicine, University of Toronto, Toronto, Ontario M5S 1A8, Canada.

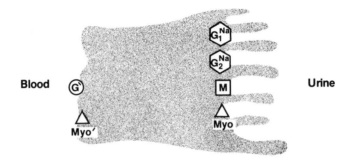

Figure 1. Diagrammatic representation summarizing luminal and contraluminal distribution of different sugar carrier mechanisms in the renal proximal tubule. Details of localization (i.e., convoluted or straight segments), specificity, and kinetic characteristics are given in the text.

A second set of high-affinity (K_m range 0.1–0.3 mM, pH 7.4, 25°C), Na^+-dependent, phlorizin-sensitive glucose carriers, G_2^{Na}, is situated in the straight portion of the proximal tubule (Moran and Turner, 1982; Barfuss and Schafer, 1981). In dog kidney the total number of G_2^{Na} sites is small, perhaps no more than 5–10% of the total number of Na^+-dependent glucose carriers. We tentatively suggest that D-mannose, D-fructose, myoinositol, and possibly 2-deoxy-D-glucose share the G_2^{Na} carrier together with D-glucose. A small component of Na^+-dependent phlorizin binding may take place at the G_2^{Na} sites. But it is not yet known whether the affinity of phlorizin for this site is the same as that for G_1^{Na}.

The model outlined above (Fig. 1) is an attempt to reconcile conflicting data in the literature relative to the importance of the C-2 hydroxyl group in the substrate–receptor interaction. We suggest that the presence of an hydroxyl group at the C-2 position of sugars is indeed critical but only for *one* of the two available Na^+-dependent glucose carriers located at the luminal membrane, i.e., for G_1^{Na} but not for G_2^{Na}. In the G_1^{Na} carrier in the proximal convoluted tubule, there is strong preference for a pyranose $^4C_1$ chair conformation with equatorial hydroxyl groups located at C-2, C-3, and C-6. For optimal interaction equatorial C-4 and C-1 hydroxyls are also necessary.

On the other hand, we postulate that the specificity of sugar interaction with the G_2^{Na} site in the straight portion of the proximal tubule is quite different. In particular, modifications at the C-2 position will be less critical and therefore mannose, 2-deoxy-D-glucose, and fructose are proposed as substrates for this carrier.

2.2 Other Sugar Carriers at the Luminal Membrane

There is a phlorizin-insensitive (probably also Na^+-independent) transport system for D-mannose somewhere along the luminal surface of the nephron. This mannose carrier (M in Fig. 1) could be situated at the level of the proximal tubule or more distally.

Myoinositol uptake in renal brush border membrane vesicles is Na^+ dependent (Takenawa *et al.*, 1977). Luminal extraction of myoinositol is partially blocked by both glucose and phlorizin loading *in vivo* (Silverman, 1974a). Therefore, myoinositol shares either the G_1^{Na} or the G_2^{Na} site. Because myoinositol exhibits little tendency to interfere with phlorizin binding (Silverman, 1976), it is tentatively included with the G_2^{Na} carrier. As there is a phlorizin- and mannose-insensitive component of luminal myoinositol uptake (Silverman, 1974a), a separate pathway must exist for myoinositol (Myo in Fig. 1) and its segmental localization remains to be determined.

3. BASOLATERAL MEMBRANE

Probably the only reliable *in vivo* approach to this problem is by use of the pulse injection indicator dilution technique (Chinard *et al.*, 1959). This procedure has been applied to investigate antiluminal uptake of sugars (Silverman *et al.*, 1970a,b; Silverman, 1974b, 1977) vitamin B_{12} (Silverman, 1979), glutamate (Silverman *et al.*, 1981), and amino acids (Foulkes and Gieske, 1973).

Recent studies using this technique have shown that the ring oxygen in pyranose sugars is essential for recognition by the antiluminal transport system (Silverman, 1980). Also, 2-deoxy-D-glucose (Silverman and Turner, 1979) is definitely taken up at the basolateral surface as originally proposed by Kleinzeller and McAvoy (1973). Table I summarizes present knowledge as to which sugar substrates interact with the antiluminal membrane of the nephron. The reasons for localizing this antiluminal interaction at the level of the proximal tubule have been given elsewhere (Silverman *et al.*, 1970a; Silverman, 1976). Each sugar substrate that exhibits an antiluminal interaction from the peritubular surface can be blocked *in vivo* by sufficiently high plasma concentration of phlorizin (Silverman, 1977), i.e., levels that are about 1000-fold higher than the concentration necessary to inhibit glucose transport at the brush border (Silverman, 1976). Complementing this evidence for a phlorizin-insensitive glucose transport system at the basolateral membrane is the finding that tracer [³H]phlorizin exhibits no demonstrable antiluminal interaction in single-pass experiments *in vivo* (Silverman, 1974b).

In order to pursue kinetic studies in greater detail, it is necessary to turn to more direct *in vitro* approaches such as the isolated perfused tubule (Schafer and Barfuss, 1980) or contraluminal membrane vesicles (Heidrich *et al.*, 1977). The use of such *in vitro* methods, however, is not without its own set of problems and limitations. Up until very recently (see Scalera *et al.*, 1980), it had been difficult to prepare "clean" antiluminal membranes in sufficient yield. Moreover, the vesicles prepared from antiluminal membranes are more heterogeneous than those obtained from the brush border surface. For example, brush border vesicles are all right side out, whereas antiluminal vesicles are approximately a 50–50 mixture of inside-out and right-side-out populations (Kinne, 1976a; Kinne and Schwartz, 1978).

Table I. Sugar Interaction with the Antiluminal Membrane of Dog Proximal Tubule from the Peritubular (Exterior) Surface

Interacting	Noninteracting
D-Glucose	L-Glucose
D-Galactose	D-Allose
D-Mannose	D-Mannitol
3-*O*-Methyl-D-glucose	D-Glucosamine
6-Deoxy-D-galactose	D-Mannosamine
D-Xylose	D-Ribose
D-Talose	2-Deoxy-D-ribose
L-Arabinose	6-Deoxy-L-galactose
D-Fructose	5-Thio-D-glucose
2-Deoxy-D-glucose	D-Arabinose
	L-Idose
	α-Methyl-D-glucose
	β-Methyl-D-glucose

Kidney antiluminal vesicles were first used to study sugar transport by Kinne and collaborators (Kinne *et al.*, 1975). These vesicles were obtained by free-flow electrophoretic separation of rat tissue. Since then, some additional new data on contraluminal uptake of sugars have been presented (Kinne, 1978). Free-flow electrophoresis has also been adopted to separate dog kidney membranes (Chant and Silverman, 1978). The canine preparation has been used for a limited number of transport studies of both sugars and anions (Turner and Silverman, 1978a,b; Grinstein *et al.*, 1980; Silverman, 1981). These vesicle studies have provided a finding of major importance: that glucose uptake at the basolateral membrane (in contrast to the brush border surface) is Na^+ independent (Kinne *et al.*, 1975; Turner and Silverman, 1978a). Similarly (as indicated above) a high-affinity, Na^+-dependent phlorizin receptor site can be readily demonstrated at the luminal surface, but there is no evidence of any such high-affinity site at the basolateral membrane (Silverman, 1981).

Also, D-glucose uptake into contraluminal vesicles is relatively insensitive to inhibition by phlorizin compared to phloretin or cytochalasin B (Silverman, 1981). Identical results were obtained in earlier studies using antiluminal vesicles from rat kidney (Kinne *et al.*, 1975) and intestine (Hopfer *et al.*, 1975). Based on these data a working hypothesis outlining the different antiluminal sugar carrier systems is indicated in Fig. 1.

Two carrier systems (at least) are postulated along the length of the proximal tubule. The G' transporter would be shared by D-glucose and nine other substrates as listed in Table I. The existence of a distinct myoinositol pathway (Myo') at the antiluminal membrane was deduced from the fact that myoinositol but not 5-thio-D-glucose (Silverman, 1980) is taken up at the peritubular surface.

At the moment it seems prudent to assume that the myoinositol transport systems at opposite poles of the tubular epithelium differ from one another.

The minimum chemical and steric determinants for the G' carrier in Fig. 1 can then be summarized as follows: in order for successful interaction to take place with the G' site, the sugar molecule must exist in the pyranose $^4C_1$ chair conformation, D-glucose configuration, and possess a hydroxyl group at the C-1 position. If OH groups are present at C-3 or C-6, they must be equatorial.

4. ENERGETICS OF SUGAR REABSORPTION

Glucose-sensitive phlorizin binding to rat kidney brush border membranes was first documented to be Na^+ dependent in 1970 (Frasch *et al.*, 1970). Later, glucose and Na^+ flux measurements in perfused proximal tubule segments of rats (Ullrich *et al.*, 1974) led to the suggestion that a Na^+/D-glucose cotransporter also exists at the luminal surface of the renal tubular epithelium. In order to dissociate transport from metabolism, further studies of this system have been pursued using isolated brush border vesicles from rat, rabbit, dog, and man (Aronson and Sacktor, 1975; Kinne *et al.*, 1975; Beck and Sacktor, 1975, 1978a,b; Turner and Silverman, 1977, 1978a,b). Detailed kinetic analyses of Na^+-dependent phlorizin binding to brush border vesicles have also been carried out (Glossmann and Neville, 1973; Aronson, 1977; Turner and Silverman, 1980a,b).

D-Glucose-dependent Na^+ transport has been studied in isolated vesicles (Hilden and Sacktor, 1979) and also investigated with electrophysiologic techniques that measure transmucosal membrane depolarization upon addition of sugars in the presence of Na^+ (e.g., see Fromter, 1979).

The essential observations can be summarized as follows:

1. The Na^+-dependent component of D-glucose transport is localized only to the luminal membrane (i.e., G_1^{Na}, G_2^{Na} in Fig. 1).

2. There is a one-to-one stoichiometric coupling between the transmucosal movement of glucose and Na^+ ions. Na^+ stimulates D-glucose uptake and D-glucose stimulates Na^+ uptake *by the same* sugar carrier.
3. The driving force for sugar transport is proportional to the transmucosal electro-chemical potential gradient for Na^+ ($\Delta\Psi_{Na}$), i.e.,

$$\text{Driving force } \alpha \ \gamma\Delta\mu_{Na} = \gamma[58 \ \ln\frac{[Na]_o}{[Na]_i} - \Delta\Psi] \tag{1}$$

where γ is a coupling coefficient between the potential gradient $\Delta\mu_{Na}$ and the transport rate. It reflects the efficiency of the coupling. (The value of RT/F at 25°C is 58. $\Delta\Psi$ is the electrical potential gradient across the membrane.)

Much evidence to date in both intestine (Schultz, 1977) and kidney (Fromter, 1979) indicates that $\gamma = 1$.

Murer and Hopfer (1974) were the first to demonstrate that stereospecific uptake of D-glucose into intestinal brush border vesicles was accelerated by two factors: (1) a trans-membrane Na^+ chemical gradient (outside $>$ inside) and (2) a transmembrane electrical po-tential gradient, inside negative. This demonstration provided important experimental ver-ification of the Na^+-gradient hypothesis and also showed that the cotransport of Na^+ and glucose involves the net transport of positive charge and is, therefore, electrogenic. Similar experiments have since been reproduced in many different laboratories using renal brush border vesicles (see above references).

These results demonstrate the validity of equation (1), but do not reveal whether the normal, *in vivo*, transmucosal Na^+ electrochemical gradient is adequate to account quanti-tatively for the observed active transport of sugars, or if an additional energy requirement is necessary derived from primary linkage to a metabolic reaction sequence.

This question has been considered (Schultz, 1977; Fromter, 1979; Kessler and Se-menza, 1979), and the consensus is that the normal transmucosal Na^+ electrochemical gradient is both necessary and sufficient to account for active transepithelial reabsorption of sugars in the intestine and kidney. A dissenting opinion has been advanced by Kimmich and Randles (1980) who have introduced evidence showing that exogenous ATP directly modifies the mucosal glucose carrier in the intestine.

5. REGULATION OF TUBULAR SUGAR REABSORPTION

Polarization of the renal proximal tubule is manifested as a unique distribution of plasma membrane constituents at opposite poles of the cell (Kinne and Kinne-Saffran, 1978; Silverman and Turner, 1979). In particular, Na^+/K^+-dependent ATPase is localized exclusively to the basolateral membrane. This enzyme system when properly activated pumps 3 Na^+ ions out of the cell in exchange for 2 K^+, and thus establishes a low in-tracellular Na^+ concentration. The Na^+ chemical gradient across the luminal membrane is $\sim$ 14 : 1 (outside : inside). At the same time, the transmucosal electrical potential is $\sim$ 60–70 mV, inside negative. The largest part of this potential is derived from activity of the Na^+ pump, but a component is also due to electrogenic extrusion of H^+ into the tubular lumen (Fromter, 1979). In any case, the overall net effect of the basolateral Na^+ pump activity is that under usual steady-state *in vivo* conditions, there is a considerable trans-mucosal Na^+ electrochemical gradient. Different luminal cotransported substrates such as amino acids, sugars, lactate, and phosphate enter the tubular cell by "sliding down" the

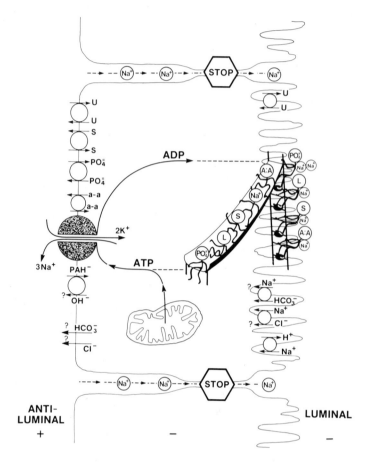

Figure 2. Schematic view depicting multiplicity of tubular ionic and nonionic transport mechanisms, and the coupling of Na$^+$-dependent carriers to cell metabolism. S, sugar; A : A (a-a), amino acid; L, lactate; U, uric acid; PAH, *p*-aminohippurate. Reprinted with permission.

transmucosal Na$^+$ electrochemical gradient as depicted schematically in Fig. 2. If 100% efficiency of energy coupling to the glucose carrier is assumed, it can be calculated that the proximal tubular cell should be able to accumulate glucose to approximately 30 times the lumen concentration. The fact that such high intracellular levels are never reached is not because of metabolic degradation (indeed the proximal tubule is gluconeogenic) (Schmidt and Guder, 1976) but because of the large passive flux of glucose across the peritubular surface, this antiluminal leak being mediated of course by the Na$^+$-independent G' carrier (Fig. 1) located at the basal surface of the cell.

6. PATHOGENESIS OF TUBULAR DISORDERS

The spectrum of renal tubular dysfunction encompasses a variety of clinical conditions ranging from aminoacidurias to the adult Fanconi syndrome. It is self-evident that a proper understanding of the pathogenesis of tubular disorders (and ultimately their clinical management) will depend upon complete knowledge of transepithelial transport physiology. Attempts have already been made to classify such tubular disorders based entirely on a functional rationale (Scriver *et al.*, 1976; Silverman and Turner, 1979). One such scheme is illustrated in Fig. 3. Two distinctly different membrane lesions at the molecular level can be postulated to rationalize tubular disease: Type I, a disorder that results from an altered gene product(s), and Type II, a disorder of membrane energization.

Type I disorders will be manifest at the clinical level by selective lesions affecting single membrane-functional proteins. By far the best known class of defects concerns those membrane proteins acting as transporters or carriers, e.g., inherited renal glucosuria.

Type II disorders are examples of complete or partial Fanconi syndrome in which the usual clinical manifestations are glucosuria, phosphaturia, aminoaciduria, decreased serum uric acid, proximal tubular acidosis, and natriuresis. It is apparent that multiple brush border, Na^+-dependent transport processes must be affected in this disorder (see Fig. 2). In many cases of the Fanconi syndrome, no pathology is visible even at the level of the electron microscope. Given such circumstances, it seems likely that the molecular defect involves one or more of the following mechanisms acting singly or in combination: (1) diminished intrinsic activity of the Na^+ pump, i.e., of Na^+/K^+-dependent ATPase, (2) decrease in the supply of ATP necessary for hydrolysis by the ATPase, (3) altered intramembrane coupling between the Na^+ electrochemical gradient and secondarily cotransported solute–carrier complexes, and (4) altered intramembrane coupling.

As an illustrative example, let us consider the experimental Fanconi syndrome induced in animals following administration of maleic acid (Berliner *et al.*, 1950). This disease model has been investigated in rats, using renal cortical slices (Rosenberg and Segal, 1964),

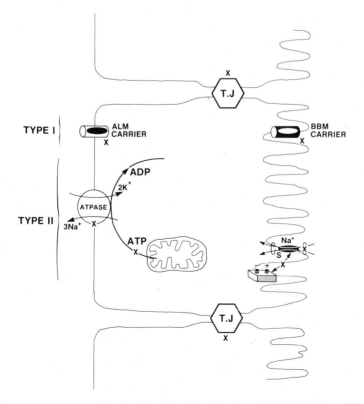

Figure 3. Diagram illustrating classification of renal tubular disorders. For details see text. TJ, tight junction; ALM, antiluminal membrane; BBM, brush border membrane. Reprinted with permission.

micropuncture (Bergeron and Vadeboncoeur, 1971), and morphologic (Rosen *et al.*, 1973; Worthen, 1963) techniques. The mechanism of glucosuria induced by maleic acid has also been studied in dogs, using the pulse injection indicator dilution method (Silverman and Huang, 1976). The cumulative data suggest that the tubular abnormality (at least for amino acid and sugar transport) results from increased backflux into the urine of normally reabsorbed substrates. Bergeron *et al.* (1976) feel that this backflux is secondary to altered permeability of the luminal surface at some point along the nephron distal to the reabsorbing site (i.e., distal to the proximal tubule). Another possibility (Silverman and Huang, 1976) is that backflux occurs at the same level as the reabsorbing site (i.e., the proximal convoluted tubule). Recent studies (Silverman, 1980) have shown that sugar uptake and phlorizin binding to vesicles prepared from maleic acid kidney are the same as normal. This means that a simple membrane effect is unlikely.

Szczepanska-Konkel and Angielski (1980) have demonstrated the protective effect of acetoacetate infusion. Their studies imply that the pathogenic factor in maleic acid nephropathy is maleyl CoA, a metabolite of maleate, and that a reduction in cell ATP is not responsible for reduced transport function. The most reasonable conclusion is that the defect in maleic acid nephropathy is somehow linked to a defect in the ability of the renal tubular cell to maintain a favorable transmucosal Na^+ electrochemical gradient despite a normal ATP supply. This focuses attention on the integrity of Na^+/K^+-dependent ATPase activity and/or on coupling between energy supply and carrier function.

7. MOLECULAR NATURE OF SUGAR CARRIER SYSTEMS IN THE KIDNEY

Because of the asymmetries inherent in epithelia, progress has been slow in the biochemical isolation and characterization of epithelial transport systems such as the one involving sugar reabsorption in the kidney. The heterogeneity of sugar carriers as a function of length along the nephron adds a further complication.

7.1. Sugar Transport System at the Antiluminal Membrane (G′ in Fig. 1)

No real attempts have been made to isolate sugar carriers from the basolateral membrane. There is some expectation (which may or may not turn out to be justified) that the contraluminal sugar transport system will function no differently at the molecular level than the sugar carrier found in red cells that have been isolated and reconstituted (Sogin and Hinkle, 1978; Goldin and Rhoden, 1978).

The erythrocyte sugar carrier behaves as an asymmetric generalized mobile carrier type with sequential translocation of sugar first at one side (external face) and then at the inside (cytoplasmic) surface (Wilbrandt, 1977; Deves and Krupka, 1978). The system is completely passive and functions as a facilitated, equilibrative mechanism. There is no Na^+ or electrical potential dependence.

7.2 Sugar Transport at the Brush Border Membrane

For Na^+/sugar cotransport systems (either G_1^{Na} or G_2^{Na} in Fig. 1), results of kinetic experiments are consistent with the behavior expected of some form of mobile carrier—either a gate or shuttle mechanism.

Turner and Silverman (1980a) have carried out an analysis of the kinetics of nontransported inhibitor binding to carrier-mediated cotransport systems. Because there is strong evidence that phlorizin is a nontransported competitive inhibitor of the Na^+-dependent sugar carrier in the renal brush border, this drug has been used to probe the operational characteristics of this mechanism (Turner and Silverman, 1980b). The results suggest that the Na^+/glucose cotransport system in the kidney brush border behaves as an iso-random bi bi mechanism. However, the Na^+ dependence of phlorizin binding is so strong that at Na^+ concentrations greater than 10–20 mM, the system effectively operates as an iso-ordered bi bi mechanism with Na^+ binding first.

Hopfer and Glasecore (1980) have investigated the kinetics of equilibrium exchange for the Na^+/glucose cotransport system in the intestine. Their data argue in favor of an iso-ordered bi bi mechanism with glide symmetry (Na^+ first on at the outside of the membrane and first off at the inside). Hopfer has also presented evidence that the translocation rate is not rate limiting compared to the association–dissociation rates of Na^+. His conclusion is that the operation of the Na^+/glucose cotransport carrier is most compatible with a gate-type mechanism. In such a scheme, the transmembrane electrical potential might have a greater effect on the conformation of the carrier binding sites rather than on the translocation rate. It should be pointed out, however, that the precise shape of the electrical potential field with respect to the distribution of transmembrane carrier protein remains completely undetermined. This makes it difficult to incorporate membrane potential effects into theoretical formulations of carrier kinetics.

There have been several reports of successful reconstitution of the Na^+-dependent glucose carrier from renal brush border membrane (Fairclough *et al.*, 1979; Crane *et al.*, 1978). Such studies are in their infancy and have so far not produced any new insights into the molecular mechanism of the transport system.

REFERENCES

Aronson, P. S. (1977). *Kidney Int.* **12**, A548.
Aronson, P. S., and Sacktor, B. (1975). *J. Biol. Chem.* **250**, 6032–6039.
Barfuss, D. W., and Schafer, J. A. (1981). *Am. J. Physiol.* **240**, F322–F332.
Beck, J. C., and Sacktor, B. (1975). *J. Biol. Chem.* **250**, 8674–8680.
Beck, J. C., and Sacktor, B. (1978a). *J. Biol. Chem.* **253**, 5531–5535.
Bergeron, M., and Vadeboncoeur, M. (1971). *Nephron* **8**, 367–374.
Bergeron, M., Dubord, L., and Hausser, C. (1976). *J. Clin. Invest.* **57**, 1181–1189.
Berliner, R. W., Hilton, J. G., Yu, T. F., and Kennedy, T. J., Jr. (1950). *J. Clin. Invest.* **29**, 396–401.
Chant, S., and Silverman, M. (1978). *Clin. Exp. Immunol.* **32**, 405–410.
Chinard, F. P., Taylor, W. R., Nolan, F., and Enns, T. (1959). *Am. J. Physiol.* **196**, 535–544.
Crane, R. K., Malathi, P., Presier, H., and Fairclough, P. (1978). *Am. J. Physiol.* **234**, E1–E5.
Deves, R., and Krupka, R. M. (1978). *Biochim. Biophys. Acta* **510**, 339–348.
Fairclough, P., Malathi, P., Presier, H., and Crane, R. K. (1979). *Biochim. Biophys. Acta* **553**, 295–306.
Foulkes, E. C., and Gieske, T. (1973). *Biochim. Biophys. Acta* **318**, 439–445.
Frasch, W., Frohnert, P. P., Bode, F., Baumann, K., and Kinne, R. (1970). *Pfluegers Arch.* **320**, 265–284.
Fromter, E. (1979). *J. Physiol.* **288**, 1–31.
Glossmann, H., and Neville, D. M. (1972). *J. Biol. Chem.* **247**, 7779–7789.
Goldin, S. M., and Rhoden, V. (1978). *J. Biol. Chem.* **253**, 2575–2853.
Grinstein, S., Turner, R. J., Silverman, M., and Rothstein, A. (1980). *Am. J. Physiol.* **238**, F452–F460.
Heidrich, H. G., Kinne, R., Kinne-Saffran, E., and Hannig, K. (1977). *J. Cell Biol.* **54**, 232–245.
Hilden, S. A., and Sacktor, B. (1979). *J. Biol. Chem.* **254**, 7090–7096.
Hopfer, U., and Glasecore, R. (1980). *J. Biol. Chem.* **255**, 4399–4402.
Hopfer, U., Sigriest-Nelson, K., and Murer, H. (1975). *Ann. N.Y. Acad. Sci.* **264**, 414–426.

Kessler, M., and Semenza, G. (1979). *FEBS Lett.* **108**, 205–208.

Kimmich, G., and Randles, J. (1980). *Am. J. Physiol.* **238**, C177–C183.

Kinne, R. (1976a). In *International Review of Physiology* (K. Thurau, ed.), pp. 169–210, University Park Press, Baltimore.

Kinne, R. (1976b). In *Current Topics in Membranes and Transport* (F. Bronner and A. Kleinzeller, eds.), pp. 209–267, Academic Press, New York.

Kinne, R. (1978). *Int. Cong. Nephrol.* **14**, 14–24.

Kinne, R., and Kinne-Saffran, E. (1978). In *Molecular Specialization and Symmetry in Membrane Function* (A. K. Solomon and M. Karnovsky, eds.), pp. 272–293, Harvard University Press, Cambridge, Mass.

Kinne, R., and Schwartz, I. L. (1978). *Kidney Int.* **14**, 547–556.

Kinne, R., Murer, H., Kinne-Saffran, E., Thees, M., and Sachs, G. (1975). *J. Membr. Biol.* **21**, 375–395.

Kleinzeller, A., and McAvoy, M. (1973). *J. Gen. Physiol.* **62**, 169–184.

Murer, H., and Hopfer, U. (1974). *Proc. Natl. Acad. Sci USA* **71**, 484–488.

Rosen, V. J., Kramer, H. J., and Gonick, H. (1973). *Lab. Invest.* **28**, 446–455.

Rosenberg, L. E., and Segal, S. (1964). *Biochem. J.* **92**, 345–352.

Scalera, V., Storelli, C., Storelli-Joss, C., Haase, W., and Murer, H. (1980). *Biochem. J.* **186**, 177–181.

Schafer, J. A., and Barfuss, D. W. (1980). *Am. J. Physiol.* **238**, F335–F346.

Schmidt, U., and Guder, W. G. (1976). *Kidney Int.* **9**, 233–342.

Schultz, S. G. (1977). *Am. J. Physiol.* **233**, E249–E254.

Scriver, C. R., Chesney, R. W., and McInnes, R. R. (1976). *Kidney Int.* **9**, 149–171.

Silverman, M. (1974a). *Biochim. Biophys. Acta* **332**, 248–262.

Silverman, M. (1974b). *Biochim. Biophys. Acta* **339**, 92–102.

Silverman, M. (1976). *Biochim. Biophys. Acta* **457**, 303–351.

Silverman, M. (1977). *Am. J. Physiol.* **232**, F455–F460.

Silverman, M. (1979). *Am. J. Physiol.* **237**, F25–F33.

Silverman, M. (1980). *Biochim. Biophys. Acta* **600**, 502–512.

Silverman, M. (1981). *Membr. Biochem.*, **4**, 63–6.

Silverman, M., and Black, J. (1975). *Biochim. Biophys. Acta* **394**, 10–30.

Silverman, M., and Huang, L. (1976). *Am. J. Physiol.* **231**, 1024–1032.

Silverman, M., and Turner, R. J. (1979). *Biomembranes* **10**, 1–50.

Silverman, M., Aganon, M. A., and Chinard, F. P. (1970a). *Am. J. Physiol.* **218**, 735–742.

Silverman, M., Aganon, M. A., and Chinard, F. P. (1970b). *Am. J. Physiol.* **218**, 743–750.

Silverman, M., Vinay, P., Shinobu, L., Gougoux, A., and Lemieux, G. (1981). *Kidney Int.* **20**, 359–365.

Sogin, D. C., and Hinkle, P. C. (1980). *Proc. Natl. Acad. Sci. USA* **77**, 5725–5729.

Szczepanska-Konkel, M., and Angielski, S. (1980). *Am. J. Physiol.* **239**, F50–F56.

Takenawa, T., Wada, E., and Tsumita, T. (1977). *Biochim. Biophys. Acta* **464**, 108–117.

Turner, R. J., and Moron, A. (1982). *Am. J. Physiol.* **242**, in press.

Turner, R. J., and Silverman, M. (1977). *Proc. Natl. Acad. Sci. USA* **75**, 2825–2829.

Turner, R. J., and Silverman, M. (1978a). *Biochim. Biophys. Acta* **507**, 305–321.

Turner, R. J., and Silverman, M. (1978b). *Biochim. Biophys. Acta* **511**, 470–486.

Turner, R. J., and Silverman, M. (1980a). *Biochim. Biophys. Acta* **596**, 272–291.

Turner, R. J., and Silverman, M. (1980b). *J. Membr. Biol.* **58**, 43–55.

Ullrich, K. J. (1976). *Kidney Int.* **9**, 134–138.

Ullrich, K. J. (1979). *Annu. Rev. Physiol.* **41**, 181–195.

Ullrich, K. J., Rumrich, G., and Baumann, K. (1974). *Pfluegers Arch.* **351**, 35–48.

Vick, H., Diedrich, D. F., and Baumann, K. (1973). *Am. J. Physiol.* **224**, 552–557.

Wilbrandt, W. (1977). In *Proceedings in Life Sciences Biochemistry of Membrane Transport* (G. Semenza and E. Carafoli, eds.), pp. 204–211, Springer-Verlag, Berlin.

Worthen, H. G. (1963). *Lab. Invest.* **12**, 791–801.

118

Renal Tubular Control of Potassium Transport

Gerhard Giebisch

1. GENERAL FEATURES OF POTASSIUM HOMEOSTASIS

Several control mechanisms regulate the body's potassium balance. The distribution of potassium ions within the body and internal potassium balance is regulated by a number of powerful and effective hormonal mechanisms that control the transfer of potassium between extracellular and cellular stores (Bia and DeFronzo, 1981; Knochel, 1977). Among these mechanisms insulin (DeFronzo *et al.*, 1974), epinephrine (DeFronzo *et al.*, 1979), mineralocorticoids (DeFronzo *et al.*, 1980), and acid–base factors (Frayley and Adler, 1977) are the best defined. Following the release of insulin, epinephrine, or aldosterone, potassium ions are effectively transferred from extracellular to intracellular potassium stores. In addition, alkalosis as well as isohydric elevation of the plasma bicarbonate level have a similar effect (Frayley and Adler, 1977). When these responses are blunted potentially life-threatening fluctuations in extracellular potassium concentration may occur.

Effective control of external potassium balance is exercised by specialized regions within the kidney and the gut. The kidneys play a dominant role in maintaining external potassium balance in vertebrates, whereas the gut and Malpighian tubules are the key sites of potassium excretion in insects. To effect efficient control of overall potassium balance, renal and intestinal mechanisms must be coordinated with the mechanisms controlling proper distribution of potassium ions within the body. A schematic summary of the distribution of potassium ions in the body and of the factors controlling internal and external potassium balance is presented in Fig. 1.

Gerhard Giebisch • Department of Physiology, Yale University School of Medicine, New Haven, Connecticut 06510.

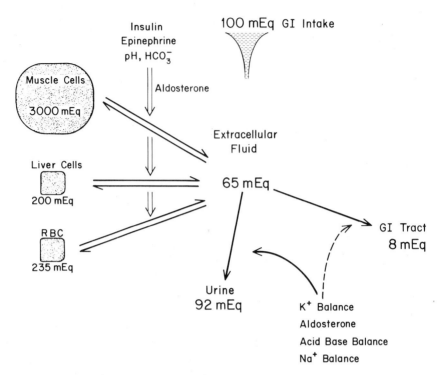

Figure 1. Distribution of potassium ions in the body. Only a small fraction (1–2%) of body potassium is within the extracellular fluid. Potassium intake occurs via the gastrointestinal tract (GI). Potassium excretion occurs via the kidney and the lower GI tract, mainly the colon. Several factors modulate the internal and external potassium balance. [Modified from Giebisch, 1979.]

2. EPITHELIAL LOCALIZATION OF POTASSIUM TRANSPORT

In vertebrates the renal tubules are ultimately charged with the excretion of the bulk of the daily potassium, although a small portion, usually less than 10% of the potassium ingested, is secreted into the colon (Giebisch, 1979; Giebisch *et al.*, 1981). Both in the kidney and in the intestine, net potassium transport occurs only across well-defined segments of these epithelial structures.

Following free filtration across the glomerulus, potassium ions are extensively reabsorbed along the "proximal" nephron, i.e., the proximal tubule and the loop of Henle. Only the epithelium of the "late" distal tubule, i.e., that of the initial and cortical collecting tubule, has the ability to respond to a variety of metabolic stimuli with dramatic and predictable changes in net potassium transport (Giebisch, 1979; Giebisch *et al.*, 1981). The most important factors modulating renal tubular potassium transport are changes in (1) potassium balance, (2) in the level of circulating adrenal steroids, (3) in acid–base balance, and (4) in the composition of distal tubular fluid with respect to flow rate, sodium and anion composition. It is by secretion or reabsorption of potassium ions across specialized regions of the "distal" nephron that regulation of renal potassium excretion is achieved.

Even along these well-defined and highly specialized epithelial structures of the "distal" nephron further localization of potassium transport may occur. Thus, it is possible that potassium transport does not take place equally in all epithelial cells. This tentative

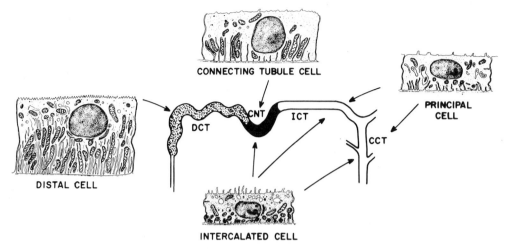

CONNECTING TUBULE CELL

PRINCIPAL CELL

DISTAL CELL

INTERCALATED CELL

Figure 2. Distribution of different cell types along superficial distal nephron of the rat. [From Stanton *et al.*, 1981.]

conclusion is based upon marked cell heterogeneity and striking relationships between function and fine structure of individual cell types in these nephron segments when changes in potassium transport occur.

Figure 2 shows that cytological heterogeneity is a feature of those renal nephron sites controlling potassium secretion (Giebisch *et al.*, 1981; Kaissling and Kritz, 1979; Rastegar *et al.*, 1980). In the mammalian "distal" tubule, potassium secretion occurs across three cytologically well-defined segments, the connecting tubule, the initial and the cortical collecting tubule. These tubular segments are made up of connecting tubule cells, principal cells, and intercalated cells. Morphometric analyses by electron microscopy has clearly shown that upon stimulation of potassium secretion, only *two* cells types, the connecting tubule cell and the principal cell, respond with transport-related alterations of their fine structure. Similar observations have also been made in the cortical and papillary collecting tubules (Rastegar *et al.*, 1980; Wade *et al.*, 1979). Only these cells show a very dramatic increase of their *basolateral* surface area when potassium secretion has been stimulated by prolonged pretreatment with a high-potassium diet. Such surface amplification is completely absent in intercalated cells. In sharp contrast, when animals are made to conserve potassium by dietary withdrawal of potassium, only intercalated cells show significant fine-structural alterations. These morphological changes are confined to the *luminal* cell membrane (Stetson *et al.*, 1980). Observations such as these are consistent with functional specialization of potassium-transporting cells suggesting that morphologically different cell types are involved in potassium secretion and potassium reabsorption. This has important implications concerning the precise localization of relevant transport mechanisms in the cells subserving potassium transport.

3. CELLULAR TRANSPORT MECHANISMS

The analysis of cellular transport mechanisms regulating renal tubular potassium secretion and reabsorption requires knowledge of the net transport rates of potassium ions under a wide variety of experimental conditions, of accurate measurements of unidirec-

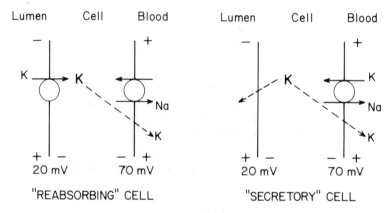

Figure 3. Cell models to account for transport and regulation of potassium ions across the initial collecting tubule ("late" distal tubule). (Top) A cell model incorporating all known transport components in *one* cell type. Inset on right demonstrates possibility of either electrically neutral or directly rheogenic operation of active peritubular Na^+–K^+ exchange. (Bottom) Hypothetical distribution of transport components in the two cell types lining the initial and cortical collecting tubule. Shown are a reabsorbing (intercalated, "dark") and a secretory (principal, "light") cell. [From Stanton and Giebisch, 1981.]

tional fluxes, of the electrochemical driving forces and specific potassium conductances of the luminal and antiluminal membranes of the transporting cells. In addition, a similar assessment of intercellular transport parameters is necessary. Such a complete analysis is not yet possible, and has been made even more difficult by the above-mentioned possibility of functional cell heterogeneity. Nevertheless, certain general features of the cellular transport mechanisms have emerged.

Figure 3 provides schematic cell models summarizing our present state of knowledge of the initial collecting tubule (Giebisch, 1979; Giebisch and Stanton, 1981). A similar model has recently also been suggested to explain the known features of potassium transport across the cortical collecting tubule (O'Neil and Boulpaep, 1979). At the top of Fig. 3, relevant transport parameters are placed in *one* cell type. The cell models on the bottom show how these processes may be distributed in two cell types, one representing a "secreting" and the other a "reabsorbing" cell.

3.1. The Peritubular (Basal) Cell Membrane

This cell membrane shares many properties of other body cells including marked electrical polarization, a relatively high potassium conductance, and a sodium–potassium exchange pump.

Evidence for *active* potassium uptake across the peritubular cell membrane has now been firmly established by measurements of transmembrane electrical potential differences and intracellular potassium activities (Oberleithner and Giebisch, 1981). Thus, the peritubular membrane potential is inadequate to account for the observed magnitude of the potassium ion activity gradient that is maintained across the peritubular cell membrane. The insets in Fig. 3 show the possibilities of either electroneutral or directly rheogenic (electrogenic) operation of the peritubular exchange mechanism. Particularly under conditions of a sharply increased luminal sodium load can it be shown that enhancement of peritubular sodium–potassium exchange directly hyperpolarizes the cell membrane (Wiederholt and Giebisch, 1974). Other properties of the peritubular active uptake of potassium ions include its saturability (Stanton and Giebisch, 1978) and the fact that several stimuli known to modulate potassium secretion, such as changes in the plasma potassium level, mineralocortiocoids, and acid–base parameters, modulate potassium uptake across the peritubular cell membrane (deMello-Aires *et al.*, 1973). A significant involvement of the peritubular sodium–potassium exchange pump in the sharp stimulation of distal tubular potassium secretion following *chronic potassium adaptation* has also been deduced from the local and sharply restricted increase of Na^+/K^+-stimulated ATPase in tubular segments of the cortical collecting tubule, a tubule with similar cell types as the initial collecting tubule (Doucet and Katz, 1980).

3.2. The Luminal (Apical) Cell Membrane

The luminal cell membrane is partly depolarized, has also a marked permeability to potassium ions, and, in addition, an active mechanism for potassium uptake from cell to lumen. It is the balance between the latter and the electrochemical potential difference of potassium ions that determines the rate of secretory potassium movement across the luminal cell membrane (Giebisch, 1979). Either elevation of *cellular* potassium activity, as observed during augmentation of peritubular potassium levels in control or potassium-adapted state (Oberleithner and Giebisch, 1981a), or lowering of the *luminal* potassium concentration by perfusion of the lumen of distal tubules with artificial solutions (Good and Wright, 1979) results in increased potassium secretion. Electrical polarization of the luminal cell membrane by an external "voltage clamp" also demonstrates the sensitivity of the luminal potassium activity to electrical driving forces (Garcia *et al.*, 1980).

The luminal cell membrane is also the site where sodium ions enter the distal tubular cell. An increased delivery of sodium ions into the distal nephron has often been shown to augment potassium secretion. Hence, the suggestion has been made that enhance sodium entry across the *luminal* cell membrane stimulates *peritubular* sodium–potassium exchange (Khuri *et al.*, 1975). However, enhancement of distal potassium secretion can also occur without an increase in net sodium reabsorption under conditions where the luminal potassium concentration falls and the electrochemical potential difference for passive potassium egress from cell to lumen is increased (Good and Wright, 1979).

The possibility that the rate of luminal active potassium uptake is subject to functional modulation has recently been explored by measuring the electrochemical potential difference and permeability of potassium ions across the luminal cell membrane (Oberleithner and

Giebisch, 1981a). Similar to the peritubular cell membrane, potassium ions are actively pumped from the lumen into the cell. This conclusion is based on the observation that the electrical potential across the luminal membrane is too small to account for this transmembrane activity difference (Oberleithner and Giebisch, 1981a). A similar double-pump model of potassium uptake across the apical *and* basal cell membrane has also been described in the isolated turtle bladder (Husted and Steinmetz, 1980). The precise nature of the active luminal potassium pump has not been fully elucidated but a number of points deserve mention. First, active luminal potassium uptake may cease when distal tubular potassium secretion is stimulated during potassium adaptation such that the passive leak of potassium ions from cell to lumen would be unopposed by active potassium reabsorption (Oberleithner and Giebisch, 1981a). A second point of interest concerns the possible interaction of potassium movement with that of other ions. Along the *early* distal tubule, a cell segment sharing some transport properties with the thick ascending limb of Henle's loop, some evidence indicates functional linkage of neutral sodium chloride cotransport to potassium reabsorption (Oberleithner and Giebisch, 1981b). The driving force for such potassium transport against a concentration gradient would be derived from the sodium and chloride activity difference across the luminal cell membrane. Arguments in favor of such coupling are based on experiments in amphibian early distal tubules. Luminal application of furosemide (a diuretic known to inhibit sodium chloride cotransport in many epithelia, including the renal tubule) or perfusion with chloride-free solutions blocks net potassium reabsorption (Oberleithner and Giebisch, 1981b). Perfusion of mammalian distal tubules with solutions in which chloride ions had been replaced by sulfate also show an increase of net potassium secretion without a change in the transepithelial potential difference (Velasquez *et al.*, 1980). Both these observations could be interpreted within the framework of a cell model in which the potassium pump–leak system in the luminal cell membrane involves (1) passive movement of potassium ions from cell to lumen and (2) a specific cotransport system that involves simultaneous electroneutral translocation of potassium with sodium chloride in the opposite direction, i.e., from the lumen into the cell. Enhanced net secretion of potassium ions would occur whenever chloride ions in the lumen are replaced by other anions or as the cotransport carrier is specifically inhibited by diuretics such as furosemide.

Another possibility is that luminal potassium reabsorption involves countertransport of potassium ions and sodium. Potassium ions could actively be pumped across the luminal cell membrane in exchange for sodium ions entering the tubular lumen. The energy for this exchange process would be derived from ATPase. Net sodium reabsorption in a direction opposite to that of active extrusion could still occur provided that there is a high passive sodium permeability across the luminal cell membrane. Such a mechanism of active potassium–sodium exchange has been suggested for the apical cell membrane of the turtle bladder, a preparation in which net potassium transport occurs under short-circuit conditions (Husted and Steinmetz, 1980).

Finally, potassium uptake from lumen to cell could also be coupled to hydrogen ion extrusion (Stetson *et al.*, 1980). It is of interest that states of potassium retention are frequently associated with enhanced hydrogen ion secretion. Stimulation of distal tubular potassium–hydrogen exchange could account for augmentation of potassium reabsorption and net hydrogen ion secretion. It is possible that several of these co- and countertransport mechanisms exist and are distributed differentially along the distal tubule.

3.3 Functional Cell Heterogeneity

The type of cell model described at the top of Fig. 3 may be an oversimplification. Based on the previously mentioned morphological evidence, it is possible that key elements of potassium transport may also be differentially distributed in the cells lining the segments of the distal tubule.

The lower panel of Fig. 3 shows one possible distribution of potassium transport components in the *luminal* membrane of "reabsorbing" and "secretory" cell types. "Reabsorbing" cells, possibly the intercalated (or dark) cells lining the initial and cortical collecting tubule, could be distinguished by low potassium permeability of the luminal membrane and presence of an active mechanism of potassium uptake. On the other hand, "secretory" cells (principal or "light" cells) would be characterized by a high luminal permeability to potassium ions and by absence of a reabsorptive potassium pump in the luminal membrane.

3.4. Paracellular Transport of Potassium Ions

In addition to the described transcellular route of potassium transport, it must also be considered that potassium movement could occur *between* cells. Such potassium movement could be driven by the lumen-negative electrical potential difference that is normally present across the initial and cortical collecting tubule (Giebisch, 1979). Direct and quantitative information on the functional role of potassium movement through a paracellular "shunt" pathway is presently not available for the mammalian distal nephron. If the normal transepithelial potential difference across the rat distal tubule is lowered to zero by application of an external current, the luminal potassium concentration remains above the level of plasma potassium (Garcia *et al.*, 1980). From this it is clear that *cellular* secretory mechanisms must participate in potassium transport. Similarly, when early distal tubules of *Amphiuma* are perfused with furosemide, the transepithelial potential difference approaches zero yet net potassium secretion is stimulated (Oberleithner and Giebisch, 1981a). Again, this evidence demonstrates participation of a transcellular secretory pathway in the overall transport of potassium. However, these results do not exclude some potassium secretion via a paracellular route. The precise magnitude and functional role of such a mode of potassium transport have yet to be defined.

ACKNOWLEDGMENT. Work in the author's laboratory was supported by NIH Grant AM17433.

REFERENCES

Bia, M. J., and DeFronzo, R. A. (1981). *Am. J. Physiol.* **240**, F257–F268.

DeFronzo, R. A., Birkhead, G., and Bia, M. (1979). *Kidney Int.* **16**, 917–925.

DeFronzo, R. A., Lee, R., Jones, A., and Bia, M. (1980). *Kidney Int.* **17**, 586–594.

deMello-Aires, M., Giebisch, G., Malnic, G., and Curran, P. F. (1973). *J. Physiol.* **232**, 47–70.

Doucet, A., and Katz, A. (1980). *Am. J. Physiol.* **238**, F380–F386.

Frayley, D. S., and Adler, S. (1977). *Kidney Int.* **12**, 354–360.

Garcia, E., Malnic, G., and Giebisch, G. (1980). *Am. J. Physiol.* **238**, F235–F246.

Giebisch, G. (1979). In *Membrane Transport in Biology* (G. Giebisch, D. C. Tosteson, and H. H. Ussing, eds.), Vol. IVa, pp. 215–298, Springer-Verlag, Berlin.

Giebisch, G., and Stanton, B. (1981). *Mineral Electrol. Met.* **5**, 100–120.

Giebisch, G., Malnic, G., and Berliner, R. W. (1981). In *The Kidney* (B. Brenner and F. C. Rector, Jr., eds.), pp. 408–439, Saunders, Philadelphia.

Good, D. W., and Wright, F. S. (1979). *Am. J. Physiol.* F192–F205.

Hiatt, N., Yamakawa, T., Davidson, M. B., and Miller, A. (1974). *Metabolism* **23**, 43–51.

Husted, R. F., and Steinmetz, P. R. (1980). Abstracts, 34th Annual Meeting of the Society of General Physiology, Abstract 62.

Kaissling, B., and Kritz, W. (1979). *Structural Analysis of the Rabbit Kidney,* Springer-Verlag, Berlin.

Khuri, R. N., Strieder, N., Wiederholt, M., and Giebisch, G. (1975). *Am. J. Physiol.* **278**, 1249–1261.

Knochel, J. P. (1977). *Kidney Int.* **11**, 443–452.

Oberleithner, H., and Giebisch, G. (1981a). In *Epithelial Ion and Water Transport* (A. D. C. MacKnight and J. Leader, eds.), pp. 97–105, Raven Press, New York.

Oberleithner, H. and Giebisch, G. (1981b). Abstracts, 8th International Congress of Nephrology, TT079.

O'Neil, R., and Boulpaep, E. L. (1979). *J. Membr. Biol.* **50**, 365–387.

Rastegar, A., Biemesderfer, D., Kashgarian, M., and Hayslett, J. P. (1980). *Kidney Int.* **18**, 293–301.

Stanton, B. A., and Giebisch, G. (1978). *Fed. Proc.* **37**, 728.

Stanton, B. A., and Giebisch, G. (1981). *Mineral and Electrolyte Metabolism* **5**, 100–120.

Stanton, B. A., Biemesderfer, D., Wade, J. B., and Giebisch, G. (1981). *Kidney Int.* **19**, 36–48.

Stetson, D., Wade, J. B., and Giebisch, G., (1980). *Kidney Int.* **17**, 45–56.

Velasquez, H., Wright, F. S., and Good D. W. (1980). Abstracts, 13th Annual Meeting of the American Society of Nephrology, p. 153A.

Wade, J. B., O'Neil, R. G., Pryor, J. L., and Boulpaep, E. L. (1979). *J. Cell Biol.* **81**, 439–445.

Wiederholt, M., and Giebisch, G. (1974). *Fed. Proc.* **33**, 387.

119

Pathways for Solute Transport Coupled to H^+, OH^-, or HCO_3^- Gradients in Renal Microvillus Membrane Vesicles

Peter S. Aronson

1. INTRODUCTION

In secondary active transport, the net transmembrane flux of a solute against its electrochemical gradient is coupled to, and therefore is secondary to, the flow across the same membrane of a second solute for which a favorable electrochemical gradient exists (Aronson, 1981). This mechanism contrasts with primary active transport, whereby the uphill flux of a solute is directly coupled to the flow of an exergonic chemical reaction. Current evidence suggests that many solutes, including sugars, amino acids, organic acids, phosphate, and sulfate, are transported across the luminal membrane of the proximal tubular cell via cotransport with Na^+ (Sacktor, 1977; Murer and Kinne, 1980). For these solutes, energization of secondary active transport is achieved by coupling solute flux to the inwardly directed electrochemical Na^+ gradient that is normally present across the luminal membrane as the result of primary active Na^+ extrusion across the basolateral membrane of the cell.

Na^+, however, is not the only ion for which an electrochemical gradient exists across the luminal membrane of the proximal tubular cell. The process of active H^+ extrusion across the luminal membrane results in the formation of an inwardly directed electrochemical H^+ gradient, an outwardly directed electrochemical OH^- gradient, and, in the presence of CO_2, an outwardly directed electrochemical HCO_3^- gradient. The purpose of this review is to briefly outline several recently described transport pathways by which solute flux across the luminal membrane of the proximal tubular cell can be coupled to electrochemical gradients of H^+, OH^-, or HCO_3^-. A pathway that mediates Na^+–H^+ exchange, by utilizing the electrochemical Na^+ gradient to energize secondary active H^+ secretion, likely

Peter S. Aronson • Departments of Medicine and Physiology, Yale University School of Medicine, New Haven, Connecticut 06510.

plays a role in generating the electrochemical gradients of H^+, OH^-, and HCO_3^-. Pathways that mediate anion–OH^- and anion–HCO_3^- exchanges may play roles in utilizing the electrochemical gradients of OH^- and HCO_3^- to energize secondary active anion absorption.

2. Na^+–H^+ EXCHANGE

In microvillus membrane vesicles prepared from the renal cortex of the rat or rabbit, uphill Na^+ accumulation results from imposition of an out < in H^+ gradient, and uphill H^+ efflux from imposition of an out > in Na^+ gradient, consistent with the presence of a transport system mediating Na^+–H^+ exchange (antiport) in the luminal membrane of the proximal tubular cell (Murer *et al.*, 1976; Kinsella and Aronson, 1980). (Although it has become customary to refer to this transport process as Na^+–H^+ exchange, it should be noted that Na^+–H^+ exchange cannot be distinguished from Na^+/OH^- cotransport on the basis of currently available evidence.) The accumulation of Na^+ by membrane vesicles in either the presence or the absence of an H^+ gradient is unaffected by experimental maneuvers that alter the transmembrane electrical potential difference, implying that the Na^+–H^+ exchange process is electroneutral with a fixed 1 : 1 stoichiometry.

The renal microvillus membrane Na^+–H^+ exchanger appears to have affinity for Li^+ and NH_4^+, in addition to Na^+ and H^+, and can likely function in multiple exchange modes involving these four cations (Kinsella and Aronson, 1981a). The rate of Na^+ uptake (K_m 6 mM) is competitively inhibited by external Li^+ (K_i 2 mM) and NH_4^+ (K_i 4 mM) but is not inhibited by external K^+, Rb^+, Cs^+, or choline. The existence of a Li^+–H^+ exchange mode is supported by the finding that uphill H^+ efflux results from imposition of an out > in Li^+ gradient. Uphill H^+ efflux also results from imposition of an out > in NH_4^+ gradient, but this may reflect entry of NH_4^+ via nonionic diffusion of NH_3 rather than via mediated NH_4^+–H^+ exchange. Nevertheless, that NH_4^+ can be transported by the system, at least in exchange for Na^+, is suggested by the observation that Na^+ efflux is accelerated by external NH_4^+. Unidirectional Na^+ efflux is also accelerated by external Na^+, indicating the presence of an Na^+–Na^+ exchange mode, but is inhibited by external Li^+. External K^+ has no effect on the efflux of either Na^+ or H^+.

Na^+ transport via this pathway is competitively inhibited by harmaline (K_i 9×10^{-5} M; Aronson and Bounds, 1980) and amiloride (K_i 2×10^{-5} M; Kinsella and Aronson, 1981b), but is not inhibited by acetazolamide, furosemide, or 4-acetamido-4'-isothiocyanostilbene-2,2'-disulfonate (SITS). Although harmaline appears to be a competitive inhibitor for the Na^+ site of several microvillus membrane transport systems, amiloride, at concentrations up to 5×10^{-4} M, has no direct effect on other Na^+-coupled transport processes, such as Na^+/glucose cotransport and Na^+/alanine cotransport. It is interesting to note that the apparent K_i for amiloride inhibition of the renal microvillus membrane Na^+–H^+ exchanger is more consistent with the relatively high concentrations of the agent (10^{-4} M) required to abolish Na^+–H^+ exchange in nonepithelial tissues (Aickin and Thomas, 1977; Johnson *et al.*, 1976) than with the apparent K_i (10^{-8}–10^{-7} M) for inhibition of conductive Na^+ channels (Na^+ uniport) in "tight" epithelia (Cuthbert and Shum, 1975).

3. ANION EXCHANGE

In microvillus membrane vesicles isolated from the dog renal cortex, imposition of an out < in pH gradient (pH_o 6.0, pH_i 7.5) markedly accelerates the uptake of urate and causes the transient accumulation of the anion above its equilibrium level of uptake (Blomstedt and Aronson, 1980). In the absence of a transmembrane pH gradient, the rate of urate uptake at pH 6.0 is not greater than at pH 7.5, suggesting that the pH gradient stimulation of urate uptake results from H^+/anion cotransport or OH^-–anion exchange rather than from transport of the undissociated weak acid per se. This pH-gradient-stimulated urate transport is saturable with respect to external urate (K_m 0.4 mM at pH_o 6.5, pH_i 7.5) and is inhibited by agents, such as salicylate, furosemide, probenecid, 4,4′-diisothiocyanostilbene-2,2′-disulfonate (DIDS), and SITS (Blomstedt and Aronson, 1980; Kahn and Aronson, 1981a), which are known to inhibit erythrocyte anion exchange (Wieth, 1970; Brazy and Gunn, 1976; Motais and Cousin, 1976; Cabantchik and Rothstein, 1972).

The same transport system appears also to have affinity for Cl^- and *para*-aminohippurate (PAH). External Cl^- inhibits pH-gradient-stimulated urate uptake and, in the absence of transmembrane pH gradients, out < in Cl^- gradients stimulate urate influx and out > in Cl^- gradients stimulate urate efflux (Kahn and Aronson, 1981b). In addition, Cl^--gradient-stimulated urate transport is sensitive to inhibition by probenecid, DIDS, and SITS. PAH inhibits pH-gradient-stimulated urate uptake and itself undergoes pH-gradient-stimulated transport that is inhibited by urate, salicylate, furosemide, probenecid, DIDS, and SITS. The simplest model accounting for the properties of this transport system is that of a 1 : 1 exchanger with affinity for multiple anions, including urate, PAH, OH^-, and Cl^-. The full range of anion specificity for the system remains to be determined. It should be emphasized that HCO_3^- rather than OH^- may well be a substrate for the exchanger, as none of these microvillus membrane vesicle studies were performed under rigorously CO_2-free conditions.

In rabbit renal microvillus membrane vesicles, imposition of an out < in pH gradient induces uphill Cl^- accumulation, suggesting the presence of a transport system mediating either H^+/Cl^- cotransport or exchange of Cl^- with OH^- or HCO_3^- (Warnock and Yee, 1981). It has not been determined to what extent this pH-gradient-stimulated Cl^- transport proceeds via the organic anion–OH^-–Cl^- exchanger described above. Likewise, the sensitivity of this pH-gradient-stimulated Cl^- transport to anion exchange inhibitors has not been reported. It is interesting to note that pH-gradient coupled Cl^- transport in rat intestinal microvillus membrane vesicles is inhibited by probenecid, furosemide, and SITS (Liedtke, 1979; Liedtke and Hopfer, 1980).

Finally, Cl^- efflux from *Necturus* renal microvillus membrane vesicles is stimulated by external Cl^- and HCO_3^-, consistent with the existence of an exchanger having affinity for these anions (Seifter *et al.*, 1981). This system is inhibited by SITS, but the effect of other potential inhibitors has not been tested. The possible interaction of organic anions with this exchanger has also not been evaluated.

4. PHYSIOLOGICAL ROLES OF THESE PATHWAYS

Luminal membrane transport pathways that couple solute flux to H^+, OH^-, or HCO_3^- gradients may play important roles in mediating active transtubular anion absorption. As discussed earlier, an inwardly directed electrochemical Na^+ gradient is normally maintained across the luminal membrane of the proximal tubular cell due to the primary active extrusion of Na^+ across the basolateral membrane. The operation of the Na^+–H^+ exchanger, as illustrated in the first panel of Fig. 1, could drive uphill H^+ secretion from cell to lumen. In the presence of luminal HCO_3^-, the net effect of this process would be to mediate uphill HCO_3^- transport into the cell, energized by the electrochemical Na^+ gradient. It is worth noting that the Na^+–H^+ exchanger could similarly mediate the absorption of the anion of any weak acid that crosses the luminal membrane by nonionic diffusion.

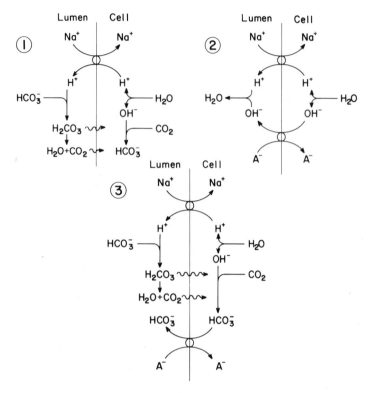

Figure 1. Possible physiological roles of ion exchangers in mediating anion absorption. See text for details.

Likely examples include glycodiazine, butyrate, propionate, and acetate (Ullrich *et al.*, 1971).

Although an amiloride-sensitive Na^+–H^+ exchanger has been demonstrated to be functional in the luminal membrane of the intact proximal tubular cell (Boron and Boulpaep, 1981; Schwartz, 1981), the importance of Na^+–H^+ exchange, as compared to primary active H^+ secretion, in mediating proximal tubular acidification has been disputed. Whereas the rate of bicarbonate absorption in the continuously perfused proximal tubule of the rabbit and rat is inhibited greater than 80% by removing Na^+ or by inhibiting primary active Na^+ transport with ouabain or K^+ removal (Burg and Green, 1977; Chantrelle *et al.*, 1979), similar maneuvers inhibit acidification only partially or not at all in the rat or hamster proximal tubule studied by stationary microperfusion (deMello-Aires and Malnic, 1979; Ullrich *et al.*, 1977). As recently discussed (Aronson, 1981), these latter findings are not necessarily inconsistent with the concept that acidification occurs predominately by secondary active H^+ secretion via Na^+–H^+ exchange, but the issue awaits final resolution.

The outwardly directed electrochemical gradient of OH^- or HCO_3^- resulting from active H^+ secretion could drive uphill anion transport across the luminal membrane via exchange for OH^- or HCO_3^-. To the extent that active H^+ secretion occurs via Na^+–H^+ exchange, the net effect of these processes, as shown in the second and third panels of Fig. 1, would be to couple uphill anion transport to the electrochemical Na^+ gradient, thereby simulating electroneutral Na^+/anion cotransport. The concept that neutral salt absorption could be accomplished by double ion exchange is at least three decades old (Berliner, 1952). A possible example of this phenomenon is the process of active Cl^-

absorption across the luminal membrane of the *Necturus* proximal tubular cell. In this cell, the chloride activity is maintained above that predicted for electrochemical equilibrium across the luminal membrane by an electroneutral process requiring the presence of both Na^+ and Cl^- in the luminal fluid (Spring and Kimura, 1978). In *Necturus* renal microvillus membrane vesicles, both a SITS-sensitive anion exchanger with affinity for Cl^- and HCO_3^- and an amiloride-sensitive Na^+–H^+ exchanger have been identified, but not a direct Na^+/Cl^- cotransport pathway (Seifter *et al.*, 1981). In contrast, the chloride activity of the rat proximal tubular cell is at the value predicted for electrochemical equilibrium across the luminal membrane (Sohtell, 1978). Hence, in the mammalian proximal tubule there is no evidence for active chloride transport across the luminal membrane and the participation of luminal membrane Cl^-–OH^- exchange in mediating transepithelial Cl^- absorption is uncertain.

The possible role of luminal membrane anion exchange in mediating transtubular urate and PAH transport depends importantly on the nature of transport processes at the basolateral membrane. For example, in the presence of a passive basolateral membrane exit step, uphill urate uptake across the luminal membrane via exchange for OH^- or HCO_3^- could drive the active transtubular urate absorption that occurs in the proximal tubules of some animals (Weiner, 1979). On the other hand, in the presence of an active urate uptake process at the basolateral membrane that is sufficiently powerful to generate an outwardly directed urate gradient across the luminal membrane in excess of the electrochemical gradient of OH^- or HCO_3^-, cell urate could exchange for luminal OH^-, HCO_3^-, or Cl^-, thereby accomplishing transtubular urate secretion. (Of course, in animals having proximal tubules that secrete urate or PAH it is quite possible that cell to lumen organic anion flux occurs entirely via passive, noncoupled diffusion.)

Clearly, the precise physiological roles for luminal membrane transport pathways that couple proximal tubular solute flux to H^+, OH^-, or HCO_3^- gradients remain to be fully elucidated. Important new information concerning these pathways should emerge over the next several years.

REFERENCES

Aickin, C. C., and Thomas, R. C. (1977). *J. Physiol.* **273**, 295–316.

Aronson, P. S. (1981). *Am. J. Physiol.* **240**, F1–F11.

Aronson, P. S., and Bounds, S. E., (1980). *Am. J. Physiol.* **238**, F210–F217.

Berliner, R. W. (1952). *Fed. Proc.* **11**, 695–700.

Blomstedt, J. W., and Aronson, P. S. (1980). *J. Clin. Invest.* **65**, 931–934.

Boron, W. F., and Boulpaep, E. L. (1981). *Kidney Int.* **19**, 233.

Brazy, P. C., and Gunn, R. B. (1976). *J. Gen. Physiol.* **68**, 583–599.

Burg, M., and Green, N. (1977). *Am. J. Physiol.* **233**, F307–F314.

Cabantchik, Z. I., and Rothstein, A. (1972). *J. Membr. Biol.* **10**, 311–330.

Chantrelle, B., Cogan, M. G., and Rector, F. C., Jr. (1979). *Kidney Int.* **16**, 809.

Cuthbert, A. W., and Shum, W. K. (1975). *Proc. R. Soc. London Ser. B* **189**, 543–575.

deMello-Aires, M., and Malnic, G. (1979). *Am. J. Physiol.* **236**, F434–F441.

Johnson, J. D., Epel, D., and Paul, M. (1976). *Nature (London)* **262**, 661–664.

Kahn, A. M., and Aronson, P. S. (1981a). *Kidney Int.* **19**, 245.

Kahn, A. M., and Aronson, P. S. (1981b). *Fed. Proc.* **40**, 463.

Kinsella, J. L., and Aronson, P. S. (1980). *Am. J. Physiol.* **238**, F461–F469.

Kinsella, J. L., and Aronson, P. S. (1981a). *Am. J. Physiol.* **241**, C220–C226.

Kinsella, J. L., and Aronson, P. S. (1981b). *Am. J. Physiol.* **241**, FR374–F379.

Liedtke, C. M. (1979). *Biophys. J.* **25**, 94a.

Liedtke, C. M., and Hopfer, U. (1980). *Fed. Proc.* **39**, 734.

Motais, R., and Cousin, J. L. (1976). *Biochim. Biophys. Acta* **419**, 309–313.

Murer, H., and Kinne, R. (1980). *J. Membr. Biol.* **55**, 81–95.

Murer, H., Hopfer, U., and Kinne, R. (1976). *Biochem. J.* **154,** 597–604.

Sacktor, B. (1977). *Curr. Top. Bioenerg.* **6,** 39–81.

Schwartz, G. J. (1981). *Am. J. Physiol.* **241,** F380–F385.

Seifter, J., Kinsella, J. L., and Aronson, P. S. (1981). *Kidney Int.* **19,** 257.

Sohtell, M. (1978). *Acta Physiol. Scand.* **103,** 363–369.

Spring, K. R., and Kimura, G. (1978). *J. Membr. Biol.* **38,** 233–245.

Ullrich, K. J., Radtke, H. W., and Rumrich, G. (1971). *Pfluegers Arch.* **330,** 149–161.

Ullrich, K. J., Capasso, G., Rumrich, G., Papavassiliou, F., and Kloss, S. (1977). *Pfluegers Arch.* **368,** 245–252.

Warnock, D. G., and Yee, V. J. (1981). *J. Clin. Invest.* **67,** 103–115.

Weiner, I. M. (1979). *Am. J. Physiol.* **237,** F85–F92.

Wieth, J. O. (1970). *J. Physiol.* **207,** 581–609.

120

Cell Culture Models to Study Epithelial Transport

Julia E. Lever

1. INTRODUCTION

Ironically, differentiated cell lines derived from transporting epithelia have been available for almost 20 years in some cases, yet the study of transepithelial transport in cell culture is a relatively recent advance. This approach promises important new possibilities to integrate biochemical, electrophysiological, cellular, and molecular approaches to study epithelial transport function and its regulation by biologically active agents (Cereijido *et al.*, 1978a,b; Handler *et al.*, 1979a).

A major difficulty encountered with the use of intact tissue preparations is the presence of more than one cell type; cell culture techniques offer the possibility of isolating each cell type as a pure culture, generating artificial epithelial sheets for electrical and transport measurements by conventional techniques. Such monolayers can be obtained even from epithelia that do not conveniently occur as sheets in nature. Increased possibilities for experimental manipulation also include control of cell proliferative, nutritional, and hormonal states (Rindler *et al.*, 1979b; Taub and Saier, 1979b). This optimism must be tempered by recognition of the increased possibility for artifacts due to malignancy, selective changes in the cell population in long-term culture, and removal of possible modulatory factors such as basement membranes and tissue architecture.

Epithelial cell lines that exhibit transepithelial transport properties are now available from whole dog kidney (MDCK) (Madin and Darby, 1958), pig kidney (LLC-PK$_1$) (Hull *et al.*, 1976), DMBA-induced rat mammary adenocarcinoma (Rama 25) (Bennett *et al.*, 1978), and toad urinary bladder (TB-M; TB6C) (Handler *et al.*, 1979b). Epithelial cell lines from human colon (McCombs *et al.*, 1976), rat, mouse, or human mammary gland (McGrath, 1975; Young *et al.*, 1978), intestine (Quaroni *et al.*, 1979), and liver (Owens *et al.*, 1974) have been isolated that exhibit certain presumptive indications of transepithe-

---

Julia E. Lever • Department of Biochemistry and Molecular Biology, University of Texas Medical School, Houston, Texas 77025.

231

lial fluid transport such as polarized morphology and focal regions of fluid accumulation in domes.

2. MORPHOLOGICAL AND GROWTH CHARACTERISTICS OF EPITHELIAL CELL CULTURES

Epithelial cells derived from transporting epithelia retain the capacity to maintain their polarized morphological orientation even after enzymatic disaggregation and plating in cell culture. This remarkable feature permits epithelial cell monolayers in culture to carry out the transepithelial transport processes characteristic of their tissue of origin.

In these cultures every cell is oriented with the basolateral plasma membrane side in contact with the solid support surface and the apical surface in contact with the growth medium. Epithelial cells must attach to a surface such as a plastic culture dish, nitrocellulose filter (Misfeldt et al., 1976), or collagen (Emerman and Pitelka, 1977); growth cannot occur in suspension although short-term survival is possible.

The mechanisms of orientation of the plasma membrane into morphologically, functionally, and enzymatically distinct opposite surfaces are not understood. Evidence for polarity of epithelial cells in culture is based on morphological studies (Leighton et al., 1969; Pickett et al., 1975), determination that [^{3}H]ouabain binding sites are confined to the basolateral surface (Mills et al., 1979; Cereijido et al., 1980a), amiloride sensitivity of apical but not basolateral surfaces of certain cell types (Cereijido et al., 1978b), polarity of budding of influenza, sendai, and vesticular stomatitis viruses in MDCK cells (Boulan and Sabatini, 1978), and by undirectional fluid transport properties (Misfeldt et al., 1976; Cereijido et al., 1978b; Bisbee et al., 1979; Bisbee, 1979). Valentich et al. (1979) demonstrated that nonproliferating suspension aggregates of MDCK cells retained morphological polarity, an observation that suggests attachment is not required to maintain polarity.

Taub et al. (1979) described a serum-free medium supplemented with insulin, hydrocortisone, transferrin, triiodothyronine, and prostaglandin E$_1$ that could support proliferation and dome formation of MDCK cells and primary kidney cultures.

3. TRANSPORT PROPERTIES OF EPITHELIAL CELL MONOLAYERS

Monolayers formed from epithelial cells in culture reconstitute the transepithelial transport properties characteristic of transporting epithelia. This fortunate characteristic derives from the maintenance in culture of morphological and functional plasma membrane polarity discussed above, as well as the formation of occluding junctions controlling permeation via the paracellular route.

Transport characterizations were preceded by early morphological evidence that epithelial cultures could transport fluid across the cell layer. Domes (Fig. 1), or focal regions of active fluid accumulation between the cell monolayer and the culture dish, were observed in a wide variety of epithelial cell types, including renal (Leighton et al., 1970), mammary (McGrath, 1975), liver (Owens et al., 1974), and urinary bladder epithelia (Handler et al., 1979a). Domes could be collapsed by ouabain (Abaza et al., 1974) or mechanical puncture. Time-lapse photography studies indicated that domes could rise, then collapse, apparently at random over the cell monolayer.

Although undirectional fluid transport from the apical to basolateral surface is clearly required to cause fluid accumulation in domes, it is not known whether specific biochemi-

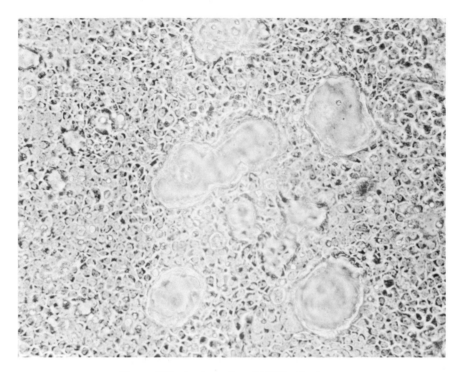

Figure 1. Domes in confluent MDCK cell cultures.

cal or functional changes accompany the appearance of domes in the monolayer. Nor is it understood why fluid accumulates in focal regions rather than uniformly under the cell monolayer. Perhaps domes represent focal regions of decreased cell–substratum adhesiveness. No morphological differences in cell–cell junctions or membrane polarity have been observed between cells in domes and those in the surrounding monolayer (Pickett *et al.*, 1975; Cereijido *et al.*, 1980b).

Misfeldt *et al.* (1976) provided the first direct measurement of transepithelial transport in cell culture. They demonstrated that MDCK cells plated on a porous support provided by nitrocellulose filters retained normal morphological polarity and epithelial intercellular junctions. Such preparations mounted in a Ussing chamber generated an apical-negative transepithelial potential of 1.42 mV, a resistance of 84 ohms·cm², catalyzed an apical to basolateral water flux of 7.3 μl/cm² per hr, and demonstrated selective permeability to Na^+ compared with Cl^-.

These results were confirmed and extended by Cereijido *et al.* (1978b) using MDCK cell monolayers supported on collagen-coated nylon mesh. Evidence supporting permeation of Na^+ via amiloride-sensitive Na^+-specific channels rather than a carrier-mediated process was obtained from electrical and tracer studies. Furthermore, the importance of Ca^{2+} in maintaining integrity of junctional complexes and high epithelial electrical resistance was demonstrated. In a subsequent study using MDCK cells, Cereijido *et al.* (1980a) measured a value of net apical to basolateral Na^+ transport of 2.6 μmoles/hr per cm² and found a large passive flux of this ion as characteristic of "leaky" epithelia. Exit of Na^+ at the basolateral surface was mediated by an ouabain-sensitive Na^+ pump, which was also localized on the basolateral surface by [³H]ouabain binding measurements. Accompanying these studies of functional polarization, additional morphological evidence for structural membrane polarization and disposition of cytoskeletal elements was presented. In a separate report, Cereijido *et al.* (1980b) found evidence, based on electric field scanning and freeze-fracture electron microscopy, for functional and morphological heterogeneity of occluding junctions in the MDCK cell population grown as a monolayer on a porous support.

By contrast, Rindler *et al.* (1979a) have proposed that Na^+ entry at the apical surface of MDCK cells is mediated by an electroneutral carrier-mediated process that catalyzes either net Na^+ uptake, $Na^+–Na^+$ exchange, or Na^+/H^+ antiport. These studies were performed using MDCK cell monolayers on plastic dishes in the presence of 0.5 mM ouabain to inhibit Na^+/K^+-ATPase-driven Na^+ efflux at the basolateral surface. Using similar approaches, Taub and Saier (1979a) proposed that Ca^{2+} can act as an allosteric regulator of Na^+ flux.

Rabito and Ausiello (1980) and Mullin *et al.* (1979) demonstrated Na^+-dependent glucose transport in LLC-PK_1 pig kidney epithelial cells. Cell monolayers grown on filter supports were used, but transepithelial transport was not directly measured. Rather, uptake into the cells was assayed by incubating the filters in labeled substrate, washing the filters, and terminating uptake in trichloroacetic acid. Sugar specificity and sensitivity to phloretin and phlorizin resembled those of renal proximal tubules.

Other hormone-responsive epithelial cell culture systems that have been utilized for transepithelial transport studies are toad urinary bladder epithelial cells (Handler *et al.*, 1979b) and mammary epithelial cells (Bisbee, 1979; Bisbee *et al.*, 1979). Although MDCK and LLC-PK_1 cell lines were derived from whole kidney, new approaches are designed to initiate cell cultures from defined anatomical regions of an epithelial tissue (Quaroni *et al.*, 1979).

As large quantities of epithelial cells can be obtained from cell cultures, isolation of plasma membrane vesicles for transport studies is feasible. This approach has been used in the past to characterize transport properties of fibroblast and tumor cell cultures, as well as membranes from renal and intestinal tissue preparations (reviewed in Lever, 1980). Hanna *et al.* (1980) have presented a preliminary account of the isolation of plasma membrane vesicles from LLC-PK_1 and MDCK cells.

4. CULTURE MODELS FOR EPITHELIAL DIFFERENTIATION

A remarkable analogy between development of domes in epithelial cell cultures and processes of cell differentiation was found (Lever, 1979a,b,c, 1981). Several categories of compounds known as inducers of differentiation in such diverse cell culture systems as Friend erythroleukemia cells (Marks and Rifkind, 1978) and neuroblastoma cells (Palfrey *et al.*, 1977) also triggered a dramatic increase over the spontaneous incidence of domes in confluent MDCK kidney cell cultures and in Rama 25 mammary cell cultures. Agents such as dimethylsulfoxide, dimethylformamide, and hexamethylene bisacetamide, purines such as inosine and adenosine, and certain fatty acids such as butyrate were effective inducers of dome formation. Also, conditions expected to elevate intracellular levels of cyclic AMP increased dome formation in both mammary and MDCK kidney cells. Similar effects of cyclic AMP on dome formation in MDCK cells were noted by Valentich *et al.* (1979).

A similar pattern of regulation of induction of dome formation emerged from both systems. Domes appeared at 7–15 hr after addition of inducer. Induction required cell confluence, serum, protein synthesis, but did not require DNA synthesis. Induced domes were collapsed after addition of ouabain or cytoskeletal inhibitors such as cytochalasin B.

Dome formation is a property specific to certain fluid-transporting epithelial cell types; many epithelial cell types (such as BSC-1 cells) and fibroblastic cell types (such as 3T3 cells) do not make domes spontaneously and are refractory to chemical induction of domes. Certain subclones were isolated from MDCK cells (Lever, 1981) that differed in response to inducers of dome formation; one group did not form domes with any inducer, other groups responded to some inducers but not to others.

The regulation of dome formation may be relevant to questions concerning aspects of secretory epithelial development and modulation. At present, this response refers to visible features that arise in these cultures; biochemical correlates of dome structure and function have not yet been identified. Dome cells could represent either selective changes in trans-epithelial fluid transport systems, alterations in cell junctions, or changes in adhesion to the substratum. Ultimately, it will be necessary to establish whether biochemical changes identified with dome formation in cell culture bear any resemblance to normal secretory epithelial development *in vivo*.

ACKNOWLEDGMENT. This work was supported by U.S. Public Health Service Grant AM-27400-01.

REFERENCES

Abaza, N. A., Leighton, J., and Schultz, S. G. (1974). *In Vitro* **10**, 172–183.
Bennett, D. C., Peachey, L. A., Durbin, H., and Rudland, P. S. (1978). *Cell* **15**, 283–298.
Bisbee, C. A. (1979). *Fed. Proc.* **38**, 1056.
Bisbee, C. A., Machen, T. E., and Bern, H. A. (1979). *Proc. Natl. Acad. Sci. USA* **76**, 536–540.
Boulan, E. R., and Sabatini, D. D. (1978). *Proc. Natl. Acad. Sci. USA* **75**, 5071–5075.
Cereijido, M., Rotunno, C. A., Robbins, E. S., and Sabatini, D. D. (1978a). In *Membrane Transport Processes* (J. F. Hoffman, ed.), p. 433, Raven Press, New York.
Cereijido, M., Robbins, E. S., Dolan, W. J., Rotunno, C. A., and Sabatini, D. D. (1978b). *J. Cell Biol.* **77**, 853–880.
Cereijido, M., Ehrenfeld, J., Meza, I., and Martinez-Palomo, A. (1980a). *J. Membr. Biol.* **52**, 147–159.
Cereijido, M., Stefani, E., and Martinez-Palomo, A. (1980b). *J. Membr. Biol.* **53**, 19–32.
Emerman, J. T., and Pitelka, D. R. (1977). *In Vitro* **13**, 316–328.
Handler, J. S., Perkins, F. M., and Johnson, J. P. (1979a). *Am. J. Physiol.* **238**, F1–F9.
Handler, J. S., Steele, R. E., Sahib, M., Wade, J. B., Preston, A. J., Lawson, N., and Johnson, J. P. (1979b). *Proc. Natl. Acad. Sci. USA* **76**, 4151–4155.
Hanna, S. D., Misfeldt, D., and Wright, E. M. (1980). *Fed. Proc.* **39**, 736.
Hull, R. N., Cherry, W. R., and Weaver, G. W. (1976). *In Vitro* **12**, 670–677.
Leighton, J., Brada, Z., Estes, L., and Justh, G. (1969). *Science* **163**, 472–473.
Leighton, J., Estes, L. W., Mansukhani, S., and Brada, Z.(1970). *Cancer* **26**, 1022–1028.
Lever, J. E. (1979a). In *Hormones and Cell Culture, Cold Spring Harbor Conferences on Cell Proliferation* (G. Sato and R. Ross, eds.), Vol. 6, p. 727, Cold Spring Harbor Laboratory, Cold Spring Harbor, N.Y.
Lever, J. E. (1979b). *Proc. Natl. Acad. Sci. USA* **76**, 1323–2437.
Lever, J. E. (1979c). *J. Supramol. Struct.* **12**, 259–272.
Lever, J. E. (1980). *CRC Crit. Rev. Biochem.* **7**, 187–246.
Lever, J. E. (1981). *Ann. N.Y. Acad. Sci.* **372**, 371–383.
McCombs, W. B., Leibovitz, A., McCoy, C. E., Stinson, J. C., and Berlin, J. D. (1976). *Cancer* **38**, 2316–2327.
McGrath, C. M. (1975). *Am. Zool.* **15**, 231–236.
Madin, S. H., and Darby, N. B. (1958). As cataloged in: *American Type Culture Collection Catalogue of Strains* (1975) Vol. 2, p. 47.
Marks, P. A., and Rifkind, R. A. (1978). *Annu. Rev. Biochem.* **47**, 419–448.
Mills, J. W., MacKnight, A.D.C., Dayer, J. M., and Ausiello, D. A. (1979). *Am. J. Physiol.* **236**, C157–C162.
Midfeldt, D. S., Hamamoto, S. T., and Pitelka, D. R. (1976). *Proc. Natl. Acad. Sci, USA* **73**, 1212–1216.

Mullin, J. M., Diamond, L., and Kleinzeller, A. (1979). *Fed. Proc.* **38,** 1058.

Owens, R. B., Smith, H. S., and Hackett, A. J. (1974). *J. Natl. Cancer Inst.* **53,** 261–266.

Palfrey, C., Kimhi, Y., Littauer, U. Z., Reuben, R. C., and Marks, P. A. (1977). *Biochem. Biophys. Res. Commun.* **76,** 937–942.

Pickett, P. B., Pitelka, D. R., Hamamoto, S. T., and Misfeldt, D. S. (1975). *J. Cell Biol.* **66,** 316.

Quaroni, A., Wands, J., Trelstad, R. L., and Isselbacher, K. J. (1979). *J. Cell Biol.* **80,** 248–265.

Rabito, C. A., and Ausiello, D. A. (1980). *J. Membr. Biol.* **54,** 31–38.

Rindler, M. J., Taub, M., and Saier, M. H. (1979a). *J. Biol. Chem.* **254,** 11431–11439.

Rindler, M. J., Chuman, L. M., Shaffer, L., and Saier, M. H. (1979b). *J. Cell Biol.* **81,** 635–648.

Taub, M., and Saier, M. H. (1979a). *J. Biol. Chem.* **254,** 11440–11444.

Taub, M., and Saier, M. H. (1979b). In *Methods in Enzymology* (W. B. Jacoby and I. H. Pastan, eds.), Vol. LVIII, pp. 552–560, Academic Press, New York.

Taub, M., Chuman, L., Saier, M. H., and Sato, G. (1979). *Proc. Natl. Acad. Sci. USA* **76,** 3338–3342.

Valentich, J. D., Tchao, R., and Leighton, J. (1979). *J. Cell Physiol.* **100,** 291–304.

Young, L. J., Cardiff, R. D., and Seeley, T. (1978). *In Vitro* **14,** 895–902.

121

The Electrochemical Proton Gradient and Catecholamine Accumulation into Isolated Chromaffin Granules and Ghosts

Robert G. Johnson, Sally E. Carty, and Antonio Scarpa

1. INTRODUCTION

Most of the catecholamines contained within the chromaffin cells of the adrenal medulla are localized to a specialized subcellular organelle, the chromaffin granule (Blaschko and Welch, 1953; Hillarp *et al.*, 1953). This single observation, which has many implications concerning the intragranular biosynthesis, catabolic degradation, and quantal release of catecholamines, has spurred intensive investigation of the molecular mechanism of catecholamine accumulation and storage. Quantitation of the amount of amines contained within the isolated granule emphasizes the enormity of the accumulation process that must be employed (Smith, 1968): the catecholamine content has been assayed at 2000 nmoles/mg protein and, in addition, nucleotides (mainly ATP) are present at 500 nmoles/mg protein. Based on the intragranular water space, the internal concentration of catecholamines and ATP would be 0.5 and 0.15 M, respectively, if each of these compounds were free in solution. These are extremely large concentrations by biological standards. Invoking principles of osmolality, one can recognize that the total population of these compounds cannot exist free in solution, and a precise interaction must exist between uptake at the granule membrane and storage in the intragranular matrix. However, despite two decades of research effort, the mechanism of catecholamine uptake across the granule membrane remains largely unsolved.

From the many experiments during that time, three molecular models for catecholamine accumulation have emerged. Two of these, the storage complex hypothesis (stating that uptake occurs purely by amine permeation down a concentration gradient maintained

Robert G. Johnson, Sally E. Carty, and Antonio Scarpa. • Department of Biochemistry and Biophysics, University of Pennsylvania School of Medicine, Philadelphia, Pennsylvania 19104.

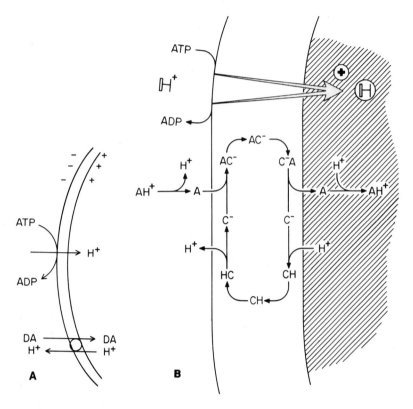

Figure 1. Models for amine accumulation. (A) Conceptualized; (B) molecular.

by strong internal binding; see Berneis *et al.*, 1969, 1970) and the active uptake hypothesis (proposing that uptake proceeds against a concentration gradient via a carrier-mediated mechanism directly coupled to ATP hydrolysis; see Kirshner, 1962; Carlsson *et al.*, 1962), are now considered untenable. Over the last few years, measurements from several laboratories interested in the physicochemical properties of the chromaffin granule membrane, the generation and maintenance of proton gradients, and the study of the electrochemical gradient for catecholamines have resulted in the H$^+$ gradient hypothesis, the basic elements of which are now generally accepted (Johnson *et al.*, 1978; Phillips, 1978; Schuldiner *et al.*, 1978; Ingebretsen and Flatmark, 1979; Njus and Radda, 1979; Johnson and Scarpa, 1979). The model, illustrated schematically in Fig. 1A, states that a proton-translocating ATPase exists in the chromaffin granule membrane, and the vectorial movement of protons from the cytosol across the membrane results in the generation of a ΔpH (inside acidic) and $\Delta\Psi$ (positive inside). Amine accumulation is thought to proceed by a separate, distinct, carrier-mediated process, dependent upon one or both components of the electrochemical proton gradient ($\Delta\bar{\mu}_{H^+}$) (Fig. 1A).

It is the purpose of this discussion to focus upon the essential features of the experimental data from which these models were constructed, which include (1) measurement of the membrane permeability, (2) properties of the ATPase, (3) measurement of the $\Delta\bar{\mu}_{H^+}$, and (4) distribution of catecholamines as a function of the $\Delta\bar{\mu}_{H^+}$. The qualitative and quantitative investigations of amine accumulation into the chromaffin granule have proven to be a very fruitful endeavor. In fact, when compared to other biological systems used for transport studies, the chromaffin granule offers several tremendous advantages, including the ability to isolate granules intact, under isotonic conditions, in high purity and large yield, and the ability to prepare ghosts devoid of all endogenous amines and nucleotides.

2. PERMEABILITY AND BUFFERING CAPACITY

The membrane of the chromaffin granule is the most impermeable to cations of any previously isolated subcellular organelle (Johnson and Scarpa, 1976a; Johnson *et al.*, 1978). For example, the conductance to protons may be one order of magnitude less than that of a mitochondrion, formerly thought to maintain the lowest permeability to protons (Scholes and Mitchell, 1970). In addition, the internal buffering capacity of the granules is quite high as well, approaching 300 μmoles H^+/pH unit per g dry wt. These two properties help to explain the experimental observation that isolated granules maintained at 4°C can maintain a ΔpH across the membrane for over 48 hr (Johnson *et al.*, 1978).

The empirical observations concerning the physicochemical properties of the chromaffin granule membrane may ultimately relate to its unique composition. For example, the lipid/protein ratio of the membrane is the highest of any cellular membrane, excepting the Schwann cell (Winkler, 1976). Moreover, the membrane contains the highest percentage of lysolecithin found within any mammalian biological membrane. Correlation of the physicochemical and biophysical measurements with the physiologic function of the chromaffin granule is just now beginning.

3. THE H^+-ATPase

While the study of the ATPase activity of the chromaffin granule membrane is still in its infancy, the majority of ATPase activity appears to be due to an H^+-translocating ATPase similar in many respects to the F_1 mitochondrial ATPase (Apps and Schatz, 1979). Activity due to other ATPases may exist, but it has yet to be identified. The existence of the H^+-translocating ATPase is based upon fulfillment of the minimal criteria that: (1) the addition of ATP to an isolated ghost preparation in which the internal and external pH values are initially identical results in generation of a ΔpH, inside acidic (Johnson *et al.*, 1979); (2) there is a fixed stoichiometry between ATP hydrolysis and H^+ translocation (Njus *et al.*, 1978; Flatmark and Ingebretsen, 1977); and (3) under appropriate conditions ATP can be synthesized at the expense of a proton gradient (Roisin *et al.*, 1980). In addition, recent attempts at isolation of the ATPase have resulted in measurements of physicochemical properties that are consistent with the presence of an H^+-translocating ATPase (Apps and Schatz, 1979), and inhibitors known to affect proton pumps demonstrate an inhibitory effect (Apps and Glover, 1978; Buckland *et al.*, 1979). The physicochemical properties of the ATPase, on the other hand, are very controversial, as measurements suffer from the lack of defined proton gradients during the time they were made. With regard to inhibitors, the mitochondrial and chromaffin granule ATPases identify closely with one another, however: both are inhibited by DCCD and alkyl tin compounds; oligomycin and aurovertin affect only the mitochondrial ATPase (Apps and Schatz, 1979; Apps and Glover, 1978; Buckland *et al.*, 1979).

4. THE EXISTENCE AND MEASUREMENT OF THE $\Delta\bar{\mu}_{H^+}$

It is generally accepted that the H^+-translocating ATPase within the chromaffin granule membrane is responsible for the development of an electrochemical proton gradient ($\Delta\bar{\mu}_{H^+}$) across the membrane, composed of interconvertible electrical and concentration components according to the chemiosmotic hypothesis (Mitchell, 1968):

$$\Delta \bar{\mu}_{H^+} = \Delta \Psi - Z\Delta pH$$

where $\Delta \Psi$ is the transmembrane potential and $Z = 2.3RT/F$.

Unfortunately, the small size of the granules precludes direct measurement of the internal pH or membrane potential with microelectrodes. Other, less direct but independent techniques are commonly used to measure these components. For measurement of the ΔpH, [^{14}C]methylamine distribution, ^{31}P NMR, and fluorescent dye distribution have been utilized (Johnson and Scarpa, 1976b; Njus *et al.*, 1978; Salama *et al.*, 1980). For measurement of the $\Delta \Psi$, the independent methods of [^{14}C]-SCN[-] and fluorescent dye distribution have been reported to yield equivalent results (Johnson and Scarpa, 1979; Holz, 1978; Salama *et al.*, 1980).

In intact chromaffin granules, suspended in isotonic sucrose at physiologic pH, the ΔpH across the membrane approaches 2 pH units, indicating an internal pH of 5.5 (Johnson and Scarpa, 1976b). This acidic internal pH is independent of the ionic composition of the media (NaCl, KCl, choline Cl) and of the presence of MgATP, indicating that it is not due to the establishment of a Donnan equilibrium.

The transmembrane potential can reach large positive or negative values, depending upon the experimental conditions (Johnson and Scarpa, 1979). In the presence of MgATP, a $\Delta \Psi$ of 80 mV, inside positive (as measured by [^{14}C]-SCN[-] distribution), is observed. This large $\Delta \Psi$ most likely exists under physiologic conditions. When FCCP, an ionophore that allows protons to distribute according to their electrochemical potential, is present, a large negative potential (measured by [^{14}C]-TPMP[+]) approaching the Nernst potential for protons is observed. Figure 2 illustrates the simultaneous measurement of the $\Delta \Psi$ and ΔpH under various conditions.

5. AMINE UPTAKE

The properties of amine accumulation are consistent with a carrier-mediated active process, based upon the following empirical evidence (Johnson and Scarpa, 1982): uptake exhibits structural specificity and stereospecificity, a Q_{10} of 4.6, dependency upon the presence of ATP and Mg, and specific inhibition by reserpine. The coupling of the inward movement of the amine to a driving force for transport remained until recently an enigma. The most important recent contribution to the study of amine uptake has been the attempt to elucidate and quantitate the contribution of the $\Delta \bar{\mu}_{H^+}$ as the driving force for amine accumulation, given the importance of proton gradients to the mechanism of substrate transport in other biological systems. The fundamental issues that were addressed by investigators can be grouped into two major questions: (1) is the driving force for amine uptake provided by the ΔpH, $\Delta \Psi$, or both together as the $\Delta \bar{\mu}_{H^+}$?; and (2) the more challenging question, what is the precise quantitative relationship between the $\Delta \bar{\mu}_{H^+}$ and the $\Delta \bar{\mu}_A$ (electrochemical gradient for amines)?

The chromaffin ghost preparation has provided the means by which to approach these questions. Utilizing the selective permeability properties of the membrane to anions, ghosts can be formed in differing ionic media so as to generate a ΔpH alone or a $\Delta \Psi$ alone upon MgATP addition. An illustrative experiment is outlined in Fig. 3, and is based on the conceptualization that activation of the H[+]-ATPase should result in the movement of H[+] vectorially inward. This movement of a positive charge, resulting in the rapid generation of a membrane potential, serves to limit further influx of protons. In the presence of an impermeable anion (such as isethionate), therefore, a large $\Delta \Psi$ and minimal ΔpH would

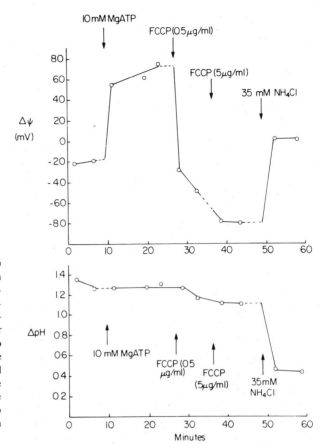

Figure 2. Measurement of ΔpH and $\Delta\psi$ under identical conditions in chromaffin granules. The reaction mixtures contained 0.27 M sucrose, 30 mM Tris–maleate (pH 6.8), and 10.8 mg chromaffin granules/ml. In one chamber [^{14}C]-SCN$^-$ and ^{3}H$_2$O were added (to measure membrane potential, positive inside), to another [^{3}H]-TPMP$^+$ and [^{14}C]dextran (to measure the membrane potential, negative inside), and to the third [^{14}C]methylamine and ^{3}H$_2$O (to measure the ΔpH). More details in Johnson and Scarpa (1979).

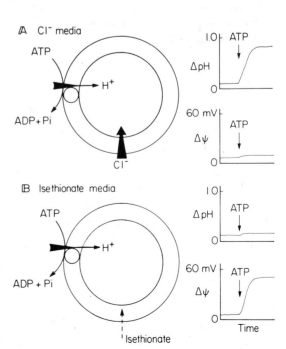

Figure 3. The generation of a ΔpH or $\Delta\psi$ in chromaffin ghosts.

be generated. On the other hand, the presence of a permeable anion (such as chloride) results in inward movement of the anion down its electrochemical gradient. The overall result, dependent upon a low internal buffering capacity in the ghosts, is the generation of a large ΔpH and minimal $\Delta \Psi$.

Recent extensive experiments clearly demonstrate that either the ΔpH or the $\Delta \Psi$, or both can drive amine accumulation, with the maximum rate and extent of accumulation occurring in the presence of both together.

The improvement of the methodology for preparation of highly purified ghosts has enabled investigation of the precise quantitative relationship between the $\Delta \tilde{\mu}_{H^+}$ and the $\Delta \tilde{\mu}_A$ under identical conditions. Their relationship is predicted by the chemiosmotic hypothesis, which states that the energy for substrate transport systems is derived from the $\Delta \tilde{\mu}_{H^+}$ (generated from the proton-pumping ATPase), and that the coupling between the transported species (amine and proton in this case) would approach equilibrium, i.e., net transport vanish, whenever (Mitchell, 1968; Rottenberg, 1979):

$$\Delta \tilde{\mu}_A - n \Delta \tilde{\mu}_{H^+} = 0$$

where n is the number of protons translocated in the opposite direction (antiport), i.e., the stoichiometry of the reaction. Thus, H^+/amine antiport is defined by nonspontaneous obligatory coupling through the existence of a putative translocator.

Using the highly purified ghost preparation and incorporating the concepts of the chemiosmotic hypothesis, the results provide unequivocal evidence for the role of the $\Delta \tilde{\mu}_{H^+}$ in the accumulation of biogenic amines and indicate that the magnitude of the amine gradient is equal to $\Delta \Psi - 2Z \Delta pH$ (Fig. 4) (Knoth *et al.*, 1980; Johnson *et al.*, 1981). The evidence to date cannot discern unequivocally the actual amine species transported (i.e., cation, anion, zwitterion, or neutral form) or the charge on the carrier molecule. The model most consistent with the evidence has been presented in Fig. 1B. The essence of the model is that a negatively charged carrier molecule is capable of binding an uncharged lipophilic catecholamine. The complex moves vectorially inward due to the presence of an inwardly directed positive potential. At the inside membrane face, dissociation of the carrier occurs,

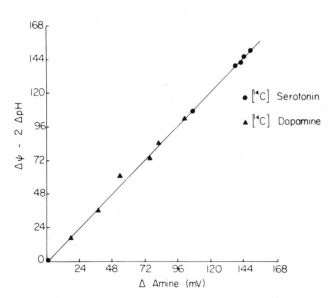

Figure 4. The relationship between the driving force and amine accumulation into isolated chromaffin ghosts. Chromaffin ghosts, formed in 185 mM Na isethionate, 20 mM ascorbate, and 4 mM Tris–maleate, were suspended (1.2 mg protein/ml) in Na isethionate media at pH 7.0 in the presence of 30 mM Tris–maleate, 8 mM MgATP, ^3H_2O, either [^{14}C]methylamine, [^{14}C]serotonin, [^{14}C]dopamine, or [^{14}C]-SCN$^-$, and various concentrations of both NaSCN (0–60 mM) and $(NH_4)_2SO_4$ (0–30 mM). Sample volumes were 1.2 ml. Labeled biogenic amines were not added to the reaction media until 10 min incubation had elapsed. After 20 min incubation, all samples were centrifuged and assayed for radioactivity. More details in Johnson *et al.* (1981).

the catecholamine becomes rapidly protonated, and, due to the high intragranular H^+ concentration and the pK_a of the carrier, protonation of the carrier also ensues. Movement of the carrier in the opposite vectorial plane then takes place, unaffected by the $\Delta\Psi$ but at the expense of the H^+ gradient.

6. CONCLUSIONS

The H^+-translocating ATPase within the chromaffin granule membrane is responsible for the generation of a large $\Delta\bar{\mu}_{H^+}$, which in turn is the driving force for amine accumulation. At present, active areas of investigation include: (1) the ionic species of catecholamine transported; (2) the properties of the amine transporter; and (3) isolation and characterization of the H^+-ATPase.

Other subcellular organelles contain amines, including the serotonin dense granules of platelets, the histamine granules of mast cells, anterior pituitary granules, insulin granules of pancreatic β cells, and peptide granules in the gut. The evidence to date strongly suggests that each of these amine-containing subcellular organelles maintains a similar mechanism for amine accumulation based upon the electrochemical proton gradient.

ACKNOWLEDGMENTS. The authors are indebted to Mr. Daniel Brannen for the preparation of the manuscript. The research described herein has been supported by Grants HL-18708 and HL-24010 from the National Institute of Health.

REFERENCES

Apps, D. K., and Glover, S. A. (1978). *FEBS Lett.* **85**, 254–258.
Apps, D. K., and Schatz, G. (1979). *Eur. J. Biochem.* **100**, 411–419.
Berneis, K. H., Pletscher, A., and DaPrada, M. (1969). *Nature* (London) **244**, 281–283.
Berneis, K. H., Pletscher, A., and DaPrada, M. (1970). *Br. J. Pharmacol.* **39**, 382–389.
Blaschko, H., and Welch, A. D. (1953). *Arch. Exp. Pathol. Pharmakol.* **219**, 17–22.
Buckland, R. M., Radda, G. K., and Wakefield, L. M. (1979). *FEBS Lett.* **103**, 323–327.
Carlsson, A., Hillarp, N.-Å., and Waldeck, B. (1962). *Med. Exp.* **6**, 47–53.
Flatmark, T., Ingebretsen, O. C. (1977). *FEBS Lett.* **78**, 53–56.
Hillarp, N.-Å., Lagerstedt, S., and Nilson, B. (1953). *Acta Physiol. Scand.* **29**, 251–263.
Holz, R. W. (1978). *Proc. Natl. Acad. Sci. USA* **75**, 5190–5194.
Ingebretsen, O. C., and Flatmark, T. (1979). *J. Biol. Chem.* **254**, 3833–3839.
Johnson, R., and Scarpa, A. (1976a). *J. Biol. Chem.* **251**, 2189–2191.
Johnson, R. G., and Scarpa, A. (1976b). *J. Gen. Physiol.* **68**, 601–631.
Johnson, R. G., and Scarpa, A. (1979). *J. Biol. Chem.* **254**, 3750–3760.
Johnson, R. G., and Scarpa, A. (1982). *Physiol. Rev.*, in press.
Johnson, R. G., Carlson, N. J., and Scarpa, A. (1978). *J. Biol. Chem.* **253**, 1512–1521.
Johnson, R. G., Pfister, D., Carty, S. E., and Scarpa, A. (1979). *J. Biol. Chem.* **254**, 10963–10972.
Johnson, R. G., Carty, S. E., and Scarpa, A. (1981). *J. Biol. Chem.* **256**, 5773–5780.
Kirshner, N. (1962). *J. Biol. Chem.* **237**, 2311–2317.
Knoth, J., Handloser, K., and Njus, D. (1980). *Biochemistry* **19**, 2938–2942.
Mitchell, P. (1968). *Chemiosmotic Coupling and Energy Transduction*, Glynn Research, Bodmin.
Njus, D., and Radda, G. K. (1979). *Biochem. J.* **180**, 579–585.
Njus, D., Sehr, P. A., Radda, G. K., Ritchie, G., and Seelig, R. J. (1978). *Biochemistry* **17**, 4337–4343.
Phillips, J. H. (1978). *Biochem. J.* **170**, 673–674.
Roisin, M. P., Scherman, D., and Henry, J. P. (1980). *FEBS Lett.* **115**, 143–146.
Rottenberg, H. (1979). *Biochim. Biophys. Acta* **549**, 225–253.
Salama, G., Johnson, R. G., and Scarpa, A. (1980). *J. Gen. Physiol.* **75**, 109–140.
Scholes, P., and Mitchell, P. (1970). *Bioenergetics* **1**, 61–72.

Schuldiner, S., Kishkes, H., and Kanner, B. I. (1978). *Proc. Natl. Acad. Sci. USA* **75**, 3713–3716.

Smith, A. D. (1968). In *The Interaction of Drugs and Subcellular Components in Animal Cells* (P. N. Campbell, ed.), pp. 239–292, Churchill, London.

Winkler, H. (1979). *Neuroscience* **1**, 65–80.

122

Nucleoside and Nucleobase Uptake in Cultured Animal Cells*

Peter G. W. Plagemann and Robert M. Wohlhueter

1. SALVAGE OF EXOGENOUS NUCLEOSIDES AND NUCLEOBASES

The salvage pathways for nucleosides and nucleobases in animals presumably serve for the utilization by cells of these substances liberated during degradation of nucleotides and nucleic acids. They are nonessential for many types of cells, and mutants deficient in salvage activity can readily be isolated from many types of animal cells in culture and have found wide use in somatic cell genetics (Siminovitch, 1976). However, they are important in the overall physiology of the animal because certain tissues are deficient in the pathways for *de novo* synthesis of purines and thus rely entirely on purines derived from food or synthesized in other tissues (Murray, 1971).

The salvage pathways for nucleosides and nucleobases, reduced to their simplest, consist of two components, a nonconcentrative transporter, which moves the substrate across the cell membrane, and an intracellular, phosphorylative enzyme, which introduces either a phosphoryl or a phosphoribosyl group, thereby trapping the substrate intracellularly and activating it for subsequent metabolic purposes. A clear understanding of the individual steps and their interplay is important for elucidating possible mechanisms of regulation of the salvage pathways, and also for a number of practical reasons, as for example, in the use of radioactively labeled nucleosides and nucleobases to specifically label nucleic acids

---

*For more extensive information and documentation the reader should consult recent reviews on this topic by the authors. One review (Plagemann and Wohlhueter, 1980) deals with the transport of nucleosides and nucleobases; the other (Wohlhueter and Plagemann, 1980) focuses on the interrelationships between transport and phosphorylation in the uptake of phosphorylatable nutrients. Because of the voluminous literature on the topic under discussion, we have limited the references in this article to those not already cited in the above reviews, other related reviews, and a few key papers. Earlier reviews of this subject are Plagemann and Richey (1974), Berlin and Oliver (1975), Hochstadt (1974), and Paterson *et al.* (1975).

Peter G. W. Plagemann and Robert M. Wohlhueter • Department of Microbiology, University of Minnesota, Minneapolis, Minnesota 55455.

of cells and viruses and to assess the rates of nucleic acid synthesis. Furthermore, many anticancer and immunosuppressive agents are nucleoside and nucleobase analogs that become activated to toxic metabolites by incorporation via salvage pathways. Knowledge of their mode of entry into cells and metabolism is important in the assessment of their mode of action, efficacy, and optimal administration and of the development of drug-resistant mutants (Sirotnak *et al.*, 1979).

2. KINETIC CHARACTERIZATION OF UPTAKE

Salvage of nucleosides and nucleobases has been measured in a variety of cultured animal cells by determining the uptake* of radioactively labeled substrate from the external medium. Upon close examination, the kinetics of uptake are seen to be biphasic: a rapid initial phase precedes a much longer and more extensive phase (generally lasting for at least several min). Because the latter phase so dominates the kinetic picture, particularly at concentrations < 50 μM, the first phase has frequently been overlooked, and the rates of isotope accumulation during the second phase have been construed as rates of transport. Some indirect lines of evidence, reviewed in detail by Plagemann and Richey (1974), supported this conclusion, and the literature is replete with kinetic studies of uptake that have been mistakenly interpreted to reflect those of the transport step. Representative kinetic parameters for the uptake of nucleosides and nucleobases by Novikoff rat hepatoma cells in suspension are summarized in Table I. Similar values have been reported for other types of cultured cells.

A clear delineation of the transport step could only be achieved with cells in which the phosphorylation step was nonfunctional due either to a genetic deficiency of the re-

Table I. Kinetic Parameters for the Transport and "Uptake" of Nucleosides and Nucleobases by Novikoff Rat Hepatoma Cells [a]

Substrate	Transport K (μM)	Transport V (pmoles/μl cell H$_2$O per sec)	Uptake K_m (μM)	Uptake V_{max} (pmoles/μl cell H$_2$O per sec)
Adenosine	103 ± 8	17 ± 0.5	6–10	2–4
Thymidine	228 ± 14	27 ± 5	0.4–0.5	0.15–0.3
Uridine	250 ± 13	25 ± 4	12–16	1.0–2.6
Deoxycytidine	626 ± 52	41 ± 6	0.8–1.6	0.04–0.08
Cytidine	$2,425 \pm 497$	48 ± 17	15–25	1–3
Hypoxanthine	349 ± 17	53 ± 10	4–8	1.5–2.2
Adenine	$3,300 \pm 524$	119 ± 11	30–50	3–4.5
Uracil	$14,200 \pm 950$	164 ± 5		

[a] Excerpt of values previously summarized by Plagemann and Wohlhueter (1980). Kinetic parameters for transport are means of at least three independent determinations ± S.E.M. or from single experiments stated ± error of the estimate. Those for uptake reflect ranges observed in at least three experiments.

* "Uptake" denotes the transfer of radioactivity from exogenous labeled substrate to intracellular space or components regardless of metabolic conversions. "Transport" denotes solely the transfer of unmodified exogenous substrate across the cell membrane as mediated by a saturable, selective carrier.

sponsible phosphorylating enzyme or to experimentally induced depletion of ATP and *P*-ribosyl-*PP*. Experiments of this sort have properly identified the kinetics of the first phase with those of transport per se, and have served to emphasize the rapidity of transport. In the first-order range of substrate concentration for the transporter, the half-time of transmembrane equilibration of nucleosides and purines at 25°C is only 5 to 30 sec, depending on the substrate.

3. KINETIC CHARACTERIZATION OF TRANSPORT

The rapidity of transport necessitated the development of techniques for measuring accurate initial time courses of substrate accumulation by cells. Wohlhueter *et al*. (1978) have developed a rapid mixing/sampling method for cell suspensions that permits the determination of changes in intracellular substrate concentration during substrate entry or exit in time intervals as short as 1 sec. Procedures for comparable resolution of substrate accumulation curves in monolayer cultures have also been developed. Even with these techniques it has proved difficult, if not impossible, to capture an initial linear segment of substrate entry; curvature is manifest within a few seconds of incubation, particularly at low substrate concentrations. This problem has been circumvented by fitting appropriate integrated rate equations based on the simple carrier model (see Eilam and Stein, 1974) to complete time courses of transmembrane equilibration of the substrate. Initial velocities are then estimated as the slope of the curves at time zero. Linearized forms of the same integrated rate equations have also been applied to estimate initial velocities (Heichal *et al*., 1979).

Time courses of equilibration of labeled nucleosides and nucleobases across the membrane of a variety of cultured cells blocked in the phosphorylation step are consistent with the operation of simple carriers. Influx and efflux when directly compared have been found to be approximately equal and not to be affected by the presence of unlabeled substrate on the *trans* side of the membrane (the side opposite to that initially exposed to labeled substrate). These results indicate that the nucleoside and nucleobase transporters of cultured cells are indifferent with respect to direction of transport and that the carriers have the same mobility whether or not loaded with substrate. Representative Michaelis–Menten constants and maximum velocities for the transport of nucleosides and nucleobases by Novikoff rat hepatoma cells are summarized in Table I. Similar kinetic parameters have been estimated for other cell lines.

A translocation of the ribosyl moiety of extracellular uridine and inosine catalyzed by membrane-bound phosphorylases, whereby the bases are released extracellularly and ribose-1-phosphate intracellularly, has been ascribed to membrane vesicles of mammalin cells (Hochstadt and Quinlan, 1976). Such processes have not been detectable in whole cells of a number of lines, or at least do not occur at a significant rate relative to the rate of nucleoside transport (Plagemann *et al*., 1981b). The same is true for a hypoxanthine phosphoribosyltransferase-catalyzed translocation and conversion of extracellular hypoxanthine to intracellular IMP, which has been postulated to occur in vesicles (Hochstadt and Quinlan, 1976). Nonmediated permeation of nucleosides and nucleobases is generally very slow in comparison to transport (2% as rapid in the first-order substrate range for transport) because of the low lipid solubility of these hydrophilic substances. Thus, it does not play a significant role in the uptake of natural nucleosides and bases, but may be more significant in the entry of analogs with increased lipid solubility.

4. MOLECULAR MECHANISM OF TRANSPORT

The actual mechanism of substrate translocation by carriers is not understood. The following observations need to be considered in this regard. Arrhenius and van't Hoff plots for the transport of thymidine in Novikoff cells are linear between 5 and 37°C ($E_A = 18.3$ kcal/mole; $\Delta H^{\circ\prime} = 9.3$ kcal/mole, respectively). Nucleoside and nucleobase transport are little affected by external pH between 6 and 8.5, except in the case of 5-fluorouracil, which dissociates in this range ($pK_a = 8.0$) and is only transported in its undissociated form (Wohlhueter *et al.*, 1980). Nucleoside and purine transport in various types of cultured cells have been found to be inhibited by sulfhydryl reactive reagents, but the uptake of nucleosides by golden hamster MCT cells has been found to be either stimulated or inhibited by different organomercurials (see Bibi *et al.*, 1978). Thus, the role of sulfhydryl groups in transport is still uncertain. The functional portions of the carriers seem to be deeply embedded in the membrane, as the removal of surface proteins and carbohydrates by treatment of cells with various hydrolytic enzymes has no effect on their transport capacity.

5. SPECIFICITY OF TRANSPORTERS

All data available at present indicate that all natural ribo- and deoxyribonucleosides are transported by a single carrier. Purine nucleosides have the highest affinity for the carrier, whereas cytidine nucleosides have the lowest. In respect to its broad substrate specificity and relatively high K_m, the nucleoside transporter of cultured cells is similar to that of human erythrocytes (Cass and Paterson, 1972; Cabantchik and Ginsburg, 1977). There are, however, differences in functional details. In contrast to the nucleoside transporters of cultured cells, that of human erythrocytes has been reported to move in either direction several times more rapidly when loaded than when empty. Furthermore, the empty carrier in erythrocytes seems to move four times more rapidly from the outside to the inside than in the opposite direction.

The number of distinct carriers involved in nucleobase transport, and the extent of overlap with nucleoside transport, is not so straightforward. The case of hypoxanthine is illustrative. The findings that the transport of nucleosides is inhibited by hypoxanthine and hypoxanthine transport is inhibited by nucleosides might suggest that these substances are transported by the same carrier. But the transport of nucleosides and hypoxanthine is distinguishable in that nucleoside transport, but not hypoxanthine transport, is strongly inhibited by nitrobenzylthioinosine ($K_i \approx 1$ nM). Furthermore, nucleoside-transport-defective mutants of S49 lymphoma cells take up purines and pyrimidines unabated (Cohen *et al.*, 1979).

The system responsible for uracil transport is also poorly defined. Uracil transport is inhibited by a number of nucleosides, but it is a difficult substrate to investigate, because of its low affinity for its carrier (Table I). The same is true for adenine. Some experimental evidence suggests that adenine and folate are transported by a single transporter (Suresch *et al.*, 1979). Cytosine does not seem to be a substrate for any of these carriers, and, indeed, its entry seems not to be mediated at all.

6. TRANSPORT AND PHOSPHORYLATION ACTING IN TANDEM

It is evident in Table I that the apparent Michaelis–Menten constants and maximum velocities of uptake as measured over time periods of minutes in cells not blocked in phosphorylation are one to two powers of ten less than those for transport of the same substrate. The reason is that both the specificity and the rate of nucleoside and nucleobase uptake reflect those of the kinase or phosphoribosyltransferase involved, and not those of the transporter.

In some cases, for example hypoxanthine, the uptake K_m corresponds closely to the K_m for the phosphoribosyltransferase as measured *in vitro*. In other cases, notably thymidine and uridine, there is a large discrepancy between the K_m of the phosphorylation reaction effective *in situ* and that measured *in vitro*. The striking preponderance of enzymatic lesions in cells selected for resistance to nucleoside and nucleobase analogs is readily understood in terms of the rate-limiting roles of the phosphorylating enzymes in uptake.

Perhaps the most meaningful kinetic comparison between transport and phosphorylation is a comparison of V_{max}/K_m ratios, because physiologically relevant concentrations (say, $\geq 10 \ \mu M$) fall within the first-order range of transport, and may approach that for phosphorylation *in situ* (see Table I). Thus, it is the relative size of the first-order rate constants that determines the pool size of intracellular, free base or nucleoside, with respect to a given exogenous concentration. Because the first-order rate constants for *in situ* phosphorylation are similar to or even higher than those for transport, the salvage pathways operate very efficiently at exogenous substrate concentrations in the physiological range.

7. UPTAKE OF NUCLEOSIDE AND NUCLEOBASE ANALOGS

Analogs generally enter cells via the same carrier transporting their natural counterparts, although often at a lower rate (see Paterson et al., 1981). But here again the conversion of the analogs to their phosphorylated intermediates is generally limited at the phosphorylation rather than the transport step. The entry of some analogs, such as 8-azaguanine, however, seems not to be mediated and does limit incorporation of the analog into the nucleotide pool and into nucleic acids (Plagemann *et al.*, 1981a).

8. REGULATION OF UPTAKE

The capacity of cultured cells to transport nucleosides and purines is not subject to regulation by hormones or cyclic nucleotides. It is, however, inhibited by an unusual array of unrelated and structurally complex substances, such as dipyridamole, which are rather nonspecific in that they also inhibit the transport of substances other than nucleosides and nucleobases.

The capacity of cultured cells to transport nucleosides and purines is little affected by culture age and quiescence, whereas the capacity to take up these substrates is decreased in stationary phase or quiescent cultures. Where investigated, this low uptake capacity of nongrowing cells has clearly been shown to be due to a decreased activity in phosphorylation, but the reasons for the decrease may vary with the salvage system or the cell type.

For example, low rates of uridine uptake in stationary-phase Novikoff cells seem to be due to a loss of uridine kinase, whereas in quiescent 3T3 cells it seems related to altered affinity of uridine kinase for ATP. Hypoxanthine uptake in stationary-phase Novikoff cells, on the other hand, is limited by a lack of P-ribosyl-PP (Wohlhueter and Plagemann, unpublished observation).

9. A VIEW TO THE FUTURE

Purification of carrier proteins, physical and chemical characterization of them, and their reconstitution in artificial membranes are tasks of the future. The strong ($K_D = 0.1-1$ nM) and specific binding of p-nitrobenzylthiopurine ribosides to the nucleoside carrier (Paterson *et al.*, 1981) raises the hope that these substances might be useful in tagging the carrier and thus making practicable its eventual isolation and characterization. A deeper understanding of the mode of inhibition by nitrobenzylthiopurine ribosides will, in itself, contribute to our understanding of transport at the molecular level. The isolation of carrier-deficient mutants will complement isolation studies and help resolve the open questions of specificity. Resolution of the discrepancies apparent between *in situ* and *in vitro* kinetics of nucleostide phosphorylation may help define the mechanisms of regulation of uptake.

REFERENCES

Berlin, R. D., and Oliver, J. M. (1975). *Int. Rev. Cytol.* **42,** 287–336.
Bibi, O., Schwartz, J., Eilam, Y., Shohami, E., and Cabantchik, Z. I. (1978). *J. Membr. Biol.* **39,** 159–183.
Cabantchik, Z. I., and Ginsburg, H. (1977). *J. Gen. Physiol.* **69,** 75–96.
Cass, C. E., and Paterson, A. R. P. (1972). *J. Biol. Chem.* **247,** 3314–3320.
Cohen, A., Ullman, B., and Martin, D. W., Jr. (1979). *J. Biol. Chem.* **254,** 112–118.
Eilam, Y., and Stein, W. D. (1974). *Methods Membr. Biol.* **2,** 283–354.
Heichal, O., Ish-Shalom, D., Koren, R., and Stein, W. D. (1979). *Biochim. Biophys. Acta* **551,** 169–186.
Hochstadt, J. (1974). *CRC Crit. Rev. Biochem.* **2,** 259–310.
Hochstadt, J., and Quinlan, D. (1976). *J. Cell. Physiol.* **89,** 839–852.
Murray, A. W. (1971). *Annu. Rev. Biochem.* **40,** 811–826.
Paterson, A. R. P., Kim, S. C., Bernard, O., and Cass, C. E. (1975). *Ann. N.Y. Acad. Sci.* **255,** 402–411.
Paterson, A. R. P., Kolassa, N., and Cass, C. E. (1981). *Pharmacol. Ther.* **12,** 515–536.
Plagemann, P. G. W., and Richey, D. P. (1974). *Biochim. Biophys. Acta* **344,** 263–305.
Plagemann, P. G. W., and Wohlhueter, R. M. (1980). *Curr. Top. Membr. Transp.* **14,** 225–330.
Plagemann, P. G. W., Marz, R., Wohlueter, R. M., Graff, J. C., and Zylka, J. M. (1981a). *Biochim. Biophys. Acta* **647,** 49–62.
Plagemann, P. G. W., Wohlhueter, R. M., and Erbe, J. (1981b). *Biochim. Biophys. Acta* **640,** 448–462.
Siminovitch, L. (1976). *Cell* **7,** 1–11.
Sirotnak, F. M., Chello, P. L., and Brockman, R. W. (1979). *Methods Cancer Res.* **16,** 381–447.
Suresch, M. R., Henderson, G. B., and Huennekens, F. M. (1979). *Biochem. Biophys. Res. Commun.* **87,** 135–139.
Wohlhueter, R. M., and Plagemann, P. G. W. (1980). *Int. Rev. Cytol.* **64,** 171–240.
Wohlhueter, R. M., Marz, R., Graff, J. C., and Plagemann, P. G. W. (1978). *Methods Cell Biol.* **20,** 211–236.
Wohlhueter, R. M., McIvor, R. S., and Plagemann, P. G. W. (1980). *J. Cell. Physiol.* **104,** 309–319.

123

Transport and Trans-Plasma-Membrane Redox Systems

Frederick L. Crane, Hans Löw, and Michael G. Clark

The source of energy for plasma membrane functions such as transport or vesicle formation and movement is generally considered to be ATP derived from the mitochondria or glycolysis in the cytoplasm. Very little attention has been paid to energy-rich redox agents such as NADH or glutathione in contact with the interior of the plasma membrane and the fact that potential oxidants such as oxygen, ferric compounds, or dehydroascorbic acid are readily available either inside or outside the surface membrane. The presence of low-redox-potential compounds and high-redox-potential compounds at the plasma membrane means that energy is available, independent of ATP, if a suitable transduction system is present in the membrane. Direct redox energization of a plasma membrane function would provide a degree of independence from a supply of ATP. It is quite clear that ATP supplies the bulk of energy for plasma membrane transport functions such as the sodium–potassium pump. The source of energy for specialized transport and for control functions in the plasma membrane is not so clear. A second source of energy input allows greater degrees of freedom in control of plasma membrane function than an unsegregated cytoplasmic ATP supply.

Several redox systems have now been recognized in plasma membranes from various eukaryotic sources (Crane *et al.*, 1979). They include NADH oxidase (Crane and Löw, 1976; Gayda *et al.*, 1977), NADH semidehydroascorbate reductase (Goldenberg, 1980), NADH transferrin reductase (Crane, unpublished), glutathione oxidase (Ormstad *et al.*, 1979), NADPH oxidase (Mukherjee and Lynn, 1977), ascorbic acid oxidase (Crane *et al.*, 1980), xanthine oxidase (Jarasch *et al.*, 1977), NADH glutathione reductase, and NADPH glutathione reductase (Tillmann *et al.*, 1975). Artificial electron acceptors have also been used to measure the activity of NADH and NADPH dehydrogenases in the plasma membrane. These acceptors include ferricyanide, cytochrome *c*, indophenol, and tetrazolium

Frederick L. Crane • Department of Biological Sciences, Purdue University, West Lafayette, Indiana 47906. **Hans Löw** • Karolinska Hospital, Stockholm, Sweden. **Michael G. Clark** • Division of Human Nutrition, CSIRO, Adelaide, South Australia.

salts, but it should be emphasized that activity measured in this way is only an assay and not a definition of the natural redox system (Crane *et al.*, 1979).

There appear to be three different types of NADH dehydrogenases in eukaryotic plasma membranes (Löw *et al.*, 1979; Wang and Alaupovic, 1978). These include (1) an enzyme that is oriented exclusively toward the inside of the cell that acts as an NADH cytochrome *c* reductase or as a ferricyanide reductase and is apparently identical with the NADH cytochrome b_5 reductase of microsomes. (2) There is a transmembrane NADH ferricyanide reductase that transmits electrons from internal NADH to external ferricyanide. (3) In addition, a low activity of NADH ferricyanide reductase often appears exclusively on the outside of the cell to catalyze the oxidation of added NADH by ferricyanide (Cherry *et al.*, 1981). For transport studies the most interesting of these enzymes should be the transmembrane NADH dehydrogenase, which has been found in animal, plant, and yeast cells (Clark *et al.*, 1981; Crane *et al.*, 1981; Craig and Crane, 1981). There is evidence from yeast and carrot cells that this enzyme can act as a proton pump to move protons out of the cells. This enzyme is also sensitive to hormones. In animal cells it is inhibited by insulin or triiodothyronine and stimulated by glucagon at physiological concentrations (Löw and Crane, 1976; Goldenberg *et al.*, 1978). By development of a transmembrane proton gradient this enzyme could energize transport across the membrane. The amino acid transport system driven by NADH in ascites cells described by Christensen (1979) and co-workers (Garcia-Sancho *et al.*, 1977; Ohsawa *et al.*, 1980) may be an example of this type of transport. As pointed out by Sachs (1977), there are also ion transport systems that have not been defined to be driven by an ATPase.

Because the trans-plasma-membrane dehydrogenase is stimulated by protonophoric uncouplers, such as dinitrophenol, and ionophores, such as gramicidin, an effect of these reagents on transport function cannot be specifically attributed to inhibition of mitochondrial ATP formation (Löw *et al.*, 1979). The same caution should be emphasized with regard to the effects of any mitochondrial inhibitor until the properties of plasma membrane redox systems are known. For example, one can see strange discrepancies in the effects of anaerobiosis and cyanide on carnitine transport that suggest energization by a cyanide-insensitive oxidase (Huth and Shug, 1980).

A different view of the significance of the plasma membrane dehydrogenases appears in the idea that the redox state or redox protonation of membrane proteins may modify transport function. This would obviously direct attention to those dehydrogenases that respond to hormones. These include the transmembrane NADH dehydrogenase described above and the NADPH oxidase of adipocytes described by Mukherjee and Lynn (1979). A similar enzyme may be the concanavalin A-stimulated, superoxide-generating NADPH dehydrogenase of leukocytes (Dewald *et al.*, 1979). It has recently been indicated that the activation of the leukocyte enzyme is related to changes in membrane potential (Miles *et al.*, 1980) and that it may be related to transport of the superoxide anion. It does appear that NADPH dehydrogenases are often related to superoxide and peroxide generation (Dewald *et al.*, 1979), which in turn is related to insulin action (May and de Haën, 1979). The appearance of NADPH oxidase in some membranes, as well as the concentration of another superoxide-producing enzyme, xanthine oxidase, in other membranes (Jarasch *et al.*, 1977) emphasizes specialization of redox function in different plasma membranes.

The mechanism of iron transport from, or attached to, transferrin through the plasma membrane is as yet unclear (Young and Aisen, 1980). If iron in transferrin is reduced outside the cell in order to be transported into the cell, then the transmembrane NADH dehydrogenase, which can also act as a transferrin reductase, is an obvious way to convert the transferrin iron to the reduced form.

Plasma membranes contain redox carriers appropriate to a redox-driven transport function. These include cytochrome b_5, cytochrome P-420, copper, nonheme iron, and thiols. Present evidence indicates rather low levels of flavin and P-450. Nine different types of plasma membrane have been found to contain from 30 to 300 pmoles cytochrome b_5/mg protein, which is consistently about one third the amount of cytochrome b_5 found in corresponding endoplasmic reticulum (Bruder *et al.*, 1980). In addition, plasma membranes contain significant quantities of a cytochrome P-420 but very small amounts of P-450. Rather high levels of nonheme iron (MacKellar and Crane, 1979) are present in rat liver or pig erythrocyte (5.8 nmoles/mg protein) plasma membranes, and reference has been made to iron in *Neurospora* plasma membrane (Christensen, 1979). The term nonheme is used to describe this iron for it is only partially removed by acid treatment as would be expected if it was an iron–sulfur protein. Significant amounts of copper are also present.

Copper content of 2.1 and 0.45 nmoles/mg protein, respectively, has been reported in liver cell and pig erythrocyte membranes (Vassiletz *et al.*, 1967; MacKellar and Crane, unpublished).

The presence of thiol groups on plasma membrane proteins has long been recognized, and proposals have been made for control of transport by redox changes in membrane thiols (Czech, 1977). Both pyridine nucleotides and glutathione are appropriate donors for reduction of membrane thiols. Vacuole acidification may also be based on a protonophoric redox system derived from plasma membrane (Segal and Jones, 1979).

In summary, there is remarkable diversity of redox functions in plasma membranes whose relation to energization and control of transport function is a relatively unexplored territory.

REFERENCES

Bruder, G., Bretscher, A., Franke, W. W., and Jarasch, E. D. (1980). *Biochim. Biophys. Acta* **600,** 739–755.

Cherry, M. J., MacKellar, W. C., Morré, D. J., Crane, F. L., Jacobsen, L. B., and Schirrmacher, V. (1981). *Biochim. Biophys. Acta* **634,** 11–18.

Christensen, H. N. (1979). *Adv. Enzymol. Relat. Areas Mol. Biol.* **49,** 41–101.

Clark, M. G., Patten, G. S., Crane, F. L., Löw, H., and Grebing, C. (1981). *Biochem J.* **200,** 565–572.

Craig, T., and Crane, F. L. (1981). *Proc. Indiana Acad. Sci.* **90,** in press.

Crane, F. L., and Löw, H. (1976). *FEBS Lett.* **68,** 153–156.

Crane, F. L., Goldenberg, H., Morré, D. J., and Löw, H. (1979). In *Subcellular Biochemistry* (D. B. Roodyn, ed.), Vol. 6, pp. 345–399, Plenum Press, New York.

Crane, F. L., MacKellar, W. C., Morré, D. J., Ramasarma, T., Goldenberg, H., Grebing, C., and Löw, H. (1980). *Biochem. Biophys. Res. Commun.* **93,** 746–754.

Crane, F. L., Roberts, H., Linnane, A. W. and Löw, H. (1982). *J. Bioenerget. Biomembr.* **14,** in press.

Czech, M. P. (1977). *Annu. Rev. Biochem.* **46,** 359–384.

Dewald, B., Baggiolini, M., Curnutte, J. T., and Babior, B. M. (1979). *J. Clin. Invest.* **63,** 21–29.

Garcia-Sancho, J., Sanchez, A., Handlogten, M. E., and Christensen, H. N. (1977). *Proc. Natl. Acad. Sci. USA* **74,** 1488–1491.

Gayda, D. P., Crane, F. L., Morré, D. J., and Löw, H. (1977). *Proc. Indiana Acad. Sci.* **86,** 385–390.

Goldenberg, H. (1980). *Biochem. Biophys. Res. Commun.* **94,** 721–726.

Goldenberg, H., Crane, F. L., and Morré, D. J. (1978). *Biochem. Biophys. Res. Commun.* **83,** 234–240.

Huth, P. J., and Shug, A. L. (1980). *Biochim. Biophys. Acta* **602,** 621–634.

Jarasch, E. D., Bruder, G., Keenan, T. W., and Franke, W. W. (1977). *J. Cell Biol.* **73,** 223–241.

Löw, H., and Crane, F. L. (1976). *FEBS Lett.* **68,** 157–159.

Löw, H., Crane, F. L., Grebing, C., Hall, K., and Tally, M. (1979). In *Diabetes 1979* (W. K. Waldhäusl, ed.), pp. 209–213, Excerpta Medica, Amsterdam.

MacKellar, W. C., and Crane, F. L. (1979). *Fed. Proc.* **38,** 356.

May, J. M., and de Haën, C. (1979). *J. Biol. Chem.* **254,** 9017–9021.

Miles, P. R., Bowman, L., and Castranova, V. (1980). *Fed. Proc.* **39,** 2100.

Mukherjee, S. P., and Lynn, W. S. (1977). *Arch. Biochem. Biophys.* **184,** 69–76.

Mukherjee, S. P., and Lynn, W. S. (1979). *Biochim. Biophys. Acta* **568,** 224–233.

Ohsawa, M., Kilberg, M. S., Kimmel, G., and Christensen, H. N. (1980). *Biochim. Biophys. Acta* **599,** 175–190.

Ormstad, K., Moldeus, P., and Orrenius, S. (1979). *Biochem. Biophys. Res. Commun.* **89,** 497–503.

Sachs, G. (1977). *Am. J. Physiol.* **233,** F359–F365.

Segal, A. W., and Jones, O. T. G. (1979). *Biochem. J.* **182,** 181–188.

Tillmann, W., Cordura, A., and Schröter, W. S. (1975). *Biochim. Biophys. Acta* **385,** 157–171.

Vassiletz, I. M., Derkatchev, E. F., and Neifakh, S. A. (1967). *Exp. Cell Res.* **46,** 419–427.

Wang, C.-S., and Alaupovic, P. (1978). *J. Supramol. Struct.* **9,** 1–14.

Young, S. P., and Aisen, P. (1980). *Biochim. Biophys. Acta* **633,** 145–153.

124

Metabolite Transport in the Nervous System

Henry Sershen and Abel Lajtha

The question can be raised whether a separate discussion (and investigation) of transport in the brain is justified, or whether the systems present in the nervous system are the same as those observed in most other tissues. The available information indicates several reasons why such studies in the nervous system would be of merit, among them, (1) there are systems that are specific for the nervous system, (2) the properties of some systems are different in brain, and perhaps most important, (3) the functional implications of transport are very different in the brain.

1. SPECIFICITY OF AMINO ACID TRANSPORT

The major amino acid tranport systems that have been identified and described in some detail in eukaryotic cells, the L, A, ASC, and L^+ systems (Christensen, 1975), are also found in brain cells. As in other tissues, there is significant overlap among the transport systems for neutral amino acids. Overlap of the three large groups, neutral, basic, and acidic amino acids, is limited; for example, glutamate moderately inhibits small neutral amino acids, but not large neutral or basic amino acids (Blasberg and Lajtha, 1965). In general, small neutral amino acids are stronger inhibitors of small than of large neutral amino acids. Specificity in these cases is described in respect to competitive inhibition studies; however, specificity also exists in differences in Na^+ dependence, other requirements, and kinetic parameters.

A comparison of cross inhibition of uptake in brain slices, where cellular uptake of the low-affinity systems is probably predominant, indicated the presence of 10, possibly more, low-affinity transport systems (Table I) (Sershen and Lajtha, 1979). It is likely that as our armamentarium of specific inhibitors useful for separating and characterizing trans-

Henry Sershen and Abel Lajtha • Center for Neurochemistry, Rockland Research Institute, Ward's Island, New York 10035.

Table I. Amino Acid Transport Systems in Brain Cells

System	Inhibitor	Substrates
Low-affinity		
A	NMA	Pro, Ser, Gly
L	BCH	Leu, Val, Trp
ASC	Cys	Ala, Ser, Thr
Pro	Pipecolinate	Pro
Gly	Gly	Gly
GABA	GABA	GABA
Tau (β)	Tau	Tau, βAla, GABA
Glu	Asp	Glu, Asp
Ly$^+$	Homoarg	Lys, Arg, Orn
Lys	Cadavarine	Lys
High-affinity		
GABA		
Glu		
Gly		
Pro		
Tau		
Trp		
Lys		
Gln		
Tyr		

port expands, so will the number of transport classes. For example, such recently described systems as system N, which serves for amino acid amides (Kilberg *et al.*, 1980), have not yet been tested in the nervous system. Because a fairly large number of amino acid transport systems are present in several cells, some highly specific for one amino acid, nervous tissue cells are not unique in this respect. Also, properties of the systems vary in different cells; not all systems are present in all cell types, and the various brain cells are not the only ones showing such differences.

Of particular importance in brain function are transport systems in cerebral capillaries, which form an element of the blood–brain barrier system and are part of the mechanism that regulates cerebral metabolic levels. Another important system is synaptosomal high-affinity transport, which may regulate neurotransmitter activity; also of importance is the transport of the amino acids that serve as neurotransmitter precursors, for the level of these compounds in the brain influences the level of their products, the neurotransmitters. The transport of tryptophan (serotonin precursor) was the subject of a separate treatise (Baumann, 1979).

2. CAPILLARY TRANSPORT

Specific transport systems are also present in cerebral capillaries. Fewer carriers for amino acids are present in capillaries than in brain cells; only three or four systems have been clearly identified (Oldendorf and Szabo, 1976; Sershen and Lajtha, 1979) that would correspond to the L, Glu, and Ly$^+$ systems in Table I. Indications for the ASC system were also found (Hwang *et al.*, 1980). No cross inhibition between acidic, basic, and neutral amino acids was found in capillary uptake, though recently Wade *et al*. (1980) suggested an interaction between glutamic acid and the L neutral amino acid transport system that was dependent on pH. Influx and efflux differences exist, which are possibly due to the heterogeneous distribution of the systems available for metabolite transport. Where influx (blood to brain) was measured, no transport via the A neutral system was observed (Wade and Katzman, 1975; Sershen and Lajtha, 1979). When transport was measured in isolated capillary preparations, transport of methylaminoisobutyric acid was detected, which, in view of the *in vivo* studies, suggests that this A neutral system is present at the outer rather than inner capillary surface and is therefore active in the brain-to-blood direction (Betz and Goldstein, 1980; Cutler and Coull, 1978).

Carrier systems also exist in capillaries for monocarboxylic acids (Oldendorf, 1973), for glucose (Crone, 1965; Pardridge and Oldendorf, 1975), for diamines (Lajtha and Sershen, 1974), for nucleotides (Cornford and Oldendorf, 1975), and for choline (Wheeler, 1979), among others. The distribution and properties of these systems are very different from those for amino acids.

Cellular transport is different quantitatively as well as qualitatively from capillary transport. Uptake of the nonessential amino acids is very active in the cells, but is mostly absent in capillaries. In capillaries the uptake of essential amino acids is more rapid and several systems (such as GABA, Tau, Pro) seem to be absent. The CSF and extracellular space concentrations of transported amino acids are normally below the K_m values at the cerebral capillary; transport rates in capillaries are therefore dependent on normal plasma levels, which are generally 10 times higher than extracellular levels. Pardridge and Oldendorf (1975, 1977) indicate that the K_m of essential amino acid transport across the blood–brain barrier approximates plasma amino acid levels (that is, the transport under physiological conditions is half-saturated). This is in contrast to the 10-fold higher K_m for transport of amino acids in liver, intestine, and kidney. Competition of amino acids occurs in brain uptake from blood but not in other tissues. Conditions such as phenylketonuria therefore confer selective vulnerability to the CNS.

3. HIGH-AFFINITY TRANSPORT

The high-affinity uptake systems for amino acids are generally considered to be associated with a transmitter pool (low K_m), and low-affinity uptake with a metabolic pool (high K_m). High-affinity systems in addition to GABA and glycine have been found with other putative neurotransmitter amino acids such as taurine (Hruska *et al*., 1978) and aspartate (Davies and Johnston, 1976). Glycine transport is present in spinal cord, where it has a transmitter role, but could not be demonstrated in cerebral cortex, cerebellum, or midbrain, where it does not have such a role (Johnston and Iversen, 1971; Bennett *et al*., 1973). Some part of high-affinity uptake may be an exchange process rather than net removal (Levi *et al*., 1976).

Synaptosomal uptake is responsible for much of neurotransmitter uptake, but neurons and glial bodies are also involved. Neurons appear to have a selective uptake mechanism for their respective neurotransmitters (Kuhar, 1973); for example, sites of high-affinity norepinephrine uptake are localized to norepinephrine neurons. High-affinity uptake, however, is not restricted to neurons and synaptosomes or to neurotransmitter amines and neurotransmitter amino acids. The selective change in glial and neuronal cell morphology in thick and thin brain slice preparations has been used to distinguish between glial high-affinity uptake of β-alanine or taurine and neuronal GABA uptake (Kaczmarek and Davison, 1972; Riddall *et al.*, 1976). High-affinity uptake of GABA is not exclusive to neurons; it has been localized also to glial cells in rat retina (Iversen and Kelly, 1975). High-affinity glutamate uptake may also be glial (East *et al.*, 1980). High-affinity uptake has also been found for amino acids that may not have neurotransmitter roles, such as tryptophan (Höckel *et al.*, 1978), lysine (Hwang and Segal, 1979), proline (Balcar *et al.*, 1976), glutamine (Balcar and Johnston, 1975), and tyrosine (Morre and Wurtman, 1981). The differences in properties (specificity, ion dependence, etc.) between high- and low-affinity systems have not been studied in detail; when examined, such as with glycine (Aprison and McBride, 1973), some differences were found.

4. DEVELOPMENTAL CHANGES IN TRANSPORT PROCESSES

The homeostasis of the developing brain is not as strongly controlled as that of the adult brain; an increase in plasma levels of metabolites results in a greater increase in young brain, indicating a weaker "blood–brain barrier." The transport systems of the barriers, however, develop early: capillary transport systems examined in immature brain are as active as in adult brain, and some are more active (Sershen and Lajtha, 1976a; Cremer *et al.*, 1976). Cellular transport, especially the high-affinity systems, develops later than capillary transport. Neurotransmitter amino acid transport increases just before birth; smaller increases are seen with other amino acids. Changes in the cellular transport system mostly involve nonessential amino acids of putative neurotransmitter function (taurine, glycine, GABA, and acidic amino acids) (Sershen and Lajtha, 1976b). Developmental changes in levels of amino acids in the cerebral free amino acid pool are extensive; in general, the nonessential amino acids increase and the essentials decrease (a large decrease in taurine) with development (Piccoli *et al.*, 1971; Lajtha and Toth, 1973). The changes in transport with development do not parallel the changes in the levels of amino acids (Levi *et al.*, 1967), just as the regional variations in transport and in amino acid levels are different (Kandera *et al.*, 1968). These findings indicate that transport may not be the only factor that determines cerebral metabolite levels, although clearly cellular transport as measured in brain slices may not be a measure of transport activity in brain *in vivo*. The developmental changes can be complex; for example, in development the synaptosomal lysine uptake V_{max} for high-affinity uptake decreases, for low-affinity uptake increases (Hwang and Segal, 1979). Some systems (such as one of the GABA systems) may be present only in developing tissue (Levi, 1973). There are differences between the neuronal and the glial developmental changes in GABA and taurine uptake (Borg *et al.*, 1980). Reduced capacity of high-affinity glutamate transport was observed in synaptosomes from aging brain (Wheeler, 1980).

5. ALTERATIONS IN CEREBRAL TRANSPORT PROCESSES

Although uptake from blood into brain is strongly restricted and is selective, changes can occur under various conditions. Osmotic effects (Rapoport, 1980; Brightman *et al.*, 1970) and hypertension (Johansson, 1980; Larsson *et al.*, 1980) cause increased entry into brain through changes in the capillary endothelial cells and their junctions. Increases in amino acid uptake under such conditions are most likely due to increase in diffusion (Zanchin *et al.*, 1976). Transport into brain is affected by changes in blood levels of competing substrates: in phenylketonuria, for example, the elevated blood phenylalanine inhibits the cerebral uptake of other amino acids belonging to the same transport class, such as tryptophan, thereby lowering brain tryptophan and serotonin levels; it also results in an increase of brain glycine (Dienel, 1981). The increased blood (and brain) level of an amino acid can have complex effects on other amino acids: it can compete in uptake (blood to brain), exit (brain to blood), and exchange (in both directions); it could stimulate or inhibit these processes, and have metabolic effects as well. Changes, especially in the uptake and levels of neurotransmitter precursors, are important, for in many cases the levels of precursors are rate limiting in neurotransmitter synthesis. Dietary influence on the uptake of such precursors, through feeding increased amounts of the precursor (and keeping competing analogs at low levels), has therapeutic significance (Morre and Wurtman, 1981). Dietary changes in transport have also been observed in protein malnutrition; in some cases such changes in transport are responsible for alterations in brain levels, for increased cerebral level and transport of histidine occur simultaneously with decreased level and transport of lysine (Toth and Lajtha, 1980). Because the properties of various transport systems differ, any alterations in temperature (Banay-Schwartz *et al.*, 1977), ions (Margolis and Lajtha, 1968), or available energy (Sershen and Lajtha, 1976a) can result in changes in transport of some compounds. Drugs also influence transport to brain (Lajtha and Toth, 1965).

6. PROPERTIES OF TRANSPORT SYSTEMS

At present there is no evidence that energy requirement, such as that for ion or proton gradients is different in brain from that in other organs and cells (Schultz and Curran, 1970; Kaback *et al.*, 1976). Of importance in brain is the specific change in transport when Na^+ levels are low (Margolis and Lajtha, 1968). The predominance of the A system in cellular, and the L system in capillary, neutral amino acid transport explains the strong Na^+- and temperature-dependence in cellular uptake and lack of it in capillary uptake (Sershen and Lajtha, 1979; Banay-Schwartz *et al.*, 1977; Emirbekov *et al.*, 1977). The fact that the A system is not totally absent but is active on the outer surface (brain side) of capillaries (Betz and Goldstein, 1980) indicates differences between properties of uptake into and exit from the brain. The heterogeneous distribution of high-affinity transport indicates that many structures contain primarily only one specific transport system. Such specificity is not restricted to a nerve-ending and its specific neurotransmitter, as shown by differences between neuronal and glial GABA transport (Iversen and Kelly, 1975).

7. CONCLUSIONS

Many diverse transport systems for metabolites are present in the brain; as many as 30 have been characterized to some extent. Many of these are present only in specific structures and play specific roles in metabolism; neurotransmission, neuromodulation, etc. Transport systems undergo complex developmental alterations and are sensitive in various degrees to numerous factors,such as ion levels, gradients, temperature, energy supply, related metabolite levels. They are also affected in a complex way (uptake or exit only) by drugs, malnutrition, anoxia, hypertension. Undoubtedly a number of additional cerebral transport systems will be recognized and characterized in the near future.

REFERENCES

Aprison, M. H., and McBride, W. J. (1973). *Life Sci.* **12**, 449–458.

Balcar, V. J., and Johnston, G. A. R. (1975). *J. Neurochem.* **24**, 875–879.

Balcar, V. J., Johnston, G. A. R., and Stephanson, A. L. (1976). *Brain Res.* **102**, 143–151.

Banay-Schwartz, M., Lajtha, K., Sershen, H., and Lajtha, A. (1977). *Neurochem. Res.* **2**, 695–706.

Baumann, P. (ed.) (1979). *Tryptophan Transport: Transport Mechanisms of Tryptophan in Blood Cells, Nerve Cells, and at the Blood–Brain Barrier*, Springer-Verlag, Berlin.

Bennett, J. P., Logan, W. J., and Snyder, S. H. (1973). *J. Neurochem.* **21**, 1533–1550.

Betz, A. L., and Goldstein, G. W. (1980). In *The Cerebral Microvasculature* (H. M. Eisenberg and R., L. Suddith, eds.), pp. 5–16, Plenum Press, New York.

Blasberg, R., and Lajtha, A. (1965). *Arch. Biochem. Biophys.* **112**, 361–377.

Borg, J., Ramaharobandro, N., Mark, J., and Mandel, P. (1980). *J. Neurochem.* **34**, 1113–1122.

Brightman, M. W., Klatzo, I., Olsson, Y., and Reese, T. S. (1970). *J. Neurol. Sci.* **10**, 215–239.

Christensen, H. N. (1975). In *Current Topics in Membranes and Transport* (F. Bronner and A. Kleinzeller, eds.), pp. 227–258, Academic Press, New York.

Cornford, E. M., and Oldendorf, W. H. (1975). *Biochim. Biophys. Acta* **394**, 211–219.

Cremer, J. E., Braun, L. D., and Oldendorf, W. H. (1976). *Biochim. Biophys. Acta* **448**, 633–637.

Crone, C. (1965). *J. Physiol.* **181**, 103–113.

Cutler, R. W. P., and Coull, B. M. (1978). In *Taurine and Neurological Disorders* (A. Barbeau and R. J. Huxtable, eds.), pp. 95–107, Raven Press, New York.

Davies, L. P., and Johnston, G. A. R. (1976). *J. Neurochem.* **26**, 1007–1014.

Dienel, G. A. (1981). *J. Neurochem.* **36**, 34–43.

East, J. M., Dutton, G. R., and Currie, D. N. (1980). *J. Neurochem.* **34**, 523–530.

Emirbekov, E. Z., Sershen, H., and Lajtha, A. (1977). *Brain Res.* **125**, 187–191.

Höckel, S. H. J., Müller, W. E., and Wollert, U. (1978) *Neurosci. Lett.* **8**, 65–69.

Hruska, R. E., Padjen, A., Bressler, R., and Yamamura, H. I. (1978). *Mol. Pharmacol.* **14**, 77–85.

Hwang, S.-M., and Segal, S. (1979). *Biochim. Biophys. Acta* **557**, 436–448.

Hwang, S.-M., Weiss, S., and Segal, S. (1980). *J. Neurochem.* **35**, 417–424.

Iversen, L. L., and Kelly, J. S. (1975). *Biochim. Pharmacol.* **24**, 933–938.

Johansson, B. B. (1980). In *The Cerebral Microvasculature* (H. M. Eisenberg and R. L. Suddith, eds.), pp. 211–219, Plenum Press, New York.

Johnston, G. A. R., and Iversen, L. L. (1971). *J. Neurochem.* **18**, 1951–1961.

Kaback, H. R., Rudnick, G., Schuldiner, S., Short, S. A., and Stroobant, P. (1976). In *The Structural Basis of Membrane Function*, pp. 107–128, Academic Press, New York.

Kaczmarek, L. K., and Davison, A. N. (1972). *J. Neurochem.* **19**, 2355–2362.

Kandera, J., Levi, G., and Lajtha, A. (1968). *Arch. Biochem. Biophys.* **126**, 249–260.

Kilberg, M. S., Handlogten, M. E., and Christensen, H. N. (1980). *J. Biol. Chem.* **255**, 4011–4019.

Kuhar, M. J. (1973). *Life Sci.* **13**, 1623–1634.

Lajtha, A., and Sershen, H. (1974). *Arch. Biochem. Biophys.* **165**, 539–547.

Lajtha, A., and Toth, J. (1965). *Biochem. Pharmacol.* **14**, 729–738.

Lajtha, A., and Toth, J. (1973). *Brain Res.* **55**, 238–241.

Larsson, B., Skärby, T., Edvinsson, L., Hardebo, J. E., and Owman, C. (1980). *Neurosci. Lett.* **17**, 155–159.

Levi, G. (1973). In *Biochemistry of the Developing Brain* (W. Himwich, ed.), pp. 187–218, Dekker, New York.

Levi, G., Kandera, J., and Lajtha, A. (1967). *Arch. Biochem. Biophys.* **119**, 303–311.

Levi, G., Poce, U., and Raiteri, M. (1976). In *Transport Phenomena in the Nervous System: Physiological and Pathological Aspects* (G. Levi, L. Battistin, and A. Lajtha, eds.), pp. 273–289, Plenum Press, New York.

Margolis, R. K., and Lajtha, A. (1968). *Biochim. Biophys. Acta* **163**, 374–385.

Morre, M. C., and Wurtman, R. J. (1981). *Life Sci.* **28**, 65–75.

Oldendorf, W. H. (1973). *Am. J. Physiol.* **224**, 1450–1453.

Oldendorf, W. H., and Szabo, J. (1976). *Am. J. Physiol.* **230**, 94–98.

Pardridge, W. M., and Oldendorf, W. H. (1975). *Biochim. Biophys. Acta* **401**, 128–136.

Pardridge, W. M., and Oldendorf, W. H. (1977). *J. Neurochem.* **28**, 5–12.

Piccoli, F., Grynbaum, A., and Lajtha, A. (1971). *J. Neurochem.* **18**, 1135–1148.

Rapoport, S. I. (1980). In *The Cerebral Microvasculature* (H. M. Eisenberg and R. L. Suddith, eds.), pp. 179–192, Plenum Press, New York.

Riddall, D. R., Leach, M. J., and Davison, A. N. (1976). *J. Neurochem.* **27**, 835–839.

Schultz, S. G., and Curran, P. F. (1970). *Physiol. Rev.* **50**, 637–718.

Sershen, H., and Lajtha, A. (1976a). *Exp. Neurol.* **53,** 465–474.
Sershen, H., and Lajtha, A. (1976b). *Neurochem. Res.* **1,** 417–428.
Sershen, H., and Lajtha, A. (1979). *J. Neurochem.* **32,** 719–726.
Toth, J., and Lajtha, A. (1980). *Exp. Neurol.* **68,** 443–452.
Wade, L. A., and Katzman, R. (1975). *J. Neurochem.* **25,** 837–842.
Wade, L. A., Brady, H. M., and Barbay, D. M. (1980). *Soc. Neurosci. Abstr.* **6,** 831.
Wheeler, D. D. (1979). *J. Neurochem.* **32,** 1197–1213.
Wheeler, D. D. (1980). *Exp. Gerontol.* **15,** 269–284.
Zanchin, G., Sershen, H., and Lajtha, A. (1976). *Exp. Neurol.* **51,** 292–303.

X

Channels, Pores, Intercellular Communication

125

Proteins Forming Large Channels in Biological Membranes*

Hiroshi Nikaido

It is generally accepted that most specific transport processes across biological membranes are carried out by intrinsic membrane proteins that function as transmembrane channels (Singer, 1974). However, in order to prevent the loss of intracellular nutrients and metabolites, these channels must remain closed normally and allow only the passage of solutes of one specific type; we propose to call them "stringent channels." We may imagine channels with contrasting properties, i.e., channels that normally remain open and are less discriminating in terms of the structure of solutes that are allowed to pass through. Although these "relaxed channels" cannot exist in the typical plasma membranes of unicellular organisms for obvious reasons, there is no reason that precludes their existence in specialized membranes. The first such channel, identified by the reconstitution approach, was the porin channel in bacterial outer membranes.

1. PORIN CHANNELS IN ESCHERICHIA COLI AND SALMONELLA TYPHIMURIUM

Gram-negative bacteria such as *E. coli* and *S. typhimurium* are surrounded by an extra membrane layer, called the outer membrane, which is located outside the cytoplasmic (plasma) membrane and the peptidoglycan layer of the cell wall. It has been shown that the outer membrane acts as a permeability barrier for the penetration of a number of compounds, for example, antibiotics that are either hydrophobic or of large size (Nikaido, 1976). On the other hand, the membrane must permit the diffusion of nutrient molecules,

---

*For more detailed description of the topics covered in this article, the reader is referred to recent, more extensive reviews (Nikaido and Nakae, 1979; Nikaido, 1979).

Hiroshi Nikaido • Department of Microbiology and Immunology, University of California, Berkeley, California 94720.

and studies with plasmolyzed cells have shown that it is nonspecifically permeable to hydrophilic solutes with molecular weights less than about 600 (Decad and Nikaido, 1976).

The component responsible for this permeability was sought by reconstituting various outer-membrane protein fractions with lipid components of the outer membrane, and by examining whether radiolabeled sucrose diffused out through the protein-containing membranes of liposomes during gel filtration (Nakae, 1976). This search resulted in the identification of a class of proteins (porins) that were extraordinarily active in producing nonspecific diffusion channels. Thus, only 10 μg of porins reconstituted with 1.5 mg of lipids was enough to permit the efflux of 99% of the sucrose molecules originally present inside the liposomes. In a more recently developed "swelling" assay (Luckey and Nikaido, 1980a), even 0.1 μg of porins was seen to produce significant increases in the rate of influx of solute molecules into liposomes made from 1.5 mg of lipids, thereby rapidly decreasing the turbidity of the suspension.

The properties of the porin channel have been studied by using intact cells (Nikaido and Rosenberg, 1981; Nikaido, Rosenberg, and Foulds, to be published), as well as liposomes (Nikaido and Rosenberg, to be published) and planar bilayer membranes (Benz *et al.*, 1979; Schindler and Rosenbusch, 1978). The results can be summarized as follows.

1. There is no discrimination of solutes based on their specific configuration, although some workers have erroneously concluded that some porin channels are more or less specific (for review see Nikaido *et al.*, 1980). However, the rate of diffusion of solutes through the porin channel is obviously influenced by the gross physical properties of the solute, as described below.
2. The size of the solute has a very strong influence on the penetration rate, even when it is well within the exclusion limit. Thus, lactose (molecular weight 342) diffuses almost 300 times more slowly than glycerol (molecular weight 90) through the channel of *E. coli* B porin (Nikaido and Rosenberg, 1981). Behavior of this type is expected theoretically for diffusion through narrow channels (Renkin, 1954), and the use of the Renkin equation suggests that the *E. coli* porin channel can be approximated as a hollow cylinder with a diameter of 1.2 nm (Nikaido and Rosenberg, 1981).
3. The more hydrophobic solutes diffuse through the porin channel more slowly (Zimmermann and Rosselet, 1977). A more detailed study has shown that, with monoanionic cephalosporins, a 10-fold increase in the 1-octanol/water partition coefficient (of uncharged form) produced a 5- to 6-fold decrease in the permeability coefficient (Nikaido, Rosenberg, and Foulds, to be published). This effect is possibly due to the strong hydrogen bonding between the water molecules in the pore as well as between these water molecules and groups on the walls of the pore.
4. Studies with black lipid membranes (Benz *et al.*, 1979) showed that channels made by *E. coli* porins prefer cations over anions. This was also confirmed by the study of penetration rates of various cephalosporins in intact cells (Nikaido, Rosenberg, and Foulds, to be published).

The porins normally present in wild-type *E. coli* (and presumably also *S. typhimurium*) thus produce very narrow channels that generally discriminate against hydrophobic and anionic molecules. This property of the channel is advantageous for these enteric organisms, which live in an environment rich in anionic and hydrophobic inhibitors, i.e., bile salts. Thus, one of the most important functions of their outer membranes is to prevent the access of these detergents to the plasma membrane.

The synthesis of porins is controlled by a variety of physiological conditions. (1) E. *coli* K_{12} produces two species of porins, la and lb, and the synthesis of the former is shut off in media of high osmotic activity (Nakamura and Mizushima, 1976). The channels produced by these two proteins are indistinguishable in their properties, except that the permeability produced by porin lb in intact cells is nearly one order of magnitude lower than that produced by la (Nikaido, Rosenberg, and Foulds, to be published). These results suggest that in intact cells, a large fraction of lb channels is in a "closed" state. The condition that produces this closing is not yet known, but it is most interesting that Schindler and Rosenbusch (1978) found that membrane potential in excess of certain threshold values resulted in the closing of *E. coli* B porin channels reconstituted into planar lipid bilayers. (2) *E. coli* K_{12} produces, under certain conditions, a "new" porin, called "E" or "e" (Foulds and Chai, 1978). We found that this protein produced channels that did not discriminate against anionic solutes (Nikaido, Rosenberg, and Foulds, to be published). Recent results showed that the protein is coregulated with the alkaline phosphatase (Argast and Boos, 1980; Tommassen and Lugtenberg, 1980). Thus, *E. coli* produces this channel only under phosphate-limited conditions, and the channel has properties especially favorable for the diffusion of inorganic phosphate as well as other phosphorylated compounds.

The chemistry of *E. coli* porin was studied before the channel-forming function of this protein became known (Rosenbusch, 1974). Its molecular weight is around 36,000; it is one of the most abundant proteins in *E. coli*; and its amino acid composition is not particularly hydrophobic. Porin in its native state was shown to exist as trimers (Palva and Randall, 1978; Nakae *et al.*, 1979), which contain an unusually high percentage of β structure and no indication of α-helix, and is unusually resistant to denaturation by sodium dodecyl sulfate (Rosenbusch, 1974). The porin trimers associate noncovalently with the underlying peptidoglycan layer, and sometimes removal of other outer-membrane components with sodium dodecyl sulfate produces a two-dimensional hexagonal crystalline array of porin trimers on planar peptidoglycan sheets. Computer enhancement of the negatively stained, electron microscopic pictures of such preparations showed that the timer has three central "pits" that take up the stain (Steven *et al.*, 1977). These pits presumably correspond to the channels, and therefore each subunit of the porin trimer produces a single channel. This conclusion was also supported by the observation that on reconstitution into planar lipid bilayers, a unit increment in conductance, presumably corresponding to the insertion of a single trimer, always consisted of three equal substeps (Schindler and Rosenbusch, 1978).

Porin from *E. coli* has been sequenced, but the sequence did not contain any long stretches of either hydrophobic or hydrophilic amino acids (Chen *et al.*, 1979). The recent crystallization of this porin in the presence of octylglucoside (Garavito and Rosenbusch, 1980) should contribute greatly to our understanding of the three-dimensional structure of these channels.

2. PORINS FROM NONENTERIC BACTERIA

Porin of *Pseudomonas aeruginosa* has a size (apparent molecular weight 35,000) similar to those of *E. coli* and *S. typhimurium*, but it produces a much larger channel, allowing the diffusion of polysaccharides of several thousand molecular weight (Hancock *et al.*, 1979). Similar, large exclusion limits were noted with porins from *Neisseria gonorrhoeae*

(Douglas, Lee, and Nikaido, 1981) and several other gram-negative, nonenteric species (Zalman, Hancock, and Nikaido, to be published).

In spite of the large size of the porin channel, the outer membrane of *P. aeruginosa* has been known to have low permeability to a variety of solutes, including antibiotics (Brown, 1975). The logical explanation is that most of the channels remain closed under laboratory conditions, but this idea remains to be proven.

3. PORINS FROM MITOCHONDRIAL OUTER MEMBRANE

Mitochondrial outer membrane must allow the diffusion of various metabolic intermediates and coenzymes, and indeed there have been several studies indicating that this membrane was permeable to low-molecular-weight compounds (for references see Zalman *et al.*, 1980). We have purified a major protein from the mitochondrial outer membranes of rat liver and mung bean seedlings, and have shown that this protein of apparent molecular weight 30,000 is a porin that produced large channels with an exclusion limit of several thousand molecular weight (Zalman *et al.*, 1980). It is interesting that "particles" with a central "pit" or pore have been demonstrated on these membranes by electron microscopy (Parsons *et al.*, 1965), and the presence of protein oligomers with central holes has also been detected by X-ray diffraction studies (Manella and Bonner, 1975). Mitochondrial porin channels apparently prefer negatively charged solutes (Colombini, 1979) in contrast to the porins of wild-type *E. coli*.

4. PHAGE LAMBDA RECEPTOR PROTEIN, AN EXAMPLE OF A SPECIFIC CHANNEL

Phage lambda receptor protein, coded for by the gene *lamB*, was shown by the pioneering work of Szmelcman and Hofnung (1975) to facilitate the diffusion of maltose across the outer membrane. Although some workers suggested that this protein also produced large nonspecific channels and therefore was a porin, our study of the solute diffusion rates in reconstituted liposomes containing purified LamB protein showed that LamB channels definitely had elements of configurational specificity (Luckey and Nikaido, 1980a). Thus, among disaccharides, maltose diffused through this channel 40 times faster than sucrose. [In contrast, the maximal difference in diffusion rates through the porin channel among disaccharides was only two- to threefold (Nikaido *et al.*, 1980)]. The configurational discrimination became even more pronounced with higher oligosaccharides, and only the oligosaccharides of the maltose series were seen to penetrate through the LamB channel with significant rates.

This configurational specificity suggests the presence of specific binding sites within the channel, and we indeed found that maltoheptaose inhibits the diffusion of glucose with an apparent K_i of 1 mM (Luckey and Nikaido, 1980b). Ferenci *et al.* (1980) also found that LamB-containing surface of *E. coli* cells bound fluorescent derivatives of amylopectin.

In the reconstituted system, the LamB channel allows rapid diffusion of very small solutes, such as serine, Tris, or inorganic ions (Luckey and Nikaido, 1980a; Boehler-Kohler *et al.*, 1979), and thus shows some properties of the relaxed channel. In intact cells, however, the channel appears to function in a much more stringent fashion (von Meyenburg and Nikaido, 1977); possible mechanisms for this phenomenon have been proposed (Luckey and Nikaido, 1980a; Henzenroeder and Reeves, 1980; Bavoil and Nikaido, 1981).

5. FUTURE PERSPECTIVES

Much of the work described has been performed with the outer membranes of gram-negative bacteria. However, the occurrence of these channel-forming proteins is not limited to the bacterial system, as is evident from the work on mitochondrial porins. We suspect that proteins forming large, relaxed channels will be found in a number of intracellular membrane systems as well as in the areas of plasma membranes located next to the surface of other cells. Indeed in the latter area, "gap junctions" are present, and they seem to have structures very similar to the oligomeric unit of porin (Unwin and Zamphighi, 1980).

The stringent channels are obviously widely distributed and catalyze the processes of "facilitated diffusion" and "active transport." However, the functions of these channels have seldom, if ever, been understood at the molecular level. It is our hope that the LamB channel will serve as a good model in this area, in view of our extensive knowledge on the structure of LamB protein.

ACKNOWLEDGMENTS. The experimental work in the author's laboratory was supported by United States Public Health Service Grant AI-09644 and American Cancer Society Grant BC-20.

REFERENCES

Argast, M., and Boos, W. (1980). *J. Bacteriol.* **143,** 152–150.

Bavoil, P., and Nikaido, H. (1981). *J. Biol. Chem.* **256,** 11385–11388.

Benz, R., Janko, K., and Läuger, P. (1979). *Biochim. Biophys. Acta* **551,** 238–247.

Boehler-Kohler, B. A., Boos, W., Dieterle, R., and Benz, R. (1979). *J. Bacteriol.* **138,** 33–39.

Brown, M. R. W. (ed.) (1975). In *Resistance of Pseudomonas aeruginosa,* pp. 71–107, Wiley–Interscience, New York.

Chen, R., Krämer, C., Schmidmayr, W., and Henning, U. (1979). *Proc. Natl. Acad. Sci. USA* **76,** 5014–5017.

Colombini, M. (1979). *Nature (London)* **279,** 643–645.

Decad, G., and Nikaido, H. (1976). *J. Bacteriol.* **128,** 325–336.

Douglas, J. T., Lee, M. D., and Nikaido, H. (1981). *FEMS Microbiol. Lett.* **12,** 305–309.

Ferenci, T., Schwentorat, M., Ullrich, S., and Vilmart, J. (1980). *J. Bacteriol.* **142,** 521–526.

Foulds, J., and Chai, T. (1978). *J. Bacteriol.* **133,** 1478–1483.

Garavito, R. M., and Rosenbusch, J. P. (1980). *J. Cell Biol.* **86,** 327–329.

Hancock, R. E. W., Decad, G. M., and Nikaido, H. (1979). *Biochim. Biophys. Acta* **554,** 323–331.

Henzenroeder, M. W., and Reeves, P. (1980). *J. Bacteriol.* **141,** 431–435.

Luckey, M., and Nikaido, H. (1980a). *Proc. Natl. Acad. Sci. USA* **77,** 167–171.

Luckey, M., and Nikaido, H. (1980b). *Biochem. Biophys. Res. Commun.* **93,** 166–171.

Manella, C. A., and Bonner, W. D., Jr. (1975). *Biochim. Biophys. Acta* **413,** 226–233.

Nakae, T. (1976). *J. Biol. Chem.* **251,** 2176–2178.

Nakae, T., Ishii, J., and Tokunaga, M. (1979). *J. Biol. Chem.* **254,** 1457–1461.

Nakamura, K., and Mizushima, S. (1976). *J. Biochem. (Tokyo)* **80,** 1411–1422.

Nikaido, H. (1976). *Biochim. Biophys. Acta* **433,** 118–132.

Nikaido, H. (1979). In *Bacterial Outer Membranes* (M. Inouye, ed.), pp. 361–407, Wiley–Interscience, New York.

Nikaido, H., and Nakae, T. (1979). *Adv. Microb. Physiol.* **20,** 163–250.

Nikaido, H., and Rosenberg, E. Y. (1981). *J. Gen. Physiol.* **77,** in press.

Nikaido, H., Luckey, M., and Rosenberg, E. Y. (1981). *J. Supramol. Struct.,* in press.

Palva, E. T., and Randall, L. L. (1978). *J. Bacteriol.* **133,** 279–286.

Parsons, D. F., Bonner, W. D., Jr., and Verboon, J. G. (1965). *Can. J. Bot.* **43,** 647–655.

Renkin, E. M. (1954). *J. Gen. Physiol.* **38,** 225–243.

Rosenbusch, J. P. (1974). *J. Biol. Chem.* **249,** 8019–8029.

Schindler, H., and Rosenbusch, J. P. (1978). *Proc. Natl. Acad. Sci. USA* **75,** 3751–3755.

Singer, S. J. (1974). *Annu. Rev. Biochem.* **43,** 805–833.

Steven, A. C., ten Heggeler, B., Muller, R., Kistler, J., and Rosenbusch, J. P. (1977). *J. Cell Biol.* **72,** 292–301.

Szmelcman, S., and Hofnung, M. (1975). *J. Bacteriol.* **124,** 112–118.

Tommassen, J., and Lugtenberg, B. (1980). *J. Bacteriol.* **143,** 151–157.

Unwin, P. N. T., and Zampighi, G. (1980). *Nature (London)* **283,** 545–549.

von Meyenburg, K., and Nikaido, H. (1977). *Biochem. Biophys. Res. Commun.* **78,** 1100–1107.

Zalman, L. S., Nikaido, H., and Kagawa, Y. (1980). *J. Biol. Chem.* **255,** 1771–1774.

Zimmermann, W., and Rosselet, A. (1977). *Antimicrob. Agents Chemother.* **12,** 368–372.

126

Gap Junction Transport Regulation

Camillo Peracchia

1. INTRODUCTION

Direct cell-to-cell communication (cell coupling) is a feature of virtually all animal tissues from the simplest invertebrates to mammals (Peracchia, 1980). The most obvious function of cell coupling is that of electrically coordinating certain cell communities, for which synchronous activities such as heart contraction, uterine contracture at labor, simultaneous firing of groups of neurons, etc. can take place. However, many other cooperative phenomena are known to depend on direct intercellular communication as not only small ions are being exchanged but also larger charged and neutral molecules including metabolites and second messengers (Peracchia, 1980). This allows synchronization of hormonal effects, equilibration of metabolic and ionic pools, exchange of signal molecules for growth control and cell differentiation, and probably many other cellular activities yet to be identified.

Cell coupling is mediated by intercellular junctions, most frequently known as gap junctions, which represent the molecular framework of narrow intercellular channels spanning the plasma membranes of two neighboring cells and the interposed extracellular space (Fig. 1). In healthy cells the channels are stable in an open state, but can be closed by certain changes in the intracellular ionic composition. This review will discuss primarily the mechanism by which the channel permeability is regulated and the ultrastructural changes that accompany channel occlusion.

2. CHANNEL PERMEABILITY REGULATION

An interesting property of intercellular channels is their ability to close off, rendering neighboring cells independent from each other. This phenomena is known as cell uncoupling. Our knowledge of cell uncoupling finds its roots in studies on the healing property of heart tissue. Since the 19th century, heart tissue has been known to survive mechanical injury to some of its cells by virtue of a property called "healing over." At first, the

Camillo Peracchia • Department of Physiology, University of Rochester Medical Center, Rochester, New York 14642.

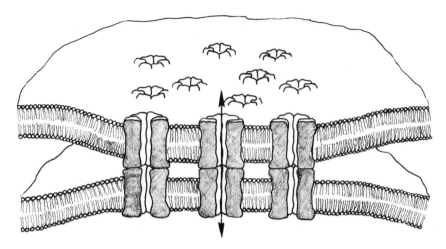

Figure 1. Model of gap junction in coupled state. Intramembrane particles composed of six main subunits span the membrane thickness and bind to similar particles of the adjacent membrane, bridging an extracellular space of 2.0–3.0 nm. The particles form a disordered array and are separated by a center-to-center distance of $\sim$10 nm (vertebrates). Channels $\sim$1.5 nm in diameter are located at the center of the particles (arrow). [From Peracchia, 1980.]

mechanism of healing over was not understood and for many years was believed to follow membrane resealing at the injured region. This hypothesis seemed supported by evidence for the necessity of a normal $[Ca^{2+}]_o$ for the occurrence of healing over (Délèze, 1965) in view of the well-known membrane-stabilizing effect of Ca^{2+}. In retrospect, these data provided the earliest evidence for the role of Ca^{2+} in cell uncoupling. In fact, it is now clear that healing over does not depend on membrane resealing but rather on the occlusion of the intercellular channels of gap junctions following an increase in $[Ca^{2+}]_i$ (Loewenstein and Rose, 1978; Peracchia, 1980; DeMello, 1980).

Evidence for a relationship between uncoupling and $[Ca^{2+}]_i$ was first produced by the observation that electrical uncoupling between perforated cells (Loewenstein et al., 1967) or cells made permeable to ions (Nakas et al., 1966) occurs only in Ca^{2+}-containing media. More recently, the effects of intracellular Ca^{2+} on cell coupling have been elegantly confirmed by studies on the changes in electrical coupling of *Chironomus* salivary gland cells simultaneous with changes in $[Ca^{2+}]_i$ monitored with aequorin (Rose and Loewenstein, 1975, 1976). Indeed, most uncoupling treatments are known to produce an increase in $[Ca^{2+}]_i$. Aside from the obvious increase in $[Ca^{2+}]_i$ following intracellular Ca^{2+} injections, cell damage, or exposure to Ca^{2+} ionophores (Loewenstein and Rose, 1978), $[Ca^{2+}]_i$ is known to increase in uncoupling resulting from inhibition of metabolism, extracellular Ca^{2+} chelation, prolonged cooling, increase in $[Na^+]_i$, stimulation to secretion (reviewed in Peracchia, 1980), and probably hypoxia (Wojtczak, 1979).

The occlusion of the intercellular channels, however, may not be a direct effect of calcium on the channel molecules, as an increase in $[Ca^{2+}]_i$ causes a variety of changes in the intracellular homeostasis. Indeed, in some cases uncoupling occurs without an aequorin-detectable increase in $[Ca^{2+}]_i$ (Rose and Loewenstein, 1976). Furthermore, an increase in $[Ca^{2+}]_i$ causes a decrease in pH_i (Meech and Thomas, 1977), and evidence for cell uncoupling, following cytoplasmic acidification to pH values lower than 6.85, has recently been produced (Turin and Warner, 1977, 1980; Bennett et al., 1978; Rose and Rick, 1978; Iwatsuki and Petersen, 1979; Weingart and Reber, 1979; Giaume et al., 1980; Rink et al., 1980; Petersen et al., 1980). On the other hand, it is possible that a decrease in pH_i causes an increase in $[Ca^{2+}]_i$ (Lea and Ashley, 1978), and thus its effect could be indirect. Rose and Rick (1978) took this stand in a study on *Chironomus* salivary gland cells in which pH_i, $[Ca^{2+}]_i$, and electrical uncoupling were monitored simultaneously during cytoplasmic acidification. However, in a recent study only a small increase in $[Ca^{2+}]_i$ (from 0.2 to 0.35 μM) was detected in uncoupling by 100% CO_2 (Rink *et al.*, 1980). In this case Ca^{2+} was believed not to be the uncoupling agent as $4–7 \times 10^{-5}M$ or higher $[Ca^{2+}]_i$ has been reported to electrically uncouple cells (Oliveria-Castro and Loewenstein, 1971). On the other hand, lower $[Ca^{2+}]_i$ is known to affect the junctions' permeability to dyes (Délèze and Loewenstein, 1976; Rose *et al.*, 1977) and to change the gap junction structure (Peracchia, 1978), and thus questions on the independent effect of Ca^{2+} and H^+ still remain.

To determine whether Ca^{2+} and/or H^+ or neither are the uncoupling agents, the most common approach, namely measuring the coupling resistance between intact cells, is not the ideal one (even when combined with pH and Ca^{2+} monitoring) because, aside from the virtual impossibility to fully evaluate the impact of all the cellular mechanisms activated by these agents, there is no way to determine the actual $[Ca^{2+}]_i$ and $[H^+]_i$ at the very junctional regions. Ideally, one would want to test the independent effect of postulated uncoupling agents on the permeability of isolated junctions. However, the methodologies for determining gap junction permeability *in vitro* are not yet available, although evidence for the capacity of gap junctions to curl into vesicles (Goodenough, 1976) suggests that eventually it might be possible to do so. Meanwhile some useful information can be obtained by a morphological approach (Peracchia, 1978).

3. GAP JUNCTION CHANGES WITH UNCOUPLING

In crayfish, reversible electrical uncoupling of lateral giant axons was found to parallel reversible changes in gap junction structure, characterized by a tightening of intramembrane particle packings into crystalline, most often hexagonal, arrays and a decrease in gap thickness and particle diameter (Peracchia and Dulhunty, 1976). Similar changes in particle crystallinity and diameter were soon demonstrated also in mammalian stomach (Figs. 2, 3)

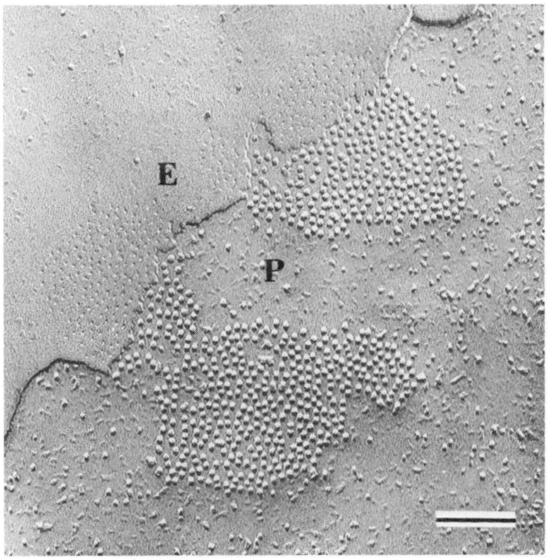

Figure 2. Freeze-fracture replica of a control gap junction from the epithelium of rat stomach fixed after a 1-hr incubation in a well-oxygenated Tyrode's solution at 37°C. The fracture plane steps down from E face (E) to P face (P). Notice that particles and complementary pits are disorderly arranged at a center-to-center distance of ~10 nm. Bar = 100 nm. [From Peracchia and Peracchia, 1980a.]

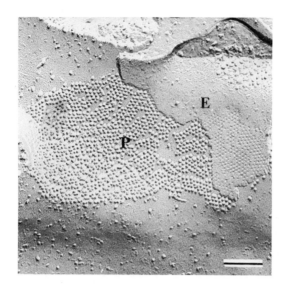

Figure 3. Freeze-fracture replica of a gap junction from the epithelium of rat stomach fixed after a 1-hr incubation at 37°C in a well-oxygenated Tyrode's solution containing 2×10^{-6}M calcium ionophore (A23187). Particles on P face (P) and pits on E face (E) are hexagonally packed into small crystalline domains separated by particle-free isles. The average center-to-center distance between neighboring particles or pits is 8.5 nm. Bar = 100 nm. [From Peracchia and Peracchia, 1980a.]

and liver cells following uncoupling treatments (Peracchia, 1977), and, more recently, gap junction crystallization with uncoupling has been confirmed in variety of tissues (Raviola *et al.*, 1978, 1980; Baldwin, 1979; Dahl and Isenberg, 1980; Meda *et al.*, 1980; Petersen *et al.*, 1980). In one study (Dahl and Isenberg, 1980), however, an increase in particle diameter was found in early uncoupling stages, followed by a decrease in diameter at later stages. In some of these studies, electrical uncoupling was monitored electrophysiologically (Dahl and Isenberg, 1980; Petersen *et al.*, 1980); however, only in crayfish axons have changes in gap junction structure been demonstrated in those very cells in which changes in junctional resistance were monitored (Peracchia and Dulhunty, 1976). The gap junction changes have been interpreted to reflect conformational rearrangements in channel protein resulting in channel occlusion (Peracchia and Dulhunty, 1976; Peracchia, 1977).

The validity of these findings has been questioned in view of the fact that glutaraldehyde fixation, used in these experiments, itself induces electrical uncoupling (Politoff and Pappas, 1972; Bennett, 1973; Ramon and Zampighi, 1980). Recently, however, the gap junction changes have been confirmed in samples freeze-fractured without previous fixation and cryoprotective treatment (Raviola *et al.*, 1978, 1980; Baldwin, 1979; Peracchia and Peracchia, 1980b); thus, glutaraldehyde is likely to uncouple with a different mechanism not involving particle crystallization.

In rat stomach, junction crystallization was also obtained by exposure to a Ca^{2+} ionophore (A23187) (Fig. 3) (Peracchia and Peracchia, 1980a). However, also in this case, it could not be unequivocally determined whether the structural change was a direct or mediated Ca^{2+} effect. Instead, gap junctions isolated from mammalian eye lens fibers provided a very useful system for studying the direct effects of divalent cations and H^+ on gap junction structure. These junctions, in fact, can be isolated with disordered particle arrays, typical of coupled junctions, in EDTA-containing media (Fig. 4) and can be induced to crystallize *in vitro* upon exposure to divalent cations (Fig. 5) or H^+ (Peracchia, 1980). Reversible crystallization occurs when the $[Ca^{2+}]$ of the medium is buffered to values of 5×10^{-7}M or greater, at pH 7 (Peracchia, 1978), when the $[H^+]$ is raised to 3×10^{-7}M in Ca^{2+}/Mg^{2+}-free media (pH 6.5 or lower) (Peracchia and Peracchia, 1980b), and when the $[Mg^{2+}]$ is brought to values greater than 10^{-3}M in Ca^{2+}-free media at pH 7

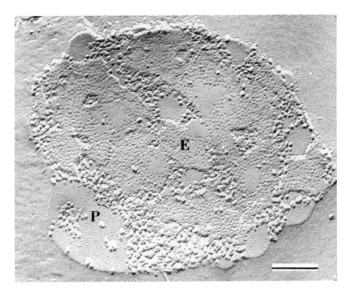

Figure 4. Freeze-fracture replica of a gap junction isolated from calf lens fibers, in the presence of EDTA, and subsequently incubated in a Ca^{2+}–EDTA buffer ($10^{-7}M$ Ca^{2+}). Notice that particles and pits form a disordered array. E, E face; P, P face. Bar = 100 nm. [From Peracchia and Peracchia, 1980a.]

(Peracchia and Peracchia, 1980a). Interestingly, these concentrations are similar to those believed to uncouple intact cells (Loewenstein and Rose, 1978; Weingart, 1977; Turin and Warner, 1977, 1980); thus, the data suggest an independent effect of divalent cations and H^+ on cell coupling. While divalent cations crystallized the gap junctions mainly into hexagonal arrays, low-pH treatment produced mostly rhombic and orthogonal particle packings. This suggested a four-subunit composition of the lens particles (Peracchia and Peracchia, 1980b).

The calcium effects on lens junctions have recently been confirmed (Alcalá *et al.*,

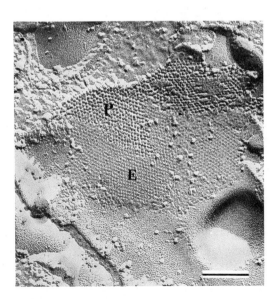

Figure 5. Freeze-fracture replica of a gap junction isolated from calf lens fibers, in the presence of EDTA, and subsequently incubated in a solution containing $10^{-4}M$ Ca^{2+} and $10^{-3}M$ Mg^{2+}. Particles and pits are packed into a crystalline array. E, E face; P, P face. Bar = 100 nm. [From Peracchia and Peracchia, 1980a.]

1979; Kistler and Bullivant, 1980b), while doubts have been raised on the low-pH results. According to Kistler and Bullivant (1980a) in fact, membranes with orthogonal arrays are individual membranes rather than junctions and can be produced by tryspin digestion at pH 5.5–8. In our study, unequivocal evidence for the junctional nature of these arrays was consistently obtained (Peracchia and Peracchia, 1980b), and thus we feel that tryspin treatment and/or the different isolation procedure employed by Kistler and Bullivant (1980a) may have caused the crystalline junctional membranes to separate. The induction of orthogonal arrays by tryspin is indeed interesting and raises the possibility that crystallinity may also result from the loss of some components of the junctional protein.

Obviously, the validity of the lens studies, in terms of uncoupling, relies on the capacity of lens fibers, like other cells, to uncouple. However, doubts in this regard have been raised on the basis of the apparent inability of lens fiber junctions to crystallize in intact cells and in isolated fractions, under certain experimental conditions (Goodenough, 1979). As reported earlier, we feel that there is now clear evidence that lens fiber junctions crystallize in isolated fractions upon exposure to divalent cations or H^+ (Peracchia, 1978; Alcalá et al., 1979; Kistler and Bullivant, 1980b; Peracchia and Peracchia, 1980a,b); moreover, recently a widespread junction crystallization has also been obtained in intact fibers by a treatment that increases cell calcium content (Peracchia et al., 1979; Bernardini and Peracchia, 1981). In addition, the capacity of lens fibers to uncouple has been strongly supported by evidence for the existence of a "healing over" mechanism in rat lens. This is suggested by the observation that in lenses loaded with $^{42}K^+$, there is a rapid return of $^{42}K^+$ efflux rate to control values, following mechanical injury (Bernardini et al., 1981). Interestingly, the $^{42}K^+$ efflux rate, greatly increased immediately after injury, decreases afterwards only if a normal $[Ca^{2+}]_o$ is maintained, which confirms the hypothesis that gap junction uncoupling is involved in lens healing over.

The mechanism of channel permeability modulation is still unclear. A hypothesis has proposed that junction crystallization and channel occlusion follows blockage of negatively charged sites on the membrane surface and conformational changes in channel protein (Peracchia et al., 1979; Peracchia, 1980). Recent data indicate that these sites may belong to calmodulin-like proteins linked to channel macromolecules. These data are based on the observation that trifluoperazine, a specific calmodulin inhibitor, prevents the Ca^{2+}- and H^+-induced crystallization of lens fiber gap junctions and strongly inhibits the electrical uncoupling of amphibian embryo cells (Peracchia et al., 1981).

Protein conformational changes, possibly resulting from the activation of a calmodulin-like protein are believed to result in channel occlusion at the cytoplasmic end (Fig. 6) (Peracchia, 1980). A similar model has recently been proposed by Unwin and Zampighi (1980). Here, however, the channels are believed to close, at the cytoplasmic end, as a result of rotation and tilt of the particle subunits. This hypothesis stems from data on changes in the architecture of isolated, negatively stained, liver junctions studied by low-dose electron microscopy. These may turn out to be very important findings. However, at present, they should be taken with caution. In fact, the junctional changes have not been obtained following exposure to known uncoupling agents such as divalent cations or H^+, but rather following a more or less extensive wash, after detergent treatment. Moreover, the changes in channel size are only presumed to take place, as the limited resolution of the technique has not allowed the investigators to visualize the channels in either of the two conformational states described.

In conclusion, present evidence suggests that direct communication between cells is mediated by small hydrophilic channels that couple electrically and metabolically adjacent cells at gap junctions. Channel permeability is regulated by intracellular free divalent cat-

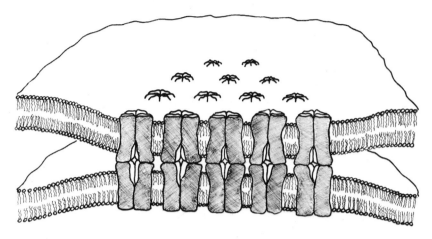

Figure 6. Tentative model of gap junction in uncoupled state. Uncoupling agents such as divalent cations and H^+ are believed to act on the junctional molecules, narrowing the intercellular channels at their cytoplasmic end and clumping the intramembrane particles into crystalline (most often hexagonal) arrays. There is evidence (Peracchia *et al.,* 1981) that both the junctional changes and the electrical uncoupling may be mediated by a calmodulin-like protein linked to the junctional macromolecules. [From Peracchia, 1980.]

ions and, probably, H^+. A small increase in cytoplasmic divalent cations or H^+ concentration is believed to induce the channels to switch from a permeable to a nonpermeable state. Channel occlusion is likely to result from conformational changes in channel protein, activated by a calmodulin-like protein.

ACKNOWLEDGMENTS. I am indebted to my wife Lillian for her excellent collaboration. This work was supported by NIH Grant GM-20113.

REFERENCES

Alcalá, J., Kuszak, J., Katar, M., Bradley, R. H., and Maisel, H. (1979). *J. Cell Biol.* **83**(2, pt. 2), 269a.

Baldwin, K. M. (1979). *J. Cell Biol.* **82**, 66–75.

Bennett, M. V. L. (1973). *Fed. Proc.* **32**, 65–75.

Bennett, M. V. L., Brown, J. E., Harris, A. L., and Spray, D. C. (1978). *Biol. Bull. (Woods Hole, Mass.)* **155**, 428a.

Bernardini, G., and Peracchia, C. (1981). *Invest. Opthalmol. Vis. Sci.* **21**, 291–299.

Bernardini, G., Peracchia, C., and Venosa, A. (1981). *J. Physiol. (London)* **320**, 187–192.

Dahl, G., and Isenberg, G. (1980). *J. Membr. Biol.* **53**, 63–75.

Délèze, J. (1965). In *Electrophysiology of the Heart* (B. Taccardi and G. Marchetti, eds.), pp. 147–148, Pergamon Press, Elmsford, N.Y.

Délèze, J., and Loewenstein, W. R. (1976). *J. Membr. Biol.* **28**, 71–86.

DeMello, W. C. (1980). In *Membrane Structure and Function* (E. E. Bittar, ed.), Vol. 3, pp. 127–170, Wiley, New York.

Giaume, C., Spira, M. E., and Korn, H. (1980). *Neurosci. Lett.* **17**, 197–202.

Goodenough, D. A. (1976). *J. Cell Biol.* **68**, 220–231.

Goodenough, D. A. (1979). *Invest. Ophthalmol. Vis. Sci.* **18**, 1104–1122.

Iwatsuki, N., and Petersen, O. H. (1979). *J. Physiol. (London)* **291**, 317–326.

Kistler, J., and Bullivant, S. (1980a). *FEBS Lett.* **111**, 73–78.

Kistler, J., and Bullivant, S. (1980b). *J. Ultrastruct. Res.* **72**, 27–38.

Lea, T. J., and Ashley, C. C. (1978). *Nature (London)* **275**, 236–238.

Loewenstein, W. R., and Rose, B. (1978). *Ann. N.Y. Acad. Sci.* **307**, 285–307.

Loewenstein, W. R., Nakas, M., and Socolar, S. J. (1967). *J. Gen. Physiol.* **50,** 1865–1891.
Meda, P., Perrelet, A., and Orci, L. (1980). *Horm. Metab. Res. Suppl.* **10,** 157–162.
Meech, R., and Thomas, R. C. (1977). *J. Physiol.* **265,** 867–879.
Nakas, M., Higashino, S., and Loewenstein, W. R. (1966). *Science* **151,** 89–91.
Oliveira-Castro, G. M., and Loewenstein, W. R. (1971). *J. Membr. Biol.* **5,** 51–77.
Peracchia, C. (1977). *J. Cell Biol.* **72,** 628–641.
Peracchia, C. (1978). *Nature (London)* **271,** 669–671.
Peracchia, C. (1980). *Int. Rev. Cytol.* **66,** 81–146.
Peracchia, C., and Dulhunty, A. F. (1976). *J. Cell Biol.* **70,** 419–439.
Peracchia, C., and Peracchia, L. L. (1980a). *J. Cell Biol.* **87,** 708–718.
Peracchia, C., and Peracchia, L. L. (1980b). *J. Cell Biol.* **87,** 719–727.
Peracchia, C., Bernardini, G., and Peracchia, L. L. (1979). *J. Cell Biol.* **83,**(2, pt. 2), 86a.
Peracchia, C., Bernardini, G., and Peracchia, L. L. (1981). *J. Cell Biol.* **91**(2, pt. 2) **91,** 124a.
Petersen, O. H., Findlay, I., Meda, P., Laugier, R., and Iwatsuki, N. (1980). In *Hydrogen Ion Transport in Epithelia* (I. Schulz, ed.), pp. 227–234, Elsevier/North-Holland, Amsterdam.
Politoff, A., and Pappas, G. D. (1972). *Anat. Rec.* **172,** 384–385.
Ramon, F., and Zampighi, G. (1980). *J. Membr. Biol.* **54,** 165–171.
Raviola, E., Goodenough, D. A., and Raviola, G. (1978). *J. Cell Biol.* **79,**(2, pt. 2), 229a.
Raviola, E., Goodenough, D. A., and Raviola, G. (1980). *J. Cell Biol.* **87,** 273–279.
Rink, T. J., Tsien, R. Y., and Warner, A. E. (1980). *Nature (London)* **283,** 658–660.
Rose, B., and Loewenstein, W. R. (1975). *Nature (London)* **254,** 250–252.
Rose, B., and Loewenstein, W. R. (1976). *J. Membr. Biol.* **28,** 87–119.
Rose, B., and Rick, R. (1978). *J. Membr. Biol.* **44,** 377–415.
Rose, B., Simpson, I., and Loewenstein, W. R. (1977). *Nature (London)* **267,** 625–627.
Turin, L., and Warner, A. E. (1977). *Nature (London)* **270,** 56–57.
Turin, L., and Warner, A. E. (1980). *J. Physiol. (London)* **300,** 489–504.
Unwin, P. N. T., and Zampighi, G. (1980). *Nature (London)* **283,** 545–549.
Weingart, R. (1977). *J. Physiol. (London)* **264,** 341–365.
Weingart, R., and Reber, W. (1979). *Experientia* **35,** 17a.
Wojtczak, J. (1979). *Circ. Res.* **44,** 88–95.

127

Transmembrane Channels Produced by Colicin Molecules

S. E. Luria and Joan L. Suit

1. INTRODUCTION

In this idiosyncratic and necessarily sketchy minireview of the action of colicins, attention is focused on a specific group of these remarkable bacterial proteins, namely on the E1-I-K group. These proteins have in common a bactericidal action that, as we shall see below, apparently results from the formation of transmembrane channels permitting passage of a variety of small molecules. In susceptible *Escherichia coli* bacteria, these colicins cause membrane depolarization followed by a variety of secondary effects: defects in active transport, loss of motility, lowering of ATP levels, and arrest of protein and nucleic acid synthesis (Luria, 1973). The mechanisms of these effects will be explored below.

Colicins are proteins coded by genes present in Col plasmids. Each colicin consists of a single polypeptide of 50,000–80,000 molecular weight. Each Col plasmid also has information that confers to the cell a specific immunity to the colicin in question. Information that regulates colicin production is also present. At present, little is known of the regulatory and immunity mechanisms for colicins of the group E1-I-K, with which this review is concerned and which will sometimes be referred to below as "our colicins."

2. BIOCHEMICAL EFFECTS OF OUR COLICINS

Loss of macromolecular syntheses was the first recognized aspect of cell damage due to the action of the colicins in question. It turned out, however, not to be the central one. A more critical effect of colicins proved to be the inhibition of specific active transport systems. The accumulation of substrates such as amino acids, β-D-galactosides, and potassium ions, which depends on a transmembrane proton gradient, was immediately blocked

S. E. Luria and Joan L. Suit • Department of Biology, Massachusetts Institute of Technology, Cambridge, Massachusetts 02139.

by our colicins (Fields and Luria, 1969). ATP levels were lowered. Accumulation of sugars such as glucose and its analogs through the phosphotransferase system was not inhibited and metabolism continued (Fields and Luria, 1969).

A key step in clarifying the mechanisms involved in colicin action was the use of *unc*A mutants of *E. coli* in which the activity of the Ca^{2+}/Mg^{2-}-ATPase is reduced to a few percent of the wild-type level. The effects of our colicins on these mutant cells were quite different from those on normal cells. There was a rise in ATP levels instead of a decrease; macromolecular syntheses continued more or less normally; but transport of amino acids, galactosides, and other substances was still abolished, as was cell motility, and the cells were killed (Plate *et al.*, 1974). ATP levels were clearly irrelevant. Membrane energy, needed for transport and movement but not for biosynthesis, stood out as the critical parameter.

Exit of labeled rubidium and potassium ions from colicin-treated cells had long been observed (Wendt, 1970). Feingold (1970) concluded that colicin allowed K^+ ions to exit from cells but did not allow entry of protons unless a proton-conducting uncoupler was present. Colicin, therefore, might resemble valinomycin in permitting selective ion losses. The situation, however, proved later to be less simple.

Experiments directly aiming at measurements of the effect of colicins on membrane potential were done first using fluorescent dyes (Brewer, 1976) or H^+/O ratios (Gould and Cramer, 1977) and then more precisely with labeled lipophilic cations (Weiss and Luria, 1978; Tokuda and Konisky, 1978a). The results obtained in two laboratories were unambiguous: the membrane potential of sensitive cells exposed to colicin K or Ia decreased rapidly from about 100 to less than 20 mV. On the other hand, measurements of intracellular pH by partition of a labeled weak acid showed no decrease in ΔpH and sometimes an increase. In terms of Mitchell's formulation: $\Delta\mu$ (mV) $= \Delta\Psi + 60\Delta pH$, only the electrical potential term ($\Delta\Psi$) appeared to be altered by colicin, not the chemical potential of the proton gradient.

3. COLICINS AND ARTIFICIAL MEMBRANES

Further analysis of our colicins' action required systems simpler than intact cells. Membrane vesicles prepared according to Kaback (Kaback, 1971) gave erratic results; presumably the receptor mechanism on the bacterial surface is readily disorganized in the preparation of vesicles. Better results have been reported from the use of vesicles frozen-thawed with colicin, reproducing the findings found on intact cells (Tokuda and Konisky, 1978b).

More informative were the effects of the colicins on artificial membranes. The first report (Schein *et al.*, 1978) was of a study of ion conductivity induced by colicin K on a planar membrane. Colicin allowed ion fluxes for several ions, but only if an electrical potential (positive on the side of colicin addition) was electrically applied. Thus, colicin action seemed to be gated, a potential difference between 10 and 30 mV being required to open the ion channels.

Studies on liposomes (Tokuda and Konisky, 1979; Kayalar and Luria, 1979) gave very different results. Liposomes preloaded with any one of a variety of substrates (Rb^+ ions, sulfate, phosphate, sucrose, glucose-6-phosphate) upon exposure to colicin exhibited an immediate efflux of substrate at rates specific for each substance. This efflux was completely independent of the membrane potential, the results being similar whether the transmembrane potential difference was set at any value between -60 and $+40$ mV by addition of valinomycin and varying potassium concentrations. Inulin and other substances of molecular weight over 1000 did not exit from liposomes.

We believe that the discrepancy between the results obtained with the two types of membrane—planar double monolayers vs. liposomes—with respect to the requirement for a membrane potential, is due to complications intrinsic in the planar membranes. Such complications may arise from the structural singularity present at the edge of planar membranes.

Colicin-treated liposomes become permeable also to protons. This was convincingly demonstrated in experiments carried out in a KC1 unbuffered medium in the absence of weak anions (Weiss and Kayalar, unpublished). Measurement of proton transport requires the absence of buffers because permeable weak acids provide a shuttle for protons. The puzzling persistence (or even rise) of ΔpH observed in colicin-treated cells (Weiss and Luria, 1978; Tokuda and Konisky, 1978a) is apparently due to the fact that the conductance of the colicin channels for protons is not sufficient to balance the steady outflow of protons through the electron transport.

We face, therefore, a simple yet puzzling picture. Our colicins, a set of very soluble proteins, can associate directly, without previous enzymatic alteration, with a phospholipid bilayer in which they generate channels for ions and small organic molecules. Two questions arise: What is the molecular nature of the channels? And how relevant are the liposome findings for the bactericidal action of colicin?

The first question remains unanswered. Measurements of the rate of exit of Rb^+ ions from liposomes as a function of the concentration of colicin reveal that the rate depends on the first power of the concentration (Kayalar, unpublished). This is in agreement with the well-known first-order rate of bacterial killing by colicins. For the sake of comparison, one may mention that channels (porins) that are integral parts of the outer membrane of gram-negative bacteria are reportedly formed by interactions among three molecules (Steven *et al.*, 1977). The channels in tight junctions between animal cells are reported to consist of a complex of six molecules (Unwin and Zampighi, 1980). Both kinds of channels have porosity limits similar to those of colicins.

The molecular structure of the channels and its relation to the colicin molecule itself remain to be explored. High-resolution electron microscopy of liposomes with associated colicin will be of help. Even more desirable are crystallization of colicin and X-ray diffraction analysis. Recent findings (Dankert *et al.*, 1982) incriminate the C-terminal segment of colicin E1 as the transmembrane channel forming portion.

4. COLICINS AND THE BACTERIAL ENVELOPE

We inquire next as to whether colicins produce in the membrane of living bacteria channels like those in liposomes. If such channels occur, they should allow exit or entry (down their own concentration gradient) of substances not usually permeating the membrane. The most convincing observations are the exit of α-methylglucoside phosphate from cells fed α-methylglucoside and treated with colicin E1, and entry of sucrose into *E. coli* cells in the presence of colicin E1 (Kayalar and Luria, 1979).

The liposome experiments suggest that the colicins here discussed act on intact bacteria by interacting with the cytoplasmic membrane, probably in intact form, without being previously degraded. Two classes of bacterial mutations can prevent colicin action, both of them affecting the outer membrane. Resistant mutants lack colicin "receptors," which are outer-membrane components that combine specifically with colicins of one or several kinds. Tolerant mutants, unresponsive to one or several colicins which they can still absorb, also show changes in outer-membrane protein. Although observations on these mutants do not yet provide a molecular interpretation of early steps in colicin action, they suggest that a series of interactions takes place between the colicin and the proteins of the outer membrane. The interactions that are abolished by tolerance mutations may be steps needed to allow the colicin molecules to reach the cytoplasmic membrane without being trapped by the phospholipids of the outer membrane.

The colicins described in this review, as well as other groups of colicins, probably act as intact molecules without prior cleavage (in contrast with certain toxins). The reported cleavage of colicins by bacterial cells or membranes (Cavard and Lazdunski, 1979) may be a side reaction without biological significance.

Inhibition of colicin A or E1 proteolysis by a protease inhibitor does not interfere with the bactericidal action of these colicins (R. N. Brey, 1982). Whole cells or outer cell membranes from mutants resistant to a colicin are still able to cleave that colicin (Bowles and Konisky, 1981). Colicin E2, the prototype of another class of colicins, is an endonuclease that degrades DNA both *in vivo* and *in vitro* (Konisky, 1978). Colicins of the E3 class cause damage to ribosomal RNA (Konisky, 1978). It is not known whether these colicins actually penetrate into the cytoplasm or remain in the cytoplasmic membrane, exposing only an active site to the cytoplasm. They probably need to reach further into the bacterial cell than colicin E1, for when E2 or E3 is immobilized on plastic beads, it can combine with cells but will not kill them (Lau and Richards, 1976). Under similar conditions colicin E1, which acts on the membrane itself, does kill bacteria.

All colicins yet studied fail to act on deenergized bacterial cells (Jetten and Jetten, 1975). Possibly the colicin molecules can reach the cytoplasmic membrane only at sites where outer and inner membranes are in functional contact. Such contact may be absent from deenergized bacteria.

5. COLICINOGENY

The production of colicins by colicinogenic cells deserves some comment. Colicinogenic cells undergo at low rates a spontaneous shift to colicin production. They may be induced by a variety of treatments (mitomycin C, UV light, etc.) to undergo massive colicin synthesis and usually also plasmid amplification. The regulatory mechanisms that control colicin production and are affected by inducing agents are still unknown. The induced cells that produce colicin are killed (Ozeki et al., 1959); for colicin E1 the events of cell death following induction involve an abrupt arrest in respiration and a deenergization of cellular metabolism (Suit, Fan, and Luria, unpublished). These events are apparently due to expression of a gene in the Col E1 plasmid (Shafferman et al., 1979) and resemble the role of comparable gene functions in bacteriophage-induced lysis.

Most of the E1 colicin synthesized after induction appears to remain for a relatively long time in the cell interior and to be released slowly as the cells become permeable (Jakes and Model, 1979). Neither E1 nor E3 is transported through the cell membrane in a precursor form: colicin synthesized *in vitro* in the presence or absence of membrane vesicles is identical to that synthesized *in vivo* (Jakes and Model, 1979). Colicin E1 extracted from inside induced cells has the same NH_2-terminal amino acid sequence as that obtained from the exterior of the cell:

Met-Glu-Thr-Ala-Val-Ala-Tyr-Tyr-Lys-Asp-Gly-Val-Pro-Tyr-Asp-Asp-Lys-Gly-Gln-Val-
(Sauer, unpublished).*

A key biological question should be considered: what is the biological significance of colicins and Col plasmids? It can be speculated that originally colicins may have been membrane-bound proteins, possibly playing a role in lethal cell-to-cell contacts. A killing defense acting only at cell contact may be valuable to the killer cells without significantly impairing survival of the sensitive strains. The genes coding for such colicins may have been on the bacterial chromosome. In the light of the newer evidence on transposable elements in bacteria (Kleckner, 1977), it is reasonable to assume that sets of genes including a colicin gene and its regulatory apparatus could have entered plasmids from the chromosome. Production of extracellular colicin may actually have arisen when the colicin gene acquired the plasmid state. It may be worthwhile to search for colicin-like proteins, possibly subject to specific inducible regulatory systems, in the membranes of noncolicinogenic bacteria, especially strains hypersensitive to colicin-inducing agents. A search for specific immunity to certain colicins among bacterial strains that can bind them may also help detect "crypto-colicinogeny."

ACKNOWLEDGMENTS. Work done in the authors' laboratory was supported by NIH Grant 5-R01-AI03038 and NSF Grant PCM-7614956.

REFERENCES

Bowles, L. K., and Konisky, J. (1981). *J. Bacteriol.* **145,** 668–671.
Brewer, G. J. (1976). *Biochemistry* **15,** 1387–1392.
Brey, R. N. (1982). *J. Bacteriol.* **149,** 306–315.
Cavard, D., and Lazdunski, C. (1979). *Eur. J. Biochem.* **96,** 525–533.
Dankert, J. R., Uratani, Y., Grabau, C., Cramer, W. A., and Hermodson, M. (1982). *J. Biol. Chem.,* in press.
Feingold, D. S. (1970). *J. Membr. Biol.* **3,** 372–386.
Fields, K. L., and Luria, S. E. (1969). *J. Bacteriol.* **97,** 64–77.
Gould, J. M., and Cramer, W. A. (1977). *J. Biol. Chem.* **252,** 5491–5497.
Jakes, K. S., and Model, P. (1979). *J. Bacteriol.* **138,** 770–778.
Jetten, A. M., and Jetten, M. E. R. (1975). *Biochim. Biophys. Acta* **387,** 12–22.
Kaback, H. R. (1971). *Methods in Enzymology* (W. B. Jacoby, ed.), Vol. 22, pp. 99–120, Academic Press, New York.
Kayalar, C., and Luria, S. E. (1979). In *Membrane Bioenergetics* (C. P. Lee, G. Schatz, and L. Ernster, eds.), pp. 297–306, Addison–Wesley, Reading, Mass.
Kleckner, N. (1977). *Cell* **11,** 11–23.
Konisky, J. (1978). In *The Bacteria* (L. N. Ornston and J. R. Sokatch, eds.), Vol. VI, pp. 71–136, Academic Press, New York.
Lau, C., and Richards, F. M. (1976). *Biochemistry* **15,** 666–671.
Luria, S. E. (1973). In *Bacterial Membranes and Walls* (L. Leive, ed.), pp. 293–320, Dekker, New York.
Ozeki, H., Stocker, B. A. D., and De Margerie, H. (1959). *Nature* (*London*) **184,** 337–339.
Plate, C. A., Suit, J. L., Jetten, A. M., and Luria, S. E. (1974). *J. Biol. Chem.* **249,** 6138–6143.
Schein, S. J., Kagan, B. L., and Finkelstein, A. F. (1978). *Nature* (*London*) **276,** 159–163.
Shafferman, A., Flashner, Y., and Cohen, S. (1979). *Mol. Gen. Genet.* **176,** 139–146.
Steven, A. C., ten Heggeler, B., Muller, R., Kistler, J., and Rosenbusch, J. P. (1977). *J. Cell Biol.* **72,** 292–301.

*This sequence differs from that published by D. H. Watson and D. J. Sherratt [*Nature* (*London*) **278,** 362 (1979)]. We are informed by Dr. Sherratt that the experimental findings reported in that paper are to be discounted.

Tokuda, H., and Konisky, J. (1978a). *Proc. Natl. Acad. Sci. USA* **75,** 2579–2583.
Tokuda, H., and Konisky, J. (1978b). *J. Biol. Chem.* **253,** 7731–7737.
Tokuda, H., and Konisky, J. (1979). *Proc. Natl. Acad. Sci. USA* **76,** 6167–6171.
Unwin, P.N.T., and Zampighi, G. (1980). *Nature (London)* **283,** 545–549.
Weiss, M. J., and Luria, S. E. (1978). *Proc. Natl. Acad. Sci. USA* **75,** 2483–2487.
Wendt, L. (1970). *J. Bacteriol.* **104,** 1236–1241.

128

On the Molecular Structure and Ion Transport Mechanism of the Gramicidin Transmembrane Channel

Dan W. Urry

1. INTRODUCTION

Elucidation of membrane transport structure–function processes can be considered in terms of (1) transport phenomenology, (2) molecular structure of the transporting unit, and (3) the detailed mechanisms of ion movement. In the case of ion transport by a channel mechanism, the transport phenomenology includes single-channel currents, ion selectivity, concentration dependence of single-channel currents with concern for saturation effects and conductance maxima, voltage dependence of single-channel currents, concentration dependence of permeability ratios, and ion competition and block (Eisenman *et al.*, 1977). The aspects of molecular structure involve the number of subunits required to form the channel and their primary, secondary, tertiary, and quaternary structural elements. The detailed mechanism for ion movement through the channel requires determination of the ion binding sites and binding constants, the rate constants for elemental ion movements, the effects of ion–ion repulsion during multiple occupancy, and changes in structural states of the channel for different occupancy states. The goal of the third aspect is to determine the complete free-energy profile for ion movements through the channel and to derive the thermodynamic basis for ion selectivity.

With respect to the gramicidin channel, the first two aspects of determining transport phenomenology and molecular structure are substantially complete and the process of detailing the ion movements within the channel is well under way. The basic problem is one of determining the unique process or processes whereby the ion(s) moves through the channel. For example, one might suppose that the dependence of single-channel conductance over wide ranges of ion concentration and transmembrane potential, with effects such as saturation and inhibition by high ion concentration, would be sufficient to identify a

---

Dan W. Urry • Laboratory of Molecular Biophysics, University of Alabama, Birmingham, Alabama 35294.

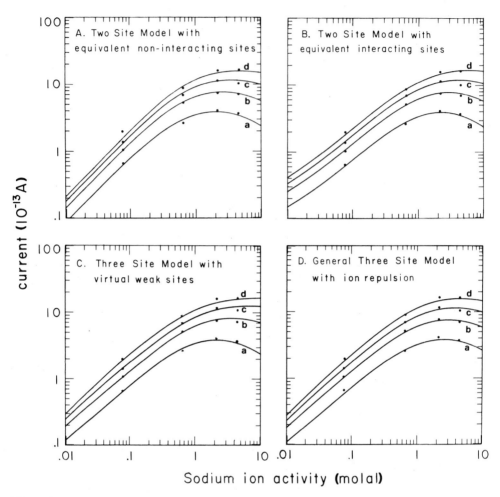

Figure 1. Calculated single-channel currents, least-squares fitted to the experimental data points. The rate constants obtained in B and C agree, to within about a factor of two, with the NMR-derived rate constants, whereas those of A and D, while fitting the experimental data equally well, are inconsistent with the NMR data. While we have found no other combinations of rate constants than these four to fit, this demonstrates the requirement for an independent determination of the rate constant in order to achieve a description of the transport process.

(A) $K_b = 1.2$ M^{-1}, $k_{off} = 6.6 \times 10^6$/sec, $k_{cb} = 1.3 \times 10^7$/sec.   (B) $K_b^t = 200$ M^{-1}, $k_{off}^t = 1.6 \times 10^5$/sec, $k_{cb} = 1.3 \times 10^7$/sec, $K_b^w = 1.3$ M^{-1}, $k_{off}^w = 7.9 \times 10^6$/sec.   (C) $K_b^t = 100$ M^{-1}, $k_{off}^t = 1.1 \times 10^5$/sec, $k_{cs} = 4.8 \times 10^6$/sec, $K_b^w = 1.0$ M^{-1}, $k_{off}^w = 5.8 \times 10^6$/sec.   (D) $K_b^t = 75$ M^{-1}, $k_{off}^t = 2.9 \times 10^7$/sec, $K_b^w = 1.0$ M^{-1}, $A = 3.1$ kcal-Å/mole. [Parts B, C, and D from Urry *et al.*, 1980.]

Eyring rate theory is utilized to introduce voltage dependence to these rate constants (Eyring and Urry, 1963; Johnson *et al.*, 1954; Parlin and Eyring, 1954).

unique mechanism. This is not the case for sodium ion transport by the gramicidin channel; there is not a unique set of binding constants and rate constants (that define a free-energy profile) for calculating the single-channel currents as a function of voltage and concentration. This is demonstrated with the malonyl gramicidin channel in Fig. 1 where the experimental single-channel currents due to NaCl at 0.1 M, 1 M, 3 M, and 5.5 M for 50 mV (a), 100 mV (b), 150 mV (c), and 200 mV (d) are given as points to be compared with calculated currents using different occupancy models and different constants. In Figs. 1A and B are two-site models with equivalent sites separated by 21 Å but in A the sites are noninteracting whereas in B the difference between the binding constant for the first ion and the second ion is due to ion repulsion, i.e., the sites are interacting. In the three-site models of Figs. 1C and D there is a single tight-binding site midway through the channel and there are equivalent weak-binding sites separated by 21 Å, but relevant constants differ with ion–ion repulsion taken into account for multiple occupancy states in Fig. 1D. The details for calculating curves of Figs. 1B, C, and D are given elsewhere (Urry *et al.*, 1980), and the relevant constants are included in the legend.

Because there is not a single set of constants that uniquely gives rise to the single-channel currents as a function of ion concentration and applied potential, independent determination of binding constants, rate constants, and site locations is sought. The recent packaging of channels in phospholipid structures at concentrations suitable for spectroscopic characterization (Urry *et al.*, 1979a,b; Masotti *et al.*, 1980) is a development that makes determination of the critical quantities possible. Cation nuclear magnetic resonance of the ion binding process can provide required binding and rate constants, and carbon-13 and possibly oxygen-17 magnetic resonance can be used to identify the locations of the binding sites as well as to provide information on the rates of ion exchange between structurally equivalent binding sites.

The determination of the binding sites, binding constants, and rate constants for the gramicidin channel gives rise to two fundamental pieces of information. One is the repulsion between monovalent cations as a function of distance, and the second is the effect of the lipid dielectric constant just beyond the single thickness of polypeptide backbone that laterally forms the first coordination shell of the ion in the channel. These quantities become determinable, in part, because of the unique twofold symmetric structure of the single-filing, single-walled gramicidin channel.

2. MOLECULAR STRUCTURE OF THE GRAMICIDIN CHANNEL

The linear gramicidin pentadecapeptides, HCO-L-Val_1-Gly_2-L-Ala_3-D-Leu_4-L-Ala_5-D-Val_6-L-Val_7-D-Val_8-L-Trp_9-D-Leu_{10}-L-ϕ_{11}-D-Leu_{12}-L-Trp_{13}-D-Leu_{14}-L-Trp_{15}-$NHCH_2CH_2OH$ where $\phi_{11} = $ Trp, Phe, Tyr at the approximate ratios of 7 : 1 : 2, respectively, and where L-Val_1 is replaced to a few percent by L-Ile_1, form the first, and so far the only, structurally characterized ion-selective transmembrane channels. As such they are models for providing insights into the ion-selective transmembrane channels, for example of muscle and nerve, particularly because these physiological channels appear to have so many transport properties in common with the gramicidin channel (Eisenman *et al.*, 1980). The primary structure of the gramicidins was determined by Sarges and Witkop (1964); the channel activity was first demonstrated by Hladky and Haydon (1970, 1972); new classes of conformations for these alternating L, D polypeptides were described by this laboratory (Urry, 1971; Urry *et al.*, 1971), by Ramachandran and Chandrasekaran (1972a,b), and by Veatch and Blout (1974); and the head-to-head (amino end to amino end) hydrogen-bonded dimer proposed in 1971 by this laboratory (Urry, 1971, 1973) to be the dominant functional channel has been substantiated by Bamberg and Läuger and their colleagues (Apell *et al.*, 1977; Bamberg *et al.*, 1977), by Veatch, Blout, and their colleagues (Weinstein *et al.*, 1979), and by this laboratory (Szabo and Urry, 1979; Urry *et al.*, 1971). Also, the crystal structure studies for the K^+ and Cs^+ complexes of Koeppe *et al.* (1979), while yet a low resolution, yield a helical structure length and diameter that are within tenths of an angstrom of those previously proposed for the channel [see table of $\beta_{3,3}^6$-helix parameters of Urry (1973) and

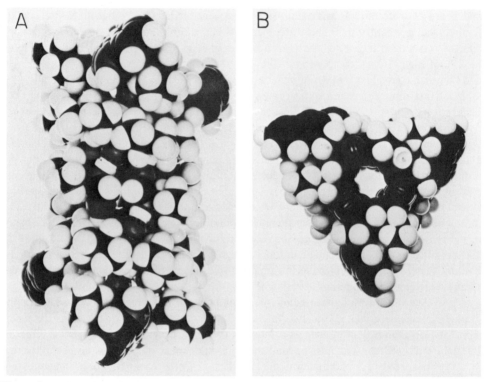

Figure 2. (A) Side view of the dimeric gramicidin channel showing the formyl head-to-head junction and the twofold symmetry. (B) Helix axis view of the gramicidin channel.

Fig. 6 of Urry *et al*. (1975)]. Accordingly, the molecular structure of this single-filing cation-selective transmembrane channel is well founded.

The single-stranded β-helical structure of the gramicidin channel is represented in Fig. 2 by space-filling models. In the side view of the channel in Fig. 2A, two molecules are seen attached head to head (formyl end to formyl end) by means of six intermolecular hydrogen bonds, giving a structure with exact twofold symmetry. This channel of two molecules can be converted to that of a single molecule by removing the formyl moiety (HCO) and tying covalently the amino ends of two molecules together by a malonyl, $-CO-CH_2-CO-$, bridge. This molecule, called malonyl gramicidin, will be used in the ion binding studies discussed below. The helix axis view in Fig. 2B shows a 4-Å-diameter channel with three peptide carbonyl oxygens pointing outward to meet an incoming ion and to replace partially its waters of hydration. Coordinated water precedes and follows the cation through the channel with lateral coordination of the ion being provided by the carbonyl oxygens. The peptide carbonyl oxygens can librate into the channel decreasing the mean channel diameter, lining the channel with negative ends of the carbonyl dipoles making the channel favorable for cation entry and unfavorable for anion entry. The energetics of libration can contribute to selectivity among monovalent cations, and is viewed as being dependent on the side chain bulk and orientation. This latter element can give rise to a multiplicity of single-channel currents for the same structure (Urry *et al*., 1981). While a single lateral coordination shell of peptide carbonyls is adequate for monovalent cations, it is not sufficient to compete with bulk water for divalent cations.

The molecular structure of the gramicidin channel replaces the yet too common perspective of ion selectivity based on radii of hydration with the hydrated ion passing through aqueous and static pores of discriminating diameters. Selectivity based on hydrated radii would have difficulty explaining the selectivity of Ca^{2+} over Na^+ in a calcium channel, whereas this is explicable on the basis of a gramicidin-like structure with dynamic and direct coordination of the cation by the channel wall (Urry, 1978a). In this regard it is essential to bear in mind that, while the dielectric constant of water is about 80, the dielectric constant of N-methylacetamide (a model of the peptide moiety) is 179. Accordingly, a polypeptide conformation that is of a form that can properly present its peptide carbonyls to a cation can readily compete with water (Urry, 1978b).

3. ION TRANSPORT MECHANISM

With the molecular structure of the channel on a firm basis, the next problem is to develop a detailed understanding of the ion movements through the channel. The questions to be answered are many. Do ions bind in the channel? If so, where are the binding sites and barriers and what are the binding constants and rate constants? Is there multiple occupancy in the channel? If so, how do the sites and constants change with changing occupancy? The answers to these questions allow for the plotting of a free-energy profile for permeation of the different permeant ions. The understanding of the energy components that comprise the free-energy profiles will provide the thermodynamic basis for ion selectivity and the answer to questions such as, how much the lipid dielectric constant, the peptide librations (torsional oscillations), the chemical nature of the lipid polar head groups, the hydrocarbon chains, and the amino acid side chains contribute to the free energies of the binding sites and barriers. One approach is to use cation nuclear magnetic resonance (NMR) to characterize the interactions of cations with phospholipid-packaged gramicidin channels (Urry *et al.*, 1979a,b; Masotti *et al.*, 1980).

3.1. Ion Binding and Rate Constants (Urry et al., 1980)

Using sodium-23 NMR, two binding constants have been observed for the micellar-packaged malonyl dimer of gramicidin, a tight binding constant (K_b^t of about 100 M^{-1}) and a weak binding constant (K_b^w of about 1 M^{-1}). Additionally, off-rate constants have also been estimated, i.e., $k_{off}^t \simeq 3 \times 10^5$/sec and $k_{off}^w \simeq 2 \times 10^7$/sec (Urry *et al.*, 1980; Venkatachalam and Urry, 1980). Because $K_b^t = k_{on}^t / k_{off}^t$ etc., this gives $k_{on}^t \simeq 3 \times 10^7$/M-sec and $k_{on}^w \simeq 2 \times 10^7$/M-sec. Thus, there appear to be at least two binding sites in the channel. Because five rate constants are required for two- and three site models and for an approximate four-site model, there is yet an undetermined rate constant which is for intrachannel ion movement. Of course, it is necessary to convert these rate constants, determined in the absence of an electric field, to voltage-dependent rate constants.

3.2. Introduction of Voltage Dependence by Eyring Rate Theory

Utilizing Eyring rate theory the equilibrium derived rate constants—k_{on}^t, k_{off}^t, k_{on}^w, and k_{off}^w—can be converted to voltage-dependent rate constants, e.g., k^1, etc., suitable for describing the steady-state rate processes in the presence of an applied electric field. For example,

$$k^1 = k_{on}^t \exp(lzFE/2dRT) \tag{1}$$

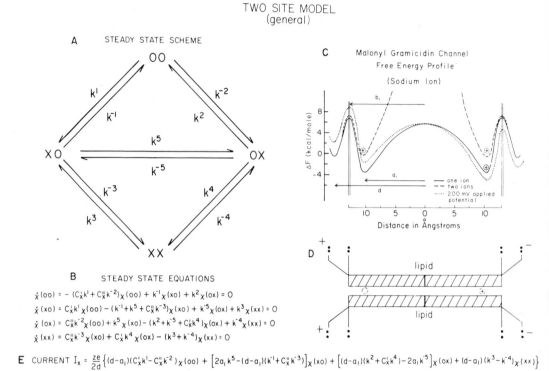

TWO SITE MODEL
(general)

A   STEADY STATE SCHEME

B   STEADY STATE EQUATIONS

$\dot{\chi}(oo) = -(C_x^l k^1 + C_x^u k^{-2})_\chi(oo) + k^{-1}{}_\chi(xo) + k^2{}_\chi(ox) = 0$

$\dot{\chi}(xo) = C_x^l k^1{}_\chi(oo) - (k^{-1} + k^5 + C_x^u k^{-3})_\chi(xo) + k^{-5}{}_\chi(ox) + k^3{}_\chi(xx) = 0$

$\dot{\chi}(ox) = C_x^u k^{-2}{}_\chi(oo) + k^5{}_\chi(xo) - (k^2 + k^{-5} + C_x^l k^4)_\chi(ox) + k^{-4}{}_\chi(xx) = 0$

$\dot{\chi}(xx) = C_x^u k^{-3}{}_\chi(xo) + C_x^l k^4{}_\chi(ox) - (k^3 + k^{-4})_\chi(xx) = 0$

C   Malonyl Gramicidin Channel
Free Energy Profile
(Sodium Ion)

one ion
two ions
200 mV applied potential

D

E   CURRENT $I_x = \frac{ze}{2d}\left\{(d-a_1)(C_x^l k^1 - C_x^u k^{-2})_\chi(oo) + \left[2a_1 k^5 - (d-a_1)(k^{-1} + C_x^u k^{-3})\right]_\chi(xo) + \left[(d-a_1)(k^2 + C_x^l k^4) - 2a_1 k^5\right]_\chi(ox) + (d-a_1)(k^3 - k^{-4})_\chi(xx)\right\}$

Figure 3

where zFE is the free-energy change on crossing from one side of the membrane to the other with z being the charge on the ion (+1 for sodium), F the Faraday (23,053 cal/mole-V.), and E the transmembrane potential, e.g., 0.1 V. The fraction $l/2d$ is the fraction of the voltage change or drop, in the case of k_{on}^t, on going from the minimum just outside the channel at the voltage of the aqueous solution with $2d$ being the total transmembrane distance over which the voltage drops (see Fig. 3C). For a linear electric field gradient, l is the transmembrane distance from the outer minimum at d to the rate-limiting barrier for entry into the channel. These lengths are defined in Fig. 3C with $l = b_1 - d$. The quantities R and T have their usual meaning with $R = 1.987$ cal/deg and T in degrees Kelvin.

3.3. Calculation of Single-Channel Currents

In order that the NMR-derived binding and rate constants for the ion interactions with the channels be demonstrable as relevant to the ion transport observed in the planar lipid bilayer experiment, the formalism is required for utilizing these constants in the calculation of current passing through the channel. Once this is achieved, then the constants can be used to describe the free-energy profile for ion permeation through the channel, and it is the comparison of the free-energy profiles for permeation by different ions that provides the thermodynamic basis for understanding ion selectivity.

The single-channel current is the flow of charge per second through one gramicidin channel. Interestingly, this quantity is directly observable; it is seen in the electrical lipid bilayer experiment as a step change in current as one channel "opens" by dimerization or "closes" by dissociation into monomers. The magnitude of the single-channel current is dependent on the concentration of ions on each side of the planar bilayer and on the transmembrane potential. Following a formalism that is fundamentally similar to that of Eisenman and colleagues (Eisenman *et al.*, 1977; Sandblom *et al.*, 1977), the single-channel current, I, may be written, in general, as

$$I = ze \sum_{i=0}^{n} \chi_i \left(\sum_j \lambda_{ij} q_{ij} \right) \qquad (2)$$

where e is the charge on an electron in coulombs, i.e., 1.6×10^{-19} C; χ_i is the probability of a given channel state, specified by the number of location of the ions in the channel; λ_{ij} is the signed fraction of channel length of the jth ion translocation beginning at i; it is positive when with the potential gradient and negative when against the gradient; and q_{ij} is the rate for the ion translocation. The units of q_{ij} are sec^{-1}. For a first-order process, i.e., for intrachannel ion translocations and an ion leaving the channel, it is the rate constant k_{ij}. For a second-order process, i.e., for an ion entering the channel, it is the product of a second-order rate constant times the concentration. In the convention used here, when an ion enters the channel from the left-hand (positive) side the rate q_{ij} is $k_{ij} C'_x$ and from the right-hand (negative) side it is $k_{ij} C''_x$.

The steady-state scheme for a two-site model is given in Fig. 3A for empty (oo), for singly occupied (xo) and (ox), and for doubly occupied (xx) states of the channel. The associated steady-state equations are listed in Fig. 3B where $\dot{\chi}$(oo), for example, is the change with time of the probability of the empty channel. The χ_i of equation (2) therefore become $\chi_0 = \chi$(oo), $\chi_1 = \chi$(xo), $\chi_2 = \chi$(ox), and $\chi_3 = \chi$(xo) for the two-site model. Using the equations of Fig. 3B, each χ_i can be solved in terms of χ(oo), i.e.,

$$\chi_i = \chi(oo) f_i \qquad (3)$$

where each of the f_i is a complicated function of the q_{ij}, except for f_0, which is 1. For a detailed treatment of each occupancy model see Urry *et al.* (1980).

What remain in the calculation of the single-channel currents are determining voltage-dependent rates, q_{ij}, and positioning of the ion in the channel, which allows evaluation of λ_{ij}. Also most fundamentally the positioning of the ion is required for the voltage-dependent rate constant as considered above.

The gramicidin channel is approximately 26 Å long. Locating the ion within that length can be achieved while the channel is within the lipid structure by observing, in the ^{13}C magnetic resonance spectra, the perturbation of systematically positioned ^{13}C-enriched carbonyl carbons as a function of ion concentration. For the present, however, the X-ray data of Koeppe *et al.* (1979) on crystalline complexes of gramicidin with K$^+$ and with Cs$^+$ will be used.

Using the definitions of a_1 and d in Figs. 3C and D and substituting for λ_{ij} and q_{ij} of equation (2), the expression for the single-channel current is seen as E in Fig. 3. The denominator of λ_{ij}, i.e., the total length of the channel ($2d$), has been factored out leaving the lengths for the individual ion movements, e.g., $(d - a_1)$ for the movement of an ion from solution to the binding site. Assuming the steady-state conditions apply, however, one need only count the ions going over a single barrier in which case the general equation reduces to

$$I = ze [C'_x k^1 \chi(oo) - k^{-1} \chi(xo) + C''_x k^4 \chi(ox) - k^{-4} \chi(xx)] \qquad (4)$$

for the left-hand barrier of Fig. 3C. Thus, in the steady-state assumption these lengths $d-a_1$ and a_1 are not required, but lengths were required in converting the NMR-derived rate constants to voltage-dependent constants (see above).

3.4. Comparison of Calculated and Experimental Single-Channel Currents

With the values for the lengths of a_1, b_1, and d it is now possible to calculate the single-channel currents of the gramicidin channel with but one parameter and that parameter, k_{cb}, should fall within the limits $10^{10} > k_{cb} > 3 \times 10^5$. Using the structurally reasonable values of $a_1 = 10.5$ Å, $b_1 = 13$Å, and $d = 15$ Å and fitting to the 100-mV curve, the set of curves for the calculated single-channel currents as a function of ion molal activity and transmembrane potential are calculated. The result is surprisingly close to the experimental values with a value for k_{cb} of 3.2×10^6/sec and using, of course, the experimentally derived rate constants $k_{off}^t = 3 \times 10^5$/sec, $k_{on}^t = 3 \times 10^7$/M-sec, $k_{off}^w = 2 \times 10^7$/sec, and $k_{on}^w = 2 \times 10^7$/M-sec.

Next, all five rate constants are allowed to vary to achieve a "best fit" to the experimental single-channel currents both as a function of ion concentration and of transmembrane potential. The values resulting from a best fit can then be compared with the NMR-derived constants. In the case of the malonyl gramicidin channel, the striking result is that none of the four constants varied by more than a factor of about two from the NMR-derived values, and the comparison of the calculated curves with the experimental data points is superb (see Fig. 1B). The best-fit values were $k_{off}^t = 1.6 \times 10^5$/sec, $k_{on}^t = 3.2 \times 10^7$/M-sec, $k_{off}^w = 7.9 \times 10^6$/sec, and $k_{on}^w = 1.0 \times 10^7$/M-sec.

3.5. Determination of Free-Energy Profiles and the Thermodynamic Basis for Ion Selectivity

Once the position of the binding sites and major barriers are located and the binding and rate constants are determined for a particular ion passing through the channel, the free-energy profile for permeation can be plotted. For the case of the sodium ion, the binding constant for the first ion in the channel, K_b^t, is approximately 100 M^{-1} giving a decrease in free energy, i.e., -2.8 kcal/mole, using the expression

$$\Delta F_b = -2.3RT \log K_b \tag{5}$$

At present, the best estimate for the locations is 2.5 Å in from the mean end of the channel. The off-rate constant for single occupancy, i.e., k_{off}^t, is 3×10^5/sec; using this value in the Eyring rate theory to define a free energy of activation, i.e.,

$$\Delta F\ddagger = 2.3RT(12.79 - \log k_{off}^t) \tag{6}$$

gives a value for the free energy of activation for leaving the channel of 9.9 kcal/mole with respect to the binding site, which becomes 7.1 ($=9.9-2.8$) kcal/mole with respect to a solution reference of zero free energy. Plotting this barrier at the mean end of the channel provides another pair of points for the free-energy profile. With the rate over the central barrier being 3.2×10^6/sec, this gives $\Delta F\ddagger_{cb} = 8.6$ kcal/mole, which when referenced to solution gives a barrier height of 5.8 kcal/mole. Placing this barrier midway through the channel and connecting the points gives the free-energy profile indicated by the solid line in Fig. 3C.

For the second ion interaction with the channel, i.e., for the case of double occupancy, the binding constant for the weak site is 1 M^{-1} and the off-rate constant is 2×10^7/sec giving $\Delta F_b = 0$ and $\Delta F\ddagger = 7.5$ kcal/mole. As two ions, one in each binding site, approach each other, repulsion causes the free energy to go to infinity such that there is no value to plot for the central barrier for double occupancy. The free-energy plot for double occupancy is seen as the dashed line in Fig. 3C. Also, of course, because the channel has twofold symmetry, this requires that the difference between ΔF_b for one ion in the channel and for two ions must be due to repulsion. Thus, for the two-site model, there would be a free energy of repulsion of 2.8 kcal/mole at a distance of 21 Å. Because the distance dependence for charge–charge repulsion is r^{-1}, this requires that a similar repulsion be sensed at 23 Å and, if the channel structure were unchanged, that the barrier for the entry of the second ion would be about 2.5 kcal/mole higher than for the first ion. This would indicate an on-rate of $k_{on}^W \simeq 2.5 \times 10^5$/sec rather than the observed $k_{,on}^W$ of 2×10^7/sec. If the two-site model is to be correct, this would require that the presence of one ion in the channel results in the selection of a conformation that is not the same as with no ion in the channel. The ion appears to choose a conformation for the channel that is a metastable state in the absence of the ion. Because at the time of entry of the second ion into the channel the first ion is in the opposite half of the channel some 23 Å away, the interconversion between the ion-selected conformation and the stable conformation in the absence of ion must be slow with respect to ion translocation between halves of the channel. The sort of conformational difference that one might imagine is one in which the carbonyl moieties are tilted inward toward the channel axis in the presence of an ion such that the second, incoming ion is met with a more favorable orientation of carbonyls during the partial dehydration step. Thus, as the details of ion transport are explored, the perspective of a dynamic channel structure emerges with subtle structural changes capable of significant changes in elemental rate processes. Once the free-energy profile is correctly determined for sodium and other ions and once the components of the free energy are dissected, then we will have a rather complete understanding of the basis for ion selectivity in the channel transport mechanism.

ACKNOWLEDGMENTS. This work was supported in part by NIH Grant GM-26898. The author wishes to thank Dr. C. M. Venkatachalam for calculating and plotting Fig. 1A.

REFERENCES

Apell, H.-J., Bamberg, E., Alpes, H., and Läuger, P. (1977). *J. Membr. Biol.* **31**, 171–188.

Bamberg, E., Apell, H.-J., and Alpes, H. (1977). *Proc. Natl. Acad. Sci. USA* **74**, 2402–2406.

Bradley, R. J., Urry, D. W., Okamoto, K., and Rapaka, R. S. (1978). *Science* **200**, 435–437.

Eisenman, G., Sandblom, J., and Neher, E. (1977). In *Metal Ligand Interactions in Organic and Biochemistry* (B. Pullman and N. Goldblum, eds.), Part 2, pp. 1–36, Reidel, Dordrecht.

Eisenman, G., Enos, B., Hägglund, J., and Sandblom, J. (1980). *Ann. N.Y. Acad. Sci.* **339**, 8–20.

Eyring, H., and Urry, D. W. (1963). *Ber. Bunsenges. Phys. Chem.* **67**, 731–740.

Hladky, S. B., and Haydon, D. A. (1970). *Nature (London)* **225**, 451–453.

Hladky, S. B., and Haydon, D. A. (1972). *Biochim. Biophys. Acta* **274**, 294–312.

Johnson, F. H., Eyring, H., and Polissar, M. I. (1954). In *The Kinetic Basis of Molecular Biology*, Chapter 14, Wiley, New York.

Koeppe, R. E., II, Berg, J. M., Hodgson, K. O., and Stryer, L. (1979). *Nature (London)* **279**, 723–725.

Masotti, L., Spisni, A., and Urry, D. W. (1980). *Cell Biophys.* **2**(2), 241–251.

Parlin, B., and Eyring, H. (1954). In *Ion Transport across Membranes* (H. T. Clark, ed.), pp. 103–118, Academic Press, New York.

Ramachandran, G. N., and Chandrasekaran, R. (1972a). *Prog. Peptide Res.* **2**, 195–215.

Ramachandran, G. N., and Chandrasekaran, R. (1972b). *Indian J. Biochem. Biophys.* **9,** 1–11.

Sandblom, J., Eisenman, G., and Neher, E. (1977). *J. Membr. Biol.* **31,** 383–417.

Sarges, R., and Witkop, B. (1964). *J. Am. Chem. Soc.* **86,** 1862–1863.

Szabo, G., and Urry, D. W. (1979). *Science* **203,** 55–57.

Urry, D. W. (1971). *Proc. Natl. Acad. Sci. USA* **68,** 672–676.

Urry, D. W. (1973). In *Conformation of Biological Molecules and Polymers—The Jerusalem Symposia on Quantum Chemistry and Biochemistry* V (E. D. Bergmann and B. Pullman, eds.), pp. 723–736, Israel Academy of Sciences, Jerusalem.

Urry, D. W. (1978a). In *Frontiers of Biological Energetics* (P. L. Dutton, J. Leigh, and A. Scarpa, eds.), Vol. 2, pp. 1227–1234, Academic Press, New York.

Urry, D. W. (1978b). *Ann. N.Y. Acad. Sci.* **307,** 3–27.

Urry, D. W., Goodall, M. C., Glickson, J. D., and Mayers, D. F. (1971). *Proc. Natl. Acad. Sci. USA* **68,** 1907–1911.

Urry, D. W., Long, M. M., Jacobs, M., and Harris, R. D. (1975). *Ann. N.Y. Acad. Sci.* **264,** 203–220.

Urry, D. W., Spisni, A., and Khaled, M. A. (1979a). *Biochem. Biophys. Res. Commun.* **88,** 940–949.

Urry, D. W., Spisni, A., Khaled, M. A., Long, M. M., and Masotti, L. (1979b). *Int. J. Quantum Chem.: Quantum Biol. Symp. No. 6,* pp. 289–303.

Urry, D. W., Venkatachalam, C. M., Spisni, A., Bradley, R. J., Trapane, T. L., and Prasad, K. U. (1980). *J. Membr. Biol.* **55,** 29–51.

Urry, D. W., Venkatachalam, C. M., Prasad, K. U., Bradley, R. J., Parenti-Castelli, G., and Lenaz, G. (1981). *Int. J. Quantum Chem.: Quantum Biol. Symp. No. 8,* pp. 385–389.

Veatch, W. R., and Blout, E. R. (1974). *Biochemistry* **13,** 5257–5264.

Venkatachalam, C. M., and Urry, D. W. (1980). *J. Magn. Reson.* **41,** 313–335.

Weinstein, S., Wallace, B., Blout, E. R., Morrow, J. S., and Veach, W. (1979). *Proc. Natl. Acad. Sci. USA* **76,** 4230–4234.

129

Ion Conduction through a Pore

Gramicidin A

S. B. Hladky

Gramicidin A is a linear pentadecapeptide that forms pores in lipid membranes (Hladky and Haydon, 1972). Because both ends of the peptide are blocked with neutral groups and the side chains are all hydrophobic (Sarges and Witkop, 1965), the only polar groups that can interact with ions and water are the carbonyl oxygens and amide hydrogens of the backbone. The accepted conformation of the pore (see e.g. Weinstein *et al.*, 1980) is a dimer of two helices joined end to end (Urry *et al.*, 1971; Ramachandran and Chandrase-karan, 1972). Except at the ends of the dimer, all the polar groups are internally hydrogen bonded as part of the lining of a right-cylindrical hole. The high currents observed for each pore, more than 10^8 ions/sec, preclude major arrangement of this structure as part of ion movement. Gramicidin thus provides an opportunity to investigate selective ion conduct-ance in a pore with particularly simple properties.

The pore is selective for monovalent cations. The conductances for the alkali cations fall in the aqueous diffusion sequence $(H) > Cs > Rb > K > Na > Li$, but differ by larger amounts. The conductances increase with ion concentration up to ca. 1.0 M, then reach a plateau, and, most clearly for cesium, actually decline at higher activities. From these data it was suggested that the pores contain water, that the ions prefer the pore to the aqueous solutions, and that ions are transported partially hydrated (Hladky, 1972; Hladky and Hay-don, 1972). They are not fully hydrated for then the pore would need to be too large to be selective against anions, divalent cations, and dimethylammonium ions (Urban, 1978). They are not bare for then the large free energy of dehydration would ensure that the pore would be rarely occupied. At concentrations below 0.1 M, as the voltage is increased the current tends to a limit determined by the rate of access of ions to the pore. At high concentrations, the current increases exponentially with potential reflecting the potential dependence of the transfer process (Hladky and Haydon, 1972; Urban *et al.*, 1980; Ander-sen and Procopio, 1980; Eisenman *et al.*, 1980a,b).

S. B. Hladky • Department of Pharmacology, University of Cambridge, Cambridge CB2 2QD, England.

It might appear from the preceding data that the pores become occupied by one ion at concentrations near 1.0 M, which could account for the plateau. If so, at higher concentrations the pores might become doubly occupied, which could lead both to a decrease in conductance and to a shift in the rate-limiting step from the access process to transfer into the occasional vacancies that occur when one of the ions leaves. However, there are good reasons for believing that the conduction process is more complicated, at least for some of the cations. (1) Ion selectivity can be estimated from the potential for zero current when different solutions of two cations are separated by the membrane. For one species on each side and concentrations above 0.01 M, this selectivity is stronger than estimated from conductances and the discrepancy increases with concentration (Myers and Haydon, 1972; Urban *et al.*, 1980). Furthermore, using mixtures with thallium, the thallium appears more strongly preferred if its concentration exceeds that of the other ion (Eisenman *et al.*, 1978). (2) Thallium and silver conductances are similar to those for potassium at concentrations above 0.01 M. Nevertheless, at low concentrations they can reduce the conductance seen with much higher concentrations of sodium or potassium (Neher, 1975; Andersen, 1975; Neher *et al.*, 1978; McBride and Szabo, 1978). (3) Conductances are greater than estimated from the flux of labeled ions, i.e., there is single-filing and the flux ratio exponent is greater than 1 (Schagina *et al.*, 1978; Procopio and Andersen, 1979). (4) Conductances below 0.01 M are higher than expected (Neher *et al.*, 1978; Urban *et al.*, 1980). The first three points imply interaction between ions within the pore even at concentrations below 0.1 M (Hladky, 1972; Eisenman *et al.*, 1978; Hladky *et al.*, 1979; Urban and Hladky, 1979). The fourth suggests that at low concentrations, pore occupancy does not drop as expected (Neher *et al.*, 1978). All of this evidence for significant occupation of the pore by ions below 0.1 M must be reconciled with the continued increase in conductance up to 1 M.

Two explanations have been proposed. In the four-ion, equilibrium-binding model (Sandblom *et al.*, 1977), ions can be bound simultaneously to any combination of four sites, two at each end of the pore. The conductance is limited by movement between the two inner sites, which become saturated only at high activities. Binding to the outer sites affects the fluxes by modulating both the binding of ions to the inner sites and the rate constants for transfer between them. This modulation varies with both the species of ion bound to the outer site and the species being transferred. If the modulation is assumed independent of the species being transferred, the model fails to explain even the qualitative features of the zero-current potential data. With species dependence it has provided no explanation for the change in shape of the current–voltage relation. Similarly, equilibrium modulation cannot explain the discrepancy between conductances and tracer fluxes. Furthermore, no basis has been proposed for the simultaneous binding of three or four cations or for the required species dependence of the modulation. The equilibrium-binding assumption has now been abandoned.

In the second model, interaction enters as a result of two assumptions: ions only move into regions of the pore that do not already contain an ion, and ions within the pore repel each other (Hladky, 1972, 1974; Aityan *et al.*, 1977; Urban, 1978; Levitt, 1978a, b; Hille and Schwarz, 1978; Urban and Hladky, 1979; Kohler and Heckmann, 1979). In a first approximation, the repulsion can be assumed independent of ion species. It could result from long-range electrostatic forces, the incompressibility of water trapped between the ions, or both. If only two ions can be inside the pore, there can be two modes of ion transport. At low concentrations, transport occurs by entry, transfer, and exit. If this sequence is the only mode, the conductance reaches a maximum when the pore is most frequently singly occupied. However, if a second ion can enter easily, entry of a second ion may promote the rapid exit of either of the ions then present. Provided transfer is sufficiently rapid, increases in concentration may therefore lead to further increases in conductance even if the pore is already singly occupied. At still higher concentrations, the pore is usually doubly occupied and whenever an ion leaves the pore it or another immediately reenters from the same side before transfer can occur, i.e., the pore is blocked.

Qualitatively, this kinetic model accounts for all the observed effects with gramicidin (Urban and Hladky, 1979). For instance, blocking of other ions by thallium is predicted if thallium conduction occurs in the two-ion mode and in mixed-occupancy pores it is usually the other ion that leaves. Less dramatic blocking can explain why the preferred ion appears to be more preferred when the selectivity is determined by competition, i.e., by zero-current potential, rather than by conductance.

Qualitatively, this model is very adjustable. Urban *et al.* (1980) have shown that electrical data for the alkali cations and ammonium can be fitted assuming the repulsion is independent of species. The general two-ion model appears capable of fitting all the published conductance and current–voltage data for fixed ions such as sodium, rubidium, and thallium (e.g., the data of Eisenman *et al.*, 1980b) provided no particular potential dependence of the rate constants is assumed in advance. Only hydrogen appears to have more complicated properties that may entail occupancy by more than two ions (Eisenman *et al.*, 1980a). This ion, however, is conducted by a different mechanism involving exchange with the hydrogen of water molecules (Hladky and Haydon, 1972; Rosenberg and Finkelstein, 1978; Levitt *et al.*, 1978).

Unfortunately, the electrical data alone do not convincingly determine the values of all the rate constants, in particular those for ions leaving singly occupied pores, and hence they do not adequately determine the first binding constants. Urban (1978; Urban *et al.*, 1978) assuming species-independent repulsion succeeded in fitting all the data above 0.1 M for the alkali cations and ammonium using values of the exit rate constants, roughly similar in size for all species, ranging from ca. 3×10^5 ions/sec to ca. 2×10^7 ions/sec. However, when the less accurate low-activity data were included, all these constants were small. If now the species-independence assumption is dropped, there may instead be a gradation downward from a large value for sodium, ca. 10^7 ions/sec. If so, repulsion is stronger for the more strongly bound ions, which might result if, for instance, stronger binding meant binding further into the pore.

The electrical properties are only one facet of the function of the pore. Others for which data exist are tracer fluxes, equilibrium binding, and water movements. If, as well as the conductances, the difference between the flux ratio exponent and one were available for a range of ion concentrations near the maximum of the exponent, that would specify all the rate constants in the model. Unfortunately, the measured flux ratios are for channels embedded in phospholipid membranes while the conductances etc. are for monoglyceride membranes. The conductances are known to change with the type of membrane (see e.g. Fröhlich, 1979). The data of Schagina *et al.* (1978) with ox brain lipids and rubidium suggest that the first binding constants are large (small rate constant for exit), while those of Procopio and Andersen (1979) with diphytanoylphosphatidylcholine and sodium and cesium suggest a range of values.

There have been two equilibrium-binding studies. Urry *et al.* (1980) assert that gramicidin incorporated into lysolecithin micelles is in the conformation of the pore. Using ^{23}Na NMR they infer that two (or possibly three) ions can bind, the first with a binding constant $K^t \sim 100$ M^{-1}, the second with $K^w \sim 3$ M^{-1}. They also calculate exchange rates for sodium between the pore and the aqueous phase. At low concentrations when the pore can be empty or can contain one ion, $k_{off}^t \sim 3 \times 10^5$ sec^{-1}, while at high concentrations when the pore can be occupied by one or more ions, $k_{off}^{W} \sim 4 \times 10^7$ sec^{-1}. These values are surprisingly close to those for sodium obtained by Urban et al. (1980), 122 M^{-1}, 0.2 M^{-1}, 5×10^5 sec^{-1}, and 3×10^8 sec^{-1}, respectively. However, because the lipids used by Urry et al. were different from those used by Urban et al., there is no reason why the constants should be the same.

Veatch and Durkin (1980) have used equilibrium dialysis to determine the extra binding of thallium to dimyristoylphosphatidylcholine vesicles when gramicidin is incorporated. Their results clearly demonstrate strong binding by thallium and show that only one ion binds to each pore with high affinity. However, in their quantitative analysis they ignored the Donnan potential. This develops across the dialysis membrane to hold free anions near the bound cations, thus preserving electroneutrality. As a consequence of this potential, free cations are partially excluded, and, particularly at low ion concentrations, the activity of tracer is lower than in the reference compartment and thus the binding constants calculated ignoring the Donnan effect are too small. When corrected, the estimate for thallium increases from 500–1000 M^{-1} to 1700–4800 M^{-1}, which is intermediate between the 9000 M^{-1} inferred by Urban et al. (1980) and the 550 M^{-1} inferred by Eisenman et al. (1978). Veatch and Durkin also sought to measure the displacement of bound thallium by sodium and rubidium. Using the small displacement they observed, which was the same for both ions, they estimated that $K_X/K_{Tl} < 0.04$ for both.

There is clearly water within the pore and ion movements can only occur if at least some of this water also moves (Rosenberg and Finkelstein, 1978; Levitt et al., 1978). Andersen and Procopio (1980) report that ions move so fast that they must be able to knock water out of the way rather than having to wait for water movements to create a vacancy. Perhaps this "knocked-on" water is just the water of partial hydration still bound to the ion. However, it would not be too surprising if even more water could be knocked-on, for an ion (with or without bound water) that has gained sufficient energy to move within the pore might transfer sufficient energy to an adjacent water molecule in a collision. It is worth noting that in the presumed structure for the pore, the water would not be strongly hydrogen bonded to the wall. Such knock-on transfer of energy and momentum between ions is much less likely so long as free water molecules separate them in the pore. The decline in conductance at high activities suggests that indeed knock-on processes between ions are not dominant. The models so far used for gramicidin have assumed the extreme case that they do not occur.

It seems probable that a coherent accumulation of data and a convincing kinetic interpretation will soon exist for gramicidin. All the evidence available at present for fixed ions points to a maximum of double occupancy for the pore as envisaged in the two-ion kinetic model. However, this description leaves unanswered many questions of interest. Thus, the nature of the interactions of ions and water with the walls of the pore are not made clear. Indeed, the description does not even reveal the number of ion binding sites, possibly more than 20, only the maximum number that can be occupied simultaneously, probably two.

REFERENCES

Aityan, S. K., Kalandadze, I. L., and Chizmadjev, Y. A. (1977). *Bioelectrochem. Bioenerg.* **4**, 30–44.
Andersen, O. S. (1975). *International Biophysics Congress, Copenhagen*, p. 369.
Andersen, O. S., and Procopio, J. (1980). *Acta Physiol. Scand. Suppl.* **481**, 7–14.
Eisenman, G., Sandblom, J., and Neher, E. (1978). *Biophys. J.* **22**, 307–340.
Eisenman, G., Enos, B., Hägglund, J., and Sandblom, J. (1980a). *Ann. N.Y. Acad. Sci.* **339**, 8–20.

Eisenman, G., Hägglund, J., Sandblom, J., and Enos, B. (1980b). *Upsala J. Med. Sci.* **85,** 247–257.
Eisenman, G., Hägglund, J., Sandblom, J., and Enos, B. (1981). *Upsala J. Med. Sci.,* in press.
Fröhlich, O. (1979). *J. Membr. Biol.* **48,** 365–383.
Hille, B., and Schwarz, W. (1978). *J. Gen. Physiol.* **72,** 409–442.
Hladky, S. B. (1972). Ph.D. dissertation, University of Cambridge, England.
Hladky, S. B. (1974). In *Drugs and Transport Processes* (B. A. Callingham, ed.), pp. 193–210, Macmillan, London.
Hladky, S. B., and Haydon, D. A. (1972). *Biochim. Biophys. Acta* **274,** 294–312.
Hladky, S. B., Urban, B. W., and Haydon, D. A. (1979). In *Membrane Transport Processes* (C. F. Stevens and R. W. Tsien, eds.) Vol. 3, pp. 89–103, Raven Press, New York.
Kohler, H.-H., and Heckmann, K. (1979). *J. Theor. Biol.* **79,** 381–401.
Levitt, D. G. (1978a). *Biophys. J.* **22,** 209–219.
Levitt, D. G. (1978b). *Biophys. J.* **22,** 221–248.
Levitt, D. G., Elias, S. R., and Hautman, J. M. (1978). *Biochim. Biophys. Acta* **512,** 436–451.
McBride, D., and Szabo, G. (1978). *Biophys. J.* **21,** 25a.
Myers, V. B., and Haydon, D. A. (1972). *Biochim. Biophys. Acta* **274,** 313–322.
Neher, E. (1975). *Biochim. Biophys. Acta* **401,** 540–544. Erratum: *Biochim. Biophys. Acta* **469,** 359.
Neher, E., Sandblom, J., and Eisenman, G. (1978). *J. Membr. Biol.* **40,** 97–116.
Procopio, J., and Andersen, O. S. (1979). *Biophys. J.* **25,** 8a.
Ramachandran, G. N., and Chandrasekaran, R. (1972). *Indian J. Biochem. Biophys.* **9,** 1–11.
Rosenberg, P. A., and Finkelstein, A. (1978). *J. Gen. Physiol.* **72,** 327–340.
Sandblom, J., Eisenman, G., and Neher, E. (1977). *J. Membr. Biol.* **31,** 383–417.
Sarges, R., and Witkop, B. (1965). *J. Am. Chem. Soc.* **87,** 2011–2020.
Schagina, L. V., Grinfeldt, A. E., and Lev, A. A. (1978). *Nature* (London) **273,** 243–245.
Urban, B. W. (1978). Ph.D. dissertation, Cambridge University, England.
Urban, B. W., and Hladky, S. B. (1979). *Biochim. Biophys. Acta* **554,** 410–429.
Urban, B. W., Hladky, S. B., and Haydon, D. A. (1978). *Fed. Proc.* **37,** 2628–2632.
Urban, B. W., Hladky, S. B., and Haydon, D. A. (1980). *Biochim. Biophys. Acta* **602,** 331–354.
Urry, D. W., Goodall, M. C., Glickson, J. S., and Mayers, D. C. (1971). *Proc. Natl. Acad. Sci. USA* **68,** 1907–1911.
Urry, D. W., Venkatachalam, C. M., Spisni, A., Läuger, P., and Khaled, M. A. (1980). *Proc. Natl. Acad. Sci. USA* **77,** 2028–2032.
Veatch, W. R., and Durkin, J. T. (1980). *J. Mol. Biol.* **143,** 411–417.
Weinstein, S., Wallace, B. A., Morrow, J. S., and Veatch, W. R. (1980). *J. Mol. Biol.* **143,** 1–19.

130

Conformational States of Gramicidin A in Solution and in the Membrane

Sergei V. Sychev and Vadim T. Ivanov

The linear polypeptide antibiotic gramicidin A (GA) is known to form alkali metal ion conducting channels in biological and artificial membranes (Hladky and Haydon, 1972; Myers and Haydon, 1972). Urry (1971) and Urry *et al.* (1971) assumed that the channels are formed by two GA molecules in $\pi_{LD}^{6.3}$-helix conformation joined in NH$_2$-terminal to NH$_2$-terminal fashion ($\overrightarrow{\pi_{LD}}\overleftarrow{\pi_{LD}}$). Veatch *et al.* (1974) suggested parallel ($\uparrow\uparrow\pi\pi_{LD}$) and antiparallel ($\uparrow\downarrow\pi\pi_{LD}$) double helices as alternative possibilities. Comparison of the calculated and experimentally observed infrared (IR) spectra coupled with circular dichroism (CD) measurements provided evidence for the presence of dominating quantities (ca. 75%) of $\underrightarrow{\uparrow\downarrow\pi}\pi_{LD}^{5.6}$ or $\underrightarrow{\downarrow\uparrow\pi}\pi_{LD}^{7.2}$ helices (right and left handed), in equilibrium with ca. 25% of $\pi_{LD}^{4.4}\pi_{LD}^{4.4}$ and $\pi_{LD}^{6.3}\pi_{LD}^{6.3}$ dimers in dioxane solution (Sychev *et al.*, 1980).

The next logical step would be a structural study of GA in the membrane environment. Fluorescence measurements of the energy transfer in gramicidin C derivatives incorporated into liposomes at a 1 : 1000 molar peptide : lipid ratio (Veatch and Stryer, 1977) have confirmed the dimeric nature of the gramicidin channel. On the other hand, GA was studied in phospholipid liposomes and micelles by ^{13}C and ^{19}F NMR (Urry *et al.*, 1979; Weinstein *et al.*, 1979) and by CD (Urry *et al.*, 1979; Masotti *et al.*, 1980) spectroscopy at much higher, 1 : 10–30 ratios. Interpretation of the results obtained (see below) was based on the a priori assumption of the dimeric state of GA incorporated into the lipid phase, in spite of the fact that according to Raman spectroscopy and calorimetry data (Weidekamm *et al.*, 1977; Chapman *et al.*, 1977) and osmotic pressure investigation of liposomes (Boehler *et al.*, 1978), the peptide : lipid ratio is an important parameter influencing the state of the GA–lipid system. With this in mind, an IR and CD spectroscopic study of GA in liposomes was carried out in a broad range of GA : lipid ratios (1 : 10– 350), temperatures (10–60°C), and with various lipids (Sychev and Ivanov, to be published).

Sergei V. Sychev and Vadim T. Ivanov • Shemyakin Institute of Bioorganic Chemistry, USSR Academy of Sciences, 117988 Moscow V-334, USSR.

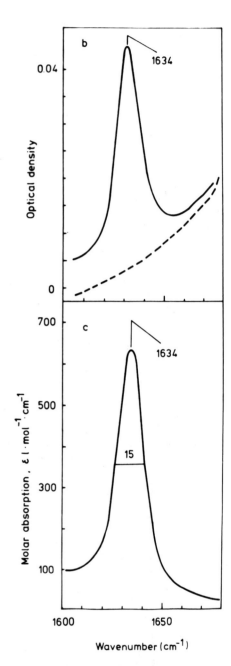

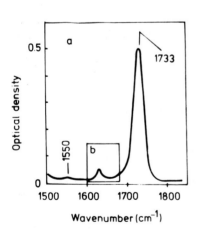

Figure 1. IR spectrum of GA–DPPC liposomes (1 : 310 molar ratio). (a) Spectrum of a liposome suspension in D_2O. (b) An analogous spectrum obtained with 20-fold higher sensitivity of the spectrophotometer (Perkin–Elmer 180). Dashed line indicates the lipid absorption. (c) Amide I band after subtraction of the lipid absorption. (a–c) Sonication and spectra measurements were made at 20°C.

Some results of this study are presented in Figs. 1 and 3 and in Table I. Figure 1a illustrates the IR spectrum obtained for GA in dipalmitoylphosphatidylcholine (DPPC) liposomes, at a GA : lipid ratio of 1 : 310 at 20°C, i.e., below the phase transition temperature. The spectrum shows an intensive phospholipid carbonyl band (1733 cm^{-1}) and weak amide bands: amide I (1634 cm^{-1}) and amide II (1550 cm^{-1}). Figure 1b displays the Amide

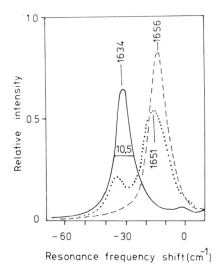

Figure 2. IR spectra in the amine I region calculated for different GA dimer models: ——, $\uparrow\downarrow\pi\pi_{LD}^{5,6}$; - - - -, $\pi_{LD}^{6,3}\pi_{LD}^{6,3}$; · · · · ·, $\uparrow\uparrow\pi\pi_{LD}^{5,6}$.

I band obtained at a higher sensitivity. Figure 1c presents the same band in a molecular extinction scale after subtraction of the phospholipid absorption. Figure 2 displays the earlier calculated IR spectra (Sychev *et al.*, 1980) for different GA dimer conformations. The observed frequency 1634 cm^{-1} is consistent with the $\uparrow\downarrow\pi\pi_{LD}^{5,6}$ helices, and somewhat less so with the $\uparrow\downarrow\pi\pi^{7,2}$ helices ($\nu = 1636$ cm^{-1}), the frequencies of the $\overrightarrow{\pi_{LD}}\overleftarrow{\pi}_{LD}$ dimers as well as of the $\uparrow\uparrow$ double helices being considerably higher ($\nu \geq 1647$ cm^{-1}). The CD spectrum 1 in Fig. 3, taken at similar conditions, shows positive Cotton effects at ca. 220 nm indicative of dominating right-handed helices, unlike GA in dioxane solution (curve 3), where left-handed helices predominate (Sychev *et al.*, 1980). Increase in the

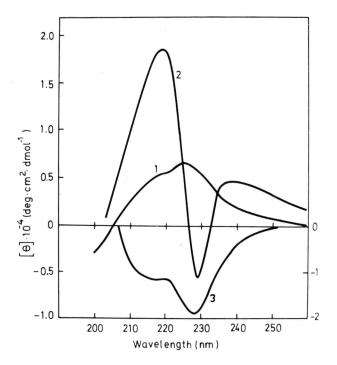

Figure 3. CD spectra of GA in DPPC liposomes at 1 : 330 (1) and 1 : 75 (2) GA : lipid ratios (molar). A very thin layer of about 10 μm was used to decrease the light scattering. The sensitivity was 1×10^{-6}. Spectrum 3 (to which the right-handed scale refers) belongs to GA dimers equilibrated in dioxane.

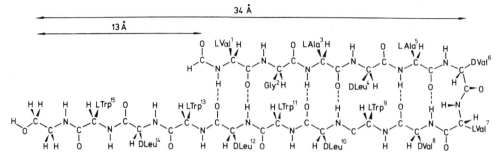

Figure 4. β_{DL} hairpin conformation of GA.

GA content up to a molar ratio of 1 : 75–150 gives rise to a drastic change in the CD spectrum (curve 2), accompanied by an increase in the intensity and by the amide I band shift from 1634 cm^{-1} to 1630 cm^{-1}, which testify to the formation of the $\uparrow\downarrow\beta$ structure (Sychev *et al.*, 1980). The $\uparrow\downarrow\beta$ structure formed by the fully extended GA molecules would have a length of more than 50 Å, which is too long for a single penetration of a bilayer membrane. Besides, in the CD spectrum the $\uparrow\downarrow\beta$ structure has a negative rather than a positive band at 216–220 nm (Ikeda *et al.*, 1979). At the same time, the recent CD spectral study of the β turns in the LD cyclopeptide (Iseli *et al.*, 1979) has shown that the positive Cotton effect at ca. 220 nm (n–π^* transition) corresponds to the DL turn, while formation of the LD turn brings about a negative effect at the same wavelength. On the basis of this observation we suggested a β_{DL} hairpin conformation with one DL turn (type II'), as illustrated in Fig. 4, for GA in DPPC liposomes, at a ratio of 1 : 75–150. Because the n–π^* Cotton effects should cancel each other in the $\uparrow\downarrow\beta$ structure with equal L and D residue contents (Brack and Spach, 1979), a main contribution to the CD of the proposed structure comes from the DL turn. A pleated sheet formed by intermolecular association of such structures is shown in Fig. 5. Formation of such aggregates on passing from 1 : 300 to 1 : 150–75 GA : lipid ratios apparently interferes with the cooperativity of the lipid phase transition detected earlier at 1 : 150–100 ratios (Weidekamm *et al.*, 1977; Boehler *et al.*, 1978).

The results obtained with GA in dimyristoylphosphatidylcholine (DMPC), regardless of temperature (20 and 52°C) and GA : lipid ratio (1 : 350, 1 : 150, and 1 : 35), are similar to those obtained with DPPC at 20°C and 1 : 150 ratio (amide I frequency 1631 cm^{-1}),

Table I. Conformational States of GA in DPPC and DMPC Liposomes (20°C)[a]

Lipid	GA: lipid ratio (molar)	Predominant conformation
DPPC	1:10–30	$\uparrow\downarrow\pi\pi_{LD}$ helix $+ \uparrow\downarrow \beta$ sheet
	1:150	$\uparrow\downarrow \beta$ sheet
	1:>300	$\uparrow\downarrow\pi\pi_{LD}$ helix
DMPC	1:35–350	$\uparrow\downarrow \beta$ sheet

[a] Incorporation of GA into the lipid zone of liposomes was followed by fluorescence measurements (blue shift of the tryptophan emission).

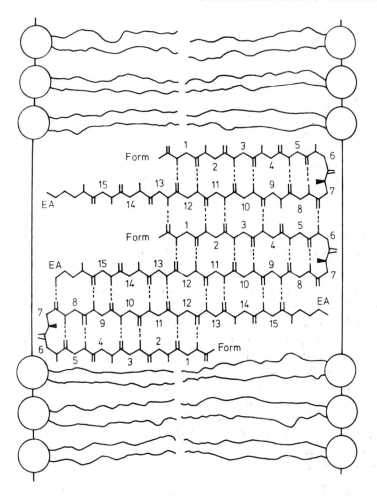

Figure 5. Antiparallel β-sheet structure of GA. Only three GA molecules are shown, though much larger associates can also form.

i.e., bear evidence of the presence of pleated sheet associates (see Table I). Sonication time (30 min) and temperature (37°C) employed for preparation of liposomes as well as the temperature of measurement (52°C), the choice of lipid (DMPC) and GA : lipid ratio accurately reproduce the conditions of Weinstein *et al.* (1979). These authors consider their data as favoring the Urry $\overrightarrow{\pi}_{LD}\overleftarrow{\pi}_{LD}$ dimeric structure of GA in the membrane. We believe that the NMR results of Weinstein and colleagues, according to which the COOH terminus of GA is localized near the membrane surface, whereas the NH$_2$ terminus is located deep in the lipid bilayer, are in full accord with the structures presented in Figs. 4 and 5, and that the above NMR data should be accordingly reinterpreted.

Let us now return to the consideration of the GA dimer state in the membrane. Bamberg *et al.* (1978, 1979) produced a number of experimental data that they consider as strongly supporting the existence of the $\overrightarrow{\pi}_{LD}\overleftarrow{\pi}_{LD}$ dimer(s) in the membrane. For instance, such dimers are expected to be less stable if negative charges appear at the junction site, and indeed

1. *N*-Pyromellityl GA was found 2–3 orders of magnitude less potent a channel former than *O*-pyromellityl (Bamberg *et al.*, 1978).

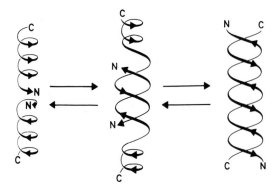

Figure 6. "Zipper" mechanism of $\uparrow\downarrow\pi\pi_{LD}$-helix formation.

2. Conductivity induced by N-succinyl GA falls by 3 orders of magnitude on passing from the nonionized (pH 4) to the ionized (pH 6) state of the carboxyl (Bamberg *et al.*, 1979).

However plausible, we believe that these facts can be explained just as well on the basis of the $\uparrow\downarrow\pi\pi_{LD}$ channel structure. As was suggested in 1975 by Urry and colleagues, and recently confirmed in the study of covalently linked GA dimers (Sychev *et al.*, in preparation), formation of antiparallel double helices occurs by intertwining of two single-stranded π_{LD} helices by a "zipper" mechanism, schematically shown in Fig. 6. In other words, the $\overrightarrow{\pi}_{LD}\overleftarrow{\pi}_{LD}$ dimer serves as an intermediate on the way to the $\uparrow\downarrow\pi\pi_{LD}$ dimer, and the introduction of NH_2-terminal charged groups causes kinetic limitations in the reaction path. Moreover, the time required for the current drop after adding KOH to the solution, which is over 5 min for the N-succinyl GA (Bamberg *et al.*, 1979) and more than 1 hr for some other N-carboxylic derivatives (Apell *et al.*, 1979), agrees with the unwinding time of the $\uparrow\downarrow$ double helix (Veatch and Blout, 1974; Sychev *et al.*, 1980) and is too slow for dissociation of the $\overrightarrow{\pi}_{LD}\overleftarrow{\pi}_{LD}$ dimer.

Finally, it should be noted that formation of the $\uparrow\downarrow\pi\pi_{LD}$ helix observed by us in DPPC at 20°C and low concentrations of the antibiotic is consistent with the fact that the same conformation predominates in a nonpolar solvent, such as dioxane or chloroform (Sychev *et al.*, 1980). According to Boheim *et al.* (1980), the cation conductivity of the GA-treated membrane remains practically invariable on cooling the system below the phase transition temperature. This allows to assume the same structure for GA in the membrane at temperatures both below and above the phase transition temperature. Whether the $\uparrow\downarrow\pi\pi_{LD}$ helix is the sole structure of the GA transmembrane channel, or other structures can be formed under various conditions, requires further study.

REFERENCES

Apell, H.-J., Bamberg, E., and Alpes, H. (1979). *J. Membr. Biol.* **50**, 271–285.

Bamberg, E., Apell, H.-J., Alpes, H., Gross, E., Morell, J. L., Harbaugh, J. F., Janko, K., and Läuger, P. (1978). *Fed. Proc.* **37**, 2633–2638.

Bamberg, E., Alpes, H., Apell, H.-J., Bradley, R., Härter, B., Quelle, M.-J., and Urry, D. W. (1979). *J. Memb. Biol.* **50**, 257–270.

Boehler, B. A., de Gier, J., and van Deenen, L. L. M. (1978). *Biochim. Biophys. Acta* **512**, 480–488.

Boheim, G., Hanke, W., and Eibl, H. (1980). *Proc. Natl. Acad. Sci. USA* **77**, 3403–3407.

Brack, A., and Spach, G. (1979). *J. Mol. Evol.* **13**, 35–46.

Chapman, D., Cornell, B. A., Eliasz, A. W., and Perry, A. (1977). *J. Mol. Biol.* **113**, 517–538.

Hladky, S. B., and Haydon, D. A. (1970). *Nature (London)* **225**, 451–453.

Ikeda, S., Fukutome, A., Imae, T., and Yoshida, T. (1979). *Biopolymers* **18**, 335–349.

Iseli, M., Wagniere, G., Brahms, J. G., and Brahms, S. (1979). *Helv. Chim. Acta* **62**, 921–931.

Masotti, L., Spisni, A., and Urry, D. W. (1980). *Cell Biophys.* **2**, 241–251.

Myers, V. B., and Haydon, D. A. (1972). *Biochim. Biophys. Acta* **274**, 313–322.

Sychev, S. V., Nevskaya, N. A., Jordanov, S., Shepel, E. N., Miroshnikov, A. I., and Ivanov, V. T. (1980). *Bioorgn. Chem.* **9**, 121–151.

Sychev, S. V., Fonina, L. A., Ivanov, V. T., and Ovchinnikov, Y. A. (1981). In preparation.

Urry, D. W. (1971). *Proc. Natl. Acad. Sci. USA* **68**, 672–676.

Urry, D. W., Goodall, M. C., Glickson, J. D., and Mayers, D. F. (1971). *Proc. Natl. Acad. Sci. USA* **68**, 1907–1911.

Urry, D. W., Long, M. M., Jacobi, M., and Harris, R. D. (1975). *Ann. N.Y. Acad. Sci.* **264**, 203–220.

Urry, D. W., Spisni, A., and Khaled, M. A. (1979). *Biochem. Biophys. Res. Commun.* **88**, 940–949.

Veatch, W. R., and Blout, E. R. (1974). *Biochemistry* **13**, 5257–5264.

Veatch, W., and Stryer, L. (1977). *J. Mol. Biol.* **113**, 89–102.

Veatch, W. R., Fossel, E. T., and Blout, E. R. (1974). *Biochemistry* **13**, 5249–5256.

Weidekamm, E., Bamberg, E., Brdiczka, D., Wildermuth, G., Macco, F., Lehmann, W. and Weber, R. (1977). *Biochim. Biophys. Acta* **464**, 442–447.

Weinstein, S., Wallace, B. A., Blout, E. R., Morrow, J. S., and Veatch, W. (1979). *Proc. Natl. Acad. Sci. USA* **76**, 4230–4234.

131

Bacteriorhodopsin

Benno Hess, Dietrich Kuschmitz, and Martin Engelhard

1. DISCOVERY OF A PURPLE MEMBRANE OF HALOBACTERIA AS A LIGHT-ENERGY CONVERTER

In 1967 Stoeckenius and Kunau (1968) and, independently, McClare (1967) described a purple fragment in the cell membrane of halobacteria. Later, its chemical nature was identified through the isolation of a polypeptide of molecular weight 26,000 with a retinal residue bound via a Schiff base linkage to a lysine side chain, and the new membrane protein was called bacteriorhodopsin (BR) (Oesterhelt and Stoeckenius, 1971).

In 1972, it was shown that upon illumination BR undergoes a thermoreversible photochemical reaction cycle coupled to a release and uptake of protons. Action spectrum as well as quantum yield were given, and a companion reaction of the protein moiety in terms of protein fluorescence changes was reported (Oesterhelt and Hess, 1973). Furthermore, evidence for the function of BR as a light-driven proton pump yielding a proton gradient across a membrane was presented (Oesterhelt, 1972; Oesterhelt and Stoeckenius, 1973). Here, on the basis of the structural composition we review the current state of knowledge of the function of BR. We do not dwell upon work carried out with vesicles, artificial layers, or cellular membranes if it does not add to the current fundamental knowledge. Furthermore, gradient-coupled reactions such as amino acid transport or ATP synthesis will not be discussed. Additional information can be obtained from earlier reviews (Henderson, 1977; Stoeckenius et al., 1979; Eisenbach and Caplan, 1979a; Stoeckenius, 1980; Ottolenghi, 1980).

2. CHEMISTRY

Recently, the complete amino acid sequence of BR became available with minor discrepancies (Ovchinnikov et al., 1979; Khorana et al., 1979). Whereas from earlier stud-

Benno Hess, Dietrich Kuschmitz, and Martin Engelhard • Max-Planck-Institut für Ernährungsphysiologie, 4600 Dortmund, West Germany.

ies the retinal was indicated to be attached to the ϵ-amino group of the lysine-41 residue (Bridgen and Walker, 1976), more recently it was also found at lysine-216 (Ovchinnikov *et al.*, 1980; Katre *et al.*, 1981; Lemke and Oesterhelt, 1981).

With the sequence data at hand, the results of studies of the crystalline array of BR in two different forms (hexagonal and orthorhombic) in the purple membrane by low-dose electron microscopy (Henderson and Unwin, 1975; Michel *et al.*, 1980) yielded a BR-model consisting of seven α-helix rods located in two parallel rows approximately perpendicular to the plane of the membrane.

The charge polarity of this arrangement is determined by the high helical content, the uneven number of helices, and the excess of carboxylate groups located at the cytoplasmic membrane surface. Within the membrane, three such molecules are arranged in a trimeric cluster with 10–11 moles of lipids per mole of BR (25% of the mass of the purple membrane). The lipid distribution is also asymmetric (Blaurock, 1975) and contributes to the higher overall negative charge density of the extracellular face (Neugebauer *et al.*, 1978).

Neutron diffraction and scattering experiments allow localization of the retinylidene isomers in the membrane plane occurring in roughly equimolar distribution as 13-*cis*- and all-*trans*-retinal in the dark-adapted form of BR. These studies suggest a position of the β-ionone ring near the center of the BR molecule and of the isoprenoid stretching between two α helices toward the next BR within the trimeric unit (King *et al.*, 1980). Linear dichroism studies yield the angle of the transition moment between the chromophore and the plane of the membrane between 19 and 31° (Bogomolni *et al.*, 1977; Heyn *et al.*, 1977; Korenstein and Hess, 1978; Acuna and Gonzalez-Rodriguez, 1979; Bamberg *et al.*, 1979; Keszthelyi, 1980). A position of the Schiff base approximately 17 Å below the nearer membrane surface has been suggested (King *et al.*, 1979). The overall orientation of the BR molecule is supported by the data obtained from circular dichroism, infrared spectroscopy, and X-ray diffraction (see Stoeckenius *et al.*, 1979).

Finally, transient dichroism studies indicate that BR is immobilized within the native purple membrane matrix and that the rotational freedom of the BR monomer, trimer, or clusters of trimers is excluded. (Razi-Naqvi *et al.*, 1973; Sherman *et al.*, 1976; Korenstein and Hess, 1978; Kouyama *et al.*, 1981). Using fluorescence depolarization, the microviscosity of the lipid domain in the purple membrane was found to be 5 poise (Korenstein *et al.*, 1976). The studies show that the immobilization of BR is mainly due to protein–protein interactions. Our current view of the overall chemistry and topology of BR is supported by reconstitution experiments (Oesterhelt and Schuhmann, 1974; Huang *et al.*, 1981).

3. PHOTOCHEMICAL REACTION CYCLES

BR is found in a dark (BR_{560}^{DA}) and in a light (BR_{570}^{LA})-adapted form. Irradiation of BR^{DA} causes a red shift of the absorption maximum with an increase in the molar extinction coefficient (Oesterhelt and Stoeckenius, 1971; Korenstein and Hess, 1977a), which slowly reverts in a thermal dark reaction. Whereas BR^{DA} consists of equimolar 13-*cis*- and all-*trans*-retinal, BR^{LA} contains only the all-*trans*-retinal isomer with an extinction of $\epsilon \sim 63{,}000$ M^{-1} cm^{-1} (Oesterhelt *et al.*, 1973). Thus, the dark–light adaptation represents a photochemical isomerization reaction.

Steady white light illumination of the BR^{LA} form results in a blue shift of the spectrum toward a maximum near 412 nm that can be stabilized by the presence of ether and salt or at lower temperature (Oesterhelt and Hess, 1973; Oesterhelt and Stoeckenius, 1973).

In the dark, this component relaxes thermally to the purple BR^{LA} form, indicating an overall thermoreversible photochemical reaction cycle that is supported by an action spectrum and an appropriate quantum yield. The process is accompanied by a release and uptake of protons (Oesterhelt and Hess, 1973).

Investigation of the kinetics of the formation and decay of the 412-nm product by flash photolysis and modulation excitation spectrophotometry demonstrated the generation of several intermediates prior to and after the 412-nm component. The following reaction sequence was obtained (Kung *et al.*, 1975; Chance *et al.*, 1975; Lozier *et al.*, 1975; Slifkin and Caplan, 1975; Dencher and Wilms, 1975; Gillbro, 1978; Applebury *et al.*, 1978; Stoeckenius *et al.*, 1979):

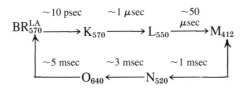

This tentative scheme neglects the excited state of BR_{570} and summarizes the thermal decay process with approximate half-times for room-temperature conditions. Different thermal stability allows to trap the intermediates at lower temperature and to obtain the optical absorption parameter of BR and the K, L, and M intermediates. The O and N intermediates are still ill-defined (Hess and Kuschmitz, 1977; Kriebel *et al.*, 1979; Stoeckenius *et al.*, 1979). Early experiments showed that the M intermediate is composed of two conformations (Slifkin and Caplan, 1975; Hess and Kuschmitz, 1977), which are in thermal equilibrium (Korenstein *et al.*, 1978), and suggested a branched reaction cycle (Korenstein *et al.*, 1979).

This photocycle can be perturbed by photochemical reactions of the K intermediate (Goldschmidt *et al.*, 1976), the L intermediate (Hurley *et al.*, 1978), and the M intermediates (Oesterhelt and Hess, 1973; Hess, 1976; Hess and Kuschmitz, 1977; Hurley *et al.*, 1978), all of which relax through separate pathways to BR.

Furthermore, BR and the photocycle intermediates are sensitive to pressure (Crespi and Ferraro, 1979; Tsuda and Ebrey, 1980), electric fields (see below), state of hydration (Korenstein and Hess, 1977b; Lazarev and Terpugov, 1980), pH (Moore *et al.*, 1978; Fischer and Oesterhelt, 1979; Mowery *et al.*, 1979; Kuschmitz and Hess, 1980; Kalisky *et al.*, 1981b), ionic strength (Kuschmitz and Hess, 1980), and organic solvents (Oesterhelt *et al.*, 1973; Eisenbach and Caplan, 1979b,c).

At low pH a blue chromophore species (absorbing at 605 nm) is formed (Moore *et al.*, 1978; Fischer and Oesterhelt, 1979; Mowery *et al.*, 1979), which also shows a photocycle upon illumination not involving the M intermediate nor pumping protons (Mowery *et al.*, 1979; Tsuji and Rosenheck, 1979b).

Molecular interpretation of the photocycle reactions is facilitated by the results of Raman spectroscopy. This technique (Callender and Honig, 1977) supports the protonated Schiff base model of BR, whereas the M intermediate is found to be deprotonated; deprotonation occurs between the L and the M intermediate and reprotonation between the M and the O intermediate (Lewis *et al.*, 1974; Aton *et al.*, 1977; Marcus and Lewis, 1978; Terner *et al.*, 1979; Stockburger *et al.*, 1979; Narva *et al.*, 1981). Also, Raman patterns of the K, L, O, and X intermediates are described by these authors. Recent studies show the M component to be an unprotonated Schiff base of 13-*cis*-retinal (Braiman and Mathies, 1980; Tsuda *et al.*, 1980; Mowery and Stoeckenius, 1981; Kuschmitz and Hess, 1982).

4. PROTON CYCLE

The basis for the physiological function of BR is a vectorial proton cycle originating from the photocycle that translocates protons across the membrane from the cytoplasmic to the extracellular space. Calculated (Hartmann *et al.*, 1977) and directly measured (Govindjee *et al.*, 1980; Bogomolni *et al.*, 1980; Renard and Delmelle, 1980) quantum efficiency yielded 0.5–0.7 protons translocated per absorbed photon. Based on a photocycle quantum efficiency of 0.25–0.35 (Goldschmidt *et al.*, 1976, 1977; Becher and Ebrey, 1977; Govindjee *et al.*, 1980), a transport stoichiometry of two protons per photocycle is observed.

In isolated purple membrane the proton cycle is a function of pH. At neutral and higher pH protons are dissociated in the light and reassociated in the dark (Oesterhelt and Hess, 1973; Chance *et al.*, 1975; Lozier *et al.*, 1975; Garty *et al.*, 1977), whereas at lower pH a reverse reaction is observed (Dencher and Wilms, 1975; Garty *et al.*, 1977). The number of protons dissociated per photocycle relative to the M_{412} intermediate is a function of the pH (Bakker and Caplan, 1978), the ionic strength (Hess and Kuschmitz, 1978; Avi-Dor *et al.*, 1979; Ort and Parson, 1979; Govindjee *et al.*, 1980), and the light intensity and illumination time (Kuschmitz, 1978), ranging from 0.2–3 up to several hundred at very low intensities (Caplan *et al.*, 1978). This behavior points to the participation of proton binding groups of the opsin moiety of BR in addition to a deprotonation/protonation reaction of the retinal Schiff base binding site in the overall proton translocation reaction (see Section 6). Furthermore, the proton cycle is also sensitive to blue light activation (Oesterhelt and Hess, 1973; Hess, 1976).

At subzero temperatures, where no O_{660} intermediate can be observed, the rates of deprotonation and protonation correlate closely with the rates of formation and decay of the M_{412} intermediate (Chance *et al.*, 1975), whereas at room temperature the deprotonation reaction lags behind the rate of M_{412} formation (Lozier *et al.*, 1976; Ort and Parson, 1978; Govindjee *et al.*, 1980) and the proton cycle correlates better with the rates of formation and decay of the O_{660} intermediate (Gillbro, 1978).

5. CHARGE CYCLES

Light-induced charge cycles are observed by appropriate electrical circuits. Conductivity changes in a purple membrane suspension at low ionic strength (oriented by a small electric field) correlate kinetically with the proton cycle (Keszthelyi and Ormos, 1980) but are slower at high ionic strength (randomly distributed purple membrane) reflecting charge reorientation (Slifkin *et al.*, 1979). Photopotentials and photocurrents due to proton transport are observed with purple membranes absorbed on or incorporated into lipid bilayers with and without membrane supports (e.g., Millipore filter, collodion), and in synthetic membranes (Drachev *et al.*, 1974; Dancshazy and Karvaly, 1976; Shieh and Packer, 1976; Block *et al.*, 1977; Eisenbach *et al.*, 1977; Herrmann and Rayfield, 1978; Bamberg *et al.*, 1979; Seta *et al.*, 1980; Korenbrot and Hwang, 1980). Similar observations were made with purple membranes in an oil–water interface provided a proton acceptor was present in the oil phase (Boguslavsky *et al.*, 1975; Hwang *et al.*, 1977).

Higher time resolution with oriented purple membranes reveals electric signals originating from charge displacement reactions within the membrane that occur prior to proton release into the medium (Drachev *et al.*, 1978; Keszthelyi and Ormos, 1980; Korenbrot and Hwang, 1980). They are attributed to the formation and decay of the photocycle inter-mediates K_{590} (instantaneously appearing and of opposite polarity), L_{550}, M_{412}, and O_{660} (Drachev *et al.*, 1978; Keszthelyi and Ormos, 1980).

Intrinsic charge displacement potentials are also observed with purple membranes ad-sorbed on electric insulator (Teflon film), capacitively coupled to the measuring device (Trissl and Montal, 1977; Hong and Montal, 1979). Again, an instantaneous appearance of a photopotential of opposite polarity (attributed to K) is observed followed by two increasing exponentials in the millisecond range and a slowly decreasing exponential in the second range. Purple membrane layers oriented between transparent metal electrodes give photopotentials following the profile of the exciting flash. Furthermore, photopoten-tials of opposite polarity are observed upon excitation of the photocycle intermediates M_{412}, N_{520}, and O_{660} (Hwang *et al.*, 1978). Finally, a quenching effect of blue light on the generation of photopotentials was described (Karvaly and Dancshazy, 1977; Nagy, 1978).

Three mechanisms are discussed for the intrinsic charge displacement reactions: dis-crete displacement of two protons through the membrane following the time course of the photocycle (Keszthelyi and Ormos, 1980); interfacial charge transfer; and reorientation of electric dipoles across the membrane (Hong, 1976). An interfacial charge cycle (change of the overall surface charge density) was also reported using spin labels responding to the surface charge density (Carmeli *et al.*, 1980; Tokutomi *et al.*, 1980).

As expected from these electrical responses, charge displacement and transfer reac-tions of BR are also observed in response to electrical field perturbation. Simple models of free retinal already display field-induced shifts of negative charge toward the carbonyl (Mathies and Stryer, 1976). Purple membrane sheets orient in low electrical fields (Shinar *et al.*, 1977; Tsuji and Rosenheck, 1979a; Keszthelyi and Ormos, 1980), and the chrom-ophore itself responds to an electric field of approximately 10 kV/cm in the absence (Shinar *et al.*, 1977; Borisevitch *et al.*, 1979) and in presence of light (Lukashev *et al.*, 1980) the latter state stimulates the formation of the M state. A response is also recorded if the reaction of BR or the apomembrane (with a functional retinal absent) is analyzed at 300 nm with a polarization angle of 55° indicating a field-induced dipole change, perhaps due to a tyrosine deprotonation (see below). With the well-known electrical gradients in bio-logical membranes, such dielectric polarization of groups or systems of groups is of sig-nificance and might play a role in the overall stabilization of the light-driven proton transfer process in BR (Hess and Korenstein, 1978).

6. PROTEIN CONFORMATION CYCLE

The involvement of the opsin moiety in the overall mechanism of the photo- and proton cycle was recognized with the discovery of the protein fluorescence cycle following illumination in a time course comparable to the formation and decay of the M intermediate (Oesterhelt and Hess, 1973; Bogomolni *et al.*, 1978). Chemical modification (Konishi and Packer, 1978) and kinetic studies of the spectral changes in the UV absorption of the protein (Bogomolni *et al.*, 1978; Rafferty, 1979; Hess and Kuschmitz, 1979) show a de-protonation of one tyrosine and reactions of two tryptophan residues; these indicate charge and proton displacements and changes of the micropolarity within the protein during the photocycle (Hess and Kuschmitz, 1979). Nanosecond laser excitation differentiates in the UV absorption changes a slow-phase spectral component, which closely follows the for-mation of M, and a faster component with a half-time of 600 nsec at 296 nm, representing the involvement of tryptophan and tyrosine also in the very early reaction of the photocycle (Efremov *et al.*, 1978; Kuschmitz and Hess, 1982). The proximity of the aromatic amino acid residues to the chromophore is substantiated by simultaneous cross excitation; excitation at 412 nm changes the kinetics in the UV range and vice versa (Hess and Kuschmitz, 1979). Energy transfer to the chromophore was found upon 265-nm excitation alone (Kalisky *et al.*, 1981a).

The involvement of proton-binding amino acid residues in the photo–proton cycle was deduced from the behavior of the proton-to-M_{412} ratio as a function of ionic strength (Hess and Kuschmitz, 1978), and shown directly by pH difference titration between the light and the dark state showing an increasing buffer capacity around neutrality and a decreasing buffer capacity at acidic pH (Hess *et al.*, 1978). The kinetics of the photocycle as a function of pH and ionic strength identify proton binding groups of pK 8.6, 7–7.5, and 4.7 at high ionic strength, which are not active in the dark, and shift up to 1.7 units higher at low ionic strength. This indicates conformational changes by which proton binding groups "buried" in the dark turn to the interface region during the photocycle. The appearance of groups of the latter two pKs lags behind M_{412} formation (Kuschmitz and Hess, 1980).

The biphasic behavior of M_{412} decay (Hess and Kuschmitz, 1977; Eisenbach *et al.*, 1976) as a function of temperature in comparison to the temperature sensitivity of the O_{660} intermediate points to a temperature-dependent conformational transition of BR (Hoffman *et al.*, 1978).

7. MECHANISM

The BR cycle catalyzes a vectorial charge separation generated by the photochemical activation of the retinal chromophore according to the general scheme:

$$
1.\ [\text{retinal}\!=\!\text{N}\!-\!\text{R}] \xrightarrow{h\nu} [\text{retinal}\ \overset{\delta+}{\text{···}}\text{N}\!-\!\overset{\delta-}{\text{R}}]^* \quad \text{excitation}
$$

$$
\begin{array}{cc}
\vdots & | \\
\text{H}^+ & \text{H}
\end{array}
$$

2. $[\text{DH}\text{···}A_1] \longrightarrow [\text{D}^-\text{···}\text{H}^+A_1]$ geminate charge pair
3. $[\text{H}^+A_1\text{···}A_2] \longrightarrow [A_1\text{···}\text{H}^+A_2]$ proton migration
4. $A_2\text{H}^+ \longrightarrow A_2 + \text{H}^+$ proton dissociation (outside)
5. $\text{D}^- + \text{H}^+ \longrightarrow \text{DH}$ proton association (inside)

The basis for such chemical mechanisms is the knowledge of the chemical structure of BR and the chromophore configuration in its ground state. Theoretical calculations and model experiments (Honig *et al.*, 1979a; Sheves *et al.*, 1979; Motto *et al.*, 1980; Nakanishi *et al.*, 1980) lead to the postulation of a point charge model with a negative charge located opposite to the C-13 and C-14 bond of the retinylidene chromophore. Also, a negative charge interaction with the β-ionone ring is discussed. These charges are in resonance with the protonated Schiff base (Honig *et al.*, 1979a). Excitation leads to an intrachromophoric charge displacement. In this transient charge recombination is prevented by two possible mechanisms:

1. Electrostatic charge stabilization (Warshel, 1978; Warshel, 1979).
2. *Trans–cis* isomerization of the C-13 bond (Schulten and Tavan, 1978; Honig *et al.*, 1979b; Warshel, 1979).

Following this event, two mechanisms of charge transfer through the membranes are discussed. A charge diffusion (proton conduction) along two hydrogen-bonded pathways has been suggested (Nagle and Morowitz, 1978), in which the retinal chromophore is regarded as an active switch, shifting a proton into one diffusion pathway leading to the external medium and accepting one proton from the cytoplasmic space through a second diffusion pathway (Kozlov and Skulachev, 1977; Stoeckenius *et al.*, 1979). Alternatively, discrete displacement reactions of protons have been suggested connecting donor–acceptor groups within the protein moiety, which change pK cyclically during the photocycle (Hess *et al.*, 1978; Keszthelyi and Ormos, 1980).

Studies for identification of amino acid residues involved in such a mechanism are under way in a number of laboratories. In order to account for the finding that two protons are transported per photocycle, a mechanism is proposed where a sequential transfer of a tyrosyl proton followed by the Schiff base proton to two protein acceptor sites is assumed. Both protons are released simultaneously to the external aqueous medium followed by a sequential reprotonation of tyrosine and the chromophore (Hess and Kuschmitz, 1979; Kalisky *et al.*, 1981b).

8. COOPERATIVITY

BR molecules are arranged in a two-dimensional array, forming an almost perfect crystal lattice of space group $P3$. Thus, three protein molecules are in direct contact, forming a trimeric cluster and imposing strong protein–protein interactions, which might result from the large contact areas between helical domains. Such rigid, immobilized structure within the bacterial membrane suggests a functional cooperativity of the BR system. Indeed, an electronic coupling between the retinal chromophores has been found by circular dichroic spectrum (Heyn *et al.*, 1975; Becher and Cassim, 1975). Evidence for structurally cooperative units in the form of trimers and multiples of trimers is given by chemical cross-linking of BR trimers (Dellweg and Sumper, 1978), by dissociation experiments (Becher and Cassim, 1977), and by regeneration experiments (Ebrey *et al.*, 1977; Rehorek and Heyn, 1979). Such a high level of protein organization also suggests a kinetic coupling among the three BR molecules. Indeed, the rate constant for the decay of the M intermediate of the photochemical cycle was found to depend nonlinearly on the light intensity, thus indicating a nearest-neighbor interaction (Korenstein *et al.*, 1979). Furthermore, a high ratio of dissociated protons per M state in the range up to 3 at low light intensities and the decrease of this ratio to unity at saturating light intensities was observed (Kuschmitz and Hess, 1981). This suggests an interaction between one molecule in the M state and its neighbors inducing proton transfer reactions. Cooperative functions were also suggested on the basis of structural changes in BR induced by electric field perturbation (Tsuji and Neumann, 1981).

The observation of cooperativity in the purple membrane might be of importance for membrane functions in general. Its physiological significance is understood in terms of the overall efficiency of coupled molecular functions in the oligomeric forms of the membrane organization. We visualize cooperativity involved in transport and electrochemical gradient formation. Indeed, the photocycle kinetics are dependent on the electrochemical gradient (Quintanilha, 1980), and are expected to play a significant role in the stabilization of vectorial proton transfer.

REFERENCES

Acuna, A. U., and Gonzalez-Rodriguez, J. (1979). *An. Quim.* **75**, 630–635.

Applebury, M. L., Peters, K. S., and Rentzepis, P. M. (1978). *Biophys. J.* **23**, 375–382.

Aton, B., Doukas, A. G., Callender, R. H., Becher, B., and Ebrey, T. G. (1977). *Biochemistry* **16**, 2995–2998.

Avi-Dor, Y., Rott, R., and Schnaiderman, R. (1979). *Biochim. Biophys. Acta* **545**, 15–23.

Bakker, E. P., and Caplan, S. R. (1978). *Biochim. Biophys. Acta* **503**, 362–379.

Bamberg, E., Apell, H. J., Dencher, N. A., Sperling, W., Stieve, H., and Laeuger, P. (1979). *Biophys. Struct. Mech.* **5**, 277–292.

Becher, B., and Cassim, J. (1975). *Biophys. J.* **15**, 66a.

Becher, B., and Cassim, J. (1977). *Biophys. J.* **19**, 285–297.

Becher, B., and Ebrey, T. G. (1977). *Biophys. J.* **17**, 185–191.

Blaurock, A. E. (1975. *J. Mol. Biol.* **93**, 139–158.

Block, M. C., Hellingwerf, K. J., and Van Dam, K. (1977). *FEBS Lett.* **76**, 45–50.

Bogomolni, R. A., Hwang, S.-B., Tseng, Y.-W., King, G. I., and Stoeckenius, W. (1977). *Biophys. J.* **17**, 98a.

Bogomolni, R. A., Stubbs, L., and Lanyi, J. K. (1978). *Biochemistry* **17**, 1037–1041.

Bogomolni, R. A., Baker, R. A., Lozier, R. H., and Stoeckenius, W. (1980). *Biochemistry* **19**, 2152–2159.

Boguslavsky, L. I., Kondrashin, A. A., Kozlov, I. A., Metelsky, S. T., Skulachev, V. P., and Volkov, A. G. (1975). *FEBS Lett.* **50**, 223–226.

Borisevitch, G. P., Lukashev, E. P., Kononenko, A. A., and Rubin, A. B. (1979). *Biochim. Biophys. Acta* **546**, 171–174.

Braiman, M., and Mathies, R. (1980). *Biochemistry* **19**, 5421–5428.

Bridgen, J. A., and Walker, I. D. (1976). *Biochemistry* **15**, 792–798.

Callender, R. H., and Honig, B. (1977). *Annu. Rev. Biophys. Bioenerg.* **6**, 33–55.

Caplan, S. R., Eisenbach, M., and Garty, H. (1978). In *Energetics and Structure of Halophilic Microorganisms* (S. R. Caplan and M. Ginzburg, eds.), pp. 49–61, Elsevier/North-Holland, Amsterdam.

Carmeli, C., Quintanilha, A. T., and Packer, L. (1980). *Proc. Natl. Acad. Sci. USA* **77**, 4707–4711.

Chance, B., Porte, A., Hess, B., and Oesterhelt, D. (1975). *Biophys. J.* **15**, 913–917.

Crespi, H. L., and Ferraro, J. R. (1979). *Biochem. Biophys. Res. Commun.* **91**, 575–582.

Dancshazy, Z., and Karvaly, B. (1976). *FEBS Lett.* **72**, 136–138.

Dellweg, H.-G., and Sumper, M. (1978). *FEBS Lett.* **90**, 123–126.

Dencher, N., and Wilms, M. (1975). *Biophys. Struct. Mech.* **1**, 259–271.

Drachev, L. A., Jasaitis, A. A., Kaulen, A. D., Kondrashin, A. A., Lieberman, E. A., Nemecek, I. B., Ostroumov, S. A., Semenov, A. Y., and Skulachev, V. P. (1974). *Nature (London)* **249**, 321–324.

Drachev, L. A., Kaulen, A. D., and Skulachev, V. P. (1978). *FEBS Lett.* **87**, 161–167.

Ebrey, T. G., Becher, B., Mao, B., Kilbride, P., and Honig, B. (1977). *J. Mol. Biol.* **112**, 377—397.

Efremov, E. S., Sotnichenko, A. I., Aldashev, A. A., Surin, A. M., and Miroshnikov, A. I. (1978). In *Proc. 2nd FRG–USSR Symp. Chem. Peptides Proteins,* Grainau-Eibsee, May 1978, pp. 71–73.

Eisenbach, M., and Caplan, S. R. (1979a). In *Current Topics in Membrane Transport* (B. Horecker and E. Stadtman, eds.), Vol. 12, pp. 165–249, Academic Press, New York.

Eisenbach, M., and Caplan, S. R. (1979b). *Biochim. Biophys. Acta* **554**, 269–280.

Eisenbach, M., and Caplan, S. R. (1979c). *Biochim. Biophys. Acta* **554**, 281–292.

Eisenbach, M., Bakker, E. P., Korenstein, R., and Caplan, S. R. (1976). *FEBS Lett.* **71**, 228–232.

Eisenbach, M., Weissmann, C., Tanny, G., and Caplan, S. R. (1977). *FEBS Lett.* **81**, 77–80.

Fischer, U., and Oesterhelt, D. (1979). *Biophys. J.* **28**, 211–230.

Garty, H., Klemperer, G., Eisenbach, M., and Caplan, S. R. (1977). *FEBS Lett.* **81**, 238–242.

Gillbro, T. (1978). *Biochim. Biophys. Acta* **504**, 175–186.

Goldschmidt, C. R., Ottolenghi, M., and Korenstein, R. (1976). *Biophys. J.* **16**, 839–843.

Goldschmidt, C. R., Kalisky, O., Rosenfeld, T., and Ottolenghi, M. (1977). *Biophys. J.* **17**, 179–183.

Govindjee, R., Ebrey, T., and Crofts, A. R. (1980). *Biophys. J.* **30**, 231–242.

Hartmann, R., Sickinger, H. D., and Oesterhelt, D. (1977). *FEBS Lett.* **82**, 1–6.

Henderson, R. (1977). *Annu. Rev. Biophys. Bioenerg.* **6**, 87–109.

Henderson, R., and Unwin, P. T. (1975). *Nature (London)* **257**, 28–32.

Herrmann, T. R., and Rayfield, G. W. (1978). *Biophys. J.* **21**, 111–125.

Hess, B. (1976). *FEBS Lett.* **64**, 26–28.

Hess, B., and Korenstein, R. (1978). In *Advances in Chemical Physics* (I. Prigogine and S. A. Rice, eds.), Vol. 34, pp. 224–227, Wiley, New York.

Hess, B., and Kuschmitz, D. (1977). *FEBS Lett.* **74**, 20–24.

Hess, B., and Kuschmitz, D. (1978). In *Frontiers of Biological Energetics* (P. L. Dutton, J. Leigh, and A. Scarpa, eds.), pp. 257–264, Academic Press, New York.

Hess, B., and Kuschmitz, D. (1979). *FEBS Lett.* **100,** 334–340.

Hess, B., Korenstein, R., and Kuschmitz, D. (1978). In *Energetics and Structure of Halophilic Microorganisms* (S. R. Caplan and M. Ginzburg, eds.), pp. 89–108, Elsevier/North-Holland, Amsterdam.

Heyn, M. P., Bauer, J. P., and Dencher, N. A. (1975). *Biochem. Biophys. Res. Commun.* **67,** 897–903.

Heyn, M. P., Cherry, R. J., and Müller, U. (1977). *J. Mol. Biol.* **117,** 607–620.

Hoffmann, W., Graca-Miguel, M., Barnard, P., and Chapman, D. (1978). *FEBS Lett.* **95,** 31–34.

Hong, F. T. (1976). *Photochem. Photobiol.* **24,** 155–189.

Hong, F. T., and Montal, M. (1979). *Biophys. J.* **25,** 465–472.

Honig, B., Dinur, U., Nakanishi, K., Balogh-Nair, V., Gawinowicz, M. A., Arnaboldi, M., and Motto, M. (1979a). *J. Am. Chem. Soc.* 7084–7086.

Honig, B., Ebrey, T., Callender, R. H., Dinur, U., and Ottolenghi, M. (1979b). *Proc. Natl. Acad. Sci. USA* **76,** 2503–2507.

Huang, K., Bayley, H., Liao, M., London, E., and Khorana, H. G. (1981). *J. Biol. Chem.* **256,** 3802–3809.

Hurley, J. B., Becher, B., and Ebrey, T. G. (1978). *Nature (Londong)* **272,** 87–88.

Hwang, S., Korenbrot, J. I., and Stoeckenius, W. (1977). *J. Membr. Biol.* **36,** 137–158.

Hwang, S., Korenbrot, J. I., and Stoeckenius, W. (1978). *Biochim. Biophys. Acta* 300–317.

Kalisky, O., Feitelson, J., and Ottolenghi, M. (1981a). *Biochemistry* **20,** 205–209.

Kalisky, O., Ottolenghi, M., Honig, B., and Korenstein, R. (1981b). *Biochemistry* **20,** 649–655.

Karvaly, B., and Dancshzy, Z. (1977). *FEBS Lett.* **76,** 36–40.

Katre, N. V., Wolber, P. K., Stoeckenius, W., and Stroud, R. M. (1981). *Biochemistry,* in press.

Katre, N. V., Wolber, P. K., Stoeckenius, W., and Stroud, R. M. (1981). *Proc. Natl. Acad. Sci. USA* **78,** 4068–4072.

Keszthelyi, L. (1980). *Biochim. Biophys. Acta* **598,** 429–436.

Keszthelyi, L., and Ormos, P. (1980). *FEBS Lett.* **109,** 189–193.

Khorana, H. G., Gerber, G. E., Herlihy, W. C., Gray, C. P., Anderegg, R. J., Nihei, K., and Biemann, K. (1979). *Proc. Natl. Acad. Sci. USA* **76,** 5046–5050.

King, G. I., Stoeckenius, W., Crespi, H. L., and Schoenborn, B. P. (1979). *J. Mol. Biol.* **130,** 395–404.

King, G. I., Mowery, P. C., Stoeckenius, W., Crespi, H. L., and Schoenborn, B. P. (1980). *Proc. Natl. Acad. Sci. USA* **77,** 4726–4730.

Konishi, T., and Packer, L. (1978) *FEBS Lett.* **92,** 1–4.

Korenbrot, J. I., and Hwang, S. (1980). *J. Gen. Physiol.* **76,** 649–682.

Korenstein, R., and Hess, B. (1977a). *FEBS Lett.* **82,** 7–11.

Korenstein, R., and Hess, B. (1977b). *Nature (London)* **270,** 184–186.

Korenstein, R., and Hess, B. (1978). *FEBS Lett.* **89,** 15–20.

Korenstein, R., Sherman, W. V., and Caplan, S. R. (1976). *Biophys. Struct. Mech.* **2,** 267–276.

Korenstein, R., Hess, B., and Kuschmitz, D. (1978). *FEBS Lett.* **93,** 266–270.

Korenstein, R., Hess, B., and Markus, M. (1979). *FEBS Lett.* **102,** 155–161.

Kouyama, T., Kimura, Y., Kinosita, K., and Ikegami, A. (1981). *FEBS Lett.* **124,** 100–104.

Kozlov, I. A., and Skulachev, V. P. (1977). *Biochim. Biophys. Acta* **463,** 29–89.

Kriebel, A. N., Gillbro, T., and Wild, U. P. (1979). *Biochim. Biophys. Acta* **546,** 106–120.

Kung, M. C., DeVault, D., Hess, B., and Oesterhelt, D. (1975). *Biophys. J.* **15,** 907–911.

Kuschmitz, D. (1978). In *Energetics and Structure of Halophilic Microorganisms* (S. R. Caplan and M. Ginzburg, eds.), pp. 356–357, Elsevier/North-Holland, Amsterdam.

Kuschmitz, D., and Hess, B. (1980). In *ICRO Course on Basic Principles and Methods in Membrane Bioenergetics, Light Energy Transduction by Bacteriorhodopsin,* Szeged, Hungary, September 1980.

Kuschmitz, D., and Hess, B. (1981a). *Biochemistry* **21,** 5950–5957.

Kuschmitz, D., and Hess, B. (1982). *FEBS Lett.* **138,** 137–140.

Lazarev, Y. A., and Terpugov, E. L. (1980). *Biochim. Biophys. Acta* **590,** 324–338.

Lemke, H.-D., and Oesterhelt, D. (1981). *FEBS Lett.* **128,** 255–260.

Lewis, A., Spoonhower, J., Bogomolni, R. A., Lozier, R. H., and Stoeckenius, W. (1974). *Proc. Natl. Acad. Sci. USA* **71,** 4462–4466.

Lozier, R. H., Bogomolni, R. A., and Stoeckenius, W. (1975). *Biophys. J.* **15,** 955–962.

Lozier, R. H., Niederberger, W., Bogomolni, R. A., Hwang, S., and Stoeckenius, W. (1976). *Biochim. Biophys. Acta* **440,** 545–556.

Lukashev, E. P., Vozary, E., Kononeko, A. A., and Rubin, A. B. (1980). *Biochim. Biophys. Acta* **592,** 258–266.

McClare, C. W. F. (1967). *Nature (London)* **216,** 766–771.

Marcus, M. A., and Lewis, A. (1978). *Biochemistry* **17,** 4722–4735.

Mathies, R., and Stryer, L. (1976). *Proc. Natl. Acad. Sci. USA* **73**, 2196–2143.

Michel, H., Oesterhelt, D., and Henderson, R. (1980). *Proc. Natl. Acad. Sci. USA* **77**, 338–342.

Moore, T. A., Edgerton, M. E., Parr, G., Greenwood, C., and Perham, R. N. (1978). *Biochem. J.* **171**, 469–476.

Motto, M. G., Sheves, M., Tsujimoto, K., Balogh-Nair, V., and Nakanishi, K. (1980). *J. Am. Chem. Soc.* **102**, 7947–7949.

Mowery, P. C., and Stoeckenius, W. (1981). *Biochemistry* **20**, 2302–2306.

Mowery, P. C., Lozier, R. H., Chae, Q., Tseng, Y-W, and Taylor, M. (1979). *Biochemistry* **18**, 4100–4107.

Nagle, J. F., and Morowitz, H. J. (1978). *Proc. Natl. Acad. Sci. USA* **75**, 298–302.

Nagy, K. (1978). *Biochem. Biophys. Res. Commun.* **85**, 383–390.

Nakanishi, K., Balogh-Nair, V., Arnaboldi, M., Tsujimoto, K., and Honig, B. (1980). *J. Am. Chem. Soc.* **102**, 7947–7949.

Narva, D. L., Callender, R. H., and Ebrey, T. G. (1981). *Photochem. Photobiol.* **33**, 567–571.

Neugebauer, D. C., Oesterhelt, D., and Zingsheim, H. P. (1978). *J. Mol. Biol.* **125**, 123–135.

Oesterhelt, D. (1972). *Hoppe-Seyler's Z. Physiol. Chem.* **353**, 1554–1555.

Oesterhelt, D., and Hess, B. (1973). *Eur. J. Biochem.* **37**, 316–326.

Oesterhelt, D., and Schuhmann, L. (1974). *FEBS Lett.* **44**, 262–265.

Oesterhelt, D., and Stoeckenius, W. (1971). *Nature New Biol.* **233**, 149–152.

Oesterhelt, D., and Stoeckenius, W. (1973). *Proc. Natl. Acad. Sci. USA* **70**, 2853–2857.

Oesterhelt, D, Meentzen, M., and Schuhmann, L. (1973). *Eur. J. Biochem.* **40**, 453–463.

Ort, D. R., and Parson, W. W. (1978). *J. Biol. Chem.* **253**, 6158–6164.

Ort, D. R., and Parson, W. W. (1979). *Biophys. J.* **25**, 341–354.

Ottolenghi, M. (1980). *Adv. Photochem.* **12**, 97–200.

Ovchinnikov, Y. A., Abdulaev, N. G., Feigina, M. Y., Kiselev, A. V., and Lovanov, N. A. (1979). *FEBS Lett.* **100**, 219–224.

Ovchinnikov, Y. A., Abdulaev, N. G., Tsetlin, V. I., Kiselev, A. V., and Zakis, V. I. (1980). *Bioorg. Khim.* **6**, 1427–1429.

Quintanilha, A. T. (1980). *FEBS Lett.* **117**, 8–12.

Rafferty, C. N. (1979). *Photochem. Photobiol.* **29**, 109–120.

Razi-Naqvi, K., Gonzalez-Rodriguez, J., Cherry, R. J., and Chapman, D. (1973). *Nature New Biol.* **245**, 249–251.

Rehorek, P. M., and Heyn, P. M. (1979). *Biochemistry* **18**, 4977–4983.

Renard, M., and Delmelle, M. (1980). *Biophys. J.* **32**, 993–1006.

Schulten, K., and Tavan, P. (1978). *Nature (London)* **272**, 85–86.

Seta, P., Ormos, P., D'Epenoux, B., and Gavach, C. (1980). *Biochim. Biophys. Acta* **591**, 37–52.

Sherman, W. V., Slifkin, M. A., and Caplan, S. R. (1976). *Biochim. Biophys. Acta* **423**, 238–248.

Sheves, M., Nakanishi, K., and Honig, B. (1979). *J. Am. Chem. Soc.* **101**, 7086–7088.

Shieh, P., and Packer, L. (1976). *Biochem. Biophys. Res. Commun.* **71**, 603–609.

Shinar, R., Druckmann, S., Ottolenghi, M., and Korenstein, R. (1977). *Biophys. J.* **19**, 1–5.

Slifkin, M. A., and Caplan, S. R. (1975). *Nature (London)* **253**, 56–58.

Slifkin, M. A., Garty, H., Sherman, W. V., Vincent, M. P., and Caplan, S. R. (1979). *Biophys. Struct. Mech.* **5**, 313–320.

Stockburger, M., Klusmann, W., Gattermann, H., Massig, G., and Peters, R. (1979). *Biochemistry* **18**, 4886–4900.

Stoeckenius, W. (1980). *Acc. Chem. Res.* **13**, 337–344.

Stoeckenius, W., and Kunau, W. H. (1968). *J. Cell Biol.* **38**, 337–357.

Stoeckenius, W., Lozier, R. H., and Bogomolni, R. A. (1979). *Biochim. Biophys. Acta* **505**, 215–278.

Terner, J., Hsieh, C.-L., Burns, A. R., and El-Sayed, M. A. (1979). *Biochemistry* **18**, 3629–3634.

Tokutomi, S., Iwasa, T., Yoshizawa, T., and Ohnishi, S. (1980). *FEBS Lett.* **114**, 145–148.

Trissl, H.-W., and Montal, M. (1977). *Nature (London)* **266**, 655–657.

Tsuda, M., and Ebrey, T. G. (1980). *Biophys. J.* **30**, 149–157.

Tsuda, M., Glaccum, M., Nelson, B., and Ebrey, Th. G. (1980). *Nature* **287**, 361–363.

Tsuji, K., and Neumann, E. (1981). *Int. J. Biol. Macromol.* **3**, 231–242.

Tsuji, K., and Rosenheck, K. (1979a). In *Electro-Optics and Dielectrics of Macromolecules and Colloids* (B. R. Jennings, ed.), pp. 77–88, Plenum Press, New York.

Tsuji, K., and Rosenheck, K. (1979b). *FEBS Lett.* **98**, 368–372.

Warshel, A. (1978). *Proc. Natl. Acad. Sci. USA* **75**, 2558–2562.

Warshel, A. (1979). *Photochem. Photobiol.* **30**, 285–290.

132

Light-Induced Aldimine Bond Migration as a Possible Mechanism for Proton Transfer in Bacteriorhodopsin

Yuri A. Ovchinnikov and Najmutin G. Abdulaev

Bacteriorhodopsin (BR) is the only protein component of the purple membrane of some halophilic microorganisms. It functions as the light-driven proton pump establishing an electrochemical gradient of hydrogen ions across the cell membrane (Stoeckenius *et al.*, 1979).

Determination of the primary structure of BR and elucidation of its topography in the native purple membrane stimulated the study of its functioning as the proton pump. In the course of this work a number of interesting features were elucidated, typical of the integral membrane transport proteins (Ovchinnikov *et al.*, 1979). In particular, it was shown that the polypeptide chain of the protein forms seven α-helical rods spanning the membrane. The chromophore attached via the aldimine bond to the Lys-41 residue is located at the second rod, nearer to the cytoplasmic surface of the membrane. The majority of basic amino acids are placed in the vicinity of the membrane surface and can stabilize the protein structure interacting with the negatively charged phosphate and sulfogroups of lipid residues. Of considerable interest is the location of the negatively charged carboxyl groups, the majority of which are in the vicinity of the cytoplasmic surface of the membrane (Fig. 1). Formation of polarizable hydrogen bonds between these carboxyl groups could mediate the uptake of protons on the cytoplasmic surface of the membrane and serve as the first step in a proton-conducting hydrogen-bonded system. It is noteworthy that practically all tryptophan residues are located nearer to the outside of the membrane. Structural asymmetry of BR in the membrane is undoubtedly an important factor providing vectorial character for the proton translocation. If BR functions as a light-driven proton pump, it must provide the ways for protons to cross the membrane. This translocation of protons can be theorized to occur in three possible ways: (1) through the aqueous pore, which should

Yuri A. Ovchinnikov and Najmutin G. Abdulaev • Shemyakin Institute of Bioorganic Chemistry, USSR Academy of Sciences, 117988 Moscow V-334, USSR.

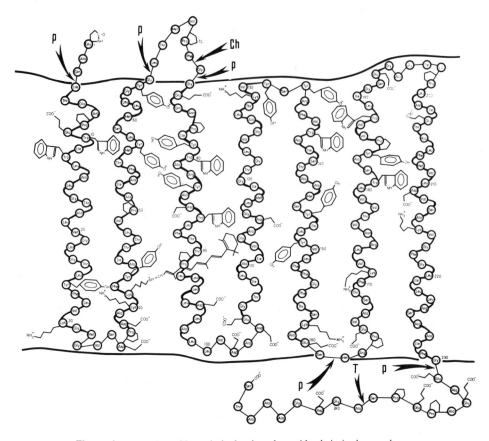

Figure 1. Disposition of bacteriorhodopsin polypeptide chain in the membrane.

include a proton-specific gating mechanism; (2) by means of the proton-binding group, which could be translocated from one membrane surface to the other by the rotation of the protein molecule or by its conformational rearrangement; and (3) by means of the proton-conducting channel situated inside the protein molecule or near its surface.

The first possibility, which could not be excluded from the available model due to its low (7 Å) resolution, was shown to be inconsistent with neutron scattering data (Zaccai and Gilmore, 1979). It is also hard to imagine rapid rotation of the BR molecule in the crystal structure of the purple membrane, for such a process is not observed even in fluid membranes.

Thus, proton transfer in the BR molecule can proceed through the proton-specific channel formed by the chain of hydrogen bonds (Nagle and Morowitz, 1978).

Dunker (Dunker and Marvin, 1978) postulated the existence of a continuous chain of hydrogen bonds capable of effective translocation of protons from the cytoplasmic side to the outer surface of the membrane. To avoid short-circuit one should assume at least one interruption in this chain. In BR this could be done by the Schiff base moiety, which is not accessible to the aqueous environment (Stoeckenius *et al.*, 1979) and in principle could function as a "shuttle" catching the proton from the cytoplasm and translocating it across the hydrophobic barrier. The discovery of aldimine bond migration during the photochemical cycle of BR provides evidence in favor of this mechanism.

We found that a migration of the aldimine bond takes place in the course of the BR photochemical cycle. It was also shown that, depending on the $NaBH_4$ reduction of BR or its modified derivatives in the light or in the dark, the retinyl moiety is attached to ϵ-amino groups of different lysine residues. Such a phenomenon is not an artifact of the reduction conditions, but is a consequence of the photochemically induced aldimine migration (Ovchinnikov *et al.*, 1980).

Our experimental approach relies upon analysis of the products formed on limited proteolysis of native (Abdulaev *et al.*, 1978) or reduced BR catalyzed by chymotrypsin or papain. By polyacrylamide gel electrophoresis two fragments are obtained in both cases: fragments 1–71 and 72–247 (248) with chymotrypsin; 3–65 and 73–230 (231) with papain [numbering according to Ovchinnikov *et al.* (1979) or—in parentheses—according to Khorana *et al.* (1980)]. In each case, these fragments are termed ''small'' and ''large,'' respectively. Intense fluorescence of the retinyl residue enabled its facile detection on electrophoretograms both in retinyl-bacterioopsin and in the above-mentioned fragments.

After $NaBH_4$ reduction of BR in the light (in all experiments a 250-W halogen lamp was used along with a 3-cm layer of 10% $CuSO_4$ and a yellow-orange glass filter) at pH 9.5 and 20°C (Schreckenbach *et al.*, 1977), retinyl moiety was found in the small and large fragments. The data obtained indicate that the ϵ-amino group of Lys-41 is not the only site for covalent attachment of retinal to the BR polypeptide chain and that an aldimine bond may be formed by a lysine residue in a small fragment (apparently Lys-41) or by a large-fragment lysine residue (Lys-215). This conclusion was confirmed by isolating the ϵ-N-retinyl derivative of H-Ala-Lys-OH from the pronase digest of fragment 72–247.

After cleavage with chymotrypsin or papain, BR preparations essentially preserve the spectral and functional properties of the native molecule, but acquire the $NaBH_4$ reducibility. We capitalized on this fact to obtain information about the aldimine localization in dark conditions. In dark preparations the retinyl moiety is associated exclusively with a small fragment. After reduction of native or proteolyzed BR in the light, fluorescence was detected in both fragments. These data testify to the $N \rightarrow N$ retinal migration in the course of photochemical transformations of BR.

After dark reduction of proteolyzed BR, previously adapted to light or dark, or after reduction of preparations regenerated from proteolyzed BR with 13-*cis*- or all-*trans*-retinal, all the fluorescence is found only in the small fragment. This fact means that in the dark, regardless of the double bond configuration, retinal is attached to Lys-41, and aldimine formation with a new lysine residue (Lys-215) in the light is not caused by retinal isomerization during the photocycle.

BR reduction in the light was also carried out under conditions in which an intermediate of the photocyle (so-called M_{412} form or its congeners) having deprotonated aldimine bond is predominant. Stabilization of such a form is achieved by adding gramicidin S (Melnik *et al.*, 1979) or by lowering the temperature (Peters *et al.*, 1976). In both cases, reduction proceeds at a considerably higher rate and results in localization of retinyl on Lys-215. It might be possible that under these conditions both Lys-41 and Lys-215 aldimines are present, but only the latter is reducible by $NaBH_4$. However, because Lys-41 aldimine in proteolyzed BR is accessible for $NaBH_4$ in the dark as well as in the light, localization of the retinyl moiety only in a large fragment when reduction is carried out in the light under conditions favoring M_{412} stabilization, is indicative of the absence of Lys-41 aldimine and of complete migration of aldimine onto Lys-215. Thus, a correlation exists between the aldimine bond deprotonation and its migration.

The migration may proceed via aldimine hydrolysis and subsequent reaction of released retinal with another lysine amino group [earlier, a photoinduced hydrolysis of the aldimine bond was discovered and a hypothesis about possible migration was discussed (Shkrob *et al.*, 1978)], or via a direct attack of aldimine by a spatially proximal amino group leading to a tetrahedral intermediate. We found that trifluoroacetylation or succinylation of the exposed amino groups in BR fails to prevent the aldimine migration, thus allowing a conclusion to be made that the Lys-215 amino group, inaccessible for modification, is a suitable acceptor for migrating retinal, probably being in a hydrophobic environment and having a low pK.

Collectively, the data on reduction of BR and its modified derivatives clearly show that migration is not a random formation of aldimines with any of the accessible amino groups induced by the pH of the medium or by the presence of NaBH$_4$. The retinyl residue occupies a chromophoric site in BR when it is bound either to Lys-41 or to Lys-215, for none of the reduced preparations regenerates with added all-*trans*-retinal. However, marked differences exist in the microenvironment of retinyl residues attached to different lysines: the characteristic fine structure and bathochromic shift of the absorption band are seen only for Lys-215 retinyl.

A preference for aldimine localization on Lys-41 in the dark and a capability of photoinduced migration onto Lys-215 are characteristic of native BR and may be lost under certain conditions affecting the relative stability of corresponding aldimines. For example, after light and dark reduction, retinyl is bound only to Lys-41 when analogs modified at the retinal C-4 position are reduced, and only to Lys-215 when proteolyzed BR is reduced at acidic pH with NaBH$_3$CN.

Aldimine bond migration reported in this communication provides evidence that at least two amino groups, Lys-41 and Lys-215 (Lys-216), which are separated in the amino acid sequence and belong to two different α-helical segments, are involved in the BR photocycle. This finding should be taken into consideration when interpreting the data on physicochemical studies of BR or the mechanisms of its function, wherein retinal aldimine is given a key role. A possibility of protonating and deprotonating different aldimines, formed by amino groups of different residues and having different microenvironments, seems very advantageous for realization of a proton transfer across the purple membrane. A photoinduced migration of the aldimine bond may be a component part of the mechanism of BR functioning as a proton pump.

REFERENCES

Abdulaev, N. G., Feigina, M. Y., Kiselev, A. V., Ovchinnikov, Y. A., Drachev, L. A., Kaulen, A. D., Khitrina, L. V., and Skulachev, V. P. (1978). *FEBS Lett.* **90**, 190–194.

Dunker, A. K., and Marvin, D. A. (1978). *J. Theor. Biol.* **72**, 9–16.

Khorana, H. G., Gerber, G. E., Harlihy, W. G., Gray, C. P., Anderegg, R. J., Nihei, K., and Biemann, K. (1980). *Proc. Natl. Acad. Sci. USA* **76**, 5046–5050.

Melnik, E. I., Chizhov, I. V., Skopinskaya, S. N., Snezhkova, L. G., and Miroshnikov, A. I. (1979). *Abstracts of the First Soviet–Swiss Symposium: Biological Membranes: Structure and Function,* Tbilisi, p. 75.

Nagle, J. F., and Morowitz, H. J. (1978). *Proc. Natl. Acad. Sci. USA* **75**, 298–301.

Ovchinnikov, Y. A., Abdulaev, N. G., Feigina, M. Y., Kiselev, A. V., and Lobanov, N. A. (1979). *FEBS Lett.* **100**, 219–224.

Ovchinnikov, Y. A., Abdulaev, N. G., Tsetlin, V. I., Kiselev, A. V., and Zakis, V. I. (1980). *Bioorg. Khim.* **6**, 1427–1429.

Peters, I., Peters, R., and Stoeckenius, W. (1976). *FEBS Lett.* **61**, 128–134.

Schreckenbach, T., Walckhoff, B., and Oesterhelt, D. (1977). *Eur. J. Biochem.* **76**, 499–511.

Shkrob, A. M., Radionov, A. V., and Ovchinnikov, Y. A. (1978). *Bioorg. Khim.* **4**, 354–359.

Stoeckenius, W., Lozier, R. H., and Bogomolni, R. A. (1979). *Biochim. Biophys. Acta* **505**, 215–278.

Zaccai, G., and Gilmore, D. J. (1979). *J. Mol. Biol.* **132**, 181–191.

133

Light-Energy Transduction in Halobacterium halobium

Thomas G. Ebrey

1. INTRODUCTION

Under conditions of low oxygen tension, *Halobacterium halobium* inserts into its plasma membrane patches of a specialized membrane, the purple membrane. Three attributes of this new type of plasma membrane make it extremely interesting. First, the purple membrane consists of only a single protein—bacteriorhodopsin—organized in a rigid fashion with high symmetry, in the manner of a two-dimensional crystal. A wide variety of physical techniques, including especially the elegant electron-density mapping of the molecule using electron microscopy, have provided a great deal of structural information about bacteriorhodopsin (Henderson and Unwin, 1975). Second, the chromophore responsible for light absorption is retinal, the chromophore of the visual pigment. Both chromophores are covalently linked to their apoproteins as protonated Schiff bases. It is believed that the action of light in both pigment systems is to cause a photoisomerization about one of the double bonds of the retinal (Honig *et al.*, 1979). Finally, bacteriorhodopsin, by acting as a light-driven proton pump, can convert light energy into stored chemical free energy in the form of a proton gradient (see review by Stoeckenius *et al.*, 1979).

Recently, a second light-energy transduction process in halobacteria has been discovered, which seems to involve the light-driven expulsion of sodium from the interior of the cells (Lindley and MacDonald, 1979; Matsuno-Yagi and Mukohata, 1980). A second retinal-based pigment, halorhodopsin, has been implicated. While it is likely that the ion and proton movements associated with this second pigment will be important in understanding the physiology of the intact halobacteria, the probable location of halorhodopsin in the "normal" plasma membrane of the cell suggests that it may have many of the characteristics of a "normal" membrane protein. I will be concerned primarily with the purple membrane of halobacteria because of the exciting possibility of eventually having detailed

Thomas G. Ebrey • Department of Physiology and Biophysics, University of Illinois, Urbana, Illinois 61801.

structural information from this unusual membrane to correlate with its light-energy-transducing properties.

2. STRUCTURAL STUDIES

So far, structural studies of the Henderson–Unwin type at 7-Å resolution have shown that bacteriorhodopsin is clustered in trimers in the membrane and that each bacteriorhodopsin molecule probably consists of seven α helices spanning the membrane (see, however, Glaeser *et al.*, 1981). The determination of the amino acid sequence of bacteriorhodopsin (Ovchinnikov *et al.*, 1979; Khorana *et al.*, 1979) has led to attempts to fold the 248 amino acids into the seven α helices seen in the electron-density maps (e.g., Engelman *et al.*, 1980). Because the protein traverses the membrane seven times, the NH_2 and COOH termini are on different sides of the membrane; the COOH terminus has been localized to the cytoplasmic side (Gerber *et al.*, 1977). A large number of other properties of the purple membrane have been correlated with either the cytoplasmic or the extracellular side of the membrane (Zingsheim *et al.*, 1978). Engelman *et al.* (1980) have tried to match the seven helical segments of the amino acid sequence with the seven helices seen in the electron-density map. Already some tests of these matches have been made by difference neutron diffraction measurements with deuterium-labeled amino acids (Engelman and Zaccai, 1980).

3. PHOTOCHEMISTRY

Light, absorbed by the chromophore of bacteriorhodopsin, causes a photochemical alteration in the pigment with the formation of the metastable primary photochemical product. This primary photoproduct then decays in the dark through a series of spectroscopically distinct intermediates until a cycle is completed and the original bacteriorhodopsin is recovered. At room temperature, this photocycle is complete in about 20 msec (pH 7.0). Several intermediates in the photocycle have been identified and studied (Lozier *et al.*, 1975). In order of appearance they are:

$$\text{bacteriorhodopsin } (BR) \overset{h\nu}{\rightarrow} K \rightarrow L \rightarrow M \rightarrow O$$

There are numerous complications in this simple scheme. For example, there is good evidence that there is more than one form of the M intermediate and that part of the L intermediate can decay directly to bacteriorhodopsin (Lozier *et al.*, 1978; Iwasa *et al.*, 1980; Kalisky *et al.*, 1981).

Raman spectroscopy has shown what is suggested by absorption spectroscopy—that bacteriorhodopsin, K, L, and O are protonated Schiff bases, while M is an unprotonated Schiff base (Lewis *et al.*, 1974; Aton *et al.*, 1977; Terner *et al.*, 1979; Stockburger *et al.*, 1979; Narva *et al.*, 1981). The primary photochemical event has been studied by low temperature (Lozier *et al.*, 1975; Hurley *et al.*, 1977) and picosecond spectroscopy (Applebury *et al.*, 1978). The similarities between the primary photochemistry of bacteriorhodopsin and visual pigments include (1) the primary photochemical product being a protonated Schiff base, (2) the absorption spectrum of the primary photoproduct being shifted to longer wavelengths than the parent pigment (bathochromically shifted), and (3) photoreversibility between the primary photoproduct and the parent pigment. This strongly suggests that the effect of light in forming the primary photochemical product in the two cases is similar—to isomerize the chromophore (Rosenfeld *et al.*, 1977). Extraction of the chromophores of the L and M intermediates of the photocycle shows that they are in the 13-*cis* conformation, suggesting that light has isomerized the chromophore to this conformation from all-*trans* in bacteriorhodopsin (Pettei *et al.*, 1977; Tsuda *et al.*, 1980; see also Braiman and Mathies, 1980). The quantum efficiency of this photoisomerization is about 0.28 ± 0.05. The sum of the efficiencies of the forward and back reactions is about 1, suggesting that bacteriorhodopsin and K reach their ground state from a common, barrierless, excited state (Hurley *et al.*, 1977).

Because the photocycle proceeds spontaneously after the formation of the primary photochemical product, the free energy to drive the photocycle and the concomitant proton translocations must reside in the primary photochemical product. Thus, the operational role of light is to take a pigment with low free energy, bacteriorhodopsin, and transform it into a pigment with high free energy, the bathointermediate K (Rosenfeld *et al.*, 1977). The molecular properties of K that distinguish it from bacteriorhodopsin and give it its high free energy are not known. One particularly simple hypothesis is that the isomerization event that characterizes the bacteriorhodopsin to K transformation leads to the increased free energy of K by a charge-separation mechanism (Honig *et al.*, 1979). The hypothesis is based on the assumption that, in the interior of a protein, a charged species like the protonated Schiff base of the chromophore would be expected to form a charge pair with a negatively charged amino acid acting as a counterion. Upon isomerization, the positive charge on the nitrogen of the chromophore would move away from the negative charge of the counterion. The resulting metastable species (K) would have higher free energy than the pigment because of the coulombic charge separation.

4. PROTON PUMPING

The actual coupling of the bacteriorhodopsin photocycle to its proton-pumping function is not at all well understood. Neither the release nor the rebinding of protons from purple membrane can be kinetically correlated with the formation or decay of any of the photocycle intermediates. For example, at low pH, the purple membrane releases its protons before significant quantities of M (the deprotonated intermediate) are formed and picks them up again more rapidly than M decays (see e.g. Ort and Parson, 1978). Moreover, it seems that both the release and the rebinding of protons follow single-exponential kinetics, while the formation and decay of the M intermediate, which occur at roughly the same time scale, are not single exponentials.

Several different preparations have been used to study the light-induced proton movements associated with the purple membrane. When intact cells of halobacteria are illuminated, a complex set of pH transients are recorded extracellularly. The major features are a transient alkalinization followed by an acidification of the external medium, which lasts as long as the light remains on (see e.g. Bogomolni *et al.*, 1976, Fig. 1a). The prolonged acidification is due to the light-induced transport of protons from the interior to the exterior of the cell. That there actually is a vectorial movement of protons due to light absorbed by the purple membrane was shown best by the experiments using cell envelope vesicles (broken cells that are resealed), where light caused an acidification of the external medium (Lozier *et al.*, 1976). When the purple membrane was incorporated into phospholipid vesicles, light caused an alkalinization of the medium. Electron microscopy showed that the side of the purple membrane facing outwards in the cell envelope vesicles was reversed so that it faced inward in the phospholipid/purple membrane reconstituted vesicles. Thus, in the cell envelope vesicles, light causes a release of protons at the extracellular surface and an uptake at the cytoplasmic surface. The bacteriorhodopsin molecules can be turned over many times so that many protons per bacteriorhodopsin can be transported. The resulting efflux of protons leads to the net acidification of the external medium. It is not yet clear what the final form of the energy gradient across the cell is, but it may be that the active proton movements are converted into a potassium gradient (Stoeckenius *et al.*, 1979).

The nature of the transient alkalinization seen upon illumination of whole cells is more problematic, and several hypotheses have been offered. It seems that the alkalinization observed in whole cells is due to two distinct processes; in standard cell envelope preparations, only one of these can be easily seen (Lindley and MacDonald, 1979). It has been suggested that the transient alkalinization seen in standard cell envelope preparations is due to a rapid change in membrane potential resulting from light-induced H^+ ejection from the cell envelope vesicles, leading to an influx of sodium through an electrogenic Na^+/H^+ antiporter, causing a net influx of protons. This would cause an alkalinization of the external medium until light-induced proton pumping predominates to yield a net acidification of the external medium (Lanyi and MacDonald, 1976; Bogomolni *et al.*, 1976). In certain carefully selected cell envelope vesicle preparations (Lindley and MacDonald, 1979) as well as whole cells, a second type of proton uptake can also be seen. This alkalinization is due to the second pigment of halobacteria, halorhodopsin, which uses light energy to actively remove sodium from the cell. The protons then flow into the cell in response to the increased negative membrane potential caused by the expulsion of the positively charged sodium (Lindley and MacDonald, 1979; Matsuno-Yagi and Mukohata, 1980). Several mutants and chemical treatments tentatively help to dissect the alkalinization that is due to passive flow of protons in response to the sodium pump, from the acidification due to light-powered proton pumping of bacteriorhodopsin. Matsuno-Yagi and Mukohata (1977) found that cells of a mutant lacking bacteriorhodopsin upon illumination showed alkalinization of the medium but no acidification. Hydroxylamine treatment of wild-type cells also could abolish the acidification, presumably by selectively destroying the bacteriorhodopsin. Moreover, heating whole cells for several hours to ca. 70°C seems to denature the sodium-pump pigment while preserving bacteriorhodopsin. In this case, the transient alkalinization was abolished and only a prolonged acidification was observed (Matsuno-Yagi and Mukohata, 1980).

Because the photocycle starts with bacteriorhodopsin as a protonated Schiff base, and during the photocycle the chromophore becomes deprotonated (the M intermediate), there was a strong hope among workers in the field that the proton(s) that was pumped was the Schiff base proton. However, if this was true it would require a stoichiometry of no more than one proton pumped per M formed in the photocycle. That this may not be true was first suggested by the studies of Ort and Parson (1979) who showed that in purple membrane sheets, the number of protons released per M formed ranges from 0.8 to 1.4 depending on salt concentration. Subsequent studies in purple-membrane-containing phospholipid vesicles and whole cells (Govindjee *et al.*, 1980; Bogomolni *et al.*, 1980; Renard and Delmelle, 1980) established unequivocally that these protons are actually pumped across the cell membrane and not just released and rebound from the same side of the membrane ("Bohr" protons, see Eisenbach and Caplan, 1979). Thus, more than one proton can be pumped per M formed under physiological conditions. This unexpected stoichiometry leaves unclear the role of the Schiff base proton in pumping and will no doubt require more complicated hypotheses to try to explain how the formation of the high-free-energy intermediate K leads to the transport of more than one proton across the cell membrane (e.g., Kalisky *et al.*, 1981).

There are a number of reports of light-induced conformational changes and amino acid side-chain alterations in bacteriorhodopsin during the photocycle (e.g., Becher *et al.*, 1978; Bogomolni *et al.*, 1978). Some models have incorporated amino acid (tyrosine) ionization as part of the pumping process (see e.g. Kalisky *et al.*, 1981). A particular role in proton pumping for the COOH-terminal end of bacteriorhodopsin (its "tail" of ca. terminal 21 amino acids) has recently been suggested (Govindjee *et al.*, 1981). These workers found that clipping off the COOH-terminal end of bacteriorhodopsin with proteolytic enzymes reduced the quantum efficiency of proton release by twofold without altering the photocycle. This suggests that the COOH-terminal tail of bacteriorhodopsin is partially responsible for the high quantum yields of proton pumping of the purple membrane.

REFERENCES

Applebury, M., Peters, K., and Rentzepis, P. (1978). *Biophys. J.* **23,** 375–382.

Aton, B., Doukas, A., Callender, R., Becher, B., and Ebrey, T. (1977). *Biochemistry* **16,** 2995–2999.

Becher, B., Tokunaga, F., and Ebrey, T. (1978). *Biochemistry* **17,** 2293–2300.

Bogomolni, R., Baker, R., Lozier, R., and Stoeckenius, W. (1976). *Biochim. Biophys. Acta* **440,** 68–88.

Bogomolni, R., Stubbs, L., and Lanyi, J. (1978) *Biochemistry* **17,** 1037–1041.

Bogomolni, R., Baker, R., Lozier, R., and Stoeckenius, W. (1980). *Biochemistry* **19,** 2152–2159.

Braiman, M., and Mathies, R. (1980). *Biochemistry* **19,** 5421–5428.

Eisenbach, M., and Caplan, S. R. (1979). *Curr. Top. Membr. Transp.* **12,** 165–247.

Engelman, D., and Zaccai, G. (1980). *Proc. Natl. Acad. Sci. USA* **77,** 5894–5898.

Engelman, D., Henderson, R., McLachlan, A., and Wallace, B. (1980). *Proc. Natl. Acad. Sci. USA* **77,** 2023–2027.

Gerber, G., Gray, C., Wildenauer, D., and Khorana, H. (1977). *Proc. Natl. Acad. Sci. USA* **74,** 5426–5430.

Glaeser, R., Jap, B., Maestre, M., and Hayward, S. (1981). *Biophys. J.* **33,** 218a.

Govindjee, R., Ebrey, T., and Crofts, A. (1980). *Biophys. J.* **30,** 231–242.

Govindjee, R., Ohno, K., and Ebrey, T. (1981). Abstracts, American Society of Photobiology.

Henderson, R., and Unwin, P. N. (1975). *Nature (London)* **257,** 28–32.

Honig, B., Ebrey, T., Callender, R., Dinur, U., and Ottolenghi, M. (1979). *Proc. Natl. Acad. Sci. USA* **76,** 2503–2507.

Hurley, J., Ebrey, T., Honig, B., and Ottolenghi, M. (1977). *Nature (London)* **270,** 540–542.

Iwasa, T., Tokunaga, F., and Yoshizawa, T. V. (1980). *Biophys. Struct. Mech.* **6,** 253–270.

Kalisky, D., Ottolenghi, M., Honig, B., and Korenstein, R. (1981). *Biochemistry* **20,** 649–655.

Khorana, M., Gerber, G., Herlihy, W., Gray, C., Anderegg, R., Nihei, K., and Biemann, K. (1979). *Proc. Natl. Acad. Sci. USA* **76,** 5046–5050.

Lanyi, J., and MacDonald, R. (1978). *Biochemistry* **15,** 4608–4614.

Lewis, A., Spoonhower, J., Bogomolni, R., Lozier, R., and Stoeckenius, W. (1974). *Proc. Natl. Acad. Sci. USA* **71,** 4462–4466.

Lindley, E., and MacDonald, R. (1979). *Biochem. Biophys. Res. Commun.* **88,** 491–499.

Lozier, R., Bogomolni, R., and Stoeckenius, W. (1975). *Biophys. J.* **15,** 955–962.

Lozier, R., Niederberger, W., Bogomolni, R., Hwang, S.-B., and Stoeckenius, W. (1976). *Biochim. Biophys. Acta* **440,** 545–556.

Lozier, R., Niederberger, W., Ottolenghi, M., Sivorinovsky, G., and Stoeckenius, W. (1978). In *Energetics and Structure of Halophilic Microorganisms* (S. R. Caplan and M. Ginzburg, eds.), pp. 123–142, Elsevier/North-Holland, Amsterdam.

Matsuno-Yagi, A., and Mukohata, Y. (1977). *Biochem. Biophys. Res. Commun.* **78,** 237–243.

Matsuno-Yagi, A., and Mukohata, Y. (1980). *Arch. Biochem. Biophys.* **199,** 297–303.

Narva, D., Callender, R., and Ebrey, T. (1981). *Photochem. Photobiol.* **33,** 567–571.

Ort, D., and Parson, W. (1978). *J. Biol. Chem.* **253,** 6158–6164.

Ort, D., and Parson, W. (1979). *Biophys. J.* **25,** 341–354.

Ovchinnikov, Y., Abdulaev, N., Feigina, M., Kiselev, A., and Lobanov, N. (1979). *FEBS Lett.* **100,** 219–224.

Pettei, M., Yudd, A., Nakanishi, K., Henselman, R., and Stoeckenius, W. (1977). *Biochemistry* **16,** 1955–1959.

Renard, M., and Delmelle, M. (1980). *Biophys. J.* **32,** 993–1006.

Rosenfeld, T., Honig, B., Ottolenghi, M., Hurley, J. and Ebrey, T. (1977). *Pure Appl. Chem.* **49,** 341–351.

Stockburger, M., Klausmann, W., Gattermann, H., Massing, G., and Peters, R. (1979). *Biochemistry* **18,** 4886–4900.

Stoeckenius, W., Lozier, R., and Bogomolni, R. (1979). *Biochim. Biophys. Acta* **505,** 215–278.

Terner, J., Hsieh, C., and El-Sayed, M. (1979). *Biophys. J.* **26,** 527–542.

Tsuda, M., Glaccum, M., Nelson, B., and Ebrey, T. (1980). *Nature (London)* **287,** 351–353.

Zingsheim, H., Neugebauer, D., and Henderson, R. (1978). *J. Mol. Biol.* **123,** 275–278.

XI

Excitable Membranes

134

Reconstitution of the Acetylcholine Receptor in Lipid Vesicles and in Planar Lipid Bilayers

M. Montal and J. Lindstrom

1. INTRODUCTION

Transmission of signals from nerve to muscle is mediated through nicotinic acetylcholine receptors (AChRs). When an action potential depolarizes a motor nerve ending, acetylcholine (ACh) is released, which, after diffusing across the synaptic cleft, binds to AChRs in the postsynaptic membrane of the muscle. ACh binding causes an increase in the permeability of the postsynaptic membrane to both sodium and potassium ions, thereby reducing the membrane potential and triggering an action potential in the muscle membrane. The relative ease of pharmacological, electrophysiological, and morphological studies of muscle, and the relative abundance of AChR in fish electric organs have made the AChR by far the best characterized neurotransmitter receptor. It provides an archetype for understanding the structure and function of signal-transducing membrane proteins.

Several recent comprehensive reviews have dealt with biophysical, biochemical, and immunochemical aspects of AChRs (e.g., Colquhoun, 1979; Changeux, 1981; Karlin, 1980; Lindstrom, 1979). Before discussing AChR reconstitution, we will briefly consider some aspects of AChR function, as analyzed in muscle cells, and AChR structure, as studied using AChR purified from fish electric organs. Biochemical (Nathanson and Hall, 1979) and immunochemical studies (Lindstrom *et al.*, 1979b) suggest that AChRs from these seemingly diverse sources are fundamentally homologous in structure as well as function.

It is generally presumed that agonist binding induces a conformation change in the AChR molecule that results in opening of a cation-specific channel. The shape of dose–response curves suggests that there are two ACh binding sites per functional unit (Dionne

M. Montal • Departments of Biology and Physics, University of California at San Diego, La Jolla, California 92093.　*J. Lindstrom* • Receptor Biology Laboratory, The Salk Institute for Biological Studies, San Diego, California 92158.

et al., 1978). The channel has a functional cross section of at least a 6.5-Å square (Watanabe and Narahashi, 1979; Dwyer *et al.*, 1980; Adams *et al.*, 1980). Thus, it allows the passage of not only Na^+, but also, to some extent, organic cations the size of agonists. Electrophysiological analysis of membrane noise indicates that open channels have high conductances (20–30 pS, i.e., $\simeq 50,000$ monovalent cations/msec) (Katz and Miledi, 1972; Anderson and Stevens, 1973; Neher and Sakmann, 1976). The channel open time is of the order of 1 msec and increases somewhat with affinity of the agonist (Sakmann *et al.*, 1980). Single-channel conductances have the form of square pulses of fixed amplitude and exponentially distributed duration (Neher and Sakmann, 1976; Sakmann *et al.*, 1980). The lifetime of the channel is voltage dependent such that the probability of channel closing varies over an *e*-fold range for every 80-mV change in membrane potential (Anderson and Stevens, 1973; Magleby and Stevens, 1972; Neher and Sakmann, 1976). With each liganding there is a probability that the AChR will relax into a desensitized conformation characterized by a closed channel and high affinity for agonists. Thus, at high agonist concentrations, there is transient activation followed by unresponsiveness. Desensitized AChRs relax over seconds to the resting state (Katz and Thesleff, 1957; Sakmann *et al.*, 1980; Neubig and Cohen, 1980).

Negatively stained AChRs visualized by electron microscopy appear from the extracellular surface to have a doughnut shape about 90 Å in diameter centered around a hydrophilic pit (Cartaud *et al.*, 1978). The AChR protrudes about 50 Å on the extracellular surface, but only about 15 Å on the cytoplasmic surface (Ross *et al.*, 1977; Klymkowsky and Stroud, 1979). Both freeze fracture (Heuser and Salpeter, 1979) and tannic acid staining (Potter and Smith, 1977) suggest that the pit runs through the center of the molecule, perpendicular to the membrane, implying that it might correspond to the cation channel. At this time the best analog for the shape of the AChR molecule appears to be a mushroom with a roughly peaked asymmetric cap on the external surface of the membrane and a stem extending across the membrane (Changeux, 1981; Karlin, 1980; Zingsheim *et al.*, 1980).

AChR from *Torpedo californica* electric organ has four kinds of glycopeptide subunits. AChR monomers (molecular weight $\simeq 250,000$) consists of two α subunits ($\simeq$ 38,000) and one each of β ($\simeq 50,000$), γ ($\simeq 57,000$), and δ ($\simeq 64,000$) (Weill *et al.*, 1974; Reynolds and Karlin, 1978; Lindstrom *et al.*, 1979a; Raftery *et al.*, 1975, 1980). The two ACh binding sites per monomer appear to be formed by α subunits (Karlin *et al.*, 1975; Weill *et al.*, 1974; Weiland *et al.*, 1978). As will be explained below, the cation channel is an integral component of the monomer (Neubig *et al.*, 1979; Wu and Raftery, 1979; Changeux *et al.*, 1979; Anholt *et al.*, 1980). Local anesthetics label the δ subunit (Saitoh and Changeux, 1980). Which subunits form the channel is not known. The three larger subunits are relatively protease sensitive, whereas the α subunits are comparatively resistant, but proteolytically nicked subunits remain intimately noncovalently associated (Lindstrom *et al.*, 1980b). For a long while this led to confusion in establishing the subunit structure of the AChR, because some laboratories observed only α subunits. Sequence homologies (Raftery *et al.*, 1980) and immunological cross-reaction (Tzartos and Lindstrom, 1980) indicate that the subunits were all derived by duplication of an ancestral gene. This suggests that, despite their at least partially independent functions, they must also exhibit some structural homologies.

Reconstitution has been an important goal since the beginning of biochemical studies of AChRs a decade ago, but it has only recently been achieved. Fortunately, it turns out that the multisubunit AChR protein initially purified on the basis of its ligand binding properties (Lindstrom and Patrick, 1974; Raftery *et al.*, 1975; Karlin *et al.*, 1975; cf. Changeux, 1981) contains both the ligand binding sites and the cation channel whose permeability they regulate. What had been lacking were techniques to preserve the native

AChR conformation and to reassemble it into model membranes. The first goal of reconstitution, to determine if all the components of the AChR are in the purified monomer, has now been achieved. The remaining goals are to determine which subunits and parts of subunits are involved in opening and closing the channel, forming its structure, and regulating desensitization. Also, reconstitution techniques can help study the role of lipid in AChR integration and function.

2. RECONSTITUTION IN LIPID VESICLES

The search for procedures to reproducibly reconstitute the agonist-stimulated cation conductance of AChR into planar lipid bilayers or lipid vesicles has been characterized by failures and unreproducible reports of success (reviewed by Montal, 1976; Heidmann and Changeux, 1978; Changeux, 1981). The critical factor was the design of a method for solubilizing and reconstituting AChRs without losing their cation channels. Epstein and Racker (1978) showed that if membranes rich in AChR were solubilized in cholate–lipid mixtures and then the cholate was removed by dialysis, vesicles were formed that exhibited a rapid, carbamylcholine-dependent influx of $^{22}Na^+$, which was inhibited by α-bungarotoxin and several other antagonists. Exposure of the vesicles to carbamylcholine before adding $^{22}Na^+$ caused no specific influx, indicating that the reconstituted AChRs exhibited desensitization. The detergents routinely used to solubilize AChRs, Triton X-100 or cholate alone, preserved ligand binding, but denatured the cation channel (Lindstrom et al., 1980a). Detergent–lipid mixtures have also been used to preserve the labile binding activities of the action potential Na^+ channel (Agnew et al., 1978; Barchi et al., 1980) and may prove a valuable approach for preserving the native conformation of many membrane proteins.

Alkaline extraction (pH 11.0) at low ionic strength of highly purified membrane fragments from *T. californica* (Wu and Raftery, 1979; Moore and Raftery, 1980) and *T. marmorata* (Changeux et al., 1979) produces membranes containing virtually only the subunits characteristic of affinity-purified AChR (Neubig et al., 1979). These vesicles were solubilized in cholate plus soybean phospholipids and then reconstituted into vesicles that displayed carbamylcholine-sensitive sodium flux (Wu and Raftery, 1979; Changeux et al., 1979). Furthermore, fast kinetics studies on the interaction of a fluorescent agonist with AChR revealed that these reconstituted vesicles expressed the low-affinity state for agonist binding as well as the slow interconversion to the high-affinity state (Heidmann et al., 1980).

AChR solubilized in cholate–lipid mixtures can also be affinity purified by any of several procedures, or even purified by conventional procedures, and then can be reconstituted into vesicles that exhibit activation and desensitization (Huganir et al., 1979; Lindstrom et al., 1980a; Heidmann et al., 1980; Sobel et al., 1980). These results as well as the reconstitution of purified membrane-bound AChR demonstrate that only the four subunits characteristic of purified AChR are required for function. Reconstitution of affinity-purified AChR further shows that no lipids or other components not intimately associated with solubilized AChRs are necessary for biological activity.

Torpedo AChRs, but not the AChRs of other species, exist primarily as dimers linked by a disulfide bond between their δ subunits (Chang and Bock, 1977; Hamilton et al., 1979). Reconstituted monomers and dimers were equally active, independent of their concentration in the vesicles, and did not appear to be noncovalently associated in the membranes. This indicates that the channel is an integral component of the monomer rather than formed at the interface between monomers (Anholt et al., 1980).

The conditions required for the stabilization of the AChR in cholate-lipid mixtures upon solubilization and during the reassembly process were characterized (Anholt *et al.*, 1981). It was found that while relatively low concentrations of lipid in the micellar solution (cholate:lipid, 20:1 molar ratio) were effective in preserving channel activity, a 10-fold excess was required during removal of the detergent. The packing density of the AChRs in the reconstituted vesicles was 5-fold lower than in native membranes. Reassembly at higher AChR/lipid ratios led to denser packing but was concomitant to the irreversible inactivation of a fraction of the AChR channels. AChRs reconstitute asymmetrically. At least 70% of the AChRs in reconstituted vesicles are oriented normally, with their toxin binding sites on the external surface of the vesicles (Anholt *et al.*, 1980; Lindstrom *et al.*, 1980a). Vesicle size and AChR distribution can be manipulated to some extent: inclusion of cholesterol (20% w/w) during the reassembly process enhances vesicle fusion during a freeze-thaw cycle and produces larger vesicles with a random orientation of AChRs (Anholt *et al.*, 1982).

Reconstituted AChR vesicles are more convenient to study some aspects of AChR function than are native AChR-rich membranes. Because of the dense packing of AChRs in native membranes, equilibration of $^{22}Na^+$ across the vesicles limits the response (Moore *et al.*, 1979; Lindstrom *et al.*, 1980a). This restriction can be partially overcome by using stopped-flow techniques (Boyd and Cohen, 1980; Neubig and Cohen, 1980; Moore and Raftery, 1980; Cash *et al.*, 1980; Hess *et al.*, 1979). Because of the low concentration of AChR in reconstituted vesicles, desensitization rather than equilibration limits the response, and the total $^{22}Na^+$ influx is directly proportional to the amount of AChR (Lindstrom *et al.*, 1980a). Correlation of the fractional occupancy of ligand binding sites by *Naja naja siamensis* toxin with inhibition of receptor function was used to demonstrate that in the reconstituted system the doubly liganded receptor prevails in controlling channel gating (Anholt *et al.*, 1982), in line with results using different cholinergic systems (Sine and Taylor, 1980). This behavior is more conspicuous as vesicle leakiness and as the limitation on the available internal volume per receptor for the agonist-induced uptake of Na^+ are reduced by using (cholesterol-supplemented) larger vesicles (Anholt *et al.*, 1982).

The initial reconstitution experiments were performed with a crude mixture of soybean phospholipids (Epstein and Racker, 1978). It was then shown that removal of neutral lipids from asolectin or replacement of asolectin with a mixture of pure phospholipids yielded vesicles that were less active in carbamylcholine-induced $^{22}Na^+$ influx. However, supplementing the purified phospholipids with α-tocopherol, coenzyme Q-10, or vitamin K during the reconstitution procedure led to vesicles that were considerably more active than those reconstituted with crude phospholipid mixtures (Kilian *et al.*, 1980). This fact appears not to be due to the antioxidant properties of α-tocopherol, for other antioxidants were ineffective. It was suggested that α-tocopherol or quinones modify the packing arrangement of the phospholipid bilayer (Kilian *et al.*, 1980).

All the studies detailed before have in common the use of cholate as the solubilizing detergent. It was recently reported that AChR could also be purified in octylglucoside, and then reconstituted with *Torpedo* lipids into vesicles by dialysis (Gonsalez-Ross *et al.*, 1980). However, others have observed that octylglucoside added to cholate–lipid solubilized AChR prevents reconstitution of carbamylcholine-induced $^{22}Na^+$ uptake (Anholt *et al.*, 1981; Huganir and Racker, personal communication). Gonsalez-Ross *et al.* (1980) used more than 80-fold higher AChR-to-lipid ratios for reconstitution than the other groups. Thus, it appears that residual amounts of active AChR remaining after exposure to octylglucoside could account for the carbamylcholine-induced $^{22}Na^+$ influx reported.

The studies with AChR reconstituted in lipid vesicles show that, in contrast with the situation several years ago (for reviews see Montal, 1976; Briley and Changeux, 1977;

Changeux, 1981), AChR reconstitution is now a reproducible, easy, and efficient process. As long as the cation channel is stabilized by lipids in cholate solution, several purification methods yield active AChR incorporated in lipid vesicles. Reconstitution into vesicles can now be used as a routine assay for studying AChR function associated to biochemical studies of AChR structure.

3. RECONSTITUTION IN PLANAR LIPID BILAYERS

The ultimate criterion for reconstitution is to mimic in detail the function of native AChR. This requires measuring the function of single AChR channels. The conductance of single AChR channels can be measured electrophysiologically both in intact muscle (Neher and Sakmann, 1976; Sakmann *et al.*, 1980) and, more recently, in small patches of excised muscle membrane (Horn and Patlak, 1980). Now this is possible using AChR reconstituted into planar lipid bilayers (Schindler and Quast, 1980; Nelson *et al.*, 1980). These reports must be distinguished from previous claims of recovery of functional activity from black lipid films made with or modified by membrane extracts prepared in ways that we now know either did not extract AChRs or that denatured the channels of the extracted AChRs (for review see Montal, 1976; Heidmann and Changeux, 1978; Karlin, 1980; Changeux, 1981).

AChR has been incorporated into planar lipid bilayers from AChRs initially in the form of vesicles. Monolayers at the air–water interface were spontaneously generated from a vesicle suspension and, thereafter, a planar lipid bilayer was assembled by apposition of two such monolayers (Montal and Mueller, 1972; Schindler and Quast, 1980; Nelson *et al.*, 1980). Two sets of preparations were used: (1) native membrane vesicles rich in AChR from *T. marmorata* (Schindler and Quast, 1980); (2) reconstituted vesicles containing AChR that had been previously solubilized in lipid–cholate mixtures and purified by toxin and concanavalin A affinity chromatography (Nelson *et al.*, 1980).

In the experiments performed with native membrane vesicles (Schindler and Quast, 1980), bilayers were formed from mixed vesicle suspensions containing native vesicles and soybean lecithin plus cholesterol vesicles in various ratios. Preservation of AChR function in the planar membrane was demonstrated by several criteria: (1) Addition of carbamylcholine caused a steady-state conductance that was competitively inhibited by *d*-tubocurarine. Preincubation of native vesicles with α-bungarotoxin prevented the conductance response to carbamylcholine. (2) Asymmetric membranes were formed that displayed complete asymmetry with respect to functional ligand binding sites. (3) The time course of the conductance after a single dose of 1 μM carbamylcholine showed a pronounced maximum. This maximum together with half-saturation of conductance increase between 50 and 100 nM carbamylcholine suggest that AChR assumed a desensitized state also in the planar membrane. (4) Carbamylcholine-induced single-channel fluctuations were resolved at low concentrations of native vesicles with respect to lipid vesicles. The single-channel conductance was 90 pS in 1 M NaCl and 20 to 25 pS in 0.25 M NaCl. Channel open times ranged between 1 and 2 msec. (5) The carbamylcholine-induced conductance was cation selective, about 7 times higher for K^+ and Na^+ than for Cl^-.

The results obtained with purified AChR (Nelson *et al.*, 1980) are fundamentally analogous to those described for planar bilayers derived from native vesicles (Schindler and Quast, 1980): (1) Addition of carbamylcholine caused a large increase in membrane conductance, which spontaneously relaxed. This decay, consistent with desensitization (Katz and Thesleff, 1957), was enhanced by addition of excess carbamylcholine at the peak of the response. (2) The carboamylcholine-induced increase in membrane conductance was inhibited by curare. (3) Asymmetric bilayers formed by apposing two monolayers composed of AChR–lipid and lipid without AChR showed asymmetric response to carbamylcholine. (4) The carbamylcholine-induced increase in membrane conductance was accompanied by an increase in the conductance noise. Fluctuation analysis indicated a single-channel conductance of 16 pS (in 0.1 M NaCl) with a mean channel open time of ∼35 msec. Single-channel conductance fluctuations of 60 pS (in 0.5 M NaCl) and open time of ∼50 msec were recorded. These initial results were followed by a detailed characterization of the single channel properties (Labarca *et al.*, 1981, 1982). Analysis of the fluctuations in membrane conductance due to the opening and closing of single AChR channels shows that the lifetime of the open state (τ_0) depends on the agonist used, following the sequence τ_0, Suberyldicholine $> \tau_0$ Ach $> \tau_0$ Carbachol. τ_0 is voltage-dependent, doubling each 80 mV of membrane hyperpolarization. Channel fluctuations occur in "bursts" represented by the sudden appearance of channel activity followed by quiescent periods. Burst duration depends on the agonist used, being longest in the presence of Suberyldicholine. The open state of the AChR channel is blocked by the local anesthetic QX-222 in analogy to its action on muscle AChR. Thus, the agonist-regulated ion channel of the purified AChR displays the pharmacological specificity expected of this receptor in native membranes.

Reconstitution of AChRs into planar bilayers is still technically laborious but it offers great potential. Although planar bilayers are not as amenable to biochemical studies as are vesicles, in planar bilayers AChR activity can be measured with a sensitivity and time resolution inaccessible in vesicles. A combined approach using both vesicles and planar bilayers emerges as potentially fruitful, especially because the planar bilayers can be formed directly from the vesicles.

4. PROSPECTS FOR THE FUTURE

Studies of AChR reconstituted into both vesicles and planar bilayers should begin to contribute significantly to studies of how the structure of AChR produces its function. The greatest advantage of reconstituted AChRs over native AChRs for studying function is that in a reconstituted system the protein, lipid, and ionic components can be completely defined. What is needed are probes for relating structure to function. An approach we are taking to this problem is to use monoclonal antibodies directed at many defined parts of the AChR macromolecule (Gullick *et al.*, 1981; Conti-Tronconi *et al.*, 1981; Lindstrom, 1979; Lindstrom *et al.*, 1981). We hope that monoclonal antibodies can be used as molecular templates to compare both the structure and the function of AChRs from electric organ and muscle.

Acknowledgments. Work from the authors' laboratories was supported by grants from the Office of Naval Research (N00014-79-C-0798 to M.M. and J.L.), the National Institutes of Health (EY 02084 to M.M. and NS 11323 to J.L.), and the Muscular Dystrophy Association (to J.L.).

REFERENCES

Adams, D. J., Dwyer, T. M, and Hille, B. (1980). *J. Gen. Physiol.* **75**, 493–510.
Agnew, W., Levinson, S. R., Brabson, J. S., and Raftery, M. A. (1978). *Proc. Natl. Acad. Sci. USA.* **75**, 2606–2608.

Anderson, C. R., and Stevens, C. F. (1973). *J. Physiol. (London)* **235**, 655–691.

Anholt, R., Lindstrom, J., and Montal, M. (1980). *Eur. J. Biochem.* **209**, 481–487.

Anholt, R., Lindstrom, J., and Montal, M. (1981). *J. Biol. Chem.* **256**, 4377–4387.

Anholt, R., Fredkin, D. R., Deerinck, T., Ellisman, M., Montal, M., and Lindstrom, J. (1982). *J. Biol. Chem.*, in press.

Barchi, R. L., Cohen, S. A., and Murphy, L. E. (1980). *Proc. Natl. Acad. Sci. USA* **77**, 306–1310.

Boyd, N., and Cohen, J. B. (1980). *Biochemistry* **19**, 5344–5353.

Briley, M. S., and Changeux, J.-P. (1977). *Int. Rev. Neurobiol.* **20**, 31–63.

Cartaud, J., Benedetti, E., Sobel, A., and Changeux, J.-P. (1978). *J. Cell Sci.* **29**, 313–337.

Cash, D., Aoshima, H., and Hess, G. (1980). *Biochem. Biophys. Res. Commun.* **95**, 1010–1016.

Chang, H. W., and Bock, E. (1977). *Biochemistry* **16**, 4513–4520.

Changeux, J.-P. (1981). *Harvey Lectures,* pp. 85–254. Academic Press, New York.

Changeux, J.-P., Heidmann, T., Popot, J. L., and Sobel, A. (1979). *FEBS Lett.* **195**, 181–187.

Colquhoun, D. (1979). In *The Receptors* (R. D. O'Brien, ed.), Vol. 1, pp. 93–142, Plenum Press, New York.

Conti-Tronconi, B., Tzartos, S., and Lindstrom, J. (1981). *Biochemistry* **20**, 2181–2191.

Dionne, V. E., Steinbach, J. H., and Stevens, C. F. (1978). *J. Physiol. (London)* **281**, 421–444.

Dwyer, T. M., Adams, D. J., and Hille, B. (1980). *J. Gen. Physiol.* **75**, 469–492.

Epstein, M., and Racker, E. (1978). *J. Biol. Chem.* **253**, 6660–6662.

Gonsalez-Ross, J. M., Paraschos, A., and Martinez-Carrion, M. (1980). *Proc. Natl. Acad. Sci. USA* **77**, 1796–1800.

Gullick, B., Tzartos, S., and Lindstrom, J. (1981). *Biochemistry* **20**, 2173–2180.

Hamilton, S. L., McLaughlin, M., and Karlin, A. (1979). *Biochemistry* **18**, 155–163.

Heidmann, T., and Changeux, J.-P. (1978). *Annu. Rev. Biochem.* **47**, 317–357.

Heidmann, T., Sobel, A., Popot, J.L., and Changeux, J.-P. (1980). *Eur. J. Biochem.* **110**, 35–55.

Hess, G. P., Cash, D. J., and Aoshima, H. (1979). *Nature (London)* **282**, 329–331.

Heuser, J., and Salpeter, S. (1979). *J. Cell Biol.* **82**, 150–173.

Horn, R., and Patlak, J. (1980). *Proc. Natl. Acad. Sci. USA* **77**, 6930–6934.

Huganir, R. L., Shell, M. A., and Racker, E. (1979). *FEBS Lett.* **108**, 155–160.

Karlin, A. (1980). In *Cell Surface Reviews* (G. Poste, G. Nicholson, and C. N. Cotman, eds.), Vol. VI, pp. 191–260, Elsevier/North-Holland, Amsterdam.

Karlin, A., Weill, C. L., McNamee, M. G., and Valderrama, R. (1975). *Cold Spring Harbor Symp. Quant. Biol.* **40**, 203–213.

Katz, B., and Miledi, R. (1972). *J. Physiol. (London)* **224**, 665–699.

Katz, B., and Thesleff, S. (1957). *J. Physiol. (London)* **138**, 63–80.

Kilian, P. L., Dunlap, C. R., Mueller, P., Schell, M. A., Huganir, R. L., and Racker, E. (1980). *Biochem. Biophys. Res. Commun.* **93**, 409–414.

Klymkowsky, W., and Stroud, R. M. (1979). *J. Mol. Biol.* **128**, 319–334.

Labarca, P., Lindstrom, J., and Montal, M. (1981). *VII Intl. Biophys. Congr. and III Pan American Biochem. Congr. Mexico.* pp. 258.

Labarca, P., Lindstrom, J., and Montal, M. (1982). *Biophys. J.* **37**, 170a.

Lindstrom, J. (1979). *Adv. Immunol.* **27**, 1–50.

Lindstrom, J., and Patrick, J. (1974). In *Synaptic Transmission and Nerve Interaction* (W. C. Bennett, ed.), pp. 191–216, Raven Press, New York.

Lindstrom, J., Merlie, J., and Yogeeswaran, G. (1979a). *Biochemistry* **18**, 4465–4470.

Lindstrom, J., Walter, B., and Einarson, B. (1979b). *Biochemistry* **18**, 4470–4480.

Lindstrom, J., Anholt, R., Einarson, B., Engel, A., Osame, M., and Montal, M. (1980a). *J. Biol. Chem.* **255**, 8340–8350.

Lindstrom, J., Gullick, W., Conti-Tronconi, B., and Ellisman, M. (1980b). *Biochemistry* **19**, 4791–4795.

Lindstrom, J., Tzartos, S., and Gullick, W. (1981). *Ann. N.Y. Acad. Sci.*, in press.

Magleby, K. L., and Stevens, C. F. (1972). *J. Physiol. (London)* **223**, 173–197.

Montal, M. (1976). *Annu. Rev. Biophys. Bioeng.* **5**, 119–175.

Montal, M., and Mueller, P. (1972). *Proc. Natl. Acad. Sci. USA* **69**, 3561–3566.

Moore, H.-P., and Raftery, M. A. (1980). *Proc. Natl. Acad. Sci. USA* **77**, 4509–4513.

Moore, H.-P., Hartig, P. R., and Raftery, M. A. (1979). *Proc. Natl. Acad. Sci. USA* **76**, 6265–6269.

Nathanson, N. M., and Hall, Z. W. (1979). *Biochemistry* **18**, 3392–3401.

Neher, E., and Sakmann, B. (1976). *Nature (London)* **260**, 799–802.

Nelson, N., Anholt, R., Lindstrom, J., and Montal, M. (1980). *Proc. Natl. Acad. Sci. USA* **77**, 3057–3061.

Neubig, G. R., and Cohen, J. B. (1980). *Biochemistry* **19**, 2770–2779.

Neubig, G. R., Krodel, E. K., Boyd, N. D., and Cohen, J. B. (1979). *Proc. Natl. Acad. Sci. USA* **76**, 690–694.

Potter, L., and Smith, O. (1977). *Tissue Cell* **9**, 585–594.

Raftery, M. A., Vandlen, R. L., Reed, K. L., and Lee, T. (1975). *Cold Spring Harbor Symp. Quant. Biol.* **40**, 193–202.

Raftery, M. A., Hunkapiller, M. W., Strader, C. D., and Hood, L. A. (1980). *Science* **208**, 1454–1457.

Reynolds, J., and Karlin, A. (1978). *Biochemistry* **17**, 2035–2038.

Ross, M., Klymkowsky, W., Agard, D., and Stroud, R. (1977). *J. Mol. Biol.* **116**, 635–659.

Saitoh, T., and Changeux, J.-P. (1980). *Eur. J. Biochem.* **105**, 51–62.

Sakmann, B., Patlak, J., and Neher, E. (1980). *Nature (London)* **286**, 71–73.

Schindler, H. G., and Quast, U. (1980). *Proc. Natl. Acad. Sci. USA* **77**, 3052–3056.

Sine, S. M., and Taylor, P. (1980). *J. Biol. Chem.* **255**, 10144–10156.

Sobel, A., Heidmann, T., Cartaud, J., and Changeux, J.-P. (1980). *Eur. J. Biochem.* **110**, 13–33.

Tzartos, S. J., and Lindstrom, J. (1980). *Proc. Natl. Acad. Sci. USA* **77**, 755–759.

Watanabe, S., and Narahashi, T. (1979). *J. Gen. Physiol.* **74**, 615–628.

Weiland, G., Freeman, D., and Taylor, P. (1978). *Mol. Pharmacol.* **15**, 213–226.

Weill, C. L., McNamee, M. G., and Karlin, A. (1974). *Biochem. Biophys. Res. Commun.* **61**, 997–1003.

Wu, W. C. S., and Raftery, M. A. (1979). *Biochem. Biophys. Res. Commun.* **89**, 26–35.

Zingsheim, H. P., Neugebauer, V.-C., Barrantes, F. J., and Frank, J. (1980). *Proc. Natl. Acad. Sci. USA* **77**, 952–956.

135

Acetylcholinesterase

The Relationship of Protein Structure to Cellular Localization

Terrone L. Rosenberry

1. INTRODUCTION

Acetylcholinesterase (EC 3.1.1.7) is associated primarily with cells involved in cholinergic synaptic transmission, but it is also found in a variety of other neuronal and a few nonneuronal cells (Nachmansohn, 1959). The catalytic properties of acetylcholinesterase have been studied intensively for more than 40 years, and details about its catalytic mechanism and about features that distinguish it from the similar enzyme cholinesterase (EC 3.1.1.8) can be found in many reviews (eg., Froede and Wilson, 1971; Rosenberry, 1975; Massoulié, 1980). This review will focus on an area of recent intense interest, the relationship of acetylcholinesterase protein structure to its cellular localization. Two contributions have significantly influenced current research directions. In 1969, Massoulié and Rieger demonstrated that fresh extracts of eel electric organ contained multiple forms of acetylcholinesterase that could be distinguished by their sedimentation coefficients. This report led to similar investigations of many nerve and muscle cells, and in extracts of most of these cells more than one sedimenting acetylcholinesterase form could be identified. In 1973, Hall found multiple forms in extracts of rat diaphragm. One of these forms, a 16 S species, was specifically associated with sections of the diaphragm that are rich in neuromuscular junctions. This 16 S form decreased to very low levels on denervation. Hall's observations were confirmed and extended in studies in other laboratories as noted below, but they were particularly significant for two reasons. They suggested that differences in the various acetylcholinesterase forms were related to their cellular localizations, and they indicated that

Terrone L. Rosenberry • Department of Pharmacology, Case Western Reserve University, Cleveland, Ohio 44106.

the levels of at least one form specifically localized at skeletal neuromuscular junctions could be maintained by innervation of the muscle.

2. ACETYLCHOLINESTERASE FORMS WITH COLLAGEN-LIKE TAIL STRUCTURES

Acetylcholine has an extremely high turnover rate with acetylcholinesterase, and this catalytic efficiency has permitted the resolution of various sedimenting forms even when they are present in minute amounts in crude cellular extracts. Examination of the protein structure of a particular sedimenting form requires the isolation of that form in milligram amounts, and it is for this reason that the acetylcholinesterase forms from electric organs of several *Torpedo* species and of the eel *Electrophorus electricus* (Nachmansohn and Lederer, 1939) are by far the most thoroughly characterized. For example, electric organs from the eel contain approximately 50–100 mg acetylcholinesterase/kg fresh tissue (the size of organs from one typical eel), while rat diaphragm contains about 0.1 mg/kg muscle. The electric organs are phylogenetically derived from muscle (see Nachmansohn, 1959; Fessard, 1958) and contain large amounts of actin and myosin that are not organized in striated structures (Amsterdam *et al.*, 1976). The unusually large quantities of acetylcholinesterase in these organs derive from the fact that an electric organ cell or electroplaque is a giant syncytium that receives thousands of nerve terminals at synapses on its innervated face, and these synapses are remarkably like skeletal neuromuscular junctions in their structure, pharmacology, and acetylcholinesterase content. Following the introduction of useful affinity chromatography resins (see Dudai and Silman, 1974a; Rosenberry, 1975; Rosenberry *et al.*, 1981), acetylcholinesterase forms from electric organs were isolated and characterized in several laboratories. The structure of the predominant form from eel electric organ (Rosenberry and Richardson, 1977; Anglister and Silman, 1978; Rosenberry *et al.*, 1980) is schematized in Fig. 1. This structure represents an 18 S molecule, in which twelve 75,000-dalton catalytic subunits (circles) are arranged in three tetrameric groups. Within each tetramer, two catalytic subunits are directly linked by a single disulfide bond while the remaining two are each covalently attached by a single disulfide bond to one "tail" subunit. Each tail subunit is about 35,000 daltons and consists of two domains: a 24,000-dalton segment that interacts with corresponding segments in the other two tail subunits to form a collagen-like triple helix, and a noncollagen-like region that includes the disulfide bonds to the catalytic subunits. About 35% of the proline residues and 70% of the lysine residues in the collagen-like domain are hydroxylated, a modification typical of collagen-like polypeptides. Disulfide bonds also cross-link the tail subunits directly to each other in the region where the two domains intersect. The precise stoichiometry of disulfide bonds between the tail subunits has not been determined, but the total n within the tail structure appears to be about 5. The triple-helical region of the tail structure resists degradation by pepsin and can be isolated free of catalytic subunit contamination (Mays and Rosenberry, 1981). In contrast, several proteases cleave in the noncollagen-like domain of the tail subunits to release 11 S catalytic subunit tetramers that retain residual tail subunit fragments (Massoulié and Rieger, 1969; Dudai and Silman, 1974b; McCann and Rosenberry, 1977; Rosenberry *et al.*, 1980).

Endogenous proteolytic activity may be partially responsible for the variety of acetylcholinesterase forms that are observed in nerve and muscle tissues. A nomenclature that permits analysis of this polymorphism has been introduced by Bon *et al.* (1979) and defines two classes of acetylcholinesterase: "asymmetric" forms A_n that are assemblies of n cat-

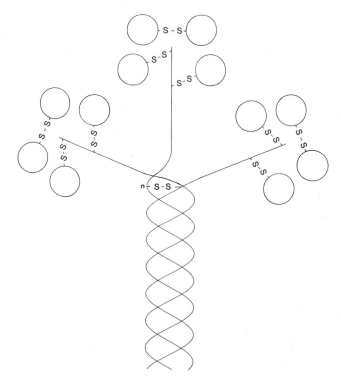

Figure 1. Schematic structure of the predominant 18 S form of acetylcholinesterase from eel electric organs.

alytic subunits associated with a collagen-like tail structure, and "globular" forms G_n in which the assemblies of n catalytic subunits are devoid of any detectable collagenlike components. The 18 S eel enzyme in Fig. 1 thus is an A_{12} form and comprises 60–70% of the total acetylcholinesterase in eel electric organ. Most of the remainder is A_8, a form that can be generated by the proteolytic removal of one G_4 from A_{12} enzyme as noted above. Careful determinations of eel A_{12}, A_8, A_4, and G_4 molecular weights together with electron micrographs of these forms (Rieger *et al.*, 1973; Dudai *et al.*, 1973; Bon *et al.*, 1976) established the relationships among these forms and provided the first clues to the A_{12} structure in Fig. 1. The A_{12} form also accounts for virtually all the acetylcholinesterase in the PLD muscle from adult chicken, particularly when extraction is conducted in the presence of protease inhibitors (Sketelj *et al.*, 1978; Silman *et al.*, 1979). In contrast, A_{12} forms constitute less than half of the total acetylcholinesterase in embryonic muscles and in most mammalian muscles.

Although acetylcholinesterases from chicken and mammalian muscles have not been isolated, several close similarities suggest that structural information about asymmetric eel electric organ forms is directly applicable to those in muscle:

1. The molecular weight relationships among A_{12}, A_8, A_4, and G_4 forms are similar (Bon *et al.*, 1979; Massoulié *et al.*, 1980).
2. Asymmetric forms are thought to be localized in extracellular basement membrane (evidence for this is presented below) and generally require high salt plus nonionic detergents for maximal solubilization. If the ionic strength of the extract is reduced by dialysis, aggregation or precipitation of these forms occurs. This aggregation,

which may be an important model for the localization of these forms *in situ*, is promoted by chondroitin sulfate and certain other polyanions and blocked by acetylation of the enzyme forms (Bon *et al.*, 1978, 1979).

3. Collagenase degrades asymmetric forms in at least two stages. Initially, less asymmetric A'_{12}, A'_8, and A'_4 forms are generated by removal of part of the triple-helical tail structure shown in Fig. 1. This cleavage increases the sedimentation coefficient of each form by about 10% while decreasing its molecular weight (as shown by many laboratories, eg., Dudai and Silman, 1974b; Johnson *et al.*, 1977; Watkins *et al.*, 1977; Anglister and Silman, 1978; Rotundo and Fambrough, 1979). Further degradation to yield at least some G_4 enzyme can occur, particularly at 37°C (Lwebuga-Mukasa *et al.*, 1976; Webb, 1978; Bon and Massoulié, 1978). Neither the A'_n nor the G_4 forms aggregate at low ionic strength.

Observations in these three areas are very consistent among asymmetric acetylcholinesterases from different species (Bon *et al.*, 1979), and they provide extremely useful criteria for the characterization of tiny amounts of native asymmetric forms in partially purified or even crude solutions.

Asymmetric acetylcholinesterases appear to be localized in the extracellular basement membrane matrix. Lwebuga-Mukasa *et al.* (1976) detected hydroxylated proline and lysine in purified asymmetric but not globular acetylcholinesterase forms from *Torpedo* electric organ and reported a *Torpedo* membrane fraction that was enriched in both acetylcholinesterase and hydroxyproline. They suggested that collagen-like subunits are responsible for an *in vivo* localization of the asymmetric forms in the extracellular basement membrane matrix at the synapse. A similar suggestion based on more tentative evidence had been made earlier (Dudai and Silman, 1974b), in part because this localization could readily explain the release of acetylcholinesterase from the neuromuscular junctions on exposure of muscles to collagenase (Hall and Kelly, 1971; Betz and Sakmann, 1973). Recent striking histochemical evidence to support this localization in frog neuromuscular junction has been presented by McMahan *et al.* (1978), who demonstrated that a significant fraction of the junctional acetylcholinesterase remains associated with residual basement membrane following destruction of the nerve and muscle cells. Asymmetric forms are associated particularly with synapses in electric organs and with skeletal neuromuscular junctions. These are locations at which acetylcholine synaptic transmission is extremely rapid and is mediated by "nicotinic" acetylcholine receptors. Asymmetric acetylcholinesterase sites are highly concentrated in such synapses and ensure that the lifetime of acetylcholine in the synaptic cleft is less than a few hundred microseconds (see Rosenberry, 1979). Asymmetric forms are present in smaller amounts in autonomic ganglia, in axons of motor nerves (Fernandez *et al.*, 1979), and in smooth and cardiac muscles and their innervating nerves (Skau and Brimijoin, 1980). The relationship of acetylcholinesterase structure to synaptic function in these systems is less clear, particularly in those where acetylcholine synaptic transmission is slower and is generally mediated by "muscarinic" receptors.

3. GLOBULAR ACETYLCHOLINESTERASE FORMS

In contrast to the asymmetric forms with their distinct collagen-like tail structures, the class of globular acetylcholinesterases is defined more ambiguously here as all other acetylcholinesterase forms that lack a collagen-like tail. The G_4 form generated by endogenous proteases from asymmetric forms in stored eel electric organ tissue was the first acetylcho-

Table I. Sedimentation Coefficients of Acetylcholinesterase Forms in T28 Cell Extracts and Culture Media [a]

Gradient buffer	Total cellular forms	Soluble cellular forms	Secreted forms
With Triton	9.7 ± 0.06	9.7 ± 0.02	10.6 ± 0.04
	3.5 ± 0.04	3.2 ± 0.07	4.4 ± 0.06
Without Triton	no distinct	10.2 ± 0.04	10.5 ± 0.07
	peaks	4.6 ± 0.04	4.3 ± 0.10

[a] Sedimentation coefficients were determined in 5–20% sucrose gradients containing 10 mM Tris–Hcl, 50 mM $MgCl_2$, and 1 M NaCl at pH 7. The tabulated values are means ± S.E.M. (where the number of observations was 4 to 8). Total cellular forms reflect primarily detergent-solubilized forms and refer to the activity in supernatants of cell homogenates prepared in the presence of Triton X-100. Soluble cellular forms refer to the activity in high-speed supernatants of cell homogenates prepared in the absence of detergent at low ionic strength. Secreted forms refer to soluble extracellular activity found in culture media following removal of floating cells. See Lazar and Vigny (1980) for further details.

linesterase to be isolated and characterized in detail (see Rosenberry, 1975), and subsequently G_2 and G_1 forms from eel were generated by degradation involving prolonged sonication in detergent or extensive autolysis in partially purified extracts (see Bon and Massoulié, 1976). However, it is clear that globular forms can also be synthesized independently of asymmetric forms because they are found in certain tissues like brain that appear to be devoid of asymmetric forms. The class of globular acetylcholinesterases can be subdivided into three, structurally distinct groups based on their interactions with non-ionic detergents (Rieger and Vigny, 1976; see Massoulié et al., 1980). This point is illustrated in Table I, which contains data obtained by Lazar and Vigny (1980) on the sedimentation properties of G_4 ($\sim$ 10 S) and G_1 ($\sim$ 4 S) acetylcholinesterase forms synthesized by a neuroblastoma × sympathetic ganglion cell hybrid cell line called T28. Part of the acetylcholinesterase associated with the tissue in these cultures can be extracted by homogenization in the absence of nonionic detergent, but the remainder requires detergent for extraction and thus presumably corresponds to integral membrane protein. In addition, acetylcholinesterase is secreted in soluble form into the medium from these cultures as it is from many other cells. According to the signal hypothesis (Blobel and Dobberstein, 1975), secreted proteins are synthesized with an NH_2-terminal "leader" sequence of 15 to 30 hydrophobic amino acids that directs the extrusion of the protein through the rough endoplasmic reticulum membrane into the lumen during synthesis. This hydrophobic signal peptide is cleaved by a specific protease during extrusion into the lumen. In a proposed variation on this hypothesis, the signal peptide may not be cleaved in the lumen but instead reinserts into the rough endoplasmic reticulum membrane to give an integral membrane protein (see Wickner, 1979). Because both integral membrane-bound and secreted acetylcholinesterases are synthesized by the T28 cultures, one may question whether their structures differ in a way related to the presence of a hydrophobic signal peptide. An experimental technique for detecting an exposed hydrophobic peptide is its binding of nonionic detergent, and Table I indicates that the acetylcholinesterase fractions extracted from T28 cultures interact with nonionic detergent during sucrose gradient sedimentation in different ways. Tissue enzyme that requires detergent for solubilization aggregates when sedimented in the absence of detergent. Thus, this detergent-solubilized fraction behaves as a typical integral phospholipid membrane protein. Soluble cellular enzyme extracted in the absence of detergent nonetheless interacts somewhat with detergent as indicated by a 0.5–1.2 S decrease when sedimented in the presence of detergent. In contrast, secreted soluble enzyme in the medium appears devoid of an exposed hydrophobic peptide, because it gives identical sedimentation coefficients in the presence or absence of detergent and thus shows no detergent interactions. In this respect the secreted enzyme is analogous both to the degradative G_n forms from eel, which do not bind radiolabeled detergent (Millar *et al.*, 1978), and indeed to asymmetric acetylcholinesterases, none of which show shifts in sedimentation coefficient in the presence of nonionic detergents (Bon *et al.*, 1978). In further contrast, none of the globular forms show ionic-strength-dependent sedimentation properties characteristic of the asymmetric forms.

No globular acetylcholinesterase that interacts with detergent has yet been isolated and structurally characterized to determine the polypeptide structure responsible for the detergent interaction and its possible relevance to a signal peptide. Nevertheless, observations in several systems suggest that it is relatively small. Integral membrane-bound forms have been purified from human and bovine erythrocytes (Ott *et al.*, 1975; Grossmann and Liefländer, 1975; Niday *et al.*, 1977; Berman, 1973), bovine brain (Chan *et al.*, 1972; Reuss *et al.*, 1976), and *Torpedo* electric organ (Viratelle and Bernhard, 1980) and partially purified from bovine superior cervical ganglion (Vigny *et al.*, 1979). In each of these cases, catalytic subunits of 75,000–80,000 daltons appeared predominantly as disulfide-linked dimers on sodium dodecyl sulfate electrophoresis gels in the absence of disulfide reducing agents. This is precisely the pattern observed for the soluble G_4 forms generated by proteolysis of asymmetric eel or *Torpedo* acetylcholinesterases, and thus no sizable hydrophobic peptides ($> 10,000$ daltons) are apparent either as extensions of the catalytic subunit polypeptides or as noncatalytic subunits that are disulfide linked to the integral membrane-bound forms in a fashion analogous to that of the collagenlike subunits in the asymmetric forms. Each of these four detergent-solubilized acetylcholinesterases aggregates in the absence of nonionic detergent, just as the integral membrane-bound form in Table I (Ott and Brodbeck, 1978; see Massoulié *et al.*, 1980), and the ability of these forms to aggregate under these conditions is blocked by certain proteases (Bon and Massoulié, 1981; Massoulié *et al.*, 1980).* This protease conversion appears to involve only a small portion of the integral membrane-bound forms, as the sedimentation coefficients of the active protease-derived forms in the absence of detergent are only slightly shifted from those of the original detergent-solubilized forms in the presence of detergent. These observations all are consistent with the presence of a small, hydrophobic peptide only in the detergent-binding acetylcholinesterase forms, but its relationship to a signal peptide must be explored in a system where enzyme biosynthesis and structure can be studied coordinately.

4. RELATIONSHIPS AMONG ACETYLCHOLINESTERASE FORMS

A fundamental question that underlies current work in many laboratories is the following: does the polymorphism of acetylcholinesterase forms observed in a single tissue reflect exclusively posttranslational processing of a single catalytic subunit gene product, or do these forms arise independently from related but distinct genes? Although several observations, including protease-mediated interconversions of forms in extracts, the similarity of catalytic subunit sizes and turnover numbers (e.g., Vigny *et al.*, 1978; Viratelle and Bernhard, 1980), and the possible indirect identification of a signal peptide, all suggest

---

*Unpublished observations on human erythrocyte acetylcholinesterase by Dutta-Choudhury and Rosenberry.

extensive posttranslational processing of acetylcholinesterase *in vivo,* very few direct genetics experiments to test this question have been conducted. In a preliminary note, Dudai (1977) reported the presence of both asymmetric and globular forms of acetylcholinesterase in *Drosophila melanogaster*, and he identified mutant flies heterozygous for a small deficiency covering the putative acetylcholinesterase structural locus. These mutants showed a normal distribution of forms but contained only one-half the activity of normal flies, an observation that Dudai noted was consistent with the assumption that all forms contain a common polypeptide. In cultured mammalian cells in which acetylcholinesterase is totally inactivated by an organophosphate, newly synthesized enzyme reappears first in the G_1 fraction followed by reappearance in G_4 and finally A_{12} (Rieger *et al.*, 1976; Koenig and Vigny, 1978). These observations suggest but by no means prove that the G_1 form is a biosynthetic precursor of the G_4 and A_{12} forms. In general, G_1 forms in mammalian cells are found predominantly in the soluble cell extract fraction as defined in Table I, whereas G_2 and particularly G_4 forms distribute primarily into the detergent-solubilized and secreted fractions (see Massoulié *et al.*, 1980; Chang and Blume, 1976). During both development *in vivo* and transition from log to stationary phase in tissue culture, the percentage of total acetylcholinesterase found as G_1 decreases severalfold. Furthermore, studies with cationic organophosphate inactivating agents that do not significantly penetrate cell membranes indicate that G_1 forms are much less accessible to the extracellular medium than G_2 and G_4 forms (Lazar and Vigny, 1980; Taylor *et al.*, 1981). Similar patterns are observed with avian cells except that the internal, cell-soluble form apparently is G_2 rather than G_1 (Rotundo and Fambrough, 1979, 1980; Taylor *et al.*, 1980). These observations suggest that intracellular soluble G_1 (or G_2 in avians), presumably containing a small, exposed hydrophobic peptide as in Table I, serves as a precursor to membrane-bound G_4 and/or secreted G_4 (Massoulié *et al.*, 1980; Gisiger and Vigny, 1977), as required by the signal hypothesis. Furthermore, experiments involving selective organophosphate inactivation and the use of protein synthesis inhibitors in cultured chick embryo muscle cells suggest that acetylcholinesterase synthesized in the intracellular pool is transported over a 2- to 3-hr period to two alternative destinations: a small fraction appears in the plasma membrane, where it slowly turns over with a half-life of about 50 hr, while the majority is secreted into the extracellular medium (Rotundo and Fambrough, 1980). Cleavage of a signal peptide from enzyme in the intracellular pool could be obligatory either for secretion or, as Bon and Massoulié (1981) have suggested, for assembly of the asymmetric forms.

In addition to regulating the subcellular distribution of globular acetylcholinesterase forms, cells also concentrate asymmetric forms in the region of nicotinic synapses, as Hall noted in 1973. The quantitative distribution of forms found in the rat diaphragm is shown in Table II (Younkin *et al.*, 1981). Globular forms are distributed rather uniformly throughout the muscle, while asymmetric forms are localized, although not exclusively, to regions containing neuromuscular junctions. A fraction that is not readily solubilized is localized even more exclusively in the junctions. Acetylcholinesterase in this fraction appears to be collagen-tailed because it is solubilized and degraded by collagenase in the pattern (noted above) typical of asymmetric forms, and this fraction may contribute significantly to the basement membrane acetylcholinesterase identified by McMahan *et al.* (1978). Denervation of the diaphragm results in loss of about one-half of the globular forms after 3 days, while the asymmetric forms, after a brief lag, fall to about one-fourth of their innervated levels (Collins and Younkin, 1981; McLaughlin and Bosmann, 1976). Furthermore, the asymmetric forms do not change uniformly over this period: A_{12} decreases by 85%; A_8, by 70%; and A_4, perhaps a degradation product of A_{12} and A_8, increases by 60% (Collins and Younkin, 1981). Denervation of other muscles also results in a rapid disappearance of

**Table II. Percentage Distribution of Junctional and Nonjunctional
Acetylcholinesterase Forms in Rat Diaphragm** [a]

Form	Junctional	Nonjunctional
All globular forms	7.3 ± 1.2	60.5 ± 1.8
G_4	2	25
$G_2 + G_1$	5	36
All asymmetric forms	11.6 ± 0.9	11.0 ± 0.8
A_{12}	8	
A_8	3	
A_4	0	
Nonextractable	7.0 ± 0.6	2.5 ± 0.3

[a] Percentages for the six major categories (globular, asymmetric, and nonextractable fractions in junctional and nonjunctional regions) are averages of 20 determinations and sum to 100%. Within categories, subdivisions among specific forms are given. Diaphragms were divided into junctional and nonjunctional segments, and enzyme activities per milligram tissue were determined. Nonjunctional activities were assumed uniform throughout the muscle and were deducted from the total activities in junctional segments to obtain the junctional activities. Segments were homogenized first with 1% Triton X-100, 10 mM sodium phosphate (pH 7) to solubilize globular forms. Particulate residues were then homogenized with the same buffer containing 1 M NaCl to solubilize asymmetric forms, and the nonextractable fractions remained in the particulate residues following extraction of the asymmetric forms. See Younkin *et al.* (1981) for further details.

asymmetric forms (Vigny *et al.*, 1976) and may be followed by an increase in the amount of cell-soluble G_1 or G_2 forms by 2 weeks postdenervation (Sketelj *et al.*, 1978; see Collins and Younkin, 1981). Reinnervation of muscles at their original junctional sites resulted in reappearance of the A_{12} form, while reinnervation at muscle sites previously free of junctions resulted in reappearance of this form both at the new sites and at the (nonreinnervated) endplates of the original junctions (Vigny *et al.*, 1976; Weinberg and Hall, 1978; Lømo and Slater, 1980). Relative distributions of acetylcholinesterase in rat and chick cultured muscle cells approximate those of denervated muscle (Koenig and Vigny, 1978; Rotundo and Fambrough, 1979), and the appearance of asymmetric forms in mixed cultures of muscle and nerve has been used as a specific biochemical marker of functional nerve–muscle contacts (Koenig and Vigny, 1978; Rubin *et al.*, 1979). A great deal of current attention is focused on the mechanisms by which nerve regulates the levels not only of A_{12} acetylcholinesterase but also of junctional acetylcholine receptor. A review of pertinent literature is far beyond the scope of this article; it suffices to state that regulation is thought to be mediated both by the activity (electrical and/or mechanical) generated in muscle by nerve and by trophic factors delivered to muscle by nerve (Davey *et al.*, 1979; Lømo and Slater, 1980).

ACKNOWLEDGMENTS. The author thanks Dr. Steven G. Younkin and Patricia L. Collins for their helpful comments and critical reading of this review.

REFERENCES

Amsterdam, A., Lamed, R., Josephs, R., Silman, I., and Gröschel-Stewart, U. (1976). *Sixth European Congress on Electron Microscopy, Jerusalem,* p. 270.
Anglister, L., and Silman, I. (1978). *J. Mol. Biol.* **125,** 293–311.
Berman, J. D. (1973). *Biochemistry* **12,** 1710–1715.
Betz, W., and Sakmann, B. (1973). *J. Physiol. (London)* **230,** 673–688.
Blobel, G., and Dobberstein, B. (1975). *J. Cell Biol.* **67,** 835–851.

Bon, S., and Massoulié, J. (1976). *FEBS Lett.* **67**, 99–103.

Bon, S., and Massoulié, J. (1978). *Eur. J. Biochem.* **89**, 89–94.

Bon, S., and Massoulié, J. (1980). *Proc. Natl. Acad. Sci. USA* **77**, 4464–4468.

Bon, S., Huet, M., Lemonnier, M., Rieger, F., and Massoulié, J. (1976). *Eur. J. Biochem.* **68**, 523–530.

Bon, S., Cartaud, J., and Massoulié, J. (1978). *Eur. J. Biochem.* **85**, 1–14.

Bon, S., Vigny, M., and Massoulié, J. (1979). *Proc. Natl. Acad. Sci. USA* **76**, 2546–2550.

Chan, S. L., Shirachi, D. Y., Bhargava, H. N., Gardner, E., and Trevor, A. J. (1972). *J. Neurochem.* **19**, 2747–2758.

Chang, C.-H., and Blume, A. J. (1976). *J. Neurochem.* **27**, 1427–1435.

Collins, P. L., and Younkin, S. G. (1982). *J. Biol. Chem.*, in press.

Davey, B., Younkin, L. H., and Younkin, S. G. (1979). *J. Physiol. (London)* **289**, 501–515.

Dudai, Y. (1977). *Isr. J. Med. Sci.* **13**, 944.

Dudai, Y., and Silman, I. (1974a). *Methods Enzymol.* **34**, 571–580.

Dudai, Y., and Silman, I. (1974b). *J. Neurochem.* **23**, 1177–1187.

Dudai, Y., Herzberg, M., and Silman, I. (1973). *Proc. Natl. Acad. Sci. USA* **70**, 2473–2476.

Fernandez, H. L., Duell, M. J., and Festoff, B. W. (1979). *J. Neurochem.* **32**, 581–585.

Fessard, A. (1958). In *Traité de Zoology* (P. P. Grasse, ed.), Vol. XIII, pp. 1143–1238, Masson. Paris.

Froede, H. C., and Wilson, I. B. (1970). In *The Enzymes* (P. D. Boyer, ed.), Vol. V, 3rd ed., pp. 87–114, Academic Press, New York.

Gisiger, V., and Vigny, M. (1977). *FEBS Lett.* **84**, 253–256.

Grossmann, H., and Liefländer, M. (1975). *Hoppe-Seyler's Z. Physiol. Chem.* **356**, 663–669.

Hall, Z. W. (1973). *J. Neurobiol.* **4**, 343–361.

Hall, Z. W., and Kelly, R. (1971). *Nature New Biol.* **232**, 62–63.

Johnson, C. D., Smith, S. P., and Russell, R. L. (1977). *J. Neurochem.* **28**, 617–624.

Koenig, J., and Vigny, M. (1978). *Nature (London)* **271**, 75–77.

Lazar, M., and Vigny, M. (1980). *J. Neurochem.* **35**, 1067–1079.

Lømo, T., and Slater, C. R. (1980). *J. Physiol. (London)* **303**, 191–202.

Lwebuga-Mukasa, J. S., Lappi, S., and Taylor, P. (1976). *Biochemistry* **15**, 1425–1434.

McCann, W. F. X., and Rosenberry, T. L. (1977). *Arch. Biochem. Biophys.* **183**, 347–352.

McLaughlin, J., and Bosmann, H. B. (1976). *Exp. Neurol.* **52**, 263–271.

McMahan, U. J., Sanes, J. R., and Marshall, L. M. (1978). *Nature (London)* **271**, 172–174.

Massoulié, J. (1980). *Trends Biochem. Sci.* **5**, 160–164.

Massoulié, J., and Rieger, F. (1969). *Eur. J. Biochem.* **11**, 441–455.

Massoulié, J., Bon, S., and Vigny, M. (1980). *Neurochem. Int.* **2**, 161–184.

Mays, C., and Rosenberry, T. L. (1981). *Biochemistry* **20**, 2810–2817.

Millar, D. B., Christopher, J. P., and Burrough, D. O. (1978). *Biophys. Chem.* **9**, 9–14.

Nachmansohn, D. (1959). *Chemical and Molecular Basis of Nerve Activity,* Academic Press, New York [revised edition, with Neumann, E. (1975)].

Nachmansohn, D., and Lederer, E. (1939). *Bull. Soc. Chim. Biol.* **21**, 797–808.

Niday, E., Wang, C. S., and Alaupovic, P. (1977). *Biochim. Biophys. Acta* **469**, 180–193.

Ott, P., and Brodbeck, U. (1978). *Eur. J. Biochem.* **88**, 119–125.

Ott, P., Jenny, B., and Brodbeck, U. (1975). *Eur. J. Biochem.* **57**, 469–480.

Reuss, K.-P., Weinert, M., and Leifländer, M. (1976). *Hoppe-Seyler's Z. Physiol. Chem.* **357**, 783–793.

Rieger, F., and Vigny, M. (1976). *J. Neurochem.* **27**, 121–129.

Rieger, F., Bon, S., and Massoulié, J. (1973). *Eur. J. Biochem.* **34**, 539–547.

Rieger, F., Favre-Bauman, A., Benda, P., and Vigny, M. (1976). *J. Neurochem.* **27**, 1059–1063.

Rosenberry, T. L. (1975). *Adv. Enzymol.* **43**, 103–218.

Rosenberry, T. L. (1979). *Biophys. J.* **26**, 263–290.

Rosenberry, T. L., and Richardson, J. M. (1977). *Biochemistry* **16**, 3550–3558.

Rosenberry, T. L., Barnett, P., and Mays, C. (1980). *Neurochem. Int.* **2**, 25–147.

Rosenberry, T. L., Barnett, P., and Mays, C. (1981). *Methods Enzymol.* **82**, 325–339.

Rotundo, R. L., and Fambrough, D. M. (1979). *J. Biol. Chem.* **254**, 4790–4799.

Rotundo, R. L., and Fambrough, D. M. (1980). *Cell* **22**, 483–594.

Rubin, L. L., Schuetze, S. M., and Fischbach, G. D. (1979). *Dev. Biol.* **69**, 46–58.

Silman, I., Lyles, J. M., and Barnard, E. A. (1979). *FEBS Lett.* **94**, 166–170.

Skau, K. A., and Brimijoin, S. (1980). *J. Neurochem.* **35**, 1151–1154.

Sketelj, J., McNamee, M. G., and Wilson, B. W. (1978). *Exp. Neurol.* **60**, 624–629.

Taylor, P., Rieger, F., and Greene, L. A. (1980). *Brain Res.* **182**, 383–396.

Taylor, P., Rieger, F., Shelanski, M. L., and Greene, L. A. (1981). *J. Biol. Chem.* **256**, 3827–3830.

Vigny, M., Koenig, J., and Rieger, F. (1976). *J. Neurochem.* **27,** 1347–1353.

Vigny, M., Bon, S., Massoulié, J., and Leterrier, F. (1978). *Eur. J. Biochem.* **85,** 317–323.

Vigny, M., Bon, S., Massoulié, J., and Gisiger, V. (1979). *J. Neurochem.* **33,** 559–565.

Viratelle, O. M., and Bernhard, S. A. (1980). *Biochemistry* **19,** 4999–5007.

Watkins, M. S., Hitt, A. S., and Bulger, J. E. (1977). *Biochem. Biophys. Res. Commun.* **79,** 640–647.

Webb, G. (1978). *Can. J. Biochem.* **56,** 1124–1132.

Weinberg, C. B., and Hall, Z. W. (1978). *Dev. Biol.* **68,** 631–635.

Wickner, W. (1979). *Annu. Rev. Biochem.* **48,** 23–45.

Younkin, S. G., Rosenstein, C. C., Collins, P L., and Rosenberry, T. L. (1982). *J. Biol. Chem.*, in press.

136

Scorpion Neurotoxins and Their Role in the Study of Sodium Channel in Electrically Excitable Membranes

Eugene V. Grishin and Yuri A. Ovchinnikov

The molecular mechanism of nerve impulse propagation is among the most intriguing problems of modern physicochemical biology. The role of the fast sodium channels in excitation phenomena can be analyzed using neurotoxins that modify some distinctive channel functions (Narahashi, 1974).

The secretion products of the scorpion poisonous gland induce the prolongation of action potentials in single nodes of Ranvier (Adam et al., 1966). This effect is due to a retardation of sodium channel inactivation that is sometimes accompanied by a substantial shift of the potential dependence of activation in the direction of hyperpolarization (Koppenhöfer and Schmidt, 1968; Mozhayeva et al., 1978). The biological activity of the scorpion venoms is due to the presence of moderately basic neurotoxins of polypeptide nature (Rochat et al., 1970). The common structural feature of these miniproteins is the presence of four disulfide bridges.

According to their biological specificity, scorpion neurotoxins can be divided into several groups of polypeptides that induce paralysis in mammalians, insects or crustaceans (Zlotkin, 1973). To date, the complete amino acid sequence of 18 mammalian neurotoxins has been determined from the venom of nine different scorpions. All these proteins are characterized by a molecular weight of about 7000 and by the absence of methionine residues. These features are well illustrated by the amino acid sequence of toxin M_{10} from the venom of the Middle Asian scorpion *Buthus eupeus* (Grishin et al., 1980d):

Val-Arg-Asp-Gly-Tyr-Ile-Ala-Asp-Asp-Lys[10]-Asp-Cys-Ala-Tyr-Phe-Cys-Gly-Arg-Asn-Ala[20]-

Tyr-Cys-Asp-Glu-Glu-Cys-Lys-Lys-Gly-Ala[30]-Glu-Ser-Gly-Lys-Cys-Trp-Tyr-Ala-Gly-Gln[40]-

Eugene V. Grishin and Yuri A. Ovchinnikov • Shemyakin Institute of Bioorganic Chemistry, USSR Academy of Sciences, 117988 Moscow V-334, USSR.

Tyr-Gly-Asn-Ala-Cys-Trp-Cys-Tyr-Lys-Leu-Pro-Asp-Trp-Val-Pro-Ile-Lys-Gln-Lys-Val-
Ser-Gly-Lys-Cys-Asn

Among these structures three toxic variants were studied from the venom of the North American scorpion *Centruoides sculpturatus,* which possess a high paralyzing activity to insects and low lethality to vertebrates (Babin *et al.,* 1974). The insectotoxin I_2 from *B. eupeus* venom also belongs to the structural group of *Centruroides* toxic variants (Grishin *et al.,* 1979). Along with I_2, Middle Asian scorpion venom also contains a number of "short" insectotoxins, which are built up of 35–36 amino acid residues cross-linked by four disulfide bridges. Thus far, insectotoxin I_1 was shown to possess the following amino acid sequence (Zhdanova *et al.,* 1977):

Met-Cys-Met-Pro-Cys-Phe-Thr-Thr-Arg-Pro-Asp-Met-Ala-Gln-Gln-Cys-Arg-Ala-Cys-
Cys-Lys-Gly-Arg-Gly-Lys-Cys-Phe-Gly-Pro-Gln-Cys-Leu-Cys-Gly-Tyr-Asp

An insectotoxin of the same structural type was shown to be present in *Androctonus mauretanicus* venom (Rochat *et al.,* 1979). The mode of action of these proteins is now under study.

Mammalian scorpion toxins are easily iodinated without loss of toxicity and are suitable for radioligand analysis of interaction with receptors. These studies, as well as voltage-clamp assays, have shown that scorpion neurotoxins interact with surface membrane receptors in equimolar ratio (Mozhayeva *et al.,* 1979), and the affinity of binding sites for the toxins decreases by depolarization of the membrane (Mozhayeva *et al.,*1979; Catterall, 1977). The potential dependence of neurotoxin dissociation may be attributed to the energy difference between the inactivated state of normal channels and channels treated with the toxin. The transition of a fraction of sodium channels into a slow inactivated state could also result in additional reduction of toxin binding (Mozhayeva *et al.,* 1978, 1980).

Divalent cations effectively suppress the specific binding of scorpion neurotoxins to the receptors, while Ca^{2+} ions increase the nonspecific binding of labeled toxin (Soldatov *et al.,* 1980; Jover *et al.,* 1980). This effect may be due to stronger nonspecific interaction of the toxin with the lipid phase, in the presence of Ca^{2+}. Phospholipase A_2 treatment effectively decreases the binding of *Leiurus* toxin to bilayer membranes (Okamoto, 1980).

Scorpion toxins have only negligible influence on model phosphatidylserine and lecithin bilayers (Faucon *et al.,* 1979; Grishin *et al.,* 1980a). On the other hand, the solubilization of synaptosomal membrane by Triton X-100 completely suppresses the effect of toxin on the receptor (Catterall *et al.,* 1979). These data demonstrate the importance of lipid environment for receptor function.

Scorpion toxins induce a saturable quenching of tryptophan fluorescence upon interaction with nerve membranes (Grishin *et al.,* 1980a). This effect is followed by a saturable decrease in the light scattering of the membrane suspension, which was used for determination of toxin–receptor interaction. The binding of iodinated toxin is also accompanied by selective quenching of iodotyrosine fluorescence. All these effects are sensitive to membrane depolarization, thus supporting the conclusion about the specificity of these spectral changes.

The radioligand assay showed that nonspecific adsorption of scorpion toxins to excitable membranes, which does not depend on the membrane potential, accounts for up to 50% of the total binding (Catterall, 1977; Okamoto, 1980; Grishin *et al.,* 1980b). So it could be suggested that toxins act by inserting their hydrophobic chain segments into the lipid phase, thereby inducing an electrostatic or steric restriction for the voltage-dependent shift of the membrane charge particle, which is coupled to the channel inactivation process.

Sea anemone toxins constitute another class of polypeptide neurotoxins (Rathmayer and Béress, 1978) that share a common receptor with scorpion toxins (Catterall and Béress, 1978; Courand *et al.*, 1978). *Anemonia sulcata* toxin II (ATX II) possesses the same mode of action as scorpion toxin (Bergman *et al.*, 1976), but the number of its binding sites is 10 times higher (Vincent *et al.*, 1980) and competitive displacement by scorpion toxin is not observed. It must be noted that competitive interactions are effective only among those scorpion toxins that share antigenic specificity (Couraud *et al.*, 1980). This would indicate the involvement of discrete but overlapping binding sites in toxin–receptor interaction with ATX II.

The use of photosensitive toxin analogs is an effective method for the direct study of their respective binding components. Thus, the monosubstituted dinitrophenylazide derivative of toxin M_{10} (DNPA-M_{10}) irreversibly interacts with sodium channels upon UV-light irradiation at wavelengths higher than 320 nm (Grishin *et al.*, 1980c) and its radioactive radicals specifically bind to a pair of synaptosomal proteins of 77,000 and 51,000 molecular weight (Grishin *et al.*, 1980b) (Fig. 1). A similar protein composition of the receptor was found in neuroblastoma cells and in crab nerve plasma membrane (Grishin *et al.*, 1980b). The crayfish nerve binding components for [^{125}I]-NAP-ATX I are of 53,000 and 86,000 molecular weight (Hucho *et al.*, 1979). The modification of two membrane components suggests their spatial proximity to the binding site of the toxin molecule in the membrane.

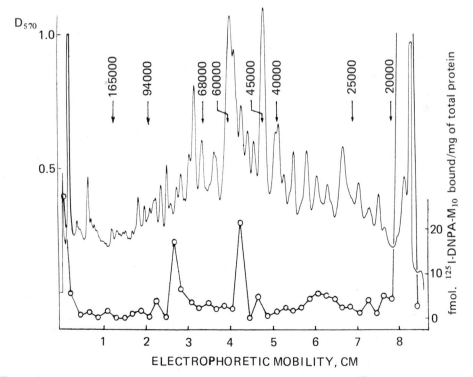

Figure 1. The densitogram and the profile of radioactivity of specifically bound ^{125}I-labeled DNPA-M_{10} after SDS-polyacrylamide gel electrophoresis of photoaffinity-modified synaptosomes (Grishin *et al.*, 1980b). The arrows show the molecular weight values for reference proteins.

The photosensitive derivative of *Leiurus* scorpion toxin labels a 250,000 membrane component in neuroblastoma cells and a 250,000 plus a 32,000 polypeptide in rat brain synaptosomes. This toxin derivative was prepared at a one-to-one molar ratio of succinyl-imidyl-5-azido-2-nitrobenzoate to total reactive amino groups of the toxin (Beneski and Catterall, 1980).

Tetrodotoxin (TTX) is known to be a specific marker for sodium channels (Narahashi, 1974) and its receptor has a molecular weight about 230,000 according to radiation inactivation of excitable membranes by accelerated electrons (Levinson and Ellory, 1973). Approximately the same molecular size was estimated for the TTX receptor component that has been detected electrophoretically in solubilized *Electrophorus electroplax* membrane (Agnew *et al.*, 1980). The separation of skeletal muscle membrane proteins in nonionic detergents resulted in purification of the saxitoxin receptor fraction composed of acidic glycopolypeptides of 53,000, 60,000, and 64,000 molecular weight (Barchi *et al.*, 1980). Finally, in a recent review Lazdunski and his colleagues (1979) reported that a photoaffinity analog of TTX modifies membrane components of about 50,000 and 34,000 molecular weight.

It is difficult to find reasons for these differences, but some thermodynamic and kinetic studies of the saxitoxin receptor provide evidence in favor of the subunit nature of the receptor (Weigele and Barchi, 1978).

In conclusion, the study of different neurotoxin receptors provides essential information about the molecular characteristics and functioning of fast sodium channels in excitable membranes. Among them, the scorpion toxin receptor is one of the most intriguing structures because of its participation in potential-dependent regulation of the fast sodium channel.

REFERENCES

Adam, K. R., Schmidt, H., Stampfli, R., and Weiss, C. (1966). *Br. J. Pharmacol.* **26,** 666–677.
Agnew, W. S., Moore, A. C., Levinson, S. R., and Raftery, M. A. (1980). *Biochem. Biophys. Res. Commun.* **92,** 860–866.
Babin, D. R., Watt, D. D., Goos, S. M., and Mlejnek, R. V. (1974). *Arch. Biochem. Biophys.* **164,** 694–706.
Barchi, R. L., Cohen, S. A., and Murphy, L. E. (1980). *Proc. Natl. Acad. Sci. USA* **77,** 1306–1310.
Beneski, D. A., and Catterall, W. A. (1980). *Proc. Natl. Acad. Sci. USA* **77,** 639–643.
Bergman, C., Dubois, J. M., Rojas, E., and Rathmayer, W. (1976). *Biochim. Biophys. Acta* **455,** 173–184.
Catterall, W. A. (1977). *J. Biol. Chem.* **252,** 8660–8668.
Catterall, W. A., and Béress, L. (1978). *J. Biol. Chem.* **253,** 7393–7396.
Catterall, W. A., Morrow, C. S., and Hartshorne, R. P. (1979). *J. Biol. Chem.* **254,** 11379–11387.
Couraud, F., Rochat, H., and Lissitzky, S. (1978). *Biochem. Biophys. Res. Commun.* **83,** 1525–1530.
Couraud, F., Rochat, H., and Lissitzky, S. (1980). *Biochemistry* **19,** 457–462.
Faucon, J. F., Dufourcq, J., Couraud, F., and Rochat, H. (1979). *Biochim. Biophys. Acta* **554,** 332–339.
Grishin, E. V., Soldatova, L. N., Soldatov, N. M., Kostetsky, P. V., and Ovchinnikov, Y. A. (1979). *Bioorg. Khim.* **5,** 1285–1294.
Grishin, E. V., Efremov, E. S., Soldatov, N. M., Podrezova, E. I., and Petrenko, A. G. (1980a). *Bioorg. Khim.* **6,** 576–584.
Grishin, E. V., Soldatov, N. M., and Ovchinnikov, Y. A. (1980b). *Bioorg. Khim.* **6,** 914–922.
Grishin, E. V., Soldatov, N. M., Ovchinnikov, Y. A., Mozhayeva, G. N., Naumov, A. P., Zubov, A. N., and Nisman, B. C. (1980c). *Bioorg. Khim.* **6,** 724–730.
Grishin, E. V., Soldatova, L. N., Shakhparonov, M. I., and Kazakov, V. K. (1980d). *Bioorg. Khim.* **6,** 714–723.
Hucho, F., Stengelin, S., and Bandini, G. (1979). In *Recent Advances in Receptor Chemistry* (F. Gualtieri, M. Giannella, and C. Melchiorre, eds.), pp. 37–58, Elsevier/North-Holland, Amsterdam.
Jover, E., Martin-Moutot, N., Couraud, F., and Rochat, H. (1980). *Biochemistry* **19,** 463–467.

Koppenhöfer, E., and Schmidt, H. (1968). *Pfluegers Arch.* **303,** 150–161.

Lazdunski, M., Balerna, M., Chicheportiche, R., Fosset, M., Jacques, Y., Lombet, A., Romey, G., and Schweitz, H. (1979). *Adv. Cytopharmacol.* **3,** 353–361.

Levinson, S. R., and Ellory, J. C. (1973). *Nature New Biol.* **245,** 122–123.

Mozhayeva, G. N., Naumov, A. P., and Nosyreva, E. D. (1978). *Mechanisms of Zootoxin Action* No. 6, pp. 9–27, Gorkiy University.

Mozhayeva, G. N., Naumov, A. P., Soldatov, N. M., and Grishin, E. V. (1979). *Biofizika* **24,** 235–241.

Mozhayeva, G. N., Naumov, A. P., Nosyreva, E. D., and Grishin, E. V. (1980). *Biochim. Biophys. Acta* **597,** 587–602.

Narahashi, T. (1974). *Physiol. Rev.* **54,** 813–889.

Okamoto, H. (1980). *J. Physiol.* **299,** 507–520.

Rathmayer, W., and Béress, L. (1978). In *Toxins: Animal, Plant and Microbial* (P. Rosenberg, ed.), pp. 549–556, Pergamon Press, Elmsford, N.Y.

Rochat, H., Rochat, C., Kupeyan, C., Miranda, F., Lissitzky, S., and Edman, P. (1970). *FEBS Lett.* **10,** 349–351.

Rochat, H., Bernard, P., and Couraud, F. (1979). *Adv. Cytopharmacol.* **3,** 325–334.

Soldatov, N. M., Kovalenko, V. A., and Grishin, E. V. (1980). In *III USSR–FRG Symposium on Chemistry of Peptides and Proteins*, Makhachkala, Abstract, p. 28.

Vincent, J. P., Balerna, M., Barhanin, J., Fosset, M., and Lazdunski, M. (1980). *Proc. Natl. Acad. Sci. USA* **77,** 1646–1650.

Weigele, J. B., and Barchi, R. L. (1978). *FEBS Lett.* **91,** 310–314.

Zhdanova, L. N., Adamovich, T. B., Nazimov, I. V., Grishin, E. V., and Ovchinnikov, Y. A. (1977). *Bioorg. Khim.* **3,** 485–493.

Zlotkin, E. (1973). *Experientia* **29,** 1453–1466.

137

Pumiliotoxins

New Tools to Investigate Calcium-Dependent Events in Nerve and Muscle

Edson X. Albuquerque, Jordan E. Warnick, Frederick C. Kauffman, and John W. Daly

1. INTRODUCTION

Advances in understanding the role of calcium in cellular function have been made through the use of various toxins and pharmacological agents. Calcium antagonists that have been particularly useful are caffeine (Endo, 1977), D-600 (Watanabe and Besch, 1974), verapamil (McLean *et al.*, 1974), and quercetin (Shoshan *et al.*, 1980). Other agents that have been used in studies of the role of calcium include: diltiazem (Nagao *et al.*, 1980), bepridil (Vogel *et al.*, 1979), quinidine (Scales and McIntosh, 1968), nefedipine (Raschak, 1976), mersalyl (MacLennan, 1970), FR7534 (Jolly and Gross, 1979), dimethylaminoethylene dioxidenes (Rahwan *et al.*, 1977), ryanodine (Hajdu and Leonard, 1961; Frank and Sleator, 1975), and various opioids (Lee and Berkowitz, 1977). Although these agents have been useful in studies of calcium-dependent events, their utility is frequently limited by the high concentrations that are required and by their lack of specificity. Purified toxins of animal origin such as tetrodotoxin (Kao, 1966; Narahashi, 1974), batrachotoxin (Albuquerque and Daly, 1976), histrionicotoxin (Albuquerque *et al.*, 1973), and α-bungarotoxin (Lee, 1971; Changeux *et al.*, 1970) have invariably been found to act at relatively low concentrations and to be highly specific in altering ionic events involved in electrical excitation, neurotrophic phenomena, and receptor-channel kinetics.

The function of toxins of animal origin is not clearly defined. However, certain alkaloids found in the skins of neotropical frogs may serve as a passive "chemical defense"

Edson X. Albuquerque, Jordan E. Warnick, Frederick C. Kauffman, and John W. Daly • Department of Pharmacology and Experimental Therapeutics, University of Maryland School of Medicine, Baltimore, Maryland 21201.

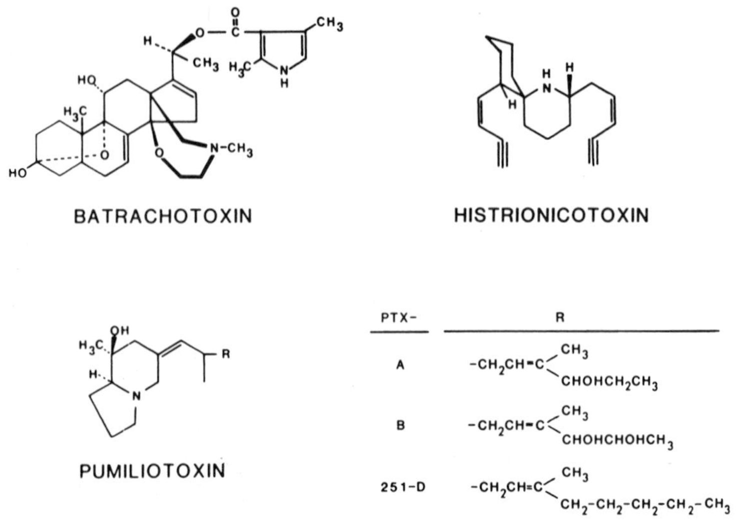

BATRACHOTOXIN

HISTRIONICOTOXIN

PUMILIOTOXIN

PTX-	R
A	$-CH_2CH=C\begin{smallmatrix}CH_3\\CHOHCH_2CH_3\end{smallmatrix}$
B	$-CH_2CH=C\begin{smallmatrix}CH_3\\CHOHCHOHCH_3\end{smallmatrix}$
251-D	$-CH_2CH=C\begin{smallmatrix}CH_3\\CH_2-CH_2-CH_2-CH_2-CH_3\end{smallmatrix}$

Figure 1. Structures of toxins found in South American frogs of the genus *Dendrobates*. Batrachotoxin (molecular weight 538) is obtained from *Phyllobates aurotaenia*; histrionicotoxin (molecular weight 268.21) is obtained from *Dendrobates histrionicus*; pumiliotoxins are obtained from *D. pumilio* (molecular weights: PTX-A, 307.26; PTX-B, 323.26; 251-D, 251.23).

against predators. Indeed, the dendrobatid frogs possess over 100 different alkaloids each with rather novel features that are not found in any other source in nature (Daly *et al.*, 1978). Most of these alkaloids are highly active pharmacological agents, and understanding their mechanism and site of action is currently the subject of great interest. For example, the complex steroidal alkaloid batrachotoxin (Fig. 1) exerts most of its effect by increasing voltage-dependent sodium permeability and thus has been used by a large number of investigators as a specific probe to study the molecular nature of the sodium channel. The toxin reacts with a site in the sodium channel different from that of tetrodotoxin (Albuquerque and Daly, 1976; Catterall, 1979). Another class of important toxins found in the dendrobatid frogs are the histrionicotoxins (Fig. 1). These azaspiro undecane compounds have been the key to an understanding of the molecular nature of the acetylcholine receptor (Albuquerque *et al.*, 1973; Eldefrawi *et al.*, 1978). The introduction in 1973 of histrionicotoxin and its analogs opened the field for biochemical and biophysical studies of the ionic channel of the acetylcholine receptor.

We recently introduced a novel series of toxins that represent two new classes of indolizidine alkaloids and were obtained from the poison frog *Dendrobates pumilio*. The structure and absolute configuration of this unique series of alkaloidal toxins from this Ecuadoran frog were revealed by X-ray crystallography. The first and simplest member of the pumiliotoxin A class was found to be 8-hydroxy-8-methyl-6-(2'-methylhexylidene)-1-azabicyclo[4.3.0]nonane (Daly *et al.*, 1980). This compound, 251-D ($C_{16}H_{29}NO$) (Fig. 1), was the first structurally defined member of the pumiliotoxin A class of 24 dendrobatid alkaloids. Further analysis of mass spectra allowed formulation of the structures of six additional members of this class of alkaloids, particularly pumiliotoxin-A (PTX-A; $C_{19}H_{33}NO_2$) and pumiliotoxin-B (PTX-B; $C_{19}H_{33}NO_3$). Mass spectra distinguished PTX-A from PTX-B, the latter having an additional hydroxyl group at C-16 (Fig. 1). The pharmacological effects of PTX-B, which have been detailed elsewhere (Albuquerque *et al.*, 1981; Tamburini *et al.*, 1981), are unique in altering Ca^{2+}-dependent events in nerve and muscle.

Recent studies in our laboratories on pumiliotoxins suggest that these compounds may be extremely useful in delineating the role of Ca^{2+} in electrogenic tissue.

2. RESULTS AND DISCUSSION

2.1 Interaction of PTX-B at the Presynaptic Nerve Terminal

After exposing a frog sartorius muscle to PTX-B (≤ 1.5 μM), single stimuli applied to the nerve induced a large number of repetitive endplate potentials (EPPs) and increased the frequency of potassium-evoked transmitter release but not the amplitude and frequency of spontaneous transmitter release (Fig. 2). The repetitive EPPs persisted throughout the presence of PTX-B and also after washout for 60–120 min. In every case, each successive

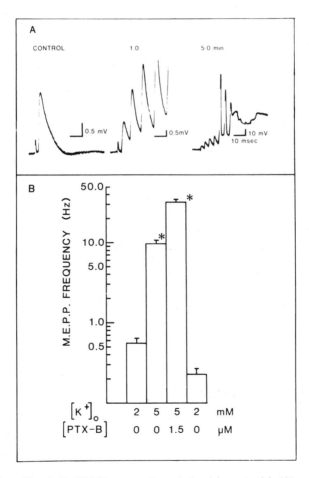

Figure 2. Effect of pumiliotoxin-B (PTX-B) on neurally evoked endplate potentials (A) and potassium-evoked transmitter release (B) in frog sartorius muscles. In (A), the external potassium calcium concentration was reduced to 0.045 mM to prolong the appearance of repetitive endplate potentials to single shocks to the nerve at 1 and 5 min after exposure to PTX-B (1.5 μM). In (B), the external potassium concentration ($[K^+]_o$) was first increased to 5 mM prior to addition of the toxin. The significant ($p < 0.001$) increases in miniature endplate potential (MEPP) frequency from control level and from 5 mM $[K^+]_o$ are indicated by an asterisk.

EPP was greater in amplitude than the preceding EPP. It should be mentioned that this effect of PTX-B on EPPs is unrelated to any effect on sodium, potassium, or chloride conductance, for the toxin does not cause membrane depolarization, does not alter the nerve or muscle action potential, delayed sodium inactivation, or delayed rectification of the muscle fiber, the latter property being ascribed to an increase in potassium conductance. Thus, the toxin's action differs markedly from agents such as tetraethylammonium (Katz and Miledi, 1966). 9-aminoacridine (Volle, 1970), germine (Detwiler, 1972), and tityustoxin (Warnick *et al.*, 1976), which also have the ability to potentiate muscle twitch.

Decreasing the external calcium concentration ($[Ca^{2+}]_o$) by 75 and 97.5% did not affect the action of the toxin on the EPP. Even at the lowest concentration of $[Ca^{2+}]_o$ examined (0.045 mM), where EPPs were less than 0.4 mV in amplitude, PTX-B-treated nerves were still able to generate EPPs. Thus, at any particular $[Ca^{2+}]_o$ between 0.045 and 1.8 mM, successive EPPs are potentiated to a greater extent than the second EPP in the train; facilitation is, however, greater at 1.8 mM $[Ca^{2+}]_o$. Such findings are in agreement with the proposed role of Ca^{2+} in the release process from the nerve terminal (Katz and Miledi, 1965, 1967a, 1968; Rahamimoff, 1968). Simultaneous extracellular recordings of nerve terminal spikes and EPPs from surface fibers of the frog sartorius muscle normally evoked one nerve terminal spike and an EPP in response to single shocks to the nerve. In the presence of PTX-B, repetitive EPPs were generated, each preceded by a nerve terminal spike. In another series of experiments we decided to verify whether or not the nerve terminal potential was involved in the action of PTX-B. Using tetrodotoxin (0.3 μM) to block the nerve terminal spike, we evoked EPPs by focal depolarization using a 2 M NaCl electrode positioned over the nerve terminal arborization in frog sartorius muscles. Because the quantal nature of the neurally evoked EPP is no longer all-or-none, an EPP of graded amplitude can be evoked by simply altering the magnitude and duration of the applied current (Katz and Miledi, 1967b). The relationship between the magnitude of the evoked EPP (in mV) and that of the current applied through the micropipet (in μA) assumes a nearly sigmoidal form. When PTX-B was applied to the muscle, the curve relating EPP amplitude to current applied, shifted to the left so that lower currents were required to evoke the EPP but without altering the shape or magnitude of the curve. Thus, the involvement of the presynaptic nerve terminal during sodium inactivation to induce the marked increase in evoked transmitter release is most unlikely. Indeed, this experiment also renders most unlikely the involvement of sodium conductance but supports the notion that the effect of the toxin on evoked EPPs and MEPPs is a calcium-dependent phenomenon. One then must consider the possibility that the toxin binds to calcium receptive sites and competitively displaces the ion, a phenomenon possibly coupled with an inhibition of Ca^{2+}-dependent ATPase. Under this condition, one might expect a progressive depression of the EPPs as is usually observed during long periods of repetitive stimulation. In these experiments, the apparent increase in the flux of calcium into the presynaptic nerve terminal or its release from binding sites within the nerve terminal has resulted in EPPs that increase in amplitude.

An additional possibility is that of calcium–sodium exchange and prolongation of the nerve terminal in frog sartorius muscles by PTX-B cannot be excluded nor can the possibility that the toxin may generate Ca^{2+}-dependent action potentials which are responsible for the repetitive firing in the nerve terminal. In crayfish muscle, relatively high concentrations of PTX-B (30 μM) were required before a significant effect on calcium-dependent potentials occurred (Albuquerque *et al.*, 1981).

As will be shown below, PTX-B alters the contraction of skeletal muscle by causing repetitive activation of the sarcolenimal membrane and thereby facilitating the release of calcium from the sarcoplasmic reticulum and by blocking the reuptake of calcium by Ca^{2+}-dependent ATPase. PTX-B may have similar actions in nerve, facilitating calcium exchange across the nerve terminal membrane; it is unlikely that this toxin is mobilizing Ca^{2+} from endoplasmic reticulum (Rahamimoff, 1968). PTX-B has no effect on Ca^{2+}-dependent ATPase of mitochondria (Albuquerque et al., 1981; Tamburini et al., 1981; E. Racker, personal communication). Although the effect of PTX-B on Ca^{2+}-dependent ATPase of the nerve terminal has not been investigated, an increase in the flux of calcium across the membrane to internal sites is essential for the release of vesicular contents and for an increase in the EPP amplitude. A sustained increase in internal Ca^{2+} caused by blockade of a Ca^{2+}-pump of the endoplasmic reticulum with PTX-B might be expected to lead to an increased probability of the release of quanta.

Finally, the blockade of evoked activity may depend on several mechanisms. First, PTX-B may saturate the sites for Ca^{2+} entry into the nerve terminal. Second, the internal Ca^{2+} may rise as a result of: (1) a blockade of the Ca^{2+} pump of the endoplasmic reticulum, and/or (2) as a consequence of the influx of Ca^{2+}. Lastly, and perhaps as a result of the foregoing, with the increase in internal Ca^{2+}, the internal membrane potential builds toward that of the equilibrium potential for Ca^{2+} (Katz and Miledi, 1967a,b), effectively terminating further influx of extracellular Ca^{2+} necessary for the evoked release.

2.2 Effect on the Contractile Properties of Skeletal Muscle

When applied either to whole frog sartorius muscle or to small bundles (15–25 fibers) of fibers from semitendinosus muscles, PTX-B (50 nM–30 μM) produces a transient concentration, and frequency-dependent potentiation and prolongation of the directly evoked muscle contraction. To eliminate any influence of the nerve and receptor effects of PTX-B, most muscles were first exposed to α-bungarotoxin (5 μg/ml) to inactivate acetylcholine receptors. Figure 3 illustrates the full concentration–response curve observed in bundles of semitendinosus fibers together with an example of a control muscle contraction and one at

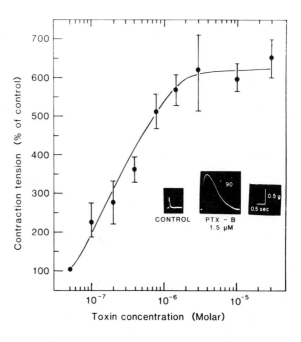

Figure 3. Dose-response relationship for the potentiation of contraction by varying concentrations of pumiliotoxin-B (PTX-B) in small bundles of frog semitendinosus fibers. Peak responses were obtained between 5 sec and 3 min after application of the toxin, depending upon the concentration. The insets show a typical control response of the 15- to 25-fiber bundles and the peak response to PTX-B at 90 sec after exposure. The muscles were stimulated repetitively at a frequency of 0.2 Hz.

peak effect of PTX-B action (1.5 μM; at 90 sec). The action of the toxin occurs in a nearly instantaneous manner upon its addition to the bath. Because the muscle was stimulated at a frequency of 0.2 Hz (i.e., every 5 sec), the first contraction upon the addition of toxin was always less than 5 sec distant. Immediately, the later part of the falling phase was protracted and the amplitude increased. Subsequently, the rising phase became longer, the prolonged falling phase increased in amplitude and combined with the increased rising phase to yield the tracing in Fig. 3 shown at the peak of toxin effect at 2 min. At lower frequencies of stimulation (i.e., 0.05 and 0.0167 Hz), the potentiation reached values in excess of 10-fold greater than control. This potentiation and its prolongation from an average 90% relaxation time of 45 msec to 3000 msec at 0.0167 Hz by 1.5 μM PTX-B extended at least over a 30-min period rather than 3–5 min at 0.2 Hz, and was completely reversible upon washing.

If the muscles were stimulated only once every 2–5 min, a peak increase in tension of 425% could be attained, contraction time increased 8- to 9-fold, half-relaxation time 15-fold, and 90% relaxation time 17-fold. When stimulated at a frequency of 10 Hz/1 sec, the bundles of semitendinosus fibers responded with single distinct twitches superimposed upon a graded increase in muscle tension reaching about 200 mg at the end of the train (Fig. 4). At 20 Hz, the amplitude of the individual responses is severely reduced but visible and tetanic fusion of the response begins. Tetanic fusion is obvious at 50 and 100 Hz. At the end of the train of pulses, tension drops to prestimulation levels within 100–150 msec. In the presence of PTX-B, tetanic stimulation at 10 Hz results in nearly complete fusion within 5 min, somewhat similar to that of the control at 50 Hz. After cessation of stimulation, the falling phase appears to be biphasic. The response at 20 Hz closely approaches maximal tension at 100 Hz. While the tension induced by 100-Hz stimulation only increases 20%, the tension generated at 10, 20, and 50 Hz rises dramatically in the presence of PTX-B to the tension developed at 100 Hz (Fig. 4). Thus, tetanic fusion occurs at lower frequencies in the presence of PTX-B, tetanic tension is increased more at 10 and 20 Hz than at 50 and 100 Hz, and an after-contraction is apparent upon cessation of tetanic stimulation.

The twitch/tetanus ratio at 100 Hz was increased in the presence of PTX-B from 0.3 to more than 1.1 msec as a result of the increase in contraction amplitude rather than the tetanic tension developed. In these experiments, single stimuli were applied intermittently (at 2- to 5-min intervals) and thus tension developed was greater than at higher stimulus frequencies (i.e., once every 2–5 min vs. once every 5 sec).

In the absence of external calcium (i.e., in the presence of EGTA), PTX-B prolonged

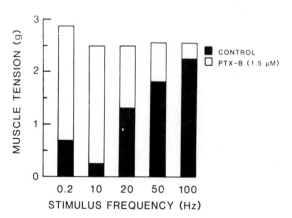

Figure 4. Muscle tension developed by small bundles of fibers of frog semitendinosus muscles before and after exposure to pumiliotoxin-B (PTX-B) at various stimulus frequencies. At 0.2 Hz the tension represents the amplitude of the single twitch. At 10 Hz, the tension is that increase in baseline tension without regard to twitches, while at 20–100 Hz the tension developed is not obscured by single responses. The muscle bundles were not stimulated continually but only every 2.5 to 5 min. The data presented for the toxin were obtained after 25–30 min. Note the change in twitch/tetanus ratio from 0.31 for control to 1.1 for PTX-B.

but did not potentiate the twitch. Restoration of calcium and simultaneous removal of toxin allowed return of normal twitch characteristics. The muscle then responded to applied PTX-B in a normal manner. Neither methoxyverapamil, a compound thought to block the calcium channel (Kohlhardt *et al.*, 1972; Baker *et al.*, 1973), nor dantrolene, which is proposed to inhibit calcium release from the sarcoplasmic reticulum (Van Winkle, 1976; Morgan and Bryant, 1977), greatly altered the response to PTX-B. Although the absolute magnitudes of the responses to PTX-B were reduced by dantrolene (8.3 μM), the percentage changes were similar to controls. This is most probably caused by the reduction in twitch amplitude by dantrolene to 20% of the controls. With methoxyverapamil (50 μM), the twitch was only reduced by 20%; PTX-B potentiated the twitch by 226% (260 ± 80% in controls) and prolonged the twitch up to 80% of the maximum in the presence of methoxyverapamil. The results with these calcium antagonists are equivocal at best primarily because their mechanisms are as yet uncertain and effective concentrations may not have been attained.

The wavelength of the action potential elicited by a 2 msec pulse in muscle was unaffected by the toxin (i.e., there was no change in amplitude, threshold, overshoot, or rate of rise). However, in the presence of toxin it began to fire repetitively at frequencies of 100–160 Hz for periods of up to 100 msec after the stimulus ceased and was superimposed on a generator-type potential. A reduction in the extracellular Cl^- concentration had no effect on this repetitive activity but a reduction of $[Ca^{2+}]_0$ or the application of EGTA or tetrodotoxin blocked repetitive activity but not the generation potential.

Certainly the evidence thus far points to the primary involvement of calcium in the action of PTX-B. We therefore examined a partially calcium-dependent action potential in crayfish skeletal muscle. To obtain the "calcium" spike, NaCl was partially replaced with $SrCl_2$ (100 mM). When exposed to PTX-B at concentrations from 1.5 to 60 μM for 1 to 2 hr, the threshold, overshoot, and amplitude of the action potential remained unchanged. At 15 μM, the rate of rise was increased 25% and the time to half-amplitude of the action potential was decreased 52%. A further decrease in the rate of rise occurred at 30 μM PTX-B. At higher concentrations, the contractions induced by such direct stimulation were so intense as to completely sever the bundle of fibers.

Thus, PTX-B causes significant increases in the amplitude and duration of the twitch and of the tetanic responses of muscle fibers. The twitch amplitude approaches that of a 100-Hz tetanus in the same muscle; the tetanic responses at low frequencies (10 to 20 Hz) increase to the greatest extent and approach those at 50 and 100 Hz, which increase to the smallest extent in the presence of PTX-B. Tetanic responses are, in addition, followed by an after-contraction. The ability of PTX-B to induce these changes in amphibian skeletal muscle, the reduction of the effect by EGTA, and the increase in the rate of rise but decrease in the duration of a Ca^{2+}-dependent action potential in crayfish skeletal muscle suggest that the actions of PTX-B are due to displacement and/or release of calcium from extracellular sites and at sites within the sarcoplasmic reticulum. A highly appealing mechanism of action is an inhibition of Ca^{2+}-dependent ATPase (*vide infra*).

2.3. Effect of Indolizidine Alkaloids on Ca^{2+}-Dependent ATPase Activity of the Sarcoplasmic Reticulum

Because both prolongation and potentiation of muscle twitch by PTX-B could be related to alteration in Ca^{2+} binding and translocation in muscle, we examined the effects of this agent and two other closely related indolizidine alkaloids on Ca^{2+}-dependent ATPase in sarcoplasmic reticulum (SR) preparations from muscles of several species (Albuquerque

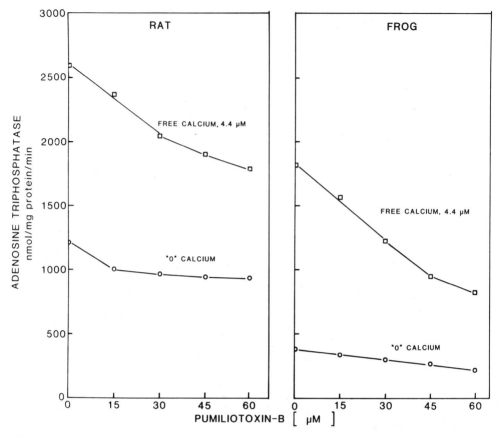

Figure 5. Inhibition of ATPase in frog sarcoplasmic reticulum. Fractions (0.05 μg protein) were incubated at 30°C for 20 min in the presence and absence of 4.4 μM free Ca^{2+} and increasing concentrations of pumiliotoxin-B. Assay conditions are as described by Tamburini *et al.* (1981). Each point is the average of three replicate samples.

et al., 1981; Tamburini *et al.*, 1981). PTX-B appears to be a remarkably potent and specific inhibitor of this activity. Figure 5 illustrates the effect of this toxin on Ca^{2+}-dependent ATPase in SR from frog hindlimb muscles. Assays were performed in the presence of 4.4 μM free Ca^{2+}. The slight inhibition noted in the absence of added calcium is probably due to small amounts of Ca^{2+}-dependent ATPase activated by residual calcium in the crude muscle preparation. Kinetic analysis of the inhibition by PTX-B indicates that this is noncompetitive with Ca^{2+} (Fig. 6). The inhibition constant (K_i) of PTX-B for Ca^{2+}-dependent ATPase in frog SR is about 20 μM. Similar inhibition was observed in SR preparations from chickens (Albuquerque *et al.*, 1981) and rats (Tamburini *et al.*, 1981). The finding that only about 40% of Ca^{2+}-dependent ATPase is inhibited by PTX-B in rat muscle while 251-D completely inhibits this activity (Table I) (Tamburini *et al.*, 1981) supports the conclusion that there is more than one form of Ca^{2+}-dependent ATPase in mammalian SR. PTX-A, which has only one hydroxyl group in the side chain, was a very weak inhibitor of Ca^{2+}-ATPase (Table I).

The remarkable difference between 251-D and the other two forms of the toxin in inhibiting Ca^{2+}-ATPase suggests that these toxins may be useful in delineating the function of various forms of SR ATPase. The action of PTX-A and PTX-B on Ca^{2+}-ATPase cor-

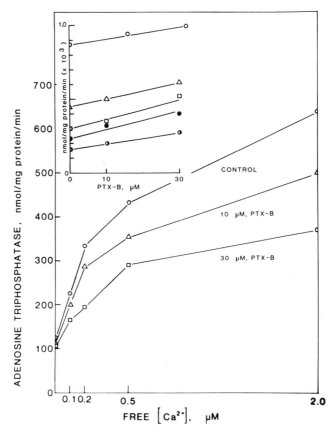

Figure 6. Inhibition of Ca^{2+}-dependent ATPase in frog sarcoplasmic reticulum as a function of Ca^{2+} concentration. Assays were performed in the presence or absence of 10 μM and 30 μM PTX-B in a reagent consisting of 20 mM imidazole HCl, pH 7.0, 100 μM KCl, 1 mM $MgCl_2$, 2 mM ATP, 0.1 mM EGTA, and various amounts of Ca^{2+} to give the following free Ca^{2+} concentrations: (o) Zero Ca^{2+}; ($\triangle$) 0.1 μM; ($\square$) 0.2 μM; ($\bullet$) 0.5 μM; (◑) 2 μM.

Table I. Effect of Pumiliotoxin-B on ATPase Activity from Sarcoplasmic Reticulum and Brain Synaptosomes [a]

	Source of ATPase		
	Sarcoplasmic reticulum		Rat brain synaptosomes
Toxin	$-Ca^{2+}$	$+Ca^{2+}$ [b]	$+Na^+$ and K^+ [c]
None	0.87 [d]	3.10	20.9
PTX-A (40 μM)	0.87	2.91	22.5
PTX-B (100 μM)	0.71	2.12	20.8
251-D (100 μM)	0.23	0.23	5.8

[a] Values are averages from three replicate samples. Sarcoplasmic reticulum and synaptosomes were isolated from hindlimb muscle and brain of adult rats (200 g). The isolation procedures and assay of ATPase activity are described in Tamburini *et al.* (1981). PTX-B only inhibited about 40% of the total Ca^{2+}-activated ATPase in rat muscle with concentrations ranging up to 100 μM.
[b] Ca^{2+} was present at 4.4 μM.
[c] Na^+ and K^+ were present at 100 and 40 μM, respectively.
[d] Values are μmoles/mg protein per min.

relates with their action on muscle contraction, PTX-A being considerably less potent than PTX-B. To date, we have not carried out an extensive study of the action of 251-D on muscle contraction, but this compound appears to be at least as active as PTX-B in prolonging muscle relaxation.

251-D differs from the two other toxins in its effects on synaptosomal ATPase activity, which consists mainly of the Na^+/K^+-ATPase and Mg^{2+}-ATPase. Neither PTX-B nor PTX-A affected this activity whereas 251-D markedly inhibited the synaptosomal ATPase. Indeed, 100 μM 251-D was a better inhibitor of synaptosomal ATPase than 100 μM ouabain, which only inhibited this activity by about 25%. Thus, 251-D appears to be less specific than PTX-A and PTX-B because it acts on Na^+/K^+-ATPase and Mg^{2+}-ATPase in addition to Ca^{2+}-ATPase.

The selective inhibition of Ca^{2+}-related events in nerve and muscle by PTX-B has provided new insight into the pathology of avian muscular dystrophy. In muscles from normal chickens, PTX-B produces effects similar to those noted in amphibian and mammalian skeletal muscle, namely an increase in the amplitude and duration of muscle contraction and enhanced transmitter release and multiple EPP discharge after a single nerve stimulus (Albuquerque *et al.*, 1981). In contrast, these effects either were not observed or were greatly attenuated in muscles from dystrophic chickens (Line 413). These effects on the contractile and electrical properties of muscle and nerve correlated with a failure of PTX-B to inhibit Ca^{2+}-ATPase in SR preparations from dystrophic chicken muscle (Albuquerque *et al.*, 1981).

ACKNOWLEDGMENTS. This work was supported in part by USPHS Grants NS-12063 and NS-14728, The Muscular Dystrophy Association of America, and US Army Grant DAAG 29-78-G-0203.

REFERENCES

Albuquerque, E. X., and Daly, J. W. (1976). In *Receptors and Recognition* (P. Cuatrecasas, ed.), Series B, Vol. 1, pp. 297–338, Chapman & Hall, London.

Albuquerque, E. X., Barnard, E. A., Chiu, T.-H., Lapa, A. J., Dolly, J. O., Jansson, S.-E., Daly, J., and Witkop, B. (1973). *Proc. Natl. Acad. Sci. USA* **70**, 979–953.

Albuquerque, E. X., Warnick, J. E., Tamburini, R., Kauffman, F. C., and Daly, J. W. (1981). In *Disorders of the Motor Unit* (D. L. Schotland, ed.), pp. 611–626, J. Wiley & Sons, New York.

Baker, R. F., Meves, H., and Ridgway, E. B. (1973). *J. Physiol. (London)* **231**, 511–526.

Catterall, W. A. (1979). *Adv. Cytopharmacol.* **3**, 305–316.

Changeux, J.-P., Kasai, M., and Lee, C. Y. (1970). *Proc. Natl. Acad. Sci. USA* **67**, 1241–1247.

Daly, J. W., Brown, G. B., Mensah-Dwumah, M., and Myers, C. W. (1978). *Toxicon* **16**, 163–168.

Daly, J. W., Tokuyama, T., Fujiwara, T., Highet, R. J., and Karle, I. L. (1980). *J. Am. Chem. Soc.* **102**, 830–837.

Detwiler, P. B. (1972). *J. Pharmacol. Exp. Ther.* **180**, 244–254.

Eldefrawi, M. E., Eldefrawi, A. T., Mansour, N. A., Daly, J. W., Witkop, B., and Albuquerque, E. X. (1978). *Biochemistry* **17**, 5474–5484.

Endo, M. (1977). *Physiol. Rev.* **57**, 71–108.

Frank, M., and Sleatos, W. W. (1975). *N. S. Arch. Pharmakol.* **290**, 35–47.

Hajdu, S., and Leonard, E. (1961). *Circ. Res.* **9**, 1291–1297.

Jolly, S. R., and Gross, G. J. (1979). *Eur. J. Pharmacol.* **54**, 289–293.

Kao, C. Y. (1966). *Pharmacol. Rev.* **18**, 997–1049.

Katz, B., and Miledi, R. (1965). *Nature (London)* **207**, 1097–1098.

Katz, B., and Miledi, R. (1966). *Nature (London)* **212**, 1242–1245.

Katz, B., and Miledi, R. (1967a). *Proc. R. Soc. London Ser. B* **167**, 8–22.

Katz, B., and Miledi, R. (1967b). *Proc. R. Soc. London Ser. B* **167**, 23–28.

Katz, B., and Miledi, R. (1968). *J. Physiol.* (*London*) **195**, 481–492.

Kohlhardt, M., Bauer, B., Krause, H., and Fleckenstein, A. (1972). *Pfluegers Arch. Eur. J. Physiol.* **335**, 309–322.

Lee, C. H., and Berkowitz, B. A. (1977). *J. Pharmacol. Exp. Ther.* **202**, 646–653.

Lee, C. Y. (1971). In *Neuropoisons: Their Pathophysiological Actions* (L. L. Simpson, ed.), pp. 21–70, Plenum Press, New York.

McLean, M. J., Shigenobu, K., and Sperelakis, N. (1974). *Eur. J. Pharmacol.* **26**, 379–382.

MacLennan, D. H. (1970). *J. Biol. Chem.* **245**, 4508–4518.

Morgan, K. G., and Bryant, S. H. (1977). *J. Pharmacol. Exp. Ther.* **201**, 138–147.

Nagao, T., Matlib, M. A., Frankin, D., Millard, R. W., and Schwartz, A. (1980). *J. Mol. Cell. Cardiol.* **12**, 29–43.

Narahashi, T. (1974). *Physiol. Rev.* **54**, 813–889.

Rahamimoff, R. (1968). *J. Physiol.* (*London*) **195**, 471–481.

Rahwan, R. G., Raust, M. M., and Witisk, D. T. (1977). *J. Pharmacol. Exp. Ther.* **201**, 126–137.

Raschak, M. (1976). *Arzneim. Forsch.* **26**, 1330–1333.

Scales, B., and McIntosh, D. A. D. (1968). *J. Pharmacol. Exp. Ther.* **160**, 249–260.

Shoshan, V., Campbell, K. P., MacLennan, D. H., Frodis, W., and Britt, B. A. (1980). *Proc. Natl. Acad. Sci. USA* **77**, 4435–4438.

Tamburini, R., Albuquerque, E. X., Daly, J. W., and Kauffman, F. C. (1981). *J. Neurochem.* **37**, 755–780.

Van Winkle, W. B. (1976). *Science* **193**, 1130–1131.

Vogel, S., Crompton, R., and Sperelakis, N. (1979). *J. Pharmacol. Exp. Ther.* **210**, 378–385.

Volle, R. L. (1970). *Life Sci.* **9**, 753–758.

Warnick, J. E., Albuquerque, E. X., and Diniz, C. R. (1976). *J. Pharmacol. Exp. Ther.* **198**, 155–167.

Watanabe, A. M., and Besch, H. R., Jr. (1974). *J. Pharmacol. Exp. Ther.* **191**, 241–251.

138

The Acetylcholine-Receptor-Regulated Ion Transport System of Skeletal Muscle and Electric Organs

Mohyee E. Eldefrawi and Amira T. Eldefrawi

The nervous system regulates body function by means of a highly specialized chemical communication system, where recipient postsynaptic cells carry receptors for specific neurotransmitters. The vertebrate motor endplate represents a synapse where the motor neuron releases acetylcholine (ACh) to transmit its commands to the skeletal muscle fiber, which carries densely packed receptors ($2 \times 10^4/\mu m^2$) at the tops of the junctional folds in juxtaposition to the release sites.

1. THE IONIC CHANNEL AS A TRANSDUCER

Unlike most other neurotransmitter and hormone receptors, the ACh receptor in muscle endplates is not coupled to a nucleotide cyclase, i.e., a strong transducer that amplifies 10^3- to 10^4-fold the signal resulting from receptor activation (Ariens and Rodrigues de Miranda, 1979). Instead it contains an ionic channel whose activation results in increased influx of Na^+ and efflux of K^+ along their chemical gradients, causing localized depolarization of the cell membrane in the endplate area (Stevens, 1980). When activated, a single ionic channel opens for approximately 1 msec and transports $\sim 10^4$ ions, resulting in ~ 0.1 μV of depolarization (Katz and Miledi, 1972). Thus, it is a weak transducer. Spontaneous release of a single quantum of ACh from the motor neuron results in activation of $\sim 10^3$ channels, causing a depolarization of ~ 1 mV or less miniature endplate potential, which is insufficient to trigger muscle contraction. The simultaneous release of 100–200 quanta of ACh as a result of nerve stimulation activates enough ionic channels ($\sim 10^5$–10^6) to produce depolarizations sufficient to trigger an action potential whose prop-

Mohyee E. Eldefrawi and Amira T. Eldefrawi • Department of Pharmacology and Experimental Therapeutics, University of Maryland School of Medicine, Baltimore, Maryland 21201.

agation along the muscle fiber leads to muscle contraction. This ionic channel is selective against anions but discriminates little among small cations. It is permeable even to organic cations as large as choline and glycine ethylester and is even more permeable to ammonium, hydroxylamine, and guanidine than to Na^+ (Dwyer *et al.*, 1980). The ionic channel lifetime is both voltage and temperature sensitive (Anderson and Stevens, 1973), and the nature of the permeant cation affects both channel lifetime and conductance (Van Helden *et al.*, 1977). The high conductance and low selectivity for metal ions are consistent with a large water-filled open pore (6.5×6.5 Å) with neighboring negative surface charges of low field strength but none of high field within the pore (Adams *et al.*, 1980), though data with ammonium solutions suggest an endplate channel containing high-field-strength neutral sites (Takeda *et al.*, 1980).

2. STRUCTURE OF THE ACh-RECEPTOR/CHANNEL MOLECULE

This ACh-receptor/channel molecule is an integral membrane glycoprotein, which is found in micromolar concentrations (1 nmole/g tissue) in the electric organs of *Torpedo*. The pure protein was isolated by affinity chromatography from detergent extracts of electric organs (Eldefrawi and Eldefrawi, 1973a). Although most biochemical evaluations were made on the electric organ receptor/channel molecule, the cross-reactivity of antireceptor antibodies raised against the electric organ receptor with mammalian skeletal muscle receptors (Lindstrom and Dau, 1980) strongly suggests that the two receptors are similar, which is also evidenced by their drug specificity (Moreau and Changeux, 1976).

The receptor/channel molecule of *Torpedo* electric organs appears to be ellipsoid, has an apparent molecular weight estimated to be 250,000–330,000 (Edelstein *et al.*, 1975; Reynolds and Karlin, 1978), is asymmetric, is 110 Å in length and tranverses the bilayer, extending some 55 Å extracellularly and 15 Å intracellularly as revealed by X-ray diffraction (Ross *et al.*, 1977). The molecule exists in *Torpedo* membranes mainly as a disulfide cross-linked dimer, but there is no functional significance for this dimeric nature since the ACh receptor of the electric eel exists as a monomer (Chang and Bock, 1977; Hamilton *et al.*, 1977). It is made up of five subunits as revealed by SDS electrophoresis ($\alpha\beta\gamma\delta$) at ratios of 2 : 1 : 1 : 1 (Raftery *et al.*, 1980). There is a considerable amount of homology among the four subunits at the NH_2 termini, such that at 11 of 54 points, all four subunits have the same amino acids (Raftery *et al.*, 1980). These subunits are exposed extracellularly because they are all glycosylated (Lindstrom *et al.*, 1979), and interact with extracellularly applied antibodies of each of the subunits (Claudio and Raftery, 1977) and lactoperoxidase (Hartig and Raftery, 1977). They also appear to be exposed on the internal surface of the membrane as shown by trypsin hydrolysis (Strader and Raftery, 1980).

The smallest subunit (α, molecular weight 40K) carries the recognition site for ACh and other cholinergic drugs (Damle and Karlin, 1978; Rübsamen *et al.*, 1978), but the function of the other three subunits (50K, 60K, and 65K) is still unclear. Reconstitution of the purified protein (Nelson *et al.*, 1980) argues strongly that the receptor and channel are one molecule. Our studies on a number of channel probes ([³H]perhydrohistrionicotoxin, [³H]phencyclidine, and [³H]imipramine) suggest that there is more than one binding site for these drugs (Eldefrawi *et al.*, 1980b), and we may have to use several of them before a final conclusion is reached regarding which subunit(s) constitutes the channel. Some of these sites may be on the 50K or 66K subunits (Oswald *et al.*, 1980; Saitoh *et al.*, 1980).

At present there is no compelling reason not to consider that all four subunits, including the 40K, may contribute to the structure of the ionic channel.

3. MULTIPLICITY OF BINDING SITES ON THE RECEPTOR/CHANNEL COMPLEX

The structure of the receptor/channel monomer suggests that there may be two sites for binding ACh, and occupation of the two sites may be necessary to open the channel as shown by voltage clamp (Stevens, 1980) and tracer ion flux (Hess *et al.*, 1979) studies. Binding of agonists and antagonists to these sites is mutually exclusive, but the two sites are not intrinsically equivalent (Sine and Taylor, 1980; Damle and Karlin, 1978). Binding of agonists to these sites is positively cooperative (Eldefrawi and Eldefrawi, 1973b). In addition to these voltage-independent ACh and snake venom toxin binding sites (identified as receptor sites), the receptor/channel molecule has voltage-dependent sites (identified as channel, allosteric, or local anesthetic sites), which bind drugs that differ widely in structure and action such as the antiviral amantadine (Tsai *et al.*, 1978), the local anesthetic piperocaine (Tiedt *et al.*, 1979), the ganglionic blocker tetraethylammonium (Adler *et al.*, 1979), the antimalarial quinacrine (Grünhagen and Changeux, 1976; Tsai *et al.*, 1979), the general anesthetic phencyclidine (Albuquerque *et al.*, 1980a), the tricyclic antidepressant imipramine (Eldefrawi *et al.*, 1981), the antispasmodic meproadifen (Krodel *et al.*, 1979), and perhydrohistrionicotoxin (Eldefawi *et al.*, 1977, 1978). There appears to be a variety of channel sites; some are extracellularly exposed, some are intracellular, while others are exposed only when the channel opens (Horn *et al.*, 1980; Albuquerque *et al.*, 1980b).

Antireceptor antibodies raised in experimental animals bind to several determinants or antigenic sites on the receptor/channel molecule. Although some antibodies bind to the ACh-receptor site, thereby blocking binding of ACh or α-bungarotoxin (Eldefrawi, 1978), the majority seem to bind to other determinants (Lindstrom and Dau, 1980) and accelerate receptor degradation or attract complement and cause membrane lysis. Some antibodies may inhibit receptor function through interference with the agonist-induced conformational changes (Karlin *et al.*, 1978; Desouki *et al.*, 1981). Recently, monoclonal antibodies were isolated that identify selective antigenic sites on all four subunits (Tzartos and Lindstrom, 1980).

The receptor/channel molecule binds Ca^{2+} possibly at the receptor site as well as at other sites (Eldefrawi *et al.*, 1975; Chang and Neumann, 1976; Rübsamen *et al.*, 1978), which affects receptor affinity for cholinergic ligands (Cohen *et al.*, 1974) and receptor desensitization. In addition, the carbohydrate moiety of the receptor/channel molecule has binding sites for lectins (Meunier *et al.*, 1974), and the proteins appear to be closely associated with certain lipids (Chang and Bock, 1979).

In the presence of ATP and an endogenous protein kinase, the receptor/channel molecule is phosphorylated (Gordon *et al.*, 1977; Teichberg *et al.*, 1977). The site of phosphorylation appears to be on the δ chain. Also, ATP is found in presynaptic vesicles along with ACh at a 1 : 4 molar ratio (Whittaker and Dowdall, 1975) and may be released with ACh. Recently, it has been shown that the ACh-receptor/channel molecule is an excellent substrate for the enzyme carboxymethylase, which is found in the electric organ (Kloog *et al.*, 1980). It is not known if the receptor/channel is methylated *in vivo*, and whether this and phosphorylation have functional significance.

4. ALLOSTERIC INTERACTIONS AND CONFORMATIONAL STATES

When the ACh-receptor/channel molecule was still a hypothetical entity, it was proposed that its ability to regulate ion transport resulted from its undergoing conformational changes upon binding of ACh (Nachmansohn, 1955). Models of receptor–drug interaction show three conformational states, and these are detected by monitoring changes in intrinsic fluorescence (Barrantes, 1976), extrinsic fluorescence probes (Grünhagen and Changeux, 1976; Quast *et al.*, 1979), and agonist affinities shown by kinetics of α-bungarotoxin binding (Weber *et al.*, 1975). In addition, binding of agonists to the receptor sites increases the affinities of the channel sites two- to fourfold for [³H]perhydrohistrionicotoxin and [³H]phencyclidine and increases their initial rate of binding up to several hundredfold (Aronstam *et al.*, 1981; Eldefrawi *et al.*, 1980a,b). Binding of antagonists (except for α-bungarotoxin) also increases the initial rate of channel drug binding, but to a lower degree than for agonists. This suggests that binding of antagonists induces a conformation of the receptor/channel complex that is nonconducting and different from the three other detected conformations (i.e., resting, active, desensitized). Conversely, binding of drugs to the ionic channel sites increases the affinity of ACh and other agonists to the receptor site (Kato and Changeux, 1976). Therefore, it is obvious that there is an allosteric link or coupling between the receptor and channel sites.

5. CALCIUM ROLE IN RECEPTOR FUNCTION

The ACh receptor has a high affinity for Ca^{2+}, such that the protein purified from *Torpedo* electric organs in Ca^{2+}-containing solutions has 4.7% of its molecular weight as bound Ca^{2+} (Eldefrawi *et al.*, 1975). Binding of activators (e.g., ACh or carbamylcholine) causes the release of bound Ca^{2+}, whereas binding of inhibitors (e.g., *d*-tubocurarine or α-bungarotoxin) does not (Rübsamen *et al.*, 1978). The question is how Ca^{2+} interacts with the protein and in what way it affects its binding and channel properties. In addition, higher Ca^{2+} in the medium is known to increase the rate of desensitization (Nastuk and Parsons, 1970) possibly through raising ionized Ca^{2+} inside the membranes (Miledi and Parker, 1980), as well as to affect the allosteric coupling of receptor and channel (Cohen *et al.*, 1974). These results raise interesting questions regarding the role of Ca^{2+} in receptor function.

6. CONCLUDING REMARKS

In short, it appears that the nicotinic receptor/channel complex is a single multifunctional asymmetric glycoprotein of four distinct homologous subunits, which binds agonists and antagonists to receptor sites, and local anesthetics and other channel drugs to ionic channel sites, and also translocates cations. It interacts with surrounding lipids and probably with intracellular filaments for localization in the membrane. The receptor/channel molecule has a high affinity for Ca^{2+}, whose role in its function and effect on its conformations is still not clear. Similarly, a functional role for ATP, which may be released with ACh, is still unknown. Reconstitution of the pure receptor/channel molecule in lipid membranes has been achieved, but improvements are still required for it to exhibit properties that are identical with the molecule's function *in vivo*. Binding of agonists to the receptor sites opens the ionic channel, allowing cations but not anions to pass through, though its cation selectivity is very low, unlike axonal K^+ and Na^+ channels. It is not known which or how many of the five subunits contribute to the channel structure, but it is becoming clear that the recognition (or receptor) portion of the protein is small, while the majority of the molecule makes the ionic channel.

ACKNOWLEDGMENTS. Research reported herein from the authors' laboratories was supported by NIH Grant NS-15261 and Army Research Office Grant DAAG 29-78-G-0203.

REFERENCES

Adams, D. J., Dwyer, T. M., and Hille, B. (1980). *J. Gen. Phsyiol.* **75**, 493–510.

Adler, M., Oliveira, A. C., Albuquerque, E. X., Mansour, N. A., and Eldefrawi, A. T. (1979). *J. Gen. Physiol.* **74**, 129–152.

Albuquerque, E. X., Tsai, M.-C., Aronstam, R. S., Witkop, B., Eldefrawi, A. T., and Eldefrawi, M. E. (1980a). *Proc. Natl. Acad. Sci. USA* **77**, 1224–1228.

Albuquerque, E. X., Adler, M., Spivak, C. E., and Aguayo, L. (1980b). *Ann. N.Y. Acad. Sci.* **358**, 204–238.

Anderson, C. R., and Stevens, C. F. (1973). *J. Physiol. (London)* **235**, 655–691.

Ariens, E. J., and Rodrigues de Miranda, J. F. (1979). In *Recent Advances in Receptor Chemistry* (Gualtieri, F., Gianella, M., and Melchiorre, C., eds.), pp. 1–37, Elsevier/North-Holland, Amsterdam.

Aronstam, R. S., Eldefrawi, A. T., Pessah, I. N., Daly, J. W., Albuquerque, E. X., and Eldefrawi, M. E. (1981). *J. Biol. Chem.* **256**, 2843–2850.

Barrantes, F. J. (1976). *Biochem. Biophys. Res. Commun.* **72**, 479–488.

Chang, H. W., and Bock, E. (1977). *Biochemistry* **16**, 4513–4520.

Chang, H. W., and Neumann, E. (1976). *Proc. Natl. Acad. Sci. USA* **73**, 3364–3368.

Claudio, T., and Raftery, M. A. (1977). *Arch. Biochem. Biophys.* **181**, 484–489.

Cohen, J. B., Weber, M., and Changeux, J.-P. (1974). *Mol. Pharmacol.* **10**, 904–932.

Damle, V., and Karlin, A. (1978). *Biochemistry* **17**, 2039–2045.

Desouki, A., Eldefrawi, A. T., and Eldefrawi, M. E. (1981). *Exp. Neurol.* **73**, 440–450.

Dwyer, T. M., Adams, D. J., and Hille, B. (1980). *J. Gen. Physiol.* **75**, 469–492.

Edelstein, S. J., Beyer, W. B., Eldefrawi, A. T., and Eldefrawi, M. E. (1975). *J. Biol. Chem.* **250**, 6101–6106.

Eldefrawi, A. T., Eldefrawi, M. E., Albuquerque, E. X., Oliveira, A. C., Mansour, N., Adler, M., Daly, J. W., Brown, G. B., Burgermeister, W., and Witkop, B. (1977). *Proc. Natl. Acad. Sci. USA* **74**, 2172–2176.

Eldefrawi, M. E. (1978). *Fed. Proc.* **37**, 2823–2827.

Eldefrawi, M. E., and Eldefrawi, A. T. (1973a). *Arch. Biochem. Biophys.* **159**, 362–373.

Eldefrawi, M. E., and Eldefrawi, A. T. (1973b). *Biochem. Pharmacol.* **22**, 3145–3150.

Eldefrawi, M. E., Eldefrawi, A. T., Penfield, L. A., O'Brien, R. D., and Van Campen, D. (1975). *Life Sci.* **16**, 925–936.

Eldefrawi, M. E., Eldefrawi, A. T., Mansour, N. A., Daly, J. W., Witkop, B., and Albuquerque, E. X. (1978). *Biochemistry* **17**, 5474–5484.

Eldefrawi, M. E., Eldefrawi, A. T., Aronstam, R. S., Maleque, M. A., Warnick, J. E., and Albuquerque, E. X. (1980b). *Proc. Natl. Acad. Sci. USA* **77**, 7458–7462.

Eldefrawi, M. E., Warnick, J. E., Schofield, G. G., Albuquerque, E. X., and Eldefrawi, A. T. (1981). *Biochem. Pharmacol.* **30**, 1391–1394.

Eldefrawi, M. E., Warnick, J. E., Schofield, G. G., Albuquerque, E. X., and Eldefrawi, A. T. (1981). *Biochem. Pharmacol.*, in press.

Gordon, A. S., Davis, G. C., and Diamond, I. (1977). *Proc. Natl. Acad. Sci. USA* **74**, 263–267.

Grünhagen, H.-H., and Changeux, J.-P. (1976). *J. Mol. Biol.* **106**, 497–516.

Hamilton, S. L., McLaughlin, M., and Karlin, A. (1977). *Biochem. Biophys. Res. Commun.* **79**, 692–699.

Hartig, P. R., and Raftery, M. A. (1977). *Biochem. Biophys. Res. Commun.* **78**, 16–22.

Hess, G. P., Cash, D. J., and Aoshima, H. (1979). *Nature (London)* **282**, 329–331.

Horn, R., Brodwick, M. S., and Dickey, W. D. (1980). *Science* **210**, 205–207.

Karlin, A., Holtzman, E., Valderrama, R., Damle, V., Hsu, K., and Reyes, F. (1978). *J. Cell Biol.* **76**, 577–592.

Kato, G., and Changeux, J.-P. (1976). *Mol. Pharmacol.* **12**, 92–100.

Katz, B., and Miledi, R. (1972). *J. Physiol.* **224**, 665–699.

Kloog, Y., Flynn, D., Hoffman, A. R., and Axelrod, J. (1980). *Biochem. Biophys. Res. Commun.* **97,** 1474–1480.

Krodel, E. K., Beckman, R. A., and Cohen, J. B. (1979). *Mol. Pharmacol.* **15,** 294–312.

Lindstrom, J., and Dau, P. (1980). *Annu. Rev. Pharmacol. Toxicol.* **20,** 337–362.

Lindstrom, J., Walter, B., and Einarson, B. (1979). *Biochemistry* **18,** 4470–4480.

Meunier, J.-C., Sealock, R., Olsen, R., and Changeux, J.-P. (1974). *Eur. J. Biochem.* **45,** 371–394.

Miledi, R., and Parker, I. (1980). *J. Physiol. (London)* **306,** 567–577.

Moreau, M., and Changeux, J.-P. (1976). *J. Mol. Biol.* **106,** 457–467.

Nachmansohn, D. (1955). *Harvey Lect.* **53/54,** 57–99.

Nastuk, W. L., and Parsons, R. L. (1970). *J. Gen. Physiol.* **56,** 218–249.

Nelson, N., Anholt, R., Lindstrom, J., and Montal, M. (1980). *Proc. Natl. Acad. Sci. USA* **77,** 3057–3061.

Oswald, R., Sobel, A., Waksman, G., Roques, B., and Changeux, J.-P. (1980). *FEBS Lett.* **111,** 29–34.

Quast, U., Schimerlik, M. I., and Raftery, M. A. (1979). *Biochemistry* **18,** 1891–1901.

Raftery, M. A., Hunkapiller, M. W., Strader, C. D., and Hood, L. E. (1980). *Science* **208,** 1454–1457.

Reynolds, J. A., and Karlin, A. (1978). *Biochemistry* **17,** 2035–2038.

Ross, M. J., Klymkowsky, M. W., Agard, D. A., and Stroud, R. M. (1977). *J. Mol. Biol.* **116,** 635–659.

Rübsamen, H., Eldefrawi, A. T., Eldefrawi, M. E., and Hess, G. P. (1978). *Biochemistry* **17,** 3818–3825.

Saitoh, T., Oswald, R., Wennogle, L. P., and Changeux, J.-P. (1980). *FEBS Lett.* **116,** 30–36.

Sine, S. M., and Taylor, P. (1980). *J. Biol. Chem.* **255,** 10144–10156.

Stevens, C. F. (1980). *Annu. Rev. Physiol.* **42,** 643–652.

Strader, C. D., and Raftery, M. A. (1980). *Proc. Natl. Acad. Sci. USA* **77,** 5807–5811.

Takeda, K., Barry, P. H., and Gage, P. W. (1980). *J. Gen. Physiol.* **75,** 589–613.

Teichberg, V. I., Sobel, A., and Changeux, J.-P. (1977). *Nature (London)* **267,** 540–542.

Tiedt, T. N., Albuquerque, E. X., Bakry, N. M., Eldefrawi, M. E., and Eldefrawi, A. T. (1979). *Mol. Pharmacol.* **16,** 909–921.

Tsai, M.-C., Mansour, N. A., Eldefrawi, A. T., Eldefrawi, M. E., and Albuquerque, E. X. (1978). *Mol. Pharmacol.* **14,** 787–803.

Tsai, M.-C., Oliveira, A. C., Albuquerque, E. X., Eldefrawi, M. E., and Eldefrawi, A. T. (1979). *Mol. Pharmacol.* **16,** 382–392.

Tzartos, S. J., and Lindstrom, J. M. (1980). *Proc. Natl. Acad. Sci. USA* **77,** 755–759.

Van Helden, D., Hamill, O. P., and Gage, P. W. (1977). *Nature (London)* **269,** 711–713.

Weber, M., David-Pfeuty, T., and Changeux, J.-P. (1975). *Proc. Natl. Acad. Sci. USA* **72,** 3443–3447.

Whittaker, V. P., and Dowdall, M. J. (1975). In *Cholinergic Mechanism* (P. G. Waser, ed.), pp. 23–43, Raven Press, New York.

139

Ion Permeation through Channels in Excitable Cells

How Many Ions per Pore?

Ted Begenisich

1. INTRODUCTION

The membranes of excitable cells are characterized by the presence of certain proteins that respond to electrical or chemical stimulation by opening pores allowing ions to flow across the membrane. These pores allow the passage of millions of ions per second and yet can discriminate among ions that differ only slightly in ionic radius. Until recently the mathematical formalism used to describe ion permeation was based on solutions of the electrodiffusion equations of Nernst (1888, 1889) and Planck (1890a,b). A very general expectation from such theories is that for the movement of a monovalent ion, X, across a membrane, the unidirectional efflux-to-influx ratio is equal to the ratio of the electrochemical activities of that ion:

$$m_e/m_i = \frac{[X]_i}{[X]_o} \exp(V_m F/RT) \tag{1}$$

where $[X]_i$ and $[X]_o$ are the internal and external activities, V_m is the transmembrane potential, and R, T, and F have their usual thermodynamic meanings. This is the Ussing (1949) flux-ratio equation and is quite general: the only assumption used in the derivation is that the ions move independent of each other.

Under more restrictive conditions, most significantly a constant transmembrane field, the current of an ion through a transmembrane pathway characterized by a permeability, P_x, to that ion, can be written as

Ted Begenisich • Department of Physiology, University of Rochester School of Medicine and Dentistry, Rochester, New York 14642.

$$I_x = \frac{V_m F^2}{RT} \; P_x \; \frac{[X]_o - [X]_i \exp(V_m F/RT)}{\exp(V_m F/RT) - 1} \tag{2}$$

This is the Goldman (1943), Hodgkin and Katz (1949) constant-field equation. If the pathway is permeable to two ions, X and Y, the zero-current or reversal potential is related to the ionic activities and permeabilities by the following:

$$V_{rev} = \frac{RT}{F} \; \ln \frac{P_x[X]_o + P_y[Y]_o}{P_x[X]_i + P_y[Y]_i} \tag{3}$$

These three equations formed the basis of our understanding of ion permeation for many years. They were fundamental to the development of the ionic hypothesis. Because these equations share the basic assumption that ion movements are independent as in free diffusion, models of ion permeation using these equations can be classified as free-diffusion models.

In 1955 Hodgkin and Keynes found that the Ussing flux-ratio equation could not describe the K^+ fluxes across *Sepia* axons unless the right-hand side was raised to a power n', called the flux-ratio exponent. A value of about 2.5 was needed to describe the results from *Sepia* axons, and later Horowicz *et al.* (1968) obtained a flux-ratio exponent of 2 for K^+ fluxes in skeletal muscle.

New theories were therefore necessary to account for these results. In these new theories the ion pathway is viewed as a sequence of energy barriers and energy wells (Danielli, 1939; Davson and Danielli, 1943); the permeating ions hop—according to a Poisson process—from well to well with a rate that decreases exponentially with the height of the intervening energy barrier (Eyring *et al.*, 1949). These models can be generally classified by how many ions can simultaneously occupy the pore. There are then one-ion pores (Läuger, 1973; Hille, 1975a,b; Begenisich and Cahalan, 1979) and multi-ion pores (Heckmann, 1965a,b, 1972; Eisenmann, G., Sandblom, J. and Neher, E. (1978). Biophys. J. 22, 307–340, 1978; Hille and Schwarz, 1978; Begenisich and Cahalan, 1979). It is the purpose of this review to summarize the types of experiments that can distinguish among the possible pore models and to determine what types of pores exist in naturally excitable membranes. This definition excludes much excellent work on channels (e.g., gramicidin and alamethacin) from various sources studied in artificial lipid bilayer systems.

Table I (after Table I in Hille and Schwarz, 1979) summarizes the predictions of free-diffusion, one-ion, and multi-ion pores. The first three entries in the left-hand column of the table list the three general types of channels. The other entries in this column are the biological channels that will be discussed in this review. The headings of the various columns list the types of experiments that can distinguish among channel types.

2. FLUX VS. TRANS-CONCENTRATION

A very simple experiment to test for independence is to measure, say, the efflux of a permeant ion (with a radioactive tracer ion) as a function of the external concentration of that ion. Free-diffusion pores will show no effect of external concentration on efflux. Raising the external concentration in one-ion and multi-ion pores will inhibit the efflux as external ions occupy sites in the pore and block the outward movement of the tracer ion.

Hodgkin and Keynes (1955) and later Begenisich and De Weer (1977) showed that K^+ efflux through the potassium channel of *Sepia* and squid axons was reduced by raising

Table I.

Pore type	Permeation properties			
	Flux vs. trans-concentration	Current vs. concentration	Concentration-dependent permeability ratio	Flux-ratio exponent
Free-diffusion	no effect	linear	no	$n' = 1$
One-ion	decrease	saturating	no	$n' = 1$
Multi-ion	decrease	up to $m - 1$ maxima and finally self-block	yes	$1 \leq n' \leq m$
Na$^+$ channel in nerve	—	saturating	yes	$n' = 1$
K$^+$ channel in nerve	decrease	—	—	$1 \leq n' \leq 3.3$
K$^+$ channel from sarco-plasmic reticulum	—	saturating	no	—
AChR channel	—	saturating	no	—
Anomalous rectifier in muscle	increase	nonlinear	—	$1 \leq n' \leq 2$
Anomalous rectifier in egg cells	—	nonlinear	yes	—

external potassium. Similar experiments on frog skeletal muscle where the K$^+$ efflux was occurring through the inward or anomalous rectifier channel showed an *increase* of K$^+$ efflux as external potassium was increased (Horowicz *et al.*, 1968; Spalding *et al.*, 1981).

3. CURRENT VS. CONCENTRATION

Measurement of current (or conductance) as a function of ion concentration provides another means for distinguishing models. In models based on free diffusion, doubling the ion concentration (at a fixed voltage, far away from the reversal potential) will double the ion current. In one-ion models, the current will increase with ion concentration at first linearly, then eventually saturate at high concentrations. In multi-ion pores (containing up to m ions), there may be up to $m - 1$ maxima in the current–concentration relationship and at very high concentrations the current will actually decrease.

Hille (1975a) and Mazhayeva *et al.* (1977) found that current through the sodium channel in frog myelinated nerve saturated as the external concentration of several different permeant cations was increased. The half-saturating concentrations for Na$^+$, NH$_4^+$, and K$^+$ were found to be about 368, 368, and 220 mM, respectively. In these experiments the concentration of permeant ions was kept below 250 mM to minimize osmotic damage to the fibers. In sodium channels of squid, Begenisich and Cahalan (1980b) used concentrations of Na$^+$, NH$_4^+$, and K$^+$ up to 550 mM and found current saturation. The data were well described by rectangular hyperbolas with half-saturating concentrations (at a membrane potential of 50 mV) of 623, 268, and 161 mM.

The conductance of anomalous rectifier channels of muscle and echinoderm egg cells is a function of V_m-V_k rather than V_m alone (Katz, 1949; Hagiwara and Takahashi, 1974). At a constant value of V_m, the conductance of starfish egg cells increases at fixed [K$^+$]$_i$ as the square root of external K$^+$, and at fixed [K$^+$]$_o$ increases as the square root of [K$^+$]$_i$ (Hagiwara and Yoshi, 1979). Ohmori (1978) has found (using fluctuation analysis) that the single-channel conductance of the anomalous rectifier of the tunicate eggs increases with the square root of $[K^+]_o$. These data can, however, also be fit with a rectangular hyperbola with a half-saturating concentration of about 45 mM. These experiments must be extended to higher concentrations to unambiguously determine whether or not the conductance saturates or continues to increase as the square root of concentration.

The currents through single acetylcholine-activated channels (called AChR channels) of rat myotubes have recently been measured as a function of internal sodium concentration (Horn and Patlak, 1980). At a transmembrane voltage of $+100$ mV, the current–concentration relation is well described by a saturating rectangular hyperbola with a K_m value of 102 mM.

Coronado *et al.* (1980) have found that the single-channel conductance of a K^+-selective channel isolated from rat sarcoplasmic reticulum vesicles saturates with ionic concentration. K_m values for Li^+, Na^+, Rb, NH_4^+, and K^+ of 19, 34, 46, 47, and 54 mM were found. These experiments covered a concentration range from 20 mM to 1–3 M, concentrations high enough to clearly demonstrate saturation.

4. REVERSAL POTENTIAL EXPERIMENTS

Measurements of current reversal potentials at known internal and external ionic concentrations can be used to compute relative ion permeabilities via equation (3). The permeability ratio should be independent of ion concentration in pores obeying independence (free-diffusion models). These ratios should also be concentration independent in one-ion pore models but can be voltage dependent. Only multi-ion pores are expected to show concentration-dependent permeability ratios.

The first measurements of concentration-dependent permeability ratios were made by Chandler and Meves (1965). They found that P_{Na}/P_K of sodium channels in squid decreased as internal potassium was reduced. In these experiments ionic strength was also allowed to change. They could not distinguish a concentration effect from an ionic strength or membrane voltage effect. Nor could they eliminate the possibility that a small chloride permeability could account for their results. Cahalan and Begenisich (1976) showed that it was the internal K^+ concentration alone that was responsible for the P_{Na}/P_K change. Later, Begenisich and Cahalan (1980a) extended these experiments to include alterations in internal and external Na^+ and NH_4^+ and found that the permeability ratio P_{Na}/P_{NH_4} depended only on internal NH_4^+ activity and not on internal or external Na^+ activity, nor on external NH_4^+ activity.

Hagiwara *et al.* (1977) showed that P_K/P_{Tl} of the anomalous rectifier in starfish eggs depended on the relative concentration of Tl^+ and K^+ bathing these cells. Coronado *et al.* (1980) found that P_{Na}/P_K of the K^+ channel from sarcoplasmic reticulum was independent of ionic concentration. Adams *et al.* (1980) found that the permeability ratio in the AChR channel was also independent of ion concentration.

5. FLUX-RATIO EXPONENT

These types of experiments produced the first results inconsistent with the simple theories of Nernst and Planck as described briefly above. Begenisich and De Weer (1980) extended the original observations of Hodgkin and Keynes to include several voltages and internal and external K^+ concentrations. They found that the flux-ratio exponent, n', was a function of both voltage and internal (but not external) K^+ concentration. The values of n' ranged from 1.5 at -4 mV and 200 mM $[K^+]_i$ to 3.3 at -38 mV and 350 mM $[K^+]_i$.

Begenisich and Busath (1980) found n' of unity, independent of membrane potential, in the sodium channel of squid axons under approximately physiological conditions (440 mM $[Na^+]_o$, 50 mM $[Na^+]_i$). In frog skeletal muscle, n' appears to be close to unity at low external K^+ (Sjodin, 1965; Spalding *et al.*, 1981) and increases to about 2 at elevated K^+ concentrations (Horowicz *et al.*, 1968; Spalding *et al.*, 1981).

6. SUMMARY

As can be seen from Table I, the types of experiments described here can be used to determine how many ions can simultaneously occupy an ionic channel. Several of these tests have been applied to some biological channels. In no case have all tests been applied to a single type of channel.

The most completely studied channel is the sodium channel in nerve. The concentration-dependent permeability ratio identifies this as a multi-ion pore. The low value of n' and the lack of self-block in the current–concentration experiments reflect the very high K_m values for Na^+ ions for these channels (368 mM in myelinated nerve and 623 mM in squid nerve). Consequently, the manifestations of the multi-ion nature of this channel are expected only at very high ionic concentrations.

The potassium channel of nerve is clearly a multi-ion pore as demonstrated by the large n' values. There may be as many as three or four ions simultaneously in these channels. It will be very interesting to see the results of current–concentration and reversal potential experiments on these channels.

The potassium channel isolated from sarcoplasmic reticulum is probably the simplest biological channel. It appears to be a one-ion pore. The current–concentration experiments were carried out to about 40 times the K_m value with no sign of self-block.

The AChR channel also appears to be a one-ion pore. Considering the large size and large conductance of this channel, it is striking that it does not obey independence. This same conclusion was reached by Lewis (1979) from experiments similar to (but more complicated than) those described above.

The anomalous rectifier channel is still anomalous. The expectations of the types of pore models described here are inconsistent with the available experimental data. One way to reconcile these differences is to assume that the anomalous rectifier pore can be blocked by some internal charged species (see Hille and Schwarz, 1978). Unfortunately, such a blocking ion has yet to be identified.

In conclusion, it appears that none of the ionic channels from excitable cells obey independence. There are both one-ion and multi-ion pores with as many as three to four ions. It will be interesting to see if this is a consistent finding, or will we find, as other channels are put to these tests, channels obeying independence? In any case, we must now start to ask: what is it about the molecular structure of these pores that gives them their independent, one-ion, or multi-ion nature? An answer to this question will also probably go a long way toward providing an understanding of what makes a pore cation selective (and not anion selective) and how a pore can discriminate among many similar ions.

REFERENCES

Adams, D. J., Dwyer, T. M., and Hille, B. (1980). *J. Gen. Physiol.* **75,** 493–510.

Begenisich, T., and Busath, D. (1980). *Fed. Proc.* **39,** 1840.

Begenisich, T., and Cahalan, M. (1979). In *Membrane Transport Processes* (C. F. Stevens and R. W. Tsien, eds.), Vol. 3, Raven Press, New York.

Begenisich, T., and Cahalan, M. D. (1980a). *J. Physiol.* **307,** 217–242.

Begenisich, T., and Cahalan, M. D. (1980b). *J. Physiol.* **307,** 243–257.

Begenisich, T., and De Weer, P. (1977). *Nature (London)* **269,** 710–711.

Begenisich, T., and De Weer, P. (1980). *J. Gen. Physiol.* **76,** 83–98.

Cahalan, M. D., and Begenisich, T. (1976). *J. Gen. Physiol.* **68,** 111–125.

Chandler, W. K., and Meves, H. (1965). *J. Physiol.* **180,** 788–820.

Coronado, C., Rosenberg, L., and Miller, C. (1980). *J. Gen. Physiol.* **76,** 425–446.

Danielli, J. F. (1939). *J. Physiol.* **96,** 2P.

Davson, H., and Danielli, J. F. (1943). *The Permeability of Natural Membranes*, pp. 310–352, Cambridge University Press, London.

Eisenman, G., Sandblom, J., and Neher, E. (1978). *Biophys. J.* **22,** 307–340.

Eyring, H., Lumry, R., and Woodbury, J. W. (1949). *Rec. Chem. Prog.* **10,** 100–114.

Goldman, D. E. (1943). *J. Gen. Physiol.* **27,** 37–60.

Hagiwara, S., and Takahashi, K. (1974). *J. Membr. Biol.* **18,** 61–80.

Hagiwara, S., and Yoshi, M. (1979). *J. Physiol.* **292,** 251–265.

Hagiwara, W., Miyasaki, S., Kragne, S., and Ciani, S. (1977). *J. Gen. Physiol.* **70,** 269–281.

Heckmann, K. (1965a). *J. Phys. Chem.* **44,** 184–203.

Heckmann, K. (1965b). *J. Phys. Chem.* **46,** 1–25.

Heckmann, K. (1972). *Biomembranes* **3,** 127–153.

Hille, B. (1975a). *J. Gen. Physiol.* **66,** 535–560.

Hille, B. (1975b). In *Membranes: A Series of Advances* (G. Eisenmann, ed.), Vol. 3, Dekker, New York.

Hille, B., and Schwarz, W. (1978). *J. Gen. Physiol.* **72,** 409–442.

Hille, B., and Schwarz, W. (1979). *Brain Res. Bull.* **4,** 159–162.

Hodgkin, A. L., and Katz, B. (1949). *J. Physiol.* **108,** 37–77.

Hodgkin, A. L., and Keynes, R. D. (1955). *J. Physiol.* **128,** 61–88.

Horn, R., and Patlak, J. (1980). *Proc. Natl. Acad. Sci. USA* **77,** 6930–6934.

Horowicz, P., Gage, P. W., and Eisenberg, R. S. (1968). *J. Gen. Physiol.* **51,** 1935S–2035S.

Katz, B. (1949). *Arch. Sci. Physiol.* **2,** 285–299.

Läuger, P. (1973). *Biochim. Biophys. Acta* **311,** 423–445.

Lewis, C. A. (1979). *J. Physiol.* **286,** 417–445.

Mozhayeva, G. N., Naumov, A. P., Negulyaev, Y. A., and Nosyveva, E. D. (1977). *Biochim. Biophys. Acta* **466,** 461–473.

Nernst, W. (1888). *Z. Phys. Chem.* **2,** 613–637.

Nernst, W. (1889). *Z. Phys. Chem.* **4,** 129–181.

Ohmori, H. (1978). *J. Physiol.* **281,** 77–99.

Planck, M. (1890a). *Ann. Phys. Chem. Nevefolge.* **39,** 161–186.

Planck, M. (1890b). *Ann. Phys. Chem. Nevefolge.* **40,** 561–576.

Sjodin, R. A. (1965). *J. Gen. Physiol.* **48,** 777–795.

Spalding, B. C., Senyk, O., Swift, J. G., and Horowicz, P. (1981). *Am. J. Physiol.* **241,** C68–C75.

Ussing, H. H. (1949). *Acta Physiol. Scand.* **19,** 43–56.

140

Identification of Na⁺ Gating Proteins in Conducting Membranes

E. Schoffeniels

Already contained in the formulation princeps of the ionic theory by Hodgkin and Huxley (1952) is the idea that Na^+ and K^+ channels are two distinct entities. A wealth of data accumulated mostly in the early 1970s (reviewed by Narahashi, 1974) indicate that Na^+ and K ions pass through the axonal membrane by specific gateways having quite distinct chemical affinities and thus most probably being different proteins spatially separated in the membrane matrix. Intramembranous charge displacements provide additional support for this idea (Swenson, 1980).

The basic pharmacological observations concerning the two ionic currents (I_{Na} and I_K) are as follows. Two toxins, tetrodotoxin (TTX) and saxitoxin (STX), completely block I_{Na} in both myelinated and unmyelinated fibers at low concentrations (0.1 μM) when applied *outside*. I_K and I_L (leakage current) are not affected even by 100-fold greater concentrations.

Tetraethylammonium ions (TEA) when applied *inside* selectively block I_K, leaving almost unaffected I_{Na} and I_L. The simultaneous treatment of nerve fibers with TTX and TEA does not influence the specific blocking effect of either agent, which further substantiates the functional independence of the two types of gateways. Single Na^+ gateway currents have indeed been observed in cultured rat muscle cells (Sigworth and Neher, 1980).

The specificity of a gateway is far from being absolute, and the relative permeability of Na^+ or K^+ gateways to other ionic species has been estimated (Ulbricht, 1974; Hille, 1975; Meves, 1975). Thus, neurotoxins that increase the rate of Na^+ influx through the sodium gateway also give rise to an extra efflux of K^+ ions through the same gateway (Jacques *et al.*, 1980a).

The kinetics of toxin membrane interaction is compatible with a one-to-one molecular interaction, which provides a possibility to estimate the surface density of sodium and potassium gateways. Thus, 12,000 sodium gateways per square micrometer of nodal mem-

E. Schoffeniels • Laboratory of General and Comparative Biochemistry, University of Liège, B-4020 Liège, Belgium.

brane are evidenced using [³H]-STX (Ritchie and Rogart, 1977a) while values of 50 and 75 are found respectively for crab and rabbit vagus nerves (Keynes *et al.*, 1971).

The nature of the molecular transconformations underlying the conductance changes of both Na^+ and K^+ gateways is far from being unambiguously defined. As shown by Jakobsson (1973, 1976), unimolecular models with classical voltage-dependent rate constants may not be able to cover all the electrophysiological data obtained with voltage-clamped axonal membranes. More specifically, the cyclic evolution of g_{Na} under voltage-clamp conditions is hard to model as only a field-dependent process. During every gating cycle there is entropy production and free-energy consumption (Hille, 1978). An agreement is far from being reached though as to the nature of the free-energy source. However, the commonly held idea that these transitions are mere displacements of electric charges solely due to the electric field cannot account for basic voltage clamp and microcalorimetric data and does not resist thermodynamic tests that point to the existence of external energy input to the axonal membrane (Margineanu and Schoffeniels, 1977; Schoffeniels, 1980c). Therefore, as in any other dissipative process found in biology, a chemical must be used up to explain the impedance variation cycle (IVC).

Bimolecular reaction steps in axonal Na^+-channel gating are indeed suggested by abstract kinetic models that can successfully simulate the evolution of g_{Na} (Dorogi and Neumann, 1980; Neumann, 1980).

It is generally agreed that the Na^+ gateway must be formed by protein(s) (Agnew *et al.*, 1978) spanning the membrane (intrinsic protein). It is indeed possible to alter the Na^+ gateway properties from both sides of the membrane: TTX, STX, scorpion or sea anemone toxins are active from outside (Lazdunski *et al.*, 1979; Neumcke *et al.*, 1980) but the channel protein can be inactivated from the inside by Pronase (Armstrong *et al.*, 1973) or the anion iodate (IO_3^-) (Neumcke *et al.*, 1980).

Moreover, on the channel protein, there exist different categories of sites. Thus, external sodium sites must be saturated to obtain a conducting form of the Na^+ gateway in the presence of any of various toxins (Ponzio *et al.*, 1980). Moreover, a synergism is observed between pyrethroids and other compounds known to affect the Na^+ gateway. Thus, at least four different binding sites in addition to the Na^+ binding sites are described: (1) one for TTX and STX (Ritchie and Rogart, 1977b); (a) one for BTX, veratridine, and dihydrograyanotoxin II (Catterall, 1975); (3) at least one site for polypeptide toxins (Vincent *et al.*, 1980); and (4) one site for pyrethroids (Jacques *et al.*, 1980b).

The temperature dependence of the electrical activity of ion channels supports the possibility that the fluidity of membrane lipids may influence the transconformation of the protein gateway (Romey *et al.*, 1980).

As to the various theories invoked to explain the action of local anesthetics (fluidization and disordering of lipids, expansion of membrane lipids, alteration of membrane electric field), results obtained with *n*-octanol are more in favor of a direct interaction with the Na^+ gateway leading to a decrease in the number of conducting channels or a decrease in the conductance of a single Na^+ channel (Swenson and Narahashi, 1980). That part of the Na^+ gateway responsible for its TTX or STX sensitivity and ionophoric properties has been successfully included in liposomes, thus opening a very promising approach to further characterizing the properties and molecular organization of the native Na^+ gateway (Malysheva *et al.*, 1980; Goldin *et al.*, 1980).

The problem of the conductance of an ion channel is obviously a matter of energy profile (Läuger, 1980): the ion moves from a binding site across an energy barrier to an adjacent site. Ample evidence exists showing that the energy barriers usually considered to be fixed are indeed time dependent for even at thermal equilibrium protein molecules may exist in various conformational states (Frauenfelder *et al.*, 1979). Therefore, the energy profile of ionic channels is subjected to individual fluctuations as demonstrated by the very existence of the conductance noise (Stevens, 1972). However, in conducting membranes, the stimulus induces a *collective* transition of a population of channels from one mean energy level to another. Therefore, the problem remains to explain: (1) how the stimulus provides the coordination of many of such elementary processes; (2) the origin of the free energy needed (a) to recruit a population of individual channels and (b) to fuel the IVC of each channel. In view of the fact that most of the proteins playing a critical role in biological processes are activated (or deactivated) through covalent alterations such as phosphorylation, glycosylation, methylation, etc., it has been suggested that the conformation of the protein(s) forming the site controlling the Na^+ ion flow during nerve activity could be related to the net state of phosphorylation.

The results obtained with intact nerves show that the phosphorylation of high-molecular-weight components (above 300,000) together with minor components (around 60,000 and 40,000 molecular weights) is differently affected depending on the experimental conditions (Schoffeniels, 1980a,b).

The results on intact nerves have also been reproduced on membranes prepared from the same species (Schoffeniels and Dandrifosse, 1980), and the high-molecular-weight fraction of proteins extracted from crab nerves (Bontemps *et al.*, 1981) can be shown to phosphorylate in the presence of $[\gamma\text{-}^{32}P]$-ATP concentration ranging from 5×10^{-8} to 10^{-7} M.

Prior to any incubation in the presence of exogenous ATP, the proteins are already phosphorylated as shown by monitoring phosphoserine + phosphothreonine, and moreover a subsequent incubation *without* ATP is accompanied by an increase in phosphoserine +

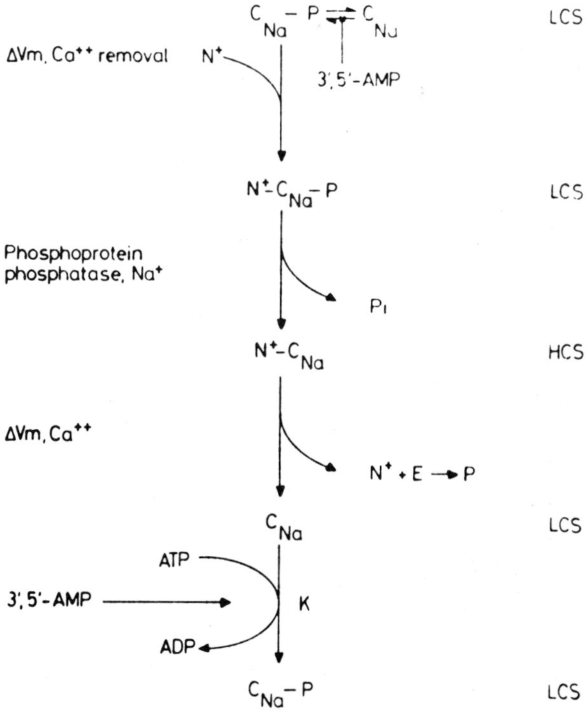

Figure 1. Biochemical cycle of impedance variation (IVC) in axonal membrane. In resting conditions, the LCS is made of a population of C_{Na} that is partially phosphorylated (C_{Na}–P). Upon electrical stimulation, i.e., an adequate change (ΔV_m) in the electric field across the membrane, a quaternary ammonium derivative (N^+) is liberated. It is only recognized by the phosphorylated form of C_{Na} (C_{Na}–P) and thus forms an LCS activated complex N^+–C_{Na}–P, the only substrate for a phosphoprotein phosphatase activated by Na^+ ions. This gives rise to the HCS complex N^+–C_{Na}. The change in electric field thus resulting favors the dissociation of the complex and C_{Na} is restored in LCS. N^+ is also available for a subsequent hydrolysis that should take place after completion of the membrane potential change. C_{Na} may again be phosphorylated, a process requiring a specific kinase stimulated by K^+ ions. Notice that all the Na^+ gateways are not phosphorylated and that to enter into the sodium activation process (in the terminology of Hodgkin and Huxley) they must be phosphorylated. [After Schoffeniels, 1980c.]

phosphothreonine corresponding to a phosphorylation 10,000 times higher than the one determined with radioactive ATP. This clearly indicates that an endogenous source of ATP (or other phosphorylating material) is present in the protein extract.

Interestingly enough, this endogenous phosphorylation is activated two- to threefold by 1.5×10^{-7} M TTX while no significant effect is observed when monitoring the ^{32}P incorporation (Bontemps *et al.*, 1981). A similar activation of the endogenous serine + threonine phosphorylation is observed in the presence of 10^{-4} M carbamylcholine.

These data have been integrated to fit in a biochemical cycle involving phosphorylation–dephosphorylation processes (Fig. 1). The protein(s) that controls the sodium conductance of the axonal membrane, the sodium conductin or C_{Na} (Schoffeniels, 1980a), can assume a so-called closed configuration or low-conductance state (LCS) and the open configuration or high-conductance state (HCS).

$3':5'$-AMP displaces the equilibrium existing in resting conditions between C_{Na}–P $\rightleftarrows$ C_{Na} in favor of C_{Na}, an effect that could well be of ecophysiological significance in the case of aquatic invertebrates (Schoffeniels, 1976, 1980c).

Much less is known of the molecular properties of the K^+ channel, mainly because of the absence of specific and high-affinity toxins.

Finally, a reversible swelling contemporary to the action potential is observed in squid giant axon and in crab nerves (Iwasa *et al.*, 1980; Iwasa and Tasaki, 1980).

The meaning of this process is not understood but could perhaps be related to the phosphorylation cycle involving some axonal proteins during activity. Though mechanical changes such as those reported by Tasaki and his colleagues bear some resemblance to swelling of polyelectrolyte gels, let us not be fooled by too simple a physicochemical analogy when dealing with such a complex object as an axon.

REFERENCES

Agnew, W. S., Levinson, S. R., Brabson, J. S., and Raftery, M. A. (1978). *Proc. Natl. Acad. Sci. USA* **75,** 2606–2610.

Armstrong, C. M., Bezanilla, F., and Rojas, E. (1973). *J. Gen. Physiol.* **62,** 375–391.

Bontemps, J., Dandrifosse, G. and Schoffeniels, E. (1981). *IRCS Med. Sci.* **9,** 70.

Catterall, W. A. (1975). *Proc. Natl. Acad. Sci. USA* **72,** 1782–1786.

Dorogi, P. L., and Neumann, E. (1980). *Proc. Natl. Acad. Sci. USA* **77,** 6582–6586.

Frauenfelder, H., Petsko, G. A., and Tsernoglu, D. (1979). *Nature (London)* **280,** 558–563.

Goldin, S. M., Rhoden, V., and Hess, E. J. (1980). *Proc. Natl. Acad. Sci. USA* **77,** 6884–6888.

Hille, B. (1975). In *Membranes: A Series of Advances 3* (G. Eisenman, ed.), Dekker, New York.

Hille, B. (1978). *Biophys. J.* **22,** 283–294.

Hodgkin, A. L., and Huxley, A. F. (1952). *J. Physiol. (London)* **116,** 449–506.

Iwasa, K., and Tasaki, I. (1980). *Biochem. Biophys. Res. Commun.* **95,** 1328–1331.

Iwasa, K., Tasaki, I., and Gibbons, R. C. (1980). *Science* **210,** 338–339.

Jacques, Y., Romey, G., and Lazdunski, M. (1980a). *Eur. J. Biochem.* **111,** 265–273.

Jacques, Y., Romey, G., Cavey, M. T., Kartalovski, B., and Lazdunski, M. (1980b). *Biochim. Biophys. Acta* **600,** 882–897.

Jakobsson, E. (1973). *Biophys. J.* **13,** 1200–1211.

Jakobsson, E. (1976). *Biophys. J.* **16,** 291–301.

Keynes, R. D., Ritchie, J. M., and Rojas, E. (1971). *J. Physiol. (London)* **213,** 235–254.

Läuger, P. (1980). *J. Membr. Biol.* **57,** 163–178.

Lazdunski, M., Balerna, M., Chicheportiche, R., Fosset, M., Jacques, Y., Lombet, A., Romey, G., and Vincent, J. P. (1979). In *Function and Molecular Aspects of Biomembrane Transport* (E. M. Klingenberg, F. Palmieri, and E. Quagliariello, eds.), pp. 25–41, Elsevier/North-Holland, Amsterdam.

Malysheva, M. K., Lishko, V. K, and Chagovetz, A. M. (1980). *Biochim. Biophys. Acta* **602,** 70–77.

Margineanu, D.-G., and Schoffeniels, E. (1977). *Proc. Natl. Acad. Sci. USA* **74,** 3810–3813.

Meves, H. (1975). *Philos. Trans. R. Soc. London Ser. B* **270,** 377–387.

Narahashi, T. (1974). *Physiol. Rev.* **54,** 813–889.

Neumann, E. (1980). *Biochem. Int.* **2,** 27–43.

Neumcke, B., Schwartz, W., and Stämpfli, R. (1980). *Biochim. Biophys. Acta* **600,** 456–466.

Ponzio, G., Jacques, Y., Frelin, C., Chicheportiche, R., and Lazdunski, M. (1980). *FEBS Lett.* **121,** 265–268.

Ritchie, J. M., and Rogart, R. B. (1977a). *Proc. Natl. Acad. Sci. USA* **74,** 211–215.

Ritchie, J. M., and Rogart, R. B. (1977b). *Rev. Physiol. Biochem. Pharmacol.* **79,** 1–50.

Romey, G., Chicheportiche, R., and Lazdunski, M. (1980). *Biochim. Biophys. Acta* **602,** 610–620.

Schoffeniels, E. (1976). *Biochem. Soc. Symp.* **41,** 179–204.

Schoffeniels, E. (1980a). In *Bioelectrochemistry: Ions, Surfaces, Membranes* (M. Blank, ed.), pp. 285–297, Adv. Chem. Ser. 188.

Schoffeniels, E. (1980b). In *Epithelial Transport in Lower Vertebrates* (B. Lahlou, ed.), pp. 269–274, Cambridge University Press, Cambridge.

Schoffeniels, E. (1980c). *Neurochem. Int.* **2,** 81–93.

Schoffeniels, E., and Dandrifosse, G. (1980). *Proc. Natl. Acad. Sci. USA* **77,** 812–816.

Sigworth, F. J., and Neher, E. (1980). *Nature (London)* **287,** 447–449.

Stevens, C. F. (1972). *Biophys. J.* **12,** 1028–1047.

Swenson, R. P. (1980). *Nature (London)* **287,** 644–647.

Swenson, R. P., and Narahashi, T. (1980). *Biochim. Biophys. Acta* **603,** 228–236.

Ulbricht, W. (1974). *Biophys. Struct. Mech.* **1,** 1–16.

Vincent, J. P., Balerna, M., Barhanin, J., Fosset, M., and Lazdunski, M. (1980). *Proc. Natl. Acad. Sci. USA* **77,** 1646–1650.

141

Swelling of Axon Membrane during Excitation

K. Iwasa and I. Tasaki

In a variety of crustacean nerve fibers, as well as in squid giant axons, the nerve membrane has been found to swell concurrently with production of an action potential. This newly discovered sign of drastic structural change in the nerve membrane and its significance are discussed in this review.

1. SURVEY OF PREVIOUS STUDIES OF MECHANICAL CHANGES IN NERVE FIBERS

Using cuttlefish axons, Hill (1950) found that the axon diameter increases slowly when rapidly repeating stimuli are applied to the axons. The existence of such a slow cummulative effect of prolonged nerve stimulation was confirmed by Lieberman (1969) and Cohen and Keynes (1971). According to Bryant and Tobias (1955), this slow increase in the axon diameter is accompanied by a small reduction in the length of the fibers. The origin of these slow mechanical changes in nerve fibers is unclear at present.

The existence of rapid mechanical changes associated with individual electric responses in the frog sciatic nerve was reported by Kayushin and Lyudkovskya (1954, 1955). The magnitude of the reported surface movement was of the order of 1 μm. The reliability of the results in this report was questioned by Sandlin et al. (1968) and Cohen and Keynes (Cohen, 1973). Recently, we made another attempt at detecting such surfaces displacements of the frog sciatic nerve during excitation; again we failed to confirm the results described in the previous report.

Using a laser interferometer, Hill et al. (1977) reported the appearance of a diameter decrease in the crayfish axon within 1 msec after the arrival of an action potential. They made no direct comparison between the time course of the mechanical changes and the

K. Iwasa and I. Tasaki • Laboratory of Neurobiology, National Institute of Mental Health, Bethesda, Maryland 20205.

action potentials. These authors did not recognize the existence of an increase in the fiber diameter during the rising phase of the action potential.

Lettvin *et al.* (1962) reported a twitch of the lobster walking leg nerve accompanied by a single action potential. They used stimulating electrodes parallel to the nerve fibers (i.e., transverse stimulation). Hence, the mechanical changes they recorded are severely distorted by the strong electric current required for stimulation.

2. RECENT FINDINGS

Rapid swelling accompanied by the action potential was first observed in crab claw nerves (Iwasa *et al.*, 1980; Tasaki *et al.*, 1980). Two independent optical methods were used for detecting surface displacements of the nerve, and the results obtained were compared with pressure changes recorded by using a piezoelectric transducer. These three methods yielded consistent results indicating that the nerve swells during the rising phase of the action potential. The surface displacement of the crab nerve was 5 to 10 nm in amplitude, and the pressure increase was about 5 dyn/cm² at the peak. Similar results were obtained in the nerves of other crustaceans, including lobsters and crayfish. The expansion of nerve fibers in the transversal direction is accompanied by a contraction in the longitudinal direction (Tasaki and Iwasa, 1980).

In the squid giant axon, the swelling observed during the rising phase of the action potential was followed by a pronounced shrinkage of the axon diameter (Iwasa and Tasaki, 1980). Using an internal electrode located near the mechanical recording site, it was shown that the swelling phase coincides with the rising (i.e., depolarizing) phase of the action potential and that the shrinkage phase is simultaneous with the falling phase. The peak value of the outward surface displacement of the axon was 0.5–1 nm. The surface displacement associated with the shrinkage of the axon is usually two to four times as large as that seen during the swelling. The pressure rise observed during the depolarizing phase of the action potential was about 2–5 dyn/cm² at the peak, and the pressure fall in the hyperpolarizing phase was roughly twice as large as the pressure rise. The increase in the axon diameter was accompanied by a rise in the tension in the longitudinal direction, and the decrease in the diameter was accompanied by a small fall in the tension.

The rapid mechanical changes observed in squid axons cannot be explained as the consequence of Na^+–K^+ ion exchange known to take place during action potentials. The quantity of the monovalent cations involved in the exchange is of the order of 10^{-11} mole/cm² per spike (Hodgkin, 1964; Tasaki, 1968). The molar volume of hydrated sodium ion is about 40 ml, and that of potassium 20 ml (e.g., Gregor, 1951). Even if we ignore the potassium outflux and take the sodium flux as an upper-bound estimation, the displacement attributable to this effect accounts for only 0.04 Å, namely, for less than 1% of the observed displacement.

3. SIGNIFICANCE OF SWELLING OF THE NERVE FIBERS

The existence of rapid mechanical changes in the nerve membrane associated with the excitation process is assumed, somewhat implicitly, in the theory of nerve excitation proposed by Loeb (1906) and Höber (1920). They proposed that ion-exchange processes involving Ca^{2+} ion and alkali metal ions underlie excitation phenomena. They based their argument on the discovery of marked changes in physicochemical properties of various biocolloids produced by such ion replacements. Höber described the resting nerve membrane to be in a Ca^{2+}-rich, compact state and the excited, or depolarized, nerve membrane to be in a loosened, i.e., swollen, state. The crucial role played by Ca^{2+} ions in the excitation process is repeatedly emphasized by Tasaki (1968). Much later, extending the results of analysis of his electrohydraulic membrane model, Teorell (1962) suggested that the process of action potential production is accompanied by swelling and shrinkage of the nerve membrane resulting from movements of water molecules.

In artificial ion-exchanger membranes, the effect of substituting univalent counterions for divalent ions on the membrane water content is well understood. As a rule, such a replacement increases the water content, and profoundly enhances the membrane conductance (see e.g. Gregor and Sollner, 1946; Lagos and Kitchener, 1960; Kertész *et al.*, 1967). Katchalsky and Zwick (1955) have demonstrated enormous swelling of a polyelectrolyte gel by replacing univalent cations with divalent cations.

We note that the peak of the pressure rise in the squid axon coincides, with accuracy of about 0.05 msec, with the peak of the action potential recorded intracellularly from the site of mechanical recording. From this fact it follows that the macromolecules involved in swelling are located either within the axolemma or within a very short distance away from the axolemma. It is also known that the membrane conductance drastically rises during the rising phase of the action potential and reaches a maximum at around the peak of the action potential (Cole and Curtis, 1939). It is therefore possible to attribute the rise in the conductance to the rise in the water content of the membrane.

Judging from the magnitude of the outward displacement of the squid axon membrane during action potentials, the number of water molecules involved in swelling is of the order of $3 \times 10^7/\mu m^2$. Therefore, many macromolecules in and near the axolemma are expected to become strongly hydrated at the peak of the action potential. Metuzals and Tasaki (1978) have shown that the inner surface of the axolemma is coated with a layer of longitudinally oriented, fibrillar elements consisting of a variety of proteins. An increase in hydration of this layer is expected to produce a reduction of the birefringence of this layer. Furthermore, fluorescent probe molecules introduced into this layer are expected to experience a rise in the dielectric constant (or polarity) of their microenvironment at the peak of the action potentials. Thus, it seems reasonable to attribute both birefringence signals (Cohen *et al.*, 1968; Sato *et al.*, 1973) and fluorescence signals of the axon (Tasaki and Warashina, 1976) to invasion of water molecules into the undercoating of the axolemma.

Further studies of the macromolecular architecture of the axolemma and its surroundings are required for a complete elucidation of the mechanism of nerve excitation.

4. SUMMARY

The existence of membrane swelling during action potentials is now firmly established. This swelling is a reflection of the invasion of water into macromolecules in and near the axon membrane associated with action potential production. There is good reason to believe that the movement of water molecules underlies the generation of birefringence and fluorescence signals, as well as the electrochemical process responsible for the large increase in the membrane conductance during action potentials.

REFERENCES

Bryant, S. H., and Tobias, J. M. (1955). *J. Cell. Comp. Physiol.* **46**, 71–95.
Cohen, L. B. (1973). *Physiol. Rev.* **53**, 373–418.

Cohen, L. B., and Keynes, R. D. (1971). *J. Physiol. (London)* **212,** 259–275.

Cohen, L. B., Keynes, R. D., and Hille, B. (1968). *Nature (London)* **218,** 438–441.

Cole, K. S., and Curtis, H. J. (1939). *J. Gen. Physiol.* **22,** 649–670.

Gregor, H. P. (1951). *J. Amer. Chem. Soc.* **73,** 642–650.

Gregor, H. P., and Sollner, K. (1946). *J. Phys. Chem.* **50,** 88–96.

Hill, B. C., Schubert, E. D., Nokes, M. A., and Michelson, R. P. (1977). *Science* **196,** 426–428.

Hill, D. K. (1950). *J. Physiol. (London)* **111,** 304–327.

Höber, R. (1920). *Pfluegers Arch. Physiol.* **182,** 104–113.

Hodgkin, A. L. (1964). *The Conduction of the Nervous Impulse,* Liverpool University Press, Liverpool.

Iwasa, K., and Tasaki, I. (1980). *Biochem. Biophys. Res. Commun.* **95,** 1328–1331.

Iwasa, K., Tasaki, I., and Gibbons, R. C. (1980). *Science* **210,** 338–339.

Katchalsky, A., and Zwick, M. (1955). *J. Polym. Sci.* **17,** 221–233.

Kayushin, L. P., and Lyudkovskaya, R. G. (1954). *Dokl. Akad. Nauk SSSR* **95,** 253–255.

Kayushin, L. P., and Lyudkovskaya, R. G. (1955). *Dokl. Akad. Nauk SSSR* **102,** 727–728.

Kertész, D., de Korosy, F., and Zeigerson, E. (1967). *Desalination* **2,** 161–169.

Lagos, A. E., and Kitchener, J. A. (1960). *Trans. Faraday Soc.* **56,** 1245–1251.

Lettvin, J. Y., Sten-Knudsen, D., and Pitts, W. H. (1962). Quarterly Progress Report, Research Laboratory of Electronics, MIT, No. 64, pp. 291–292.

Lieberman, E. M. (1969). *Acta Physiol. Scand.* **75,** 513–517.

Loeb, J. (1906). *The Dynamics of Living Matter,* Columbia University Press, New York.

Metuzals, J., and Tasaki, I. (1978). *J. Cell Biol.* **78,** 597–621.

Sandlin, R., Lerman, L., Barry, W., and Tasaki, I. (1968). *Nature (London)* **217,** 575–576.

Sato, H., Tasaki, I., Carbone, E., and Hallett, M. (1973). *J. Mechanochem. Cell Motil.* **2,** 209–217.

Tasaki, I. (1968). *Nerve Excitation,* Thomas, Springfield, Ill.

Tasaki, I., and Iwasa, K. (1980). *Biochem. Biophys. Res. Commun.* **94,** 716–720.

Tasaki, I., and Warashina, A. (1976). *Photochem. Photobiol.* **24,** 191–207.

Tasaki, I., Iwasa, K., and Gibbons, R. C. (1980). *Jpn. J. Physiol.* **30,** 897–905.

Teorell, T. (1962). *Biophys. J.* **2** (Suppl.), 27–52.

142

The Physical States of Nerve Myelin

C. R. Worthington

1. INTRODUCTION

The structure of nerve myelin in a direction at right angles to the surface of the myelin layers has been studied extensively by X-ray diffraction. X-Ray studies on nerve myelin invariably refer to peripheral nervous system (PNS) myelin because of its well-defined swelling property. The method of swelling has been used in the structure analysis of nerve myelin, and as a consequence of advances in methods of X-ray analysis (Worthington *et al.*, 1973), the electron density profile was, in 1974, uniquely determined at a resolution of 17 Å (McIntosh and Worthington, 1974a). Possible profiles at a higher resolution of 7 Å have also been presented (Worthington and McIntosh, 1974). But an interpretation of the electron density profiles in terms of an assembly of molecular components has not been accomplished. It would appear that molecular labeling experiments are needed in order to provide additional structural information. Any interpretations of the molecular labeling experiments will, however, be dependent on the nature of the physical and chemical treatment during the process of labeling for it is known from X-ray diffraction that the myelin layers of nerve can exist in different physical states depending on the kind of treatment. It is the purpose of this review to briefly describe the nature and properties of the different physical states of nerve myelin.

Historically, Finean and associates (Elkes and Finean, 1949, 1953; Finean, 1961; Finean and Millington, 1957; Joy and Finean, 1963) in the period of 1949 to 1963 made attempts to gain structural information on nerve myelin by studying modified nerve tissue. Low-angle X-ray diffraction patterns of modified nerves were obtained by exposing the nerve specimens to various physical treatments (Elkes and Finean, 1949; Joy and Finean, 1963) such as air-drying, rehydration, freezing and thawing and to various chemical treatments (Elkes and Finean, 1953; Finean and Millington, 1957). At the time when these X-ray patterns were first recorded, any structural interpretations were tentative and incomplete (Finean, 1961). As it is presently possible, using improved methods of structure analysis

C. R. Worthington • Department of Biological Sciences and Physics, Carnegie-Mellon University, Pittsburgh, Pennsylvania 15213.

(Worthington *et al.*, 1973), to interpret modified nerve patterns, many new studies of modified nerves have since been reported. These X-ray patterns mainly refer to PNS myelins although a few X-ray studies on the swelling (Lalitha and Worthington, 1975) and on the action of dehydration solutions (Inouye, 1979; Melchior *et al.*, 1979) on central nervous system (CNS) myelins have been reported.

At this time, four principal physical states of nerve myelin have been identified by X-ray diffraction. These are the normal (N), swollen (S), condensed (C), and separated (E) states. The S, C, and E states can be further subdivided into a number of separate states depending on the asymmetry of the myelin membranes and on the number of membranes within the unit cell. The various physical states of nerve myelin are listed in Table I. Note that the swollen state includes the possibility of shrinking as well as swelling; for example, the subnormal state SN2 is included as a subclass of the swollen state even though the radial repeat period (d) for nerve myelin in SN2 (Worthington and Blaurock, 1969) is somewhat less than the d value for N2.

The unit cell of nerve myelin in the normal state contains two asymmetric membranes separated by fluid layers: there is a cytoplasmic fluid layer of width c and an extracellular fluid layer of width e between the membrane pair. The X-ray evidence for the presence of these two kinds of fluid layers has been documented elsewhere (Worthington, 1976). The nerve myelin membrane is slightly asymmetric about its center with a higher electron density on the cytoplasmic side (McIntosh and Worthington, 1974a). From X-ray studies (McIntosh and Worthington, 1974b; Worthington and McIntosh, 1976), it is evident that the basic difference between the various physical states N2, S2, C2, and E2 (and also between C1 and E1) is the difference in the widths of the fluid layers. This information on c, e, and d is listed in Table II for states N2, S2, C2, and E2.

Table I. Physical States of Nerve Myelin [a]

Principal state	Symbol	Substates
Normal	N	N2
Swollen	S	S2, PS2, SN2, AS4
Condensed	C	C2, C1, IC2
Separated	E	E2, E1, IE2

[a] The numerals refer to the number of membranes per unit cell. PS, partially swollen; SN, subnormal; AS, anomalous swollen; IC, intermediate condensed; IE, intermediate separated.

Table II. Comparison of the Widths of the Cytoplasmic (c) and Extracellular (e) Fluid Layers and Radial Repeat Distances (d) for the Physical States (N2, S2, C2, and E2) [a]

Physical state	c	e	d (Å)
Normal, N2	0	0	171
Swollen, S2	0	+	200–300
Condensed, C2	−	−	124–128
Separated, E2	+	+	190–210

[a] d refers to the radial repeat period of frog sciatic nerve; 0 refers to the c and e values for normal nerve; + refers to an increase in width relative to normal nerve; − refers to a decrease in width relative to normal nerve.

2. PHYSICAL STATES REVERSIBLE WITH THE NORMAL STATE

The physical states (S2, C2, and E2) are reversible with normal nerve (N2) on reimmersion of the modified nerve in Ringer's solution.

2.1. Swollen State

The swollen state S2 is reversible with N2, that is, $N2 \rightleftharpoons S2$. The S2 state is obtained by immersing the normal PNS nerve in hypotonic solutions (Finean and Millington, 1957; McIntosh and Worthington, 1974a). The normal state N2 is regained on reimmersion in Ringer's solution. The subnormal state SN2 is a transition state (Worthington and Blaurock, 1969) that is obtained in the reversal $S2 \rightarrow SN2 \rightarrow N2$. The d value for frog sciatic nerve in SN2 is 164–168 Å. This decrease in the radial repeat distance is accounted for by a decrease in c, the width of the cytoplasmic fluid layer. The partially swollen state PS2, which is obtained by immersing the nerve specimen in moderate acid or alkaline solutions (Worthington, 1979), is also reversible with N2.

2.2 Condensed State

The condensed state C2 is reversible with N2 (i.e., $N2 \rightleftharpoons C2$). The C2 state is obtained by immersing the normal PNS nerve in dehydration solutions, namely, acetone (Elkes and Finean, 1953; Inouye, 1979), ethanol (Elkes and Finean, 1953), dimethylsulfoxide (Hashizume and Masaki, 1978; Inouye, 1979; Kirschner and Caspar, 1975), and $CaCl_2$ solutions (Worthington and McIntosh, 1976). The normal state is regained on reimmersion in Ringer's solution. To ensure that the myelin layers are in C2 (and not IC2), it is important that the time of treatment is not prolonged and that the concentration of the dehydration solutions is not above a critical level.

2.3. Separated State

The separated state E2 is reversible with N2 (i.e., $N2 \rightleftharpoons E2$). The E2 state is obtained by immersing the normal PNS nerve in hypertonic solutions, namely, hypertonic NaCl and hypertonic Ringer's solution (Finean and Millington, 1957; Joy and Finean, 1963; Worthington and Blaurock, 1969; Worthington and McIntosh, 1976). The transformation $N2 \rightarrow E2$ is seldom complete in that hypertonic treatment (Worthington and McIntosh, 1976) produces a mixture of E2 and C2 states. To ensure that the myelin layers are in E2 (and C2), the time of treatment should not be prolonged and the ionic strength of the hypertonic solution should be above a critical value; otherwise, the myelin layers will move into the El (and Cl) state. The normal state is regained on reimmersion in Ringer's solution.

3. PHYSICAL STATES IRREVERSIBLE WITH THE NORMAL STATE

The condensed states IC2 and Cl and the separated states IE2 and El are irreversible with the normal states. The irreversible condensed and separated states are derived from the reversible C2 and E2 states as a result of extended treatment (several days) and increasing the concentration of the solutions above a critical level (Kirschner and Caspar, 1975; Worthington and McIntosh, 1976), C2→IC2→C1 and E2→IC2→E1 as a function of time and concentration. The condensed and separated states C1 and E1 are reversible according to C1→E1 on immersion in Ringer's solution while E1→C1 on immersion in CaCl$_2$ solution (Worthington and McIntosh, 1976).

The irreversible states IC2, C1, IE2, and E1 can be obtained as a result of a variety of chemical and physical treatments.

3.1. Freezing

Nerve myelin when frozen at a temperature below −8°C moves irreversibly into the IC2 or the C1 state (Joy and Finean, 1963). The frozen nerve myelin on thawing moves into the IE2 or the E1 state; on the other hand, if the freezing is carried out above −8°C, the frozen state resembles the normal state and, on thawing, the normal state is regained (Joy and Finean, 1963).

3.2. Air-Drying

The nerve myelin membrane on air-drying separates into at least four separate phases (Elkes and Finean, 1949; Finean, 1961). The diffraction pattern of air-dried frog sciatic nerve shows a 149-Å series (a two-membrane repeat), a 61-Å series (a single-membrane repeat with the myelin layers possibly in C1), a pseudohexagonal lipid phase with $d \approx 44$ Å, and a cholesterol phase with $d \approx 34.5$ Å. On rehydration in Ringer's solution, the air-dried nerve moves into the E1 state (Elkes and Finean, 1949; Joy and Finean, 1963; McIntosh and Worthington, 1974b).

3.3. pH Treatment

Nerve myelin in Ringer's solution maintained at acid pH below 1.8 moves irreversibly into the C1 state, whereas in Ringer's solution maintained at alkaline pH above 12.8 the nerve moves irreversibly into the E1 state (Worthington, 1979). The C1 and E1 states are reversible with each other according to C1→E1 by immersion in Ringer's solution at neutral pH while E1→C1 by immersion in Ringer's solution maintained at acid pH below 1.8.

3.4. Enzyme Treatment

The digestion of nerve with enzymes was carried out after the nerve was in the swollen state S2 in order to facilitate the penetration of the enzymes into the myelin sheath (Worthington *et al.*, 1980). After treatment with trypsin, the modified nerve moved into the E1 state, whereas after treatment with Pronase, the modified nerve moved into a mixture of E1 and C1 states (Worthington *et al.*, 1980).

4. ANOMALOUS SWOLLEN STATE

The swollen state of CNS myelin is unusual in that the swelling occurs in units of four membranes (Lalitha and Worthington, 1975); that is, if the swelling occurs at *e*, then only alternate *e* layers swell. This swollen state is referred to as the anomalous swollen state AS4. The AS4 state of CNS myelin is reversible with N2. The AS4 state of CNS myelin has since been observed by electron microscopy (McIntosh and Robertson, 1976).

In an X-ray study of the pH property of frog sciatic nerve (PNS myelin), the AS4 state was obtained by immersing the nerve in Ringer's solution in the pH range 1.8–2.5. This AS4 state of frog sciatic nerve was irreversible with N2 but reversible with C1 on decreasing the pH below 1.8 and reversible with E1 on immersing the modified nerve in Ringer's solution maintained at neutral pH.

In a recent study of the effect of heat on frog sciatic nerve (Worthington and Worthington, 1981), it was found that the myelin layers moved into the AS4 state at temperatures of 58°C or greater. This AS4 state of frog sciatic nerve was irreversible with the normal state.

5. DISCUSSION

The molecular structure of the individual myelin layers in states N2, S2, C2, and E2 is either the same or only slightly modified, for whatever the changes (if any) in the structure might be in the various states, these changes are reversible. The main difference between these four reversible states resides in the widths of the fluid layers (Worthington and McIntosh, 1976). The key factor determining the physical state (N2, S2, C2, or E2) no doubt depends on the surface charge of the outer layers of the nerve membrane and on the ionic composition of the fluid layers as previously noted (Worthington and Blaurock, 1969; Worthington and McIntosh, 1976).

The myelin layers in states C1 and E1 are irreversibly changed in that the normal asymmetric membrane (in N2) has become symmetrical about its center. The X-ray evidence (McIntosh and Worthington, 1974b; Worthington and McIntosh, 1976) suggests that there is a rearrangement of protein components within the membrane. The gel electrophoresis patterns (Brown and Worthington, 1980) do not detect any loss or breakdown of protein components when PNS myelin is transformed to C1, E1, or AS4.

It is well established that PNS and CNS myelins can be transformed into the anomalous swollen state AS4 with four membranes per repeat. Why anomalous swelling occurs in units of four membranes within the myelin sheath is unknown. The fact that AS4 and N2 of CNS myelin is reversible and AS4 of PNS myelin is reversible with C1 and E1 on varying the pH of the Ringer's solution indicates a definite molecular mechanism exists for the formation of the anomalous swollen state.

It is important to note that molecular labeling-type experiments on nerve myelin will fail to indicate any difference between the outer surfaces facing *c* or *e* when the nerve is in either the C1 or the E1 state for the myelin membrane is now symmetric about its center. For instance, storing the nerve specimens at a temperature below −8°C ensures that the asymmetry of the membranes will be destroyed on thawing. Information on sidedness can only be obtained using mild treatments in order that the asymmetry of the myelin layers is maintained.

The different physical states of nerve myelin may eventually prove to be important in elucidating the molecular structure of the nerve myelin membrane at higher resolution. X-Ray analysis is presently based on the X-ray data from nerve in states N2 and S2 and, so far, the correct structure at higher resolution has not been unequivocally determined. If the membrane profile is the same in the four reversible states, then structure analysis of the X-ray data from nerve in these states can in principle provide the correct structure. Similarly, the molecular structure of the modified symmetric membrane in states C1 and E1 may be derived on the basis that the symmetric membrane profile is the same for both C1 and E1.

ACKNOWLEDGMENT. This work was supported by a grant from the U.S. Public Health Service.

REFERENCES

Brown, W. E., and Worthington, C. R. (1980). *Arch. Biochem. Biophys.* **201,** 437–444.
Elkes, J., and Finean, J. B. (1949). *Discuss. Faraday Soc.* **6,** 134–140.
Elkes, J., and Finean, J. B. (1953). *Exp. Cell Res.* **4,** 82–95.
Finean, J. B. (1961). *Int. Rev. Cytol.* **12,** 303–334.
Finean, J. B., and Millington, P. F. (1957). *J. Biophys. Biochem. Cytol.* **3,** 89–94.
Hashizume, H., and Masaki, N. (1978). *Arch. Biochem. Biophys.* **186,** 275–282.
Inouye, H. (1979). Ph.D. thesis, Kyoto University, Japan.
Joy, R. T., and Finean, J. B. (1963). *J. Ultrastruct. Res.* **8,** 264–282.
Kirschner, D. A., and Caspar, D. L. D. (1975). *Proc. Natl. Acad. Sci. USA* **72,** 3513–3517.
Lalitha, S., and Worthington, C. R. (1975). *J. Mol. Biol.* **96,** 625–639.
McIntosh, T. J., and Robertson, J. D. (1976). *J. Mol. Biol.* **100,** 213–217.
McIntosh, T. J., and Worthington, C. R. (1974a). *Biophys. J.* **14,** 363–386.
McIntosh, T. J., and Worthington, C. R. (1974b). *Arch. Biochem. Biophys.* **162,** 523–529.
Melchior, V., Hollingshead, C. J., and Caspar, D. L. D. (1979). *Biochim. Biophys. Acta* **554,** 204–226.
Worthington, C. R. (1976). In *The Enzymes of Biological Membranes* (A. Martonosi, ed.) Vol. 1, Plenum Press, New York.
Worthington, C. R. (1979). *Int. J. Biol. Macromolecules* **1,** 157–164.
Worthington, C. R., and Blaurock, A. E. (1969). *Biochim. Biophys. Acta* **173,** 427–435.
Worthington, C. R., and McIntosh, T. J. (1974). *Biophys. J.* **14,** 703–729.
Worthington, C. R., and McIntosh, T. J. (1976). *Biochim. Biophys. Acta* **436,** 707–718.
Worthington, C. R., and Worthington, A. R. (1981). *Int. J. Biol. Macromolecules* **3,** 159–164.
Worthington, C. R., King, G. I., and McIntosh, T. J. (1973). *Biophys. J.* **13,** 480–494.
Worthington, C. R., McIntosh, T. J., and Lalitha, S. (1980). *Arch. Biochem. Biophys.* **201,** 429–436.

143

Vertebrate Rod Outer Segment Membranes

Frans J. M. Daemen and Willem J. De Grip

1. INTRODUCTION

During the last 10 years the discussion on the excitation mechanism of vertebrate photo-receptor cells has broadened considerably. The concept of a diffusible transmitter in rod cells, mediating between rhodopsin photolysis on the disk membranes and the decrease in sodium conductance of the outer segment plasma membrane, has survived. However, the "calcium-as-the-internal-transmitter" hypothesis has been challenged seriously (Fatt, 1979), and a "cyclic-GMP-as-negative-transmitter" hypothesis has received considerable support (for a review see Hubbell and Bownds, 1979). Because visual excitation in rod cells is initiated in the rhodopsin-containing membranes, these membranes have remained a matter of current interest. This short review is intended to summarize our present knowledge of vertebrate rod outer segment (ROS) membranes, following the lines of an earlier review of the same title (Daemen, 1973). Due to its limited size, the topics had to be selected more or less arbitrarily. The literature references have been chosen to maximize further orientation.

2. CHEMICAL COMPOSITION

Developments in the methods to isolate bovine ROS membranes, and their overall composition, have recently been reviewed (De Grip *et al.*, 1980). The techniques described apply equally well to other vertebrate rod photoreceptor membranes.

Whereas 10 years ago all attention was focused on rhodopsin as the main protein constituent of (often exhaustively washed) ROS membrane preparations, more subtle iso-lation and analysis methods have since revealed the presence of a number of minor pro-teins. Most of them are probably peripheral membrane proteins, which can be removed

Frans J. M. Daemen and Willem J. De Grip • Department of Biochemistry, University of Nijmegen, Nijmegen, The Netherlands.

from the membrane by low ionic strength and/or in the absence of divalent cations. Interestingly, the binding of some of these proteins is profoundly influenced by the photolytic state of rhodopsin, in parallel to the activation of enzyme activities, which are considered to play a role in visual transduction (Hubbell and Bownds, 1979). The changes upon illumination are often reversible, which suggests that an intermediate in the photolysis sequence, but not rhodopsin or opsin, acts as an allosteric activator in these processes (Kühn, 1980).

Table I is an attempt to critically summarize present ideas on the relatively well-characterized protein components of ROS membranes. Besides rhodopsin, the large intrinsic membrane protein is the only other glycoprotein recognized thus far. The long-known, almost certainly intrinsic membrane protein retinol dehydrogenase has not yet been identified on (for example) SDS gels. Guanylate cyclase, previously suggested to be associated with rod outer segments, has recently been found to be concentrated in axonemes (Fleischman *et al.*, 1980). This observation has to remind us of the as yet unresolved question of contamination of disk membrane preparations with closely connected structures like cilia and plasma membranes. Due to their different function, as compared to disk membranes, they can be expected to have a different chemical composition as well.

The most recent data on the (phospho)lipid composition of bovine and other vertebrate ROS membranes from different laboratories are in excellent agreement and do not differ greatly from earlier values. A summary is presented in Table II. There is no positive evidence for glycolipids. The degree of unsaturation of the ROS membranes, reported in

Table I. Well-Characterized Polypeptide Chains of Rod Outer Segment Membranes

	Apparent molecular weight $(\times 10^{-3})$	Reference
Large protein	270 (frog)	Papermaster *et al.* (1979)
	220 (cattle)	Molday and Molday (1979)
cGMP phosphodiesterase[a]	120, 110 (frog)	Miki *et al.* (1975)
	88, 83, 13 (cattle)	Baehr *et al.* (1979)
Protein kinase[a,b]	68 (cattle)	Kühn (1978)
Rhodopsin	34, 35, 36 (frog)	Molday and Molday (1979)
	36 ± 1 (cattle)	De Grip *et al.* (1980)
G-protein[a,b]	37, 35, 6 (cattle)	Godchaux and Zimmerman (1979), Kühn (1980)

[a] Binding to membrane dependent on ionic strength and/or divalent cations.
[b] Binding to membrane dependent on photolytic state of rhodopsin.

Table II. Lipid Composition of Bovine Rod Outer Segment Membranes (Moles/Mole Rhodopsin ± SD)

Phospholipids		Other lipids	
Phosphatidylcholine	23 ± 1	11-*cis*-Retinal	≤ 1
Phosphatidylethanolamine	27 ± 1	Diglycerides	2 ± 1
Phosphatidylserine	10 ± 1	Free fatty acids	4 ± 1
Phosphatidylinositol	1	Cholesterol (ester)	5 ± 1
Sphingomyelin	< 1	Tocopherol	0.1
Total	62 ± 2		

Table III. Fatty Acid Composition of Bovine Rod Outer Segment Membranes [a]

Fatty acids	Overall outer segments	Phosphatidyl-choline	Phosphatidyl-ethanolamine	Phosphatidyl-serine
C16 : 0	19.6	30.6	12.6	4.1
C18 : 0	21.8	19.4	25.0	21.0
C18 : 1ω9	3.2	4.5	4.2	1.5
C18 : 2ω6	0.4	0.9	0.9	<0.1
C20 : 4ω6	4.7	2.7	2.4	4.3
C22 : 4ω6	1.6	0.4	0.8	3.0
C22 : 5ω6	2.2	0.9	1.5	1.6
C22 : 5ω3	1.9	1.4	1.4	3.3
C22 : 6ω3	42.4	35.9	50.2	48.1
C24 : 4 ⎫	1.2	<0.1	<0.1	3.9
C24 : 5 ⎭		<0.1	<0.1	9.3

[a]Values are in mole% of total fatty acids. Fatty acids making up less than 0.3% in any fraction have been omitted.

the last few years, is distinctly higher than in older publications. This largely reflects the higher purity of the present ROS membrane preparations and the more extensive precautions taken against lipid oxidation. Very recent fatty acid analyses of bovine ROS membranes and their major phospholipids (Table III; Drenthe *et al.*, 1981) agree fairly well with the overall data of Stone *et al.* (1979) for bovine, rat, and frog ROS membranes and excellently with values for individual phospholipids of bovine membranes, calculated from Miljanich *et al.* (1979). The latter authors document the interesting finding that about 24% of ROS phosphatidylcholine and phosphatidylethanolamine, and about 43% of ROS phosphatidylserine molecules contain two polyunsaturated fatty acyl chains.

3. TOPOLOGY OF ROS MEMBRANES

More refined use of diffraction techniques (see e.g. Dratz *et al.*, 1979) has not altered previous ideas on the architecture of disk-stacking in rod outer segments. Neutron-scattering analysis of intact frog retina yields a disk repeat distance of 29.5 nm, an interdisk distance of 16 nm, and an intradisk space of about 3.5 nm (Yeager *et al.*, 1980). The arrangement of rhodopsin in the membrane, as studied by electron microscopy, has recently been reviewed (Olive, 1980). Upon freeze-cleavage, rhodopsin remains with the cytoplasmic (P) face. However, the size (5–18 nm) as well as the density ($5–10 \times 10^3/\mu m^3$) of the supposed rhodopsin particles vary considerably for different species (and authors). If monomeric, a rhodopsin concentration in the rod outer segment of 3 mM would result in 20×10^3 particles/μm^3. Yet, concrete biochemical evidence for the occurrence of rhodopsin oligomers is lacking (Fatt, 1981).

Considerable advance has been made over the last 10 years with respect to the organization of the lipids in the ROS membrane. The bilayer model was confirmed (see Yeager *et al.*, 1980), and recent ^{31}P NMR studies show that the bilayer (lamellar) phase occurs exclusively, irrespective of the photolytic state of rhodopsin (De Grip *et al.*, 1979). The membrane phospholipids appear to be distributed symmetrically over the two faces of the bilayer, both with respect to head groups and to fatty acyl chains (Drenthe *et al.*, 1981).

The extremely blurred image of the rhodopsin molecule in the ROS membrane could be focused considerably, since compelling evidence has accumulated for its transbilayer disposition from a variety of techniques (see Hubbell and Bownds, 1979; Olive, 1980). Proteolytic (Fung and Hubbell, 1978) and diffraction studies (see Yeager *et al.*, 1980) indicate that rhodopsin is asymmetrically inserted in the membrane, the protein body extending into the cytoplasmic space being considerably larger than that present in the intradisk space. Elegant proteolytic, chemical modification, lectin-labeling, and sequence studies (see Hubbell and Bownds, 1979; Hargrave *et al.*, 1980) show that the NH_2-terminal part (bearing the two oligosaccharide residues) resides in the intradisk space, while the COOH-terminal part (containing a number of serines and threonines that become phosphorylated upon illumination) is accessible at the cytoplasmic side of the disk. CD and IR spectroscopy on oriented samples (Chabre, 1978; Rothschild *et al.*, 1980) strongly suggest a high α-helical content (60–80%) positioned largely perpendicular to the bilayer plane. Thus, rhodopsin α-helical fragments cross the membrane an odd (up to 9) number of times. This probably will pose very specific demands on its insertion in the membrane during biosynthesis, in particular because no evidence for an NH_2-terminal signal peptide has been found (Schechter *et al.*, 1979).

Rhodopsin–lipid interactions have been studied by means of ESR and NMR spectroscopy. Isolated ROS membrane lipids do not appear to present a correct frame of reference for studying the ROS membrane, for in the absence of rhodopsin they adopt other phases next to the bilayer (De Grip *et al.*, 1979). ^{13}C NMR of reconstituted systems might indicate preferential interaction of rhodopsin with the saturated acyl chain of phospholipids (Zumbulyadis and O'Brien, 1979). No clear preference for a specific phospholipid could be detected so far, although a slight selectivity for phosphatidylserine may exist in the ROS membrane (Watts *et al.*, 1979). According to ESR studies, rhodopsin appears to slow down the mobility of its annulus lipids only moderately (Favre *et al.*, 1979; Watts *et al.*, 1979). Thus, in general there are no indications for a rigid lipid annulus around rhodopsin in the native membrane. However, at low lipid-to-rhodopsin ratios (<20), the lipids become strongly immobilized (Davoust *et al.*, 1980). A first lipid shell around rhodopsin of 20–25 molecules would support a monomeric distribution of rhodopsin in the membrane.

4. SOME FUNCTIONAL ASPECTS OF ROS MEMBRANES

Like other excitable membranes, photoreceptor membranes are biochemically characterized by the presence of a highly specific receptor protein (rhodopsin), which undergoes conformational changes upon stimulation and returns to its original conformation for renewed excitability. The functional relation between phospholipids and rhodopsin has been extensively studied by gradual removal of phospholipids from bovine ROS membranes, in rhodopsin preparations completely depleted of phospholipids, and in reconstituted phospholipid–rhodopsin vesicles (Applebury *et al.*, 1974; O'Brien *et al.*, 1977; Van Breugel *et al.*, 1978). Many properties, but not the absorbance spectrum in the visual region, change upon lipid depletion, typically starting when 20 of the 62 phospholipids have been removed. Upon reincorporation of lipid-free rhodopsin into lipid bilayer membranes, these properties are restored to those observed in the native membrane. Most clearly, the photochemical functionality of rhodopsin depends on the presence of an appropriate microenvironment. Many indications suggest that the photolytic sequence of rhodopsin is a transient process: localized unfolding of the protein upon illumination, culminating at the metarhodopsin I to II transition, followed by a refolding that is largely completed in the free opsin stage. While the unfolding phase seems largely independent of the microenvironment of the rhodopsin molecule, the refolding phase seems to require embedding in a partially hydrophobic environment that allows sufficient thermal motion, e.g., a semifluid biomembrane (Daemen and De Grip, 1980).

Young's concept of a constantly renewing rod outer segment has proven to be very stimulating. Studies on the biosynthesis and transport to the outer segment have led to the isolation and (partial) *in vitro* translation of the RNA messenger of bovine opsin (Schechter *et al.*, 1979). The discovery that rods preferentially shed membranes from the tips of the outer segments shortly after the beginning of the day (LaVail, 1976) has opened up an entire field, in which cyclic patterns (chemical rhythmicity) have been recognized as important variables in retinal chemistry (Basinger and Hollyfield, 1980; Young, 1980). Impressive advances have also been made with regard to our knowledge of the biosynthesis and metabolism of ROS phospholipids. In contrast to the rhodopsin molecule, which remains to the end in the disk in which it was originally incorporated, the phospholipids exhibit *in vivo* a high mobility between disks. Whereas rhodopsin is always asymmetrically inserted in the membrane, phospholipids may well be prone to vigorous transbilayer movements over the membrane (see e.g. Anderson *et al.*, 1980). The physiological relevance of a dense membrane system with a relatively static major protein and very dynamic phospholipids is as yet unknown.

The extremely high degree of unsaturation of the disk membranes remains an enigma. The presence of unsaturated lipids seems required for proper photolytic behavior, but one double bond per acyl chain may be sufficient (O'Brien *et al.*, 1977). However, up to 55% of the fatty acids are polyunsaturated with an average of more than 5.5 double bonds per chain. In spite of the essential fatty acid nature of the polyunsaturated chains and the protection mechanisms required to prevent oxidative damage (Daemen, 1973; Farnsworth and Dratz, 1976), the photoreceptor cells (and the retina) very efficiently defend their high unsaturation, even under extreme conditions (see e.g. Tinoco *et al.*, 1977), which seems to emphasize its importance. The fact that the electrical response of rat retina appears to be a function of position and total number of double bonds in the diet (Wheeler *et al.*, 1975) seems to suggest a role of fatty acid unsaturation during visual transduction. Whether this could concern ion mobility, (reversible) postphotolytic protein–protein interactions, or membrane modulation phenomena is as yet unclear. Recent proposals that rhodopsin photolysis might open up an ion channel (Montal, 1979) lack, in our opinion, the specificity, velocity, and reversibility one would expect, but have not been tested with respect to the influence of fatty acid unsaturation of the membrane.

5. *RECENT DEVELOPMENTS*

The expected different chemical composition of discs vs. plasma membranes seems to be substantiated (Kamps *et al.*, 1982). NMR-studies present additional evidence that rhodopsin only moderately influences the dynamics of its surrounding lipids (Deese *et al.*, 1981; Fisher and Levy, 1981) but seems preferentially to interact with highly unsaturated lipids. The phase behavior of the ROS membrane lipids may however be more complex than originally assumed (Deese *et al.*, 1981). One of the primary excitation–transduction mechanisms has been further unravelled. Upon light-capture, rhodopsin transiently activates a GTP-binding protein (G-protein), which is subsequently released and in its turn activates a c-GMP phosphodiesterase. Shut-off mechanisms include rhodopsin-phosphorylation and GTP-hydrolysis (cf. Fung *et al.*, 1981; Shinozawa and Bitensky 1981; Kohnken *et al.*, 1981). Interaction between rhodopsin and G-protein was studied in further detail by light-scattering (Kühn *et al.*, 1981).

REFERENCES

Anderson, R. E., Maude, M. B., and Kelleher, P. A. (1980). *Biochim. Biophys. Acta* **620**, 236–246.

Applebury, M. L., Zuckerman, D. M., Lamola, A. A., and Jovin, T. M. (1974). *Biochemistry* **13**, 3448–3457.

Baehr, W., Devlin, M. J., and Applebury, M. L. (1979). *J. Biol. Chem.* **254**, 11669–11677.

Basinger, S. F., and Hollyfield, J. G. (1980). *Neurochem. Int.* **1**, 81–92.

Chabre, M. (1978). *Proc. Natl. Acad. Sci. USA* **75**, 5471–5474.

Daemen, F. J. M. (1973). *Biochim. Biophys. Acta* **300**, 255–288.

Daemen, F. J. M., and De Grip, W. J. (1980). *Neurochem. Int.* **1**, 539–550.

Davoust, J., Bienvenue, A., Fellmann, P., and Devaux, P. F. (1980). *Biochim. Biophys. Acta* **596**, 28–42.

Deese, A. J., Dratz, E. A., and Brown, M. F. (1981). *FEBS Lett.* **124**, 93–99.

Deese, A. J., Dratz, E. A., Dahlquist, F. W., and Paddy, M. R. (1981). *Biochemistry* **20**, 6420–6427.

De Grip, W. J., Drenthe, E. H. S., Van Echteld, C. J. A., De Kruijff, B., and Verkleij, A. J. (1979). *Biochim. Biophys. Acta* 558, 330–337.

De Grip, W. J., Daemen, F. J. M., and Bonting, S. L. (1980). *Methods Enzymol.* **67**, 301–320.

Dratz, E. A., Miljanich, G. P., Nemes, P. P., Gaw, J. E., and Schwartz, S. (1979). *Photochem. Photobiol.* **29**, 661–670.

Drenthe, E. H. S., Klompmakers, A. A., Bonting, S. L., and Daemen, F. J. M. (1981). *Biochim. Biophys. Acta* **641**, 377–385.

Farnsworth, C. C., and Dratz, E. A. (1976). *Biochim. Biophys. Acta* **443**, 556–570.

Fatt, P. (1979). *Nature (London)* **280**, 355–356.

Fatt, P. (1981). *Exp. Eye Res.* **33**, 31–46.

Favre, E., Baroin, A., Bienvenue, A., and Devaux, P. F. (1979). *Biochemistry* **18**, 1156–1162.

Fisher, T. H. and Levy, G. C. (1981). *Chem. Phys. Lip.* **28**, 7–23.

Fleischman, D., Denisevich, M., Raveed, D., and Pannbacker, R. G. (1980). *Biochim. Biophys. Acta* **630**, 176–186.

Fung, B. K., and Hubbell, W. L. (1978). *Biochemistry* **17**, 4403–4410.

Fung, B. K., Hurley, J. B., and Stryer, L. (1981). *Proc. Natl. Acad. Sci. USA* **78**, 152–156.

Godchaux, W., III, and Zimmerman, W. F. (1979). *Exp. Eye Res.* **28**, 483–500.

Hargrave, P. A., Fong, S. L., McDowell, J. H., Mas, M. T., Curtis, D. R., Wang, J. K., Juszczak, E., and Smith, D. P. (1980). *Neurochem.* **1**, 231–244.

Hubbell, W. L., and Bownds, M. D. (1979). *Annu. Rev. Neurosci.* **2**, 17–34.

Kamps, K. M. P., De Grip, W. J., and Daemen, F. J. M. (1982). *Biochim. Biophys. Acta,* in press.

Kohnken, R. E., Eadie, D. M., and McConnell, D. G. (1981). *J. Biol. Chem.* **256**, 12510–12516.

Kühn, H. (1978). *Biochemistry* **17**, 4389–4395.

Kühn, H. (1980). *Nature (London)* **283**, 587–589.

Kühn, H., Bennet. M., Michel-Villaz, M., and Chabre, M. (1981). *Proc. Natl. Acad. Sci. USA* **78**, 6873–6877.

LaVail, M. M. (1976). *Science* **194**, 1071–1074.

Miki, N., Baraban, J. M., Keirns, J. J., Boyce, J. J., and Bitensky, M. W. (1975). *J. Biol. Chem.* **250**, 6320–6327.

Miljanich, G. P., Sklar, L. A., White, D. L., and Dratz, E. A. (1979). *Biochim. Biophys. Acta* **552**, 294–306.

Molday, R. S., and Molday, L. L. (1979). *J. Biol. Chem.* **254**, 4653–4660.

Montal, M. (1979). *Biochim. Biophys. Acta* **559**, 231–257.

O'Brien, D. F., Costa, L. F., and Ott, R. A. (1977). *Biochemistry* **16**, 1295–1303.

Olive, J. (1980). *Int. Rev. Cytol.* **64**, 107–169.

Papermaster, D. S., Schneider, B. G., Zorn, M. A., and Kraehenbuhl, J. P. (1978). *J. Cell Biol.* **78**, 415–425.

Rothschild, K. J., De Grip, W. J., and Sanches, R. (1980). *Biochim. Biophys. Acta* **596**, 338–351.

Schechter, I., Burstein, Y., Zemell, R., Ziv, E., Kantor, F., and Papermaster, D. S. (1979). *Proc. Natl. Acad. Sci. USA* **76**, 2654–2658.

Shinozawa, T., and Bitensky, M. W. (1981). *Biochemistry* **20**, 7068–7074.

Stone, W. L., Farnsworth, C. C., and Dratz, E. A. (1979). *Exp. Eye Res.* **28**, 387–397.

Tinoco, J., Miljanich, P., and Medwadowksi, B. (1977). *Biochim. Biophys. Acta* **486**, 575–578.

Van Breugel, P. J. G. M., Geurts, P. H. M., Daemen, F. J. M., and Bonting, S. L. (1978). *Biochim. Biophys. Acta* **509**, 136–147.

Watts, A., Volotovski, I. D., and Marsh, D. (1979). *Biochemistry* **18**, 5006–5013.

Wheeler, T. G., Benolken, R. M., and Anderson, R. E. (1975). *Science* **188**, 1312–1314.

Yeager, M., Schoenborn, B., Engelman, D., Moore, P., and Stryer, L. (1980). *J. Mol. Biol.* **137**, 315–348.

Young, R. W. (1980). *Neurochem. Int.* **1**, 123–142.

Zumbulyadis, N., and O'Brien, D. F. (1979). *Biochemistry* **18**, 5427–5432.

144

Sensitivity of Biomembranes to Their Environment

Role of Stochastic Processes

Franklin F. Offner

1. INTRODUCTION

An essential characteristic of biomembranes is their sensitivity to their environment: e.g., ionic concentrations, transmembrane potential differences. An aspect of such sensitivity that has been widely studied is the electrically excitable membrane. An extreme example of such sensitivity is found in certain electrically sensitive fish, which can respond to electric fields as low as 0.1 μV/cm (Murray, 1965). The Na^+ conductance of the axonal membrane is a less extreme example: Hodgkin and Huxley (1952) pointed out that this increases 6 times as rapidly as can be explained by the direct action of a small stimulus acting on a single electron charge.

2. STATISTICAL STATEMENT OF THE PROBLEM

According to Boltzmann's principle, in a system of elements that can exist in two states, p and q, differing by a free energy ΔG, the equilibrium population ratio of the system will be

$$\bar{P}/\bar{Q} = \exp(-\Delta G/kT) \tag{1}$$

ΔG is positive if free energy must be added to shift an element from its q to p state. $\bar{P}$ is the fraction of all the elements in the p state, and $\bar{Q}$ the fraction in the q. Alternatively, $\bar{P}$

Franklin F. Offner • Biomedical Engineering Center, Northwestern University, Evanston, Illinois 60201.

may be considered to be the fraction of the time any element is in the p state, and similarly for $\bar{Q}$.

The problem is that the free-energy change that could result from, for example, the change in the electric field produced directly by a stimulus on an element would often appear to be too small to be detectable in the thermal background noise; i.e., it would be obscured by Brownian movement. Thus, 26 mV acting on one electron charge has an energy equal to kT at 300°K; to obtain the factor e^6 found by Hodgkin and Huxley, classical theory would require the *full* stimulus voltage to act on elements carrying six electron charges. In the electrically sensitive fish, many thousands of charges acting together would appear to be required.

We now find, however, that the error has been to associate the value of ΔG *directly* with the magnitude of the stimulus. Biomembranes exist in systems that are far from electrochemical equilibrium, and it will be shown that under suitable conditions a small stimulus can permit the membrane to make use of this far larger source of energy.

3. EFFECT OF STOCHASTIC FLUCTUATIONS ON MEMBRANE SENSITIVITY

In the axon, as in many other cells, the internal medium has a high K^+ concentration relative to the external. With an internal potential ≈ -80 mV, the membrane is close to electrochemical equilibrium for K^+. However, the Na^+ concentration is in almost inverse ratio, so that Na^+ is ~ 150 mV from equilibrium. Inflow of Na^+ through conducting channels is blocked in the resting (polarized) state by "gates" of some form. One gating mechanism is almost surely the adsorption of Ca^{2+} ions (Frankenhaeuser and Hodgkin, 1957; Offner, 1970, 1972; Offner and Kim, 1976), but conformational changes are probably also involved; we will discuss the theory in terms of Ca^{2+} adsorption.

When a channel is open (p state, in our treatment), ions flow through the channel resistance, with a voltage drop dispersed along the channel. With the gate closed, however, the voltage difference across the membrane is concentrated at the gate. With a cation gate (Ca^{2+}), the increased negative interface potential in the q state tends to hold the gate closed, while when the gate is open the binding energy is decreased, favoring the open (p) state of the channel.

These changes of potential will not occur instantaneously, but only as the ions within the channel relax toward their new steady-state distribution; this may be with a time constant of tens to hundreds of microseconds (Meves, 1975). The open–close cycle of the gate due to thermal agitation will be 10^{-7} sec or less.

How this rapid stochastic opening-and-closing of the gates reacts with the much slower change in gate potential to increase the membrane sensitivity will now be shown: first graphically, then analytically. A more complete analysis will be found in Offner (1980).

4. A GRAPHICAL, INTUITIVE EXPLANATION

If a gate were to be held closed a long time, the voltage across the gating region would relax toward a maximum value, V_{max}, as shown in Fig. 1A; similarly, with an open gate the voltage would relax toward V_{min}. The mean time a gate is closed will increase exponentially with the energy holding it closed [equation (1)], i.e., with V. This is shown

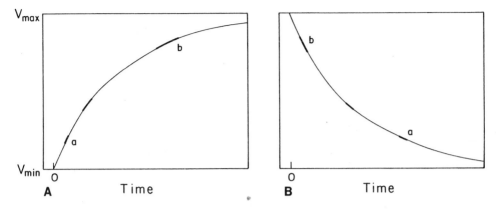

Figure 1. Graphical explanation of effect of stochastic fluctuations on membrane sensitivity. A and B show the relaxation of gate potential toward higher and lower values, with closed and open gate, respectively. Heavy segments show upward and downward mean fluctuations; these are equal at a and at b; these are thus regions of possible steady state. See text.

in Fig. 1A by the increasing duration of the heavy line segments, representing the closed time of a gate at various voltages across the gate. The change in voltage during a period of closure does not increase as rapidly as the duration of closure, however, because of the asymptotic approach toward V_{max}.

Similarly, in Fig. 1B is plotted the fall in voltage with gate open; here the open time is assumed independent* of V, but the change in voltage during an open period decreases with a decrease in voltage because of the asymptotic approach toward V_{min}.

A channel would tend to remain in a region of gate voltage where the voltage upswings and downswings are equal.† In Fig. 1 there are two such regions shown, at a and b; this occurs with simple exponential relaxations of V. We will find that a small change in a membrane parameter, e.g., the voltage across the membrane, can result in a shift in the population between these two points, resulting in a large change in conductance. Where there is only one such point of equal swings, the steady state will be in the region of such point; the gate voltage corresponding to that point will in general change much more rapidly than the membrane voltage, again resulting in high sensitivity.

5. ANALYTICAL TREATMENT

If all the gate voltages in the membrane were equal to V_{max}, we would have, by equation (1), $\overline{P/Q} = k_e$, where $k_e = \exp(-\Delta G/kT)$, ΔG being the free-energy change with the gate voltage equal to V_{max}. At any other gate voltage, $\Delta G/kT$ will be reduced by some energy γ (measured in units of kT), so that

$$P/Q = k_e e^{\gamma} \tag{2}$$

---

*This assumption is purely for convenience. The results would be unchanged if the energy change were to be divided between the two states.

†The distribution would be concentrated exactly at such points only if the p–q transition rate were infinitely fast and the relaxation curves were strictly deterministic.

P and Q are functions of γ, and are the probability densities of channels in the open and closed states, respectively, at energy coordinate γ. The fraction of all the channels with gates open is then $\bar{P} = \int_{-\infty}^{\infty} P d\gamma$, and $\bar{Q} = 1 - \bar{P}$.

Because the gate voltage relaxes with gate state, we write, assuming first-order relaxations,

$$d\gamma/dt = \alpha_p(\Gamma - \gamma) \tag{3a}$$

in the p (gate-open) state, and

$$d\gamma/dt = -\alpha_q \cdot \gamma \tag{3b}$$

in the q (closed) state. Γ is the maximum value of γ, i.e., its value when $V = V_{min}$; α_p and α_q are the rate constants of relaxation in the p and q states, respectively.

Let $S = P + Q$, i.e., the probability density of channels at γ; $\int_{-\infty}^{\infty} S d\gamma = 1$. Then from (2)

$$P/S = k_e/(k_e + e^{-\gamma}) \tag{4a}$$
$$Q/S = 1/(1 + k_e e^{\gamma}) \tag{4b}$$

Equations (4) give the fractions of the time that corresponding equations (3) are applicable to the system to give the change in the distribution of S with time. This would give a complete description of the system if equations (3) were strictly applicable, that is, if γ exactly followed these equations in a deterministic manner. In fact, due to Brownian movement of ions within the channel, the gate voltage, and thus γ, will have a random component; in electronics this is familiar as Nyquist or Johnson noise.

This randomicity of γ results in a quasi-diffusion term, $-dDs/d\gamma$, similar to Fick's law of diffusion. D is Einstein's diffusion constant, as derived for Brownian movement. Its magnitude may be estimated from channel resistance and the relaxation rate constants. Combining this term with the relaxation terms gives the Fokker–Planck partial differential equation for the system (Wang and Uhlenbeck, 1945):

$$\frac{\partial S}{\partial t} = \frac{\partial}{\partial \gamma} [S(\frac{P}{S}\alpha_p(\Gamma - \gamma) - \frac{Q}{S}\alpha_q \cdot \gamma) - \frac{\partial DS}{\partial \gamma}] \tag{5}$$

The steady state is found by letting $\partial S/\partial t = 0$ in (5), and solving the resultant ordinary differential equation numerically, after substituting for P/S and Q/S from (4). The time-dependent solution may then be found, starting with the steady-state distribution as the initial condition in solving (5) numerically.

6. RESULTS OF CALCULATION

A change in V_m, the voltage across the membrane, will change both k_e and Γ: $\delta\Delta G = -ze\delta V_m$, ze being the charge carried by a gate. Then k_e is multiplied by the factor $\exp(\delta\Delta G/kT)$, and Γ is reduced by $\delta\Delta G/kT$. These assumptions are made in the calculated results.

The steady-state calculations of Fig. 2 are for $\alpha_p = 10\alpha_q$. This is based both on Meves (1975) and on the expected dynamics of relaxation in the open and closed states. $\Gamma = 6$

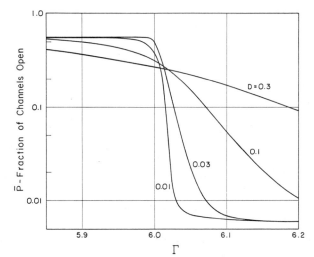

Figure 2. Change of relative membrane conductance, $\bar{P}$, as function of T, the energy available to the gate from departure of system from equilibrium, in units of kT. A change of one unit in T would be produced by a change of approximately 25 mV across the membrane. D parameter depends upon fluctuations in gate voltage; D increases directly with channel resistance.

corresponds to a change of 75 mV binding a Ca^{2+} ion, and k_e was chosen to place the conduction change in the vicinity of $\Gamma = 6$. The figures on the curves are the values of D/α_p; the range 0.03–0.3 corresponds to various estimates of channel density and conductance.

Figure 3 shows the time course of the conductance change for a step from $\Gamma = 6.1$ to 5.5; based on the assumed relationship between Γ and V_m, this would correspond to a voltage clamp step of 15 mV. The time unit of the plot is $1/\alpha_p$. If $1/\alpha_p = 10$ μsec, time $= 10^3$ in the figure would equal 10 msec.

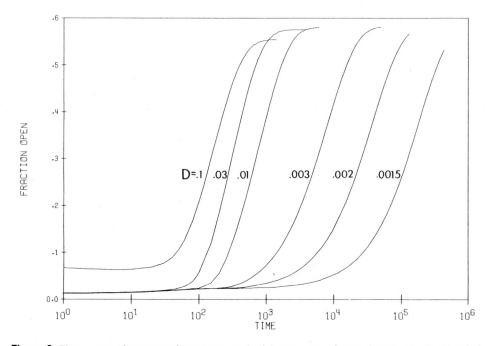

Figure 3. Time course of response of membrane conductivity to a step of approximately 15 mV. Time is in units of $1/\alpha_p$, which is of the order of 10 μsec.

7. DISCUSSION

The above results are based on only the simplest model: a single gating process, and simple exponential relaxation. The results are modified when, for example, two gating processes are considered, acting at the same time; for example, a singly charged conformational change, as well as a doubly charged Ca^{2+} gate. Such a model resembles the Hodgkin–Huxley Na^+ activation process; inactivation has not yet been included in my calculations.

8. CONCLUSIONS

We now find that the high sensitivity of membrane conductance, or permeability, to the environment is a direct and simple consequence of the stochastic nature of the control process. This should facilitate the analysis of many biological problems, possibly ranging over such a broad spectrum as from the central nervous system to kidney function.

REFERENCES

Frankenhaeuser, B., and Hodgkin, A. L. (1957). *J. Physiol. (London)* **137,** 218–244.
Hodgkin, A. L., and Huxley, A. F. (1952). *J. Physiol. (London)* **117,** 500–544.
Meves, H. (1975). *Philos. Trans. R. Soc. London* **270,** 493–500.
Murray, R. W. (1965). *Cold Spring Harbor Symp. Quant. Biol.* **30,** 233–243.
Offner, F. F. (1970). *J. Gen. Physiol.* **56,** 272–296.
Offner, F. F. (1972). *Biophys. J.* **12,** 1583–1629.
Offner, F. F. (1980). *J. Phys. Chem.* **84,** 2652–2662.
Offner, F. F., and Kim, S. H. (1976). *J. Theor. Biol.* **61,** 113–127.
Wang, M. C., and Uhlenbeck, G. E. (1945). *Rev. Mod. Phys.* **17,** 323–342.

XII

The Structure and Permeability of Blood Cell Membranes

145

Molecular Associations of the Erythrocyte Cytoskeleton

Jonathan M. Tyler, Carl M. Cohen, and Daniel Branton

The erythrocyte is a remarkable cell, strong enough to resist the potentially damaging effects of turbulent cardiac flow, yet flexible enough to negotiate splenic channels and capillaries whose diameters are less than one third its own. The red blood cell is also distinguished by its tendency to assume a defined shape. These unique characteristics are generally ascribed to the existence of an extensive cytoskeletal network of proteins that underlies and is attached to the cytoplasmic surface of the red cell membrane. Defined operationally as the shell of extensively cross-linked proteins remaining after extraction of a cell with nonionic detergents such as Triton X-100 (Yu *et al.*, 1973), the appearance of the cytoskeleton has been described by Tilney and Detmers (1975) as ''. . . an anastomosing framework like a net woven by a myopic fisherman.'' Here we review the discoveries that have elucidated the interactions that link the cytoskeleton to the membrane and that have made it possible to visualize the molecular interactions among its components. These discoveries are the result of a coordinated structural and biochemical approach that has revealed a complex set of specific interactions whose existence emphasizes the fundamental continuity between the erythrocyte membrane and the erythrocyte cytoskeleton (see Branton *et al.*, 1981, for a more detailed review).

The cytoskeleton of the erythrocyte consists of several classes of polypeptides interacting in a dynamic equilibrium to form a supramolecular complex bound specifically to a limited number of lipid-associated sites in the plasma membrane. The major extrinsic proteins comprising this complex can be separated by SDS-PAGE and consist of bands 1 and 2 (collectively, spectrin), band 2.1 (ankyrin), band 4.1, and band 5 (actin). In addition, a number of minor elements including band 4.9 are present and may have important structural or regulatory functions in the cytoskeletal complex.

Spectrin is the major extrinsic protein of the red cell, accounting for approximately

Jonathan M. Tyler and Daniel Branton • The Biological Laboratories, Harvard University, Cambridge, Massachusetts 02138.   ***Carl M. Cohen*** • Department of Research, Division of Hematology/Oncology, St. Elizabeth's Hospital, Boston, Massachusetts 02146. *Present address of J. M. T.:* Department of Genetics, University of Alberta, Edmonton, Alberta, Canada T6G-2E9.

75% of the cytoskeletal protein mass. Although the simplest form of the molecule is the α,β heterodimer of the two polypeptide chains, bands 1 and 2 (molecular weight 460,000), it is known to self-associate to form the $(\alpha,\beta)_2$ tetramer (molecular weight 920,000), which may be the native form of the molecule on the membrane. The interconversion of these two forms in solution is governed by simple equilibrium thermodynamics (Ungewickell and Gratzer, 1978). The two polypeptide subunits are structurally similar (for reviews see Steck, 1974; Kirkpatrick, 1976; Marchesi, 1979), but the smaller is multiply phosphorylated by an endogenous cAMP-independent protein kinase (Avruch and Fairbanks, 1974) at four closely spaced sites near its COOH-terminal end (Anderson, 1979; Harris and Lux, 1980). Evidence from a variety of physical methods including electron microscopy (Shotton *et al.*, 1979), light scattering (Elgsaeter, 1978; Mikkelsen and Elgsaeter, 1978), and hydrodynamic studies (Ralston, 1976; Elgsaeter, 1978) indicates that spectrin is an elongated, flexible molecule.

Spectrin's role in controlling erythrocyte membrane properties through its interactions with other polypeptides of the cytoskeletal network was first elucidated by Bennett and Branton (1977) in a quantitative study of spectrin–membrane binding. Endogenously labeled [^{32}p]spectrin was purified and its reassociation with spectrin-depleted membrane vesicles was measured as a function of time, concentration, and varying buffer conditions. This binding, characterized by relatively high affinity ($K_D = 10^{-7}$ M at pH 7.6), could be abolished by pretreating the spectrin-depleted membranes with trypsin under mild conditions, a finding that suggested that the membrane-associated "receptor" for spectrin is itself a protein. A 72,000-molecular-weight polypeptide that was released from the trypsin-treated membranes inhibited the reassociation of spectrin to unproteolyzed membrane vesicles and was capable of forming a stable complex with spectrin in solution (Bennett, 1978). This indicated that the 72,000-molecular-weight polypeptide has a binding site for spectrin. This polypeptide is now known to be a proteolytic fragment of a high-molecular-weight phosphoprotein, band 2.1 (Bennett and Stenbuck, 1979; Luna *et al.*, 1979; Yu and Goodman, 1979).

Tyler *et al.* (1979, 1980a) purified band 2.1 and characterized its interactions with spectrin in solution, demonstrating that spectrin heterodimers form a 1 : 1 stoichiometric complex of high affinity with band 2.1. In addition, electron microscopy was used to map the position of the band 2.1 binding site along the 100-nm length of the spectrin heterodimer. The localization of the band 2.1 binding site at regions 80 nm from the extreme distal ends of the spectrin tetramer arms has been confirmed and extended by detailed chemical mapping studies (Speicher *et al.*, 1980; Morrow *et al.*, 1980).

With band 2.1 identified as the major spectrin-binding protein of the cytoskeleton, a search was made for the band 2.1–membrane binding site. Again, this proved to be a protein component. The major integral membrane protein, band 3, was found to bind specifically and with high affinity to band 2.1 (Bennett and Stenbuck, 1980; Hargreaves *et al.*, 1980). Thus, the interactions between spectrin and band 2.1, between band 2.1 and band 3, and finally between band 3 and the membrane lipids, explain how this complex is firmly anchored to the red cell plasmalemma. It cannot, however, account for the cross-linked nature of the cytoskeleton that remains intact after detergent extraction eliminates the lipid continuum. Other cytoskeletal interactions must be postulated to account for these properties, and studies of the reassociation of purified cytoskeletal protein in solution show that both actin and band 4.1 play key roles in maintaining the integrity of the cytoskeleton.

Direct association between actin and spectrin has been demonstrated using [^{32}p]spectrin and F-action in a sedimentation assay (Brenner and Korn, 1979). This association has been characterized by electron microscopy to show that specific domains at the two distal ends of the spectrin tetramer can cross-link actin filaments (Cohen *et al.*, 1980).

The dense and convoluted nature of the cytoskeletal network makes the localization of its individual components difficult, and direct visualization of actin filaments in erythrocyte cytoskeletons has not been forthcoming. Rather, reports from a number of laboratories (Cohen and Branton, 1979; Pinder *et al.*, 1979; Brenner and Korn, 1979, 1980; see also Branton *et al.*, 1981, for review) suggest that the majority of red cell actin may exist as short filament seeds or protofilaments possibly stabilized by their associations with other cytoskeletal components. Although the proportion of actin present as oligomers is uncertain, it appears unlikely that red cell actin exists in the type of dynamic G–F equilibrium thought to characterize more complex, motile cells.

The spectrin–actin interaction is further complicated by the influence of another major cytoskeletal protein, band 4.1. Tyler *et al.* (1980b) have demonstrated the existence of a high-affinity association between the phosphoprotein, band 4.1, and spectrin in solution. Using biochemical and morphological analyses, they determined that saturation of available band 4.1 binding sites occurs at a stoichiometric ratio of 2 : 1 (band 4.1 monomer : spectrin heterodimer), and have localized the band 4.1 binding sites at the distal ends of the spectrin tetramer arms. By electron microscopy, the position on spectrin of these band 4.1 binding sites are indistinguishable from the position of the actin binding sites. This close association of binding domains is particularly interesting in light of evidence from a number of laboratories that band 4.1 is capable of modulating the spectrin–actin interaction (Cohen *et al.*, 1980; Fowler and Taylor, 1980; Cohen and Korsgren, 1981). Cross-linked complexes formed in the presence of band 4.1 have a consistently higher viscosity and an increased ability to survive high-shear conditions. These results suggest that band 4.1 may play a crucial role in making the erythrocyte such a remarkably plastic cell.

One of the most important results to emerge from the studies of protein interactions in the erythrocyte is the characterization of specific linkages between many of the major membrane polypeptides. Together with previous studies that have examined the interactions of band 4.2 and band 6 (Yu and Steck, 1975), the set of associations that have been characterized by the binding studies and the electron microscopic analyses reviewed here lead to a view of the erythrocyte membrane (Fig. 1) that specifies interactions among

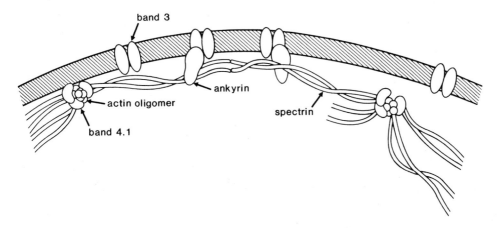

Figure 1. Associations of erythrocyte cytoskeletal proteins. [From Branton *et al.*, 1981.]

nearly all of the major membrane polypeptides and that identifies band 3, the major transmembrane protein, as a key intermediate between the lipid bilayer and the erythrocyte cytoskeleton. However, the means by which these interactions are controlled to produce morphological effects at the level of the cell membrane remain uncertain. Numerous investigations in which phosphate levels on the spectrin molecule were manipulated in an effort to determine whether phosphorylation exerts control over spectrin's known interactions have met with negative results (Ungewickell and Gratzer, 1978; Anderson and Tyler, 1979, 1980). The erythrocyte expends significant amounts of energy in maintaining phosphate levels on spectrin as well as on bands 2.1, 4.1, and 4.9 (Fairbanks and Avruch, 1974; Hosey and Tao, 1976), but whether these or other functional domains regulate association states among cytoskeletal components remains to be determined. The manner in which band 4.1 modulates spectrin–actin interaction and the possibility that similar relationships exist among other, as yet uncharacterized polypeptides, are important lines of investigation.

Although the molecular organization of the erythrocyte membrane appears to be unique, studies of its cytoskeletal protein interactions have provided us with techniques and approaches that will be generally useful in exploring the membrane structures of more complex eukaryotic cells. Techniques developed for exploring structural (Tyler and Branton, 1980) and functional (Fowler *et al.*, 1981) properties of elongate, flexible cross-linking molecules have already proven useful in examining such proteins in other cell types (Luna *et al.*, 1981; Hartwig and Stossel, 1981; Ungewickell and Branton, 1981). Certainly actin (Pollard and Weihing, 1974) and ankyrin (Bennett, 1979) are common to many cells, and spectrin, itself unique to the red cell (Hiller and Weber, 1977), appears to have structurally and functionally related analogs (actin-binding proteins) in many motile eukaryotic cells (Tyler *et al.*, 1980a). Detailed investigations of interactions linking shape-determining or motility-related proteins to the plasma membranes of these cells remain important avenues of research.

REFERENCES

Anderson, J. M. (1979). *J. Biol. Chem.* **254**, 939–944.
Anderson, J. M., and Tyler, J. M. (1979). In *Normal and Abnormal Red Cell Membranes* (S. E. Lux, V. T. Marchesi, and C. F. Fox, eds.), pp. 531–534, Liss, New York.
Anderson, J. M., and Tyler, J. M. (1980). *J. Biol. Chem.* **255**, 1259–1265.
Avruch, J., and Fairbanks, G. (1974). *Biochemistry* **13**, 5507–5514.
Bennett, V. (1978). *J. Biol. Chem.* **253**, 2292–2299.
Bennett, V. (1979). *Nature (London)* **281**, 597–599.
Bennett, V., and Branton, D. (1977). *J. Biol. Chem.* **252**, 2753–2763.
Bennett, V., and Stenbuck, P. J. (1979). *J. Biol. Chem.* **254**, 2533–2541.
Bennett, V., and Stenbuck, P. J. (1980). *J. Biol. Chem.* **255**, 6424–6432.
Branton, D., Cohen, C. M., and Tyler, J. M. (1981). *Cell* **24**, 24–32.
Brenner, S. L., and Korn, E. D. (1979). *J. Biol. Chem.* **254**, 8620–8627.
Brenner, S. L., and Korn, E. D. (1980). *J. Biol. Chem.* **255**, 1670–1676.
Cohen, C. M., and Branton, D. (1979). *Nature (London)* **279**, 163–165.
Cohen, C. M., and Korsgren, C. (1981). *Biochem. Biophys. Res. Commun.* **97**, 1429–1435.
Cohen, C. M., Tyler, J. M., and Branton, D. (1980). *Cell* **21**, 875–883.
Elgsaeter, A. (1978). *Biochim. Biophys. Acta* **536**, 235–244.
Fairbanks, G., and Avruch, J. (1974). *Biochemistry* **13**, 5514–5521.
Fowler, V., and Taylor, D. L. (1980). *J. Cell Biol.* **85**, 361–376.
Fowler, V. M., Luna, E. J., Hargreaves, W. R., Taylor, D. L., and Branton, D. (1981). *J. Cell Biol.* **88**, 388–395.

Hargreaves, W. R., Giedd, K. N., Verkleij, A., and Branton, D. (1980). *J. Biol. Chem.* **255**, 11965–11972.

Harris, H. W., and Lux, S. E. (1980). *J. Biol. Chem.* **255**, 11512–11520.

Hartwig, J. H., and Stossel, T. P. (1981). *J. Mol. Biol.* **145**, 563–581.

Hiller, G., and Weber, K. (1977). *Nature (London)* **266**, 181–183.

Hosey, M. M., and Tao, M. (1976). *Biochim. Biophys. Acta* **482**, 348–357.

Kirkpatrick, F. H. (1976). *Life Sci.* **19**, 1–18.

Luna, E. J., Kidd, G. H., and Branton, D. (1979). *J. Biol. Chem.* **254**, 2526–2532.

Luna, E. J., Fowler, V. M., Swanson, J., Branton, D., and Taylor, D. L. (1981). *J. Cell Biol.* **88**, 396–409.

Marchesi, V. (1979). *J. Membr. Biol.* **51**, 101–131.

Mikkelsen, A., and Elgsaeter, A. (1978). *Biochim. Biophys. Acta* **536**, 245–251.

Morrow, J. S., Speicher, D. W., Knowles, W. J., Hsu, C. J., and Marchesi, V. T. (1980). *Proc. Natl. Acad. Sci. USA* **77**, 6592–6596.

Pinder, J. C., Ungewickell, E., Calvert, R., Morris, E., and Gratzer, W. B. (1979). FEBS *Lett.* **104**, 396–400.

Pollard, T. D., and Weihing, R. R. (1974). *CRC Crit. Rev. Biochem.* **2**, 1–65.

Ralston, G. B. (1976). *Biochim. Biophys. Acta* **455**, 163–172.

Shotton, D. M., Burke, B. E., and Branton, D. (1979). *J. Mol. Biol.* **131**, 303–329.

Speicher, D. W., Morrow, J. S., Knowles, W. J., and Marchesi, V. T. (1980). *Proc. Natl. Acad. Sci. USA* **77**, 5673–5677.

Steck, T. L. (1974). *J. Cell Biol.* **62**, 1–19.

Tilney, L. G., and Detmers, P. (1975). *J. Cell Biol.* **66**, 508–520.

Tyler, J. M., and Branton, D. (1980). *J. Ultrastruct. Res.* **71**, 95–102.

Tyler, J. M., Hargreaves, W. R., and Branton, D. (1979). *Proc. Natl. Acad. Sci. USA* **76**, 5192–5196.

Tyler, J. M., Reinhardt, B. N., and Branton, D. (1980a). *J. Biol. Chem.* **255**, 7034–7039.

Tyler, J. M., Anderson, J. M., and Branton, D. (1980b). *J. Cell Biol.* **85**, 489–495.

Ungewickell, E., and Branton, D. (1981). *Nature (London)* **289**, 420–422.

Ungewickell, E., and Gratzer, W. B. (1978). *Eur. J. Biochem.* **88**, 379–385.

Yu, J., and Goodman, S. (1979). *Proc. Natl. Acad. Sci. USA* **76**, 2340–2344.

Yu, J., and Steck, T. L. (1975). *J. Biol. Chem.* **250**, 9176–9184.

Yu, J., Fischmann, D. A., and Steck, T. L. (1973). *J. Supramol. Struct.* **1**, 233–248.

146

Spectrin and the Red Cell Membrane Cytoskeleton

G. B. Ralston

The shape and deformability of the red cell are controlled by a "membrane cytoskeleton" composed of peripheral proteins arranged in a network on the cytoplasmic face of the membrane. The major proteins of this network are spectrin (bands 1 and 2 in the nomenclature of Steck, 1972) and actin (band 5, Tilney and Detmers, 1975). The peripheral proteins can be solubilized, leading to fragmentation of the membranes into small vesicles, by incubation at low ionic strength (Marchesi and Steers, 1968). Conversely, the nonionic detergent Triton X-100 (Yu *et al.*, 1973) solubilizes most of the lipids and integral proteins, leaving behind an insoluble residue of peripheral proteins, arranged in an anastomosing network of fibrils, shown beautifully in the scanning electron microscope study of Hainfeld and Steck (1977).

Spectrin, actin, and bands 4.1 and 4.9 are components of the "Triton shells" produced at very low ionic strength, or after washing with hypertonic KC1 (Sheetz, 1979), and thus appear to form the core of the structure.

The Triton shell does not appear to be an artifactual structure produced only in the presence of the detergent. Similar structures have been seen in fixed and sectioned membranes (Tsukita *et al.*, 1980). Localization of spectrin at the immediate surface of the cytoplasmic face of the erythrocyte membrane has been verified by labeling of fixed, thin sections of whole red cells with ferritin-labeled antispectrin antibodies (Ziparo *et al.*, 1978).

The coextraction of actin with spectrin (Tilney and Detmers, 1975), the coelution of spectrin and actin in a very-high-molecular-weight complex on agarose gel filtration (Dunbar and Ralston, 1978), and the occurrence of spectrin and actin in Triton shells (Yu *et al.*, 1973) suggest a spectrin–actin interaction that may be responsible for maintaining the integrity of the red cell membrane (Ralston, 1978). However, the details of the interaction of spectrin with actin are poorly understood. Chemical cross-linking studies have indicated that band 1 and actin may be close enough on the membrane to interact (Liu *et al.*, 1977).

G. B. Ralston • Department of Biochemistry, University of Sydney, Sydney, New South Wales, 2006, Australia.

There is good evidence for spectrin binding to F-actin, mediated either by magnesium ions (Cohen *et al.*, 1980) or by component 4.1 (Ungewickell *et al.*, 1979), and interaction of spectrin with G-actin as well as with F-actin has been reported by Cohen *et al.* (1980). The dimer of spectrin appears to be monovalent toward actin, while the tetramer seems to have two actin-binding sites and is capable of cross-linking actin filaments (Brenner and Korn, 1979).

Purified spectrin can reassociate with inside-out membrane vesicles at a specific binding site that displays high affinity ($K_D \approx 10^{-8}$ M; Bennett and Branton, 1977). This site has been identified with protein "ankyrin" (band 2.1) (Bennett and Stenbuck, 1979). Band 2.1 has now been purified, and its interaction with spectrin has been studied in solution (Tyler *et al.*, 1980). The interaction shows the same affinity as is seen on association of spectrin with inverted vesicles, and stoichiometry indicates that there is one binding site for 2.1 per dimer of spectrin.

Other binding sites for spectrin are likely to exist on the membrane. Band 4.1 is also found in Triton shells, and in the very-high-molecular-weight complex extracted at low ionic strength (Dunbar and Ralston, 1978). Band 4.1 presumably also interacts strongly with other components of the membrane, as approximately 80% of band 4.1 remains membrane bound after extraction of the water-soluble proteins (Lux, 1979). Band 4.1 has been demonstrated to bind directly with spectrin (Tyler *et al.*, 1980), and the site of its interaction is indistinguishable from that of actin (Cohen *et al.*, 1980).

Spectrin can be isolated in several different oligomeric states. Much of the early work on spectrin was confused by the failure to recognize that the water-soluble extracts are heterogeneous (Ralston, 1975), and that because of the enormous size of spectrin, small mass concentrations of lower-molecular-weight proteins could represent molar concentrations comparable with that of the spectrin. Gel filtration on Sephadex G-200 is totally inadequate for the purification of spectrin. The most appropriate medium for its isolation seems to be 4% agarose.

The two polypeptides of spectrin, bands 1 and 2, are normally associated to form a heterodimer of the form $(1 + 2)$, which is itself able to dimerize reversibly to form the tetramer $(1 + 2)_2$ in a reaction that is strongly dependent on temperature and salt concentration (Ungewickell and Gratzer, 1978). In low-ionic-strength solutions the dimer is favored, while in higher salt concentrations the tetramer is the more stable state. The equilibrium is rapid only above 30°C.

The tetramer appears to be the basic state of spectrin in the membrane *in vivo*, a conclusion strengthened by recent cross-linking experiments (Ji *et al.*, 1980). Alteration of the tetramer : dimer ratio in red cell ghosts by manipulation of the temperature and ionic strength has recently revealed that the tetramer–dimer equilibrium may be important for controlling the stability of the red cell membrane (Liu and Palek, 1980).

Dissociation of the spectrin dimer to individual polypeptide chains does not appear to be measurable at the concentrations of protein normally examined. Nevertheless, the two polypeptide chains appear to be held together by non-covalent interactions only, and can be dissociated by relatively low concentrations of urea (Calvert *et al.*, 1980). Dissociation seems to precede unfolding of the proteins.

In spite of earlier suggestions that the individual polypeptides might be composed of smaller "subunits," more recent research seems to confirm the fact that the two components of spectrin (bands 1 and 2) are, in fact, extremely large linear polypeptides of molecular weight 250,000 and 220,000, respectively (Pinder *et al.*, 1976; Dunn *et al.*, 1978). Peptide maps of both polypeptides display a marked similarity, although there are many peptides peculiar either to band 1 or to band 2 (Dunn *et al.*, 1978; Anderson, 1979). The similarities that may exist between bands 1 and 2, however, are not sufficiently close for immunological cross-reactivity (Kirkpatrick *et al.*, 1978).

The very broad urea concentration-dependence of spectrin unfolding (Calvert *et al.*, 1980) indicates that spectrin is not folded as a single cooperative unit. The CD spectrum shows significant changes at very low urea concentrations (Calvert *et al.*, 1980), indicating that even the native state contains labile structures. Thus, spectrin's tertiary structure may be comprised of separate folded domains, linked by relatively flexible regions. Approximately 20% of the polypeptide chain of spectrin undergoes significant segmental mobility, as evidenced by the presence of sharp peaks in the NMR spectrum (Calvert *et al.*, 1980). This part of the molecule is presumably the flexible regions between folded domains.

The ability of both intact spectrin and tryptic fragments from purified bands 1 and 2 to undergo rapid refolding after urea denaturation (Knowles *et al.*, 1979) is also consistent with the existence of discrete, independent folding domains in spectrin.

Spectrin is phosphorylated only on band 2, and recent work indicates that there are four sites for phosphorylation located in a cluster within a 20,000-molecular-weight peptide at the COOH-terminal end of the band 2 polypeptide (Harris and Lux, 1980). ^{32}P-labeled phosphate groups on spectrin in intact red cells (Smith and Moore, 1980) turn over with a half-life of the order of 40 hr, while in the ghosts the half-time is only 4 hr.

The role of phosphorylation is unknown: phosphorylation of spectrin does not influence the tetramer–dimer equilibrium (Dunbar and Ralston, 1978; Ungewickell and Gratzer, 1978), the binding of spectrin to everted vesicles (Bennett and Branton, 1977; Anderson and Tyler, 1980), or the conformational stability of spectrin (Calvert *et al.*, 1980). However, ATP has already been implicated directly in a number of processes. ATP is associated with the spectrin–actin complexes in water-soluble extracts (Dunbar and Ralston, 1978), and Low and Brandts (1978) found that ADP and AMP–PNP (a nonhydrolyzable ATP analog) both reduce the extent of the temperature-dependent transition of spectrin in red cell membranes near 50°C. These effects were not seen with isolated spectrin, and may reflect the stabilization of spectrin by its interaction with other cytoskeletal components, an interaction that may be potentiated by ATP.

It is now generally accepted that spectrin does not closely resemble myosin, although the two proteins share similarities. Moreover, recent theories for the biconcave shape of the red blood cell do not require the operation of a contractile system (Ralston, 1978). The hydrodynamic properties of spectrin suggest a flexible, expanded molecule, rather than the rigid rod model applicable for myosin (Ralston, 1976).

Low-angle platinum shadowing (Shotton *et al.*, 1979) reveals the dimer as a long wormlike molecule, about 1000 Å long, in which the two subunits are aligned side-by-side. Individual molecules exhibit great variety in their detailed shape, indicating considerable flexibility along the length of the molecule. The tetramer appears to be formed by a head-to-head association of dimers, to produce a molecule of exactly twice the contour length of the dimer (194 and 97 nm, respectively). No linear association states larger than the tetramer were seen, suggesting that the tetramerization was of the head-to-head type. However, Tyler *et al.* (1980) have reported occasional cyclic hexamers, consistent with the concepts of reciprocal band 1–band 2 interactions. The relative rarity of these structures may reflect steric constraints on their formation.

The structures seen in shadowed micrographs of spectrin are consistent with the known hydrodynamic data. Spectrin shows a strong polyelectrolyte effect: the frictional properties of the molecule are strongly salt-dependent (Ralston and Dunbar, 1980), and suggest neither a compact globular structure nor a rigid rod. The radius of gyration of the dimer in moderate salt concentrations has been determined from light scattering to be approximately

22 nm, but increases markedly as the salt concentration is lowered (Elgsaeter, 1978). This increase in R_G is inconsistent with a rigid rod model, but is consistent with a flexible rod or wormlike model, with a contour length greater than 76 nm. A contour length greater than 40–50 nm is also indicated from the birefringence relaxation times (Mikkelsen and Elgsaeter, 1978). The ratio of the intrinsic viscosity for spectrin tetramer and dimer (approx. 0.95) is not consistent with the rigid rod or random coil models, but is consistent with the flexible rod (Dunbar and Ralston, 1981).

The polyelectrolyte effect of spectrin is also seen in the salt-dependent contraction of the cytoskeletons prepared with Triton X-100 (Johnson *et al.*, 1980). This contraction is detectable even at the isoelectric plant, indicating that considerable intramolecular charge interactions occur.

The Stokes radius of the tetramer of spectrin (approx. 20 nm, Ralston, 1976) is sufficient for spectrin to form a monolayer over the membrane surface (Ralston, 1978), given the number of copies of spectrin per cell. However, the contour length of spectrin is considerably greater than 40 nm, and therefore the molecule *in situ* must be bent and kinked. This fact may underlie the irregularity of the cytoskeleton seen in electron micrographs, and in addition, the flexibility of spectrin may have important implications for the elasticity and extensibility of the red cell membrane.

Now that we have some knowledge of the molecular shape and size of spectrin and the identity of the important proteins with which it interacts, the next step in expanding our understanding of the cytoskeleton will be the unraveling of the detailed interactions between the cytoskeletal proteins—their stoichiometry, mutual interaction, strength of binding.

REFERENCES

Anderson, J. (1979). *J. Biol. Chem.* **254**, 939–944.
Anderson, J., and Tyler, J. M. (1980). *J. Biol. Chem.* **255**, 1259–1265.
Bennett, V., and Branton, D. (1977). *J. Biol. Chem.* **252**, 2753–2763.
Bennett, V., and Stenbuck, P. J. (1979). *J. Biol. Chem.* **254**, 2533–2541.
Brenner, S. L., and Korn, E. D. (1979). *J. Biol. Chem.* **254**, 8620–8627.
Calvert, R., Ungewickell, E., and Gratzer, W. B. (1980). *Eur. J. Biochem.* **107**, 363–371.
Cohen, C. M., Tyler, J. M., and Branton, D. (1980). *Cell* **21**, 875–884.
Dunbar, J. C., and Ralston , G. B. (1978). *Biochim. Biophys. Acta* **510**, 283–291.
Dunbar, J. C., and Ralston, G. B. (1981). *Biochim. Biophys. Acta* **667**, 177–184.
Dunn, M. J., Kemp, R. B., and Maddy, A. H. (1978). *Biochem. J.* **173**, 197–205.
Elgsaeter, A. (1978). *Biochim. Biophys. Acta* **536**, 235–244.
Hainfeld, J. F., and Steck, T. L. (1977). *J. Supramol. Struct.* **6**, 301–311.
Harris, H. W., and Lux, S. E. (1980). *J. Biol. Chem.* **255**, 11512–11520.
Ji, T. H., Kiehm, D. J., and Middaugh, C. R. (1980). *J. Biol. Chem.* **255**, 2990–2993.
Johnson, R. M., Taylor, G., and Mayer, D. B. (1980). *J. Cell Biol.* **86**, 371–376.
Kirkpatrick, F. H., Rose, D. J., and La Celle, P. (1978). *Arch. Biochem. Biophys.* **186**, 1–8.
Knowles, W., Speicher, D., Morrow, J., and Marchesi, V. T. (1979). *J. Cell Biol.* **83**, 272a.
Liu, S.-C., and Palek, J. (1980). *Nature (London)* **285**, 586–588.
Liu, S.-C., Fairbanks, G., and Palek, J. (1977). *Biochemistry* **16**, 4066–4074.
Low, P. S., and Brandts, J. F. (1978). *Arch. Biochem. Biophys.* **190**, 640–646.
Lux, S. E. (1979). *Nature (London)* **281**, 426–429.
Marchesi, V. T., and Steers, E. (1968). *Science* **159**, 203–204.
Mikkelsen, A., and Elgsaeter, A. (1978). *Biochim. Biophys. Acta* **536**, 245–251.
Pinder, J. C., Tidmarsh, S., and Gratzer, W. B. (1976). *Arch. Biochem. Biophys.* **172**, 654–660.
Ralston, G. B. (1975). *Aust. J. Biol. Sci.* **28**, 259–266.
Ralston, G. B. (1976). *Biochim. Biophys. Acta* **455**, 163–172.

Ralston, G. B. (1978). *Trends Biochem. Sci.* **3,** 195–198.

Ralston, G. B., and Dunbar, J. C. (1980). *Biochim. Biophys. Acta* **579,** 20–30.

Sheetz, M. P. (1979). *Biochim. Biophys. Acta* **557,** 122–134.

Shotton, D. M., Burke, B. E., and Branton, D. (1979). *J. Mol. Biol.* **131,** 303–329.

Smith, J. E., and Moore, K. (1980). *J. Lab. Clin. Med.* **95,** 808–815.

Steck, T. L. (1972). *J. Mol. Biol.* **66,** 295–305.

Tilney, L. G., and Detmers, P. (1975). *J. Cell Biol.* **66,** 508–520.

Tsukita, S., Tsukita, S., and Ishikawa, H. (1980). *J. Cell Biol.* **85,** 567–576.

Tyler, J. M, Reinhardt, B. N., and Branton, D. (1980). *J. Biol. Chem.* **255,** 7034–7039.

Ungewickell, E., and Gratzer, W. B. (1978). *Eur. J. Biochem.* **88,** 379–385.

Ungewickell, E., Bennett, P. M., Calvert, R., Ohanian, V., and Gratzer, W. B. (1979). *Nature (London)* **280,** 811–814.

Yu, J., Fischmann, D. A., and Steck, T. L. (1973). *J. Supramol. Struct.* **1,** 233–248.

Ziparo, E., Lemay, A., and Marchesi, V. T. (1978). *J. Cell Sci.* **34,** 91–101.

147

Molecular Features of the Cytoskeleton of the Red Cell Membrane

Vincent T. Marchesi, Jon S. Morrow, David W. Speicher, and William J. Knowles

Isolated red blood cell membranes are stable structures composed of lipids and proteins. The lipids form a single bilayer that is supported by two types of proteins. One type is tightly integrated within the lipid bilayer and is composed primarily of transmembrane glycoproteins. The remaining proteins are more loosely associated with the bilayer; most of these are attached to the inner surface of the membrane and appear to form a submembranous cytoskeleton.

On the basis of what is now known about the erythrocyte cytoskeleton we can offer the following generalizations: Red cell cytoskeletons are macromolecular protein complexes that form flexible, submembranous networks that are attached to the inner surfaces of the membrane and extend over the entire surface. Although these cytoskeletal complexes are stable structures, they also have a dynamic quality in that their shape is sensitive to metabolites such as ATP, 2,3-diphosphoglycerate, cations, and possibly other factors. This protein cytoskeleton is believed to be responsible for the plastic deformation of red blood cells and is also thought to control the distribution of integral membrane proteins. The erythrocyte cytoskeleton is composed predominantly of a single protein molecule called spectrin. Recent evidence suggests that spectrin may act in concert with other membrane proteins including an erythrocyte form of actin. Additional proteins are thought to provide links between this spectrin–actin complex and the membrane proper.

Spectrin has a number of properties that make it ideally suited to serve as the backbone of a cytoskeletal network (Marchesi, 1979). Spectrin is composed of two subunits, referred to as α and β on the basis of their molecular weights (250,000 and 225,000, respectively). Spectrin molecules exist in solution in a number of different forms depending upon how they are extracted from ghost membranes. If spectrin is extracted from red cell membranes in low-ionic-strength media at 37°C, the predominant species is an $\alpha\beta$ dimer

Vincent T. Marchesi, Jon S. Morrow, David W. Speicher, and William J. Knowles • Department of Pathology, Yale University School of Medicine, New Haven, Connecticut 06510.

(Marchesi and Steers, 1968). Depending upon the concentration of the spectrin and other factors, spectrin dimers can associate to form tetramers $(\alpha_2\beta_2)$ (Ungewickell and Gratzer, 1978) or even higher oligomers (Morrow and Marchesi, 1981). The capacity of spectrin to self-associate may be one of the principal mechanisms by which cytoskeletal dynamism is maintained. Both spectrin dimers and tetramers bind to the inner surface of the red cell membrane through a high-affinity association with another membrane protein called ankyrin (Bennett and Branton, 1977; Litman et al., 1980; Bennett and Stenbuck, 1979a). Ankyrin is also a relatively large polypeptide chain (subunit structure approximately 200,000), and it is believed to be attached to the membrane by noncovalent associations with band 3 (Bennett and Stenbuck, 1979b), one of the major transmembrane glycoproteins.

In addition to binding to ankyrin, spectrin can also associate with another red cell membrane protein, operationally designated as band 4.1 (Ungewickell et al., 1979; Tyler et al., 1980). Spectrin can also associate with the fibrous form of actin (Brenner and Korn, 1979). The conditions for these associations have only been defined in virto, and it is not known whether spectrin–actin or spectrin–4.1 associations actually take place in vivo.

These potential associations between spectrin, ankyrin, and band 3 might account for spectrin's capacity to influence the distribution of intramembranous particles (IMPs) of the red cell membrane. As these IMPs are believed to be composed predominantly of the transmembrane glycoproteins, largely band 3, it is conceivable that spectrin's capacity to influence IMP topography is mediated by a spectrin–ankyrin–band 3 linkage.

Some investigators have also suggested that spectrin is able to bind directly to polar groups of certain membrane phospholipids. Spectrin binds weakly to liposomes in vitro (Mombers et al., 1977), and chemical cross-linking studies of intact membranes suggest that spectrin is in close proximity to both phosphatidylethanolamine (PE) and phosphatidylserine (PS) (Marinetti and Crain, 1978). It has been further postulated that spectrin's capacity to interact with PE and PS may explain the asymmetric distribution of PE and PS, which occur predominantly on the inner leaflet of the lipid bilayer (Haest et al., 1978).

The chemical homogeneity of spectrin and its constituent subunits has been the subject of controversy for many years. Early investigators suggested that each of its subunits was composed of multiple polypeptide chains; more recent data show that each spectrin subunit is a single unique polypeptide chain. Experiments employing limited proteolytic cleavage indicate that each spectrin subunit is composed of multiple domains that can be defined by their sensitivity to proteolytic cleavage and their capacity to undergo reversible denaturation under appropriate conditions (Speicher et al., 1980). Specific functional sites have also been identified on certain chemical domains of each of the spectrin subunits (Morrow et al., 1980). For example, an 80,000-molecular-weight peptide derived from the terminal end of the α chain appears to be involved in dimer–tetramer associations and in the formation of higher oligomers. A 50,000-molecular-weight peptide isolated from the β chain is involved in ankyrin binding, and several peptide fragments have been identified that mediate noncovalent associations between the two subunits (Morrow et al., 1980). A schematic model of the spectrin dimer is shown in Fig. 1.

Studies on the structure and function of spectrin have led to much speculation as to how a dynamic cytoskeletal structure can exist and function in the mature red cell. Early theories suggested that spectrin might undergo a directed polymerization reaction to produce long cablelike structures that buttressed the lipid bilayer. This theory was based on the then-prevalent notion that spectrin molecules polymerized in neutral salt solutions. Early electron microscopic studies of spectrin using negative staining provided some supporting evidence for this idea. However, recent studies indicate that spectrin is not particularly sensitive to salt solutions, nor does it necessarily form aggregates in the presence of diva-

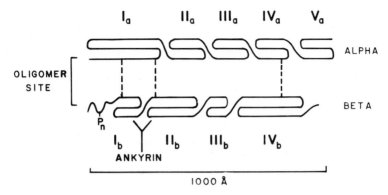

Figure 1. This diagram shows in schematic form how the two spectrin subunits may be associated to form the rodlike structures shown in Fig. 2. Each of the spectrin subunits appears to be made up of proteolytically resistant domains that have been defined on the basis of limited proteolytic digestion studies (Speicher *et al.*, 1980). Parallel functional studies have also demonstrated that spectrin oligomers are formed by chain–chain associations at the end of the molecule designated the oligomer site. This region is also close to the location of the phosphorylated segment of the β chain and reasonably close to the ankyrin binding site (Morrow *et al.*, 1980). The two subunits also appear to interact noncovalently at the sites designated by the dashed lines. [Taken from Morrow *et al.*, 1980.]

lent cations. The binding of spectrin to the inner surface of the red cell membrane does require a minimal amount of salt, but calcium or magnesium seems to play no specific role (Litman *et al.*, 1980). Furthermore, electron microscopic studies of spectrin using new rotary shadowing techniques indicate that spectrin molecules are relatively short 1000-Å flexible rods (Shotton *et al.*, 1979). It is now believed that the long fibrous structures of spectrin seen by negative staining are probably artifacts. The appearance of rotary shadowed spectrin is shown in Fig. 2.

On the basis of results from many laboratories, there is now good evidence that spectrin can form specific macromolecular complexes *in vitro* with actin, band 4.1, and possibly other membrane proteins. One hint that such a spectrin–actin–band 4.1 complex might exist was derived from studies of red cell membranes following extraction by Triton X-100 (Hainfeld and Steck, 1977; Sheetz, 1979). This detergent extracted much of the lipid and essentially all of the integral membrane proteins but left behind a shell-like structure composed almost entirely of the proteins listed above. These "Triton shells" appear as reticulated networks when examined by scanning electron microscopy and are thought by many to be reasonably faithful representations of the cytoskeletal network of the ghost membrane. Recent electron microscopic studies of red cell membranes following tannic acid–glutaraldehyde fixation also provide convincing evidence in favor of an anastomosing submembranous network immediately beneath the lipid bilayer of intact red cells (Tsukita *et al.*, 1980).

Studies from our laboratory provide yet another possible mechanism to explain the cytoskeletal form and its function. When spectrin is purified free of other membrane proteins by repeated recycling gel filtration and then concentrated by vacuum dialysis, it is possible to prepare pure spectrin solutions in excess of 24 mg/ml in isotonic salt media. Under these conditions, spectrin does not aggregate nonspecifically but rather undergoes specific oligomerization reactions that are both temperature and concentration dependent. By examining spectrin under these conditions, we have found that pure spectrin dimers have the capacity to form higher oligomeric units that represent incremental additions of

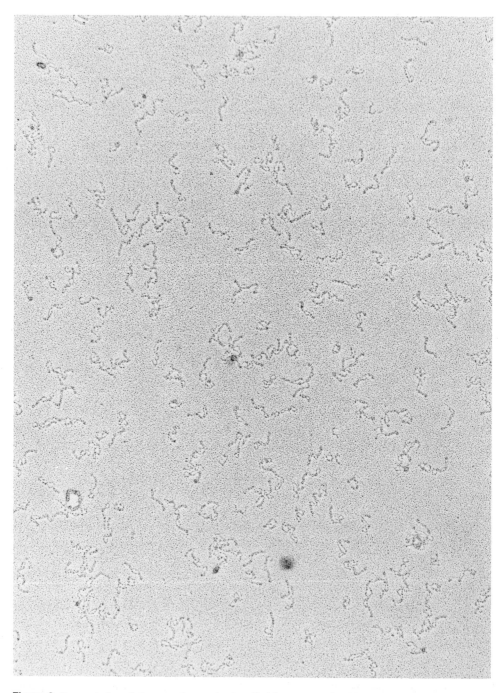

Figure 2. Rotary shadowed electron micrograph of purified human spectrin. The molecules in this preparation appear either as $\alpha\beta$ dimers of approximately 1000 Å in length, or $\alpha_2\beta_2$ tetramers of twice that length. Magnification $\times$ 80,000.

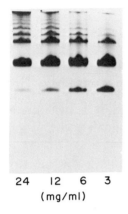
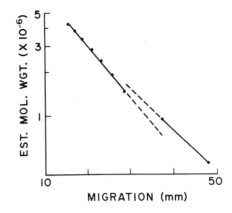

Figure 3. Oligomeric forms of spectrin can be demonstrated by nondenaturing polyacrylamide gel electrophoresis as described by Morrow and Marchesi (1981). The fastest migrating band on these gels is the spectrin dimer ($\alpha\beta$), followed by the spectrin tetramer, hexamer, and higher forms, each with an apparent incremental weight of 450,000 [Taken from Morrow and Marchesi, 1981.]

dimer forms (Morrow and Marchesi, 1981). These oligomeric structures can be demonstrated by nondenaturing gel electrophoresis (Fig. 3), sucrose gradients, gel filtration, and rotary shadowing electron microscopy. On the basis of these observations, we have postulated that spectrin can undergo a concentration-dependent oligomerization reaction through specific noncovalent associations mediated by the same peptides that are involved in the dimer–tetramer equilibrium. We have devised a simple scheme that explains the capacity of spectrin alone to form large oligomers and even branching structures by this mechanism (Morrow and Marchesi, 1981).

On the basis of these new findings and previous data on the capacity of spectrin to form complexes with other proteins, it is possible to conceive of a relatively simple model for the form and function of the erythrocyte cytoskeleton based on the capacity of spectrin to form higher oligomers, which may then be further complexed by actin and possibly band 4.1. Together, these components could provide a dynamic cytoskeleton for the red blood cell.

REFERENCES

Bennett, V., and Branton, D. (1977). *J. Biol. Chem.* **252,** 2753–2763.
Bennett, V., and Stenbuck, P. J. (1979a). *J. Biol. Chem.* **254,** 2533–2541.
Bennett, V., and Stenbuck, P. J. (1979b). *Nature (London)* **280,** 468–473.
Brenner, S. L., and Korn, E. D. (1979). *J. Biol. Chem.* **254,** 8620–8627.
Haest, C. W. M., Plasa, G., Kamp, D., and Deuticke, B. (1978). *Biochim. Biophys. Acta* **509,** 21–32.
Hainfeld, J. F., and Steck, T. L. (1977). *J. Supramol. Struct.* **6,** 301–311.
Litman, D., Hsu, C. J., and Marchesi, V. T. (1980). *J. Cell Sci.* **42,** 1–22.
Marchesi, V. T. (1979). *J. Membr. Biol.* **51,** 101–131.
Marchesi, V. T., and Steers, E. J. (1968). *Science* **159,** 203–204.
Marinetti, G. V., and Crain, R. C. (1978). *J. Supramol. Struct.* **8,** 191–213.
Mombers, C., Van Dijck, P. W. M., Van Deenen, L. L. M., De Gier, J., and Verkleij, A. J. (1977). *Biochim. Biophys. Acta* **470,** 152–160.
Morrow, J. S., and Marchesi, V. T. (1981). *J. Cell Biol.* **88,** 463–468.
Morrow, J. S., Speicher, D. W., Knowles, W. J., Hsu, C. J., and Marchesi, V. T. (1980). *Proc. Natl. Acad. Sci. USA* **77,** 6592–6596.

Sheetz, M. P. (1979). *Biochim. Biophys. Acta* **557,** 122–134.

Shotton, D., Burke, B., and Branton, D. (1979). *J. Mol. Biol.* **131,** 303–329.

Speicher, D. W., Morrow, J. S., Knowles, W. J., and Marchesi, V. T. (1980). *Proc. Natl. Acad. Sci, USA* **77,** 5673–5677.

Tsukita, S., Tsukita, S., and Ishikawa, H. (1980). *J. Cell Biol.* **85,** 567–576.

Tyler, J. M., Reinhardt, B. N., and Branton, D. (1980). *J. Biol. Chem.* **255,** 7034–7039.

Ungewickell, E., and Gratzer, W. (1978). *Eur. J. Biochem.* **88,** 379–385.

Ungewickell, E., Bounett, P. M., Calvert, R., Ohanian, V., and Gratzer, W. B. (1979). *Nature (London)* **280,** 811–814.

148

Structural Variation in Human Erythrocyte Sialoglycoproteins

David J. Anstee, William J. Mawby, and Michael J. Tanner

1. INTRODUCTION

The human erythrocyte membrane contains four distinct sialic acid-rich glycoproteins. These sialoglycoproteins can be resolved using the discontinuous sodium dodecyl sulfate-polyacrylamide gel electrophoresis system of Laemmli (1970) and visualized using the periodic acid–Schiff (PAS) stain (Fig. 1). Different laboratories have used a variety of nomenclatures for these PAS-staining bands and some of these nomenclatures are shown in Fig. 1. A complicated pattern of bands is observed because two of the components (α and δ) form dimeric complexes with themselves and with each other (Marton and Garvin, 1973; Tuech and Morrison, 1974; Dahr *et al.*, 1976a; Mueller *et al.*, 1976).

An erythrocyte contains approximately 10^6 copies of α (Gahmberg *et al.*, 1979) and approximately 2.5×10^5 copies of δ (Tanner *et al.*, 1980), and these molecules contribute 60 and 15% respectively of the total surface sialic acid (Gahmberg *et al.*, 1976; Taliano *et al.*, 1980; Dahr *et al.*, 1977a). The complete amino acid sequence of α has been determined (Tomita *et al.*, 1978), but only partial sequence data are available for δ (Furthmayr, 1978a; Dahr *et al.*, 1980a,b). Amino acid changes at residues 1 and 5 of α determine the expression of blood group M or N antigen activity on this molecule (Fig. 2; Dahr *et al.*, 1977b; Wasniowska *et al.*, 1977; Blumenfeld and Adamany, 1978; Furthmayr, 1978a). Sialoglycoprotein δ invariably carries the blood group N antigen (Hamaguchi and Cleve, 1972), and amino acid differences at residue 29 determine the S or s antigen activity of this molecule (Fig. 2; Dahr *et al.*, 1980a,b).

The α and δ sialoglycoproteins have an unusual sequence relationship. They have identical amino acid sequences for the NH$_2$-terminal 26 residues (although δ does not have an oligosaccharide at Asn-26), the remaining nine residues of the known sequence of δ

David J. Anstee • Southwestern Regional Blood Transfusion Centre, Southmead, Bristol BS10 5ND, England. ***William J. Mawby and Michael J. Tanner*** • Department of Biochemistry, University of Bristol, Bristol BS8 1TD, England.

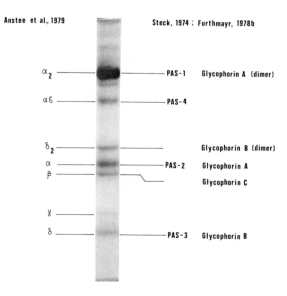

Figure 1. Sialoglycoproteins of the human erythrocyte membrane. SDS-polyacrylamide gel (10% acrylamide) of normal erythrocyte membranes stained with the PAS stain.

(which contains the Ss antigenic determinants) show homology with residues in the region 56–67 of α (Fig. 2) (Marchesi *et al.*, 1976; Furthmayr, 1978a; Dahr *et al.*, 1980ab). Little is known about the structure of the other sialoglycoproteins β and γ (Furthmayr, 1978b; Owens *et al.*, 1980; Dahr *et al.*, 1980b).

Several types of sialoglycoprotein variants involving α and δ are known and many of these are associated with altered blood group MN and Ss antigens (see Anstee, 1980, for review of serological aspects). Because the serologically based nomenclature used for the variant erythrocytes does not often reflect the nature of the primary defect in the cells, in the following discussion they have been classified according to their primary defect.

```
         M Ser    .    .  Gly          1Ọ  .   .   .   .   .              20    .        .
α  (Glycophorin A)     -Ser-Thr-Thr-    -Val-Ala-Met-His-Thr-Thr-Thr-Ser-Ser-Ser-Val-Ser-Lys-Ser-Tyr-Ile-Ser-Ser-Gln-Thr-
         N Leu           Glu
         ‡                30                       .           40                 .          .          5Ọ
         Asn-Asp-Thr-His-Lys-Arg-Asp-Thr-Tyr-Ala-Ala-Thr-Pro-Arg-Ala-His-Glu-Val-Ser-Glu-Ile-Ser-Val-Arg-Thr-

                                      60
         Val-Tyr-Pro-Pro-Glu-Glu-Glu-Thr-Gly-Glu-Arg-Val-Gln-Leu-Ala-His-His-
```

```
                     .    .    .                   1Ọ  .   .   .   .   .              20    .        .
δ  (Glycophorin B)  Leu-Ser-Thr-Thr-Glu-Val-Ala-Met-His-Thr-Ser-Thr-Ser-Ser-Ser-Val-Thr-Lys-Ser-Tyr-Ile-Ser-Ser-Gln-Thr-
                                                       (Thr)                        (Ser)
                            S Met 30
                     Asn-Gly-Glu-    -Gly-Gln-Leu-Val-His-Arg-
                            s Thr    (Glu)
```

Figure 2. Partial amino acid sequence of α and δ. Data are taken from Tomita *et al.* (1978) and Dahr *et al.* (1980b). Amino acid polymorphisms at positions 1 and 5 of α and 29 of δ give rise to the M, N, and Ss blood groups, respectively. Furthmayr (1978a) has reported a partial amino acid sequence for δ that differs from that of Dahr *et al.* (1980b) in some respects. Some of these differences are indicated by the residues in parentheses. Asterisks denote *O*-linked sialotetrasaccharides; ‡ denotes the *N*-linked oligosaccharide.

2. VARIANTS INVOLVING α

The rare red blood cells of type En(a−) completely lack α, and the absence of the entire molecule has been established by a variety of techniques (Dahr *et al.*, 1976b; Gahmberg *et al.*, 1976; Tanner and Anstee, 1976; Furthmayr, 1978a). In these cells there is a concomitant increase in the carbohydrate content of the anion transport protein (band 3; Tanner *et al.*, 1976; Gahmberg *et al.*, 1976), which results from the addition of approximately equimolar amounts of galactose and N-acetylglucosamine to the oligosaccharide chain of the protein. This probably simply represents an increase in the number of repeating linear galactose $\beta(1\rightarrow4)$ N-acetylglucosamine disaccharide units and/or branching by the same units. Fukuda *et al* (1979) have shown that these are the predominant structural units of the carbohydrate chain of the anion transport protein of normal erythrocytes. Similar changes in band 3 occur in all the α-deficient cells so far studied (see below), and this increased glycosylation appears to be secondary to the absence of α. As glycosylation of membrane proteins occurs during transit from the rough endoplasmic reticulum to the plasma membrane, this implies that the synthesis of the anion transport protein and α occurs at the same time, and that they may be closely associated at some stage in their biosynthesis. Gahmberg *et al.* (1978) have shown that α is synthesized between the pronormoblast and reticulocyte stages of erythrocyte maturation, suggesting that the anion transport protein is also synthesized at this point. The biosynthesis of the anion transport protein has been shown to occur between these stages in the rabbit (Foxwell and Tanner, 1981).

A rare structural variant of α occurs in M^g erythrocytes and this appears to be a simple allele of α (as are the M and N antigens). In this case the sialoglycoprotein has an apparent molecular weight that is approximately 2000 less than α, and a reduced sialic acid content. This probably reflects the absence of a small number of the O-linked sialic acid-rich oligosaccharides as a result of changes in the amino acid sequence of the polypeptide (Anstee and Tanner, 1978).

3. VARIANTS INVOLVING δ

Oriental populations contain a relatively high proportion (5–10%) of an allelic form of δ called $\delta^{Mi.III}$ (Anstee *et al.*, 1979). This molecule has an apparent molecular weight that is 15,000 greater than that of δ, contains more sialic acid, and is more accessible to cell surface radioiodination than δ. These differences imply that $\delta^{Mi.III}$ has an amino acid sequence that is distinct from that of normal δ. The $\delta^{Mi.III}$ sialoglycoprotein carries s blood group antigen activity, while an alternative form of this protein, designated $\delta^{Mi.IV}$, carries blood group S antigen activity.

The erythrocytes of some individuals apparently completely lack δ (Dahr *et al.*, 1975; Tanner *et al.*, 1977; Anstee *et al.*, 1979) and consequently lack blood group Ss antigens (S−s−). This variant is relatively common (up to 30%) in negroid populations (Lowe and Moores, 1972). The cells contain an unusual component of slightly higher molecular weight than δ that shows some of the properties of δ, but does not appear to be glycosylated. This unusual component may be an altered form of δ that is deficient in carbohydrate, and has a polypeptide chain that is substantially increased in size.

4. VARIANTS INVOLVING BOTH α AND δ

A unique family containing two apparently healthy individuals homozygous for the absence of both α and δ has been found (homozygous M_k, Tokunaga *et al.*, 1979). Other studies on unrelated heterozygous M^k individuals support this conclusion (Dahr *et al.*, 1977c; Furthmayr, 1978a; Anstee and Tanner, 1978). These cells appear to have an altered glycosylation of the anion transport protein similar to that found in other α-deficient cells.

Erythrocytes with interesting hybrid molecules containing portions of both α and δ are also known. Miltenberger Class V erythrocytes contain a hybrid sialoglycoprotein with an NH_2 terminus that has characteristics of α and a COOH terminus with characteristics of δ [denoted $(\alpha-\delta)^{Mi.V}$, Anstee *et al.*, 1979]. The chromosome that gives rise to the $(\alpha-\delta)^{Mi.V}$ component does not give rise to normal α or δ, so that individuals homozygous for this condition lack both normal α and δ (Vengelen-Tyler *et al.*, 1981). The $(\alpha-\delta)^{Mi.V}$ sialoglycoprotein can have the NH_2 terminus of either blood group M or N forms of α, but is always associated with s blood group antigen activity. However, a homozygous individual is known who appears to have an analogous hybrid molecule expressing S rather than s blood group activity (Langley *et al.*, 1981). In contrast, a hybrid molecule that appears to have the NH_2 terminus of δ and the COOH terminus of α, $(\delta-\alpha)^{Ph}$, occurs in Ph erythrocytes (Tanner *et al.*, 1980; Mawby *et al.*, 1981). In this case the chromosome expressing $(\delta-\alpha)^{Ph}$ retains the gene for normal α but lacks the gene for normal δ. A similar hybrid sialoglycoprotein $(\delta-\alpha)^{St^a}$ of slightly lower molecular weight than $(\delta-\alpha)^{Ph}$ is associated with the St^a antigen (a rare MNSs-related antigen) but in this case both the normal α and δ genes are present (D. J. Anstee *et al.*, 1982).

All of these hybrids probably result from unequal crossover events between misaligned chromosomes and arise by mechanisms similar to that proposed for the origin of the Lepore and anti-Lepore hemoglobins (Bunn *et al.*, 1977; Weatherall and Clegg, 1977). The $(\alpha-\delta)$ and $(\delta-\alpha)$ hybrids have apparent molecular weights that are intermediate between those of α and δ. However, from a straightforward crossover between two normal chromosomes, one would expect $(\delta-\alpha)$ products to have the same molecular weight as α and $(\alpha-\delta)$ products the same molecular weight as δ. This would result from the identity in the NH_2-terminal 26 amino acids of α and δ as the crossover must occur in a region of sequence homology (Fig. 3a). Dahr *et al.* (1980b) have pointed out that residues 27–35 of δ (Fig. 2) are homologous with two overlapping regions of α (residues 56–64 and 59–67, Fig. 2). This suggests the additional possibility of crossing-over events between the internal homology regions of normal α and δ with a concomitant looping out of the region between 27 and 55 or 58 on the α gene (Fig. 3b). Therefore, the $(\delta-\alpha)$ hybrids would have a lower molecular weight than α because of the absence of the looped-out region (residues 27–55/58 of α), which contains four O-linked oligosaccharides as well as one N-linked oligosaccharide. This would result in the reduction of approximately 10,000 to the apparent molecular weight compared with α. Such a reduction is consistent with the observed decrease in apparent molecular weight of the $\delta-\alpha$ components relative to α. The $(\alpha-\delta)$ hybrids would be larger than δ for they would retain this looped-out region (Fig. 3b). The difference in molecular weight between $(\delta-\alpha)^{Ph}$ and $(\delta-\alpha)^{St^a}$ may result from different alignments of the two internal homology regions of α with that of δ. Similarly, different $(\alpha-\delta)$ hybrids may occur. These crossovers would lead to new amino acid sequences at the crossover site that might well give rise to novel antigens.

The events depicted in Fig. 3a would give products that would be difficult to distinguish from normal sialoglycoproteins. In the case of the Lepore-type hybrids, these would

(a)

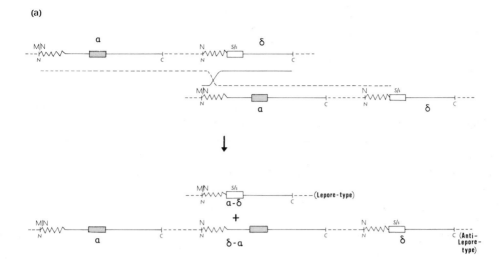

(b)

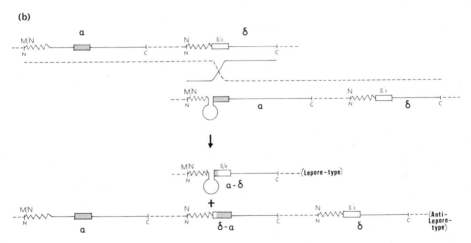

Figure 3. Schematic representation of the origin of hybrid sialoglycoproteins. The figure illustrates hypothetical crossovers between two chromosomes carrying normal α and δ genes. The crossovers shown by solid lines give rise to Lepore-type hybrids; the broken lines depict the events leading to the anti-Lepore-type hybrids. In Fig. 3a, crossover occurs in the region of NH$_2$-terminal identity ($\wedge\wedge\wedge$) between α and δ. In Fig. 3b, crossover occurs in the internal homology regions of α ($\blacksquare$) and δ ($\square$). M, N, S, and s refer to blood group antigens characteristic of the sialoglycoproteins. **N** and **C** denote the NH$_2$ terminus and COOH terminus of the individual sialoglycoproteins.

be distinguished by the absence of normal α and the presence of δ with either M or N blood group antigen activity. Cells homozygous for this condition would be phenotypically identical to En(a−) erythrocytes. In fact, a blood group M active δ sialoglycoprotein is found in one En(a−) donor (Anstee *et al.*, 1977; Dahr *et al.*, 1978) and this is most probably a Lepore-type hybrid. The anti-Lepore-type hybrids produced by the mechanisms shown in Fig. 3a would be almost indistinguishable from normal erythrocytes except for a possible increase in the α-containing bands on SDS-polyacrylamide gel electrophoresis. It follows from the mechanisms shown in Fig. 3 that the genes coding for α and δ are closely linked on the same chromosome and in the order α–δ.

5. CONCLUSIONS

It is clear from the above discussion that considerable variation occurs in the structure of both α and δ in normal individuals. The complete absence of α and δ does not cause any apparent impairment of red cell production and function and therefore the role of α and δ in normal erythrocytes is somewhat puzzling. However, the variants lacking α and δ are sufficiently rare to suggest that there is some selective advantage for erythrocytes to retain these proteins. The nature of this advantage is unknown but may be related to the carbohydrates carried by these proteins (Tanner, 1978). It may be relevant that in all the α-deficient variants there is increased glycosylation of the anion transport protein. This may contribute to the viability of these cells. However, it should be noted that the oligosaccharide structures added to the anion transport protein are different from those that are normally present on α. This suggests that the oligosaccharides carried on α and δ have no essential function.

It is interesting that no variant involving β and γ has yet been established. However, as it is not clear that these sialoglycoproteins have any marker antigens, it may simply be very difficult to pick them out. Alternatively, this conservation of their structure may reflect their more crucial function in the red blood cell.

ACKNOWLEDGMENTS. The authors' work was supported in part by grants from the Medical Research Council and the Wellcome Trust.

REFERENCES

Anstee, D. J. (1980). In *Immunobiology of the Erythrocyte* pp. 67–98, Liss, New York.

Anstee, D. J., and Tanner, M. J. A. (1978). *Biochem. J.* **175,** 149–157.

Anstee, D. J., Barker, D. M., Judson, P. A., and Tanner, M. J. A. (1977). *Brit. J. Haematol.* **35,** 309–320.

Anstee, D. J., Mawby, W. J., and Tanner, M. J. A. (1979). *Biochem. J.* **183,** 193–203.

Anstee, D. J., Mawby, W. J., Parsons, S., Tanner, M. J. A., and Giles, C. M. (1982). *J. Immunogenet.,* in press.

Blumenfeld, O. O., and Adamany, A. M. (1978). *Proc. Natl. Acad. Sci. USA* **75,** 2727–2731.

Bunn, H. F., Forget, B. G., and Ranney, H. M. (1977). *Human Haemoglobins,* pp. 151–154, Saunders, London.

Dahr, W., Uhlenbruck, G., Issitt, P. D., and Allen, F. H. (1975). *J. Immunogenet.* **2,** 244–251.

Dahr, W., Uhlenbruck, G., Janssen, E., and Schmalisch, R. (1976a). *Blut* **32,** 171–181.

Dahr, W., Uhlenbruck, G., Leikola, J., Wagstaff, W., and Landfried, K. (1976b). *J. Immunogenet.* **3,** 329–346.

Dahr, W., Issitt, P. D., and Uhlenbruck, G. (1977a). In *Human Blood Groups, 5th International Convocation on Immunology,* pp. 197–205, Karger, Basel.

Dahr, W., Uhlenbruck, G., Janssen, E., and Schmalisch, R. (1977b). *Humangenetik* **35,** 385–343.

Dahr, W., Uhlenbruck, G., and Knott, H. (1977c). *J. Immunogenet.* **4,** 191–195.

Dahr, W., Uhlenbruck, G., Leikola, J., and Wagstaff, W. (1978). *J. Immunogenet.* **5,** 117–127.

Dahr, W., Gielen, W., Beyreuther, K., and Kruger, J. (1980a). *Z. Physiol. Chem.* **361,** 145–152.

Dahr, W., Beyreuther, K., Steinbach, H., Gielen, W., and Kruger, J. (1980b). *Z. Physiol. Chem.* **361,** 895–906.

Foxwell, B., and Tanner, M. J. A. (1981). *Biochem. J.* **195,** 129–137.

Fukuda, M., Fukuda, M. N., and Hakomori, S. (1979). *J. Biol Chem.* **254,** 3700–3703.

Furthmayr, H. (1978a). *Nature (London)* **271,** 519–524.

Furthmayr, H. (1978b). *J. Supramol. Struct.* **9,** 79–95.

Gahmberg, C. G., Myllyla, G., Leikola, J., Pirkola, A., and Nordling, S. (1976). *J. Biol. Chem.* **251,** 6108–6116.

Gahmberg, C. G., Jokinen, M., and Andersson, L. C. (1978). *Blood* **52,** 379–387.

Gahmberg, C. G., Jokinen, M., and Andersson, L. C. (1979). *J. Biol. Chem.* **254,** 7442–7448.

Hamaguchi, H., and Cleve, H. (1972). *Biochim. Biophys. Acta* **278,** 271–280.

Laemmli, U. K. (1970). *Nature (London)* **227,** 680–681.

Langley, J. W., Issitt, P. D., Anstee, D. J., McMahan, M., Smith, N., Pavone, B. G., Tessel, J. A., and Carlin, M. A. (1981). *Transfusion* **21,** 15–24.

Lowe, R. F., and Moores, P. (1972). *Human Hered.* **22,** 344–350.

Marchesi, V. T., Furthmayr, H., and Tomita, M. (1976). *Annu. Rev. Biochem.* **45,** 667–698.

Marton, L. S. G., and Garvin, J. E. (1973). *Biochem. Biophys. Res. Commun.* **52,** 1457–1462.

Mawby, W. J. Anstee, D. J., and Tanner, M. J. A. (1981). *Nature (London)* **291,** 161–162.

Mueller, T. J., Dow, A. W., and Morrison, M. (1976). *Biochem. Biophys. Res. Commun.* **72,** 94–99.

Owens, J. W., Mueller, T. J., and Morrison, M. (1980). *Arch. Biochem. Biophys.* **204,** 247–254.

Steck, T. L. (1974). *J. Cell Biol.* **66,** 295–305.

Taliano, V., Guevin, R. M., Hebert, D., Daniels, G. L., Tippett, P., Anstee, D. J., Mawby, W. J., and Tanner, M. J. A. (1980). *Vox Sang.* **38,** 87–93.

Tanner, M. J. A. (1978). *Curr. Top. Membr. Transp.* **11,** 279–325.

Tanner, M. J. A., and Anstee, D. J. (1976). *Biochem. J.* **153,** 271–277.

Tanner, M. J. A., Jenkins, R. E., Anstee, D. J., and Clamp, J. R. (1976). *Biochem. J.* **155,** 701–703.

Tanner, M. J. A., Anstee, D. J., and Judson, P. A. (1977). *Biochem. J.* **165,** 157–161.

Tanner, M. J. A., Anstee, D. J., and Mawby, W. J. (1980). *Biochem. J.* **187,** 1493–1500.

Tokunaga, E., Sasakawa, S., Tamaka, K., Kawamata, H., Giles, C. M., Ikin, E. W., Poole, J., Anstee, D. J., Mawby, W. J., and Tanner, M. J. A. (1979). *J. Immunogenet.* **6,** 383–390.

Tomita, M., Furthmayr, H., and Marchesi, V. T. (1978). *Biochemistry* **17,** 4756–4770.

Tuech, J. K., and Morrison, M. (1974). *Biochem. Biophys. Res. Commun.* **59,** 352–360.

Vengelen-Tyler, V., Anstee, D. J., Issitt, P. D., Pavone, B. G., Ferguson, S. J., Mawby, W. J., Tanner, M. J. A., Blachman, M. A., and Lorque, P. (1981). *Transfusion* **21,** 1–14.

Wasniowska, K., Drzeniek, Z., and Lisowska, E. (1977). *Biochem. Biophys. Res. Commun.* **76,** 385–390.

Weatherall, D. J., and Clegg, J. B. (1977). *Cell* **16,** 467–479.

149

Functional Structure of Band 3, the Anion Transport Protein of the Red Blood Cell, as Determined by Proteolytic and Chemical Cleavages

Aser Rothstein

Band 3, an abundant intrinsic protein of the red blood cell membrane, has been identified as the anion transporter involved in physiological exchanges of Cl^- and HCO_3^- (Cabantchik and Rothstein, 1974a; Passow *et al.*, 1975). The structural and functional arrangement of the peptide in the membrane has been assessed largely by the use of proteolytic enzymes, chemical cleaving agents, and inhibitory chemical probes, some of which react covalently. The literature on the subject is extensive and has been generously reviewed (Cabantchik *et al.*, 1978; Koziarz *et al.*, 1978; Knauf, 1979; Tanner *et al.*, 1980; Drickamer, 1980). To avoid redundancy we will be particularly concerned in this review with those features of the band 3 architecture that are functionally important.

1. EFFECTS OF PROTEOLYTIC CLEAVAGES ON ANION TRANSPORT FUNCTION

Chymotrypsin applied to the intact cell cleaves band 3 into two membrane-bound segments of 60,000 and 35,000 daltons (P_1-N_t and P_1-C_t in Fig. 1) without inhibitory effect (Cabantchik and Rothstein, 1974b; Passow *et al.*, 1977). Chymotrypsin or trypsin applied to the cytoplasmic face of the membrane in resealed ghosts (Lepke and Passow, 1976) or inside-out vesicles (Steck *et al.*, 1976) cleaves at P_2. A segment of 55,000 daltons (P_2-C_t) is membrane-bound, and the remaining 40,000 daltons of peptide (P_2-N_t) is solubilized. Nevertheless, function is maintained (Lepke and Passow, 1976; Grinstein *et al.*, 1978). Nor is function reduced in vesicles containing only the 17,000- and 35,000-dalton

Aser Rothstein • Research Institute, The Hospital for Sick Children, Toronto, Canada.

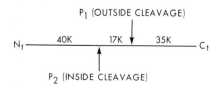

Figure 1. Location of outside and inside chymotryptic cleavages of band 3 in intact membranes.

segments, following cleavage at P_1 and P_2 (Grinstein *et al.*, 1978). Thus, the soluble cytoplasmic P_2–N_t segment is not essential for transport, a conclusion supported by the findings that the binding of glyceraldehyde-3-phosphate dehydrogenase (Rice and Steck, 1976) or of specific antibodies (England *et al.*, 1980) to that part of band 3 does not influence the transport.

In leaky ghosts treated with high concentrations of chymotrypsin, additional cleavages occur. A 15,000-dalton segment is derived from the 17,000-dalton segment, and a 9000-dalton (Ramjeesingh *et al.*, 1980a) and a carbohydrate-containing fragment of as yet undefined molecular weight (Ramjeesingh and Rothstein, unpublished observations) are derived from the 35,000-dalton segment. The capacity of chymotrypsin to produce the "extra" cleavages in ghosts is dependent on the ionic strength (Rothstein *et al.*, 1981), but probably also depends on an altered conformation of the protein in ghosts, for no "extra" cleavages are produced during chymotrypsin treatment of cells at low ionic strength. At least one of the "extra" cleavages is inhibitory, for the rate of specific anion transport by vesicles containing increasing amounts of 15,000- and 9000-dalton segments is diminished and is absent when proteolysis is complete.

In contrast to chymotryptic treatment of cells, which is noninhibitory (Cabantchik and Rothstein, 1974b), exposure to Pronase (Passow, 1971; Knauf and Rothstein, 1971) or to papain (Passow *et al.*, 1977; Jennings and Passow, 1979) is inhibitory. In the case of Pronase, a 60,000-dalton segment is produced, but the fate of the 35,000-dalton segment is not clear (Cabantchik and Rothstein, 1974b; Passow *et al.*, 1977). In the case of papain, a 60,000-dalton segment is also produced, but other cleavages occur so that the COOH-terminal segment is about 25,000 rather than 35,000 daltons (Jennings and Passow, 1979).

2. LOCATION OF IRREVERSIBLE TRANSPORT INHIBITORS IN THE 17,000-DALTON TRANSMEMBRANE SEGMENT OF BAND 3

Several covalent inhibitors of anion transport are highly localized in band 3 including NAP-taurine [*N*-(4-azido-2-nitrophenyl)-2-aminoethylsulfonic acid], and derivatives of sulfanilic acid or disulfonic stilbenes (Cabantchik *et al.*, 1978; Knauf, 1979; Drickamer, 1980). In the case of the disulfonic stilbene, DIDS (4,4'-diisothiocyanostilbene-2,2'-disulfonic acid), the binding in relation to inhibition is linear, with complete inhibition associated with a 1 : 1 mole ratio of DIDS to band 3 (Lepke *et al.*, 1976; Ship *et al.*, 1977). The conclusion that each band 3 monomer has one inhibitory site to which DIDS specifically binds is supported by the recent observations of Jennings and Passow (1979).

Within band 3, certain covalent probes have been further localized in the 17,000 transmembrane segment. These include IBS (1-isothiocyano-4-benzene sulfonic acid) and DASA (diazosulfonic acid) (Drickamer, 1980), DIDS (Grinstein *et al.*, 1978), and NAP-taurine (Knauf *et al.*, 1978a). The latter two compounds have been demonstrated to be affinity probes, binding reversibly largely through interactions of their sulfonic acid groups, followed by the covalent reaction (Cabantchik *et al.*, 1978; Knauf, 1979). A kinetic analysis of reversible inhibition of the probes applied to the outside of cells indicates that DIDS competes with Cl^- for binding to the transport site (Shami *et al.*, 1978), whereas NAP-taurine binds to a modifier site (Knauf *et al.*, 1978b). These findings would imply that the transport and modifier sites, defined by the kinetic behavior of the transport, are located in the 17,000-dalton segment. This conclusion is not, however, completely justified. The sulfonic acid ligands of the probes interact with positive groups in the functional sites, whereas the covalent reaction involves interaction through azido or thiocyano groups that are located in the probes as much as 10 Å from the sulfonic acid groups. It can be firmly concluded that the covalent binding sites in the 17,000-dalton segment must be close neighbors to the positive ligands of the functional sites within the tertiary structure of the protein, although they may be at some distance in the primary structure.

The location of some of the probes within the primary structure of the 17,000-dalton segment has been determined by the use of chemical cleaving agents. IBS is found largely within 7000 daltons of the P_1 site (Fig. 1). A small fraction of DASA is found in the same location, but most of it is between 7000 and 11,000 daltons of P_1 (Drickamer, 1980). DIDS has been localized in a 2000-dalton fragment between 7000 and 9000 daltons of P_1, probably at 9000 daltons (Ramjeesingh *et al.*, 1980b). Based on the fact that the probes, as used, are nonpenetrating and that they react with the inhibitory site in band 3 from the outside, the locations of the probes are consistent with a looped structure involving three crossings of the bilayer (Drickamer, 1980; Ramjeesingh *et al.*, 1980b) as illustrated in Fig. 2. A looped structure is also consistent with data derived from use of iodination with lactoperoxidase from the two sides of the membrane, followed by peptide mapping (Tanner *et al.*, 1980).

3. THE ARRANGEMENT OF THE 35,000-DALTON COOH-TERMINAL SEGMENT OF BAND 3

The 35,000-dalton COOH-terminal segment of band 3 (P_1–C_t, Fig. 1) produced by chymotryptic treatment of cells contains the carbohydrate of band 3, exposed to the external medium (Steck *et al.*, 1976; Tanner *et al.*, 1980; Drickamer, 1980). It has been concluded (Rao, 1979) that this segment crosses the bilayer because its two sulfhydryl groups can be labeled with permeant, but not impermeant, sulfhydryl agents. On the other hand, the impermeant agent can label the groups in "leaky" ghosts (Rao, 1979) or in inside-out

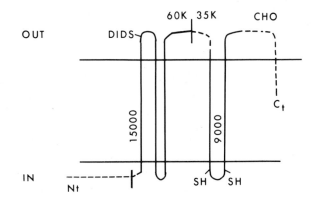

Figure 2. A possible arrangement of band 3 in the bilayer. The outside and inside chymotryptic cleavages are indicated by vertical bars. The 15,000- and 9000-dalton intrinsic segments are indicated by solid lines.

vesicles (Ramjessingh and Rothstein, unpublished observations). The conclusion that the 35,000-dalton segment crosses the bilayer is supported by peptide mapping of the 35,000-dalton segment after iodination with lactoperoxidase from the two sides of the membrane (Markowitz and Marchesi, personal communication).

As pointed out previously, in ghosts treated with high concentrations of chymotrypsin, the 35,000-dalton segment undergoes additional cleavages. The products are a 9000-dalton segment (Ramjeesingh *et al.*, 1980a) and another of undetermined molecular weight containing the carbohydrate (Ramjeesingh and Rothstein, unpublished observations). The 9000-dalton segment has no carbohydrate, but it does contain the two cytoplasmic sulfhydryl groups of the parent 35,000-dalton segment (Rothstein *et al.*, 1981). After chemical cleavage of the 9000-dalton segment with nitrocyanobenzoic acid at the two cysteine residues, three fragments are recovered of 4500, 4000, and 1000 daltons. It has been proposed the 9000-dalton segment forms a loop across the membrane, with two crossing strands as illustrated in Fig. 2 (Rothstein and Ramjeesingh, 1981; Rothstein *et al.*, 1981). The peptide to which the carbohydrate is attached must also be hydrophobically associated with the bilayer, but it is not known whether it crosses the bilayer.

4. THE STRUCTURE OF THE INTRINSIC MEMBRANE-CROSSING ELEMENTS OF BAND 3

Several sets of information are relevant in considering the relationships of the 15,000- and 9000-dalton membrane-crossing elements of band 3 in the bilayer: (1) The 15,000-dalton segment appears to cross the membrane three times and the 9000-dalton segment two times for a total of five crossings (as discussed previously. (2) The amino acid composition of the 15,000- and 9000-dalton segments, or of fragments of those segments produced by chemical cleavages, is not particularly hydrophobic (Rothstein and Ramjeesingh, 1981; Rothstein *et al.*, 1981). The percentage of hydrophobic residues ranges from 25 to 45%, indicating that their hydrophobic behavior is probably not due to the presence of long runs of hydrophobic amino acid residues in their primary structure, but rather to their tertiary structure. (3) The parent segments of the 15,000- and 9000-dalton segments (17,000- and 35,000-dalton segments) are closely associated. In Triton X-100 extracts, they cocentrifuge (Reithmeier and Rao, 1979), and they can be cross-linked by DIDS under certain conditions in the intact membrane (Jennings and Passow, 1979). The 15,000- and 9000-dalton segments coextract and copurify in Triton X-100 extracts (Ramjessingh *et al.*, 1981. (4) Band 3 seems to be present in the membrane in the form of dimers, but transport apparently occurs independently via each monomer, based on an analysis of kinetic behavior, and on the fact that each monomer has one DIDS-binding site that contributes equally to the inhibitory effect. (See reviews of Cabantchik *et al.*, 1978; Knauf, 1979.)

The above observations can be rationalized by proposing that the membrane crossing strands of band 3 (Fig. 2) are present in the bilayer in the form of an assembly arranged so that its outer surface is largely composed of hydrophobic residues in direct contact with the side chains of the phospholipids, and with its hydrophilic residues largely internalized to form an aqueous interior that provides a pathway through which anion transport occurs (Rothstein *et al.*, 1976; Cabantchik *et al.*, 1978; Knauf, 1979; Rothstein *et al.*, 1981). The aqueous pathway cannot be a completely open channel because the anion transport involves a one-for-one electroneutral exchange of high electrical resistance, with carrierlike kinetic behavior. It has been proposed that a barrier to anion diffusion is present near the outer surface of band 3 and that passage of anions across this barrier involves a local spontaneous conformational change such that an anion-binding site (the transport site) that was topologically out becomes topologically in (and vice versa). Knauf *et al.* (1980) have elaborated on the model, and Passow *et al.* (1980) have also proposed a conformational allosteric model involving sites in the 17,000- and 35,000-dalton segments. The changes in conformation may involve a local shift in the relationship of membrane-crossing strands with respect to each other near the permeability barrier of the proposed assembly. The crossing strands would be close neighbors accounting for the close associations of various segments of band 3. The functional sites might involve several such strands with outside loops, consistent with the observation that interactions of inhibitors such as DIDS involve multiple interactions, ionic, hydrophobic, covalent, and nucleophilic (Cabantchik *et al.*, 1978; Knauf, 1979). Assemblies of membrane-crossing strands with hydrophilic interiors have been proposed as the basic architecture of other membrane proteins that are not particularly hydrophobic in composition (Wickner, 1980). Such assemblies may represent a general type of structure of transport proteins within the bilayer.

ACKNOWLEDGMENT. This work was supported by Grant MT 4665 from the Medical Research Council of Canada.

REFERENCES

Cabantchik, Z. I., and Rothstein, A. (1974a). *J. Membr. Biol.* **15**, 207–226.

Cabantchik, Z. I., and Rothstein, A. (1974b). *J. Membr. Biol.* **14**, 227–248.

Cabantchik, Z. I., Knauf, P. A., and Rothstein, A. (1978). *Biochim. Biophys. Acta* **515**, 239–302.

Drickamer, L. K. (1980). *Ann. N.Y. Acad. Sci.* **341**, 419–432.

England, B. J., Gunn, R. B., and Steck, T. L. (1980). *Biochim. Biophys. Acta* **623**, 171–182.

Grinstein, S., Ship, S., and Rothstein, A. (1978). *Biochim. Biophys. Acta* **507**, 294–304.

Jennings, M. L., and Passow, H. (1979). *Biochim. Biophys. Acta* **554**, 498–519.

Knauf, P. A. (1979). *Curr. Top. Membr. Transp.* **12**, 248–263.

Knauf, P. A., and Rothstein, A. (1971). *J. Gen. Physiol.* **58**, 190–210.

Knauf, P. A., Breuer, L., McCulloch, L., and Rothstein, A. (1978a). *J. Gen. Physiol.* **72**, 632–649.

Knauf, P. A., Ship, S., Breuer, W., McCulloch, L., and Rothstein, A. (1978b). *J. Gen. Physiol.* **72**, 607–630.

Knauf, P. A., Tarshis, T., Grinstein, S., and Furuya, W. (1980). In *Membrane Transport in Erythrocytes* (U. V. Lassen, H. H. Ussing, and J. O. Wieth, eds.), pp. 389–408, Munksgaard, Copenhagen.

Koziarz, J. J., Kohler, H., and Steck, T. L. (1978). *J. Supramol. Struct. Suppl.* **2**, 215–221.

Lepke, S., and Passow, H. (1976). *Biochim. Biophys. Acta* **455**, 353–370.

Lepke, S., Fasold, H., Pring, M., and Passow, H. (1976). *J. Membr. Biol.* **29**, 247–277.

Passow, H. (1971). *J. Gen. Physiol.* **6**, 233–258.

Passow, H., Fasold, H., Zaki, L., Schuhmann, B., and Lepke, S. (1975). In *Biomembranes: Structure and Function* (G. Gárdos and I. Szász, eds.), pp. 197–214, North-Holland, Amsterdam, and Publishing House of the Hungarian Academy of Sciences, Budapest.

Passow, H., Fasold, H., Lepke, S., Pring M., and Schuhmann, B. (1977). In *Membrane Toxicity* (M. W. Miller and A. Shamoo, eds.), pp. 353–377, Plenum Press, New York.

Passow, H., Kampmann, L., Fasold, H., Jennings, M., and Lepke, S. (1980). In *Membrane Transport in Erythrocytes* (U. V. Lassen, H. H. Ussing, and J. O. Wieth, eds.), pp. 345–371, Munksgaard, Copenhagen.

Ramjeesingh, M., Gaarn, A., and Rothstein, A. (1980a). *Biochim. Biophys. Acta* **599**, 127–139.

Ramjeesingh, M., Grinstein, S., and Rothstein, A. (1980b). *J. Membr. Biol.* **57**, 95–102.

Rao, A. (1979). *J. Biol. Chem.* **254**, 3503–3511.

Reithmeier, R. A. F., and Rao, A. (1979). *J. Biol. Chem.* **254**, 3054–3060.

Rice, W., and Steck, T. L. (1976). *Biochim. Biophys. Acta* **433**, 39–53.

Rothstein, A., and Ramjeesingh, M. (1980). *Ann. N.Y. Acad. Sci.* **358**, 1–12.

Rothstein, A., Cabantchik, Z. I., and Knauf, P. (1976). *Fed. Proc.* **35**, 3–10.

Rothstein, A., Ramjeesingh, M., and DuPre, A. (1981). In *Structure and Function of Blood Cell Membranes* (B. Sarkadi, I. Szász, and G. Gárdos, eds.), pp. 263–274, Publishing House of the Hungarian Academy of Sciences, Budapest.

Shami, Y., Rothstein, A., and Knauf, P. A. (1978). *Biochim. Biophys. Acta* **308,** 357–363.

Ship, S., Shami, Y., Breuer, W., and Rothstein, A. (1977). *J. Membr. Biol.* **33,** 311–324.

Steck, T. L., Ramos, B., and Strapazon, E. (1976). *Biochemistry* **14,** 1154–1161.

Tanner, M. J. A., Williams, D. G., and Jenkins, R. E. (1980). *Ann. N.Y. Acad. Sci.* **342,** 455–464.

Wickner, W. (1980). *Science* **210,** 861–868.

150

Kinetic Asymmetry of the Red Cell Anion Exchange System

Philip A. Knauf

In the past 10 years our concept of red cell anion transport has been radically altered by the realization that the anion exchange is not a simple diffusional process, but rather a very tightly coupled one-for-one exchange that is mediated by a specific and abundant red cell membrane protein known as band 3 (for review see Knauf, 1979). Recent developments have centered around the question of precisely how band 3 mediates anion exchange, a question of importance not only for understanding the red cell anion transport system but also for gaining insight into the operation of other protein-mediated transport systems. This review focuses on recent developments in the kinetics of the transport mechanism. For other aspects of the system, particularly the band 3 structure and the transport-related conformational change, the reader is referred to chapters by Rothstein and Passow in this volume.

1. NEW FEATURES OF THE TITRATABLE CARRIER MODEL

The titratable carrier model, first proposed by Gunn in 1972, has provided the framework for interpreting most kinetic studies of the system. According to this model, monovalent anions cross the membrane by first combining with a transport site, designated E in Fig. 1 by analogy with an enzyme catalytic site. Titration with H^+ converts this site to a form that transports divalent anions such as sulfate. Several predictions of this model have been verified. In particular, the chloride or sulfate flux saturates with increasing anion concentration (Gunn *et al.*, 1973; Schnell *et al.*, 1977), there is competition among various anions (Dalmark, 1976), and most inhibitors have parallel effects on sulfate and chloride fluxes (Ku *et al.*, 1979). Also, proton fluxes consistent with an exchange of Cl^- for SO_4^{2-}

Philip A. Knauf • Department of Radiation Biology and Biophysics, University of Rochester School of Medicine and Dentistry, Rochester, New York 14642.

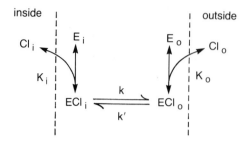

inside   outside

Figure 1. Kinetic model of the anion exchange system. E_i and E_o refer to forms of the system in which the transport site faces inward (toward the cytoplasm) or outward, respectively. K_i and K_o represent the dissociation constants for chloride at the inside and outside of the membrane, respectively. The rate constant for the conformational change from inward-facing to outward-facing is designated k, and the rate constant for the reverse conformational change is k'.

and a proton (Jennings, 1976) and with proton/Cl^- cotransport (Jennings, 1978) have been observed, as predicted by the model.

Several modifications or extensions of the model have been necessitated by recent experimental evidence:

1. In addition to the transport site, there is a modifier site with a much lower affinity for substrate anions (Dalmark, 1976), located at the extracellular surface of the membrane (Schnell *et al.*, 1978). When anions are bound to this site, translocation of the substrate site–anion complex is inhibited. This may affect the physiological rate of anion exchange (Wieth and Brahm, 1980).

2. Binding of H^+ to a site at the cytoplasmic surface of the membrane with a pK_a of 6.2 at 0°C inhibits chloride exchange (Gunn and Fröhlich, 1980; Wieth *et al.*, 1980).

3. Sulfate can bind to the monovalent carrier, E, thereby inhibiting chloride exchange (Milanick, 1980). After dansylation of the membrane, such complexes may even be able to transport SO_4^{2-} across the membrane (Legrum *et al.*, 1980).

4. Chloride exchange is not affected by increasing the pH from 7.1 to 11 in ghosts (Funder and Wieth, 1976), indicating that if a positively charged amino acid binds anions, it must be one with a very high pK_a, such as arginine (Wieth *et al.*, 1980; Knauf, 1979). An arginine-selective reagent inhibits transport (Zaki, 1981).

5. In addition to inorganic anions, the system can transport cations as carbonate complexes (Funder, 1980), superoxide radical (Lynch and Fridovich, 1978), and large organic anions (Motais, 1977), including even cyclic AMP (Kury and McConnell, 1975). The hydrophobic portion of such molecules can be large, providing that the polar head group does not contain charges spaced more than about 4 Å apart (Motais, 1977).

Although the original model was framed in terms of a diffusible carrier, a more realistic version of this model, consistent with the fact that the band 3 protein is asymmetrically arranged in the membrane and does not function as a diffusible carrier, proposes that the transport event involves a localized conformational change in the band 3 protein that changes the transport site from an inside-facing (E_i) to an outside-facing (E_o) form, or vice versa (Gunn, 1977; Knauf, 1979; Passow *et al.*, 1980a; Passow, this volume).

2. MECHANISM OF COUPLING

The anion transport system is a very tightly coupled exchange process, which permits net flow of chloride at less than 1/10,000th the rate of anion exchange (Knauf, 1979). This tight coupling could be explained by either of two mechanisms, shown in Fig. 2. According to the simultaneous mechanism (Fig. 2A), there are two anion transport sites, one on each side of the membrane, and transport occurs when the anions bound to these sites exchange sides. This provides a simple mechanical explanation for the tight one-for-one coupling of anion influx and efflux. A second model, the ping-pong model (Fig. 2B), proposes that there is only a single anion transport site, which can exist in either an inward-facing (E_i) or an outward-facing (E_o) form. To explain the tight coupling with this model, it must be assumed that the conformational change from inward-facing to outward-facing or vice versa can only occur if the transport site is loaded with an anion.

2.1. Evidence from Chloride Flux Measurements

The ping-pong and sequential models predict different effects of the internal chloride concentration on the concentration of external chloride required to half-saturate the transport system and on the maximal transport rate at high external chloride. The observed effects are most consistent with a ping-pong mechanism, but could also be explained by a sequential mechanism in which the internal site is half-saturated with chloride at less than 1 mM (Gunn and Fröhlich, 1979, 1980).

2.2. Transmembrane Effects of Probes

Several experiments have demonstrated that binding of a probe at one side of the membrane decreases the number of sites available at the opposite side of the membrane (Rothstein *et al.*, 1976; Passow *et al.*, 1980b; Passow and Zaki, 1978; Grinstein *et al.*, 1979; reviewed by Passow, this volume). Because the ping-pong model, but not the sequential model, predicts such *trans* effects, these data support the ping-pong model and particularly the concept that the transport site can alternately face inward or outward. Such effects, however, could be explained by the sequential model if binding of the probes at one side of the membrane causes a change in protein conformation at the opposite side. This possibility is not unlikely, for probes such as extracellular DIDS (4,4'-diisothiocy-

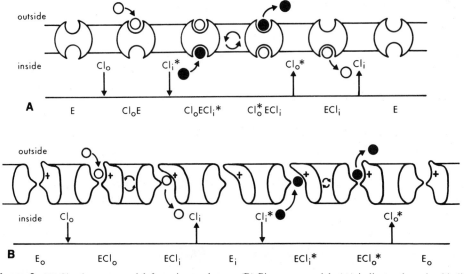

Figure 2. (A) Simultaneous model for anion exchange. (B) Ping-pong model. (+) indicates the anion-binding site.

anostilbene-2,2'-disulfonate) can affect binding of hemoglobin and glycolytic enzymes to the cytoplasmic portion of band 3 (Salhany and Gaines, 1981), as well as NAP-taurine [*N*-(4-azido-2-nitrophenyl)-2-aminoethanesulfonate] binding to proteins other than band 3 (Grinstein *et al.*, 1979).

2.3. Transmembrane Effects of Cl⁻ on Probe Binding

One way to avoid such perturbations is to make use of the substrate itself to reorient the transport system. As first shown by Dalmark (1975), for a ping-pong model, a chloride gradient across the membrane will cause an asymmetric distribution of the outward- and inward-facing unloaded forms of the carrier, E_o and E_i, such that

$$\frac{E_o}{E_i} = A\frac{Cl_i}{Cl_o} \text{ where } A = \frac{kK_o}{k'K_i}. \tag{1}$$

A is an asymmetry factor (Knauf *et al.*, 1980) related to the differences in the chloride dissociation constants at the inside and outside of the membrane, K_i and K_o, and the rate constants for conversion of the system from inside-facing to outside-facing (k) and vice versa (k'), as shown in Fig. 1.

Even for a perfectly symmetric carrier, with $k = k'$ and $K_i = K_o$, the chloride gradient will be reflected in an asymmetric distribution of E_o and E_i. Changes in the relative amount of E_o can be monitored by making use of a reversible competitive inhibitor, such as external H_2DIDS (4,4'-diisothiocyano-dihydrostilbene-2,2'-disulfonate), which will only bind to the E_o form. By the law of mass action, if E_o increases, less H_2DIDS will be required to form the inhibited complex, E–DIDS, and so the inhibitory potency of H_2DIDS will increase. With constant Cl_o, a fivefold chloride gradient (with $Cl_i > Cl_o$) causes a more than twofold increase in the inhibitory potency of H_2DIDS (Knauf *et al.*, 1980). If the chloride gradient is set up in the opposite direction ($Cl_i < Cl_o$), the inhibitory potency decreases (Furuya, 1980). Experiments in which the pH gradient is not permitted to equilibrate with the chloride gradient, or in which the membrane potential is altered by means of valinomycin (Knauf and Law, unpublished data), demonstrate that the effect is caused by changes in the chloride gradient itself rather than by the changes in membrane potential and pH gradient that usually accompany changes in the chloride distribution.

These results, which are difficult to explain by the sequential model, provide strong evidence for the ping-pong model and demonstrate that chloride gradients can be used to change the orientation of the transport site.

3. DIFFERENCES BETWEEN INWARD- AND OUTWARD-FACING CONFORMATIONS

If it is possible to orient the transport system into either the inward-facing or the outward-facing form by means of chloride gradients or chemical probes, it should be possible to examine the differences between these conformations.

3.1. Orientation by Means of Chemical Probes

Probes that bind preferentially to the system when it is in either the inward- or the outward-facing conformation and that inhibit transport should trap the system in one form. Even partial inhibitors can have a similar effect by changing the parameters k, k', K_i, and

K_o (Fig. 1), thereby favoring some forms over others. This change in conformation can then be probed by physical, chemical, or enzymatic techniques. For example, external DIDS traps the system in the outward-facing form, and causes a change in position of tryptophan residues in band 3 (Kleinfeld *et al.*, 1980), as well as the transition temperature of a putative highly-ordered lipid region adjacent to band 3 (Snow *et al.*, 1978). Passow and co-workers (Passow *et al.*, 1980a,b; Passow, this volume) have monitored the conformation of the system by measuring the rate of reaction of a particular lysine residue (Lys a) with fluorodinitrobenzene. Some inhibitors enhance the reactivity (*cis* form) while other inhibitors decrease the reactivity (*trans* form). Chloride also increases the reaction rate. In all of these cases, however, it is not yet certain whether the alterations caused by these probes are related to orientation of the transport site inward or outward (with inward being *trans* and outward *cis*) or whether they may represent direct effects of the probes or chloride on the protein conformation.

3.2. NAP-Taurine Binding

It would be expected that changes in substrate site orientation might affect the binding of noncompetitive inhibitors to other sites. Chloride gradient experiments similar to those described in Section 2.3 reveal that external NAP-taurine, which seems to bind to the modifier site (Knauf *et al.*, 1978), does so only (or at least preferentially) when the substrate site faces outward (Knauf *et al.*, 1978, 1980). This suggests a very strong interaction between the substrate and modifier sites, which is not unexpected in view of the mutual competition between inhibitors that bind to these sites (Cabantchik *et al.*, 1976; Macara and Cantley, 1981). Because NAP-taurine binds to the outward-facing form and inhibits transport, it must trap the system in this form.

3.3. Niflumic Acid Binding

Niflumic acid is a potent noncompetitive inhibitor of chloride exchange, that is, it binds equally well to loaded and unloaded transport sites (Cousin and Motais, 1979). Nevertheless, it competes with disulfonic stilbenes for binding, suggesting a close proximity to or strong interaction with the transport site. Chloride gradient experiments show that niflumic acid binds preferentially to the system when the transport site faces outward (Knauf *et al.*, 1981). Thus, niflumic acid must lock the transport system in the outward-facing conformation.

Because the inhibitory potency of niflumic acid is not changed by increasing the chloride concentration well above the level required to half-saturate the modifier site (Cousin and Motais, 1979), it appears that chloride binding to the modifier site has no effect on the orientation of the transport system. This in turn implies that external chloride can bind to the modifier site regardless of the conformation of the transport site. This is not inconsistent with the finding that external NAP-taurine binding is strongly affected by transport site conformation, for the noncovalent interactions that enhance NAP-taurine binding affinity are probably different from those that determine chloride binding.

4. INTRINSIC ASYMMETRY IN THE DISTRIBUTION OF INWARD- AND OUTWARD-FACING FORMS

If the transport site can exist in two forms, corresponding to two different conformations of the band 3 protein, there would seem to be no reason why the dissociation constants for chloride of the two forms (K_i and K_o) or the rate constants for their interconversion (k and k') should be the same. As equation (1) shows, differences in either or both of these factors could lead to an asymmetric distribution of the unloaded forms of the system, even when $Cl_i = Cl_o$, that is, $A \neq 1$. Various methods, involving substrate gradients and chemical probes, have been used to determine this asymmetry.

4.1. Half-Saturation Concentrations for Cl⁻ at Inside and Outside

For a ping-pong model, it can be readily shown (Furuya, 1980) that

$$A = \frac{K_{1/2}^o}{K_{1/2}^i} \tag{2}$$

where $K_{1/2}^o$ is the concentration of external chloride required to half-saturate the transport system when internal chloride is maximal, and $K_{1/2}^i$ is the corresponding value for internal chloride. From such measurements, Schnell et al. (1978) obtained results corresponding to an A value of 2.5, while Gunn and Fröhlich (1979) obtained an A value of 0.064. Lambert and Lowe's (1978) data also suggest that $A < 1$.

4.2. Effects of Chloride Gradients on Binding of Probes

Quantitative analysis of the effects of chloride gradients on the inhibitory potency of probes that bind preferentially to the system when the transport site is in the outward-facing form provides an independent way for setting limits on the value of A. If, for example, most of the sites face the outside ($A > 1$), reorientation of the sites toward the outside with chloride gradients ($Cl_i > Cl_o$) will have little effect on the number of outward-facing sites and hence little effect on inhibitory potency. If, on the other hand, most of the sites face inward ($A < 1$), chloride gradients will have a very large effect. Results with H_2DIDS and NAP-taurine (Knauf et al., 1980; Furuya, 1980) show that A is significantly less than 1, in agreement with the results of Gunn and Fröhlich (1979). Schnell (personal communication) has reevaluated his earlier data and now agrees with this conclusion. Thus, there are many more inward-facing than outward-facing unloaded transport sites, probably 15 times as many.

4.3. Nature of the Asymmetry

As equation (1) shows, asymmetry in the distribution of unloaded transport sites could be due to differences in the rate constants, k and k', or to differences in the dissociation constants, K_i and K_o. For the loaded sites, however, because at equilibrium, $k \cdot ECl_i = J_o = J_i = k' \cdot ECl_o$,

$$L = \frac{ECl_o}{ECl_i} = \frac{k}{k'} \tag{3}$$

That is, the asymmetry ratio of the loaded sites, L, depends only on the rate constants for the conformational changes from inward-facing to outward-facing and vice versa (Furuya, 1980; Passow et al., 1980a; Passow, this volume).

The noncompetitive inhibitor, niflumic acid, binds preferentially to the outward-facing loaded or unloaded forms (E_o or ECl_o). Its inhibitory potency at low chloride concentrations, when most of the sites are unloaded, will depend on the ratio of the unloaded forms, A. At high chloride concentrations, when most of the sites are loaded, the inhibitory potency will depend on the asymmetry ratio of the loaded sites, L. If $A = L$, there will be no change in inhibitory potency as the chloride concentration is changed, as long as $Cl_i = Cl_o$. For a system with $L = 1$ and $A = 0.064$, a more than threefold increase in ID_{50} (concentration of inhibitor required to cause 50% inhibition) would be expected when chloride is changed from 150 mM to 10 mM. In contrast, within the limits of statistical error there is no change in inhibitory potency (Cousin and Motais, 1979; Knauf, Mann, and Law, unpublished data), so $A = L$, and thus $K_i = K_o$ (because $A/L = K_o/K_i$). The asymmetry seems to be entirely determined by the rate constants, suggesting that for chloride the ECl_i form is more stable than the ECl_o form. For both loaded and unloaded forms, most of the sites face inward (by a factor of about 15). Jennings (1980) using a different method has observed a similar asymmetry for the sulfate form of the carrier.

5. RELATIONSHIP TO THE STRUCTURE OF BAND 3

The kinetic data can be combined with structural information about the band 3 protein (Rothstein, this volume) in the conceptual model shown in Fig. 3. Because the polar portion of the transport site is small (about 4 Å diameter) relative to the thickness of the bilayer, and because the turnover number of the system is so large (5×10^4 ions/sec at 38°C; Brahm, 1977), the conformational change involved in transport must involve a small region of the band 3 protein. Anions might gain access to the transport site by diffusing through an aqueous channel formed by the band 3 protein's transmembrane segments (see Rothstein, this volume; Passow, this volume; Knauf, 1979). Although the change is localized, it can affect the binding of probes to the modifier site and to other sites in the protein.

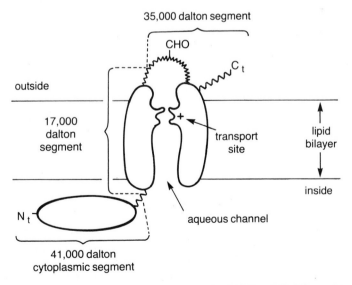

Figure 3. Conceptual model of the anion exchange system. The 17,000 and 35,000 membrane-spanning segments of band 3 form a hydrophilic channel through the membrane. This channel contains a transport site, which mediates anion exchange by means of a conformational change such as that depicted in Fig. 2B. N_t and C_t are the NH_2 terminus and COOH terminus, respectively, of the band 3 polypeptide chain; CHO is the site of carbohydrate attachment (see Rothstein, this volume).

With chloride bound to the system, the inward-facing form is preferred over the outward-facing form by a factor of perhaps 15. This indicates that the two conformations differ in free energy, and also suggests that it should be possible to further examine the conformational change by using probes and chloride gradients to orient the system toward the outside, away from its normal inside-facing conformation.

ACKNOWLEDGMENTS. This work was supported by NIH Grant 1R01 AM 27495 and in part by Contract DE-AC02-76EV03490 with the U.S. Department of Energy at the University of Rochester Department of Radiation Biology and Biophysics and has been assigned Report No. UR-3490-2054.

REFERENCES

Brahm, J. (1977). *J. Gen. Physiol.* **70**, 283–306.
Cabantchik, Z., Knauf, P., Ostwald, T., Markus, H., Davidson, L., Breuer, W., and Rothstein, A. (1976). *Biochim. Biophys. Acta* **455**, 526–537.
Cousin, J. L., and Motais, R. (1979). *J. Membr. Biol.* **46**, 125–153.
Dalmark, M. (1975). *J. Physiol. (London)* **250**, 39–64.
Dalmark, M. (1976). *J. Gen. Physiol.* **67**, 223–234.
Funder, J. (1980). *Acta Physiol. Scand.* **108**, 31–37.
Funder, J., and Wieth, J. (1976). *J. Physiol. (London)* **262**, 679–698.
Furuya, W. (1980). M.Sc. thesis, University of Toronto.
Grinstein, S., McCulloch, L., and Rothstein, A. (1979). *J. Gen. Physiol.* **73**, 493–514.
Gunn, R. B. (1972). In *Oxygen Affinity of Hemoglobin and Red Cell Acid–Base Status* (M. Rørth and P. Astrup, eds.), pp. 823–827, Munksgaard, Copenhagen.
Gunn, R. B. (1977). *Proc. Int. Union Physiol. Sci.* (Paris) **12**, 122.
Gunn, R. B., and Fröhlich, O. (1979). *J. Gen. Physiol.* **74**, 351–374.
Gunn, R. B., and Fröhlich, O. (1980). In *Membrane Transport in Erythrocytes* (U. V. Lassen, H. H. Ussing, and J. O. Wieth, eds.), pp. 431–442, Munksgaard, Copenhagen.
Gunn, R., Dalmark, M., Tosteson, D., and Wieth, J. (1973). *J. Gen. Physiol.* **61**, 185–206.
Jennings, M. L. (1976). *J. Membr. Biol.* **28**, 187–205.
Jennings, M. L. (1978). *J. Membr. Biol.* **40**, 365–391.
Jennings, M. L. (1980). In *Membrane Transport in Erythrocytes* (U. V. Lassen, H. H. Ussing, and J. O. Wieth, eds.), pp. 450–463, Munksgaard, Copenhagen.
Kleinfeld, A. M., Matayoshi, D. E., and Solomon, A. K. (1980). *Fed. Proc.* **39**, 1714.
Knauf, P. A. (1979). *Curr. Top. Membr. Transp.* **12**, 249–363.
Knauf, P. A., Ship, S., Breuer, W., McCulloch, L., and Rothstein, A. (1978). *J. Gen. Physiol.* **72**, 607–630.
Knauf, P. A., Tarshis, T., Grinstein, S., and Furuya, W. (1980). In *Membrane Transport in Erythrocytes* (U. V. Lassen, H. H. Ussing, and J. O. Wieth, eds.), pp. 389–403, Munksgaard, Copenhagen.
Knauf, P. A., Mann, N., and Law, F.-Y. (1981). *Biophys. J.* **33**, 49a.
Ku, C. P., Jennings, M. L., and Passow, H. (1979). *Biochim. Biophys. Acta* **553**, 132–141.
Kury, P. G., and McConnell, H. M. (1975). *Biochemistry* **14**, 2798–2803.
Lambert, A., and Lowe, A. G. (1978). *J. Physiol. (London)* **275**, 51–63.
Legrum, B., Fasold, H., and Passow, H. (1980). *Hoppe-Seyler's Z. Physiol. Chem.* **361**, 1573–1590.
Lynch, R. E., and Fridovich, I. (1978). *J. Biol. Chem.* **253**, 4697–4699.
Macara, I. G., and Cantley, L. C. (1981). *Biochemistry* **20**, 5695–5701.
Milanick, M. A. (1980). *Fed. Proc.* **39**, 1715.
Motais, R. (1977). In *Membrane Transport in Red Cells* (J. C. Ellory and V. L. Lew, eds.), pp. 197–220, Academic Press, New York.
Passow, H., and Zaki, L. (1978). In *Molecular Specialization and Symmetry in Membrane Function* (A. K. Solomon and M. Karnovsky, eds.), pp. 229–250, Harvard University Press, Cambridge, Mass.
Passow, H., Kampmann, L., Fasold, H., Jennings, M., and Lepke, S. (1980a). In *Membrane Transport in Erythrocytes* (U. V. Lassen, H. H. Ussing, and J. O. Wieth, eds.), pp. 345–367, Munksgaard, Copenhagen.
Passow, H., Fasold, H., Gartner, E. M., Legrum, B., Ruffing, W., and Zaki, L. (1980b). *Ann. N.Y. Acad. Sci.* **341**, 361–383.

Rothstein, A., Cabantchik, Z., and Knauf, P. (1976). *Fed. Proc.* **35,** 3–10.

Salhany, J. M., and Gaines, K. C. (1981). *Trends Biochem. Sci.* **6,** 13–15.

Schnell, K. F., Gerhardt, S., and Schoppe-Fredenburg, A. (1977). *J. Membr. Biol.* **30,** 319–350.

Schnell, K. F., Besl, E., and Manz, A. (1978). *Pfluegers Arch. Gesamte Physiol. Menschen Tiere* **375,** 87–95.

Snow, J. W., Brandts, J. F., and Low, P. S. (1978). *Biochim. Biophys. Acta* **512,** 579–591.

Wieth, J. O., and Brahm, J. (1980). In *Membrane Transport in Erythrocytes* (U. V. Lassen, H. H. Ussing, and J. O. Wieth, eds.), pp. 467–482, Munksgaard, Copenhagen.

Wieth, J. O., Brahm, J., and Funder, J. (1980). *Ann. N.Y. Acad. Sci.* **341,** 394–418.

Zaki, L. (1981). *Biochem. Biophys. Res. Commun.* **99,** 243–251.

151

Anion-Transport-Related Conformational Changes of the Band 3 Protein in the Red Blood Cell Membrane

Hermann Passow

1. THE ANION TRANSPORT PROTEIN

The lipid bilayer that surrounds the red blood cell is virtually impermeable to hydrophilic anions. There exists, however, a specific transport protein that facilitates the penetration of monovalent and divalent inorganic anions and of many organic anions including sulfonates, phosphates, and dicarboxylates. The protein also accepts monocarboxylates as substrates, although in the red cells of humans and rodents these anions are predominantly transported by a separate transport system (Deuticke, 1980).

The anion transport protein is called "band 3 protein" according to its location on SDS-polyacrylamide gel electropherograms of the red cell membrane, where it represents the third major band from the top. It has a molecular weight of 96,000 and constitutes some 25% of the total membrane protein (Steck, 1978).

The participation in anion transport of the band 3 protein was demonstrated by radioactive labeling with specific inhibitors. It was shown independently in three laboratories that 4,4'-diisothiocyano-dihydrostilbene-2,2'-disulfonate (H_2DIDS) (Cabantchik and Rothstein, 1974), 4-acetamido-4'-isothiocyanostilbene-2,2'-sulfonate (SITS) and 4,4'-diacetamidostilbene-2,2'-disulfonate (DAS) (Passow et al., 1975), and m-isothiocyano-4-benzene sulfonate (IBS) (Ho and Guidotti, 1975) bind exclusively or preferentially to band 3. There exists a linear relationship between inhibition of transport and binding of the inhibitors to the protein (Lepke et al., 1976; Ship et al., 1977). Inhibition is complete when 10^6 molecules of inhibitor are bound per cell (Zaki et al., 1975; Lepke et al., 1976; Ship et al., 1977). This number is close to the only existing estimate of band 3 molecules per cell (Steck, 1978). It is assumed, therefore, that the stoichiometry of inhibition is 1 : 1. This is corroborated by studies of the intramolecular cross-linking of the band 3 molecule.

Hermann Passow • Max-Planck-Institut für Biophysik, Frankfurt am Main, West Germany.

After cleavage by externally applied chymotrypsin, virtually all of the band 3 protein is split into fragments of 60,000 and 35,000 molecular weight. These fragments can be quantitatively cross-linked by the two isothiocyanate groups of H_2DIDS (Jennings and Passow, 1979).

The band 3 protein exists in the membrane in the form of dimers and tetramers (Steck, 1978; Margaritis *et al.*, 1977) that are possibly at equilibrium with the monomers (Dorst and Schubert, 1979). The cross-linking experiments of Jennings and Passow suggest that the monomers accomplish anion transport as independently operating units.

Efforts have been made to purify band 3 *in situ* by enzymatic (Lepke and Passow, 1976) and chemical (Wolosin *et al.*, 1977) removal from the isolated membrane of proteins that are not involved in transport. In addition, reconstitution of the transport system by incorporation of more or less impure band 3 preparations into liposomes has been attempted. Many features of the original transport system could be reestablished (Cabantchik *et al.*, 1980, Köhne *et al.*, 1981), but the complete reconstitution of all functions from pure components has still to be achieved.

2. TRANSPORT BY CONFORMATIONAL CHANGES OF THE BAND 3 PROTEIN

2.1. General Considerations

The peptide chain of the band 3 protein traverses the lipid bilayer several times. It extends its carbohydrate moiety to the outer membrane surface and a large hydrophilic segment near the NH_2 terminus towards the inner membrane surface (cf. Tanner *et al.*, 1980; Drickamer, 1980). For this reason, the molecule can neither undergo translational nor rotational diffusion across the membrane (Cherry *et al.*, 1977) and hence is incapable of mediating the exchange like a diffusible carrier. Moreover, the electrical conductance of the membrane is much lower than calculated from the rate of charge transfer that is associated with anion exchange. Hence, the protein does not simply form an aqueous pore (Harris and Pressman, 1967; Scarpa *et al.*, 1970; Hunter, 1971). One must conclude, therefore, that the anion exchange is brought about by conformational changes of the stationary transport protein.

The conformational changes that lead to anion transport are initiated by anion binding to a limited number of amino acid residues that form the so-called "transfer site." The bound anion is then translocated by a reorientation of the complex-forming amino acid residues. It is unknown whether a single set of complex-forming amino acid residues is involved or if the bound anion is transferred from one set of complexing sites to the next until it arrives at the other membrane surface. The turnover number of Cl^--Cl^- exchange is rather high (2×10^4/sec, Lepke *et al.*, 1976; Brahm, 1977). This would suggest that the complex formation and the *cis–trans* isomerization of the transfer site are restricted to a narrow gate composed of a single set of complex-forming amino acid residues that reside at one end of an aqueous channel. Such a channel could be formed by parallel transmembrane segments of the peptide chain that seem to traverse the lipid bilayer perpendicular to the surface (Lesslauer, 1980).

The isomerization reaction at the transfer site and any structural changes associated with the passing of an anion down a channel are likely to require a reorientation of other portions of the transport molecule. Conversely, these other portions of the transport molecule can be expected to affect the isomerization of the transfer site. Specific identifiable regions that in this way are linked to the transfer site are called "modifier sites." Thus, the study of mutual interactions between transfer and modifier sites and amongst the modifier sites themselves should enable one to construct a map of functional interrelationships between the various amino acids in specific regions of the transport protein. An elementary introduction is presented below on the current attempts to provide a theoretical guideline for the establishment of such a map and to obtain pertinent experimental data.

The discussion will proceed at two different levels: first, we shall consider the different conformational states that can be defined by biochemical methods and we shall try to relate them to distinct situations of the kinetics of transport. For this purpose, there is no need for a knowledge of the specific details of the structural differences between the different conformers. It is only necessary to possess criteria that enable one to distinguish clearly between the distinct conformers. Second, we shall look at the details of the conformational changes in terms of changes of the relative locations of the specific amino acid residues that constitute transfer and modifier sites.

2.2 Transport and Isomerization of the Band 3 Protein

2.2.1. Theoretical Aspects

The simplest reaction scheme that preserves the essential features of a transport by conformational change is represented in Fig. 1. The anion combines at the inner surface with the transfer site of the *trans* conformer r of the transport protein. This results in the formation of the complex ar. The subsequent isomerization of ar leads to the formation of the *cis* conformer as, in which the transfer site is exposed to the outer surface. After release of a, the unloaded protein s can pick up another anion for the journey in the opposite direction. The isomerization of the loaded transport protein seems to take place much faster than that of its unloaded form. Thus, anion exchange can probably be described as a catalytic enhancement of the *cis–trans* isomerization of the transport protein, with the substrate as a catalyst (Passow and Fasold, 1981).

The slow isomerization of the unloaded form of the transport protein is called "slippage." The rate of net anion transport via the transport protein under the influence of an electrochemical potential difference is limited by the rate of slippage. However, the net rate of anion efflux is not a direct measure of slippage for some net anion transport takes place across a parallel, as yet unidentified "conductance pathway." In contrast to the electrically "silent" anion exchange, both slippage and diffusion through this conductance pathway contribute to the electrical conductance of the membrane. In accord with the observation that the rate of anion exchange far exceeds the rate of slippage and diffusion through the conductance pathway, the electrical resistance of the red blood cell membrane

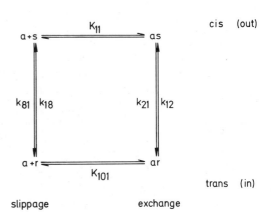

Figure 1. Minimum requirements for description of band 3-mediated anion transport. Models of this type have been discussed in the literature on transport proteins by Patlak (1957), Jacquez (1964), Stein and Lieb (1974), and others, and in the literature on enzyme kinetics by Cleland (1964) and others. For discussion, see text.

is many orders of magnitude higher than expected on the basis of the rapid anion exchange (for a recent challenge of this view, see Stone and Kregenow, 1980).

For the special case of equal concentrations of substrate a on both surfaces of the membrane, the model in Fig. 1 yields for the transport-protein-mediated flux j_a ("equilibrium exchange")

$$j_a = k_{12} \cdot \overline{RS} \cdot \frac{a}{K_{101} + a} \cdot p \tag{1}$$

where $\overline{RS} = \overline{ar} + r + as + s$ is equal to the sum of all forms (mass conservation) and $p = (ar + r)/\overline{RS}$ represents that fraction of all conformers whose transfer site r is in contact with the *trans* surface of the membrane. p can be calculated by applying the mass law, and taking into account the mass conservation as well as the establishment of a steady state:

$$p^{-1} = 1 + \frac{k_{12}}{k_{21}} \cdot \frac{a + K_{101}(k_{18}/k_{12})}{a + K_{101}} \cdot \frac{a + K_{11}}{a + K_{11}(k_{81}/k_{21})} \tag{2}$$

The two mass law constants K_{101} and K_{11} reflect differences of the properties of the transfer sites of *cis* and *trans* conformers.

If the isomerization of the substrate-loaded transport protein is very fast compared to that of the unloaded protein ($k_{18}/k_{12} = k_{81}/k_{21} = 0$), one obtains

$$j_a = V_{max} \cdot \frac{a}{K_{1/2} + a}$$

where

$$V_{max} = \frac{k_{12} \cdot k_{21}}{k_{12} + k_{21}} \text{ and } K_{1/2} = \frac{k_{12} \cdot K_{11}}{k_{12} + k_{21}} + \frac{k_{21} \cdot K_{101}}{k_{12} + k_{21}} \tag{3}$$

Equation (3) predicts simple saturation kinetics with a Hill coefficient of 1.0 for all concentrations of a at which the rate of "slippage" is small compared to the rate of isomerization of the loaded form of the transport protein. Down to the lowest substrate concentrations that could be established experimentally, the Hill coefficient was indeed 1.0 (for a review see Knauf, 1979). At high substrate concentrations, self-inhibition takes place (Dalmark, 1976; Schnell et al., 1977). The work of Gunn and Fröhlich (1979) indicates that for Cl^- ions $k_{12} \cdot K_{11}/(k_{12} + k_{21})$ is about 3.0 mM, $k_{21} \cdot K_{101}/(k_{12} + k_{21})$ about 60 mM (pH 7.8, temperature 0°C).

Attempts have been made to render the theoretical treatment more realistic by the incorporation of two characteristic features of anion transport: substrate inhibition (Passow et al., 1980b) and pH dependence (Legrum et al., 1980; Lepke and Passow, 1981). The transport of both monovalent and divalent anions is inhibited by the protonation of a modifier site that is located at the inner membrane surface and allosterically linked to the transfer site (Wieth et al., 1980; Gunn and Fröhlich, 1980). In contrast to monovalent anions, divalent anions are cotransported with a proton (Jennings, 1976). Thus, the deprotonation of the modifier site leads to an increase of monovalent anion transport with increasing pH until a plateau is reached that extends from pH 7.4 to at least 9.5 (Funder and Wieth, 1976). The rate of divalent anion transport also increases with increasing pH, but then passes through a maximum around pH 6.3 where the release of the inhibitory proton from the modifier site is compensated by the dissociation of the cotransported proton. The theoretical interpretation of the observed kinetics on the basis of the refined calculations quoted above indicates that proton and anion binding take place essentially independently but that the capacity of the transport protein to isomerize and hence to execute transport depends on the simultaneous presence of a bound proton and a bound anion (Milanik, 1980; Legrum *et al.*, 1980; Knauf, 1979).

2.2.2. Biochemical Investigations

The simplifying equation (1) shares an important feature with equations that follow from the more refined attempts to describe anion transport. It shows the existence of a relationship between the transport process (as expressed by j_a) and the distribution ratio of conformers with the transfer site oriented towards the inner (*trans*) or outer (*cis*) surface (as expressed by p). For the demonstration of the predicted transport-related conformational changes, attempts have been made to vary the distribution ratio p and to measure the ensuing variations of the susceptibility of the transfer and modifier sites to chemical modification.

A simple method of varying the ratio between *cis* and *trans* conformers consists of changing the substrate concentration on both surfaces equally from 0 to saturation of the transport system [equation (2)]. In the absence of substrates [$a = 0$ in equation (2)], the distribution of the conformers is entirely determined by k_{18}/k_{81}, which represents the distribution ratio of the *cis* and *trans* conformers of the unloaded protein. When the protein is partially saturated, the distribution becomes dependent on the distribution ratio for the substrate-loaded forms k_{12}/k_{21}, and the relative affinities of the transfer site in *cis* (K_{11}) or *trans* (K_{101}) conformation. Finally, when the transfer site is saturated in *cis and trans* conformation, the distribution ratio of the loaded transport protein k_{12}/k_{21} becomes the sole determinant [equation (2), $a \to \infty$].

Superimposed on the effects of the substrate are those of inhibitors. Penetrating inhibitors that compete with the substrate will arrest the transport protein with a distribution ratio of *cis* and *trans* conformers that reflects the relative affinities to the transfer site in *cis* and *trans* conformation (*s* and *r*, respectively). When the inhibitor combines with a modifier site (noncompetitive inhibition), one needs to take into account that the properties of that site are most likely to be different when the transfer site is in contact with the *cis* or *trans* surface, even though the modifier site always remains exposed to the same surface. Hence, as a rule, the immobile site will exist in two conformational states with different affinities to the inhibitor. The modifier sites on the *cis* and *trans* conformers will compete for the inhibitor. The result of this competition will determine the distribution of the inhibitor between *cis* and *trans* conformers and hence the ratio between the two conformers at which a given degree of inhibition of the transport is achieved (cf. Passow and Fasold, 1981).

Transport-related conformational changes have been investigated by determining the accessibility of specific binding sites on band 3 to the radioactively labeled inhibitory H$_2$DIDS, 2,4-dinitrofluorobenzene (N$_2$ph-F) (Passow *et al.*, 1980a), *N*-(4-azido-2-nitrophenyl)-2-aminoethanesulfonate (NAP-taurine) (Grinstein *et al.*, 1979; Knauf *et al.*, 1980), phenylisothiocyanate (Kempf *et al.*, 1979), and dansyl chloride (Legrum *et al.*, 1980).

Much work has been done concerning the dinitrophenylation of a specific lysine residue (Lys *a*) that is located on an outward-facing portion of the 17,000-molecular-weight segment of the transport protein and that is involved in covalent H_2DIDS binding. This residue may exist in two different states that are characterized by different accessibilities to N_2ph-F and that may be called "exposed" and "buried." The proportion of band 3 molecules in which the lysine residue is in either one of the two states can be determined by measuring the rate at which it can be dinitrophenylated (Passow *et al.*, 1980a).

Increasing the concentration of the substrate Cl^- enhances drastically the rate of dinitrophenylation of Lys *a*, indicating an increase of the fraction of band 3 molecules with exposed Lys *a* residues at the expense of the fraction of buried residues. We attribute this to a shift of the equilibrium between *cis* and *trans* conformers in favor of the *cis* conformers as indicated by equation (2) (Passow *et al.*, 1980a): In the absence of Cl^-, the transfer site of most of the transport protein molecules is oriented towards the inner, *trans*, membrane surface. With increasing Cl^- concentration, a gradually increasing fraction changes its orientation towards the outer, *cis*, surface. The divalent substrate SO_4^{2-} does not produce this shift, suggesting that different substrates affect the equilibrium between *cis* and *trans* conformers differently (Passow *et al.*, 1980b). The results with Cl^- are in accord with the observations of Knauf *et al.* (1980) who used NAP-taurine in place of N_2ph-F and of Gunn and Fröhlich (1979) who found that the apparent affinity of Cl^- to the *cis* conformer exceeds that to the *trans* conformer considerably.

Reversibly acting inhibitors can be classified according to whether they arrest the transport system with Lys *a* exposed or buried (Passow *et al.*, 1980a). In particular, externally applied phlorizin and positively charged furosemide derivatives produce inhibition when Lys *a* is exposed, whereas intracellular Ca^{2+} and 2-(4'-aminophenyl)-6-methylbenzenethiazol-3;7-disulfonate (APMB) inhibit when Lys *a* is buried. As Ca^{2+} and, most likely, *internal* APMB do not combine with the transfer site, it is clear that the interactions between modifiers and their respective modifier sites at the inner membrane surface are transmitted all the way across the membrane to the H_2DIDS binding site at the outer surface where the location of Lys *a* is altered. This is reminiscent of an observation of Salhany *et al.* (1980). They found a change of hemoglobin binding to the intracellular 40K segment of the band 3 protein when an H_2DIDS molecule becomes attached to the outward-facing portion of the 17K segment of that protein.

Experiments on the sidedness of action of inhibitors that penetrate very slowly in comparison to the substrate have received considerable interest. They were designed to demonstrate the movements of the substrate binding site from one surface of the membrane to the other. Passow and Zaki (1978) observed that Lys *a* becomes nearly inaccessible to dinitrophenylation when exposure to N_2ph-F is performed in the presence of the nonpenetrating disulfonic acid APMB (Zaki *et al.*, 1975). This compound inhibits anion transport at either surface of the membrane and renders Lys *a* inaccessible to dinitrophenylation regardless of whether it is present at the inner or outer membrane surface (Passow and Zaki, 1978). Grinstein *et al.* (1979) used NAP-taurine, a photo-label that inhibits anion transport after noncovalent binding in the dark and after covalent bond formation after photolysis. They could demonstrate that the label reacts on band 3 with the H_2DIDS binding site (or a binding site that is allosterically linked to it), regardless of whether the agent is applied to the outer or the inner membrane surface. Both the experiments with APMB and NAP-taurine suggest that the agents are capable of trapping the transfer site on the outer as well as on the inner membrane surface, pointing to a movement of this site across the permeability barrier towards that surface to which the inhibitors are applied. Although suggestive, this interpretation is not conclusive for it rests on the as yet unproven assumption that the binding site at the inner and outer membrane surface is indeed the same amino acid residue. At least for APMB, the available evidence points to a combination with different sites in the outer and inner surface: with the transfer site at the outer surface and with a modifier site at the inner surface. When APMB is bound to the inner surface, the effect of binding is transmitted across the membrane to Lys a, which results in a removal of that residue from the outer surface without exposing it to the inner membrane surface (Passow *et al.*, 1980a). In contrast, Grinstein *et al.* (1979) suggest that the NAP-taurine binding amino acid residue that is a constituent of the transfer site is indeed capable of getting in contact with the media at both membrane surfaces. Further work along these lines will be of great value to decide the question whether the translocation of the substrate is brought about by the movement of a single set of complex-forming amino acid residues or if it is transferred from one set to one or several others along the way towards the other surface.

In summary, the observations described in this section indicate that the binding of substrates and inhibitors to the transport protein are accompanied by conformational changes. These changes can be interpreted, to a first approximation, within the theoretical framework underlying the derivation of equations (1) and (2). The allosteric interactions between noncompetitive, reversibly acting inhibitors and the H_2DIDS binding site as well as the substrate-dependent changes of Lys a, both indicate that the conformational changes are not restricted to the amino acid residues that participate directly in substrate binding and translocation. They involve others that may be located far away from the transfer site. Thus, the transfer site does not seem to operate in a rigid frame, but in a frame that itself undergoes easily detectable deformations when the transfer site is modified by substrate binding.

2.3. The Structure of Transfer and Modifier Sites

Noncompetitive inhibition and acceleration of anion transport point to alterations of modifier sites. Competitive inhibition suggests an effect at the transfer site but does not necessarily prove the case. This uncertainty is due to the possible existence of allosteric competition, when inhibitors that bind to a modifier site affect the substrate binding to the transfer site and vice versa (see above and Passow *et al.*, 1980b).

Among the inhibitors studied so far, the reversible binding of H_2DIDS that precedes covalent bond formation and its noncovalently binding analog 4_1-$4'$-dinitrostilbene-2_1-$2'$-disulfonate have been shown to compete with substrate anions. Hence, they are bound to the transfer site itself (Shami *et al.*, 1978; Gunn and Fröhlich, 1981) or to a site that is allosterically linked to it (see the discussion of the subject in Lassen *et al.*, 1980, pp. 404–408). However, the reversibly binding H_2DIDS analog tetrathionate shows a mixed type of inhibition (Deuticke *et al.*, 1978) and niflumic acid is capable of preventing stilbene binding but does not compete with Cl^-. These somewhat contradictory observations throw a shade of doubt on the assumption that the H_2DIDS binding site is identical with the transfer site. It seems, however, a good guess that the amino acid residues involved in H_2DIDS binding encompass at least some of those that participate in substrate binding (see below).

The properties of the binding sites for a large variety of inhibitors including H_2DIDS have been explored by studies of the relationship between their chemical structure and efficiency. Work along these lines was initiated by Motais and Cousin (1978) and Motais *et al.*, (1978) who demonstrated that inhibitors belonging to different chemical classes produce an effect that cannot be related to simple physicochemical parameters of these molecules. This suggests that they are bound to different sites. However, chemicals belonging to the same class show a variation of their potency that can be related to binding to the same site whereby the efficiency with which they produce their effects depends on both their respective Hammett constant (a measure of the capacity to exchange electrons) and their hydrophobicity. This rule also applies to the binding of H_2DIDS and its analogs (Barzilay *et al.*, 1979), indicating that this site, like the others, forms a fairly hydrophobic niche that contains some tyrosine and/or some histidine residues that are capable of participating in electron exchange.

The structure of the H_2DIDS binding site could be studied in some more detail by following the covalent bond formation that takes place after the preceding noncovalent binding. As has already been pointed out, the lysine residues involved in the intramolecular cross-linking by H_2DIDS are contributed by the 17K and 35K segments of the protein. After H_2DIDS binding, which may possibly lead to the establishment of an induced fit, they are about 20 Å apart. The kinetics of dinitrophenylation of Lys *a* indicate an abnormally low pK value, which suggests a location in a positively charged environment. Lys *a* does not participate in substrate binding, for it becomes more accessible to dinitrophenylation when Cl^- is bound to the substrate site. Nevertheless, a modification of Lys *a* by dinitrophenylation leads to an inhibition, demonstrating a close relationship to the transfer site. Thus, the H_2DIDS binding site may encompass the substrate binding site but is not simply identical with it (Passow *et al.*, 1980a). It would seem feasible that transport-related arginine residues that have been postulated to exist by Zaki (1981) and Wieth (personal communication) are located near Lys residues *a* and *b*. They could possibly be involved in substrate binding and the neutralization of the negative charge of the sulfonate groups of H_2DIDS.

2.4. A Model of Protein-Mediated Anion Transport

The information provided may be pulled together in an entirely hypothetical model of the operation of the transfer site (Fig. 2). According to this model (Passow *et al.*, 1980a; Passow and Fasold, 1981), amino acid residues that reside at the outward-facing opening of a narrow pore form a gate. When the gate is in the *trans* conformation, an anion coming from the inner surface will be entrapped to form the complex *ar* (Fig. 1). This complex formation may be induced by electrostatic attraction between the anion and charged amino acid residues, although the reaction with uncharged groups cannot be excluded. The complex formation at the mouth of the pore prohibits a contribution to the conductance. A slight conformational change that includes a movement of the complex-forming amino acid residues of the gate over a short distance (perhaps of the order of 1 Å or so) transforms the *trans* conformer into the *cis* conformer and lifts the anion across the rate-limiting bar-

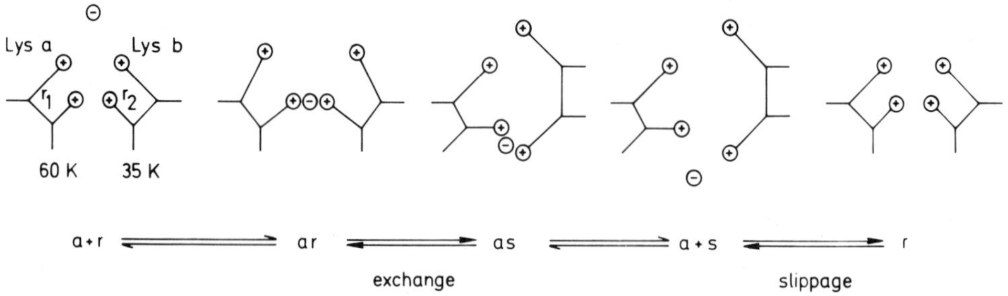

Figure 2. The low-conductance gate (Passow *et al.*, 1980b). The symbols *a*, *s*, *as*, and *ar* have the same meaning as in Fig. 1 and in the equations in the text. Occasionally, an anion *a* may diffuse across the open gate (*r* or *s*) before the formation of *as* or *ar* takes place.

rier. *ar* in Fig. 1 has turned into *as*. After the release of *a*, *s* is ready to accept another anion for the journey in the opposite direction.

If we assume that two or more of the amino acid residues that constitute the gate carry a net positive charge, one could stipulate that electrostatic repulsion between these charges would largely prevent the return of the *trans* into the *cis* conformation (slippage) without an "encapsulated" anion. Thus, the gate would form an effective insulating barrier that allows a rapid anion exchange with a minimum of slippage and hence with little contribution to the membrane conductance.

Nevertheless, in the uncombined states *r* and *s*, an approaching anion could occasionally pass through the open gate before the complex can be formed. This would constitute ionic diffusion that, in contrast to the anion exchange and similar to the slippage, would contribute to the conductance of the membrane. Thus, in addition to the predominant electrically silent exchange and similar to the independent conductance pathway, the band 3-mediated anion transport would give rise to a small conductance.

The changes at the gate that are associated with the binding or release of an anion and with the isomerization of the transport protein are allosterically linked to other portions of the transport protein, e.g., to Lys *a*, whose position can be monitored by following the rate of dinitrophenylation (Fig. 2).

ACKNOWLEDGMENTS. I thank Drs. V. Rudloff, D. Schubert, and Ph. Wood for their comments on the manuscript.

REFERENCES

Barzilay, M., Ship, S., and Cabantchik, Z. I. (1979). *Membr. Biochem.* **2**, 227–254.

Brahm, J. (1977). *J. Gen. Physiol.* **70**, 283–306.

Cabantchik, Z. I., and Rothstein, A. (1974). *J. Membr. Biol.* **15**, 207–226.

Cabantchik, Z. I., Volsky, D. J., Ginsburg, H., and Loyter, D. (1980). *Ann. New York Acad. Sci.* **341**, 444–454.

Cherry, R. J., Bürkli, A., Busslinger, M., and Schneider, G. (1977). In *Biochemistry of Membrane Transport* (G. Semenza and E. Carafoli, eds.), pp. 86–95, Springer-Verlag, Berlin.

Cleland, W. W. (1963). *Biochim. Biophys. Acta* **67**, 104–137.

Cousin, J. L., and Motais, R. J. (1979). *J. Membr. Biol.* **46**, 125–153.

Dalmark, M. (1976). *Prog. Biophys. Mol. Biol.* **31**, 145–164.

Deuticke, B. (1980). In *Membrane Transport in Erythrocytes* (U. V. Lassen, H. H. Ussing, and J. O. Wieth, eds.), pp. 539–551, Munksgaard, Copenhagen.

Deuticke, B., von Bentheim, M., Beyer, E., and Kamp, D. (1978). *J. Membr. Biol.* **44**, 135–158.

Dorst, H.-J., and Schubert, D. (1979). *Hoppe-Seyler's Z. Physiol. Chem.* **360**, 1605–1618.

Drickamer, K. (1980). *Ann. N.Y. Acad. Sci.* **341**, 419–432.

Funder, J., and Wieth, J. O. (1976). *J. Physiol.* **262**, 679–698.

Grinstein, S., McCulloch, L., and Rothstein, A. (1979). *J. Gen. Physiol.* **73**, 493–514.

Gunn, R. B., and Fröhlich, O. (1979). *J. Gen. Physiol.* **74**, 351–374.

Gunn, R. B., and Fröhlich, O. (1980). *Ann. N.Y. Acad. Sci.* **341**, 384–393.

Gunn, R. B., and Fröhlich, O. (1981). In press.

Harris, E. J., and Pressman, B. C. (1967). *Nature (London)* **216**, 918–920.

Ho, M. K., and Guidotti, G. (1975). *J. Biol. Chem.* **250**, 675–683.

Hunter, M. J. (1971). *J. Physiol.* **218**, 49P–50P.

Jacquez, J. A. (1964). *Biochim. Biophys. Acta* **79**, 318–328.

Jennings, M. L. (1976). *J. Membr. Biol.* **28**, 187–205.

Jennings, M. L., and Passow, H. (1979). *Biochim. Biophys. Acta* **554**, 498–519.

Kempf, C., Sigrist, H., and Zahler, P. (1979). *Experientia* **35**, 937.

Knauf, P. A. (1979). *Curr. Top. Membr. Transp.* **12**, 249–363.

Knauf, P. A. Tarshis, T., Grinstein, S., and Furuya, W. (1980). In *Membrane Transport in Erythrocytes* (U. V. Lassen, H. H. Ussing, and J. O. Wieth, eds.), pp. 389–403, Munksgaard, Copenhagen.

Kohne, W., Haest, C. W. M., and Deuticke, B. (1981). *Biochim. Biophys. Acta* **664,** 108–120.

Lassen, H. H. Ussing, and J. O. Wieth, eds.), pp. 389–403, Munksgaard, Copenhagen.

Lassen, U. V., Ussing, H. H., and Wieth, J. O. (eds.) (1980). In *Membrane Transport in Erythrocytes,* pp. 404–408, Munksgaard, Copenhagen.

Lepke, S., and Passow, H. (1976). *Biochim. Biophys. Acta* **455,** 353–370.

Legrum, B., Fasold, H., and Passow, H. (1980). *Hoppe Seyler's Z. Physiol. Chem.* **361,** 1573–1590.

Lepke, S., and Passow, H. (1981). *J. Physiol. (London),* in press.

Lepke, S., Fasold, H., Pring, M., and Passow, H. (1976). *J. Membr. Biol.* **29,** 147–177.

Lesslauer, W. (1980). *Biochim. Biophys. Acta* **600,** 108–116.

Margaritis, L. H., Elgsaeter, A., and Branton, D. (1977). *J. Cell Biol.* **72,** 47–56.

Milanik, M. A. (1980). American Society of Biological Chemistry and Biophysics Society, Joint Meeting in New Orleans, June 1–5, 1980, Abstract 585.

Motais, R., and Cousin, J. L. (1978). In *Cell Membrane Receptors for Drugs and Hormones* (R. W. Straub and L. Bolis, eds.), pp. 219–225, Raven Press, New York.

Motais, R., Sola, F., and Cousin, J. L. (1978). *Biochim. Biophys. Acta* **510,** 201–207.

Passow, H., and Fasold, H. (1981). In *Adv. Physiol. Sci. 6* (S. R. Hollán, G. Gárdos, and B. Sarkadi, eds.), pp. 249–261, Pergamon Press, New York.

Passow, H., and Zaki, L. (1978). In *Molecular Specialization and Symmetry in Membrane Function* (A. K. Solomon and M. Karnovsky, eds.), pp. 229–252, Harvard University Press, Cambridge, Mass.

Passow, H., Fasold, H., Zaki, L., Schuhmann, B., and Lepke, S. (1975). In *Biomembranes: Structure and Function* (G. Gárdos and I. Szász, eds.), pp. 197–214, North-Holland, Amsterdam, and Publishing House of the Hungarian Academy of Sciences, Budapest.

Passow, H., Fasold, H., Gärtner, E. M., Legrum, B., Ruffing, W., and Zaki, L. (1980a). *Ann. N.Y. Acad. Sci.* **341,** 361–383.

Passow, H., Kampmann, L., Fasold, H., Jennings, M., and Lepke, S. (1980b). In *Membrane Transport in Erythrocytes* (U. V. Lassen, H. H. Ussing, and J. O. Wieth, eds.), pp. 345–367, Munksgaard, Copenhagen.

Patlak, C. S. (1957). *Bull. Math. Biophys.* **19,** 209–235.

Salhany, J. M., Cordes, K. A., and Gaines, E. D. (1980). *Biochemistry* **19,** 1447–1454.

Scarpa, A., Cecchetto, A., and Azzone, G. F. (1970). *Biochim. Biophys. Acta* **219,** 179–188.

Schnell, K. F., Gerhardt, S., and Schöppe-Fredenburg, A. (1977). *J. Membr. Biol.* **30,** 319–350.

Shami, Y., Rothstein, A., Knauf, P. A., and McCulloch, L. (1978). *Biochim. Biophys. Acta* **508,** 357–363.

Ship, S., Shami, Y., Breuer, W., and Rothstein, A. (1977). *J. Membr. Biol.* **33,** 311–323.

Steck, T. L. (1978). *J. Supramol. Struct.* **8,** 311–324.

Stone, L. C., and Kregenow, F. M. (1980). *J. Gen. Physiol.* **76,** 455–478.

Tanner, M. J. A., Williams, D. G., and Jenkins, R. E. (1980). *Ann. N.Y. Acad. Sci.* **341,** 455–464.

Wieth, J. O., Brahm, J., and Funder, J. (1980). *Ann. N.Y. Acad. Sci.* **341,** 394–418.

Wolosin, S. M., Ginsburg, H., and Cabantchik, Z. I. (1977). *J. Biol. Chem.* **252,** 2419–2427.

Zaki, L. (1981). *Biochem. Biophys. Res. Comm.* **99,** 243–251.

Zaki, L., Fasold, H., Schuhmann, B., and Passow, H. (1975). *J. Cell Physiol.* **86,** 471–494.

152

Water Transport in Red Blood Cells

Robert I. Macey

Water transport through the red cell membrane involves at least two parallel paths: (1) through channels presumably formed by integral proteins, and (2) via diffusion through the background lipid bilayer. The primary focus of this report is on efforts to separate and to model the two paths. More general aspects of water transport in red cells have been reviewed by Sha'afi (1977), Deuticke (1977), and Macey (1979). Advances in methods for diffusional permeability measurements that use small amounts of blood include a greatly improved continuous-flow device (Brahm, 1977) and new NMR methods (Mathur-DeVré, 1979; Pirkle *et al.*, 1979). A white ghost preparation with unaltered water permeability is described by Bjerrum (1979). The anomalous osmotic equilibrium properties of red cells have been clarified by Freedman and Hoffman (1979), Hladky and Rink (1978), and Wolf (1980).

1. TWO PARALLEL PATHS

A significant pathway for water diffusion through the lipid bilayer can be anticipated from theoretical calculations and from measurements of water permeability in artificial bilayers (Finkelstein and Cass, 1968). Evidence for the existence of channels also comes from comparisons of red cells with lipid bilayers. In particular: (1) Osmotic permeability, P_f, in red cells is approximately 10 times larger than P_f of "tight" phosphatidylcholine cholesterol bilayers. (2) The activation energy, E_a, for both P_f and diffusional permeability, P_d, in red cells is in the range of 4–6 kcal/mole (consistent with E_a for self-diffusion and viscous flow of water) while E_a for lipid bilayers is in the range of 10–14 kcal/mole (consistent with dissolution and diffusion of water in fluid hydrocarbons). (3) In red cells, the ratio $g \equiv P_f/P_d > 3$, in lipid bilayers $g = 1$. This finding of $g > 1$ in red cells (Paganelli and Solomon, 1957), perhaps the strongest evidence for the existence of water channels, has been contested because of uncertainties introduced by unstirred layers. If the finding $g > 1$ is erroneous, then either the measured P_f is too high or the P_d is too low, or both.

Robert I. Macey • Department of Physiology–Anatomy, University of California, Berkeley, California 94720.

Unstirred layer errors cause underestimates of the true permeability. Hence, the error would be an underestimated P_d. However, recent investigations do not support this contention. Theoretical studies of rapid-mix devices (Rice, 1980) as well as experimental measurements [of butanol transport, which is assumed to be rate limited by unstirred layers (Brahm, personal communication)] both indicate unstirred layers of only 1–4 μm. Further, NMR estimates (which are free from external unstirred layers) suggest that, if anything, THO exchange data *overestimate* P_d (Pirkle *et al.*, 1979).

2. MERCURIALS CLOSE CHANNELS IN AN ENTROPY-DRIVEN REACTION

Red cell water permeability is depressed by mercurial sulfhydryl reagents. When saturating doses (e.g., 2 mM PCMBS) are applied, evidence for channels disappears. Thus, P_f decreases 10-fold, E_a increases to 11.5 kcal/mole, and $g = 1$ (Macey *et al.*, 1972). Water permeability in the cell and in lipid bilayers become indistinguishable. These changes are rapidly reversed with cysteine.

The simplest interpretation is that the mercurials close water channels. Further, chicken red cell studies (Brahm and Wieth, 1977) suggest that the reagents do not influence the lipid path. Water permeability of chicken cells, which apparently have no channels (P_f is 10-fold smaller than human cells, $E_a = 11.4$ kcal/mole, and $g = 1$), is not influenced by PCMBS. Assuming, then, that PCMBS acts only on channels, it can be utilized for separating the permeability into component parts. Permeability measured in the presence of saturating doses of PCMBS represents P_1, the constant permeability of the lipid path. The permeability of the channels is obtained under any other condition by subtracting P_1 from the measured value.

Using these definitions, it follows that normally 90% of the osmotic flow takes place via channels while the remaining 10% passes through the lipids. Further, after subtracting P_1 from P_f and P_d, (Macey *et al.*, 1979) indicates that the ratio $g = (P_f - P_1)/(P_d - P_1)$ remains invariant as permeability is progressively reduced by higher and higher concentrations of PCMBS. This implies that the reagent closes the channels in an all-or-none way. Further, if the channel sizes are distributed, the result implies that the reagent attacks them at random.

Assuming that channel closure is all-or-none prompts the following model:

$$nC + M(\text{open}) \overset{K}{\rightleftharpoons} C_nM(\text{closed})$$

where C is the reagent, M(open) represents a membrane receptor and associated open channel, C_nM(closed) represents the reagent–receptor complex associated with a closed channel, and n is the reaction stoichiometry. Detailed study of this model (Macey *et al.*, 1979) shows that $n = 1$, $K = 0.2$ mM and that $\Delta H° = 25$ kcal/mole and $\Delta S° \approx 100$ e.u. Thus, although channel closure is energetically unfeasible, it is entropically driven. This large positive $\Delta S°$ is characteristic of hydrophobic bond formation and is usually interpreted in terms of randomization of water structures associated with the reaction.

3. SOLUTES ARE EXCLUDED FROM THE MAJOR WATER CHANNELS

Concentrations of PCMBS required to block water channels increase rather than decrease cation permeability by an order of magnitude (Sutherland *et al.*, 1967). Further, the reagent has no effect on SO_4^{2-} exchange (Passow and Schnell, 1969) nor on net Cl^- transport (Knauf, 1979). Therefore, it is unlikely that ions are transported via water channels.

When small polar nonelectrolytes are tested, it is found that permeability of glycols is insensitive to PCMBS and only small amides appear to be serious candidates for the channel transport (Macey *et al.*, 1972; Naccache and Sha'afi, 1974). However, inescapable evidence is mounting that these solutes traverse special facilitated pathways, that their transport inhibition by PCMBS is only incidental, and that they are not transported via the water channels. This case is more detailed for urea, which shows saturation kinetics and competition between the various amides and diamides (Wieth *et al.*, 1974; Mayrand and Levitt, 1980). Separation of water channels and urea transport can be demonstrated with several inhibitors (e.g., phloretin, nitrophenols) that dramatically reduce urea permeability but do not appreciably change water permeability (Macey *et al.*, 1972; Macey, 1979; Maynard and Levitt, 1980). Further, after correcting for light-scattering artifacts, Levitt (personal communication) found the reflection coefficient of urea to be close to unity; i.e., there is no evidence for water–urea coupling. In addition, the fortuitous nature of the PCMBS inhibition of urea transport is shown by the lack of correspondence in both the time and concentration dependence of PCMBS inhibition of urea and water transport (Chien and Macey, unpublished; Levitt, personal communication). Comparative studies (Wieth and Brahm, 1977) also support the concept that water and urea paths are separate. Duck red cells have a high water permeability, but a low bilayer-type urea permeability. *Amphiuma* cells, on the other hand, have a high urea permeability, but a low bilayer-type water permeability.

Lastly, there appears to be a physiological role for fast urea transport in mammalian red cells. It confers osmotic stability in the renal circulation. As blood reaches the deeper regions of the renal medulla, it is exposed to a very hypertonic NaCl/urea media and is threatened with severe shrinkage to levels that are damaging; upon emerging from the medulla, the urea-loaded cell is faced with the opposite problem—swelling to intolerable levels. Simulation studies (Macey, unpublished) show that these large volume fluctuations in cells with lipid bilayer-type urea permeabilities are greatly minimized in cells with urea permeabilities similar to those found in humans. The fact that chicken cells do not possess a fast urea transport is consonant with their inability to produce a concentrated urine.

4. SINGLE-FILE WATER TRANSPORT

Features of the water channel are illustrated in Fig. 1. The fact that urea (molecular radius ~ 2 Å) is excluded from the channel implies that the channel radius lies somewhere between 1.5 Å (molecular radius of water) and 2 Å. It follows that water molecules cannot slide past each other; they move in a single file. The ratio g for a single-file transport channel is equal to the number of water molecules occupying the channel (Dick, 1966; Levitt, 1974). After correcting for the permeability of the lipid path, we find that $g = 5$. Therefore, five water molecules are placed in the channel.

5. COMPARISON WITH GRAMICIDIN CHANNEL

Up to this point, properties of the red cell channel are similar to a gramidicin channel. The gramicidin channel excludes urea, has a 2-Å radius, and $g = 5.3$ (Rosenberg and Finkelstein, 1978). However, there are also striking differences. The gramicidin channel freely

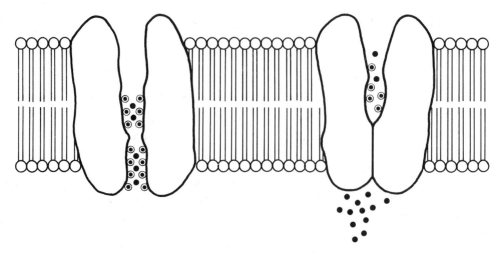

Figure 1. Schematic diagram of open (left) and closed (right) water channel. Mobile water moving in single file is denoted by filled circles. Immobile (structured) water is denoted by filled circles with halos.

admits Na^+, having a permeability ratio $P_{Na+}/P_{H_2O} = 8.4$* (Finkelstein, 1973). In red cells, Na^+ is excluded: $P_{Na+}/P_{H_2O} \sim 10^{-7}$. Further, gramicidin channels are very permeable to H^+. The high ratio $P_{H+}/P_{H_2O} = (P_{H+}/P_{Na+})(P_{Na+}/P_{H_2O}) \sim 120$ (Hladky and Haydon, 1972), together with the lack of H^+/H_2O transport coupling (Levitt *et al.*, 1978), is interpreted in terms of a Grotthus mechanism where protons jump along a continuous row of water molecules. This high H^+ permeability is not apparent in red cells. From pH equilibration kinetics in CO_2-free media, Tosteson *et al.* (1973) estimated a conductance of 10^{-7} ohms/cm², which presumably is composed of both OH^- and H^+ conductances. Using this figure as though the conductance were entirely due to H^+ gives an upper estimate (at pH 6.43) for P_{H+} of about 3×10^{-4} cm/sec. P_{H+} for the water channel is probably much lower (there are multiple paths for pH equilibration). Using a figure $P_d = P_{H_2O} = 2 \times 10^{-3}$ cm/sec yields $P_{H+}/P_{H_2O} < 0.2$ and it appears that H^+ ions are virtually excluded from the channel. The exclusion would follow if the channel were positively charged, but this would now require another assumption to explain the low electrogenic anion permeability. Alternatively, the poor H^+ conduction could result from isolation of water molecules within the channel so that protons cannot jump from one to the next. If the water molecules are separated by regions of relatively low dielectric, it may also explain ion exclusion from the channel. Isolation of water molecules has been suggested by Farrington (1979) to explain the 7–9 orders of magnitude difference in proton conduction in two very similar crystals (beta alumina and beta″ alumina).

Finally, the large entropy change suggests that the membrane becomes less hydrated when the channels close. Although there is no a priori reason to place these membrane-bound molecules in any particular position, it is tempting to associate them with the channel. Accordingly, parts of the channel have been lined with structured water that is displaced when the channel closes. This water is assumed to be relatively immobile and is considered part of the channel wall in discussions of channel dimensions.

---

*P_{H_2O} refers to diffusional water permeability, P_d. The original $P_{Na+}/P_{H_2O} = 1.4$ is corrected by a factor of 6 to correspond to newer measurements (Rosenberg and Finkelstein, 1978).

6. THE LIPID PATH

Transport through the lipid path is presumably similar to transport through artificial lipid bilayers. The classical lipid bilayer model used to predict water permeability properties views the membrane as a thin slice of bulk liquid hydrocarbon. As diffusion through the membrane is assumed rate limiting, the permeability can be predicted from measured solubility and diffusion coefficients in the appropriate solvent. Using hexadecane as the model solvent, Ohrbach and Finkelstein (1980) successfully predicted the permeabilities of water and a small number of nonelectrolytes. Specific limitations of the liquid hydrocarbon model have been pointed out, e.g., by Simon (1977) and by Petersen (1980).

Alternatives to the liquid hydrocarbon model generally introduce more rigid structures. The most explicit model (Trauble, 1971) emphasizes mobile structural defects or kinks in the hydrocarbon chains (e.g., the 2gl kink) formed by thermally induced rotations about C–C bonds. The concentration of these kinks is high (1–2 kinks per CH_2 chain), their mobility corresponds to a diffusion coefficient of 10^{-5} cm^2/sec, and approximately 50% of the free volume formed by kinks is occupied by water. Trauble proposes that water transport is simply due to movement of these water-filled kinks and he calculates a water permeability that falls within the measured range.

REFERENCES

Bjerrum, P. J. (1979). *J. Membr. Biol.* **48**, 43–67.
Brahm, J. (1977). *J. Gen. Physiol.* **70**, 283–306.
Brahm, J., and Wieth, J. O. (1977). *J. Physiol.* **266**, 727–749.
Deuticke, B. (1977). *Rev. Biochem. Pharmacol.* **78**, 1–97.
Dick, D. A. T. (1966). *Cell Water,* Butterworths, London.
Farrington, G. C. (1979). In *Membrane Transport Processes* (C. F. Stevens and R. W. Tsien, eds.), Vol. 3, pp. 43–71, Raven Press, New York.
Finkelstein, A. (1973). In *Drugs and Transport Processes: A Symposium* (B. A. Callingham, ed.), pp. 241–250, University Park Press, Baltimore.
Finkelstein, A., and Cass, A. (1968). *J. Gen. Physiol.* **52**, 145s–172s.
Freedman, J. C., and Hoffman, J. F. (1979). *J. Gen. Physiol.* **74**, 157–185.
Hladky, S. B., and Haydon, D. A. (1972). *Biochim. Biophys. Acta* **274**, 294–312.
Hladky, S. B., and Rink, T. J. (1978). *J. Physiol.* **274**, 437–446.
Knauf, P. A. (1979). *Curr. Top. Membr. Transp.* **12**, 249–363.
Levitt, D. G. (1974). *Biochim. Biophys. Acta* **373**, 115–131.
Levitt, D., Elias, S., and Hautman, J. (1978). *Biochim. Biophys. Acta* **512**, 436–451.
Macey, R. I. (1979). In *Membrane Transport in Biology* (G. Giebisch, D. C. Tosteson, and H. H. Ussing, eds.), Vol. 2, pp. 1–57, Springer-Verlag, Berlin.
Macey, R. I., Karan, D. M., and Farmer, R. E. L. (1972). In *Biomembranes* (F. Kreuzer and J. F. G. Slegers, eds.), Vol. 3, pp. 331–340, Plenum Press, New York.
Macey, R., Chien, D., Moura, T., and Karan, D. (1979). *Biophys. J.* **25**, 102a.
Mathur-DeVré, R. (1979). *Prog. Bophys. Mol. Biol.* **35**, 103–134.
Maynard, R., and Levitt, D. G. (1980). *Fed. Proc.* **39**, 957.
Naccache, P., and Sha'afi, R. I. (1974). *J. Cell Physiol.* **83**, 449–456.
Ohrbach, E., and Finkelstein, A. (1980). *J. Gen. Physiol.* **75**, 427–436.
Paganelli, C. V., and Solomon, A. K. (1957). *J. Gen. Physiol.* **41**, 259–277.
Passow, H., and Schnell, K. F. (1969). *Experientia* **25**(5), 460–467.
Petersen, D. C. (1980). *Biochim. Biophys. Acta* **600**, 666–677.
Pirkle, J. L., Ashley, D. L., and Goldstein, J. H. (1979). *Biophys. J.* **25**, 389–406.
Rice, S. A. (1980). *Biophys. J.* **29**, 65–80.
Rosenberg, P. A., and Finkelstein, A. (1978). *J. Gen. Physiol.* **72**, 327–350.

Sha'afi, R. I. (1977). In *Membrane Transport in Red Cells* (J. C. Ellory and V. L. Lew, eds.), pp. 221–256, Academic Press, New York.

Simon, S. A. (1977). *J. Gen. Physiol.* **70,** 123–127.

Sutherland, R. M., Rothstein, A., and Weed, R. (1967). *J. Cell Physiol.* **69,** 185–198.

Tosteson, D. C., Gunn, R. B., and Wieth, J. O. (1973). In *Erythrocytes, Thrombocytes, Leukocytes* (E. Gerlach, K. Moser, E. Deutsch, and W. Wilmanns, eds.), pp. 62–66, Thieme, Stuttgart.

Trauble, H. (1971). *J. Membr. Biol.* **4,** 193–208.

Wieth, J. O., and Brahm, J. (1977). *Proc. Int. Union Physiol. Sci.* **XII,** 126.

Wieth, J. O., Funder, J., Gunn, R. B., and Brahm, J. (1974). In *Comparative Biochemistry and Physiology of Transport* (L. Bolis, K. Block, S. E. Luria, and F. Lynen, eds.), North-Holland, Amsterdam.

Wolf, M. B. (1980). *J. Theor. Biol.* **83,** 687–700.

153

Acetyl Glyceryl Ether Phosphorylcholine: A Biologically Active Phosphoglyceride

Donald J. Hanahan and R. Neal Pinckard

1. BACKGROUND

Alkyl acetyl glyceryl phosphorylcholine (AGEPC) is a newly discovered potent lipid chemical mediator released from rabbit basophils *in vitro* and into the plasma of IgE-sensitized rabbits after administration of antigen, e.g., horseradish peroxidase or bovine serum albumin (Hanahan *et al.*, 1980; Pinckard *et al.*, 1979). Prior to its structural identification (Hanahan *et al.*, 1980), the released mediator was termed *platelet activating factor* (PAF) (Benveniste *et al.*, 1972; Henson, 1970). However recent evidence (Pinckard *et al.*, 1980) detailing its wide range of biological activities in the mammal has mandated that the latter term be dropped in favor of a more precise chemical one. At some appropriate time in the future, perhaps a more convenient term can be developed to replace the somewhat cumbersome chemical one. For the present, however, AGEPC will refer to the potent lipid chemical mediator released from IgE-sensitized rabbit basophils after antigen stimulation.

A particular characteristic of IgE systemic anaphylaxis in the rabbit is the rapid and significant decrease in circulating platelets, neutrophils, and basophils (Pinckard *et al.*, 1979). Intravenous infusion of synthetic AGEPC into rabbits produced exactly the same response as the injection of antigen into IgE-sensitized rabbits (McManus *et al.*, 1980). Under *in vitro* conditions, AGEPC can induce a shape change, aggregation, and release of serotonin in washed rabbit platelets, and this behavior, in essence, forms the basis of the assay system described below. The potency of AGEPC is revealed by its high activity toward platelets at a molar concentration of 10^{-10}. This compound has also been shown to have hypotensive activity (Blank *et al.*, 1979).

Donald J. Hanahan • Department of Biochemistry, The University of Texas Health Science Center, San Antonio, Texas 78284.   ***R. Neal Pinckard*** • Department of Pathology, The University of Texas Health Science Center, San Antonio, Texas 78284.

2. ISOLATION AND CHARACTERIZATION

Essentially, the procedure for isolation of naturally occurring AGEPC follows that outlined in detail by Pinckard *et al.* (1979). The basic approach in this technique is to obtain blood from the ear artery of IgE rabbits, sensitized to horseradish peroxidase or bovine serum albumin, and through careful centrifugation and washings, to isolate the "buffy coat" fraction. The latter is incubated with either horseradish peroxidase or bovine serum albumin (depending on the antigen used in sensitization) for approximately 20 to 30 min at 37°C. The cells are then removed by centrifugation and the supernatant containing the AGEPC bound to albumin treated with 95 volumes of absolute methanol. The precipitated proteins are removed by centrifugation at 2000g for 20 min. To 1 volume of the soluble fraction is added 0.95 volume of chloroform and 0.8 volume of water. This mixture is shaken vigorously and then placed in a separatory funnel. The chloroform-rich lower phase, which contains all of the functional platelet activating factor activity (AGEPC), is removed, washed once with methanol/water, 10/9 (v/v), and the lower phase again removed and stored at −25°C until needed.

At this point in the isolation procedure, all of the functional platelet stimulating activity can be recovered but is still contaminated to a significant degree with other lipids, particularly phospholipids. Also, though the functional activity levels are high in these extracts, the actual weight of the active compound is calculated to be in the nanogram range, i.e., 8–10 ng/100 ml plasma. Structural studies on this level of material can only be achieved through examination of the effect of certain reagents on retention of activity in comparison to its thin-layer chromatographic behavior. Thus, it is important to purify the platelet stimulating activity as much as possible, and this can be achieved through preparative thin-layer chromatography, using a solvent system of chloroform/methanol/water, 65/35/6 (v/v), in which the desired active substance migrated with an R_f of 0.22, which places it just below sphingomyelin ($R_f = 0.28$) and above lysolecithin ($R_f = 0.14$). Subsequent to development of the preparative plate in the above solvent, sections of the silica gel, usually between 0.20 and 0.50 cm in height, starting from the origin up to a total of 5 cm, are removed by scraping and then extracted with a solvent system of chloroform/methanol/water, 1/2/0.8 (v/v). The clear extract is mixed with chloroform and water to allow phase separation and the lower chloroform-rich layer tested for functional activity toward washed rabbit platelets. The most active platelet stimulating active extract, usually found at $R_f = 0.22$ to 0.24, is then subjected to chromatography on silica gel G in a solvent system of methanol/water, 2/1 (v/v). In this system, sphingomyelin, the most probable contaminant, remained at the origin and the desired active component migrated with an R_f of 0.40 to 0.50 on this plate, allowing high purification of the naturally occurring AGEPC.

The structure of this naturally occurring material was determined with a total available sample of 15 to 20 μg through the combined use of gas–liquid chromatography and mass spectrometry (Hanahan *et al.*, 1980). Verification of the structure of this naturally occurring material was greatly facilitated by the earlier discovery (Demopoulos *et al.*, 1979) that a semisynthetic AGEPC, prepared as below, behaved identically to the naturally derived substance on thin-layer chromatography, in reactions with various chemical reagents, and in its biological activity, both *in vitro* and *in vivo*.

3. DIRECT CHEMICAL SYNTHESIS

The most expedient route to preparation of large quantities of AGEPC is through a semisynthetic procedure, which produces a product identical in all respects to that of the naturally occurring substance. Basically this method is relatively simple, entailing only isolation of the choline plasmalogens (vinyl ether-containing phosphatidylcholine fraction) from bovine heart muscle (Pugh *et al.*, 1977). This latter fraction is hydrogenated to the saturated ether, deacylated by mild alkali, and the resulting lyso compound then subjected to acetylation to yield the desired product. These reactions can best be depicted by the following series of reactions:

$$
\text{bovine heart} \xrightarrow[\text{extraction}]{\text{solvent}} \text{total lipid extract} \xrightarrow{\text{acetone}} \text{phospholipids (insoluble)} + \text{neutral lipids (soluble)}
$$

aluminum oxide +
silicic acid chromatography

"lecithin fraction"
(plasmalogen rich) $\xrightarrow{H_2}$

where $R_1 = 14:0$ and $16:$ alkyl chain. Yields are generally about 50%.

AGEPC

A more sophisticated synthesis, in which the alkyl chain can be a single species, would involve starting with a pure 1-*O*-hexadecyl-*sn*-glycerol (chimyl alcohol), and converting it to the 3-*O*-trityl derivative followed by acetylation to yield the 2-acetyl derivative. The latter is treated with boric acid to remove the trityl group, with subsequent reaction with phosphoryl chloride and choline toluenesulfonate to give 1-*O*-hexadecyl-2-acetyl-*sn*-glyceryl-3-phosphorylcholine.

4. BIOLOGICAL ASSAY FOR AGEPC

The best system for assay of AGEPC (or its analogs or other compounds) is that of following [^{3}H]serotonin release from freshly prepared washed platelets that previously have been loaded with [^{3}H]serotonin. AGEPC is a potent stimulator of rabbit platelets, inducing a calcium-independent shape change and calcium-dependent aggregation and secretion of granular constituents. In our assay, a unit of activity (induced by AGEPC) is defined as the (molar) amount of material required to induce the release of 50% of platelet (250,000 platelets/μ1) serotonin in 60 sec at 37°C. The percent of released serotonin is determined by liquid scintillation spectroscopy relative to that of a control sample of platelets treated with 2.5% Triton X-100.

Although the platelet serotonin release assay is the preferred one, aggregation profiles can provide some qualitative guides as to the mode of action of certain compounds. Thus, the shape of the aggregation curve and whether it is reversible or nonreversible can provide some general insights into the general mode of interaction of AGEPC and its analogs with platelets and can be useful for comparative purposes with other stimulators.

5. BIOLOGICAL ACTIVITIES OF AGEPC

AGEPC has been shown to have multiple activities in experimental animals and in man. For example, it can cause irreversible aggregation and serotonin release at 2×10^{-7} M in human platelets (actually in platelet-rich plasma). Further, AGEPC in the range from 10^{-10} to 10^{-6} M can cause a dose-dependent aggregation in human neutrophils and in the presence of cytochalasin B can induce the release of specific (lysozyme and lactoferrin) and azurophilic (β-glucuronidase, myeloperoxidase) granule enzymes. In addition, AGEPC can cause contraction of guinea pig ileum at 10^{-7} to 10^{-6} M. Finally, intradermal injection of AGEPC into human volunteers induced skin blanching, following by an overt erythema together with pseudopodia formation at a dosage of 52 to 520 pg. In all the above, the deacetylated, or lyso, GEPC possessed little or no biological activity.

6. CONCLUDING REMARKS

AGEPC is a remarkably potent lipid chemical mediator, which represents one of the first examples of a biologically active phosphoglyceride. Under normal circumstances, AGEPC is not found in cells such as the basophil or neutrophil unless these cells are stimulated and this catalyzes the formation and release of this phosphoglyceride in a very short period of time (i.e., 30 to 60 sec at the most).

REFERENCES

Benveniste, J., Henson, P. M., and Cochrane, C. G. (1972). *J. Exp. Med.* **136**, 1356–1377.
Blank, M. L., Snyder, F., Byers, L. W., Brooks, B., and Muirhead, E. E. (1979). *Biochem. Biophys. Res. Commun.* **90**, 1994–1200.

Demopoulos, C. A., Pinckard, R. N., and Hanahan, D. J. (1979). *J. Biol. Chem.* **254,** 9355–9358.

Hanahan, D. J., Demopoulos, C. A., Liehr, J., and Pinckard, R. N. (1980). *J. Biol. Chem.* **255,** 5514–5516.

Henson, P. M. (1970). *J. Exp. Med.* **131,** 287–306.

McManus, L. M., Hanahan, D. J., Demopoulos, C. A., and Pinckard, R. N. (1980). *J. Immunol.* **124,** 2919–2924.

Pinckard, R. N., Farr, R. S., and Hanahan, D. J. (1979). *J. Immunol.* **123,** 1847–1857.

Pinckard, R. N., McManus, L. M., Demopoulos, C. A., Halonen, M., Clark, P. O., Shaw, J. O., Knicker, W. T., and Hanahan, D. J. (1980). *J. Reticuloendothelial Soc.* **28**(Suppl.), 95–103.

Pugh, E. L., Kates, M., and Hanahan, D. J. (1977). *J. Lipid Res.* **18,** 710–716.

154

The Effect of Intracellular Transglutaminase on Membrane Structure

Laszlo Lorand

1. NATURE OF THE REACTION

This article examines the proposition that the action of a γ-glutamine : ϵ-lysine transferase, commonly referred to as transglutaminase, can play a significant role in modifying the physical properties of cells by virtue of catalyzing a specific fusion of membrane and/or cytoskeletal proteins through γ-glutamyl-ϵ-lysine peptide bridges:

$$P\stackrel{\gamma}{\diagdown}CONH_2 + H_2N\stackrel{\epsilon}{\diagdown}\diagup\diagdown\diagup\diagdown P' \xrightarrow[Ca^{2+}]{E} P\stackrel{\gamma}{\diagdown}CONH\diagdown\stackrel{\epsilon}{\diagup}\diagdown\diagup\diagdown P' + NH_3 \quad (1)$$

The illustration applies to a simple dimerizing system between proteins P and P', but the general reaction clearly provides for forming large linear as well as branched polymers (see Lorand *et al.*, 1979a).

Plasticity of the membrane, which is such a critical attribute from the point of view of changing of cell shape, is thought to depend on the transient reversible nature of associations of intramembrane protein constituents with those in the membrane infrastructure and in the cytoskeleton. As deduced from *in vitro* studies of binary associations, in the specific case of human erythrocytes, these interactions extend to the cluster of proteins shown in Fig. 1. The major transmembrane anion carrier (i.e., band 3 protein) is linked through ankyrin (band 2.1) to spectrin (bands 1 and 2), which in turn is attached to actin through the band 4.1 protein (see e.g. Lux, 1979).

Metabolites such as ATP or 2,3-diphosphoglycerol have been shown to weaken associations within this ensemble so that the integral membrane proteins acquire a greater lateral mobility (Sheetz and Casaly, 1980).

It can be seen that a large-scale covalent polymerization of the transamidating type

Laszlo Lorand • Department of Biochemistry, Molecular and Cell Biology, Northwestern University, Evanston, Illinois 60201.

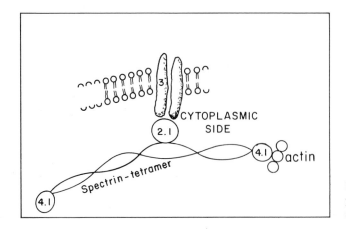

Figure 1. Topological outline of protein interactions in the membrane skeleton of the red blood cell.

given in (1) could easily lead to a significant change in membrane structure. If the reaction proceeded in a reversible manner,* the reaction could serve as a means for transmembrane modulation. In practice, however, the transglutaminase-catalyzed cross-links encountered in various systems appear to be very difficult to reverse. One may assume that, in addition to the protonation of the leaving group ($NH_3 + H^+ \rightleftharpoons NH_4^+$), some phase change occurring when the proteins become cross-linked by γ-glutamyl-ϵ-lysine bridges is responsible for the difficulty of obtaining reversal. Transglutaminase-catalyzed cross-linking thus appears to be an instrument of cell rigidification, and studies on human erythrocytes (Lorand *et al.*, 1979a), on tissue culture keratinocytes (Rice and Green, 1977, 1979) and on cataract of the eye (Lorand *et al.*, 1981) would seem to suggest a general role for this enzyme in cell senescence. Therefore, examination of the biological controls and molecular details of the following intracellular cascade assumes considerable significance:

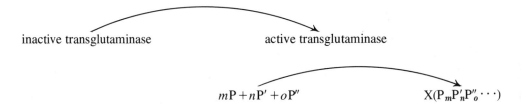

One must also keep in mind that, in addition to protein polymerization by γ-glutamyl-ϵ-lysine cross-links (X) upon activation of the latent transglutaminase, there is also a chance of incorporating a variety of primary amines (H_2NR) into the reactive γ-glutamine functionalities of protein monomers or oligomers:

$$P\overset{\gamma}{\curlywedge}CONH_2 + H_2NR \xrightarrow[Ca^{2+}]{E} P\overset{\gamma}{\curlywedge}CONHR + NH_3 \tag{2}$$

2. REGULATION OF TRANSGLUTAMINASE-CATALYZED CROSS-LINKING IN CELLS

One of the important biochemical issues is to define the controls that regulate the conversion and the activity of transglutaminase. Transglutaminases are Ca^{2+}-dependent en-

---

*As written in (1), there would be a negligible change in free energy.

zymes, and substantial conversion from latent to active state can be demonstrated in the 10–100 μM Ca^{2+} range (see Siefring *et al.*, 1978). Experimentally, Ca^{2+} loading can be obtained by exposing cells to the cation in the presence of ionophore A23187. Accumulation of concentrations of intracellular Ca^{2+} of about 100 μM has also been observed under natural conditions in old red blood cells and in sickle cells (Eaton *et al.*, 1973; Palek, 1973). In the latter case, failing of the outward-directed Ca^{2+} pump seems to be responsible (Lew *et al.*, 1980). In the keratinocytes, the disintegration of the mitochondria may give rise to the elevation of intracellular Ca^{2+}. The situation appears to be even more complex in the lens because, in this tissue, there seems to be sufficient Ca^{2+} *in toto* (Fagerholm, 1979) to cause a complete activation of transglutaminase. Of course, the latent enzyme and Ca^{2+} might well be sequestered from one another and activation might be triggered only when the two compartments merge. If on the other hand they coexisted in the same phase, the known competition between Zn^{2+} and Ca^{2+} for the active site might serve as a brake on enzyme activity. For one such enzyme, we obtained an apparent K_i of 6×10^{-7} M Zn^{2+} (Credo *et al.*, 1976), indicating that very low concentration of Zn^{2+} could be effective. It is interesting that in senile cataract there seems to be a considerable decline in the concentration of Zn^{2+} and an increase in that of Ca^{2+} (Kuck, 1970). Would such an inverse change in the composition of the lens in regard to these two antagonists tip the balance for activating transglutaminase?

Because all known transglutaminases function through a cysteine-containing catalytic site, there is always a possibility of inactivating the enzyme by direct oxidation or by disulfide exchange. In human erythrocytes, enzyme activity appears to be significantly diminished after a few hours of energy depletion (Coetzer and Zail, 1979). Cystamine was found to be a good inhibitor of transglutaminase action, mainly on account of its reaction with the active center cysteine (Lorand *et al.*, 1979a; Svahn, Credo, and Lorand, unpublished results).

Some transglutaminases respond to certain organic solutes in special ways. For example, the activity of a nonsecretory transglutaminase found in the anterior lobe of guinea pig prostate is enhanced from immeasurably low to very high velocities by tosylglycine, without any effect on the K_m for substrate (Lorand and Tong, unpublished results). Perhaps compounds such as tosylglycine mimic a metabolite that might appear at the right moment of the cell cycle to enhance the activity of the enzyme.

Naturally occurring polyamines (e.g., spermine, spermidine, putrescine) are known to be rather good substrates for transglutaminase (Lorand *et al.*, 1979b) and could have an effect on protein cross-linking by competing against the reactive ϵ-lysine residues, i.e., reaction (2) versus reaction (1). In the presence of these bifunctional amines and Ca^{2+}, transglutaminase can become membrane-bound (Griffin *et al.*, 1981a), which, by itself, might serve as an interesting regulatory feature.

Yet another control, operating at the substrate level, has been observed in lens homogenates, where the addition of glycerol specifically enhanced the reactivity of intrinsic proteins (i.e., of β-crystallin) without having much effect on a casein test protein (Lorand *et al.*, 1981).

3. STRATEGIES FOR PROBING THE ACTION OF INTRACELLULAR TRANSGLUTAMINASE

Approaches toward the intracellular reactions are very similar to those developed earlier for fibrin cross-linking (Lorand *et al.*, 1968). After demonstrating that a given physiological event is Ca^{2+} dependent, it is recommended that suitable primary amines be em-

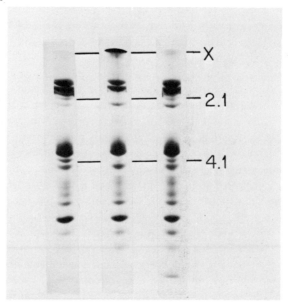

Figure 2. Changes in the SDS electrophoretic profile of membrane proteins upon incubating human erythrocytes [37°C, 3 hr, 50% hematocrit, in a buffer containing 0.1 M KCl, 0.06 M NaCl, 10 mM glucose, and 5 mM Tris–HCl (pH 7.4)] with 1.5 mM Ca^{2+} in the presence of 10 μM ionophore A23187 (center gel). In the experiment represented by the gel on the right, the cells were preincubated for 30 min with 20 mM aminoacetonitrile and this amine was also present during Ca^{2+} loading. It can be seen that this prevented the loss of bands 2.1 and 4.1 as well as the formation of the polymer, marked X. The left-hand gel shows a control where the incubation mixture contained 1.5 mM Mg^{2+} instead of Ca^{2+} and no aminoacetonitrile.

ployed as inhibitors [i.e., using reaction (2) vs. reaction (1)]. In their isotopically or fluorescently labeled forms, the amines can also serve as specific probes for marking the transglutaminase-reactive γ-glutamine-containing proteins of the cell. In order to properly evaluate inhibition, appropriate compounds (usually the $^{\alpha}N$-methyl or $^{\alpha}N$-dimethyl analogs of amines) must be used as controls. Stability as well as permeability considerations dictate the choice of the cross-linking inhibitors and of the controls. The following serve as useful examples: glycine methylester vs. sarcosine methylester; histamine vs. $^{\alpha}N$-dimethylhistamine; dansylcadaverine vs. $^{\alpha}N$-dimethylated dansylcadaverine, i.e., $(CH_3)_2N(CH_2)_5NH.dansyl$.

Profound alteration of the membrane protein profile observed in Ca^{2+}-loaded erythrocytes is illustrated in Fig. 2. The intracellular accumulation of Ca^{2+} caused a total disappearance of bands 4.1, 2.1, and the partial loss of band 3 and spectrin, together with the appearance of high-molecular-weight polymers (marked X) on top of the reduced SDS electrophoretic gels. The presence of aminoacetonitrile (H_2NCH_2CN) or histamine during Ca^{2+} loading prevented all these changes, whereas control compounds, such as $^{\alpha}N$-dimethylhistamine, were ineffective (see Lorand et al., 1979a, 1980).

With the use of a radioactive amine ([^{14}C]histamine), labeling of spectrin, band 3, and the growing polymer regions (X) could be demonstrated (Siefring et al., 1978), showing that these proteins contained transglutaminase-reactive amine-incorporating sites.

Conventional subunit analysis cannot be applied to decipher the protein composition of γ-glutamyl-ϵ-lysine cross-linked polymeric structures. However, the constituents may retain some of their original antigenic determinants such that a judicious application of immunological techniques may still provide significant information regarding composition. Bjerrum et al. (1981) recently showed that the membrane polymer produced in Ca^{2+}-loaded erythrocytes contains the antigenic sites of spectrin, band 3, 2.1, and 4.1. Acetylcholinesterase and glycophorin determinants were absent from the polymer. One may thus note that the transglutaminase-reactive membrane constituents belong to the interacting cluster of proteins shown in Fig. 1, and this study certainly represents one of the most direct methods for nearest-neighbor analysis. The question as to whether actin participates in the formation of the enzymatically cross-linked polymer is still open.

Demonstrating the existence of γ-glutamyl-ε-lysine cross-links is an essential aspect of proof for the functioning of transglutaminase. The most direct method for quantitating the frequency of this isopeptide consists of digesting the protein by the sequential addition of proteases that hydrolyze the conventional backbone α-peptides without, however, attacking the γ-peptide. As such, the protein becomes degraded essentially to free amino acids and, among these and a few miscellaneous oligopeptides remaining, there is also the γ-glutamyl-ε-lysine of interest. It is then necessary to isolate the latter from all other components of the digest and appropriately quantitate the dipeptide recovered (see Siefring *et al.*, 1978). We have recently developed a rapid and sensitive high-pressure liquid chromatographic procedure for this purpose (M. Griffin, J. Wilson, and L. Lorand, unpublished).

4. CONSEQUENCES OF TRANSGLUTAMINASE ACTION

The structural consequences of protein polymerization through γ-glutamyl-ε-lysine linkages vary from cell type to cell type. In the case of human erythrocytes, cross-linking involves some transmembrane proteins (band 3) and some (2.1, 4.1, and spectrin) from the membrane skeleton. In keratinocytes, a circularly arranged cornifying envelope forms just beneath the plasma membrane. However, the result in both cases is a rigidification of the cell and an irreversible fixation of shape.

In human erythrocytes, the Ca^{2+}-induced fixation of shape change was shown to be specifically prevented by the same amines that inhibited the production of protein-to-protein γ-glutamyl-ε-lysine bridges (Lorand *et al.*, 1979a). These compounds also prevented the concomitant change in the elastic properties of the membrane (Smith *et al.*, 1981). Furthermore, when red cells were loaded with Ca^{2+}, a significant increase in membrane density was found which could be related to the formation of the polymer of the type discussed (Griffin *et al.*, 1981b). This Ca^{2+}-induced increase in membrane density could be inhibited by histamine but not by $^\alpha N$-dimethylhistamine.

ACKNOWLEDGMENTS. This work was aided by a USPHS Research Career Award (5K06HL-03512) and by a grant from the National Institutes of Health (AM-25412).

REFERENCES

Bjerrum, O. J., Swanson, P. E., Griffin, M., and Lorand, L. (1981). *J. Supramol. Struct.* **16**, 289.
Coetzer, T. L., and Zail, S. S. (1979). *Biochem. J.* **43**, 375.
Credo, R. B., Stenberg, P., Tong, Y. S., and Lorand, L. (1976). *Fed. Proc.* **35**, 1630 (Abstr. 1390).
Eaton, J. W., Shelton, T. G., Swofford, H. S., Kolpin, C. E., and Jacob, H. S. (1973). *Nature* **246**, 105.
Fagerholm, P. P. (1979). *Exp. Eye Res.* **28**, 211.
Griffin, M., Barnes, R. N., and Walton, P. L. (1981a). *Int. J. Cancer,* in press.
Griffin, M., Swanson, P. E., Bjerrum, O.J., and Lorand, L. (1981b). *Biophys. J.* **33**, 177a (Abstr. T-AM-P068).
Kuck, J. R. R., Jr. (1970). In *Biochemistry of the Eye* (C. N. Graymore, ed.), p. 319, Academic Press, New York.

Lew, V. L., Bookchin, R. M., Brown, A. M., and Ferreira, H. G. (1980). In *Membrane Transport in Erythrocytes* (U. V. Lassen, H. H. Ussing, and J. O. Wieth, eds.), p. 196, Munksgaard, Copenhagen.

Lorand, L., Rule, N. G., Ong, H. H., Furlanetto, R., Jacobsen, A., Downey, J., Oner, N., and Bruner-Lorand, J. (1968). *Biochemistry* **7,** 1214.

Lorand, L., Siefring, G. E., Jr., and Lowe-Krentz, L. (1979a). *J. Supramol. Struct.* **9,** 427.

Lorand, L., Parameswaran, K. N., Stenberg, P., Tong, Y. S., Velasco, P. T., Jonsson, N. A., Mikiver, L., and Moses, P. (1979b). *Biochemistry* **18,** 1756.

Lorand, L., Siefring, G. E., Jr., and Lowe-Krentz, L. (1980). In *Membrane Transport in Erythrocytes* (U. V. Lassen, H. H. Ussing, and J. O. Wieth, eds.), p. 285, Munksgaard, Copenhagen.

Lorand, L., Hsu, L. K.-H., Siefring. G. E., Jr., and Rafferty, N. S. (1981). *Proc. Natl. Acad. Sci. USA* **78,** 1356.

Lux, S. E. (1979). *Nature (London)* **281,** 426.

Palek, J. (1973). *Blood,* **42,** 988 (Abstract).

Rice, R. H., and Green, H. (1977). *Cell* **11,** 417.

Rice, R. H., and Green, H. (1979). *Cell* **18,** 681.

Sheetz, M. P., and Casaly, J. (1980). *J. Biol. Chem.* **255,** 9955.

Siefring, G. E., Jr., Apostol, A. B., Velasco, P. T., and Lorand, L. (1978). *Biochemistry* **17,** 2958.

Smith, B. D., LaCelle, P. L., Siefring, G. E., Jr., Lowe-Krentz, L., and Lorand, L. (1981). *J. Membr. Biol.* **61,** 75.

XIII

Properties of Cell Surfaces. Recognition and Interaction of Cells

155

Structure, Dynamics, and Signaling across Cell Membranes

Garth L. Nicolson

A large number of signals and extracellular communications must be selectively received by cells, so it is fitting that there are probably a variety of different ways in which cells can transmit signals from their exterior to their interior environments. The ways in which cells handle these signals and transmit information in the proper discrimination, sequences, or amplification are probably important in determining a variety of physiologically critical cellular responses (Nicolson, 1982). Cells appear to have evolved a complex system of receptor control at the level of the cell membrane to transmit information; therefore, elucidating membrane structure and determining how cells control the displays and dynamics of their surface receptors are important in understanding the ways in which certain extracellular communications are processed.

A most important link in this communication system is the cell membrane. General consensus has been reached over the last few years that the structures of cell membranes generally conform to a number of basic principles (reviews: Singer, 1971, 1974; Singer and Nicolson, 1972; Bretscher, 1973; Wallach, 1975; Nicolson, 1976; Nicolson *et al.*, 1977; Nicolson, 1979). Briefly these are: (1) the major membrane lipids, such as the phospholipids, are arranged in a planar, bilayer configuration that is predominantly in a fluid state under physiological conditions; (2) the lipid bilayers are discontinuous and interrupted by numerous proteins and glycoproteins that are intercalated at various depths into the bilayer, some penetrating across the membrane; (3) the arrangement of lipids, glycolipids, proteins, and glycoproteins is asymmetric with respect to the distribution and arrangement of specific molecules in the inner and outer halves of the bilayer; and (4) the proteins and glycoproteins of cell membranes are quite heterogeneous and can be operationally divided into two main classes—integral (or intrinsic) and peripheral (or extrinsic) membrane proteins (Capaldi and Green, 1972; Singer and Nicolson, 1972). An example of this molecular arrangement is the positioning of the integral and peripheral proteins and glycoproteins in

---

Garth L. Nicolson • Department of Tumor Biology, University of Texas System Cancer Center, M. D. Anderson Hospital and Tumor Institute, Houston, Texas 77030.

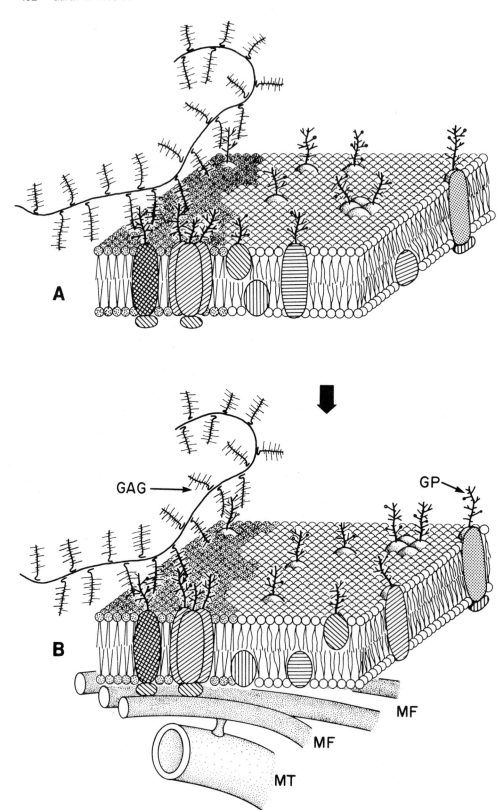

the model depicted in Fig. 1. In this proposal, integral transmembrane glycoproteins are stabilized by hydrophobic interactions with membrane lipids and also by interactions of peripheral membrane proteins at the inner surface. Also shown in this scheme are interactions with extracellular components such as glycosaminoglycans.

Evidence exists that certain integral membrane proteins are present in oligomeric complexes. These complexes can be constructed from identical or nonidentical integral membrane protein subunits, and oligomeric complexes have also been found between integral and peripheral classes of membrane proteins at the inner (Nicolson and Painter, 1973; Elgsaeter and Branton, 1974; Elgsaeter *et al.*, Pinto da Silva and Nicolson, 1974) or the outer (Hynes *et al.*, 1976; Singer, 1979) membrane surface. The dynamic formation or disaggregation of oligomeric complexes in the membrane may be important in regulating the topographic display of individual membrane components, and perhaps in regulating the mobilities of specific surface receptors.

The basic structure of the cell membrane has been described as a two-dimensional solution of a mosaic of integral membrane proteins and glycoproteins embedded in a predominantly fluid lipid bilayer matrix (Singer and Nicolson, 1972). Obviously, cell membranes are much more complex, and in addition to this fundamental structure, numerous peripheral membrane components are probably bound to both membrane surfaces. Also, transiently bound membrane-associated cytoskeletal assemblages (microfilaments and perhaps microtubules) are, on occasion, attached at the inner membrane surface (Fig. 1) (Nicolson, 1976; Loor, 1977; Nicolson *et al.*, 1977).

The notion that biological membranes are predominantly fluid in nature has important implications. First, this type of structural arrangement allows rapid lateral diffusion within the plane of the membrane (reviews: Edidin, 1974; Singer, 1974, 1976; Nicolson, 1976, 1979; Nicolson *et al.*, 1977) permitting rapid and reversible changes in the topographic display of specific surface components. Certain cell surface components appear free to undergo random diffusion in the plane of the membrane, while others move much more slowly. The movement of glycoproteins in this latter class of membrane components is thought to be restrained by a variety of mechanisms yielding a graded hierarchy of mobilities for different receptors on the cell surface and even within specific regions on the surfaces of cells. These same restraining mechanisms may also control the mobility and topographic display of cytoplasmic membrane components at the inner surface. Thus, rapid and reversible changes in the topographic display of membrane components could occur in response to stimuli originating outside or inside the cell resulting in changes in the pattern or aggregation of membrane components that could be important as a signaling event in transmembrane-mediated communication (Nicolson, 1979).

The fluid mosaic membrane structure also permits membrane components to organize in an asymmetric fashion across the membrane, allowing specific molecules to be localized to the inner or outer half of the bilayer, such that they are exposed exclusively at the inner or outer membrane surfaces. Integral membrane glycoproteins, proteins, and glycolipids exhibit striking asymmetry, and these molecules appear to selectively reside or they are

←_____

Figure 1. Modified version of the fluid mosaic model of membrane structure showing transmembrane control over the distribution and mobility of cell surface receptors by peripheral and membrane-associated cytoskeletal components. In this hypothetical model the mobility of integral glycoprotein complexes is controlled by outer surface peripheral glycosaminoglycans and also by membrane-associated cytoskeletal elements at the inner surface. (A) Some glycoprotein complexes are shown to be sequestered into a specific lipid domain indicated by the shaded area, while others exist in a free or unaggregated state and are capable of free lateral motion. (B) Aggregation of certain glycoproteins has stimulated attachment of membrane-associated cytoskeletal elements. MF, microfilaments; MT, microtubules. [From Nicolson, 1979, with permission.]

oriented specifically at one surface of the membrane. For example, in every plasma membrane that has been isolated and investigated thus far, the oligosaccharides reside only at the outer surface of the membrane (review: Nicolson *et al.*, 1977). The asymmetry of the plasma membrane dictates that components oriented to either the inner or the outer surface of the membrane should not be able to rotate at appreciable rates from one side of the membrane to the other (Singer and Nicolson, 1972). This lack of appreciable transmembrane rotation (Bretscher, 1973; Nicolson and Singer, 1974) may be due to the tremendous thermodynamic barriers that tend to prevent hydrophilic molecules from passing through the hydrophobic matrix of the membrane, thus maintaining proper asymmetry (Singer and Nicolson, 1972; Singer, 1974).

Cellular membranes are predominantly formed from a matrix of fluid-state lipids that have a viscosity similar to light machine oil (Edidin, 1974). There appears to be a considerable degree of lipid acyl chain motion in biological membranes under physiological conditions, and this probably determines their bulk fluidity (Horwitz *et al.*, 1972; Levine *et al.*, 1972; McFarland, 1972). Bulk membrane fluidity in many systems seems to be controlled by cholesterol (Emmelot and Van Hoeven, 1975). Cholesterol can preferentially interact with certain fatty acids and sphingolipids (Brockerhoff, 1974), resulting in a decrease in lateral molecular spacing and flexibility of phospholipid acyl chains (Hubbell and McConnell, 1971; Oldfield and Chapman, 1972). The presence of cholesterol in certain biological membranes could be responsible for modulating phospholipid acyl chain mobility and flexibility and may be important in segregating certain lipids into specific membrane domains. Cholesterol serves to create an intermediate condition between the fluid or solid phases in pure phospholipid systems (Oldfield and Chapman, 1972), and addition of cholesterol to biological membranes in the fluid state can result in decreased phospholipid acyl chain mobility and flexibility to a point intermediate between the fluid and solid states. In the cell membranes that have been analyzed, bulk fluidity characteristics are certainly determined by the membrane lipids, because artificial lipid bilayers constructed from extracted membrane lipids are similar in viscosity to the membranes from which they were derived (reviews: Edidin, 1974; Nicolson *et al.*, 1977). A variety of studies have estimated biological membrane viscosity to be a few poise with lipid diffusion constants varying from 1×10^{-8} to 12×10^{-8} cm^2/sec depending upon the type of lipid probe (Edidin, 1974; Nicolson *et al.*, 1977).

In model lipid membranes, as well as some biological membranes, both solid- and fluid-phase membrane regions have been identified using freeze-fracture electron microscopy. The fluid membrane regions and replicas appear smooth after freeze-fracture, while the solid-phase lipid fracture surfaces have ridges that give them a banded appearance. The surface areas of smooth or banded lipid regions in the fracture face appear to be directly proportional to the amount of fluid- versus solid-phase lipid as determined by paramagnetic probes (Grant *et al.*, 1974). Biological membranes also contain numerous intercalations or particles that are thought to be protein or glycoprotein–lipid complexes. Extracted membrane lipids do not contain the particles when observed by freeze-fracture techniques; however, reconstitution of biological membranes results in fracture surfaces with numerous particles (Hong and Hubbell, 1972; Segrest *et al.*, 1974; Vail *et al.*, 1974). Cholesterol can dampen the extent of lipid-phase separation when present in concentrations greater than 20 mole%, as well as cause differential partitioning of proteins and glycoproteins into fluid phases (Kleemann and McConnell, 1976). Local lipid domain formation can lead to protein clustering in membranes with broad lipid-phase transitions, and these events may be important in modifying the topographic display of cell surface receptors. This could be important, because many membrane-bound enzymes are sensitive to the phase state of lipid immediately surrounding them (Bennett *et al.*, 1978), and their activities may be controlled by modulation of the membrane physical state (Kimelberg, 1977).

Integral membrane proteins and glycoproteins are much less mobile in the plane of the membrane compared to phospholipids and have diffusion constants in the range of 4–10×10^{-9} cm²/sec (Poo and Cone, 1974), and some glycoproteins have very low rates of lateral diffusion ($D < 3 \times 10^{-12}$ cm²/sec) (Peters *et al.*, 1974). There appears to be heterogeneity in the abilities of membrane glycoproteins to diffuse laterally in the lipid matrix of the membrane (Schlessinger *et al.*, 1976). In some cases, peripheral protein components at the inner surface may control the distribution and mobility of transmembrane glycoproteins (Nicolson and Painter, 1973; Elgsaeter and Branton, 1974; Elgsaeter *et al.*, 1976).

The topographic distribution of most cell surface receptors is essentially random. However, the binding of multivalent ligands to surface receptors induces, in many cases, lateral movement and topographic redistribution of the ligand–receptor complexes (reviews: Unanue and Karnovsky, 1973; Edidin, 1974; Loor, 1977; Nicolson, 1976; Nicolson *et al.*, 1977). However, in some cases, cell surface receptors show nonrandom distributions before the binding of multivalent ligands (Wartiovaara *et al.*, 1974; Rostas and Jeffrey, 1975). Often this occurs in regionally specialized areas or specific membrane domains (Nicolson and Yanagimachi, 1974; Koehler, 1975).

Because cell surface receptors can be distributed in unique displays and/or possess differing lateral mobilities, controlling mechanisms must exist that govern cell surface dynamics and receptor topographic displays. The mobilities of cell surface receptors can be modulated or restricted by a variety of different mechanisms, which have been reviewed in detail elsewhere (Nicolson, 1976, 1979; Nicolson *et al.*, 1977). Briefly these are: planar associations or aggregations with other molecules, the formation of specific lipid or protein domains leading to exclusion or inclusion of specific components with respect to these domains, control by peripheral membrane components, and control by membrane-associated (cytoskeletal) systems. In these latter examples, transmembrane linkages are thought to be involved in controlling the display and mobility of cell surface receptors at the inner surface of the membrane. As mentioned, the lateral mobility of the major glycoproteins of the human erythrocyte membrane appears to be under restraint by an inner surface peripheral protein network involving spectrin, actin, and possibly other components. Transmembrane control by membrane-associated cytoskeletal elements is perhaps one of the most important mechanisms of cell surface receptor control. Cytoskeletal control appears to be transient, and its formation and reversal may be one of its most unique features.

Membrane-associated cytoskeletal control appears to be involved in regulating the topography, display, and dynamics of cell surface receptors in a variety of systems (reviews: Unanue and Karnovsky, 1973; Loor, 1977; Nicolson, 1976, 1979; Nicolson *et al.*, 1977). One of the best studied examples of cytoskeletal transmembrane control is the cellular control of surface-immunoglobulin molecules on lymphocytes (Taylor *et al.*, 1971; Yahara and Edelman, 1972; Loor *et al.*, 1972). Binding of anti-immunoglobulin to B lymphocytes results in surface-Ig receptor redistribution to form clusters, patches, and eventually caps, and microfilaments appear to supply the cytoskeletal contractile system necessary to drive cap formation. On the other hand, microtubules appear to also play a role in modulating the direction and mobilities of cell surface receptors (Edelman, 1974, 1976) by directing translocation of clusters and patches toward a specific region of the cell (Loor, 1977; de Petris, 1977; Nicolson *et al.*, 1977). Thus, the dynamics of surface-Ig receptor complexes is probably dependent upon interplay between microfilament and mi-

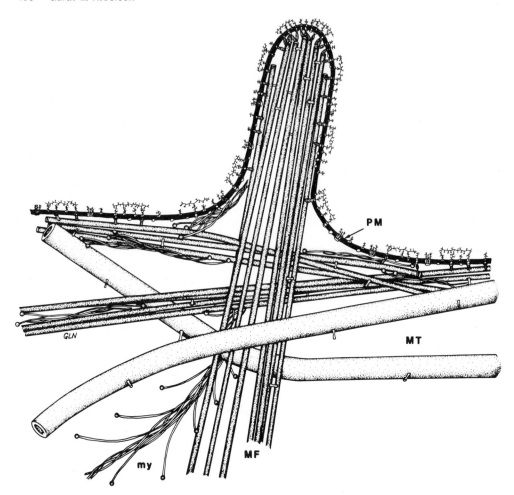

Figure 2. Hypothetical interactions between membrane-associated microtubule (MT) and microfilament (MF) systems involved in transmembrane control over cell surface receptor mobility and distribution. This model envisages opposite but coordinated roles for the microfilaments (contractile) and microtubules (skeletal and directional) and suggests that they are linked to one another or to the same plasma membrane (PM) inner surface components. This linkage may occur through myosin molecules or through cross-bridging molecules such as α-actinin. In addition, peripheral membrane component linked at the inner or outer plasma membrane surface may extend this control over specific membrane domains. [From Nicolson *et al.*, 1977, with permission.]

crotubular systems of the cell (Loor, 1977; Nicolson *et al.*, 1977). In addition, these systems may be linked to an intermediate (100 Å) filament system as described by Albertini and Anderson (1977).

Currently the basic structure of the plasma membrane is envisioned as an elaboration of the fluid mosaic membrane model (Singer and Nicolson, 1972) with added peripheral membrane-associated components (Fig. 2) (Nicolson *et al.*, 1977). On the extracellular surface, glycosaminoglycans and mucopolysaccharides are probably associated with integral membrane glycoproteins, although certain portions of their structures may actually anchor hydrophobically in the membrane (Kraemer, 1975). Integral membrane proteins and glycoproteins or their oligomeric complexes are shown asymmetrically oriented in the membrane. Various arrays of cytoskeletal elements interact with the membrane in particular regions where they are attached by bridging molecules (Mooseker and Tilney, 1975; Hepler and Palevitz, 1974). At any one time only certain classes of membrane-linked surface receptors are probably under cytoplasmic control by systems such as the membrane-associated cytoskeletal assemblages, because at any one time only some of the cell surface receptors are linked to transmembrane controlling systems or to other membrane components through peripheral interactions at either membrane surface. To a large degree the formation of transmembrane linkages to cytoskeletal systems is stimulated by the aggregation of cell surface receptors through binding of multivalent ligands. Ligand-induced transmembrane linkage of surface receptors to cytoskeletal systems has been seen in several systems (Ash and Singer, 1976; Ash *et al.*, 1977). These and other studies have shown that the microtubular system is not directly involved in cytoskeletal membrane attachment; however, microtubules may be indirectly linked to membranes, possibly through microfilaments (Nicolson *et al.*, 1977; de Petris, 1977; Loor, 1977).

In this brief essay it is impossible to discuss at length all of the different ways in which cells receive signals and forward them into the cell interior. Thus, only models of transmembrane communication-discrimination based upon membrane dynamics and receptor displays have been considered.

REFERENCES

Albertini, D. F., and Anderson, E. (1977). *J. Cell Biol.* **73**, 111–127.

Ash, J. F., and Singer, S. J. (1976). *Proc. Natl. Acad. Sci. USA* **73**, 4574–4579.

Ash, J. F., Louvard, D., and Singer, S. J. (1977). *Proc. Natl. Acad. Sci. USA* **74**, 5584–5588.

Bennett, J. P., McGill, K. A., and Warren, G. B. (1978). *Nature (London)* **274**, 823–825.

Bretscher, M. S. (1973). *Science* **181**, 622–629.

Brockerhoff, H. (1974). *Lipids* **9**, 645–650.

Capaldi, R. A., and Green, D. E. (1972). *FEBS Lett.* **25**, 205–209.

de Petris, S. (1977). In *Dynamic Aspects of Cell Surface Organization* (G. Poste and G. L. Nicolson, eds.), Vol. 3 of *Cell Surface Reviews*, pp. 643–728, North-Holland, Amsterdam.

Edelman, G. M. (1974). In *Control of Proliferation in Animal Cells* (B. Clarkson and R. Baserga, eds.), Vol. 1 of *Cold Spring Harbor Conferences on Cell Proliferation*, pp. 357–377, Cold Spring Harbor Laboratory, New York.

Edelman, G. M. (1976). *Science* **192**, 218–226.

Edidin, M. (1974). *Annu. Rev. Biophys. Bioeng.* **3**, 179–201.

Elgsaeter, A., and Branton, D. (1974). *J. Cell Biol.* **63**, 1018–1036.

Elgsaeter, A., Shotton, D. M., and Branton, D. (1976). *Biochim. Biophys. Acta* **426**, 101–122.

Emmelot, P., and Van Hoeven, R. P. (1975). *Chem. Phys. Lipids* **14**, 236–246.

Grant, C. W. M., Wu, S. H.-W., and McConnell, H. M. (1974). *Biochim. Biophys. Acta* **363**, 151–158.

Hepler, P. K., and Palevitz, B. A. (1974). *Annu. Rev. Plant Physiol.* **25**, 309–362.

Hong, K., and Hubbell, W. L. (1972). *Proc. Natl. Acad. Sci. USA* **69**, 2617–2621.

Horwitz, A. F., Horsley, W. J., and Klein, M. P. (1972). *Proc. Natl. Acad. Sci. USA* **69**, 590–593.

Hubbell, W. L., and McConnell, H. M. (1971). *J. Am. Chem. Soc.* **93**, 314–326.

Hynes, R. O., Destree, A. T., and Mautner, V. (1976). In *Membranes and Neoplasia* (V. T. Marchesi, ed.), pp. 189–202, Liss, New York.

Kimelberg, H. K. (1977). In *Dynamic Aspects of Cell Surface Organization* (G. Poste and G. L. Nicolson, eds.), Vol. 3 of *Cell Surface Reviews*, pp. 205–293, North-Holland, Amsterdam.

Kleemann, W., and McConnell, H. M. (1976). *Biochim. Biophys. Acta* **219**, 206–222.

Koehler, J. K. (1975). *J. Cell Biol.* **67**, 647–659.

Kraemer, P. M. (1975). In *Mammalian Cells: Probes and Problems* (C. R. Richmond, D. F. Petersen, P. F. Mullaney, and E. C. Anderson, eds.), pp. 242–245, USERDA Technical Information Center, Oak Ridge, Tenn.

Levine, Y. K., Birdsall, N. J. M., Lee, A. G., and Metcalfe, J. G. (1972). *Biochemistry* **11**, 1416–1421.

Loor, F. (1977). *Prog. Allergy* **23**, 1–153.

Loor, F., Forni, L., and Pernis, B. (1972). *Eur. J. Immunol.* **2**, 203–212.

McFarland, B. G. (1972). *Chem. Phys. Lipids* **8,** 303–313.

Mooseker, M. S., and Tilney, L. G. (1975). *J. Cell Biol.* **67,** 725–743.

Nicolson, G. L. (1976). *Biochim. Biophys. Acta* **457,** 57–108.

Nicolson, G. L. (1979). *Curr. Topics Dev. Biol.* **13,** 305–338.

Nicolson, G. L. (1982). In *Hormone Action* (R. F. Goldberger, ed.), Vol. 3A of *Biological Regulation and Development: A Comprehensive Treatise*, Plenum Press, New York, in press.

Nicolson, G. L., and Painter, R. G. (1973). *J. Cell Biol.* **59,** 395–406.

Nicolson, G. L., and Singer, S. J. (1974). *J. Cell Biol.* **60,** 236–248.

Nicolson, G. L., and Yanagimachi, R. (1974). *Science* **184,** 1294–1296.

Nicolson, G. L., Poste, G., and Ji, T. H. (1977). In *Dynamic Aspects of Cell Surface Organization* (G. Poste and G. L. Nicolson, eds.), Vol. 3 of *Cell Surface Reviews*, pp. 1–73, North-Holland, Amsterdam.

Oldfield, E., and Chapman, D. (1972). *FEBS. Lett.* **23,** 285–297.

Peters, R., Peters, J., Tews, K. H., and Bahr, W. (1974). *Biochim. Biophys. Acta* **367,** 282–294.

Pinto da Silva, P., and Nicolson, G. L. (1974). *Biochim. Biophys. Acta* **363,** 311–319.

Poo, M., and Cone, R. A. (1974). *Nature (London)* **247,** 438–440.

Rostas, J. A. P., and Jeffrey, P. L. (1975). *Neurosci. Lett.* **1,** 47–53.

Schlessinger, J., Koppel, D. E., Axelrod, D., Jacobson, K., Webb, W. W., and Elson, E. L. (1976). *Proc. Natl. Acad. Sci. USA* **73,** 2409–2413.

Segrest, J. P., Gulik-Krzywicki, T., and Sardet, C. (1974). *Proc. Natl. Acad. Sci. USA* **71,** 3294–3298.

Singer, I. I. (1979). *Cell* **16,** 675–685.

Singer, S. J. (1971). In *Structure and Function of Biological Membranes* (L. I. Rothfield, ed.), pp. 145–222, Academic Press, New York.

Singer, S. J. (1974). *Annu. Rev. Biochem.* **43,** 805–833.

Singer, S. J. (1976). In *Surface Membrane Receptors: Interface between Cells and Their Environment* (R. A. Bradshaw, W. A. Frazier, C. R. Merrell, D. I. Gottlieb, and R. A. Hogue-Angeletti, eds.), pp. 1–24, Plenum Press, New York.

Singer, S. J., and Nicolson, G. L. (1972). *Science* **175,** 720–731.

Taylor, R. B., Duffus, W. P. H., Raff, M. C., and de Pétris, S. (1971). *Nature New Biol.* **233,** 225–229.

Unanue, E. R., and Karnovsky, M. J. (1973). *Transplant. Rev.* **14,** 184–210.

Vail, W. J., Papahadjopoulos, D., and Moscarello, M. A. (1974). *Biochim. Biophys. Acta* **345,** 463–467.

Wallach, D. F. H. (1975). *Membrane Molecular Biology of Neoplastic Cells*, North-Holland, Amsterdam.

Wartiovaara, J., Linder, E., Ruoslahti, E., and Vaheri, A. (1974). *J. Exp. Med.* **140,** 1522–1533.

Yahara, I., and Edelman, G. M. (1972). *Proc. Natl. Acad. Sci. USA* **69,** 608–612.

156

Growth Control Mediated by Cell–Cell Interactions

Luis Glaser and Richard P. Bunge

In the last 10 or 15 years, evidence has been obtained in a number of systems that cell–cell contact, either between homologous or heterologous cells, can affect cell growth and differentiation. These effects are believed to represent direct consequences of cell–cell contact and not to reflect the release by the cells of soluble agents. Recent work in our laboratories has resulted in the development of methods that should permit the isolation of the molecules on the cell surface responsible for these phenomena and thus facilitate studies designed to understand the mechanisms responsible for these effects.

Table I lists a few selected examples of situations in cells derived from higher eukaryotes where evidence suggests that cell–cell contact is involved in the regulation of cell growth or in the regulation of a metabolic function. The data are not equally compelling in each case, but the examples listed are fairly representative of the cases described to date.

It is impossible to obtain a detailed understanding of the molecules involved in cell–cell interaction of the ligands involved can only be assayed while present on live cells. A considerable advance in our understanding of these systems was made when it could be shown that cell–cell adhesion could be reproduced (at least in part) by measurements of the ability of cells to bind appropriate plasma membrane vesicles (Merrell and Glaser, 1973; Santala *et al.*, 1977). It followed from these observations that if the physiological consequences of cell–cell contact occurred as a result of the binding of complementary ligands on adjacent cells to each other, then binding of membrane vesicles to cells could in principle elicit the same physiological response. If this assumption is correct, it would then become possible to purify the relevant ligands from the membrane by suitable fractionation procedures. This approach has been successful in the two systems that have been investigated extensively in our laboratories (Bunge *et al.*, 1979). Before describing these

Luis Glaser and Richard P. Bunge • Departments of Biological Chemistry, and Anatomy and Neurobiology, Division of Biology and Biomedical Sciences, Washington University School of Medicine, St. Louis, Missouri 63110.

**Table I. Examples of Physiological Effect of Cell–Cell Contact
on Cells in Tissue Culture**

Cell type	Physiological consequences of cell–cell adhesion	Reference
Fibroblasts	Cessation of cell growth, decrease in the rate of solute uptake	[a]
Schwann cells, neurons	Initiation of cell proliferation	[a]
C-6 glioma cells	Induction of synthesis of S-100 protein	[b]
Smooth muscle cells, BC$_3$H1, an established smooth muscle-like cell line	Induction of synthesis of muscle-specific proteins, e.g., creatine kinase and acetylcholine receptors	[c]
Myoblasts	Cell fusion, induction of synthesis of a variety of proteins characteristic of differentiated muscle	[d]

[a] This review and references cited.
[b] Gysin *et al.* (1980) and references therein.
[c] Munson *et al.* (1982), Chamley-Campbell *et al.* (1979), and references therein.
[d] Bischoff (1978).

in detail, it seems reasonable to discuss some of the technical problems that complicate such an endeavor.

1. *Availability of material.* Although some of the cells listed in Table I are established cell lines, many are primary cells and all show anchorage dependence of growth. Either novel techniques applicable to the isolation of very small quantities of material will be required, or whole animal sources for these molecules must be found.

2. *Stability of ligand.* The isolation and identification of these molecules requires that they be stable under conditions of membrane isolation and after dispersal in detergent.

3. *Fractionation of hydrophobic molecules and removal of detergents.* Methods for fractionating small quantities of hydrophobic proteins are still in their infancy. Hydrophobic proteins are extracted from various surface membrane fractions with detergents and usually can be fractionated in the presence of detergents. For assay of their biological activities, these proteins must be presented to live cells in the absence of detergent, conditions under which they may become insoluble.

4. *Complexity of the biological assay.* The best studied systems depend on the assay of complex biological events such as cell growth, which can be influenced by many variables. Simpler systems, where the induction or repression of a single protein could be measured, would be much more desirable as tools for the study of cell–cell interactions.

One of the systems we have investigated in detail is the cessation of growth of "normal" (as compared to malignant) fibroblasts in cell culture. These cells grow to confluency and cease to grow when they reach monolayer density. The precise density reached is a function of the concentration of mitogenic compounds in serum (primarily the platelet-derived growth factor), and the system behaves as if the mitogenic signal from serum could compete with negative signals derived from cell–cell contact. It is beyond the scope of this review to consider the evidence that density-dependent inhibition of growth is a contact phenomenon, but this subject has been considered in detail in a recent article (Lieberman and Glaser, 1981).

We have demonstrated that a plasma membrane-enriched fraction can be prepared from 3T3 cells that inhibits the growth of sparse 3T3 cells in a manner reminiscent of that

Table II. Comparison of Inhibition of Growth by High Density and Membranes[a]

	Membranes	Purified GIP	High cell density
Fraction of cells arrested per generation	50%	50%	50%
Position in cell cycle	G_0	G_0	G_0
Medium depletion of mitogens	No	—	No
Reversibility	Yes	Yes	Yes
Competition with mitogens (EGF, PDGF, FGF)	Yes	Yes	Yes
Occlusion of receptors on cell surface	No	—	No
Inhibition of uptake of low-molecular-weight solutes			
α-Aminoisobutyrate and uridine	Yes	No	Yes
Glucose, P_i	No	No	Yes
Sensitivity to heat	Yes	Yes	—
periodate oxidation	Yes	Yes	—
Unique to 3T3 cells	No	No	No

[a] The table compares the characteristics of 3T3 cells derived from Swiss mice arrested at high cell density or at sparse density $(3 \times 10^3$ cells/cm^2) by the addition of surface membranes from 3T3 cells or partially purified GIP. For references see Whittenberger and Glaser (1977), Whittenberger *et al.* (1978, 1979), Lieberman *et al.* (1979, 1981), Raben *et al.* (1981).

observed at high cell density (Whittenberger and Glaser, 1977; Whittenberger *et al.*, 1978, 1979; Lieberman *et al.*, 1979, 1980). The inhibition appears to be due to a membrane protein or proteins abbreviated GIP (for growth inhibitory proteins), which have been partially purified by a sequence of steps involving extraction of extrinsic membrane proteins with dimethylmaleic anhydride, selective solubilization with the nonionic detergent octylglucoside, and precipitation at acid pH. The characteristics of the inhibition of growth of 3T3 cells observed at high cell density or after addition of membranes or partially purified GIP to 3T3 cells are summarized in Table II and experimental details can be found in the references cited. Several general points regarding these observation should be made.

The experimental approach is not confined to 3T3 cells, a well-established cell line, but can also be used with 1MR91 cells, a human cell line (Lieberman *et al.*, 1981). The growth inhibitory protein present in the membranes of both cells cross-react, indicating a strong tendency for conservation of the structure of these proteins during evolution.

While many characteristics of the events that occur at high cell density are reproduced by the addition of membranes to cells, some are not. Most notably, addition of membranes does not decrease the rate of uptake of glucose or P_i by cells, and purified GIP when added to cells no longer decreases the activity of the A system for amino acid transport. These observations suggest that some of the alterations in solute transport observed at high cell density are independent of the cessation of cell growth. Conversely, the rate of amino acid uptake in sparse cells already growth-arrested in plasma (lacking the platelet-derived growth factor) can be further blocked by the addition of membranes (Moya and Glaser, 1980).

One of the serious concerns regarding growth inhibition experiments of the type described is the consequence of two possible types of artifacts. The first is that membranes deplete the medium of essential nutrients; the second is that membranes, or partially purified GIP, prevent access of essential molecules such as mitogenic factors (hormones) to the cell surface. The first is easily ruled out by the fact that membranes in the presence of cells do not deplete the medium of growth factors. The second can be ruled out by two observations, namely that the rate of uptake of some small molecules (P_i, glucose) is unaffected by addition of membranes and that the binding of one defined mitogenic compound (epidermal growth factor) to the surface of cells is unaffected by the addition of membranes that themselves cannot bind epidermal growth factor under conditions where they block the mitogenic action of epidermal growth factor.*

Growth control in which cell–cell contact promotes rather than inhibits cell proliferation is readily demonstrable in the interactions of Schwann cells and nerve fibers (axons) during the development of the peripheral nervous system. This control appears designed to provide adequate numbers of Schwann cells for ensheathment and myelination as the nerve fibers elongate during development. These interactions may be studied in cultures of fetal sensory ganglia employing collagen substrates and media enriched with serum and embryo extract (Wood, 1976; Salzer and Bunge, 1980). The axonal outgrowth from these sensory neurons is richly populated with Schwann cells that surround (ensheath) the small unmyelinated nerve fibers and in time provide segments of myelin along the larger axons (Wood, 1976). Methods are available to suppress the fibroblast component in this outgrowth zone; thus, if the neuronal cell bodies within the ganglion are removed by microsurgical excision, the remaining outgrowth will (after Wallerian degeneration of the axons) consist of a pure Schwann cell population (Wood, 1976). With axonal removal, the substantial ongoing Schwann cell proliferation (characteristic of the intact growing culture) ceases and a "quiescent" and quite stable cell population is retained. If axons are again allowed to grow among these Schwann cells, proliferation resumes, but only in those regions directly invaded by axons (Wood and Bunge, 1975). Direct contact between the axon and Schwann cell appears essential; if axons and Schwann cells are grown on opposite sides of a diaphragm of reconstituted collagen, no stimulation of proliferation is seen (Salzer *et al.*, 1980a). If the axonal surface is perturbed by gentle aldehyde treatment, Schwann cells seeded onto those axons are not stimulated to divide (or to recognize and align along the axonal shaft as they normally do) (Salzer *et al.*, 1980a). If living axons are treated with trypsin immediately prior to the addition of Schwann cells, the proliferative signal is absent, but will be restored within 48 hr, presumably by replacement of damaged membrane surface components (Salzer *et al.*, 1980a). Cultures of sensory neurons spawning axonal outgrowth may be treated to suppress Schwann cell populations and pure axonal material can then be harvested from the outgrowth zone (Wood, 1976). If this axonal material is homogenized and separated into soluble and particulate fractions, it can be shown that the capability to engender proliferation is concentrated in the particulate fraction (Salzer *et al.*, 1980b). Such fractions prepared from axons previously treated with trypsin are ineffective; the activity is also removed by heating to 60°C for 10 min (Salzer *et al.*, 1980b). Activity of axonal particulate fractions is not dependent on serum containing media for it occurs in defined medium (Cassel *et al.*, 1981). Efforts to prepare axonal membranes in large quantities for further characterization have involved separation of a myelinated axon preparation from calf brain in which the myelin and axolemmal components can subsequently be separated (DeVries, 1981; DeVries *et al.*, 1981). These axolemmal fractions are mitogenic for quiescent Schwann cells, but unlike the preparation of cultured neurites this interaction is serum dependent (Cassel *et al.*, 1982; DeVries *et al.*, 1982). It is possible that this difference derives from the fact that axolemmal preparations are derived from tissues where axonal ensheathment has already been established prior to axolemmal harvest. Schwann cells establishing ensheathment are known to be secretory and may produce some product that masks the axolemmal mitogen as ensheathment progresses (Bunge *et al.*, 1980; Moya *et al.*, 1980). Such masking agents could explain why Schwann cell proliferation ceases as more intimate and extensive contacts with axons are established. An axolemmal prepara-

---

*The utility of NR-6 for these experiments was first pointed out by Fox *et al.* (1979).

tion may retain this masking agent and require a serum factor for unmasking of the mitogen.

Schwann cell proliferation has also been observed in a variety of other conditions. Both *in vivo* and *in vitro* observations indicate that Schwann cells participating in digestion of axonal and myelin debris during Wallerian degeneration undergo proliferation (for discussion see Salzer and Bunge, 1980). Studying Schwann cells prepared by a different method, Brockes *et al.* (1979, 1980) have obtained a potent Schwann cell mitogen from bovine pituitary gland; other Schwann cell mitogens are also known (Raff *et al.*, 1978).

The litany between Schwann cell and axon includes a variety of other known interactions. Schwann cell ensheathment has been shown to be necessary for maintenance of normal axon diameter (Aguayo *et al.*, 1979). Data from invertebrates indicate that the cells ensheathing axons in this species actively transfer proteins to the enclosed axons (Lasek *et al.*, 1977). Axonal contact appears to be required for the production of the basal lamina that surrounds each Schwann cell–axonal unit. Because the production of this basal lamina material takes place in cultures containing no fibroblasts, it is apparently produced by the Schwann cell (Bunge *et al.*, 1980), and is considered evidence that the Schwann cell has secretory capacity. Recent observations suggest that Schwann cells that fail to secrete normally are not able to relate normally to axons (Moya *et al.*, 1980). It is known that the axon directs the activity of the Schwann cell in myelin production, for Schwann cells taken from unmyelinated nerves will form myelin transplanted into nerves containing axons competent to induce myelin formation (Aguayo *et al.*, 1976). Finally, it should be noted that the factors that control the proliferation and functional expression of the central myelinating cell—the oligodendrocyte—appear to be quite different from those influencing the Schwann cell (for discussion see Bunge, 1981).

ACKNOWLEDGMENTS. Work in our laboratories was supported by NSF Grant PCM 7715972 and NIH Grants GM 18405, NS 09923, and GM 28002.

REFERENCES

Aguayo, A., Charron, L., and Bray, G. (1976). *J. Neurocytol.* **5**, 565–573.

Aguayo, A., Bray, G., and Perkins, S. (1979). *Ann. N.Y. Acad. Sci.* **317**, 512–531.

Bischoff, R. (1978). In *Membrane Fusion* (G. Poste and G. L. Nicolson, eds.), pp. 127–129, Elsevier/North-Holland, Amsterdam.

Brockes, J., Fields, K., and Raff, M. (1979). *Brain Res.* **165**, 105–118.

Brockes, J., Lemke, D., and Balzer, J. (1980). *J. Biol. Chem.* **255**, 8374–8377.

Bunge, M. B., Williams, A. K., Wood, P. M., Uitto, J., and Jeffrey, J. J. (1980). *J. Cell Biol.* **84**, 184–202.

Bunge, R. P. (1982). In *Dahlem Workshop Report on Neuronal–Glial Cell Interrelationships: Ontogeny, Maintenance, Injury, Repair* (S. Bernhard, ed.), pp. 115–130, Dahlem Konferenzen, Berlin.

Bunge, R., Glaser, L., Lieberman, M. A., Raben, D., Salzer, J., Whittenberger, B., and Woolsey, T. (1979). *J. Supramol. Struct.* **11**, 175–187.

Cassel, D., Wood, P. M., Bunge, R. P., and Glaser, L. (1982). *J. Cellular Biochemistry*, in press.

Chamley-Campbell, J., Campbell, G. R., and Ross, R. (1979). *Physiol. Rev.* **59**, 1–61.

DeVries, G. H. (1981). In *Research Methods in Neurochemistry*, Vol. 5, in press.

DeVries, G. H., Salzer, J. L., and Bunge, R. P. (1982). *Brain Res.* **3**, 295–299.

Fox, C. F., Vale, R., Peterson, S. W., and Das, M. (1979). *Hormones and Cell Culture* (G. Sato and R. Ross, eds.), pp. 143–158, Cold Spring Harbor Laboratory, New York.

Gysin, R., Moore, B. W., Proffitt, R. T., Deuel, T. F., Caldwell, K., and Glaser, L. (1980). *J. Biol. Chem.* **255**, 1515–1520.

Lasek, R., Gainer, H., and Barkes, V. L. (1977). *J. Cell Biol.* **74**, 501–523.

Lieberman, M. A., and Glaser, L. (1981). *J. Membr. Biol.* **63,** 1–11.

Lieberman, M. A., Raben, D. M., Whittenberger, B., and Glaser, L. (1979). *J. Biol. Chem.* **254,** 6357–6361.

Lieberman, M. A., Rothenberg, P., Raben, D. M., and Glaser, L. (1980). *Biochim. Biophys. Acta* **92,** 696–702.

Lieberman, M. A., Raben, D. M., and Glaser, L. (1981). *Exp. Cell Res.* **133,** 413–419.

Merrell, R., and Glaser, L. (1973). *Proc. Natl. Acad. Sci. USA* **70,** 2794–2798.

Moya, F., and Glaser, L. (1980). *J. Biol. Chem.* **255,** 3258–3260.

Moya, F., Bunge, M. B., and Bunge, R. P. (1980). Proc. Natl. Acad. Sci. USA **77,** 6902–6906.

Munson, R., Caldwell, K. L., and Glaser, L. (1982). *J. Cell Biol.* **92,** 350–356.

Raben, D., Lieberman, M., and Glaser, L. (1981). *J. Cell Physiol.* **108,** 35–45.

Raff, M., Abney, E., Brockes, J., and Hornby-Smith, A. (1978). *Cell* **15,** 813–822.

Salzer, J. L., and Bunge, R. P. (1980). *J. Cell Biol.* **84,** 739–752.

Salzer, J. L., Bunge, R. P., and Glaser, L. (1980a). *J. Cell Biol.* **84,** 767–778.

Salzer, J. L., Williams, A. K., Glaser, L., and Bunge, R. P. (1980b). *J. Cell Biol.* **84,** 753–766.

Santala, R., Gottlieb, D. I., Littman, D., and Glaser, L. (1977). *J. Biol. Chem.* **252,** 7625–7634.

Whittenberger, B., and Glaser, L. (1977). *Proc. Natl. Acad. Sci. USA* **74,** 2251–2255.

Whittenberger, B., Raben, D., Lieberman, M. A., and Glaser, L. (1978). *Proc. Natl. Acad. Sci. USA* **75,** 5457–5461.

Whittenberger, B., Raben, D., and Glaser, L. (1979). *J. Supramol. Struct.* **10,** 307–327.

Wood, P. M. (1976). *Brain Res.* **115,** 361–375.

Wood, P. M., and Bunge, R. P. (1975). *Nature (London)* **256,** 662–664.

157

Function and Structure of Basal Lamina at the Neuromuscular Junction

Joshua R. Sanes

1. INTRODUCTION

The neuromuscular junctions (NMJs) formed by motor axons on skeletal muscle fibers are the best studied of all synapses. Their accessibility and relative simplicity have long recommended them to neurobiologists, and advances in electron microscopy, intracellular recording, tissue culture, and biochemical analysis have made it possible to acquire, during the past three decades, detailed knowledge of their structure and function (Kuffler and Nicholls, 1976; Zacks, 1973). More recently, many investigators have used this foundation to study, *in vitro* and *in vivo*, factors involved in the development and maintenance of the NMJ, in an attempt to understand how nerve and muscle interact to form a highly organized and specialized synapse at their point of apposition.

In vertebrates, each muscle fiber is ensheathed by a continuous layer of basal lamina (BL; sometimes called basement membrane; Fig. 1a, left). This sheath of BL extends between nerve and muscle at the NMJ, and thus comprises a large fraction of the synaptic cleft material of this synapse. The main purpose of this article is to review evidence that the synaptic portion of the BL interacts with both nerve and muscle membranes to play roles in the function, formation, and maintenance of the NMJ. This summary is preceded by a short description of the structure of the NMJ and is followed by a review of current knowledge about the composition of muscle fiber BL.

2. THE NEUROMUSCULAR JUNCTION

Portions of three cells—motor neuron, muscle fiber, and Schwann cell—make up the NMJ (Fig. 1b). The motor axon, whose cell body lies in the spinal cord, loses its myelin

Joshua R. Sanes • Department of Physiology and Biophysics, Washington University Medical Center, St. Louis, Missouri 63110.

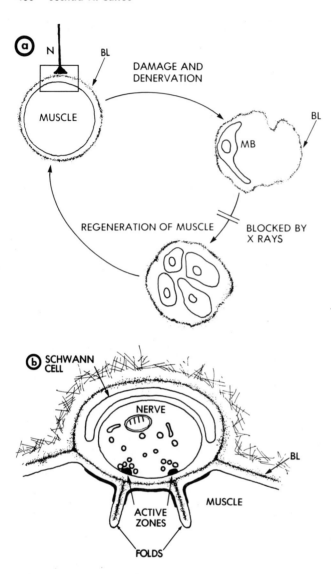

Figure 1. (a) Nerve (N) and muscle degenerate and are phagocytized after they are damaged, but their sheaths of basal lamina (BL) survive. Myoblasts (MB) within sheaths divide and differentiate to form new muscle fibers, but this process can be blocked by X-irradiation. Boxed area at upper left is enlarged in (b). (b) Cross section of a neuromuscular junction; sketch from an electron micrograph. The nerve terminal, capped by a Schwann cell process, is separated from the muscle fiber surface by a 50-nm-wide synaptic cleft. Synaptic vesicles in the terminal cluster at active zones that lie opposite mouths of functional folds in the muscle fiber surface. Acetylcholine receptors are concentrated in the thickened membrane at the crests of the folds. BL ensheaths muscle fiber and Schwann cell, passes through the synaptic cleft, and extends into junctional folds. A layer rich in collagen fibrils lies outside the BL, but does not extend through the synaptic cleft.

sheath as it approaches the muscle fiber and terminates in an array of short branches on the muscle fiber surface. The nerve terminals are capped by processes of a glial cell, the Schwann cell, and are separated from the muscle fiber by a 50-nm-wide synaptic cleft; the muscle fiber's BL sheath extends through this cleft. The Schwann cell also has a BL, and the BLs of Schwann and muscle cells fuse at the edges of the nerve terminal. In most vertebrate skeletal muscles, each muscle fiber is innervated by a single axon and has a single NMJ.

Both nerve and muscle are specialized in their region of contact. Nerve terminals, for example, bear numerous 50-nm-diameter membrane-bound vesicles, which contain the neurotransmitter, acetylcholine. Some of the vesicles are focused near thickened patches on the presynaptic membrane; these are "active zones" at which vesicles fuse with the plasma membrane to release acetylcholine into the synaptic cleft (Heuser *et al.*, 1979). Preterminal portions of the motor axon contain relatively few synaptic vesicles and no active zones, but are enriched in cytoskeletal elements such as microtubules and neurofilaments.

Muscle fibers are also specialized at the synaptic site. The subsynaptic membrane of the muscle fiber is invaginated at intervals to form 1- to 2-μm-deep junctional folds, whose mouths lie directly opposite the active zones in the nerve terminal; BL extends into the folds. Acetylcholine receptors, with which released neurotransmitter combines to activate the muscle, are concentrated in the membrane at the crests (but not the depths) of the folds: there are about 20,000 receptors/μm^2 in synaptic membrane, and less than 20/μm^2 extrasynaptically (Fambrough, 1979). The enzyme acetylcholinesterase (AChE), which inactivates acetylcholine to terminate transmitter action, is also concentrated at the NMJ. The concentration difference between synaptic and extrasynaptic areas is about 1000-fold (Hall, 1973); thus, when muscles are treated with a histochemical stain for AChE, reaction product fills the synaptic cleft but is virtually absent extrasynaptically.

3. FUNCTIONS OF BASAL LAMINA AT THE NEUROMUSCULAR JUNCTION

Our main approach to investigating the role of synaptic BL has been to remove the cellular components of the NMJ *in vivo* under conditions that preserve the BL. Characteristics of the surviving BL or the behavior of cells that repopulate it can then be studied. When skeletal muscle is damaged, the *muscle cells* degenerate and are phagocytized, but their sheaths of BL persist (Mauro, 1979; Vracko, 1974; Fig. 1a). Cells within the BL, called satellite cells, normally survive, divide, fuse, and differentiate to form new myotubes inside the old BL tubes, but this regeneration can be blocked by X-irradiation (Sanes *et al.*, 1978). *Nerve terminals* can be removed by cutting or crushing the parent axons. The distal portions of the axons, thus separated from their cell bodies, degenerate and are phagocytized. Axons regenerate from the site of damage to reinnervate muscle and form new synapses, but reinnervation can be prevented by removing a long segment of the nerve. *Schwann cells* persist at former synaptic sites after denervation, but they can be removed by keeping the muscle denervated: after several weeks, the Schwann cells migrate away (Zacks, 1973). Thus, by appropriate combinations of muscle damage, denervation, and X-irradiation, one can study BL in the absence of nerve, muscle, and Schwann cell. Furthermore, one can identify synaptic portions of the BL even in the absence of cellular elements, either (1) by using a histochemical stain for AChE (see below), or (2) by finding the struts of BL that extend into the lumen of the sheath, where they once lined junctional folds (Fig. 1b). When nerve or muscle regenerate, one can ask whether they "recognize" former synaptic sites. Results of such studies, along with other experiments, show that synaptic BL serves a number of functions that had previously been ascribed to the presynaptic and postsynaptic plasma membranes.

1. *Inactivation of released transmitter.* Because AChE is a particulate enzyme, and because it persists in denervated muscle, it had generally been thought that it was attached to the plasma membrane of the muscle fiber at the NMJ. In 1971, however, Hall and Kelly showed that treatment with collagenase, which removes the BL (Betz and Sakmann, 1973), releases AChE from the NMJ without destroying either the enzyme itself or the postsynaptic membrane. Moreover, AChE purified from electric organs of *Torpedo* or *Electrophorus* was found to consist of globular catalytic subunits attached to a fibrillar collagenlike tail (Anglister *et al.*, 1979; Lwebuga-Mukasa *et al.*, 1976; Rosenberry and Richardson,

1977; see also Bon *et al.*, 1979). Because BLs are collagenous (see below), these results suggested that AChE might be anchored by its tail to the BL.

Observation of BL sheaths that survived muscle damage, X-irradiation, and prolonged denervation confirmed this suggestion. Weeks after the nerve terminal, muscle fiber, and Schwann cell had been removed, high levels of histochemically detectable AChE persisted at original synaptic sites in muscle fiber BL (McMahan *et al.*, 1978). Thus, at least some fraction (but possibly not all; see Bon and Massoulie, 1980; Viratelle and Bernhard, 1980) of the AChE at the NMJ is contained in or tightly connected to muscle fiber BL. To date, AChE is the only enzyme known to be associated with a BL in any tissue.

2. *Attachment of nerve to muscle.* Attachment of cells to BL or BL-derived components has been documented in several nonneural systems (Kleinman *et al.*, 1981). Indirect evidence suggests that adhesion of nerve terminals to muscle fibers may also be mediated by components of the BL. When muscles are treated with proteases that digest BL but not plasma membrane, nerve terminals lose their firm attachment to the endplate and can be easily pulled away (Peper and McMahan, 1972; Betz and Sakmann, 1973). Furthermore, when muscle is damaged (and phagocytized) but not denervated, nerve terminals remain at their original synaptic sites on the BL for at least a week after the muscle cell has been removed (Sanes *et al.*, 1978). Thus, components associated with synaptic BL may be involved in maintaining the structural integrity of the NMJ.

3. *Topographic specificity during reinnervation of denervated muscle.* Axons reinnervating denervated skeletal muscle show a remarkable preference for original synaptic sites. Although synapses form at completely new sites under some circumstances, preferential reinnervation of patches of original postsynaptic membrane has been documented in mammalian, avian, and amphibian muscles reinnervated by their own or by foreign nerves (Bennett and Pettigrew, 1976). In frog muscle, over 95% of the axons that form close contacts with muscle fibers do so at original synaptic sites (Letinsky *et al.*, 1976; Sanes *et al.*, 1978).

The factors that account for the precise reinnervation of original sites are not known, but components of the BL have emerged as leading candidates. Tubes of perineurial connective tissue survive axotomy and axons often regenerate through them to reach original sites (Zacks, 1973). However, axons growing beyond or outside of perineurial tubes also prefer to reinnervate original sites (Bennett *et al.*, 1973; Letinsky *et al.*, 1976); thus, there must be factors at the site itself that account for its precise reinnervation. The Schwann cell is unlikely to be or to contain the critical factor, for reinnervation is topographically precise even when denervation is prolonged to ensure that Schwann cells are absent from original sites (Letinsky *et al.*, 1976). Experiments on denervated, damaged, X-irradiated muscle have now shown that the presence of the myofiber itself is not required for precise reinnervation of original synaptic sites (Marshall *et al.*, 1977; Sanes *et al.*, 1978). Axons regenerated to contact BL sheaths of muscle fibers after the muscle cells had degenerated and been removed. Over 95% of the axonal processes that contacted the BL did so at original synaptic sites; thus, reinnervation was as topographically precise in the absence of myofibers as in their presence. This result raises the possibility that the BL plays a role in influencing the growth of regenerating axons. Axons might, for example, adhere preferentially to molecules in synaptic BL or be repelled by molecules in extrasynaptic BL (Sanes *et al.*, 1978).

4. *Differentiation of regenerating axons into nerve terminals.* The precision of reinnervation of skeletal muscle is manifested not only in the growth of regenerating axons but also in their differentiation. Only portions of the axons within about 0.1 μm of the muscle fiber surface acquire concentrations of synaptic vesicles, and new active zones form di-

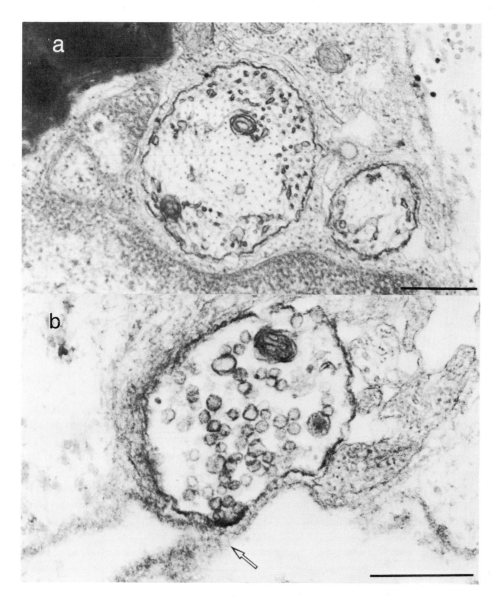

Figure 2. Preterminal and terminal portions of axons reinnervating BL sheaths in damaged, denervated, irradiated muscle. (a) Preterminal portions of axons in the nerve trunk are rich in neurofilaments and microtubules but poor in synaptic vesicles. (b) Nerve terminal, opposed to muscle fiber BL, contains numerous vesicles, some of which are focused on an active zone that lies opposite an intersection (arrow) of synaptic cleft and junctional fold BL. Bar = 0.5 μm. [From Sanes *et al.*, 1978.]

rectly above the mouths of surviving original junctional folds. Thus, factors near the muscle fiber surface must guide the transformation of a regenerating axon into a nerve terminal.

Examination of "reinnervated" BL sheaths showed that at least some of these factors are associated with synaptic BL (Sanes *et al.*, 1978). Axons that contacted BL sheaths from which muscle fibers had been removed acquired active zones and concentrations of synaptic vesicles (Fig. 2b); preterminal portions of the axons had few vesicles and no active zones (Fig. 2a). Furthermore, active zones in the terminals were preferentially associated with points where struts of BL that once lined the folds intersected the BL of the synaptic cleft—i.e., where the mouths of junctional folds had been (compare Figs. 1b and 2b). From this correspondence, it is clear that molecules in synaptic BL trigger or organize the differentiation of regenerating axons into nerve terminals.

5. *Differentiation of the postsynaptic membrane.* The stability of the postsynaptic membrane is not dependent on the continued presence of the nerve terminal. For example, although new acetylcholine receptors are synthesized and inserted into extrasynaptic membrane when muscles are denervated, the acetylcholine receptors of the postsynaptic membrane remain laterally immobile and highly concentrated compared to extrasynaptic receptors (Fambrough, 1979). It is reasonable to suppose that some cytoskeletal or exoskeletal elements induce or maintain these receptor-rich patches, and recently synaptic BL has been implicated in this process (Burden *et al.*, 1979; Bader, 1981; Sanes, unpublished). Muscles were damaged and denervated and myofibers allowed to regenerate in the absence of nerve or Schwann cells. Receptors clustered in patches of new myofiber membrane that lay beneath synaptic BL; comparably receptor-rich patches were not found elsewhere. Thus, components of synaptic BL can organize the differentiation of postsynaptic as well as presynaptic membrane.

4. COMPOSITION OF SYNAPTIC AND EXTRASYNAPTIC MUSCLE FIBER BL

The demonstration that the BL of the synaptic cleft plays important roles at the NMJ motivates biochemical study of its components and comparison of synaptic and extrasynaptic portions of the BL sheath. Unfortunately, several obstacles stand in the way of realizing these goals: (1) Biochemical analysis of all BLs has been hampered by the insolubility of these structures in neutral buffers, cold weak acids or bases, or detergents. (2) Muscle fiber BL has not yet been isolated free of other extracellular elements such as collagen fibrils, blood vessel and nerve BL, and the connective tissue sheaths that encircle fascicles of muscle fibers (perimysium) and whole muscles (epimysium). (3) Synaptic sites occupy only about 0.1% of the muscle fiber surface, and no methods for purifying synaptic regions are yet available. To circumvent the first obstacle, easily isolated BLs—notably lens capsule and glomerular BL—have been subjected to digestion with proteases (usually pepsin or collagenase) or reduction and alkylation in the presence of a denaturant, and material solubilized in these ways has been characterized (Kefalides *et al.*, 1979; Spiro, 1978). More recently, tumor cell lines have been found that secrete but do not polymerize BL-like material; analysis of these secretions has provided new insights into the composition of BL (Chung *et al.*, 1979; Hassell *et al.*, 1980; Rohde *et al.*, 1979; Timpl *et al.*, 1978). The second and third problems have not yet been solved, and so knowledge of muscle fiber BL derives mostly from immunohistochemical studies that use antibodies raised against components of nonmuscle BLs.

The nonmuscle BLs that have been characterized are rich in glycoproteins but contain no lipid, DNA, RNA, or phosphate. Although intact BLs are not fibrillar, they are largely collagenous as judged by amino acid composition, sensitivity to collagenase, physical characteristics, and immunochemical properties. Two genetically distinct collagens, Types IV and V, are associated with BL, and it is likely that there are other collagens in BL as well (Bornstein and Sage, 1980). BLs also contain noncollagenous glycoproteins, as shown by fractionation of reduced, alkylated extracts or collagenase digests. A noncollagenous protein, laminin (Rohde *et al.*, 1979; also called GP-2 by Chung *et al.*, 1979), has been purified from material secreted by cultured cells and characterized; immunohistochemical methods show that laminin is a component of authentic BLs. Fibronectin, a noncollagenous glycoprotein first purified from non-BL sources, has also been found associated with BLs. Most recently, the glycosaminoglycan heparan sulfate has been shown to be a component of BL (Kanwar and Farquhar, 1979); it is probably part of a BL-specific proteoglycan (Hassell *et al.*, 1980). A major question that has not yet been addressed is how the various components of BLs are linked to each other and rendered insoluble; neither proteolytic digests nor tumor cell secretions are well suited for studies of the supramolecular organization of BL.

Although muscle fiber BL has yet to be characterized biochemically, a combination of light and electron microscopic immunohistochemical methods has been used to probe its structure. These methods are of sufficiently high resolution to determine whether particular antigens are present in synaptic or extrasynaptic areas. We have found that some antigens are confined to one or the other of these domains, while other antigens are present in both. Collagen IV, fibronectin, and laminin, all previously known to be present in muscle (Timpl *et al.*, 1978; Chen, 1977; Rohde *et al.*, 1979), are present in both synaptic and extrasynaptic portions of muscle fiber BL (Sanes, 1982). Collagen V (also previously shown to be present in muscle; Duance *et al.*, 1977) and other still-uncharacterized proteins isolated from muscle connective tissue are associated with extrasynaptic but not synaptic regions of the muscle fiber surface (Sanes, 1982). The "extrasynaptic" proteins seem to be concentrated in a layer just external to the BL; thus, although closely associated with BL, they may not be bona fide components of BL. Perhaps most interesting, in light of the functional studies described above, has been the finding that four antisera prepared against BL-related components contain antibodies that stain synaptic but not extrasynaptic BL (Sanes and Hall, 1979). These include anti-AChE, anti-lens capsule BL, and a minor antibody in antisera raised to muscle basement membrane collagen. Adsorption experiments and differential susceptibility of the staining to proteases show that the sera recognize at least three and probably four different determinants in synaptic BL. Thus, these antibodies reveal major (i.e., intensely staining) chemical differences between synaptic and extrasynaptic BL. With the exception of AChE, the functions of these antigens are unknown, but the availability of antibodies that define and recognize them should facilitate further studies of the structure and function of the BL at the NMJ.

ACKNOWLEDGMENTS. I thank Dale Purves and Jeanette Cheney for advice and assistance.

REFERENCES

Anglister, L., Tarrab-Hazdai, R., Fuchs, S., and Silman, I. (1979). *Eur. J. Biochem.* **94**, 25–29.

Bader, D. (1981). *J. Cell Biol.* **88**, 338–345.

Bennett, M. R., and Pettigrew, A. G. (1976). *Cold Spring Harbor Symp. Quant. Biol.* **40**, 409–424.

Bennett, M. R., McLachlan, E. M., and Taylor, R. S. (1973). *J. Physiol.* **233**, 481–500.

Betz, W., and Sakmann, B. (1973). *J. Physiol.* **230**, 673–688.

Bon, S., and Massoulie, J. (1980). *Proc. Natl. Acad. Sci. USA* **77**, 4464–4468.

Bon, S., Vigny, M., and Massoulie, J. (1979). *Proc. Natl. Acad. Sci. USA* **76**, 2546–2550.

Bornstein, P., and Sage, H. (1980). *Annu. Rev. Biochem.* **49**, 957–1003.

Burden, S. J., Sargent, P. B., and McMahan, U. J. (1979). *J. Cell Biol.* **82**, 412–425.

Chen, L. B. (1977). *Cell* **10**, 393–400.

Chung, A. E., Jaffe, R., Freeman, I. L., Vergnes, J.-P., Braginski, J. E., and Carlin, B. (1979). *Cell* **16**, 277–287.

Duance, V. C., Restall, D. J., Beard, H., Bourne, F. J., and Bailey, A. J. (1977). *FEBS Lett.* **79,** 248–252.

Fambrough, D. M. (1979). *Physiol. Rev.* **59,** 165–227.

Hall, Z. W. (1973). *J. Neurobiol.* **4,** 343–361.

Hall, Z. W., and Kelley, R. B. (1971). *Nature New Biol.* **232,** 62.

Hassell, J. R., Robey, P. G., Barrach, H.-J., Wilczek, J., Rennard, S. I., and Martin, G. R. (1980). *Proc. Natl. Acad. Sci. USA* **77,** 4494–4498.

Heuser, J. E., Reese, T. S., Dennis, M. J., Jan, Y., Jan, L., and Evans, L. (1979). *J. Cell Biol.* **81,** 275–300.

Kanwar, Y. S., and Farquhar, M. G. (1979). *Proc. Natl. Acad. Sci. USA* **76,** 1303–1307.

Kefalides, N., Alper, R., and Clark, C.C. (1979). *Int. Rev. Cytol.* **61,** 167–228.

Kleinman, H. K., Klebe, R. J., and Martin, G. R. (1981). *J. Cell. Biol.* **88,** 473–485.

Kuffler, S. W., and Nicholls, J. G. (1976). *From Neuron to Brain,* Sinauer, Sunderland, Mass.

Letinsky, M. K., Fischbeck, K. H., and McMahan, U. J. (1976). *J. Neurocytol.* **5,** 691–718.

Lwebuga-Mukasa, J. S., Lappi, S., and Taylor, P. (1976). *Biochemistry* **15,** 1425–1434.

McMahan, U. J., Sanes, J. R., and Marshall, L. M. (1978). *Nature (London)* **271,** 172–174.

Marshall, L. M., Sanes, J. R., and McMahan, U. J. (1977). *Proc. Natl. Acad. Sci. USA* **74,** 3073–3077.

Mauro, A. (ed.) (1979). *Muscle Regeneration*, Raven Press, New York.

Peper, K., and McMahan, U. J. (1972). *Proc. R. Soc. London Ser. B* **181,** 431–440.

Rohde, H., Wick, G., and Timpl, R. (1979). *Eur. J. Biochem.* **102,** 195–201.

Rosenberry, T. L., and Richardson, J. M. (1977). *Biochemistry* **16,** 3550–3558.

Sanes, J. R. (1982). *J. Cell Biol.*, in press.

Sanes, J. R., and Hall, Z. W. (1979). *J. Cell Biol.* **83,** 357–370.

Sanes, J. R., Marshall, L. M., and McMahan, U. J. (1978). *J. Cell Biol.* **78,** 176–198.

Spiro, R. G., (1978). *Ann. N.Y. Acad. Sci.* **312,** 106–121.

Timpl, R., Martin, G. R., Bruckner, P., Wick, G., and Wiedemann, H. (1978). *Eur. J. Biochem.* **84,** 43–52.

Viratelle, O. M., and Bernhard, S. A. (1980). *Biochemistry* **19,** 4999–5007.

Vracko, R. (1974). *Am. J. Pathol.* **77,** 314–338.

Zacks, S. I. (1973). *The Motor Endplate*, Krieger, Huntington, N.Y.

158

Lectin Receptors on Lymphocytes

Their Use for Cell Identification and Separation

Nathan Sharon

1. INTRODUCTION

Receptors on the surface of lymphocytes are attracting a great deal of attention, mainly for two reasons. First, they play a key role in the complex cell–cell and cell–molecule interactions that are essential for the operation of the immune system. Included in these interactions are the specific recognition by lymphocytes of soluble and cellular antigens, lymphocyte activation and differentiation, and modulation of these and other activities by immune cells or their soluble products. Second, surface receptors are extremely useful markers for the identification, enumeration, and separation of the many cell subpopulations such as B and T cells, or the various subsets of the latter (e.g., helper, suppressor, or killer T cells) that populate the immune system. These markers are of two types: cell surface antigens that combine with specific antibodies, and cell surface saccharides that bind lectins.

The discovery, in the late 1960s, that mouse T cells have a specific surface marker, the Thy-l (or θ) antigen, was a major development: it made it possible for the first time to identify these cells directly, to determine their distribution in the lymphoid organs of the mouse and their development in and migration from the embryonic thymus gland (Raff, 1976). Studies of T-cell functions were also greatly facilitated.

Many other antigenic surface markers have since been identified. Although their functions are unknown, their presence on unique cell subsets or at restricted times in development suggests their association with specific roles. Best characterized are the markers on murine lymphocytes (Benacerraf and Unanue, 1979). In addition to Thy-1, they include the histocompatibility antigens H-2 (Strominger 1980), the TL and Lyt 123 markers for T-lymphocyte subsets (Boyse and Old, 1978), the T-200 or T-170 glycoprotein characteristic

---

Nathan Sharon • Department of Biophysics, The Weizmann Institute of Science, Rehovoth, Israel.

of T cells (Omary *et al.*, 1980; Hoessli and Vassalli, 1980), the recently discovered marker for natural killer (NK) T cells, the ganglioside asialo GM$_1$ (Kasai *et al.*, 1980a; Young *et al.*, 1980), and immunoglobulins (IgG, IgM) present only on B cells. Antigenic markers have also been found on human lymphocyte subpopulations both by conventional techniques (Moretta *et al.*, 1979) and by the use of monoclonal antibodies (Reinherz and Schlossman, 1980).

Knowledge of lymphocyte surface antigens has played a major role in the explosive advance of cellular immunology in the past decade. It is apparent, however, that for further progress in this area many more surface markers for lymphocytes must be identified. In particular, surface markers are required that can be used for the investigation of pathways of lymphocyte differentiation, both in normal and in pathological conditions, and for the separation of different cell subpopulations. Such separations should preferably be simple and afford cells in good recovery, high purity, and without alterations of their biological functions.

2. RECEPTORS FOR PEANUT AGGLUTININ AND SOYBEAN AGGLUTININ

Work in our laboratory during the last 5 years has been concerned with lectin receptors on murine and human lymphocyte subpopulations. Lectins (Goldstein and Hayes, 1978; Lis and Sharon, 1977, 1981; Pereira and Kabat, 1979) are excellent surface probes in that they attach tightly to specific carbohydrate moieties on cell surfaces and can be readily removed by suitable monosaccharides. In this respect they offer an important advantage over antibodies to cell surface constitutents, as it is not always possible to remove the latter from cells to which they are bound.

In 1976 we discovered that subpopulations of T cells carry distinct lectin receptors: we have found that cortical (immature) thymocytes bind peanut agglutinin (PNA) whereas medullary (mature) cells do not bind the lectin (Reisner *et al.*, 1976a). A procedure for the separation of the two murine thymocyte subpopulations was worked out. It consists of selective agglutination of the cortical cells by the lectin, followed by separation of the agglutinated (PNA$^+$) cells from the unagglutinated (PNA$^-$) cells by unit gravity sedimentation in fetal calf serum (or solutions of bovine serum albumin). Dissociation of the clumps is effected by D-galactose, a sugar that binds to PNA. The separated cells are obtained in high yield and are fully (>90%) viable. Before the development of this method, it was very difficult to isolate the cortical thymocytes, and the medullary cells could be obtained only from cortisone-treated or irradiated animals.

We have also shown that murine B splenocytes bind much larger amounts of soybean agglutinin (SBA) than do T splenocytes, and that these two cell subpopulations can be separated by selective agglutination with SBA (Reisner *et al.*, 1976b). In all cases when receptors for PNA or SBA were absent, they appeared after treatment of the cells with sialidase (neuraminidase).

Fractionation of lymphocytes by the method described offers many advantages over other techniques: it is inexpensive, simple, rapid, reproducible, efficient, and can be applied to large numbers of cells. Furthermore, it permits recovery of both the unagglutinated and the agglutinated cell populations in high yields (Sharon, 1979, 1980; Reisner and Sharon, 1980).

The difference between the lectin receptors on the lymphocyte subpopulations examined could be rationalized on the basis of the specificity of the lectins used, and the limited information available on the structure of carbohydrates present on lymphocyte surfaces. PNA is specific for D-galactosyl $\beta(1\rightarrow3)$-N-acetyl-D-galactosamine [Gal $\beta(1\rightarrow3)$GalNAc], but may also combine with nonreducing terminal D-galactose residues in other compounds (Lotan *et al.*, 1975; Pereira *et al.*, 1976). The disaccharide Gal $\beta(1\rightarrow3)$GalNAc is present in many soluble glycoproteins and has also been identified in the few membrane glycoproteins examined (e.g., in glycophorin), in membrane glycoplipids, and on lymphocytes (Sharon and Lis, 1982). In general, however, sialic acid residues are attached to the disaccharide, so that its interaction with PNA is precluded. SBA, in addition to interacting with D-galactose, binds more strongly to N-acetyl-D-galactosamine residues (Lis *et al.*, 1970; Pereira *et al.*, 1974), so that the two lectins do not necessarily combine with the same carbohydrate structures in glycoconjugates. It has also been reported that cortisone-resistant thymocytes and spleen T cells are more negatively charged and have a higher content of sialic acid than unfractionated thymocytes (Despont *et al.*, 1975).

We have therefore made the following tentative assumptions and conclusions (Sharon and Reisner, 1979; Sharon, 1980):

1. Cell surface receptors for PNA and SBA change in an orderly manner during murine lymphocyte differentiation and maturation, as evidenced by the fact that immature thymocytes are PNA⁺SBA⁺, mature thymocytes are PNA⁻SBA⁺, and T splenocytes are PNA⁻SBA⁻.
2. The change most probably involves the attachment of sialic acid residues to D-galactose and N-acetyl-D-galactosamine residues on the cell surface.
3. The PNA receptor is a marker for undifferentiated, immature cells.
4. Hemopoietic stem cells that are devoid of graft versus host (GvH) activity may be PNA⁺SBA⁺; such cells may be isolated with the aid of lectins.
5. Results of experiments with mice may be applicable to man.

Further work in our laboratory and elsewhere has provided strong support for the above assumptions and conclusions, and has established PNA and SBA as important and widely used tools for the investigation of murine lymphocytes.

Thus, treatment of cryostat sections of murine thymus with a horseradish peroxidase derivative of PNA revealed stained cells in the cortex and not in the medulla (Rose *et al.*, 1980a). Following treatment of mice with cortisone or radiation, primarily the PNA⁺ cells disappeared from the thymus, and these cells reappeared upon thymic regeneration (London *et al.*, 1978; Dumont and Nardelli, 1979). PNA⁻ and PNA⁺ cells were shown to differ in their Lyt 123 and Lyt 6 and TL surface markers (Zeicher *et al.*, 1979; Betel *et al.*, 1980; London, 1980; London and Horton, 1980), and rate of synthesis of H-2 (Rothenberg, 1980). Marked differences between enzyme levels of the two subpopulations were reported (Bauminger, 1978; Dornand *et al.*, 1980; Weinstein and Berkovich, 1981), among them being terminal deoxynucleotidyltransferase (TdT), an enzyme characteristic of immature lymphocytes, which was indeed present predominantly in the PNA⁺ cells (Rothenberg, 1980).

Correlating well with the various functional differences found by us (Reisner *et al.*, 1976a) are the findings on differences in suppressor and helper activity between PNA⁻ and PNA⁺ thymocytes (Umiel *et al.*, 1978; Eisenthal *et al.*, 1979). Similar differences in suppressor and helper activity were also found in murine spleen (Imai *et al.*, 1979; Nakano *et al.*, 1980a,b) and embryonic liver (Rabinovich *et al.*, 1979). Very significantly, when PNA⁻ thymocytes were incubated with concanavalin A and a supernatant from concanavalin A-treated lymphoid cultures (Irlé *et al.*, 1978), with thymic epithelial supernatant (Kruisbeek and Astaldi, 1979; Kruisbeek *et al.*, 1980), thymic humoral factor, a hormone known to potentiate lymphocyte differentiation (Trainin *et al.*, 1980; Umiel *et al.*, 1979), or a synthetic pentapeptide, thymopoietin 32–36 (Nash *et al.*, 1981), they acquired properties of mature cells, including the loss of the PNA receptor. These findings strongly support the model of a differentiation pathway in which the PNA$^+$ thymocytes are the precursors of the PNA$^-$ thymocytes. Examination of mice inoculated with the radiation-induced virus variant D-RadLV revealed that preleukemic lymphocytes were PNA$^+$ whereas leukemic end stage cells were PNA$^-$ (Reisner *et al.*, 1980a).

PNA has also been used to study the ontogeny of the lymphoid system (London *et al.*, 1979; Roelants *et al.*, 1979), the generation of cytotoxic lymphocytes in the thymus (Cooley and Schmitt-Verhulst, 1979; Kruisbeek *et al.*, 1980; Wagner *et al.*, 1980a,b; Umiel *et al.*, 1981), and for the detection of immature T or B cells in different murine organs (London and Horton, 1980; Rose *et al.*, 1980b; Newman and Boss, 1980; Raedler *et al.*, 1981). Last but not least, we have shown that as predicted, fractionation of murine splenocytes by SBA and PNA affords a PNA$^+$SBA$^+$ fraction enriched in pluripotential stem cells and devoid of GvH activity, and that these cells can be used for successful bone marrow transplantation across histocompatibility barriers (Reisner *et al.*, 1978).

The use of SBA has, as yet, been more limited, but mention should be made of its application to the separation of splenocytes into B and T cells for the examination of changes that occur in membranes of these cells during mitogenic stimulation (Rosenfelder *et al.*, 1979; Bödeker *et al.*, 1980) and for the isolation of a spleen cell fraction enriched in stem cells (Reisner *et al.*, 1978, 1980b).

PNA and SBA have also been used for the investigation of human lymphocytes. With the aid of PNA we have identified and isolated for the first time an immunodeficient lymphocyte subpopulation from human thymus (Reisner *et al.*, 1979) and cord blood (Lis *et al.*, 1979; Rosenberg and Sharon, 1981). As in the case of murine thymus, the PNA$^+$ cells in human thymus reside in the cortex: when frozen sections of human thymuses were treated with rhodamine-conjugated PNA, fluorescent cells were present in the cortex but not in the medulla (Christensson *et al.*, 1981). Examination of human peripheral blood lymphocytes has shown that they are all PNA$^-$; there was, however, a pronounced increase in the number of PNA$^+$ cells in peripheral blood lymphocytes of a large proportion of patients with acute lymphatic leukemias, whereas only small numbers of such cells were found in remission (Reisner *et al.*, 1979; Levin *et al.*, 1980) or in chronic lymphatic leukemia (Reisner *et al.*, 1979). Separation of antibody helper and antibody suppressor T cells in human peripheral blood by SBA has been achieved (Reisner *et al.*, 1980c). A direct outgrowth of the above findings is the promising work now in progress at the Sloan Kettering Cancer Center, on isolation from human bone marrow of stem cells devoid of GvH reactivity which are suitable for bone marrow transplantations from unrelated donors (Reisner *et al.*, 1980d).

Only little work has been done on lectin receptors on lymphocytes of animals other than mouse or man. Binding of PNA to 45% of bovine peripheral blood leukocytes was reported (Pearson *et al.*, 1979), and it was suggested that PNA$^+$ lymphocytes are T cells. The PNA receptor appears also to be an efficient marker for presumptive T cells in young and adult sheep (Fahey, 1980). In the chicken, B-cell precursors in both the bursa of Fabricus and thymus were PNA$^+$ (Schauenstein *et al.*, 1981). Considerably higher percentages of PNA$^+$ cells (50–70%) compared with those reported in mammals were found in peripheral lymphoid organs of chicken. Moreover, PNA$^-$ and PNA$^+$ chick splenocytes contained suppressor cells of different target specificity.

3. RECEPTORS FOR OTHER LECTINS

Many other lectins have been examined for differential binding to or agglutination of lymphocyte subpopulations, and in attempts to identify new subpopulations (reviewed by Sharon, 1979, 1980; Sharon and Reisner, 1979; for a recent study see Fowlkes *et al.*, 1980); satisfactory results have been obtained only in a limited number of cases. The lectin from the snail *Helix pomatia*, specific for α-*N*-acetyl-D-galactosamine, has been shown to be a useful reagent for the identification and isolation of T cells in several species, including mouse (Hammarström *et al.*, 1978). Treatment of the cells with sialidase is, however, required to unmask the receptors for this lectin. A single membrane glycoprotein, with an apparent molecular weight of 130,000, expressed on both normal and malignant mouse T lymphocytes but not on B cells, appears to be responsible for the bulk of the binding of the *Helix pomatia* lectin to sialidase-treated lymphocytes (Axelsson *et al.*, 1978).

Wheat germ agglutinin (WGA, a lectin specific for *N*-acetylglucosamine and *N*-acetylneuraminic acid) reacts selectively with murine B splenocytes, and these cells were separated from the T splenocytes by selective agglutination with this lectin (Bourguignon *et al.*, 1979). Evidence was obtained for specific interaction of a glycoprotein (designated as T 145) characteristic of murine cytotoxic T lymphocytes, with a lectin specific for *N*-acetyl-D-galactosamine, which was isolated from the seeds of *Vicia villosa* (Kimura *et al.*, 1979; Conzelmann *et al.*, 1980). Very recently it was reported that the receptor for the lectin from *Dolichos biflorus* (specific for α-linked *N*-acetylgalactosamine) is a differentiation marker selectively expressed on fetal mouse thymocytes (Kasai *et al.*, 1980b) as well as on a spontaneous murine leukemia (Muramatsu *et al.*, 1980).

4. CONCLUDING REMARKS

In this article I have reviewed data showing that cell surface carbohydrates, detectable by lectins, are characteristic markers of lymphocyte subpopulations, and that such subpopulations can in certain cases be readily separated by lectins. The structures of the molecules carrying the carbohydrates that serve as lectin receptors are not known. In attempting to isolate the receptors, it should be borne in mind that membranes may contain many lectin-reactive glycoconjugates, only some of which are expressed on the cell surface. Methods of isolation that do not require prior membrane disruption should therefore be used.

Structural characterization of the receptors will undoubtedly provide a basis for the understanding of the changes that occur on the surface of lymphocytes during differentiation and maturation of these cells, and in the course of malignant transformation. The results of such studies may also contribute greatly to clarify the role of cell surface sugars in the complex mechanisms functioning in the immune system.

REFERENCES

Axelsson, B., Kimura, A., Hammarström, S., Wigzell, H., Nilsson, K., and Mellstedt, H. (1978). *Eur. J. Immunol.* **8,** 757–764.

Bauminger, S. (1978). *Prostaglandins* **16,** 351–355.

Benacerraf, B., and Unanue, E. R. (1979). *Textbook of Immunology*, pp. 76–98, Williams & Wilkins, Baltimore.

Betel, I., Mathieson, B. J., Sharrow, S. O., and Asofsky, R. (1980). *J. Immunol.* **124,** 2209–2217.

Bödeker, B. G. D., Van Eijk, R. V. W., and Muhlradt, P. F. (1980). *Eur. J. Immunol.* **10,** 702–707.

Bourguignon, L. Y. W., Rader, R. L., and McMahon, J. T. (1979). *J. Cell Physiol.* **99**, 95–99.
Boyse, E. A., and Old, L. J. (1978). *Harvey Lect.* **17**, 23–53.
Christensson, B., Biberfeld, P., Matell, G., Smith, C. I. E., and Hammarstrom, L. (1981). *Ann. N.Y. Acad. Sci.*, in press.
Conzelmann, A., Pink, R., Acuto, O., Mach, J.-P., Dolivo, S., and Nabholz, M. (1980). *Eur. J. Immunol.* **10**, 860–868.
Cooley, M. A., and Schmitt-Verhulst, A. M. (1979). *J. Immunol.* **123**, 2328–2336.
Despont, J. P., Abel, C. A., and Grey, H. M. (1975). *Cell. Immunol.* **17**, 487–494.
Dornand, J., Bonnafous, J.-C., and Mani, J.-C. (1980). *FEBS Lett.* **118**, 225–228.
Dumont, F., and Nardelli, J. (1979). *Immunology* **37**, 217–224.
Eisenthal, A., Nachtigal, D., and Feldman, M. (1979). *Transplant. Proc.* **11**, 904–906.
Fahey, K. J. (1980). *Aust. J. Exp. Biol.* **58**, 557–569.
Fowlkes, B. J., Waxdal, M. J., Sharrow, S. O., Thomas, C. A., III, Asofsky, R., and Mathieson, B. J. (1980). *J. Immunol.* **125**, 623–630.
Goldstein, I. J., and Hayes, C. E. (1978). *Adv. Carbohydr. Chem. Biochem.* **35**, 127–340.
Hammarström, S., Hellstrom, U., Dillner, M.-L., Perlmann, P., Perlmann, H., Axelsson, B., and Robertsson, E.-S. (1978). In *Affinity Chromatography* (O. Hoffmann-Ostenhof, M. Breitenback, F. Koller, D. Kraft, and O. Scheiner, eds.), pp. 273–286, Pergamon Press, Elmsford, N.Y.
Hoessli, D. C., and Vassalli, P. (1980). *J. Immunol.* **125**, 1758–1763.
Imai, Y., Oguchi, T., Nakano, T., and Osawa, T. (1979). *Immunol. Commun.* **8**, 495–503.
Irlé, C., Piquet, P.-F., and Vassalli, P. (1978). *J. Exp. Med.* **148**, 32–45.
Kasai, M., Iwamori, M., Nagai, Y., Okumura, K., and Tada, T. (1980a). *Eur. J. Immunol.* **10**, 175–180.
Kasai, M., Ochiai, Y., Habu, S., Muramatsu, T., Tokunaga, T., and Okumura, K. (1980b). *Immun. Lett.* **2**, 157–158.
Kimura, A., Wigzell, H., Holmquist, G., Ersson, B. O., and Carlsson, P. (1979). *J. Exp. Med.* **149**, 473–484.
Kruisbeek, A. M., and Astaldi, G. C. B. (1979). *J. Immunol.* **123**, 984–991.
Kruisbeek, A. M., Zijlstra, J. J., and Krose, T. J. M. (1980). *J. Immunol.* **125**, 995–1002.
Levin, S., Russell, E. C., Blanchard, D., McWilliams, N. B., Maurer, H. M., and Mohanakumar, T. (1980). *Blood* **55**, 37–39.
Lis, H., and Sharon, N. (1977). In *The Antigens* (M. Sela, ed.), Vol. IV, pp. 429–529, Academic Press, New York.
Lis, H., and Sharon, N. (1981). In *The Biochemistry of Plants* (A. Marcus, ed.), Vol. 6, pp. 371–447, Academic Press, New York.
Lis, H., Sela, B., Sachs, L., and Sharon, N. (1970). *Biochim. Biophys. Acta* **211**, 582–585.
Lis, H., Garnett, H., Rotter, V., Reisner, Y., and Sharon, N. (1979). Abstracts, Meeting of Society of Complex Carbohydrates, Toronto, July.
London, J. (1980). *J. Immunol.* **125**, 1702–1707.
London, J., and Horton, M. A. (1980). *J. Immunol.* **124**, 1803–1807.
London, J., Berrih, S., and Bach, J.-F. (1978). *J. Immunol.* **121**, 438–443.
London, J., Berrih, S., and Papiernik, M. (1979). *Dev. Comp. Immunol.* **3**, 343–352.
Lotan, R., Skutelsky, E., Danon, D., and Sharon, N. (1975). *J. Biol. Chem.* **250**, 8518–8523.
Moretta, I., Mingari, M. C., and Moretta, A. (1979). *Immunol. Rev.* **45**, 163–193.
Muramatsu, T., Muramatsu, H., Kasai, M., Habu, S., and Okumura, K. (1980). *Biochem. Biophys. Res. Commun.* **96**, 1547–1553.
Nakano, T., Imai, Y., Naiki, M., and Osawa, T. (1980a). *J. Immunol.* **125**, 1928–1932.
Nakano, T., Oguchi, Y., Imai, Y., and Osawa, T. (1980b). *Immunology* **40**, 217–222.
Nash, L., Good, R. A., Hatzfield, A., Goldstein, G., and Incefy, G. S. (1981). *J. Immunol.* **126**, 150–153.
Newman, R. A., and Boss, M. A. (1980). *Immunol.* **40**, 193–200.
Omary, M. B., Trowbridge, I. S., and Scheid, M. P. (1980). *J. Exp. Med.* **151**, 1311–1316.
Pearson, T. W., Roelants, G. E., Lundin, L. B., and Mayor-Withey, K. S. (1979). *J. Immunol. Methods* **26**, 271–282.
Pereira, M. E. A., and Kabat, E. A. (1979). *CRC Crit. Rev. Immunol.* **1**, 33–78.
Pereira, M. E. A., Kabat, E. A., and Sharon, N. (1974). *Carbohydr. Res.* **37**, 89–102.
Pereira, M. E. A., Kabat, E. A., Lotan, R., and Sharon, N. (1976). *Carbohydr. Res.* **51**, 107–118.
Rabinovich, H., Umiel, T., Reisner, Y., Sharon, N., and Globerson, A. (1979). *Cell. Immunol.* **47**, 347–355.
Raedler, A., Raedler, E., Arndt, R., and Thiele, H.-G. (1981). *Immunol. Lett.* **2**, 335–338.
Raff, M. C. (1976). *Sci. Am.* **234**(5), 30–39.
Reinherz, E. L., and Schlossman, S. F. (1980). *Cell* **19**, 821–827.
Reisner, Y., and Sharon, N. (1980). *Trends Biochem. Sci.* **5**, 29–31.

Reisner, Y., Linker-Israeli, M., and Sharon, N. (1976a). *Cell. Immunol.* **25**, 129–134.

Reisner, Y., Ravid, A., and Sharon, N. (1976b). *Biochem. Biophys. Res. Commun.* **72**, 1585–1591.

Reisner, Y., Itzicovitch, L., Meshorer, A., and Sharon, N. (1978). *Proc. Natl. Acad. Sci. USA* **75**, 2933–2936.

Reisner, Y., Biniaminov, M., Rosenthal, E., Sharon, N., and Ramot, B. (1979). *Proc. Natl. Acad. Sci. USA* **76**, 447–451.

Reisner, Y., Sharon, N., and Haran-Ghera, N. (1980a). *Proc. Natl. Acad. Sci. USA* **77**, 2244–2246.

Reisner, Y., Ikehara, S., Hodes, M. Z., and Good, R. A. (1980b). *Proc. Natl. Acad. Sci. USA* **77**, 1164–1168.

Reisner, Y., Pahwa, S., Chiao, J. W., Sharon, N., Evans, R. L., and Good, R. A. (1980c). *Proc. Natl. Acad. Sci. USA* **77**, 6778–6782.

Reisner, Y., O'Reilly, R. J., Kapoor, N., and Good, R. A. (1980d). *Lancet* **2**, 1320–1324.

Roelants, G. E., London, J., Mayor-Withey, K. S., and Serrano, B. (1979). *Eur. J. Immunol.* **9**, 139–145.

Rose, M. L., Wallis, V., and Davies, A. J. S. (1980a). Abstract 3.5.35, 4th International Congress of Immunology, Paris, July.

Rose, M. L., Birbeck, M. S. C., Wallis, V. J., Forrester, J. A., and Davies, A. J. (1980b). *Nature (London)* **284**, 364–366.

Rosenberg, M., and Sharon, N. (1981). Abstracts, 9th Katzir Conference on Carbohydrate–Protein Interactions, Kiryat Anavim, Israel, March.

Rosenfelder, G., Van Eijk, R. V. W., and Muhlradt, P. F. (1979). *Eur. J. Biochem.* **97**, 229–237.

Rothenberg, E. (1980). *Cell* **20**, 1–9.

Schauenstein, K., Rosenberg, M., Globerson, A., and Sharon, N. (1981). In preparation.

Sharon, N. (1979). In *Affinity Chromatography and Molecular Interactions: Biochemical and Biomedical Applications* (J. Egly, ed.), Colloq. Inserm Vol. 86, pp. 197–206.

Sharon, N. (1980). In *Immunology 80* (M. Fougereau and J. Dausset, eds.), pp. 254–278, Academic Press, New York.

Sharon, N., and Lis, H. (1982). In *The Proteins* (H. Neurath and R. L. Hill, eds.), Vol. V, 3rd ed., pp. 1–144, Academic Press, New York.

Sharon, N., and Reisner, Y. (1979). In *Molecular Mechanisms of Biological Recognition"* (M. Balaban, ed.), pp. 95–106, Elsevier/North-Holland, Amsterdam.

Strominger, J. L. (1980). In *Immunology 80* (M. Fougereau and J. Dausset, eds.), pp. 541–554, Academic Press, New York.

Trainin, N., Umiel, T., and Yakir, Y. (1980). In *Thymus, Thymic Hormones and T-Lymphocytes* (F. Aiuti and H. Wigzell, eds.), pp. 201–211, Academic Press, New York.

Umiel, T., Linker-Israeli, M., Itzchaki, M., Trainin, N., Reisner, Y., and Sharon, N. (1978). *Cell. Immunol.* **37**, 134–141.

Umiel, T., Kleir, I., Sharon, N., and Trainin, N. (1979). Meeting on Immunological Aspects of Experimental and Clinical Cancer, Israel.

Umiel, T., Klein, B., Droge, W., Sharon, N., and Trainin, N. (1981). In preparation.

Weinstein, Y., and Berkovich, Z. (1981). *J. Immunol.* **126**, 998–1002.

Wagner, H., Hardt, C., Bartlett, R., Rollinghoff, M., and Pfizenmaier, K. (1980a). *J. Immunol.* **125**, 2532–2538.

Wagner, H., Rollinghoff, M., Pfizenmaier, K., Hardt, C., and Johnscher, G. (1980b). *J. Immunol.* **124**, 1058–1067.

Young, W. W., Jr., Hakomori, S., Durdik, J. M., and Henney, C. S. (1980). *J. Immunol.* **124**, 199–201.

Zeicher, M., Mozes, E., Reisner, Y., and Lonai, P. (1979). *Immunogenetics* **9**, 119–124.

159

Transfer of Membrane Constituents between Cells

A Novel Approach to Study the Involvement of Lymphocyte Plasma Membrane in the Immune Response

Aya Jakobovits and Nathan Sharon

One of the main issues in immunology deals with the role of lymphocyte plasma membrane components in the control and regulation of immune functions. Both humoral and cell-mediated immune phenomena depend upon the existence of specific membrane components and receptors located at the surface of lymphocytes. Interaction with these receptors is crucial to the ability of the cell to recognize and respond to extracellular stimuli. However, our knowledge of the relationship between the composition and the structure of lymphocyte plasma membranes and the associated cell functions is still very limited. A new and promising approach to this problem involves the transfer of membrane components from one cell type (the donor) to another (the acceptor) in order to study the effect of the inserted components on the ability of the modified cells to respond to extracellular signals and thus to determine if and to what extent the response is regulated by components that reside and act at the plasma membrane.

Present techniques for the transfer of membrane components into the plasma membranes of eukaryotic cells involve cell–cell and membrane–cell fusion using fusogenic agents such as sendai virus or polyethylene glycol (Poste and Pasternak, 1978; Papahadjopoulos *et al.*, 1979). These methods have been used to transfer receptors for hormones (Schramm *et al.*, 1977; Schramm, 1979) and asialoglycoproteins (Doyle *et al.*, 1979) to tissue culture cells as well as to transfer membrane fragments from highly metastatic cells to the plasma

Aya Jakobovits and Nathan Sharon • Department of Biophysics, The Weizmann Institute of Science, Rehovoth, Israel.

membrane of poorly metastatic sublines (Poste and Nicolson, 1980). Transfer of membrane components between primary cells such as lymphocytes was not achieved, presumably because of the inability of these cells to undergo efficient cell–cell fusion by sendai virus (Poste and Pasternak, 1978) and their poor viability after treatment with polyethylene glycol (Papahadjopoulos *et al.*, 1979).

New techniques are therefore required for efficient transfer of membrane components into the plasma membranes of intact lymphocytes so that the acceptor cells will remain fully viable and can be examined for the effect of the modified membrane composition on their functions. This has become feasible with the recent findings that reconstituted sendai virus envelopes, containing the two fusogenic viral glycoproteins, the neuraminidase-hemagglutinin glycoprotein (NH) and the fusion glycoprotein (F), can fuse efficiently with cell membranes (Volsky and Loyter, 1978). Such viral envelopes seemed to be suitable agents to serve as vehicles for intercellular transfer of isolated membrane proteins. Indeed, as first demonstrated by Volsky *et al.* (1979), detergent-solubilized sendai virus envelopes and isolated human erythrocyte band 3, the putative erythrocyte anion channel, were reconstituted into detergent-free vesicles that fused efficiently with Friend erythroleukemic cells that are devoid of anion transport activity. Insertion of band 3 molecules into the surface membrane of the Friend erythroleukemic cells was demonstrated by the specific stimulation of anion exchange in the acceptor cells.

We have used the fusogenic properties of sendai virus envelopes to confer on isolated lymphocyte plasma membranes the ability to fuse with primary lymphocytes (Fig. 1) (Prujansky-Jakobovits *et al.*, 1980). Co-reconstitution of solubilized plasma membranes isolated from murine lymphocytes with solubilized sendai virus envelopes afforded vesicles containing both membrane components and viral NH/F glycoproteins (VPM vesicles), at a 1 : 1 ratio, based on protein content. When these vesicles were incubated with lymphocytes at 4°C, they attached to the cell surface through the viral glycoproteins. Both membrane components and viral glycoproteins were detected on the bound vesicles. Upon incubation at 37°C, fusion of the attached vesicles with the acceptor plasma membranes occurred with subsequent spreading of the inserted components on the cell surface. This two-stage process could be followed by electron microscopy using VPM vesicles prepared from plasma membranes of lymphocytes labeled with biotin, which are readily labeled with ferritin-conjugated avidin (Bayer *et al.*, 1976). The integration of donor membrane components into the acceptor cells was further demonstrated by chemical and immunological methods (Prujansky-Jakobovits *et al.*, 1980). Thus, when VPM vesicles were prepared from lymphocytes, surface-labeled either on their protein or carbohydrate moieties (by lactoperoxidase-catalyzed iodination or galactose oxidase-NaB^{3}H$_4$, respectively), over 40% of the label was incorporated into the membranes of the acceptor cells. Also, when VPM vesicles prepared from mouse T-cell membranes were fused with mouse B cells, over 50% of the acceptor cells acquired the surface Thy-1 antigen, a characteristic marker for mouse T cells, as indicated by their lysis by anti-Thy-1 antiserum and complement. The transfer of functional major histocompatibility complex (MHC) antigens from cells of one mouse strain to those of another was detected by the appropriate anti-H-2 antiserum in combination with fluorescein-labeled protein A (which binds specifically to the bound antibody). Capping of the inserted histocompatibility antigens by the antibodies was observed, strongly suggesting that they are functionally integrated into the membranes of the acceptor cells. This was further substantiated by the use of the mixed lymphocyte reaction in which only cells that differ in their MHC antigens can stimulate each other. For this purpose, membranes from H-2^b splenocytes were transferred into H-2^d splenocytes. Insertion of the allogeneic MHC antigens into the acceptor cells endowed them with the ability to stimulate autologous

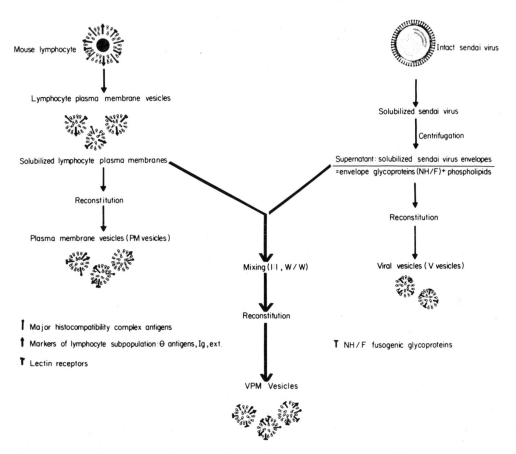

Figure 1. Scheme for preparation of fusogenic membrane vesicles (VPM) by co-reconstitution of solubilized lymphocyte plasma membranes and sendai virus envelopes.

lymphocytes in a primary one-way mixed lymphocyte reaction, indicating that the allogeneic H-2^b antigens are functionally integrated into the acceptor cell membranes.

The transfer technique described above was successfully applied to human cells as well. The receptors for sheep red blood cells, a characteristic surface marker of human T lymphocytes, were transferred to peripheral human B lymphocytes as shown by the ability of the modified acceptor cells to form regular rosettes (at 4°C) with sheep red blood cells to the same extent as T lymphocytes. Moreover, only B cells modified by fusion with thymocyte membranes formed stable rosettes at 37°C, a property that distinguishes thymocytes from peripheral T cells (Jakobovits *et al.*, 1981a).

In a similar manner, Volsky *et al.* (1980) transferred receptors for Epstein–Barr virus (EBV) from human B-cell lines into cells devoid of such receptors (human and murine T-cell derived lines), using vesicles reconstituted from sendai virus envelopes and membrane fragments rich in EBV receptors. [^{3}H]-Thymidine-labeled EBV bound efficiently to the implanted receptors, and 50–75% of the viral DNA was found inside the cells 24 hr after infection. The viral genome was functionally active in the infected cells as shown by expression of EBV-determined nuclear, early, and viral capsid antigens.

We used this approach to investigate the role of membrane constituents, encoded by the MHC genes, in the process of antigen presentation to T lymphocytes (Jakobovits *et*

Table I. Inserted H-2 Gene Membrane Products Influence Presentation of Beef Insulin to T Lymphocytes from Primed C3H.SW (H-2^b) Mice [a]

Insulin-presenting spleen cells	Origin of inserted membrane	Proliferative response (cpm × 10^{-3})		
		Without insulin	With insulin	Net response to insulin
C3H.SW (H-2^b)	None	2	28	26
	C3H.DiSn	27	46	19
C3H.DiSn (H-2^k)	None	43	50	7
	C3H.SW	65	127	62

[a] Lymph node lymphocytes were taken from C₃H.SW (H-2^b) responder mice that had been immunized against beef insulin. Antigen-presenting cells originated from C₃H.SW responder or from C₃H.DiSn (H-2^k) nonresponder splenocytes. Antigen-presenting splenocytes (fused with VPM vesicles or untreated) were incubated with or without beef insulin, irradiated to inhibit their synthesis of DNA, and added to the primed lymph node lymphocytes for 96 hr in culture. The proliferative response of lymph node lymphocytes was measured by [³H]thymidine uptake. The net response to insulin was calculated by subtracting the thymidine uptake in cultures without insulin-pulsed antigen-presenting splenocytes from that in cultures with insulin-pulsed antigen-presenting cells. (Data from Jakobovits *et al.*, 1981b.)

al., 1981b). T lymphocytes bearing specific receptors recognize and respond to the corresponding antigens not directly, but through the agency of antigen-presenting cells. Genetic studies led to the conclusion that the process of antigen presentation is controlled by products of genes in the MHC, and requires that both presenting and responding cells will share alleles within the MHC (Yano *et al.*, 1977). It has also been found that the ability to respond to a particular antigen is associated with immune response (*Ir*) genes in the MHC. For example, mice of H-2^b genotype respond well to beef insulin in contrast to H-2^k mice, which respond only poorly. Lymph node T lymphocytes, from H-2^b mice immunized with beef insulin, will be stimulated by this antigen to secondary proliferation *in vitro* only when this antigen is presented to them by H-2^b splenocytes (Keck, 1977; Cohen and Talmon, 1980). However, the nature of the MHC gene products required for antigen presentation is not known nor do we know the location of these products or their mechanism of action.

We have studied the involvement of structural membrane products of the MHC in antigen presentation by the transfer technique described. Membrane components isolated from splenocytes of H-2^b mice (strain C3H.SW) were inserted into the membranes of H-2^k splenocytes (C3H.DiSn). Whereas the unmodified H-2^k splenocytes were incapable of presenting beef insulin to primed H-2^b T splenocytes, the modified H-2^k cells acquired the ability to present beef insulin (Table I). This effect must be due to specific products of MHC genes because the donor and the acceptor cells were congenic, differing only at their H-2 alleles. Thus, structural products of the MHC genes, expressed and active at the plasma membranes, are sufficient to endow the cells with the capacity to process and present insulin to its responding T cells. Using isolated and purified H-2 gene products will make it possible to study the specific molecular requirements for these cellular functions.

A similar approach was employed by Hale *et al.* (1980) to study the molecular requirements for recognition and lysis of virally infected cells ("target cells") by cytotoxic T lymphocytes. It is known that specific recognition of viral antigens on the target cell surface requires the interaction of the cytotoxic cells with both the viral antigens and the target cell surface antigens encoded by the MHC, but the identity of the virus-coded products recognized by the cytotoxic cells is not known with certainty. Using fusogenic vesicles containing NH/F glycoproteins from sendai virus and purified H-2K^k antigen (isolated on affinity columns of monoclonal anti-H-2k^k), Hale *et al.* (1980) have succeeded in implanting both viral and H-2K^k antigens into target cells that were not susceptible to lysis by anti-H-2K^k or anti-sendai-virus cytotoxic cells and to render them susceptible to such cells. This susceptibility was acquired only with fusogenic vesicles containing NH glycoprotein, active F glycoprotein, and H-2K^k antigen, suggesting that integration of all these constituents in the acceptor cells is required. It was also shown that allogeneic cytotoxic cells (anti-H-2K^k) require the presence of membrane antigens encoded by the H-2K region in the MHC, while anti-viral cytotoxic cells require the presence of both H-2K gene products and viral antigens (NH/F) for effective lysis of the target cells.

The new approach to membrane modification described above opens novel possibilities for the investigation of the involvement of lymphocyte plasma membrane components in the immune response. The technique can also be applied in other cell systems to elucidate the structure–function relationship of their membrane constituents.

REFERENCES

Bayer, E. A., Skutelsky, E., Wynne, D., and Wilchek, M. (1976). *J. Histochem. Cytochem.* **24**, 933–939.

Cohen, I. R., and Talmon, J. (1980). *Eur. J. Immunol.* **10**, 284–289.

Doyle, D., Hou, E., and Warren, R. (1979). *J. Biol. Chem.* **254**, 6853–6856.

Hale, A. H., Ruebush, M. J., Lyles, D. S., and Harris, D. T. (1980). *Proc. Natl. Acad. Sci. USA* **77**, 6105–6108.

Jakobovits, A., Rosenberg, M., and Sharon, N. (1981a). *Eur. J. Immunol.* **11**, 440–442.

Jakobovits, A., Frenkel, A., Sharon, N., and Cohen, I. R. (1981b). *Nature* **291**, 666–668.

Keck, K. (1977). *Eur. J. Immunol.* **7**, 811–816.

Papahadjopoulos, D., Poste, G., and Vail, W. J. (1979). In *Methods in Membrane Biology* (E. D. Korn, ed.), Vol. 10, pp. 1–121, Plenum Press, New York.

Poste, G., and Nicolson, G. L. (1980). *Proc. Natl. Acad. Sci. USA* **77**, 399–403.

Poste G., and Pasternak, C. A. (1978). In *Cell Surface Reviews: Membrane Fusion* (G. Poste and G. L. Nicolson, eds.), Vol. 5, pp. 305–367, North-Holland, Amsterdam.

Prujansky-Jakobovits, A., Volsky, D. J., Loyter, A., and Sharon, N. (1980). *Proc. Natl. Acad. Sci. USA* **77**, 7247–7251.

Schramm, M. (1979). *Proc. Natl. Acad. Sci. USA* **76**, 1174–1178.

Schramm, M., Orly, J., Eimerl, S, and Korner, M. (1977). *Nature (London)* **268**, 310–313.

Volsky, D. J., and Loyter, A. (1978). *FEBS Lett.* **92**, 190–194.

Volsky, D. J., Cabantchik, Z. I., Beigel, M., and Loyter, A. (1979). *Proc. Natl. Acad. Sci. USA* **76**, 5440–5444.

Volsky, D. J., Shapiro, I. M. and Klein, G. (1980). *Proc. Natl. Acad. Sci. USA* **77**, 5453–5457.

Yano, A., Schwartz, R. H., and Paul, W. E. (1977). *J. Exp. Med.* **146**, 828–843.

160

Receptor-Mediated Segregation and Transport of Lysosomal Enzymes

William S. Sly

Studies of the transport of acid hydrolases in mammalian cells have generated great interest recently and produced several surprises. The acid hydrolases are glycoproteins that localize selectively in lysosomes where they participate collectively and often sequentially in the degradation of complex macromolecules (Neufeld *et al.*, 1975). They appear to be products of 40–50 unlinked genes. On the basis of one well-studied example, Pompe's disease, which is due to α-glucosidase deficiency, Hers (1965) elaborated the general concept that lysosomal storage diseases result from a deficiency of one of these enzymes. Early studies (Novikoff *et al.*, 1964; DeDuve and Wattiaux, 1966) suggested that lysosomes bud off the Golgi apparatus or from GERL, a smooth tubular membrane system closely associated with the Golgi apparatus and the endoplasmic reticulum (Novikoff, 1976). A recent proposal expanded this idea to include receptor-mediated transport of enzymes in these vesicles (Sly and Stahl, 1978; Fishcer *et al.*, 1980b). Specifically, it was suggested (Sly and Stahl, 1978; Sly *et al.*, 1981) that lysosomes form after newly synthesized enzymes bind to 6-phosphomannosyl-enzyme receptors on pre-Golgi membranes, and receptor-bound enzymes collect into specialized regions of the Golgi apparatus or GERL to bud off as primary lysosomes.

However, an alternate route for phosphorylated enzymes to reach lysosomes is uptake from the extracellular milieu by adsorptive pinocytosis (Kaplan *et al.*, 1977a). The exciting studies by Neufeld and co-workers on enzyme recognition and uptake (Neufeld *et al.*, 1975) caused this pathway to dominate our thinking for much of the past decade. Adsorptive pinocytosis of lysosomal enzymes by fibroblasts was discovered through studies of the uptake of "corrective factors" for enzyme-deficient fibroblasts (Neufeld and Cantz, 1971). The corrective factors proved to be lysosomal enzymes, and the uptake process was found to be specific and saturable (von Figura and Kresse, 1973). These findings implied a recognition marker on the enzymes and a specific receptor on the cell surface of fibroblasts

William S. Sly • Departments of Pediatrics, Genetics, and Medicine, Washington University School of Medicine, St. Louis, Missouri 63110, and St. Louis Children's Hospital, St. Louis, Missouri 63110.

(Neufeld *et al.*, 1977). Hickman and Neufeld (1972) postulated that different lysosomal enzymes share a "common recognition marker" for this receptor on the basis of the biochemical findings in I-cell disease. This disease results from a single gene defect that affects the posttranslational processing of most acid hydrolases (Hickman and Neufeld, 1972; Hasilik and Neufeld, 1980b).

The common recognition marker on lysosomal enzymes was eventually shown to reside in their oligosaccharide chains, and to consist of 6-phosphomannose residues (Natowicz *et al.*, 1979). Kaplan *et al.* (1977a) first proposed that mannose 6-phosphate (Man 6-P) was an essential feature of the recognition marker on the basis of competition experiments. Man 6-P and mannans containing Man 6-P were found to be strong competitive inhibitors of pinocytosis of β-glucuronidase from human platelets. Furthermore, treatment of human β-glucuronidase with alkaline phosphatase destroyed its susceptibility to uptake by fibroblasts. Until this report, there was no precedent for Man 6-P in mammalian glycoproteins. However, these findings were soon extended to other hydrolases (Kaplan *et al.*, 1977b; Sando and Neufeld, 1977; Ullrich *et al.*, 1978). The generality of these findings implicated Man 6-P in the common recognition marker for uptake.

Direct evidence to support this proposal was provided by several laboratories (Natowicz *et al.*, 1979; Distler *et al.*, 1979; von Figura and Klein, 1979; Hasilik and Neufeld, 1980b). Man 6-P was shown to be present on several "high-uptake" acid hydrolases, to be present in the recognition marker for uptake, and to be present on high-mannose oligosaccharide chains that were released when lysosomal enzymes were treated with endoglycosidase H.

The next surprise was the finding by Tabas and Kornfeld (1980) that the phosphate on newly synthesized lysosomal enzymes was sensitive to phosphatase only if the oligosaccharides were first treated with mild acid. This finding suggested that the phosphate was present in a phosphodiester. The blocking sugar released by mild acid was shown to be *N*-acetylglucosamine (Tabas and Kornfeld, 1980; Hasilik *et al.*, 1980b). These findings led Tabas and Kornfeld (1980) to postulate a second processing enzyme that would cleave the phosphodiester bond to expose Man 6-P groups on the oligosaccharides. Such an enzyme was demonstrated in membranes from rat liver (Varki and Kornfeld, 1980a). Then, phosphorylation of oligosaccharides on acid hydrolases was shown to involve transfer of *N*-acetylglucosamine 1-phosphate to the 6 position of mannose residues on high-mannose oligosaccharide chains (Hasilik *et al.*, 1981; Reitman *et al.*, 1981).

One can infer the following sequence for the transport of newly synthesized acid hydrolases to lysosomes. Lysosomal enzymes are made on membrane-bound ribosomes and gain access to the cisternal space of the endoplasmic reticulum by a conventional "signal sequence" (Erickson and Blobel, 1979). Their high-mannose oligosaccharides are modified enzymatically by transfer of *N*-acetylglucosamine 1-phosphate from UDP-*N*-acetylglucosamine (Hasilik *et al.*, 1981; Reitman *et al.*, 1981), after which a phosphodiesterase cleaves the phosphodiester bond releasing *N*-acetylglucosamine (Varki and Kornfeld, 1980a). Enzymes bind through the exposed Man 6-P recognition markers to receptors (Fischer *et al.*, 1980b), and receptor-bound enzymes collect into vesicles that bud off the Golgi apparatus or GERL to form lysosomes. The subcellular sites of the phosphorylation of the acid hydrolases and the uncovering reaction by the phosphodiesterase have not yet been determined. However, both enzymes are enriched in smooth membrane fractions from rat liver that are also enriched for galactosyltransferase (Varki and Kornfeld, 1980a; Hasilik *et al.*, 1981).

The intracellular pathway is probably the major pathway for enzyme transport to lysosomes. However, the findings from I-cell disease initially suggested that the pinocytotic pathway may be the major pathway for enzyme transport. I-Cell fibroblasts fail to phosphorylate their enzymes (Hasilik and Neufeld, 1980b; Hasilik *et al.*, 1981) and produce recognition-defective enzymes that are secreted and that are not susceptible to pinocytosis (Hickman and Neufeld, 1972). These observations led Neufeld *et al.* (1977) to propose that secretion and receptor-mediated recapture might be an obligatory route for enzymes to reach lysosomes in fibroblasts. However, once Man 6-P recognition was discovered (Kaplan *et al.*, 1977a), one could quantitate the importance of this pathway by growing cells in the presence of Man 6-P concentrations that completely inhibit the pinocytotic pathway. When this experiment was done (Sly and Stahl, 1978; von Figura and Weber, 1978; Vladutiu and Rattazzi, 1979; Sly *et al.*, 1981), Man 6-P failed to alter the distribution of lysosomal enzymes between cells and medium significantly. Thus, while some enzyme is secreted by normal fibroblasts and some enzyme may be "recaptured," most of the lysosomal enzymes produced by fibroblasts reach lysosomes without leaving the cells. The effects of the I-cell mutation can be explained by failure to segregate enzymes into lysosomes intracellularly, which results secondarily in abnormally high levels of enzyme secretion.

Lysosomal enzyme transport in normal fibroblasts is perturbed by treating cells with lysosomotropic amines which raise intralysosomal pH (Gonzalez-Noriega *et al.*, 1980). Enzyme pinocytosis continues for several hours in the absence of protein synthesis at rates that appear to require reutilization of cell surface receptors. However, when normal cells are treated with chloroquine or NH_4Cl, enzyme pinocytosis is rapidly arrested (Sando *et al.*, 1979; Gonzalez-Noriega *et al.*, 1980). In addition, amines induce a time-dependent, ligand-dependent reduction in cell surface receptors for lysosomal enzymes (Gonzalez-Noriega *et al.*, 1980). These effects have been attributed to impairment of receptor recycling by amines and can be explained by the requirement for acid pH-dependent dissociation of enzyme from enzyme receptors following internalization. Receptor reutilization would require enzyme release following internalization to regenerate free receptors. Release of enzyme from receptors was shown to occur very slowly above pH 6.0 (Gonzalez-Noriega *et al.*, 1980; Fischer *et al.*, 1980a).

The same amines have a dramatic affect on the intracellular transport pathway as well, causing most newly synthesized enzymes to be secreted rather than segregated into lysosomes (Gonzalez-Noriega *et al.*, 1980). In fact, the levels of enzyme secreted by amine-treated fibroblasts are comparable to levels secreted by I-cell fibroblasts. However, the enzyme secreted by amine-treated cells is phosphorylated enzyme with high affinity for enzyme receptors. To explain these effects of amines, we have suggested that the intracellular receptors that deliver enzyme from the Golgi apparatus to lysosomes must be reutilized following enzyme delivery to allow continued segregation of newly synthesized enzymes (Gonzalez-Noriega *et al.*, 1980; Fischer *et al.*, 1980b). Amines could impair this process in the same way they impair the function of the pinocytotic pathway, i.e., by impairing the pH-dependent release of enzyme from receptors. Once all of the intracellular receptors were charged with ligand in amine-treated cells, there would be no free receptors for subsequently synthesized enzymes to bind. These enzymes would remain soluble, would fail to be segregated into lysosomes, and would be secreted despite the fact that they bear normal recognition markers for segregation.

Phosphomannosyl enzyme receptors are not restricted to fibroblasts. In fact, they have been found in every organ examined (Fischer *et al.*, 1980b). The level in each organ correlates with the lysosomal enzyme content of that organ. Most of the receptors are found on intracellular membranes. The subcellular distribution of occupied receptors is compatible with the proposed role for the receptor in delivery of enzymes from the endoplasmic reticulum to lysosomes. The functional role of the pinocytotic pathway for lysosomal enzymes is unclear. The pinocytotic pathway does not appear to be important to all cell types, for not all cell types express the receptor for the pinocytotic pathway on their cell surface (Hasilik *et al.*, 1980a).

Two forms of enzyme processing occur after enzymes are delivered to lysosomes. Many enzymes undergo proteolytic cleavage to lower-molecular-weight forms in lysosomes (Skudlarek and Swank, 1979; Hasilik and Neufeld, 1980a). These smaller lysosomal enzyme forms are catalytically active and, for some enzymes, cleavage may be required for activation (Neufeld, 1981). The other form of processing involves removal of the recognition marker, presumably by acid phosphatase (Glaser *et al.*, 1974). This step may be important to trap enzymes in lysosomes and prevent their removal by recycling receptors. Once the recognition marker has been removed, the enzymes no longer bind enzyme receptors (Fischer *et al.*, 1980a) and are no longer susceptible to pinocytosis by fibroblasts (Glaser *et al.*, 1974).

Macrophages have intracellular Man 6-P receptors, but also a mannosyl-glycoprotein pinocytosis receptor on their cell surfaces (Achord *et al.*, 1978; Stahl *et al.*, 1978). This receptor mediates uptake of dephosphorylated lysosomal enzymes, and this uptake system mediates the clearance from plasma of infused lysosomal enzymes. The mannosyl-glycoprotein receptor is restricted to macrophages, is completely distinct from the phosphomannosyl uptake system, and its function is not known. However, it may be important for clearance of lysosomal enzymes released into the circulation by cell death and cell lysis.

It should be apparent that recent progress in elucidating the mechanism for segregation of acid hydrolases and effecting their transfer to lysosomes has been dramatic. The nature of the recognition marker is known (Varki and Kornfeld, 1980b). Two novel enzymes involved in its biosynthesis have been demonstrated and are being purified. The receptor that mediates enzyme delivery has been isolated and characterized (Sahagian *et al.*, 1980). However, a new question raised by these studies involves cellular recognition at another level. How does the cell distinguish which of its gene products should be phosphorylated for targeting to lysosomes? Presumably, the nascent polypeptide precursors for lysosomal enzymes share something in common that earmarks them for this unique form of posttranslational processing. Decoding the signal for this level of recognition is the next major challenge in the study of transport of lysosomal enzymes in mammalian cells.

REFERENCES

Achord, D. T., Brot, F. E., Bell, C. E., and Sly, W. S. (1978). *Cell* **15**, 269–278.
DeDuve, C., and Wattiaux, R. (1966). *Annu. Rev. Physiol.* **28**, 435–492.
Distler, J., Hieber, V., Sahagian, G., Schmickel, R., and Jourdian, G. W. (1979). *Proc. Natl. Acad. Sci. USA* **76**, 4235–4239.
Erickson, A., and Blobel, G. (1979). *J. Biol. Chem.* **254**, 11771–11774.
Fischer, H. D., Gonzalez-Noriega, A., and Sly, W. S. (1980a). *J. Biol. Chem.* **255**, 5069–5074.
Fischer, H. D., Gonzalez-Noriega, A., Sly, W. S., and Morre, D. J. (1980b). *J. Biol. chem.* **255**, 9608–9615.
Glaser, J. H., Roozen, K. H., Brot, F. E., and Sly, W. S. (1974). *Arch. Biochem. Biophys.* **166**, 536–542.
Gonzalez-Noriega, A., Grubb, J. H., Talkad, V., and Sly, W. S. (1980). *J. Cell Biol.* **85**, 839–852.
Hasilik, A., and Neufeld, E. F. (1980a). *J. Biol. Chem.* **255**, 4937–4945.
Hasilik, A., and Neufeld, E. F. (1980b). *J. Biol. Chem.* **255**, 4946–4950.
Hasilik, A., Voss, B., and von Figura, K. (1980a). *Physiol. Chem.* **361**, 262.
Hasilik, A., Klein, U., Waheed, A., Strecker, G., and von Figura, K. (1980b). *Proc. Natl. Acad. Sci. USA* **77**, 7074–7078.
Hasilik, A., Waheed, A., and von Figura, K. (1981). *Biochem. Biophys. Res. Commun.* **98**, 761–767.

Hers, H. G. (1965). *Gastroenterology* **48**, 625–633.

Hickman, S., and Neufeld, E. F. (1972). *Biochem. Biophys. Res. Commun.* **49**, 992–999.

Kaplan, A., Achord, D. T., and Sly, W. S. (1977a). *Proc. Natl. Acad. Sci. USA* **74**, 2026–2030.

Kaplan, A., Fischer, D., Achord, D., and Sly, W. S. (1977b). *J. Clin. Invest.* **60**, 1088–1093.

Natowicz, M. R., Chi, M. M. Y., Lowry, O. H., and Sly, W. S. (1979). *Proc. Natl. Acad. Sci. USA* **76**, 4322–4326.

Neufeld, E. F. (1981). In *Lysosomes and Lysosomal Storage Diseases* (J. W. Callahan and J. A. Lowden, eds.), pp. 115–130, Raven Press, New York.

Neufeld, E. F., and Cantz, M. J. (1971). *Ann. N.Y. Acad. Sci.* **179**, 580–587.

Neufeld, E. F., Lim, T. W., and Shapiro, L. J. (1975). *Annu. Rev. Biochem.* **44**, 357–376.

Neufeld, E. F., Sando, G. N., Garvin, A. J., and Rome, L. H. (1977). *J. Supramol. Struct.* **6**, 95–101.

Novikoff, A. B. (1976). *Proc. Natl. Sci. USA* **73**, 2781–2787.

Novikoff, A. B., Essner, E., and Quintana, N. (1964). *Fed. Proc.* **23**, 1010–1022.

Reitman, M. L., Varki, A., and Kornfeld, S. (1981). *J. Clin. Invest.* **67**, 1574–1579.

Sahagian, G., Distler, J., and Jourdian, G. W. (1980). *Fed. Proc.* **39**, 1968.

Sando, G. N., and Neufeld, E. F. (1977). *Cell* **12**, 619–627.

Sando, G. N., Titus-Dillon, P., Hall, C. W., and Neufeld, E. F. (1979). *Exp. Cell Res.* **119**, 359–364.

Skudlarek, M. D., and Swank, R. T. (1979). *J. Biol. Chem.* **254**, 9939–9942.

Sly, W. S., and Stahl, P. (1978). In *Transport of Macromolecules in Cellular Systems,* Life Science Research Report 11 (S. Silverstein, ed.), pp. 229–245, Dahlem Konferenzen, Berlin.

Sly, W. S., Natowicz, M., Gonzalez-Noriega, A., Grubb, J. H., and Fischer, H. D. (1981). In *Lysosomes and Lysosomal Storage Diseases* (J. W. Callahan and J. A. Lowden, eds.), pp. 131–146, Raven Press, New York.

Stahl, P. D., Rodman, J. S. Miller, M. J., and Schlesinger, P. H. (1978). *Proc. Natl. Acad. Sci. USA* **75**, 1399–1403.

Tabas, I., and Kornfeld, S. (1980). *J. Biol. Chem.* **255**, 6633–6639.

Ullrich, K., Mersmann, G., Weber, E., and von Figura, K. (1978). *Biochem. J.* **170**, 643–650.

Varki, A., and Kornfeld, S. (1980a). *J. Biol. Chem.* **255**, 8398–8401.

Varki, A., and Kornfeld, S. (1980b). *J. Biol. Chem.* **255**, 10847–10858.

Vladutiu, G. D., and Rattazzi, M. (1979). *J. Clin. Invest.* **63**, 595–601.

von Figura, K., and Kresse, H. (1973). *J. Clin. Invest.* **53**, 85–90.

von Figura, K., and Weber, E. (1978). *Biochem. J.,* **176**, 943–956.

von Figura, K., and Klein, U. (1979). *Eur. J. Biochem.* **94**, 347–354.

XIV

The Control of Cell Surfaces. Growth Regulation, Hormones, Hormone Receptors, and Transformation

161

Growth Regulation by Transport into Mammalian Cells

Arthur B. Pardee

The question to be discussed in this review is whether proliferation of animal cells in culture is controlled by rates of entry of nutrients. Proliferation depends on extracellular factors that are principally low-molecular-weight nutrients, the growth factors generally provided in serum, and free space on a suitable substratum. Cells stop proliferating if any of these factors is inadequate. For instance, cells may stop proliferating when they become confluent on their substratum. Such conditions regulate growth specifically, apparently affecting a control mechanism that determines during the G_1 part of the cell cycle whether or not a cell will proceed toward DNA synthesis and division (Pardee *et al.*, 1978). But also blocking any of the many metabolic "housekeeping" processes required for duplication of a cell's components stops proliferation. Cell proliferation can be arrested by many unphysiological agents: any powerful inhibitor of macromolecular synthesis such as hydroxyurea or cycloheximide, or of mitosis such as colcemid. These cells are not arrested cycle specifically, in contrast to cells arrested at high density or in low serum, which stop growth after failing to transit the G_1 part of the cell cycle, and enter a condition of quiescence (G_0).

The distinction between physiological arrest and arrest due to blocking some housekeeping function can be seen in the ability of tumorigenic and transformed cells to continue growth under conditions that arrest untransformed cells in G_0. In contrast, housekeeping inhibitors block both kinds of cells and in various parts of the cell cycle. We consider that physiologically significant mechanisms of growth regulation, including regulation by rates of nutrient transport, act to arrest nontransformed cells within G_1 and are less effective for some transformed cells. But although conditions that arrest only nontumor cells in G_0 may affect the physiological controls, yet they may never be experienced by cells *in vivo* or in conventional culture, and hence would not be physiologically significant.

A limiting intracellular supply of nutrients evidently can restrict growth, and this lim-

Arthur B. Pardee • Sidney Farber Cancer Institute, and Department of Pharmacology, Harvard Medical School, Boston, Massachusetts 02115.

itation could be imposed by slow rates of transport. Nutrient transport thus might be important for growth regulation. The defective growth regulation of cancer cells could be due to altered membranes (Pardee, 1964) that permit rapid transport of nutrients (Holley, 1972). Foster and Pardee (1969), Hatanaka *et al.* (1969), Isselbacher (1972), and Griffiths (1972) demonstrated severalfold more rapid rates of transport of some amino acids and sugars (but not of others) by transformed cells than by untransformed cells. Cunningham and Pardee (1969) further showed that conditions that arrest proliferation reduce transport rates of P_i and uridine into 3T3 cells. Thus, some transport activities correlate with growth rates, a conclusion supported by many subsequent studies. (For reviews see Kalckar, 1976; Parnes and Isselbacher, 1978; Rozengurt, 1979; and Bhargava, 1977, for a detailed model.)

Cell proliferation can be limited, with specific G_0 arrest, following nutrient deprivation. Ley and Tobey (1970) stopped CHO cells in isoleucine- or glutamine-deficient media. Holley and Kiernan (1974) demonstrated that lowering glucose, P_i, or amino acid concentrations to about 1% of normal made 3T3 cells quiescent; total deprivations arrested noncycle specifically. Thus, partial nutrient deficiencies can specifically arrest growth.

Growth depends on the nutrient concentrations that are available. On the one hand, when nutrient concentrations are low enough to inhibit growth, addition of either more dialyzed serum or growth factors can stimulate proliferation (Kamely and Rudland, 1976; Stiles *et al.*, 1979; McKeehan and McKeehan, 1980). Of course, addition of nutrients also can stimulate growth. These results are consistent with control by transport; a modest stimulation of nutrient transport by serum adequately feeding the cells. Transformed cells brought to high density stop growth because they deplete their nutrients, and to a degree where addition of serum alone will not stimulate them (Moses *et al.*, 1980).

The other situation is obtained when cells are supplied with high nutrient concentrations but inadequate serum factors, either because the latter were omitted (e.g., in medium with suboptimal serum; Kamely and Rudland, 1976) or because the factors became exhausted as with untransformed cells left at confluence (Moses *et al.*, 1980). These cells can be stimulated to proliferate by addition of serum, but not by addition of nutrients. An excess of nutrients thus cannot replace serum. Transport of Ca^{2+} and K^+ ions might be exceptions, at moderate serum concentrations (McKeehan and McKeehan, 1980). This might be anticipated because the transport systems are already nearly saturated, so that nutrient entry rates would not be increased by simply increasing concentrations. These results do not rule out control of growth by transport whose rate (V_{max}) depends on serum concentration.

Considerable literature has emerged during the past decade regarding this question of transport and growth control. We will attempt a brief summary here. There do not appear to be any results that force the conclusion that rates of transport limit growth of cells, under conditions of serum insufficiency or of high cell density. These conditions can reduce transport rates, as cited above. But these transport changes could as well be consequences of shutting down general metabolism, a part of the set of "pleiotypic responses" to physiological growth arrest (Hershko *et al.*, 1971). But lowering transport rates seems specific, not the consequence to any sort of growth arrest; certain mutants of SV40 virus-transformed 3T3 cells stop in media containing inadequate serum or at high density, but not in G_0, and their transport activities are not decreased (Dubrow *et al.*, 1978).

We need to know the growth-limiting rate of transport in order to decide whether an observed correlation between lowered transport and inhibited growth is a cause–effect relationship. To illustrate (Fig. 1), the rate of α-aminoisobutyrate (AIB) transport by 3T3 cells decreased continually as cells approached confluence and become growth arrested. This decrease did not occur with SV40-transformed cells. In order to be the cause of

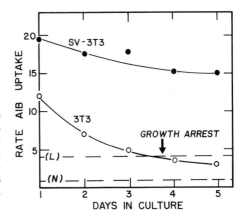

Figure 1. Changes of AIB transport in 3T3 cells. Cells were plated at $2.5 \times 10^3/cm^2$. On five successive days thereafter cells were assayed for AIB uptake, after exposure to 2 mM [^{14}C]-AIB for 5 min. Protein and DNA synthesis were monitored to determine cell growth (see Dubrow *et al.*, 1978). o, 3T3 cells; •, transformed 3T3 (SV101). The dashed lines indicate two alternative theoretical growth-limiting rates of transport.

growth arrest, on day 3, the growth-limiting transport rate would have to be at the level marked L (limiting) in Fig 1. If the actual limiting rate were (N) and was not reached, some property other than lowered transport must be responsible for the proliferation arrest. The problem simply is that we do not know the true growth-limiting rate, and so cannot decide whether the lower transport observed is the limiting factor. According to a simple calculation, an essential amino acid must be supplied at about 0.2 nmole/min per mg protein to permit protein synthesis at the rate seen in growing cells. To meet this requirement, the rate of transport under the usual conditions of assay must be about 3 nmoles/min per mg protein. This value is similar to the rate observed at confluence for AIB transport (Fig. 1). Bhargave *et al.* (1979) have provided evidence for a protein that inhibits amino acid transport into liver cells, consistent with their hypothesis regarding control of growth by inhibition of transport (Bhargava, 1977).

Some comments are in order on technical problems of studying transport, as the way experiments are performed can considerably affect results and hence conclusions (see Parnes and Isselbacher, 1978). An intact cell is very complex, and we must be assured that transport is actually being measured when we determine how rapidly a (radioactive) substrate is accumulated. After a substrate enters a cell it may be further metabolized, and thereby its ability to escape from the cell can be affected. Thus, the quantity of compound remaining after a given time depends on both the passage rates across the membrane in both directions and also on metabolic trapping, as for example by phosphorylation of deoxyglucose or uridine (see Rozengurt *et al.*, 1978). Changes of total uptake rates after transformation, or quiescence can reflect changes of actual transport or of subsequent metabolism; either can be rate limiting for accumulation. Rapid sampling alleviates some of these problems (Wohlhueter *et al.*, 1976), as do substrates that are not metabolized such as AIB or 3-*O*-methylglucose. Transport can be measured with membrane vesicles, which are incapable of further metabolism of substrates (for reviews see Hochstadt *et al.*, 1979; Parnes and Isselbacher, 1978; Lever, 1980). Some substrates, particularly amino acids, are transported by more than one system, with consequent complexities of interpretation (see Guidotti *et al.*, 1978; Oxender *et al.*, 1977). Finally, for comparison under transformed and untransformed conditions, secondary differences can be largely avoided by using a temperature-sensitive transformed line.

We now turn to results obtained with various systems. Transport has been studied mainly with a few kinds of substrates, each reviewed in detail: glucose (Hatanaka, 1974), amino acids (Guidotti *et al.*, 1978), nucleosides (Plagemann and Richey, 1974; Rozengurt,

1979), ions (Rozengurt, 1979; Leffert, 1980). A few studies have appeared on transport of other compounds; for example, putrescine is transported at least 10 times faster into growing cells (Pohjanpelto, 1976). Choline transport increases when proliferation is initiated, by some procedures (Rubin and Koide, 1975). Absence of studies on transport–growth makes conclusions related to many other nutrients impossible. Perhaps the technically closest to a study of roles of all nutrients in a complete medium is an investigation of growth rates of human cells measured when each nutrient's concentration is independently varied at several near-optimal serum concentrations (McKeehan and McKeehan, 1980). Conclusions regarding the role of transport in growth regulation unfortunately are not possible from these data.

Uridine uptake (but not adenosine transport) is highly responsive to serum (Cunningham and Pardee, 1969; Jimenez de Asua *et al.*, 1974). Uptake depends on both transport and phosphorylation, which change at different times after stimulation of quiescent cells (Rozengurt *et al.*, 1978). In any event, transport of uridine, and of other nonessential compounds that do not have to be added to the medium, cannot limit growth because their entry is not necessary for proliferation.

Transport of glucose (and of its analogs 2-deoxyglucose or 3-*O*-methylglucose) is highly dependent on growth state or transformation. The rate of glucose transport is also highly dependent on availability of sugars in the medium, and can be greatly increased through "deprivation derepression" (see Kalckar, 1976). Very low glucose puts 3T3 cells into a quiescent state (Holley and Kiernan, 1974), and can limit the glycolytic rate (Bissell, 1976). Stimulation of quiescent cells increases glucose uptake in two stages, the first being independent and the second dependent on protein synthesis (Kletzien and Purdue, 1976). Transformation with viruses can enhance glucose transport (Hatanaka, *et al.*, 1969; Hatanaka, 1975), as can chemical transformation (Oshiro and DiPaolo, 1974). Many data relate sugar transport to transformation (see Kalckar, 1976). But comparisons with untransformed cells must take into account growth state because, as for AIB uptake (Fig. 1), uptake of glucose by untransformed cells is very sensitive to cell density. A great part of the difference seen after transformation is eliminated if cell types are compared at very low densities (Bose and Zlotnick, 1973; Dubrow *et al.*, 1978).

The rate of glucose uptake does not seem to limit growth of density-inhibited cells, in spite of high dependence on growth conditions. Strong inhibition by cytochalasin B of glucose uptake did not inhibit growth of 3T3 cells (Brownstein *et al.*, 1975). Growth stimulation of 3T3 cells using glucocorticoids did not increase glucose or P_i transport. And lowering the glucose concentration in the medium so that its transport rate into growing cells fell below the rate of confluent, quiescent cells did not arrest growth (Naiditch and Cunningham, 1976).

Similar to glucose, the rate of transport of P_i is rapidly and highly sensitive to serum and to transformation (Cunningham and Pardee, 1969; Jimenez de Asua *et al.*, 1974). But data obtained on growth at very low P_i concentrations showed that transport rates were below those of quiescent cells, but did not arrest growing cells and did not lower P_i pools in these cells (Barsh *et al.*, 1977).

The roles of amino acid transport are more difficult to assess, in part because there are a dozen essential amino acids. These are transported by four main systems; some amino acids are transported by more than one system (Guidotti *et al.*, 1978; Oxender *et al.*, 1977). The A system is quite responsive to serum factors, changing its V_{max} by about fourfold (Foster and Pardee, 1969; Isselbacher, 1972). The L (and Ly) systems are not as responsive to serum concentrations. Interestingly, like the K_m for transport, the concentrations of amino acids half-optimal for growth are not affected by slightly suboptimal serum concentrations (McKeehan and McKeehan, 1980). Rates of AIB transport (by the A system) correlated with specific growth arrest in G_0 for several mutated 3T3 cell lines (Dubrow *et al.*, 1978).

Growth can be arrested at very low amino acid concentrations, about 1/100 of the concentrations usual in media (Ley and Tobey, 1970; Holley and Kiernan, 1974; Kamely and Rudland, 1976; Stiles *et al.*, 1979). Growth can be reinitiated by adding serum, and therefore under these conditions, transport rates may well be important. But in media complete except for low serum, or at high cell density, it seems very unlikely that the transport enhancement by added serum is responsible for growth initiation. Shionagi 115 cells initiate growth after testosterone is provided, but the A system is not stimulated (Robinson and Smith, 1976). Amino acid pools of serum-arrested rodent cells were stated to be higher than in growing cells (Oxender *et al.*, 1977), but were lower in density-arrested human diploid cells (Griffiths, 1972). The transport rate decrease seen when cells reach quiescence is only about 4-fold (see Fig. 1), likely far too small to arrest cells, for the reduction in external amino acid concentration (about 100-fold) that is required to arrest growth should decrease transport rates by at least 10-fold. Castor (1978) has shown that the valine concentration in MEM medium had to be decreased by only about 5-fold (to 0.05 mM) to appreciably decrease the rate of protein synthesis in WI-38 cells. Decreasing the serum concentration from 10% to 0.3% arrested growth, but did not decrease intracellular free valine or its rate of incorporation, during a 2-hr period. These results suggest for these cells that growth limitation by low serum is not due to an inadequate valine supply.

Cation transport rates depend on growth conditions (Rozengurt, 1979). They have been proposed to be important in growth regulation (see Kaplan, 1978). Growth of human cells was reciprocally dependent on both serum and cations, particularly K^+ and Ca^{2+}, and less so for Mg^{2+} (McKeehan and McKeehan, 1980). The role of Mg^{2+} in growth control has been particularly stressed (Rubin and Sanui, 1979). These ions are involved in many biochemical reactions. For example, Na^+ is a cofactor of the A amino acid transport system, and changes in this system are suggested to depend on membrane potential (Villareal and Cook, 1978). Many studies have been made on the role in growth and cell function of ouabain-inhibitable Na^+/K^+-ATPase (see Kaplan, 1978). However, the ability of serum to stimulate cell growth in a low concentration of one of the ions, or vice versa, does not prove that transport is the key property; internal processes could also be affected. For example, high serum reduces the half-optimal pyruvate concentration in the medium to zero (McKeehan and McKeehan, 1980); this may well be because the stimulated cells produce adequate pyruvate through glycolysis. The role of some cation's transport in limiting growth rates appears to remain an interesting possibility.

In summary, altered transport rates have been observed as functions of growth states of cells, and of tumorigenic transformation. These changes are generally a few fold in magnitude, and they are seen for some nutrients but not others. The attractive idea that such changes actually control growth, by limiting the supply of some essential nutrient, has not received any direct confirmation. Transport rates of most major nutrients—sugars, amino acids, purines, pyrimidines, and P_i—appear not to be growth limiting for untransformed cells, except when supplied at very low concentrations. It is still possible that transport of a cation or of some untested nutrient could become rate limiting under physiological conditions, though there is no direct evidence for these possibilities.

A great variety of membrane changes are seen after arrest or transformation (Nicolson, 1976; Wallach, 1975). Could one of these changes other than in transport be responsible for growth arrest? Possibilities that come to mind include alterations in growth factor receptors, binding of cytoskeleton to membrane, membrane potential, surface shape, and interactions between cells' surfaces.

ACKNOWLEDGMENT. This work was aided by a DHEW Grant GM 24571.

REFERENCES

Barsh, G. S., Greenberg, D. B., and Cunningham, D. D. (1977). *J. Cell Physiol.* **92,** 115–128.
Bhargava, P. M. (1977). *J. Theor. Biol.* **68,** 101–137.
Ghargava, P. M., Dwarakanath, V. N., and Prased, K. S. N. (1979). *Cell. Mol. Biol.* **25,** 85–94.
Bissell, M. J. (1976). *J. Cell Physiol.* **89,** 701–710.
Bose, S. K., and Zlotnick, B. J. (1973). *Proc. Natl. Acad. Sci. USA* **70,** 2374–2378.
Brownstein, B. L., Rozengurt, E., Jimenez de Asua, L., and Stoker, M. (1975). *J. Cell Physiol.* **85,** 579–585.
Castor, L. N. (1978). *J. Cell Physiol.* **92,** 457–468.
Cunningham, D. D., and Pardee, A. B. (1969). *Proc. Natl. Acad. Sci. USA* **69,** 1049–1056.
Dubrow, R., Pardee, A. B., and Pollack, R. (1978). *J. Cell Physiol.* **95,** 203–212.
Foster, D. O., and Pardee, A. B. (1969). *J. Biol. Chem.* **244,** 2675–2681.
Griffiths, J. B. (1972). *J. Cell Sci.* **10,** 515–524.
Guidotti, G. G., Borghetti, A. F., and Gazzola, G. C. (1978). *Biochim. Biophys. Acta* **515,** 329–366.
Hatanaka, M. (1974). *Biochim. Biophys. Acta* **355,** 77–104.
Hatanaka, M., Huebner, R. J., and Gilden, R. V. (1969). *J. Natl. Cancer Inst.* **43,** 1091–1096.
Hershko, A. P., Mamont, R., Shields, R., and Tomkins, G. M. (1971). *Nature New Biol.* **232,** 206–211.
Hochstadt, J., Quinlan, D. C., Owen, A. J., and Cooper, K. O. (1979). In *Hormones and Cell Culture* (G. H. Sato and R. Ross, eds.), pp. 751–771, Cold Spring Harbor Laboratory, New York.
Holley, R. W. (1972). *Proc. Natl. Acad. Sci. USA* **69,** 2840–2841.
Holley, R. W., and Kiernan, J. A. (1974). *Proc. Natl. Acad. Sci. USA* **71,** 2942–2945.
Isselbacher, K. J. (1972). *Proc. Natl. Acad. Sci. USA* **69,** 585–589.
Jimenez de Asua, L., Rozengurt, E., and Dulbecco, R. (1974). *Proc. Natl. Acad. Sci. USA* **71,** 96–98.
Kalckar, H. M. (1976). *J. Cell Physiol.* **89,** 503–516.
Kamely, D., and Rudland, P. S. (1976). *Nature (London)* **260,** 51–56.
Kaplan, J. G. (1978). *Annu. Rev. Physiol.* **40,** 19–41.
Kletzien, R. F., and Purdue, J. F. (1976). *J. Cell Physiol.* **89,** 723–728.
Leffert, H. L. (ed.) (1980). *Growth Regulation by Ion Fluxes*, New York Academy of Sciences, New York.
Lever, J. E. (1980). *CRC Crit. Rev. Biochem.* **7,** 187–246.
Ley, K. D., and Tobey, R. A. (1970). *J. Cell Biol.* **47,** 453–459.
McKeehan, W. L., and McKeehan, K. A. (1980). *Proc. Natl. Acad. Sci. USA* **77,** 3417–3421.
Moses, H. L., Proper, J. A., Volkenant, M. E., and Swartzendruber, D. E. (1980). *J. Cell Physiol.* **102,** 367–378.
Naiditch, W. P., and Cunningham, D. D. (1976). *J. Cell Physiol.* **92,** 319–332.
Nicolson, G. L. (1976). *Biochim. Biophys. Acta* **458,** 1–72.
Oshiro, Y., and DiPaolo, J. A. (1974). *J. Cell Physiol.* **83,** 193–202.
Oxenser, D. L., Lee, M., and Ceccini, G. (1977). *J. Biol. Chem.* **252,** 2680–2683.
Pardee, A. B. (1964). *Natl. Cancer Inst. Monogr.* **14,** 7–14.
Parnes, J. R., and Isselbacher, K. J. (1978). *Prog. Exp. Tumor Res.* **22,** 79–122.
Plagemann, P. G. W., and Richèy, D. P. (1974). *Biochim. Biophys. Acta* **344,** 263–305.
Pohjanpelto, P. (1976). *J. Cell Biol.* **68,** 512–520.
Robinson, J. H., and Smith, J. A. (1976). *J. Cell Physiol.* **89,** 111–121.
Rozengurt, E. (1979). In *Surfaces of Normal and Neoplastic Cells* (R. O. Hynes, ed.), pp. 323–353, Wiley, New York.
Rozengurt, E., Mierzejewski, K., and Wigglesworth, N. (1978). *J. Cell Physiol.* **97,** 241–252.
Rubin, H., and Koide, T. (1975). *J. Cell Physiol.* **86,** 47–58.
Rubin, A. H., and Sanui, H. (1979). In *Hormones and Cell Culture* (G. H. Sato and R. Ross, eds.), pp. 741–750, Cold Spring Harbor Laboratory, New York.

Stiles, C. D., Isenberg, R. R., Pledger, W. J., Antoniades, H. N., and Scher, C. D. (1979). *J. Cell Physiol.* **99,** 395–406.

Villareal, M. L., and Cook, J. S. (1978). *J. Biol. Chem.* **253,** 8257–8262.

Wallach, D. H. F. (1975). *Membrane Molecular Biology of Neoplastic Cells*, Elsevier, Amsterdam.

Wohlhueter, R. M., Marz, R., Graff, J. C., and Plagemann, P. G. W. (1976). *J. Cell Physiol.* **89,** 605–612.

162

Hepatic Metabolism of Serum Asialoglycoproteins

Clifford J. Steer and Gilbert Ashwell

1. INTRODUCTION

Studies on the mechanism of receptor-mediated uptake by mammalian cells have experienced an exponential growth in the last few years. Contributing in part to this development was the observation, made more than a decade ago, that enzymatic removal of the sialic acid residues of ceruloplasmin resulted in the rapid disappearance of this glycoprotein from the circulation of rabbits (Morell *et al.*, 1968). Subsequent studies showed this phenomenon to be general for asialo-derivatives of plasma glycoproteins under conditions where the fully sialylated protein survived for days (Morell *et al.*, 1971). Over the ensuing years, it became apparent that the mammalian hepatocyte possesses a unique receptor that is capable of recognizing exposed galactosyl residues of glycoproteins. Subsequent work resulted in the purification and characterization of the asialoglycoprotein lectin (Hudgin *et al.*, 1974; Kawasaki and Ashwell, 1976). As a consequence of the multitude of *in vivo* and *in vitro* studies, the asialoglycoprotein receptor proved to provide an excellent model system to examine molecular details of receptor-mediated endocytosis.

Previous reviews have provided detailed accounts of the origin and early developments of work on the asialoglycoprotein receptor (Ashwell and Morell, 1974; Neufeld and Ashwell, 1980). The present chapter seeks to summarize some of the newer information that has become available on receptor-mediated endocytosis of asialoglycoproteins. Growing interest in and utilization of the perfused liver, isolated hepatic cell systems, electron microscopy, and sophisticated biophysical tools have provided further details on the mechanism of this receptor-mediated process.

Clifford J. Steer and Gilbert Ashwell • Laboratory of Biochemistry, National Institute of Arthritis, Metabolism and Digestive Diseases, National Institutes of Health, Bethesda, Maryland 20205.

2. PHYSICAL PROPERTIES OF THE ASIALOGLYCOPROTEIN
RECEPTOR

Subsequent to studies *in vivo* implicating the liver in clearance of asialoglycoproteins, it was shown that the hepatocyte plasma membrane was the locus of binding for galactose-terminated glycoproteins (Pricer and Ashwell, 1971). A sensitive inhibition assay was then developed to quantitate the binding characteristics of various asialoglycoproteins by the plasma membranes (Van Lenten and Ashwell, 1972). It became evident that the binding process exhibited an absolute dependence on calcium as well as endogenous plasma membrane sialic acid. Treatment of the membranes with neuraminidase resulted in complete loss of binding activity, reflecting the ability of the binding protein to form stable complexes with its own galactose residues (Pricer and Ashwell, 1971). Resialylation, or removal of exposed galactosyl residues with β-galactosidase, restored binding activity to test substances (Stockert *et al.*, 1977; Paulson *et al.*, 1977).

Solubilization of the binding protein with Triton X-100 was found to have no effect on specific binding activity in solution (Morell and Scheinberg, 1972). Subsequently, Hudgin and co-workers (1974) were able to isolate and purify the hepatic receptor by affinity chromatography on asialo-orosomucoid covalently linked to Sepharose. The purified rabbit lectin was characterized as a water-soluble glycoprotein in which 10% of the dry weight consisted of sialic acid, galactose, *N*-acetylglucosamine, and mannose in molar ratios of $1 : 1 : 2 : 2$. The purified receptor in aqueous solution exhibited a high degree of aggregation resulting from self-associating properties. This tendency toward self-association was rapidly reversed in the presence of Triton X-100 with the appearance of a single oligomeric protein with an apparent molecular weight of 250,000 (Kawasaki and Ashwell, 1976). Treatment with SDS permitted the identification of two subunits with molecular weights of 40,000 and 48,000. Amino acid and carbohydrate analyses revealed that both subunit structures were glycoproteins. Using either of the neoglycoproteins, D-galactosyl-bovine serum albumin or D-glucosyl-bovine serum albumin immobilized on Sepharose, Stowell and Lee (1978) isolated material from rabbit liver indistinguishable from the rabbit hepatic asialoglycoprotein receptor. The data obtained suggested that the hepatic binding protein was incapable of distinguishing between D-galacto- and D-gluco- configurations.

A hepatic receptor capable of binding galactose-terminated glycoproteins has been identified and isolated from other mammalian species including rat (Tanabe *et al.*, 1979) and human (Baenziger and Maynard, 1980). The method of purification employed by Tanabe and co-workers to isolate the analogous rat liver binding protein followed closely the procedure described by Hudgin and co-workers (1974). However, polyacrylamide gel electrophoresis in the presence of SDS revealed a single major component with an estimated molecular weight of 47,000 as well as two lesser subunit structures of 54,000 and 57,000, respectively. Subunit structures of similar but not identical molecular weights have been reported by several laboratories. Peptide mapping studies of the subunits suggest marked similarity in the primary structures of the three bands (Warren and Doyle, 1981), a conclusion in agreement with the immunoprecipitation of all bands by a monoclonal antibody to the purified receptor (Schwartz *et al.*, 1981). The nature of the variability among laboratories in assigning molecular weights to the various rat subunit structures remains unclear. This laboratory has recently examined the minimal size of the hepatic receptor that is capable of binding galactose-terminated glycoproteins. Using high-energy irradiation and target size analysis, it appears that a receptor molecular weight of 104,000 *in situ* within the plasma membrane is necessary to recognize and bind asialoglycoproteins (Steer *et al.*,

1981). It should be noted that Sawamura *et al.*, (1980) have reported a single band of molecular weight 52,000 on SDS gels of their preparation of rat liver binding protein.

Baenziger and Maynard (1980) have recently reported the isolation and extensive characterization of the comparable human hepatic lectin. The latter protein consists of a single subunit with an apparent molecular weight of 41,000 and similar amino acid composition to each of the subunits of the rabbit lectin as well as the rat lectin. The human hepatic binding protein was also shown to be a glycoprotein with a carbohydrate composition similar to that of the rabbit (Kawasaki and Ashwell, 1976) and rat (Tanabe *et al.*, 1979).

3. THE FATE OF CIRCULATING ASIALOGLYCOPROTEINS

The rapid removal of numerous glycoproteins from circulating blood plasma involves carbohydrate recognition and occurs primarily in the liver. The first attempt to study the uptake of desialylated serum glycoproteins at the microscopic level was that recorded by Morell and co-workers (1968), who identified the hepatocyte as the site of accumulation of circulating [³H]asialoceruloplasmin. The resolution obtained was inadequate to allow localization of the ligand to intracellular organelles. By the utilization of electron microscopic autoradiography, it became possible to monitor more closely the subcellular migration of injected labeled asialoglycoproteins.

In a series of elegant studies employing this technique, Hubbard *et al.*, (1979) investigated the mechanism of internalization of variously tagged asialoglycoproteins by rat hepatocytes. A quantitative analysis of the distribution and concentration of [¹²⁵I]asialoglycoprotein revealed that greater than 90% of the molecules were associated with hepatocytes after rapid clearance from the circulation. Conversely, tagged glycoproteins terminating in *N*-acetylglucosamine or mannose were localized to the endothelial and Kupffer cells lining the sinusoidal tracts. As early as 1–2 min after infusion, [¹²⁵I]asialoglycoproteins were localized predominantly along the sinusoidal border of hepatocytes. Approximately 40–60% of the autoradiographic grains could be ascribed to the plasmalemma, 25–35% to the peripheral cytoplasm, and 12–25% to the intermediate cytoplasm. Between 4 and 15 min after intravenous administration, there was a dramatic redistribution of grains from peripheral regions to those of the lysosome–Golgi region.

In an attempt to improve resolution and more precisely define the mechanism of adsorptive endocytosis, Wall and co-workers (1980) examined two other electron microscope tracers: asialo-orosomucoid covalently coupled to either horseradish peroxidase (ASOR-HRP) or lactosaminated ferritin (Lac-Fer). Both ligands were cleared rapidly from the circulation and accumulated primarily in the liver. Internalization across the hepatocyte plasma membrane occurred via coated pits and coated vesicles of approximately 1000-Å diameter. A similar observation was noted by Stockert *et al.*, (1981) using ligand coupled to cytochemically demonstrable enzymes. They found that as early as 1 min after injection, reaction products of the enzyme markers were seen in coated pits, coated vesicles, and elongated pinocytotic channels and pleomorphic vesicles adjacent to the sinusoidal front. Numerous pinocytotic structures were also seen near the lateral surfaces of hepatocytes, suggesting that the membrane receptor is not limited to the sinusoidal pole of the hepatocyte.

Wall noted that ASOR-HRP and Lac-Fer began to accumulate in a complex of larger, smooth-surface vesicles and tubular structures adjacent to the sinusoidal periphery of the hepatocyte within 30 sec to 2 min after injection. The particulate tracer Lac-Fer was found scattered throughout the lumen of the larger vesicles, suggesting possible dissociation from the binding protein. In contrast, the Lac-Fer was closely apposed to the membrane of both coated pits and coated vesicles. There is, however, no convincing evidence to show that the receptor is associated with either coated vesicles, smooth-surface vesicles, or tubular structures. At 5 min, the tracers began to appear in the lysosome–Golgi regions, and by 15 min, fusion of the vesicles with lysosomes was prominent. Biochemical evidence for proteolysis of internalized [125I]asialoglycoproteins after incorporation into lysosomes was obtained when mono[125I]iodotyrosine was found in hepatocytes at times later than 15 min (Hubbard and Stukenbrok, 1979).

The hepatic lectin for asialoproteins provides a unique model to investigate heterophagy of serum glycoproteins. Work by Gregoriadis *et al.*, (1970) with [3]H- and ^{64}Cu-labeled asialoceruloplasmin provided early evidence that hepatocyte lysosomes were the major site of asialoglycoprotein catabolism. This initial observation was corroborated and pursued extensively by LaBadie and co-workers (1975) who followed the liver metabolism of [125I]asialofetuin. In order to characterize in more detail the heterophagic function of lysosomes, Dunn and co-workers (1980) studied the effect of temperature change on the catabolism of [125I]asialofetuin by the perfused rat liver. Uptake and catabolism of the ligand were progressively slowed as the temperature decreased from 35°C to 21°C. At 20°C, lysosomal degradation ceased completely despite continued endocytosis of [125I]asialofetuin and movement of pinocytotic vesicles from the peripheral cytoplasm to the lysosome–Golgi region. *In situ* electron microscope autoradiography revealed that temperatures at or below 20°C inhibited fusion of pinocytotic vesicles with lysosomes and therefore inhibited complete heterophagy of the ingested asialoglycoprotein. Rewarming to physiologic temperature resulted in fusion between the pinocytotic vesicles and lysosomes. However, the half time for such fusion, 7 min, was the slowest metabolic event involved in the catabolism of asialofetuin by the liver *in vivo* and in the perfused organ.

In a similar model system, Dunn and co-workers (1979) showed that treatment of a rat or perfused liver with leupeptin, an inhibitor of thiol proteinases, resulted in marked inhibition of asialofetuin degradation. Although leupeptin had no effect on clearance or ligand internalization, the rate at which radioactive products left the intact liver decreased to negligible levels. The amassing of radioactive products within lysosome-rich fractions of the liver homogenate after leupeptin treatment presumably resulted from the inactivation of one or more of the thiol-requiring cathepsins.

4. CHARACTERIZATION OF THE ASIALOGLYCOPROTEIN RECEPTOR ON ISOLATED CELLS

The binding, internalization, and catabolism of asialoglycoproteins in isolated hepatocytes have proved to be a convenient model system for the study of receptor-mediated adsorptive endocytosis. The use of a homogeneous cell population has certain advantages foremost of which include control of media composition, temperature, and ligand concentration. The first detailed study of asialoglycoprotein metabolism using isolated hepatocytes was reported by Tolleshaug and co-workers in 1977. Using a modification of the collagenase perfusion method described by Berry and Friend (1969), rat hepatocytes were isolated and studied for their ability to internalize and degrade [125I]asialofetuin. Several important points emerged from that early study: (1) Enzymatically prepared hepatocytes (collagenase perfusion) retained their ability to metabolize asialoglycoproteins; (2) calcium was an absolute requirement for binding; and (3) lysosomes were probably the site of asialoglycoprotein degradation. Studies using isopycnic and differential centrifugation as well as known inhibitors of lysosomal proteolysis (ammonium chloride and chloroquine) substantiated the primary role of lysosomes in the breakdown of asialoglycoproteins (Tolleshaug *et al.*, 1979).

Subsequent investigations performed in Berg's laboratory concluded that chloroquine (Berg and Tolleshaug, 1980), colchicine, cytochalasin B (Kolset *et al.*, 1979), concanavalin A (Tolleshaug *et al.*, 1980), and leupeptin (Berg *et al.*, 1981) were all capable of retarding fusion between endocytotic vesicles and lysosomes and thereby preventing degradation of asialoglycoproteins.

Several recent studies have concentrated on the kinetics of asialoglycoprotein binding and internalization by isolated hepatocytes. The hepatic lectin appears to provide an excellent model system to study these aspects of receptor-mediated endocytosis. Rate constants for surface binding and dissociation of radiolabeled ligands to hepatocytes have been reported by several groups. Weigel (1980) found that at 4°C the binding of [^{3}H]asialo-orosomucoid was rapid ($K_{on} \gtrsim 1.8 \times 10^4$ M^{-1} sec^{-1}) whereas dissociation of bound ligand was extremely slow ($K_{off} \lesssim 0.9 \times 10^{-5}$ sec^{-1}). The apparent association constant obtained from equilibrium binding experiments ($K_a = 2.4 \times 10^9$ M^{-1}) agreed well with the association constant calculated from the above data ($K_a = 2.0 \times 10^9$ M^{-1}). Tolleshaug (1981) studied the binding of asialofetuin and asialo-orosomucoid to the hepatocyte receptor at 10°C. He determined that the rate constant for association was 7.4×10^4 M^{-1} sec^{-1} for asialo-orosomucoid and 2.9×10^4 M^{-1} sec^{-1} for asialofetuin. The rate of dissociation for both proteins was 6.0×10^{-6} sec^{-1}. At 37°C, the rate constant for internalization was 3.7×10^{-3} sec^{-1}, corresponding to a half-life of 3 min for the occupied receptor on the cell surface. A similar residency time was reported by this laboratory calculated from the number of surface membrane receptors and the amount of asialo-orosomucoid internalized and degraded during a specific time interval (Steer and Ashwell, 1980).

A number of investigators have attempted to determine the number of asialoglycoprotein receptor sites per hepatocyte. However, quantification using conventional binding studies has produced a wide range of values ranging from 50,000 to 500,000 receptors per hepatocyte surface membrane (Schwartz *et al.*, 1980). The disparity in the literature may be the result of parameters that explain the diversity of equilibrium constants and Scatchard plots, i.e., incubation medium, perfusion technique, ligand preparation, etc. In a timely investigation, Schwartz and co-workers reported the difficulties in quantification of asialoglycoprotein receptors on the rat hepatocyte. It appears that the presence of serum and/or BSA reduce the saturation binding and uptake of [^{125}I]asialo-orosomucoid by isolated rat hepatocytes. The nature of the inhibitor(s) is presently uncertain, although certain fatty acids may be involved in this inhibition. Of note was the finding that, in all cases, the affinity constant for binding was approximately 7.0×10^{-9} M and Scatchard plots of the saturation curves yielded straight lines, suggesting a single class of binding sites (Schwartz *et al.*, 1980). It is most probable that the hepatocyte cell surface contains between 200,000 and 250,000 asialoglycoprotein receptor sites (Weigel, 1980).

The rat liver receptor, originally identified on plasma membrane, was subsequently isolated from a variety of subcellular organelles including the Golgi complex, smooth microsomes, and lysosomes (Pricer and Ashwell, 1976). More recently, examination of the topological distribution of the hepatic lectin revealed it to be a transmembrane protein (Harford and Ashwell, 1981). Several investigators have observed invariance of the receptor turnover rate during periods of protracted asialoglycoprotein catabolism, suggesting the existence of a reutilization mechanism whereby the receptor is spared lysosomal destruction (Tanabe *et al.*, 1979; Regoeczi *et al.*, 1979). Further corroboration of receptor reutilization has come from studies with isolated rat hepatocytes devoid of protein synthesis as a result of incubation with cycloheximide (Steer and Ashwell, 1980; Tolleshaug, 1981). These cells were still capable of internalizing and degrading more ligand than could be accounted for by the total cellular content of functional receptor.

The physiological significance of the intracellular pool of receptor (approximately 90% of receptor activity is subcellular) is presently unknown. However, Stockert and co-workers, using various techniques including antireceptor antibodies, provided strong evidence that the plasma membrane receptor remains segregated from the relatively larger internal pool of receptor (Stockert *et al.*, 1980). In fact, if the surface membrane receptor is internalized during heterophagy of asialoglycoproteins, it remains an enigma as to how it escapes degradation. In an innovative study, Blumenthal and co-workers (1980) have shown a translocation of the receptor across the black lipid membrane in the presence of asialo-ligand. This observation provides a unique conceptual mechanism for sparing of the receptor from degradation. Alternatively, by fusing rat liver membranes with mouse L cells, Doyle *et al.*, (1979) have shown that a preexisting pool of internal receptors is not essential for successful endocytosis and metabolism of asialoglycoproteins.

REFERENCES

Ashwell, G., and Morell, A. G. (1974). *Adv. Enzymol.* **41**, 99–128.
Baenziger, J. U., and Maynard, Y. (1980). *J. Biol. Chem.* **255**, 4607–4613.
Berg, T., and Tolleshaug, H. (1980). *Biochem. Pharmacol.* **29**, 917–925.
Berg, T., Ose, T., Ose, L., and Tolleshaug, H. (1981). *Int. J. Biochem.* **13(3)**, 253–259.
Berry, M. N., and Friend, D. S. (1969). *J. Cell Biol.* **43**, 506–520.
Blumenthal, R., Klausner, R. D., and Weinstein, J. N. (1980). *Nature (London)* **288**, 333–338.
Doyle, D., Hou, E., and Warren, R. (1979). *J. Biol. Chem.* **254**, 6853–6856.
Dunn, W. A., LaBadie, J. H., and Aronson, N. N., Jr. (1979). *J. Biol. Chem.* **254**, 4191–4196.
Dunn, W. A., Hubbard, A. L., and Aronson, N. N., Jr. (1980). *J. Biol. Chem.* **255**, 5971–5978.
Gregoriadis, G., Morell, A. G., Sternlieb, I., and Scheinberg, I. H. (1970). *J. Biol. Chem.* **245**, 5833–5837.
Harford, J., and Ashwell, G. (1981). *Proc. Natl. Acad. Sci. USA* **78**, 1557–1561.
Hubbard, A. L., and Stukenbrok, H. (1979). *J. Cell Biol.* **83**, 65–81.
Hubbard, A. L., Wilson, G., Ashwell, G., and Stukenbrok, H. (1979). *J. Cell Biol.* **83**, 47–64.
Hudgin, R. L., Pricer, W. E., Jr., Ashwell, G., Stockert, R. J., and Morell, A. G. (1974). *J. Biol. Chem.* **249**, 5536–5543.
Kawasaki, T., and Ashwell, G. (1976). *J. Biol. Chem.* **251**, 1296–1302.
Kolset, S. O., Tolleshaug, H., and Berg, T. (1979). *Exp. Cell Res.* **122**, 159–167.
LaBadie, J. H., Chapman, K. P., and Aronson, N. N., Jr. (1975). *Biochem. J.* **152**, 271–279.
Morell, A. G., and Scheinberg, I. H. (1972). *Biochem. Biophys. Res. Commun.* **48**, 808–815.
Morell, A. G., Irvine, R. A., Sternlieb, I., Scheinberg, I. H., and Ashwell, G. (1968). *J. Biol. Chem.* **243**, 155–159.
Morell, A. G., Gregoriadis, G., Scheinberg, I. H., Hickman, J., and Ashwell, G. (1971). *J. Biol. Chem.* **246**, 1461–1467.
Neufeld, E. F., and Ashwell, G. (1980). In *The Biochemistry of Glycoproteins and Proteoglycans* (W. J. Lennarz, ed.), pp. 241–266, Plenum Press, New York.
Paulson, J. C., Hill, R. L., Tanabe, T., and Ashwell, G. (1977). *J. Biol. Chem.* **252**, 8624–8628.
Pricer, W. E., Jr., and Ashwell, G. (1971). *J. Biol. Chem.* **246**, 4825–4833.
Pricer, W. E., Jr., and Ashwell, G. (1976). *J. Biol. Chem.* **251**, 7539–7544.
Regoeczi, E., Taylor, P., Debanne, M. T., März, L., and Hatton, M. W. C. (1979). *Biochem. J.* **184**, 399–407.
Sawamura, T., Nakada, H., Fujii-Kuriyama, Y., and Tashiro, Y. (1980). *Cell Struct. Funct.* **5**, 133–146.
Schwartz, A. L., Rup, D., and Lodish, H. F. (1980). *J. Biol. Chem.* **255**, 9033–9036.
Schwartz, A. L., Marshak-Rothstein, A., Rup, D., and Lodish, H. F. (1981). *Proc. Natl. Acad. Sci. USA* **78**, 3348–3352.
Steer, C. J., and Ashwell, G. (1980). *J. Biol. Chem.* **255**, 3008–3013.

Steer, C. J., Kempner, E. S., and Ashwell, G. (1981). *J. Biol. Chem.* **256,** 5851–5856.

Stockert, R. J., Morell, A. G., and Scheinberg, I. (1977). *Science* **197,** 667–668.

Stockert, R. J., Howard, D. J., Morell, A. G., and Scheinberg, I. H. (1980). *J. Biol. Chem.* **255,** 9028–9029.

Stockert, R. J., Haimes, H. B., Morell, A. G., Novikoff, P. M., Novikoff, A. B., Quintana, N., and Sternlieb, I. (1981). *Lab. Invest.* **43,** 556–563.

Stowell, C. P., and Lee, Y. C. (1978). *J. Biol. Chem.* **253,** 6107–6110.

Tanabe, T., Pricer, W. E., Jr., and Ashwell, G. (1979). *J. Biol. Chem.* **254,** 1038–1043.

Tolleshaug, H. (1981). *Int. J. Biochem.* **13,** 45–51.

Tolleshaug, H., Berg, T., Nilsson, M., and Norum, K. R. (1977). *Biochim. Biophys. Acta* **499,** 73–84.

Tolleshaug, H., Berg, T., Frölich, W., and Norum, K. R. (1979). *Biochem. Biophys. Acta* **585,** 71–84.

Tolleshaug, H., Abdelnour, M., and Berg, T. (1980). *Biochem. J.* **190,** 697–703.

Van Lenten, L., and Ashwell, G. (1972). *J. Biol. Chem.* **247,** 4633–4640.

Wall, D. A., Wilson, G., and Hubbard, A. L. (1980). *Cell* **21,** 79–93.

Warren, R., and Doyle, D. (1981). *J. Biol. Chem.* **256,** 1346–1355.

Weigel, P. H. (1980). *J. Biol. Chem.* **255,** 6111–6120.

163

Regulation of Insulin Receptor Metabolism

Michael Krupp and M. Daniel Lane

1. INTRODUCTION

Insulin serves the vital role in animal cells of regulating the synthetic phase of energy metabolism by promoting glucose and amino acid uptake, glycogen synthesis, lipogenesis, and protein synthesis. The first point at which insulin can act on a target cell to elicit these effects is through the binding interaction with its specific receptors on the plasma membrane. The magnitude of the biochemical response triggered by this interaction will depend upon several factors including the concentration of the hormone, the number of cell surface receptors, and the degree of coupling of these receptors of the response system.

Compelling evidence (Bar *et al.*, 1979; Kahn and Roth, 1976) now indicates that the level of functional insulin receptor in the plasma membrane is rigorously controlled. For example, cells respond to prolonged exposure to insulin by reducing the number of cell surface receptors, thereby decreasing their sensitivity to insulin. As the regulation of receptor level must be exerted on a specific step(s) in the synthesis, processing, or turnover of the receptor, it will be instructive to identify the sequence of steps involved. A tentative pathway for insulin receptor metabolism, consistent with current findings for this and other receptors, is shown in Fig. 1.

The total number of insulin receptors a cell possesses ultimately depends both on the rates of receptor synthesis (Step 1, Fig. 1) and receptor degradation (Step 7, Fig. 1). Not all cellular receptors, however, may be functional at a given time. An intracellular pool of inactive or active receptors may exist that becomes functional upon activation and/or insertion into the plasma membrane. Recruitment of such internal or reserve receptors could be regulated. Moreover, the insulin receptor, like certain other receptors (Steer and Ashwell, 1980; Goldstein *et al.*, 1979), may recycle (Step 8, Fig. 1), returning to the cell surface following internalization. Thus, intracellular receptors would be partitioned into

Michael Krupp and M. Daniel Lane • Department of Physiological Chemistry, The Johns Hopkins University School of Medicine, Baltimore, Maryland 21205.

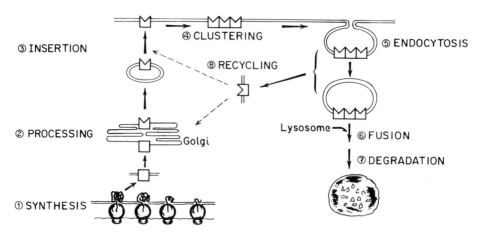

Figure 1. Pathway of insulin receptor metabolism.

either the recycling or the degradation pathway; this branch-point could be an effective site for regulation.

2. RATES OF RECEPTOR SYNTHESIS AND TURNOVER

The total number of receptors a cell possesses in the steady state will be determined by the rate of receptor synthesis relative to the rate of receptor degradation as described by the equation

$$R = k_S/k_D$$

where R is the number of receptors per cell, k_S is the zero-order rate constant for receptor synthesis, and k_D is the first-order rate constant for receptor degradation.

Two approaches have been employed to measure the rates of synthesis and degradation of the insulin receptor and to ascertain which of these rates is altered and responsible for changes in receptor level induced by physiological alterations. These approaches have led to different, but interesting, results. In theory, it should be possible to block receptor synthesis with an inhibitor of protein synthesis, such as cycloheximide, and then follow the rate of degradation of receptor through the loss of surface or total insulin-binding activity. Knowing k_D and receptor level, R, in the uninhibited state, k_S can be calculated. Several recent studies have used this approach to measure the turnover kinetics of the receptor. Kosmakos and Roth (1980), Karlsson *et al.* (1979), and Rosen *et al.* (1979) have determined the rate of receptor depletion from the cell surface in the presence of cyclo-heximide and found the half-life to be from 24 to 40 hr ($k_D = 0.017$ to 0.029 hr^{-1}). This differs markedly from the half-life of the receptor determined recently in our laboratory (Reed and Lane, 1980a; Reed *et al.*, 1981; Krupp and Lane, 1981) by the heavy-isotope density-shift method.

Density-labeling of the receptor provides an unambiguous basis for distinguishing between "old" and newly synthesized receptor in the absence of inhibitors of protein synthesis. In practice, cells are shifted from a medium containing "light" (^1H, ^{12}C, ^{14}N) amino acids to a medium containing "heavy" ($> 90\%$ ^2H, ^{13}C, and ^{15}N) amino acids. The incorporation of heavy amino acids into receptor protein substantially increases the density of newly synthesized insulin receptor. Hence, new heavy and old light receptor, solubilized with detergent, can be resolved by isopycnic banding on CsCl density gradients. By quantitating light and heavy receptor on gradients at various times after the shift to heavy medium, the rate of formation of newly synthesized heavy insulin receptor and the rate of degradation of old light receptor can be determined concomitantly. Using the density-shift technique, we have determined that the half lives of the insulin receptor of 3T3-L1 adipocytes (Reed and Lane, 1980a), 3T3-C2 fibroblasts (Reed *et al.*, 1981a), and chick liver cells (Krupp and Lane, 1981) in culture are 6.7, 8.6, and 10 hr, respectively. Insulin had no significant effect on the rate of receptor degradation (Krupp and Lane, 1981).

Importantly, the use of inhibitors of protein synthesis, which may afffect functional components of the receptor turnover system, is avoided by the density-shift procedure. This accounts for the fact that previous investigators (cited above) had obtained far longer half lives for the receptor when cycloheximide was present. Using the density-shift method, we (Reed *et al.*, 1981b) demonstrated that cycloheximide and puromycin markedly decrease the rate of receptor degradation in 3T3-L1 adipocytes, increasing its half-life from 7 to about 25 hr. Thus, continued synthesis of a short-lived protein appears to be necessary for normal insulin receptor turnover.

3. PATHWAY OF INSULIN RECEPTOR SYNTHESIS, PROCESSING, AND TURNOVER

The available evidence indicates that newly synthesized insulin receptor follows the typical pathway (Fig. 1, Steps 1–3) for other plasma membrane-associated proteins (Devreotes *et al.*, 1976; Fambrough and Devreotes, 1978). Posner and colleagues (Bergeron *et al.*, 1978; Posner *et al.*, 1978a) have detected significant levels of insulin receptor in Golgi cell fraction. Others (Horvat *et al.*, 1975) have reported that insulin receptors are present in the membranes of the endoplasmic reticulum. Presumably, active completed receptor found in the membranes of these organelles constitutes receptor en route to the plasma membrane. Insulin binding activity has also been found associated with the nuclear fraction (Goldfine and Smith, 1976; Goldfine *et al.*, 1977) of the cell; however, the origin and role of this receptor remain obscure.

One of the essential cotranslational and posttranslational processing steps in the synthesis of the insulin receptor appears to be glycosylation. The importance of the oligosaccharide moiety to insulin receptor function became apparent when Cuatrecasas' group (Cuatrecasas, 1972a,b) showed that the insulin binding capacity of the receptor was markedly decreased by stepwise treatment with neuraminidase and β-galactosidase. Thus, the presence of the capping sugars, i.e., sialyl and β-galactosyl residues, appears to be essential for insulin binding. Nevertheless, a contradictory report has been published (Clark *et al.*, 1978).

Further insight into the role of glycosylation in formation of functional insulin receptor and its turnover has come from studies using tunicamycin, a potent inhibitor of core oligosaccharide addition in the synthesis of asparagine *N*-linked glycoproteins. Treatment of 3T3-L1 adipocytes (Rosen *et al.*, 1979; Reed and Lane, 1980b; Reed *et al.*, 1981b) or IM-9 lymphocytes (Keefer and DeMeyts, 1980) with tunicamycin leads to the rapid depletion of cell surface insulin-binding activity; similarly, total cellular detergent-extractable receptor is lost in tunicamycin-treated 3T3-L1 adipocytes (Reed *et al.*, 1981b). In order to determine whether this decrease in cellular insulin receptor level is a consequence of an altered rate of receptor synthesis or degradation, Reed *et al*. (1981b) made use of the heavy-isotope density-shift technique. Tunicamycin totally blocked the formation of active, newly synthesized heavy receptor in cells shifted to medium containing heavy amino acids and yet the rate or receptor degradation was, if anything, decreased. Thus, glycosylation of the receptor during or subsequent to translation appears to be required for its activation.

Recent evidence (Ronnett and Lane, 1981) reveals that an inactive aglyco-form of the receptor accumulates in cells exposed to tunicamycin. Upon removal of tunicamycin from cells completely depleted of active receptor, insulin binding activity is largely restored within 24 hr. If, at the time tunicamycin is removed, the cells are shifted to medium containing heavy (^{15}N, ^{13}C, ^{2}H) amino acids with or without cycloheximide, it can be demonstrated that the previously synthesized inactive light receptor becomes active. Thus, the inactive aglyco-form of the receptor must accumulate in the endoplasmic reticulum or elsewhere in the cell in the presence of tunicamycin and then reenter the glycosylation pathway when the inhibitor is removed.

This raises the question of the intracellular site of core oligosaccharide addition, for it is believed that this process occurs only cotranslationally. Another possible explanation is that deglycosylation, followed by reglycosylation, occurs during the life cycle of the receptor, perhaps as part of a recycling pathway (Fig. 1, Step 8).

It is widely believed that under physiological conditions, insulin binds to its receptor, the insulin–receptor complex is internalized, and the ligand is degraded via the lysosomal pathway (Marshall and Olefsky, 1979; Gorden *et al*., 1980). Employing fluorescent conjugates of insulin, Schlessinger and colleagues (1978; Maxfield *et al*., 1978) have observed that insulin binds initially to diffusely distributed mobile receptors in the plasma membrane and these hormone–receptor complexes cluster in a temperature-sensitive process to form immobile patches over the coated regions of the membrane. These patches are believed to pinch off and form coated vesicles. Schlessinger *et al*. (1978) suggest that preceding "global" clustering of receptors in coated pits (a slow process requiring seconds to minutes), microclustering of a few ligand–receptor complexes occurs. This microaggregation is rapid (0.05–1 sec), and postulated to have a role in generating the signal for the biological response.

On the basis of the observation that inhibitors of transglutaminase block global clustering of fluorescent hormone on the cell surface (Davies, *et al*., 1980), it has been suggested that covalent cross-linking between receptors or between receptor and a protein in the coated pit may occur. Transglutaminase catalyzes the formation of ϵ-(γ-glutaminyl)lysyl cross-linkages. Insulin-induced receptor "down-regulation" (discussed in a later section), which involves endocytosis, is inhibited in fibroblasts by dansylcadaverine, a transglutaminase inhibitor (Davies *et al*., 1980; Levitzki *et al*., 1980; Baldwin *et al*., 1980). Thus, an intramolecular cross-linking step may be involved in the down-regulation process.

Despite extensive efforts, it has not yet been possible to directly trace the pathway of the insulin receptor after it leaves the plasma membrane. Willingham and Pastan (1980) have postulated a general pathway for receptor metabolism based on studies with fluorescent hormone derivatives and electron microscopy. They suggest that receptors with their ligands are endocytosed in coated pits by a process similar to that for the low-density lipoprotein receptor (Goldstein *et al*., 1979). The endocytotic vesicle is then uncoated and becomes a "receptosome," moving to the Golgi–GERL region of the cell where the ligand and receptor are separated. Presumably, the receptor would then be recycled to the plasma membrane via the Golgi secretory pathway while the ligand would be shunted to the lysosome for degradation. More definitive studies, including the isolation and characterization of the putative receptosome, will be needed to validate this hypothesis.

The question of whether the insulin receptor recycles back to the cell surface (Fig. 1, Steps 8, 2, and 3) after internalization has not been resolved. If the assumptions are made that insulin degradation is receptor mediated (Terris and Steiner, 1975) and that the insulin–receptor complex is endocytosed (Gorden *et al.*, 1980), a kinetic argument can be made that the insulin receptor must undergo recycling. It can be calculated from the data of Terris *et al.* (1979) that the rate of receptor-mediated insulin degradation by the liver cell is about 7–10 molecules of insulin per receptor per hour. Yet the rate of insulin receptor degradation by the liver cell is only 0.07 per hour (Krupp and Lane, 1982). Hence, the receptor must be endocytosed and returned to the cell surface at least 100 times during the lifetime of the receptor.

Recent studies in our laboratory indicate that receptor degradation is more complex than originally believed (Krupp and Lane, 1982). Treatment of cultured chick hepatocytes with the lysosomotropic agent chloroquine fails to alter the turnover of the receptor, but slows the rate of degradation of insulin. These data suggest either that there is an inactivation of the receptor prior to interaction with lysosomes or that insulin receptors are degraded at some site other than the lysosome. These results are in contrast to those with the acetylcholine receptor (Libby *et al.*, 1980) and suggest that different receptors follow distinct degradative pathways.

4. CONTROL OF RECEPTOR ACTIVITY: INSULIN-INDUCED RECEPTOR DOWN-REGULATION

Ligand-induced lowering of cell surface receptor level, referred to as "down-regulation," is a property shared by many types of cell surface receptors including the insulin receptor. Gavin *et al.* (1974) were the first to show *in vitro* that a prolonged high insulin concentration lowers cell surface insulin receptor level. After a 16-hr exposure to 10^{-6} M insulin, nearly half of the surface receptors of IM-9 lymphocytes are lost. Kosmakos and Roth (1980) have extended this work and demonstrated that the process is temperature dependent, insulin concentration dependent, and affected by insulin analogs in direct proportion to their biological activity. It has been found (Mott *et al.*, 1979; Baldwin *et al.*, 1980; Marshall and Olefsky, 1980) that physiological levels of insulin, i.e., as low as 5 ng/ml, can induce down-regulation. Under cell culture conditions, down-regulation of receptor level is maximal after a 6- to 9-hr exposure to a saturating level of insulin (Baldwin *et al.*, 1980; Marshall and Olefsky, 1980). Down-regulation of receptors has been observed in adipocytes (Livingston *et al.*, 1978; Marshall and Olefsky, 1980), rat hepatocytes (Blackard *et al.*, 1978; Petersen *et al.*, 1978), human fibroblasts (Mott *et al.*, 1979), 3T3-C2 fibroblasts (Reed *et al.*, 1981a), and chick hepatocytes (Krupp and Lane, 1981) in culture.

Although details of the mechanism by which down-regulation occurs are lacking, several important facts bearing on this process are known. Insulin appears to induce a redistribution of insulin receptors from the external surface of the plasma membrane to an intracellular site not accessible to exogenous insulin (Krupp and Lane, 1981). Thus, exposure of chick liver cell monolayers to insulin causes a 60% decrease in the number of cell surface receptors without altering the total number of receptors per cell, i.e., receptor extractable from total cellular membranes with Triton X-100. As a result, the number of intracellular receptors, i.e., those not accessible to exogenous insulin, increases about twofold during down-regulation of surface receptors. Amatruda and co-workers (personal communication) showed that phospholipase C treatment of down-regulated rat hepatocytes does not reverse down-regulation. Thus, chronic exposure to insulin fails to increase the "cryptic" pool of receptors which normally become accessible when cells are treated with phospholipase C (Amatruda, personal communication).

Consistent with the view that down-regulation solely involves translocation of receptors from the cell surface to the intracellular compartment, neither the rate of synthesis nor the rate of degradation of cellular insulin receptors is altered (Krupp and Lane, 1981). Heavy-isotope density-shift experiments in which newly synthesized receptors were labeled with ^{15}N, ^{13}C, and ^2H-containing amino acids and separated from previously synthesized light (^{14}N, ^{12}C, ^1H) receptors by density gradient centrifugation revealed that the rate of synthesis of heavy receptor and the rate of decay of old light receptor were unaffected by insulin-induced down-regulation. Consistent with these results, Desbuquois and colleagues (1979) found that insulin treatment of the intact rat leads to a decreased insulin receptor level in rat liver plasma membranes, but an increased receptor level in the Golgi cell fraction. Posner *et al.* (1978b) have reported similar results with the *db/db* mouse.

Two mechanisms could account for the insulin-induced redistribution of cellular receptors. Insulin could activate the rate of receptor endocytosis-internalization or could slow the rate of receptor recycling to the plasma membrane after internalization. The available evidence does not distinguish between these mechanisms. It is interesting in this connection that inhibitors of protein synthesis, such as cycloheximide, block insulin-induced down-regulation (Baldwin *et al.*, 1980). Similarly, the degradation of insulin receptors (Reed *et al.*, 1981a), a process that also requires endocytosis, is blocked by cycloheximide. These findings indicate that the continued synthesis of a short-lived protein is essential for down-regulation, as well as for the degradation of the receptor. This rapidly turning-over protein may be a functional component of the receptor internalization system, for both processes involve receptor endocytosis as a common step.

A number of agents, capable of eliciting the biological responses induced by insulin intact cells, have the capacity to cause down-regulation of the insulin receptor. Caro and Amatruda (1980) showed that these agents, i.e., anti-insulin receptor antibody, hydrogen peroxide, wheat germ agglutinin, vitamin K_5, and spermine, which mimic insulin action, cause a decrease in cell surface insulin receptor level. This observation suggests that these agents induced a common structural change in the receptor, perhaps a conformational change, which triggers "insulin action" and targets the receptor for endocytosis.

At present the physiological role of insulin-induced down-regulation of its receptor can only be speculated upon. It may constitute a mechanism by which cells can be partially desensitized to high circulating insulin level by negative feedback. Thus, the number of functional, i.e., cell surface, receptors is reduced to compensate for a prolonged elevation of plasma insulin concentration. Alternatively, receptor down-regulation may represent a mechanism by which the action of bound insulin can be terminated. Regardless of its function, the insulin-induced down-regulation process observed with cells in culture appears to reflect a true physiological process, as it can also be observed in the intact animal (Harrison *et al.*, 1976; Bar *et al.*, 1976; Kahn, 1980).

5. CONTROL OF RECEPTOR ACTIVITY: DIFFERENTIATION-INDUCED ALTERATIONS IN INSULIN RECEPTOR LEVEL

In contrast to the loss of cell surface insulin receptors that occurs during ligand-induced down-regulation, elevated levels of cell surface insulin receptors are expressed in certain cases where cell function is altered by differentiation. For example, 3T3-L1 pre-adipocytes exhibit a dramatic increase in the number of cell surface insulin receptors during differentiation (Reed *et al.*, 1977; Rubin *et al.*, 1978; Chang and Polakis, 1978; Lane *et al.*, 1980; Reed and Lane, 1980a), and thymus-derived lymphocytes acquire insulin receptors upon activation by mitogens and alloantigens (Helderman and Strom, 1977, 1979; Helderman *et al.*, 1979).

3T3-L1 fibroblasts, cloned by Green and Kehinde (1974, 1975, 1976) from mouse late-embryo cells, differentiate in culture into cells that exhibit morphological and biochemical characteristics of adipocytes (Mackall *et al.*, 1976; Green and Kehinde, 1976). Accompanying differentiation, these cells exhibit a 10- to 20-fold rise in insulin receptor level, fully differentiated 3T3-L1 adipocytes acquiring 200,000 receptors per cell (Reed and Lane, 1980b). This alteration is specific because the number of cell surface receptors for certain other polypeptide ligands, e.g., epidermal growth factor and choleragen, either remain constant or decrease, respectively (Reed *et al.*, 1977).

One of the best documented differentiated characteristics of adipocytes is the stimulation of glucose uptake and oxidation by acute insulin treatment. Prior to differentiation, acute exposure of 3T3-L1 cells to insulin has virtually no effect on the rates of [^{14}C]glucose oxidation (to $^{14}CO_2$) or glucose uptake. During differentiation, however, 3T3-L1 cells acquire the capacity to respond acutely to insulin (Reed and Lane, 1980b). Thus, brief exposure of 3T3-L1 "adipocytes" to insulin markedly stimulates both [^{14}C]glucose oxidation to $^{14}CO_2$ (Reed and Lane, 1980b) and glucose uptake rates (Karlsson *et al.*, 1979). Karlsson *et al.* (1979) have presented evidence that this increase of insulin-stimulated glucose transport during differentiation is, at least in part, due to the acquisition of coupling of the insulin receptor to the glucose transport system. These investigators showed that the fractional occupancy of receptors required for maximal stimulation of glucose transport by insulin decreased substantially during the course of differentiation. Like normal adipocytes, fully differentiated 3T3-L1 adipocytes require only 5–10% receptor occupancy with insulin for maximal activation of glucose transport. Thus, it appears that the increased sensitivity of 3T3-L1 cells during differentiation is due both to an increase in the number of insulin receptors and their coupling to the response system(s).

Using the heavy-isotope density-shift approach, it has been possible to characterize the alterations in insulin receptor metabolism responsible for the dramatic differentiation-induced rise in receptor level (Reed and Lane, 1980a). Both undifferentiated 3T3-L1 preadipocytes and 3T3-L1 adipocytes that possessed 15-fold higher insulin receptor levels were shifted from medium containing light (^{14}N, ^{12}C, ^{1}H) amino acids to medium containing heavy (^{15}N, ^{13}C, ^{2}H) amino acids. By quantitating the change in the amount of each density class of receptors with time after the shift to "heavy" medium, the effects of differentiation on rates of receptor synthesis and degradation were assessed. The differentiation-induced rise in steady-state receptor level was shown to be due solely to a 20- to 40-fold increase in the rate of receptor synthesis. A 2-fold increase in the first-order rate constant for degradation, k_D, also accompanied differentiation. Hence, the rise in insulin receptor level induced by differentiation in preadipocytes is solely the result of an increased rate of receptor synthesis.

Recent work with thymus-derived lymphocytes (T cells) has demonstrated that activation of these cells stimulates the appearance of insulin receptors (Helderman and Strom, 1977, 1979). Subsequent studies using inhibitors of protein, RNA, and DNA synthesis indicate that activation initiates receptor synthesis. It is interesting to note that Helderman *et al.* (1979) showed that the presence of insulin receptor on the T cell is crucial for the expression of cellular function, i.e., cytotoxicity.

6. SUMMARY

Recent studies on the synthesis, processing, and turnover of the insulin receptor reveal that the receptor is in a highly dynamic state. In a variety of cell types, total cellular insulin receptor turns over with a half-life of 7–8 hr; moreover, cell surface receptor can down-regulate to a new steady-state level within 9–10 hr. This enables the cell to rapidly adjust its functional insulin receptor level and thereby alter its responsiveness to insulin. For example, during the differentiation of 3T3-L1 preadipocytes into adipocytes, a large increase in receptor level and sensitivity to insulin is brought about by a marked increase in the rate of receptor synthesis. In a somewhat different physiological context, down-regulation of insulin receptor level induced by its ligand is brought about by translocation of cell surface receptors into an intracellular compartment not accessible to insulin; this process occurs without changes in the rates of receptor synthesis or degradation. A major objective of future work will be to determine the mechanisms by which these regulatory changes are brought about.

ACKNOWLEDGMENTS. Some of the work cited in this review was supported by grants from the National Institutes of Health (AM 14574), the American Heart Association (78822), and the Kroc Foundation. M.K. is a recipient of a postdoctoral fellowship from the Juvenile Diabetes Foundation.

REFERENCES

Baldwin, D., Jr., Prince, M., Marshall, S., Davies, P., and Olefsky, J. M. (1980). *Proc. Natl. Acad. Sci. USA* **77,** 5975–5978.

Bar, R. S., Gorden, P., Roth, J., Kahn, C. R., and DeMeyts, P. (1976). *J. Clin. Invest.* **58,** 1123–1135.

Bar, R. S., Harrison, L. C., Muggeo, M., Gorden, P., Kahn, C. R., and Roth, J. (1979). In *Advances in Internal Medicine* (G. H. Stollerman, ed.), pp. 23–52, Year Book Medical Publishers, Chicago.

Bergeron, J. J. M., Posner, B. I., Josefsberg, A., and Sikstrom, T. (1978). *J. Biol. Chem.* **253,** 4058–4066.

Blackard, W. G., Guzehan, P. S., and Small, M. E. (1978). *Endocrinology* **103,** 548–553.

Caro, J., and Amatruda, J. M. (1980). *Science* **210,** 1029–1030.

Chang, T., and Polakis, S. E. (1978). *J. Biol. Chem.* **253,** 4693–4696.

Clark, S., DeLuise, M., Larkins, R. G., Melick, R. A., and Harrison, L. C. (1978). *Biochem. J.* **174,** 37–43.

Cuatrecasas, P. (1972a). *J. Biol. Chem.* **247,** 1981–1991.

Cuatrecasas, P. (1972b). In *Insulin Action* (I. B. Fritz, ed.), pp. 137–169, Academic Press, New York.

Davies, P. J. A., Davies, D., Levitzki, A., Maxfield, F. R., Milhaud, P., Willingham, M., and Pastan, I. (1980). *Nature (London)* **283,** 162–167.

Desbuquois, B., Willeput, J., and Heut de Froberville, A. (1979). *FEBS Lett.* **106,** 338–344.

Devreotes, P. N., Gardner, J. M., and Fambrough, D. M. (1976). *Cell* **10,** 365–373.

Fambrough, D. M., and Devreotes, P. N. (1978). *J. Cell Biol.* **76,** 237–244.

Gavin, J. R., III, Roth, J., Neville, D. M., Jr., DeMeyts, P., and Buell, D. N. (1974). *Proc. Natl. Acad. Sci. USA* **71,** 84–88.

Goldfine, I. D., and Smith, G. J. (1976). *Proc. Natl. Acad. Sci. USA* **73,** 1427–1431.

Goldfine, I. D., Smith, G. J., Wang, K. Y., and Jones, A. L. (1977). *Proc. Natl. Acad. Sci. USA* **74,** 1368–1372.

Goldstein, J. L., Anderson, R. G., and Brown, M. S. (1979). *Nature (London)* **276,** 679–685.

Gorden, P., Carpenter, J.-L., Freychet, P., and Orci, L. (1980). *Diabetologia* **18,** 263–274.

Green, H., and Kehinde, O. (1974). *Cell* **1,** 113–116.

Green, H., and Kehinde, O. (1975). *Cell* **5,** 19–27.

Green, H., and Kehinde, O. (1976). *Cell* **7,** 105–113.

Harrison, L. C., Martin, F. I. R., and Meleck, R. A. (1976). *J. Clin. Invest.* **58,** 1435–1441.

Helderman, J. H., and Strom, T. B. (1977). *J. Clin. Invest,* **59,** 338–344.

Helderman, J. H., Strom, T. B., and Dupuy-D'Angeac, A. (1979). *Cell. Immunol.* **46,** 247–258.

Horvat, A., Li, E., and Katsoyannes, P. G. (1975). *Biochim. Biophys. Acta* **382,** 609–620.

Kahn, C. R. (1980). *Metabolism* **29,** 455–466.

Kahn, C. R., and Roth, J. (1976). In *Hormone Receptor Interaction: Molecular Aspects* (G. S. Levey, ed.), pp. 1–29, Dekker, New York.

Karlsson, F. A., Greenfeld, C., Kahn, C. R., and Roth, J. (1979). *Endocrinology* **104,** 1383–1392.

Keefer, L. M., and DeMeyts, P. M. (1980). *Diabetologia* **19,** 288.

Kosmakos, F., and Roth, J. (1980). *J. Biol. Chem.* **255,** 9860–9869.

Krupp, M. N., and Lane, M. D. (1981). *J. Biol. Chem.* **256,** 1689–1694.

Krupp, M. N., and Lane, M. D. (1982). *J. Biol. Chem.* **257,** 1372–1377.

Lane, M. D., Reed, B. C., and Clements, P. R. (1980). *ICN–UCLA Symposium on Control of Cellular Division and Development*, in press.

Levitzki, A., Willingham, M., and Pastan, I. (1980). *Proc. Natl. Acad. Sci. USA* **77,** 2706–2710.

Libby, P., Bursztajn, S., and Goldberg, A. L. (1980). *Cell* **19,** 481–491.

Livingston, J. N., Purvis, B. J., and Lockwood, D. H. (1978). *Metabolism* **27**(Suppl. 2), 2009–2014.

Mackall, J. C., Student, A. K., Polakis, S. E., and Lane, M. D. (1976). *J. Biol. Chem* **251,** 6462–6464.

Marshall, S., and Olefsky, J. (1979). *J. Biol. Chem.* **254,** 10153–10160.

Marshall, S., and Olefsky, J. M. (1980). *J. Clin. Invest.* **66,** 763–772.

Maxfield, F. R., Schlessinger, J., Schecter, Y., Pastan, I., and Willingham, M. C. (1978). *Cell* **14,** 805–810.

Mott, D. M., Howard, B. V., and Bennett, P. H. (1979). *J. Biol. Chem.* **254,** 8762–8767.

Petersen, B., Beckner, S., and Blecher, M. (1978). *Biochim. Biophys. Acta* **542,** 470–485.

Posner, B. I., Josefsberg, A., and Bergeron, J. J. M. (1978a). *J. Biol. Chem.* **253,** 4067–4073.

Posner, B. I., Raguidan, D., Josefsberg, A., and Bergeron, J. J. M. (1978b). *Proc. Natl. Acad. Sci. USA* **75,** 3302–3306.

Reed, B. C., and Lane, M. D. (1980a). *Proc. Natl. Acad. Sci. USA* **77,** 285–289.

Reed, B. C., and Lane, M. D. (1980b). *Adv. Enzyme Regul.* **18,** 97–117.

Reed, B. C., Kaufman, S. H., Mackall, J. C., Student, A. K., and Lane, M. D. (1977). *Proc. Natl. Acad. Sci. USA* **74,** 4876–4880.

Reed, B. C., Ronnett, G. V., Clements, P. R., and Lane, M. D. (1981a). *J. Biol. Chem.* **256,** 3719–3725.

Reed, B. C., Ronnett, G. V., and Lane, M. D. (1981b). *Proc. Natl. Acad. Sci. USA* **78,** 2908–2912.

Ronnett, G. V., and Lane, M. D. (1981). *J. Biol. Chem.* **256,** 4704–4707.

Rosen, O. M., Chia, G. H., Fung, C., and Rubin, C. S. (1979). *J. Cell Physiol.* **99,** 37–41.

Rubin, C. S., Hirsch, A., Fung, C., and Rosen, O. M. (1978). *J. Biol. Chem.* **253,** 7510–7578.

Schlessinger, J., Schecter, Y., Willingham, M., and Pastan, I. (1978). *Proc. Natl. Acad. Sci. USA* **75,** 2659–2663.

Steer, C. J., and Ashwell, G. (1980). *J. Biol. Chem.* **255,** 3008–3013.

Terris, S., and Steiner, D. F. (1975). *J. Biol. Chem.* **250,** 8389–8398.

Terris, S., Hofmann, C., and Steiner, D. F. (1979). *Can. J. Biochem.* **57,** 459–468.

Willingham, M. C., and Pastan, I. (1980). *Cell* **21,** 67–77.

164

Translocation Hypothesis of the Insulin-Dependent Activation of Glucose Transport in Fat Cells

Tetsuro Kono

1. INTRODUCTION

It is well established that insulin lowers the blood glucose concentration by stimulating glucose transport across the plasma membrane of certain cell types, e.g., muscle and fat cells (Pilkis and Park, 1974; Czech, 1977). However, the underlying mechanism of this hormone action is still obscure.

In this review article, we will consider a new working hypothesis on the mechanism of insulin action on glucose transport recently advanced by Cushman and Wardzala (1980) and Suzuki and Kono (1980). According to this hypothesis, insulin stimulates glucose transport in fat cells by facilitating translocation of the glucose transport mechanism to the plasma membrane from an intracellular storage site, rather than stimulating a glucose transport mechanism already associated with the plasma membrane. Interestingly, while Cushman and Wardzala (1980) and Suzuki and Kono (1980) independently reached basically identical conclusions, their experimental approaches are entirely different, as described below.

2. SUZUKI AND KONO'S APPROACH: DETERMINATION OF GLUCOSE TRANSPORT ACTIVITY IN RECONSTITUTED LIPOSOMES

In the work of Suzuki and Kono (1980), subcellular structures of fat cells were fractionated by differential and sucrose density gradient centrifugations. The fractionated subcellular structures were solubilized with sodium cholate, and the transport activities reconstituted into egg lecithin liposomes by sonication, freezing-and-thawing, and a second brief

Tetsuro Kono • Department of Physiology, Vanderbilt Medical School, Nashville, Tennessee 37232.

sonication. The glucose transport activity in the reconstituted liposomes was estimated by measuring the uptake of D-[^{3}H]glucose in the presence or absence of 20 μM cytochalasin B. As this agent is a potent inhibitor of glucose transport (Kletzien and Perdue, 1973; Czech, 1976a), the difference in the rates of glucose uptake observed in the presence and absence of cytochalasin B was assumed to show the carrier-mediated transport activity, as previously discussed by Kasahara and Hinkle (1976) and Shanahan and Czech (1977).

Suzuki and Kono (1980) found that there were two peaks of glucose transport activities in subcellular vesicles fractionated by sucrose density gradient centrifugation. One of the peaks (Peak A) coincided with the peak of 5'-nucleotidase, a marker enzyme of the plasma membrane (Emmelot et al., 1964), while the location of the other peak (Peak B) was identical to that of the peak of UDP galactose : N-acetylglucosamine galactosyltransferase, a marker enzyme of the Golgi apparatus (Fleischer, 1974). The height of Peak A in the basal form of fat cells was 13 ± 2 units (nmole glucose/min per mg protein in Peak A) and in the insulin-treated cells 26 ± 3 units. In contrast, the height of Peak B in the basal cells was 115 ± 11 units and in the insulin-treated cells 69 ± 4 units. In other words, insulin apparently increased the height of Peak A while decreasing that of Peak B. These effects of the hormone were statistically significant ($p < 0.01$). However, the increase in the height of Peak A ($+13$ units) was considerably smaller than the decrease in the height of Peak B (-46 units). The cause of this discrepancy is not clear. Apparently, insulin did not alter the activity of galactosyltransferase (the Golgi marker) in Peak B. The apparent K_m values of glucose transport mediated by the activities solubilized from Peaks A and B were indistinguishable: 11.2 ± 1.5 mM for Peak A and 13.1 ± 0.6 mM for Peak B.

3. CUSHMAN AND WARDZALA'S APPROACH: CYTOCHALASIN B BINDING ASSAY

Cushman and Wardzala (1980) prepared the plasma membrane and microsomal fractions of fat cells by differential centrifugation and subjected them to the assay for the glucose-inhibitable cytochalasin B-binding capacity. Although the binding of cytochalasin B is not specific to the glucose transport mechanism, it was previously found by Wardzala et al. (1978) that the nonspecific portion of the cytochalasin B binding was largely blocked by cytochalasin E, which does not inhibit glucose transport, and that the specific part was competitively inhibited by high concentrations of D-glucose. Therefore, Wardzala et al. (1978) and Cushman and Wardzala (1980) estimated the specific binding of cytochalasin B to the glucose transport mechanism by incubating a sample with different concentrations of [^{3}H]cytochalasin B (40–20,000 nM), in the presence of 2 μM unlabeled cytochalasin E and either in the presence or absence of 500 mM D-glucose. Subsequently, they estimated the number of specific binding sites by the Scatchard analysis.

Cushman and Wardzala (1980) found that both plasma membrane and microsomal fractions had specific cytochalasin B-binding activities, and that insulin increased the binding activity in the plasma membrane fraction while decreasing the activity in the microsomal fraction. By correcting their data for recoveries of protein, 5'-nucleotidase [a marker enzyme of the plasma membrane (Emmelot et al., 1964)], and NADH dehydrogenase [an enzyme rich in the endoplasmic reticulum (Fleischer and Kervine, 1974; Makino and Kono, 1980)], Cushman and Wardzala (1980) concluded that insulin increased the number of specific binding sites for cytochalasin B in the plasma membrane from 0.35 ± 0.21 units (sites/cell $\times 10^{-1}$) to 2.08 ± 0.50 units while decreasing the number in the microsomal membrane from 2.81 ± 0.74 units to 1.09 ± 0.14 units. In other words, insulin appeared to increase the number of plasma membrane-bound glucose transport sites 1.73 units while decreasing the number in the microsomal membrane 1.72 units. It should be noted, however, that the data of Suzuki and Kono (1980) suggest that the intracellular glucose transport activity is associated with certain Golgi-like vesicles rather than the endoplasmic reticulum.

4. OTHER DATA CONSISTENT WITH THE TRANSLOCATION HYPOTHESIS

It was previously reported by Kono and Colowick (1961), Kono *et al.* (1977a), Chandramouli *et al.* (1977), and Yu and Gould (1978) that either ATP or metabolic energy appears to be involved in the action of insulin on glucose transport. In addition, it was noted by Kono *et al.* (1977b) and Vega *et al.* (1980) that ATP or metabolic energy may also be involved in reversing the insulin effect on glucose transport after insulin-treated cells are washed with a hormone-free buffer.

Years ago, Wohltmann and Narahara (1966) reported that while insulin binds to frog muscle at 0°C, the effect of the hormone on glucose transport does not develop at 0°C. However, when the temperature is elevated to 19°C, the effect of the hormone begins to appear after an absolute lag period of 20 min. Later, Kono *et al.* (1977a) observed that insulin fails to stimulate glucose transport in fat cells at 15°C. At this temperature, the internalization of the cell-bound hormone is also blocked (Suzuki and Kono, 1979), and so are the endocytotic reactions in general (Silverstein *et al.*, 1977).

Even at 37°C, Ciaraldi and Olefsky (1979) detected a short delay in the development of the insulin effect. This delay, or lag period, is increased to 2 min by the addition of 1–5 mM vinblastine or colchicine (Häring *et al.*, 1979), which inhibit microtubule activity (Snyder and McIntosh, 1976).

Although it was once controversial, most investigators now seem to agree that insulin stimulates glucose transport by increasing the V_{max} value without changing the apparent K_m value (Czech, 1977). The K_m value for glucose transport reported by Loten *et al.* (1976) was 13 mM and that by Whitesell and Gliemann (1979), 7 mM. These values are not very different from the values obtained by Suzuki and Kono (1980) on the reconstituted system (10–15 mM).

Finally, Wardzala *et al.* (1978) have reported that insulin increases the specific binding of cytochalasin B to the plasma membrane of fat cells.

All of the results mentioned above appear to be consistent with the view that insulin initiates translocation of the transport mechanism by a temperature-sensitive, energy-requiring, and microtubule-dependent process. Although the mechanism of translocation is yet to be ascertained, it was previously suggested by Palade (1975) that certain subcellular structures may be recycled between the plasma membrane and the Golgi apparatus, possibly being mediated by endocytotic and exocytotic reactions. Incidentally, endocytosis can be a very rapid reaction as seen in the secretion of certain neurotransmitters, e.g., secretion of acetylcholine at the neuromuscular junction.

5. DATA DIFFICULT TO EXPLAIN BY THE TRANSLOCATION HYPOTHESIS

It has been shown by Czech (1976b), Vega and Kono (1979), and Whitesell and Gliemann (1979) that fat cells have different temperature coefficients for basal and insulin-stimulated glucose transport activities. Therefore, it may be postulated that glucose transport in intact cells is regulated not only by translocation but also by an additional mechanism.

In fat cells, insulin has acute regulatory effects on glucose transport, phosphodiesterase, glycogensynthase, phosphorylase, and pyruvate dehydrogenase (Pilkis and Park, 1974; Czech, 1977). However, no translocation appears to be involved in the last four hormone actions. Therefore, it may be suggested that the effects of insulin on certain enzymes are mediated by different mechanisms or by a single basic mechanism that is branched out into multiple secondary mechanisms at certain early steps in the reaction sequence.

ACKNOWLEDGMENTS. I am grateful to Mrs. Frances W. Robinson and Miss Teresa L. Blevins for their help in preparation of this manuscript.

REFERENCES

Chandramouli, V., Milligan, M., and Carter, J. R., Jr. (1977). *Biochemistry* **16**, 1151–1158.
Ciaraldi, T. P., and Olefsky, J. M. (1979). *Arch. Biochem. Biophys.* **193**, 221–231.
Cushman, S. W., and Wardzala, L. J. (1980). *J. Biol. Chem.* **255**, 4758–4762.
Czech, M. P. (1976a). *J. Biol. Chem.* **251**, 2905–2910.
Czech, M. P. (1976b). *Mol. Cell. Biochem.* **11**, 51–63.
Czech, M. P. (1977). *Annu. Rev. Biochem.* **46**, 359–384.
Emmelot, P., Bos, C. J., Benedetti, E. L., and Rümke, P. H. (1964). *Biochim. Biophys. Acta* **90**, 126–145.
Fleischer, B. (1974). *Methods Enzymol.* **29**, 180–191.
Fleischer, S., and Kervina, M. (1974). *Methods Enzymol.* **31**, 6–39.
Häring, H. U., Kemmler, W., and Hepp, K. D. (1979). *FEBS Lett.* **105**, 329–332.
Kasahara, M., and Hinkle, P. C. (1976). *Proc. Natl. Acad. Sci. USA* **73**, 396–400.
Kletzien, R. F., and Perdue, J. F. (1973). *J. Biol. Chem.* **248**, 711–719.
Kono, T., and Colowick, S. P. (1961). *Arch. Biochem. Biophys.* **93**, 514–519.
Kono, T., Robinson, F. W., Sarver, J. A., Vega, F. V., and Pointer, R. H. (1977a). *J. Biol. Chem.* **252**, 2226–2233.
Kono, T., Vega, F. V., Rains, K. B., and Shumway, S. J. (1977b). *Fed. Proc.* **36**, 341.
Loten, E. G., Regen, D. M., and Park, C. R. (1976). *J. Cell Physiol.* **89**, 651–659.
Makino, H., and Kono, T. (1980). *J. Biol. Chem.* **255**, 7850–7854.
Palade, G. (1975). *Science* **189**, 347–358.
Pilkis, S. J., and Park, C. R. (1974). *Annu. Rev. Pharmacol.* **14**, 365–388.
Shanahan, M. F., and Czech, M. P. (1977). *J. Biol. Chem.* **252**, 8341–8343.
Silverstein, S. C., Steinman, R. M., and Cohn, Z. A. (1977). *Annu. Rev. Biochem.* **46**, 669–722.
Snyder, J. A., and McIntosh, J. R. (1976). *Annu. Rev. Biochem.* **45**, 699–720.
Suzuki, K., and Kono, T. (1979). *J. Biol. Chem.* **254**, 9786–9794.
Suzuki, K., and Kono, T. (1980). *Proc. Natl. Acad. Sci. USA* **77**, 2542–2545.
Vega, F. V., and Kono, T. (1979). *Arch. Biochem. Biophys.* **192**, 120–127.
Vega, F. V., Key, R. J., Jordan, J. E., and Kono, T. (1980). *Arch. Biochem. Biophys.* **203**, 167–173.
Wardzala, L. J., Cushman, S. W., and Salans, L. B. (1978). *J. Biol. Chem.* **253**, 8002–8005.
Whitesell, R. R., and Gliemann, J. (1979). *J. Biol. Chem.* **254**, 5276–5283.
Wohltmann, H. J., and Narahara, H. T. (1966). *J. Biol. Chem.* **241**, 4931–4939.
Yu, K. T., and Gould, M. K. (1978). *Am. J. Physiol.* **234**, E407–416.

165

Hybridization and Implantation of Membrane Components by Fusion

Hormone Receptors and Associated Molecules

Michael Schramm

1. RATIONALE

The stringent limitations of conventional procedures for the study of membrane systems prompted us to seek a new approach. It had earlier been demonstrated that the mobility of proteins in the cell membrane (Singer and Nicolson, 1972; Harris, 1974; Frye and Edidin, 1970) leads to intermixing of the components when two cells are fused with each other. We therefore reasoned that the function of a certain protein in a membrane could be studied by interacting it through membrane fusion with other components in a membrane of different properties (Orly and Schramm, 1976). Intermixing of components would thus be achieved without having to resort to solubilization by detergents, which might irrevocably perturb the system. Although this rationale has so far been applied mainly in research of hormone-activated adenylate cyclase systems, its more general applicability deserves to be emphasized first (Fig. 1).

Suspected inhibitors or activators in one cell membrane could be revealed by their effect, after fusion, on the systems contributed by the other membrane. Deficiencies in a multicomponent membrane system created by disease, mutation, or intentionally, by a specific enzyme inhibitor or other chemical agent, could be analyzed by membrane hybridization. Homologous dissociable multicomponent systems would form new hybrid systems upon fusion of the two membranes. Similarly, subunit interchange between oligomeric membrane proteins would also produce new hybrid molecules. The study of such hybrids can yield important and unique information on the properties of the system. Furthermore, a solubilized purified membrane component, which by itself may have no measurable ac-

Michael Schramm • Department of Biological Chemistry, The Hebrew University of Jerusalem, Jerusalem, Israel.

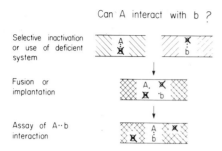

Can A interact with b ?

Selective inactivation
or use of deficient
system

Fusion or
implantation

Assay of A··b
interaction

Figure 1. Scheme describing the principle of analysis by membrane hybridization. A–B and a–b are shown as two homologous systems. However, b could also be an inhibitor or activator or a lipid that is being studied for its possible effect on A in the other membrane.

tivity (e.g., a receptor), might be reinsolubilized and implanted by fusion to test whether it is still functional.

2. ANALYSIS OF HORMONE RECEPTOR–ADENYLATE CYCLASE SYSTEMS

Membrane hybridization as outlined above was initiated (Orly and Schramm, 1976) to answer a key question: Are the various hormone receptors that activate adenylate cyclase in different cells independent units (Robison *et al.*, 1971; Cuatrecasas, 1974; Rodbell, 1975)? To tackle this problem, turkey erythrocytes, which possess an epinephrine-activated adenylate cyclase in their membrane, were treated with N-ethylmaleimide. This reagent irreversibly inactivated the catalytic activity of the adenylate cyclase. Selective heat treatment achieved the same purpose. Subsequently, the erythrocytes were fused with cells from a clone of Friend erythroleukemia that possessed adenylate cyclase but no epinephrine receptor. Efficient fusion was produced by addition of sendai virus (Loyter *et al.*, 1975; Poste and Nicolson, 1978). Membranes prepared from the fused cells demonstrated epinephrine-induced adenylate cyclase activity. Thus, fusion of the two cells produced a hybrid adenylate cyclase system consisting of the epinephrine receptor from turkey erythrocytes coupled to the adenylate cyclase of the Friend cells. The foreign transferred receptor was functional also *in vivo* in the intact fused cells (Schulster *et al.*, 1978). It was therefore concluded that a cell can be made to respond to a hormone that it does not normally recognize. In further studies, the following hormone receptors were hybridized with heterologous adenylate cyclase systems: β-adrenergic receptor of turkey erythrocytes, C-6 glioma cells, and S-49 lymphoma cyc⁻ cells; PGE₁ receptor of Friend erythroleukemia cells (Schramm *et al.*, 1977; Neufeld *et al.*, 1980); vasoactive intestinal peptide receptor of HT 29 human colon carcinoma cells (Laburthe *et al.*, 1979); glucagon receptor of isolated liver membranes (Schramm, 1979). There seemed to be no species or tissue specificity barriers with respect to receptor–adenylate cyclase hybrid formation. With the help of biochemical hybridization it was also possible to distinguish the intrinsic properties of the receptor from those of the membrane in which the receptor resides (Orly and Schramm, 1976). Along the same line, experiments suggested that the β-adrenergic receptor determines the affinity to the agonist but that the intrinsic activity of the system is determined by the adenylate cyclase components with which the receptor is hybridized (Pike *et al.*, 1979). It was further shown that a desensitized receptor apparently remains in this state even when hybridized with a foreign adenylate cyclase (Pike and Lefkowitz, 1980).

It should be emphasized that these hybrids were formed solely through the stoichiometric interaction of components that existed prior to fusion. No nucleic acid or protein synthesis was required, and the hybrids should therefore be defined as biochemical hybrids of membrane components in contradistinction to genetic hybrids.

Taken together, these findings led to some major conclusions regarding the nature of hormone and neurotransmitter receptors coupled to adenylate cyclase. (1) The receptors exist in the native cell membrane as independent units freely dissociable from the adenylate cyclase. (2) The various receptors, each specific for a different hormone, originating from different tissues and animal species, are all freely interchangeable in their coupling to different adenylate cyclase systems. (3) These receptors should be visualized as having two active loci, a binding site that is highly specific for a certain hormone or neurotransmitter and a "universal joint" by which the receptor is coupled to the regulatory unit of the adenylate cyclase and which therefore is presumably identical in all the receptors. Because of these considerations, it seems quite likely that the structure of the different receptors that activate adenylate cyclase will turn out to be quite similar, but for the limited area of the hormone-binding site, which is unique for each receptor (Schramm *et al.*, 1977).

3. METHODOLOGY

As with any new approach, once the rationale is formulated the rate of development of methodology usually determines the rate of application to actual problems. Because the activity of the biochemical hybrids depends on the stoichiometric interaction of receptor and adenylate cyclase, a high efficiency of fusion is required. Sendai virus was used as the fusogen in the earlier experiments with intact tissue culture cells. However, the virus failed to fuse isolated membranes from tissues or even turkey erythrocyte ghosts. Polyethylene glycol had been in use for genetic hybridization, which does not require a very high efficiency of fusion (Davidson and Gerald, 1976). Its application to hybridization of adenylate cyclase systems had also been demonstrated (Schwarzmeier and Gilman, 1977). A highly efficient chemical fusion procedure was therefore developed, based on polyethylene glycol but with additions of phospholipids, ATP, and Mg^{2+} (Schramm, 1979; Neufeld *et al.*, 1980). The latter supplements were essential for high fusion efficiency. In contrast to the method using virus, there seemed to be no limitations to the type of cell or even isolated membrane that could be fused by the new procedure. Cell-to-cell, membrane-to-cell, or membrane-to-membrane transfers of hormone receptors became feasible (Schramm, 1979, 1980). Even the implantation of solubilized reprecipitated hormone receptor in the membrane of a cell proceeded successfully (Eimerl *et al.*, 1980). It should, however, be noted that while the procedure preserves the activity of adenylate cyclase, the cells mostly lyse soon after fusion. In experiments where a low efficiency of fusion suffices, milder conditions which are compatible with a high viability of the fused cells can sometimes be used (Doyle *et al.*, 1979). Recently, solubilized sendai virus envelopes were employed to trap solubilized membrane components by removal of detergent. The reconstituted envelopes containing the relevant components were fused with intact cells (Volsky *et al.*, 1979). The findings with this procedure for implantation of membrane components are reviewed elsewhere in this volume (Chapter 158).

4. MECHANISM OF ACTION OF THE HORMONE RECEPTOR STUDIED BY HYBRIDIZATION AND IMPLANTATION

Three components of the adenylate cyclase system are presently recognized: R, the hormone receptor; G, the GTP-binding regulatory protein (Rodbell *et al.*, 1971; Pfeuffer and Helmreich, 1975; Cassel *et al.*, 1977; Vigne *et al.*, 1978; Howlett *et al.*, 1979); and C, the catalytic unit of the adenylate cyclase (see Ross and Gilman, 1980), for a review and Neufeld *et al.*, 1980, for a scheme illustrating the coordinated function of the three components). Development of an efficient method for membrane fusion permitted manipulation of G and C on the inner side of the membrane prior to fusion. Mild tryptic digestion of erythrocyte membranes totally inactivated C but left R essentially intact as demonstrated by hybridization (Neufeld *et al.*, 1980).

Further experiments showed that hormone activation of adenylate cyclase proceeds in two independent steps that can be separated in time and space (Neufeld *et al.*, 1980). In the presence of hormone, R activated G by facilitating the binding of GTP or its hydrolysis-resistant analog, Gpp(NH)p (Cassel and Selinger, 1978). This reaction readily proceeded in erythrocyte membranes even after inactivation of C by *N*-ethylmaleimide. Subsequently, the activated G component containing Gpp(NH)p (G*) was transferred by fusion to another membrane containing C, producing the G*C complex of high spontaneous adenylate cyclase activity. Thus, R in the membrane may not recognize C at all because it activates it indirectly, via G (Neufeld *et al.*, 1980).

In determining the relative amount of G by hybridization, the S-49 lymphoma variant, which lacks functional G (Howlett *et al.*, 1979), proved most useful. The experiments also demonstrated that R, G, and G* [activated by Gpp/NH)p] can each be hybridized with components of the adenylate cyclase system in another membrane. The amount of each transferred, measured as activation of adenylate cyclase, was directly proportional to the amount in the fusion system. The procedure therefore served as a quantitative functional microassay of the above components.

Reconstitution of a detergent-solubilized hormone receptor by conventional procedures had hitherto failed, probably because of the inherent requirement that all the components of the adenylate cyclase system must survive the detergent treatment. In contrast, reconstitution by hybridization or implantation requires only that the receptor survive solubilization in a compatible detergent. After purification or other desired manipulations, the receptor is insolubilized by removal of the detergent and is implanted in a native cell membrane where all the other components of the adenylate cyclase system are supplied in excess in their native state. Thus, reconstitution is performed in the water-insoluble state to which these membrane components are naturally adapted. Employing this approach, solubilization and reconstitution of β-adrenergic receptor function was indeed successful (Eimerl *et al.*, 1980).

These experiments paved the way for the reconstitution of the primary action of the β-adrenergic receptor by combining solubilized R with solubilized G at any desired ratio (Citri and Schramm, 1980). After removal of detergent, addition of hormone and Gpp(NH)p produced activated G*. The latter was measured by implantation in the S-49 lymphoma variant in which it joined C to form the highly active G*C adenylate cyclase. Thus, a hormone-responsive system of considerable efficiency was reconstituted. Kinetic studies showed that R acted catalytically in activating G. The rate-limiting step was not the encounter of R with G in the reconstituted membrane, but rather the subsequent step of activation of G by R in the R–G complex.

Recent studies with the nicotinic acetylcholine receptor indicate that hybridization might be a practical approach even in a system of tightly associated subunits. Experiments indicated that the ionophore in a receptor-blocked preparation and the receptor in an ionophore-blocked preparation might hybridize to form a functional system (Schiebler and Hucho, 1980).

5. CONCLUSIONS AND FUTURE PROSPECTS

Membrane hybridization readily demonstrated that the hormone receptors coupled to adenylate cyclase exist in the cell membrane as independent, dissociable units and that the different receptors, each specific for a different hormone, are all interchangeable. The ability to implant a solubilized receptor in an intact cell membrane that contained the other components of the system in their native state led to reconstitution of the primary action of the hormone receptor. Implantation by fusion could therefore serve to test whether purified membrane receptors have retained their ability to function. This criterion is essential because our present experience shows that a receptor may still demonstrate specific ligand binding despite losing its ability to activate the adenylate cyclase. It should be reiterated that analysis by membrane hybridization might be profitably adapted to any system in biological membranes. It is a relatively simple procedure that can often be performed with the intact membrane or even intact cell. It serves to gain information and answer questions about specific components in the membrane that cannot readily be obtained by conventional approaches. In the future, membrane hybridization might perhaps be performed on small amounts of biopsy material from patients in order to locate specific membrane defects.

ACKNOWLEDGMENTS. Work from our laboratory reviewed in this chapter is supported by NIH Grant AM-10451-15 and the USA–Israel Binational Science Foundation.

REFERENCES

Cassel, D., and Selinger, Z. (1978). *Proc. Natl. Acad. Sci. USA* **75**, 4155–4159.
Cassel, D., Levkovitz, H., and Selinger, Z. (1977). *J. Cycl. Nucl. Res.* **3**, 393–406.
Citri, Y., and Schramm, M. (1980). *Nature (London)* **287**, 297–300.
Cuatrecasas, P. (1974). *Annu. Rev. Biochem.* **43**, 202–214.
Davidson, R. L., and Gerald, P. S. (1976). *Somat. Cell Genet.* **2**, 165–176.
Doyle, D., Hou, E., and Warren, R. (1979). *J. Biol. Chem.* **254**, 6853–6856.
Eimerl, S., Neufeld, G., Korner, M., and Schramm, M. (1980), *Proc. Natl. Acad. Sci. USA* **77**, 760–764.
Frye, C. D., and Edidin, M. (1970). *J. Cell Sci.* **7**, 319–335.
Harris, H. (1974). *Nucleus and Cytoplasm* (3rd ed.), pp. 109–141, Oxford University Press (Clarendon), London.
Howlett, A. C., Sternweis, P. C., Macik, B. A., Van Arsdale, P. M., and Gilman, A. G. (1979). *J. Biol. Chem.* **254**, 2287–2295.
Laburthe, M., Rosselin, G., Rousset, M., Zweibaum, A., Korner, M., Selinger, Z., and Schramm, M. (1979). *FEBS Lett.* **98**, 41–43.
Loyter, A., Zakai, N., and Kulka, R. G. (1975). *J. Cell Biol.* **66**, 292–304.
Neufeld, G., Schramm, M., and Weinberg, N. (1980). *J. Biol. Chem.* **255**, 9268–9274.
Orly, J., and Schramm, M. (1976). *Proc. Natl. Acad. Sci. USA* **73**, 4410–4414.
Pfeuffer, T., and Helmreich, E. J. M. (1975). *J. Biol. Chem.* **250**, 867–876.
Pike, L. J., and Lefkowitz, R. J. (1980). *Biochim. Biophys. Acta* **632**, 354–365.
Pike, L. J., Limbird,, L. E., and Lefkowitz, R. J. (1979). *Nature (London)* **280**, 502–504.
Poste, G., and Nicolson, G. L. (eds.) (1978). *Membrane Fusion,* North-Holland, Amsterdam.
Robison, G. A., Butcher, R. W., and Sutherland, E. W. (eds.) (1971). In *Cyclic AMP,* pp. 73–75, Academic Press, New York.
Rodbell, M. (1975). *J. Biol. Chem.* **250**, 5826–5834.
Rodbell, M., Birnbaumer, L., Pohl, S. L., and Krans, M. J. (1971). *J. Biol. Chem.* **246**, 1877–1882.
Ross, E. M., and Gilman, A. G. (1980). *Annu. Rev. Biochem.* **49**, 533–564.
Schiebler, W., and Hucho, F. (1980). *Biochim. Biophys. Acta* **597**, 626–630.
Schramm, M. (1979). *Proc. Natl. Acad. Sci. USA* **76**, 1174–1178.
Schramm, M. (1980). In *Membrane Bioenergetics* (C. P. Lee, G. Schatz, and L. Ernster, eds.), pp. 349–359, Addison–Wesley, Reading, Mass.

Schramm, M., Orly, J., Eimerl, S., and Korner, M. (1977). *Nature (London)* **268,** 310–313.

Schulster, D., Orly, J., Seidel, G., and Schramm, M. (1978). *J. Biol. Chem.* **253,** 1201–1206.

Schwarzmeier, J. D., and Gilman, A. G. (1977). *J. Cycl. Nucl. Res.* **3,** 227–238.

Singer, S. J., and Nicolson, G. L. (1972). *Science* **175,** 720–731.

Vigne, J. N., Johanson, G. L., Bourne, H. R., and Coffino, P. (1978). *Nature (London)* **272,** 720–722.

Volsky, D. J., Cabantchik, Z. I., Beigel, M., and Loyter, A. (1979). *Proc. Natl. Acad. Sci. USA* **76,** 5440–5444.

166

Hormone-Activated Adenylate Cyclase and Protein Biosynthesis: A Similar Mechanism of Regulation by Guanine Nucleotides

Dan Cassel and Zvi Selinger

1. INTRODUCTION

The role that guanine nucleotide binding factors play in protein biosynthesis is now well established. In each step of protein synthesis, binding of GTP promotes association of the guanine nucleotide binding factor with the ribosome to form an active complex, whereas hydrolysis of the bound GTP results in the dissociation of the complex (Kaziro, 1978). Each step thus generates a GTPase cycle.

Recently, a regulatory GTPase cycle was discovered for the hormone-activated adenylate cyclase system (Cassel *et al.*, 1977). Extensive studies in many laboratories are revealing an interesting mechanistic similarity between the GTP binding component in the adenylate cyclase system and the elongation factors of protein synthesis. Here we wish to describe this similarity and to discuss its possible physiological and evolutionary implications.

2. ROLE OF GUANINE NUCLEOTIDES IN ADENYLATE CYCLASE

The adenylate cyclase system in the plasma membrane is now known to be composed of at least three components: a hormone receptor, a catalytic moiety, and a guanine nucleotide binding protein (see Ross and Gilman, 1980, for recent review). This basic structure of the adenylate cyclase is apparently highly conserved for it is possible to hybridize

Dan Cassel and Zvi Selinger • Department of Biological Chemistry, The Hebrew University of Jerusalem, Jerusalem, Israel.

a variety of receptors with catalytic components across a wide range of species barriers (Schramm *et al.*, 1977). The guanine nucleotide binding protein functions as an essential "transducer," which transfers the signal produced by the activated hormone receptor to the catalytic enzyme (Pfeuffer, 1977). GTP is required for this process, and its binding to the regulatory component results in the activation of the enzyme (Rodbell *et al.*, 1971). The guanine nucleotide binding component also functions in the removal of the activation signal by hydrolyzing the bound GTP (Cassel and Selinger, 1976). The GTPase reaction generates GDP which remains tightly associated with the regulatory site. Hormonal activation of the cyclase is due to the fact that the activated hormone receptor facilitates the exchange of the bound GDP for GTP (Cassel and Selinger, 1978). Studies of the molecular size of adenylate cyclase components in detergent extracts (Limbird *et al.*, 1980) indicate that the hormone promotes the formation of a stable complex between the hormone receptor and the guanine nucleotide binding protein and that GTP dissociates this complex. It has also been shown (Pfeuffer, 1979) that binding of GTP to the guanine nucleotide binding protein greatly increases its affinity to the catalytic unit. In contrast, when charged with GDP, the guanine nucleotide binding protein barely interacts with the catalytic component. The transduction of the hormonal signal in the adenylate system can thus be ascribed to a shuttle of the guanine nucleotide binding protein between the receptor and the cyclic AMP synthesizing catalytic component. A model that incorporates all these data is presented in Fig. 1.

3. COMPARISON BETWEEN ADENYLATE CYCLASE AND PROTEIN BIOSYNTHESIS

The role of guanine nucleotides in polypeptide chain elongation by the prokaryotic factors T_u and T_s is depicted in Fig. 2. Apparently, polypeptide chain elongation has been conserved through evolution, for an analogous mechanism of action was found for the eukaryotic elongation factor, EF-1 (Kaziro, 1978). The similarity between this mechanism

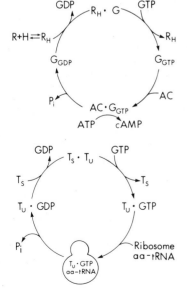

Figure 1. Regulation of adenylate cyclase activity by hormones and guanine nucleotides. R-receptor, H-hormone, G-guanine nucleotide binding protein, AC-catalytic unit of adenylate cyclase.

Figure 2. Regulation of the first elongation step in protein biosynthesis by factors T_s and T_u. T_s and T_u are elongation factors, aa-tRNA is amino acyl transfer RNA.

and that operating in the adenylate cyclase system is indeed striking. An essential feature of both systems is the presence of a guanine nucelotide binding protein that changes its reactivity toward other components, depending on whether it binds GTP or GDP. As depicted in Figs. 1 and 2, both the adenylate cyclase and the elongation step in protein synthesis operate as cyclic systems driven by the energy of GTP hydrolysis. In both systems the activation is due to an enhancement of the nucleotide exchange reaction, and activation is terminated by the GTPase reaction. Perhaps the most astonishing facet of similarity is that the hormone receptor in the plasma membrane activates the adenylate cyclase (Cassel and Selinger, 1978) in a way that is closely analogous to the action of the cytoplasmic factor T_s in protein biosynthesis (Kaziro, 1978). In the cyclase system, however, the hormone provides an additional regulatory feature. While T_s in itself facilitates the exchange reaction, the action of the receptor in the adenylate cyclase is dependent on hormone binding.

Another analogy between the systems is manifested by the action of certain bacterial toxins that catalyze a specific ADP ribosylation of the GTP binding proteins. Diphtheria toxin causes ADP ribosylation of the eukaryotic elongation factor 2 (Gill and Pappenheimer, 1971), whereas cholera toxin catalyzes ADP ribosylation of the guanine nucleotide binding protein of the cyclase system (Cassel and Pfeuffer, 1978; Gill and Meren, 1978; Johnson *et al.*, 1978b). In both cases, ADP ribosylation leads to an inhibition of the GTPase activity. Inhibition of the GTPase results in an enhancement of adenylate cyclase activity due to a decrease in the "off" rate (Cassel and Selinger, 1977). On the other hand, prevention of GTP hydrolysis results in stabilization of the EF-2 ribosome complex. This in turn leads to an inhibition of protein biosynthesis because this process is dependent upon continuous recycling of the ribosome-bound elongation factors. The effects of the toxins in both systems are mimicked by hydrolysis-resistant analogs of GTP. These analogs cause persistent activation of adenylate cyclase (Schramm and Rodbell, 1975) and an inhibition of protein biosynthesis (Yokosawa *et al.*, 1975).

4. PHYSIOLOGICAL ADVANTAGES OF THE GTPase CYCLE

The main advantage of the GTPase cycle is that it provides equally effective activation and turn-off reactions within one system. This is of particular importance for the adenylate cyclase system, which must change its catalytic activity in response to rapid fluctuations in hormone concentrations. Because the adenylate cyclase constitutes only a minute fraction of the cell membrane, effective activation requires high affinity of interaction between the adenylate cyclase components. On the other hand, such high affinity could result in practically irreversible activation and inability to turn the system off upon removal of the hormonal signal. The GTPase cycle solves this problem by coupling the turn-off reaction to hydrolysis of GTP, which supplies the free energy required to drive the turn-off reaction. Concurrent operation of the activation and turn-off reactions results in a steady state of enzyme activity that can readily shift its equilibrium in response to regulatory signals. The overall effect is that the adenylate cyclase functions as a useful biological switch that can be turned on and off with great efficiency.

Inspection of Fig. 1 reveals that the adenylate cyclase resembles in its regulatory mechanism a cyclic cascade of an interconvertible enzyme. The resemblance is due to the fact that both activation and the counteracting turn-off reaction are practically irreversible (Schramm and Rodbell, 1975; Cassel and Selinger, 1976). These reactions thus substitute for cyclic covalent modification and demodification reactions, which have not been found in the adenylate cyclase system. As discussed in depth by Stadman and Chock, cyclic cascades are endowed with extraordinary characteristics that make them highly versatile regulatory systems. Such systems can serve as amplifiers with respect to both signal response and catalytic potential (Stadman and Chock, 1977). That the GTPase cycle confers such properties on the adenylate cyclase is demonstrated by the interaction of the system with cholera toxin. Modulation of the adenylate cyclase by the toxin results in an enhancement of the catalytic potential as well as an increased sensitivity to low concentrations of hormone (Field, 1974; Bennett *et al.*, 1975; Johnson *et al.*, 1978a). It was found that cholera toxin penetrates into the cell and uses cellular NAD to cause ADP ribosylation of the guanine nucleotide binding protein (Cassel and Pfeuffer, 1978; Gill and Meren, 1978; Johnson *et al.*, 1978b). This modification is apparently not confined to bacterial toxins as a recent report describes an NAD-dependent activation of adenylate cyclase by cytosolic factor from turkey erythrocytes (Moss and Vaughan, 1978). Such modulations might operate at different stages of the cell cycle or other conditions in which adenylate cyclase activity is changed in the absence of external hormones. Taken together, these studies show that the GTPase cycle of the adenylates cyclase provides sites for regulation by both external and internal signals, thus making the adenylate cyclase a highly versatile regulatory system.

5. EVOLUTIONARY CONSIDERATIONS

Inspection of Figs. 1 and 2 reveals that the hormone receptor and the guanine nucleotide binding protein of the adenylate cyclase both have an analogous component among the elongation factors of protein synthesis. The interaction of the components with each other and with guanine nucleotides are indeed remarkably similar in the two systems. To account for this striking similarity, which is too close to be merely the result of chance, it is tempting to speculate that components of the adenylate cyclase and the elongation factors of protein synthesis have evolved from a common ancestral gene. In support of this hypothesis are the findings that in *E. coli* the elongation factor T_u is an abundant protein and its major part is associated with the cell membrane (Jacobson and Rosenbusch, 1976). Likewise, the corresponding eukaryotic factor EF-1 appears in a light and a heavy form and the latter is associated with phospholipids and cholesterol. As hormone-stimulated adenylate cyclase appeared in evolution much later than protein synthesis, one can envisage that the frequent encounter of the elongation factors with the cell membrane made it possible to test variants of these molecules for their ability to function in membrane reactions. The physiological advantage offered by the GTPase cycle conceivably acted as a selective pressure that finally led to adoption of a variant elongation factor for transduction of the hormonal signal. The acquisition of a novel physiological capability by the elongation factors of protein synthesis is not a new idea. It is well known that when *E. coli* is infected with the phage Qβ, the elongation factors T_u and T_s become subunits in the enzyme responsible for the viral RNA replication (Blumenthal *et al.*, 1972). It should be pointed out, however, that it is unlikely that the elongation factor of eukaryotes, in its present form, can substitute for the guanine nucleotide binding protein of the adenylate cyclase. This is inferred from the presence of S-49 cell variant that lacks a functional guanine nucleotide binding protein in its adenylate cyclase and yet can synthesize proteins as effectively as the wild type. Recently, the guanine nucelotide binding protein of the adenylate cyclase has been purified to homogeneity (Northup *et al.*, 1980). It is thus expected that in the near future it will be possible to test the hypothesis of a common ancestral origin by direct comparison of the physical and chemical properties of components from the adenylate cyclase and protein synthesis systems.

REFERENCES

Bennett, V., O'Keefe, E., and Cuatrecasas, P. (1975). *Proc. Natl. Acad. Sci. USA* **72**, 33–37.

Blumenthal, T., Landers, T. A., and Weber, K. (1972). *Proc. Natl. Acad. Sci. USA* **69**, 1313–1317.

Cassel, D., and Pfeuffer, T. (1978). *Proc. Natl. Acad. Sci. USA* **75**, 2669–2673.

Cassel, D., and Selinger, Z. (1976). *Biochim. Biophys. Acta* **452**, 538–551.

Cassel, D., and Selinger, Z. (1977). *Proc. Natl. Acad. Sci. USA* **74**, 3307–3311.

Cassel, D., and Selinger, Z. (1978). *Proc. Acad. Sci. USA* **75**, 4155–4159.

Cassel, D., Levkovitz, H., and Selinger, Z. (1977). *J. Cycl. Nucl. Res.* **3**, 393–406.

Field, M. (1974). *Proc. Natl. Acad. Sci. USA* **71**, 3299–3303.

Gill, D. M., and Meren, R. (1978). *Proc. Natl. Acad. Sci. USA* **75**, 3050–3054.

Gill, D. M., and Pappenheimer, A. M. (1971). *J. Biol. Chem.* 1492–1495.

Jacobson, G. R., and Rosenbusch, J. P. (1976). *Nature (London)* **261**, 23–26.

Johnson, G. L., Harden, K., and Perkins, J. P. (1978a). *J. Biol. Chem.* **253**, 1465–1471.

Johnson, G. L., Kaslow, H. R., and Bourne, H. R. (1978b). *Proc. Natl. Acad. Sci. USA* **75**, 3113–3117.

Kaziro, Y. (1978). *Biochim. Biophys. Acta* **505**, 95–127.

Limbird, L. E., Gill, D. M., and Lefkowitz, R. J. (1980). *Proc. Natl. Acad. Sci. USA* **77**, 775–779.

Moss, J., and Vaughan, M. (1978). *Proc. Natl. Acad. Sci. USA* **75**, 3621–3624.

Northup, J. K., Sternweis, P. C., Smigel, M. D., Schleifer, L. S., Ross, E. M., and Gilman, A. G. (1980). *Proc. Natl. Acad. Sci. USA* **77**, 6516–6520.

Pfeuffer, T. (1977). *J. Biol. Chem.* **252**, 7227–7234.

Pfeuffer, T. (1979). *FEBS Lett.* **101**, 85–89.

Rodbell, M., Birnbaumer, L., Pohl, S. L., and Krans, H. M. J. (1971). *J. Biol. Chem.* **246**, 1877–1882.

Ross, E. M., and Gilman, A. G. (1980). *Annu. Rev. Biochem.* **49**, 533–564.

Schramm, M., and Rodbell, M. (1975). *J. Biol. Chem.* **250**, 2232–2237.

Schramm, M., Orly, J., Eimerl, S., and Korner, M. (1977). *Nature (London)* **268**, 310–313.

Stadman, E. R., and Chock, P. B. (1977). *Proc. Natl. Acad. Sci. USA* **74**, 2761–2765.

Yokosawa, H., Kawakita, M., Arai, K., Inoue-Yokosawa, N., and Kaziro, Y. (1975). *J. Biochem. (Tokyo)* **77**, 719–728.

167

Liponomic Control of Membrane Protein Function

A Possible Model of Hormone Action

Howard Rasmussen and Toshio Matsumoto

1. INTRODUCTION

There is an increasing interest in the relationship between membrane lipids and membrane proteins. A wealth of data attest to the fact that if the lipid environment of a membrane protein is modified *in vitro*, the function of that protein is often altered (Sandermann, 1978; Jain and White, 1977; Cullis and DeKruijff, 1979). Many of these modifications of lipid composition are rather extreme. Hence, it is not immediately evident that subtle changes in membrane lipid structure could serve as a means of regulating cell function *in situ*. For instance, it is possible to alter membrane protein function by altering the environmental temperature (Brasitus *et al.*, 1979). Although this may be relevant to the behavior of certain microorganisms and unicellular eukaryotes that can adapt to and survive at different temperatures (Thompson and Nozawa, 1977; Sinensky, 1974), it is clearly irrelevant to the situation of cells in many higher organisms, which function throughout their lifetime at a relatively fixed temperature. Nonetheless, an analysis of the lipid composition of the different membranes in such a cell reveals a striking difference from one organelle to another. A most striking example of this lipid polymorphism is the difference seen in the two separate domains of the plasma membrane of an intestinal epithelial cell. In this polar cell, the basolateral (BLM) and luminal plasma membrane (or brush border membrane, BBM) are morphologically and functionally distinct. They are also quite distinct in terms of their lipid composition (Kawai *et al.*, 1974). The differences include differences in the molar ratios of glycolipids to phospholipids; differences in the relative proportion of major phospholipid classes, e.g., in the BLM the phosphatidylcholine (PC) represents 51% and the

Howard Rasmussen and Toshio Matsumoto • Departments of Medicine and Biology, Yale University School of Medicine, New Haven, Connecticut 06510

phosphatidylethanolamine (PE) 27% but in the BBM these figures are 25 and 49%, respectively; and differences in the molar ratio of fatty acid components.

A consideration of these data in the light of the results of *in vitro* manipulation of membrane lipid composition leads to the conclusion that these large differences in membrane lipid composition seen *in vivo* could alter the function of proteins in these membranes, i.e., if the same protein existed in the two domains of the plasma membrane of an enterocyte, its functional properties could be quite different in the two locations. For example, Engelhard *et al.* (1978) have shown that both basal and PGE_1-stimulated adenylate cyclase activity of the plasma membrane of the LM cell change in response to changes in the relative proportion of phospholipid polar head groups in the membrane and to the degree of unsaturation of the fatty acid in the phospholipids. The magnitude of these alterations is similar to the magnitude of the difference between the PE/PC ratio in the various domains of the mouse intestinal mucosa cell. Hence, it is possible that lipid polymorphism is of importance in determining the functional properties of membrane proteins. However, two philosophic concepts have restricted interest in this possibility. The first is the problem of specificity, the second the problem of the fluid mosaic model of membrane structure (Singer and Nicolson, 1972).

One of the more hallowed concepts of cell biology is that it is proteins, and proteins alone, that have the structural uniqueness and fidelity of structure required of the determinants of highly specific changes in cell function. Thus, the possibility that changes in membrane lipid composition, which by their very nature are conceived as general changes in membrane structure, could alter the behavior of one membrane enzyme or transport system without influencing the behavior of others has not received serious consideration. This is, in part at least, due to the widespread acceptance of the fluid mosaic model of membrane structure as a valid representation of a biological membrane. The sea of lipid envisioned in this model has been assumed to be rather homogeneous, and thus unlikely to consist of discretely different microdomains. However, recent work from a variety of experimental approaches has led to the recurrent proposal that microdomains of lipids exist within membranes, and that these separate domains exist for a sufficient time to be considered geographically (and therefore presumably functionally) distinct entities with their (presumably) distinct associated proteins (Jain and White, 1977; Cullis and DeKruijff, 1979; Klausner *et al.*, 1980).

If these latter proposals as to the nature of membrane lipid structure more closely approximate the situation in a biological membrane than does the original fluid mosaic model, then the possibility exists that a change in lipid composition could lead to a change in the function of only a few of the many proteins in a given membrane. This type of regulation of the function of a specific membrane protein(s) by an alteration in the nature of the lipid domain in which it functions has been given the name *liponomic* regulation to indicate that the change in lipid environment is the primary determinant of the change in protein structure and function (Fontaine *et al.*, 1981; Rasmussen *et al.*, 1981).

2. LIPONOMIC REGULATION AS A MECHANISM OF HORMONE ACTION

Recent studies in several laboratories have shown that changes in membrane lipid structure may be an important mechanism by which either steroid (Goodman *et al.*, 1971, 1975), sterol (Rasmussen *et al.*, 1981), peptide (Pilch *et al.*, 1980), or amine (Hirata and Axelrod, 1980) hormones regulate cell function.

Our initial studies were concerned with the mechanism of action of aldosterone in the toad urinary bladder. This steroid hormone stimulates the rate of transcellular Na^+ transport (Sharp and Leaf, 1966; Edelman and Fimogmari, 1968). This effect is not immediate. There is a delay of 60–90 min before a change in Na^+ transport occurs. The hormone increases the rate of Na^+ entry into the cell across its luminal membrane. Because of the time delay, the fact that inhibitors of RNA and protein synthesis inhibit the action of the hormone, the fact that aldosterone is apparently taken up into the nucleus of target cells, and the fact that the apparent number of Na^+ channels in this membrane increase after hormone action, the generally accepted model of its action is one in which the hormone regulates the synthesis of new cell proteins, one of which is the luminal Na^+ channel or a modulator of that channel. However, our own work has shown that the earliest metabolic effects of the hormone are on phospholipid metabolism (Goodman *et al.*, 1971, 1975). In particular, the hormone stimulates the turnover of fatty acids in membrane phospholipids, resulting in a significant increase in the weight percentage of long-chain polyunsaturated fatty acids in membrane phospholipids. Inhibitors of RNA and protein synthesis block these metabolic effects of the hormone (Lien *et al.*, 1976). In addition, an inhibitor of fatty acid synthesis blocks these effects and simultaneously blocks the action of the hormone on transcellular Na^+ transport even though this inhibitor does not block protein or RNA synthesis (Lien *et al.*, 1975). These results led us to conclude that it was possible that the primary action of this hormone is that of altering membrane lipid structure with this change leading to a change in rate of Na^+ entry into the cell.

For technical reasons it is difficult to obtain separate luminal and basolateral plasma membranes from this tissue, hence a more direct evaluation of this model has not been possible, but our more recent work concerned with 1,25-dihydroxy-vitamin D_3 action has allowed its evaluation in another tissue.

With the purpose of exploring the possibility that this mechanism of hormone action might be more general than just the action of aldosterone in the toad bladder, we have, over the past 6 years, explored the action of 1,25-dihydroxy-vitamin D_3 [1,25 $(OH)_2D_3$] on the transcellular transport of calcium in the chick intestinal mucosal cell (Max *et al.*, 1978; Rasmussen *et al.*, 1979; Matsumoto *et al.*, 1980, 1981; Fontaine *et al.*, 1981). The currently accepted model of this transport process is quite similar to that of Na^+ transport in the toad bladder: a passive entry of calcium into the cell across the luminal membrane, followed by the active extrusion of calcium across the basolateral membrane. Furthermore, a major site of action of $1,25(OH)_2D_3$ is thought to be at the point of calcium entry into the cell across the luminal membrane. In analogy with the situation in the toad bladder, there is a 1- to 2-hr delay between the time the cell takes up the hormone and the time when changes in calcium entry and transcellular calcium transport are seen. There is a substantial literature showing that this sterol hormone, just like many steroid hormones, binds to a cytosolic receptor and a considerable proportion is then translocated to the nucleus. These observations have led to the conclusion that this hormone [$1,25(OH)_2D_3$] acts by regulating the synthesis of one or more specific proteins that function to regulate calcium entry at the luminal membrane of the cell (Haussler and McCain, 1977). However, there is a major difficulty with this conclusion: Inhibitors of protein or RNA synthesis *do not* inhibit the action of $1,25(OH)_2D_3$ on luminal calcium entry or on transcellular calcium transport (Bikle *et al.*, 1978; Rasmussen *et al.*, 1979). These observations led us to explore the possibility that $1,25(OH)_2D_3$ regulates membrane function by modifying lipid structure.

The first evidence showing that this sterol hormone might influence the phospholipid composition of the brush border membrane was the observation (Max *et al.*, 1978) that hormone administration caused an increase in total lipid phosphorus of the membrane

(6.1 ± 0.5 to 6.9 ± 0.4 μg/mg protein). More recent studies (Matsumoto *et al.*, 1981) have shown a small but consistent increase in the weight percentage of PC (12.8 ± 1.8 to 15.8 ± 2.1) and a decrease of PE (48.0 ± 1.6 to 44.8 ± 2.6). From these observations, one can calculate the absolute change in phospholipid composition. These calculations show that PC increases from 0.80 ± 0.10 to 1.10 ± 0.05 mg P/mg protein but that PE remains the same (3.07 ± 0.40 to 3.09 ± 0.39). Thus, $1,25(OH)_2D_3$ administration leads to a 35% increase in the PC content of the brush border membrane. This conclusion is supported by the observation that sterol administration leads to: (1) an increase in glycerol incorporation into PC and a decrease into PE; (2) an increase in choline incorporation into PC but no change in ethanolamine incorporation into PE; and (3) no change in the rate of methylation of PE to PC (Matsumoto *et al.*, 1981). Thus, it appears that $1,25(OH)_2D_3$ stimulates the *de novo* synthesis of PC and its incorporation into the BBM of the chick duodenal mucosal cell.

Studies by Gilmore *et al.* (1979) in which the phospholipid composition of the membranes of LM cells grown in tissue culture were altered by use of defined media have shown that an increase in PC content increases the fluidity of the various membranes in this cell. However, the change in PE/PC ratio in the membranes of these cells was larger than that seen after $1,25(OH)_2D_3$ action in the intestinal cell. Hence, the change in fluidity of the BBM induced by $1,25(OH)_2D_3$ action is probably not large. However, the hormone also stimulates the turnover of fatty acids in the PC fraction and increases phospholipase A_2 activity (O'Doherty, 1979), so that this fraction contains an increased percentage of long-chain polyunsaturated fatty acids. This change in composition is known to increase membrane fluidity.

The relationship of these changes in membrane fluidity to changes in calcium transport was explored in two ways: (1) by comparing the time courses of the changes; and (2) by altering membrane fluidity *in vitro* and assessing its effect on calcium transport.

A comparison of change in PC synthesis and of fatty acid incorporation into PC to the time course of change in calcium transport showed that all occurred with a similar time course, indicating that the altered lipid structure could be responsible for the altered transport rate (Matsumoto *et al.*, 1981).

To gain further insight into this possibility, the effect of altering membrane fluidity *in vitro* on calcium transport was studied by incubating BBM vesicles obtained from vitamin D-deficient or $1,25(OH)_2D_3$-treated chicks with the methyl esters of *cis*-vaccenic acid (MCVA) and *trans*-vaccentic acid (MTVA) (Fontaine *et al.*, 1981). The first of these has been shown to increase membrane fluidity, the second to decrease it. Incubation of BBM vesicles from vitamin D-deficient chicks, but not those from treated chicks, with 0.12 mM MCVA led to an increase in the rate of calcium uptake to a value comparable to that seen after $1,25(OH)_2D_3$ administration *in vivo*. Conversely, incubation of BBM vesicles from $1,25(OH)_2D_3$-treated chicks, but not those from vitamin D-deficient animals, with 0.12 mM MTVA led to a decrease in the rate of calcium transport to a value comparable to that seen in vesicles from vitamin D-deficient animals. Incubations of either type of vesicle with either MCVA or MTVA led to no change in the rate of Na^+-dependent glucose transport.

These data led to the hypothesis that $1,25(OH)_2D_3$ acts primarily to alter the lipid structure of the BBM of the enterocyte. This change in lipid structure by modifying the function of a specific membrane protein(s) leads to an enhanced rate of calcium flux across this membrane. The change in protein function could result either from an enhanced efficiency (increased turnover number) of a fixed number of calcium carrier proteins, or from a recruitment of cryptic calcium carriers already present in the membrane.

During the same period of time that this work on $1,25(OH)_2D_3$ action was proceeding, Hirata, Axelrod, and co-workers (Hirata *et al.*, 1979; Hirata and Axelrod, 1978a,b, 1980; Ishizaka *et al.*, 1980) described an effect of amine hormones and IgE upon membrane PC synthesis. In their studies the effects seen were immediate (seconds to minutes), and were due to the activation of a specific methylation pathway catalyzing the successive methylation of PE to PC (trimethyl PE) within the plasma membrane of responsive cells. This change in membrane phospholipid composition was shown to lead to an increase in membrane fluidity, and was implicated both in the regulation of adenylate cyclase in the red blood and other cells, and in the phenomenon of calcium influx in the mast cell. It is noteworthy that an increase in fatty acid turnover in the PC fraction (activation of phospholipase A_2) always accompanied this PE-to-PC conversion. Hirata and Axelrod (1980) propose that these changes in lipid structure induced by amine hormones anc certain immunoglobulins regulate the function of membrane enzymes, hormone receptors, and membrane transport proteins in a highly specific manner.

3. CONCLUSION

The analogy between the results of Hirata and Axelrod (1980) in the mast cell and our results in the enterocyte is striking even though the time course of change and the basic process by which a change in lipid structure is produced are quite different. From this analogy, it is possible to propose that regulation of both PC content and the fatty acid composition of the PC is a common means by which hormones regulate cell function Fig. 1); that these hormonally induced changes in membrane lipid structure can be brought

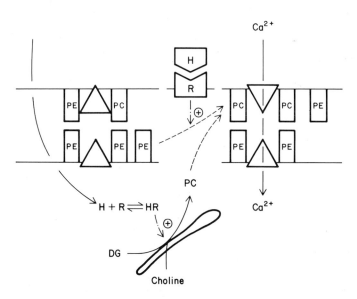

Figure 1. A schematic representation of the processes by which a hormone (H) can interact either with an intracellular receptor (R), e.g., a sterol hormone, or with a surface receptor ⊠ , e.g., an amine hormone, to lead to a change in the PC content and hence to a change in membrane fluidity that can bring about a specific change in membrane function, in this case an increase in Ca^{2+} flux. The sterol hormone achieves this effect by regulating the *de novo* synthesis of PC, which takes hours to be expressed, whereas the amine hormone achieves it by catalyzing the methylation of PE to PC, which is accomplished in seconds to minutes.

about either by hormones acting within the cell (sterol and/or steroids) to increase the *de novo* synthesis of PC, or by hormones acting at the cell surface (amines and/or proteins) to increase the conversion of PE to PC within the plasma membrane; and that these hormonally induced changes in phospholipid structure lead to an increase in membrane fluidity that in turn alters the function of one or more specific proteins within the membrane. Thus, it is postulated that these hormonally induced changes in membrane lipid structure are not epiphenomena related to the action of the particular hormone, but are one of the principal means by which the hormone regulates cell function. The data already at hand demonstrate that these hormonally induced changes in membrane lipid structure can bring about quite specific changes in the activity of one membrane protein while having little influence on that of another. Hence, they appear to have the requisite specificity to serve as a physiologic means of regulating cell function.

ACKNOWLEDGMENT. The work from this laboratory was supported by Grants AM 20570 and AM 19813 from the National Institute for Arthritis, Metabolic and Digestive Diseases of the United States Public Health Service.

REFERENCES

Bikle, D. D., Zolock, D. T., Morrissey, R. L., and Herman, R. H. (1978). *J. Biol. Chem.* **253**, 484–488.

Brasitus, T. A., Schachter, D., and Manouneas, T. G. (1979). *Biochemistry* **18**, 4136–4144.

Cullis, P. R., and DeKruijff, B. (1979). *Biochim. Biophys. Acta* **599**, 399–420.

Edelman, I. S., and Fimogmari, G. S. (1968). *Rec. Prog. Horm. Res.* **24**, 1–44.

Engelhard, V. H., Glaser, M., and Storm, D. R. (1978). *Biochemistry* **17**, 3191–3200.

Fontaine, O., Matsumoto, T., Goodman, D. B. P., and Rasmussen, H. (1981). *Proc. Natl. Acad. Sci. USA* **78**, 1751–1754.

Gilmore, R., Cohn, N., and Glaser, M. (1979). *Biochemistry* **18**, 1042–1049.

Goodman, D. B. P., Allen, J. E., and Rasmussen, H. (1971). *Biochemistry* **10**, 3825–3831.

Goodman, D. B. P., Wong, M., and Rasmussen, H. (1975). *Biochemistry* **14**, 2803–2809.

Haussler, M. R., and McCain, T. A. (1977). *N. Engl. J. Med.* **297**, 974–983, 1041–1050.

Hirata, F., and Axelrod, J. (1978a). *Proc. Natl. Acad. Sci. USA* **75**, 2348–2352.

Hirata, F., and Axelrod, J. (1978b). *Nature (London)* **275**, 219–220.

Hirata, F., and Axelrod, J. (1980). *Science* **209**, 1082–1090.

Hirata, F., Strittmatter, W. J., and Axelrod, J. (1979). *Proc. Natl. Acad. Sci. USA* **76**, 368–372.

Ishizaka, T., Hirata, F., Ishizaka, K., and Axelrod, J. (1980). *Proc. Natl. Acad. Sci. USA* **77**, 1903–1906.

Jain, M. K., and White, H. B. III. (1977). *Adv. Lipid Res.* **15**, 1–60.

Kawai, K., Fujita, M., and Nakao, M. (1974). *Biochim. Biophys. Acta* **369**, 222–233.

Klausner, R. D., Kleinfeld, A. M., Hoover, R. L., and Karnovsky, M. J. (1980). *J. Biol. Chem.* **255**, 1286–1295.

Lien, E. L., Goodman, D. B. P., and Rasmussen, H. (1975). *Biochemistry* **14**, 2749–2754.

Lien, E. L., Goodman, D. B. P., and Rasmussen, H. (1976). *Biochim. Biophys. Acta* **421**, 210–217.

Matsumoto, T., Fontaine, O., and Rasmussen, H. (1980). *Biochim. Biophys. Acta* **599**, 13–23.

Matsumoto, T., Fontaine, O., and Rasmussen, H. (1981). *J. Biol. Chem.* **256**, 3354–3360.

Max, E., Goodman, D. B. P., and Rasmussen, H. (1978). *Biochim. Biophys. Acta* **511**, 224–239.

O'Doherty, P. J. A. (1979). *Lipids* **14**, 75–77.

Pilch, P. F., Thompson, P. A., and Czech, M. P. (1980). *Proc. Natl. Acad. Sci. USA* **77**, 915–928.

Rasmussen, H., Fontaine, O., and Matsumoto, T. (1981). *Ann. N.Y. Acad. Sci.* **372**, 518–529.

Rasmussen, H., Fontaine, O., Max, E. E., and Goodman, D. B. P. (1979). *J. Biol. Chem.* **254**, 2993–2999.

Sandermann, H., Jr. (1978). *Biochim. Biophys. Acta* **515**, 209–237.

Sharp, G. W. G., and Leaf, A. (1966). *Physiol. Rev.* **46**, 593–633.

Sinensky, M. (1974). *Proc. Natl. Acad. Sci. USA* **71**, 522–525.

Singer, S. J., and Nicolson, G. L. (1972). *Science* **175**, 720–731.

Thompson, G. A., and Nozawa, Y. (1977). *Biochim. Biophys. Acta* **472**, 55–92.

168

Structure of the Oligosaccharide Residues of Membrane Glycoproteins

Mary Catherine Glick

The carbohydrate structures of membrane glycoproteins have been implicated in many diverse biological phenomena. These include virus transformation, embryogenesis, metastasis, adhesion, intracellular and extracellular recognition, differentiation, and the formation of functional proteins. All of these subjects have been reviewed (Atkinson and Hakimi, 1980; Culp et al., 1979; Frazier and Glaser, 1979; Glick and Flowers, 1978; Glick et al., 1980; Neufeld and Ashwell, 1980; Vaheri and Mosher, 1978). Several recent reports provide additional support to the concept that glycoconjugate structures participate in metastasis (Bhavanandan et al., 1980; Pearlstein et al., 1980), human tumors (Santer and Glick, 1980), intracellular recognition (Tabas and Kornfeld, 1980), neural (Giovanni et al., 1981; Littauer et al., 1980) and embryonic (Carson and Lennarz, 1979; Prujansky-Jakobovits et al., 1979) differentiation, cell adhesion (Carter et al., 1981; Rauvala et al., 1981; Rauvala and Hakomori, 1981), and protein tertiary structure (Gibson et al., 1981). In spite of these studies, very few structures of the oligosaccharide units that are presumably responsible have been elucidated. In the most well-defined system, the structure of the hepatic receptor (Fig. 1A) for serum asialoglycoproteins has been proposed (Kawasaki and Ashwell, 1976).

The inability to elucidate the carbohydrate structure of membrane glycoproteins has been due largely to the small amounts of membrane material available. Many of the analytical procedures currently in use require quantities of purified glycoproteins or oligosaccharides not available as yet from membranes. Attempts to scale down many of these procedures to nanogram amounts have not been successful. To circumvent this, many investigators have turned to glycopeptide analysis with prior purification and separation of the glycopeptides by a variety of methods including size, charge (Glick, 1979), and lectin-affinity columns (Lotan and Nicolson, 1979). Until recently, the glycopeptides have been obtained by exhaustive protease digestion. Oligosaccharides from calf thymocyte mem-

Mary Catherine Glick • Department of Pediatrics, University of Pennsylvania School of Medicine and Children's Hospital of Philadelphia, Philadelphia, Pennsylvania 19104.

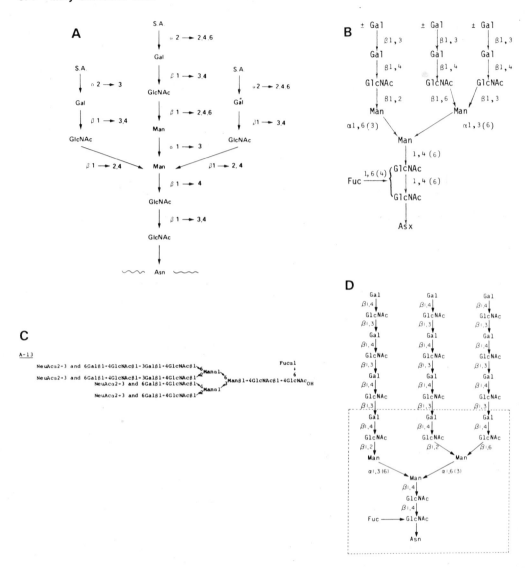

Figure 1. Partial structures proposed for glycopeptides isolated from surface membranes. (A) Glycopeptide from rat liver cell membranes (Kawasaki and Ashwell, 1976); (B) one of several glycopeptides from calf thymocyte membranes (Kornfeld, 1978); (C) an acidic oligosaccharide released from plasma membrane glycoproteins of calf thymocytes by hydrazinolysis (Yoshima *et al.*, 1980); (D) an unusual glycopeptide from CHO cells (Li *et al.*, 1980).

branes were obtained in this manner and the structures elegantly elucidated (Kornfeld, 1978). The oligosaccharide structure for one of the major glycopeptides is given in Fig. 1B. Recently, the technique of hydrazinolysis has been modified and used successfully to release the oligosaccharides from a variety of glycoproteins including those from calf thymocyte membranes (Yoshima *et al.*, 1980). One of the more unusual structures reported is shown in Fig. 1C. Enzymatic hydrolysis of the asparagine-linked oligosaccharide bond will circumvent any chemical hydrolysis during the hydrazinolysis procedure and would be a major contribution to structural analyses. Such an enzyme has been reported recently (Plummer and Tarentino, 1981). Enzymatic release of the oligosaccharide units is possible for *O*-glycosidic linkage to serine or threonine (Umemoto *et al.*, 1977). Other enzymes are extremely valuable in structural studies and can be used to obtain the sequence and anomeric linkage of small amounts of oligosaccharides (Flowers and Sharon, 1979; Kornfeld and Kornfeld, 1980). Enzymes have been used along with methylation to propose the unusual structure (Fig. 1D) of a glycopeptide from CHO cells (Li *et al.*, 1980).

Two recently developed techniques may simplify obtaining data on the structures of the oligosaccharide units of membrane glycoproteins: (1) the use of nuclear magnetic resonance spectroscopy and (2) monoclonal antibodies. In the past, 360-MHz ^{1}H-NMR spectroscopy required that the oligosaccharide units be available in milligram quantities and this is usually not the case for glycoproteins extracted or otherwise obtained from membranes. The structures of many oligosaccharides isolated from urine, for example, have been described (Montreuil, 1980), and one glycon from rat liver membranes has been reported (Debray *et al.*, 1981). Recently, the application of 500-MHz high-resolution ^{1}H NMR spectroscopy has lowered the amount of carbohydrate that can be analyzed to 25 nmoles (Vliegenthart *et al.*, 1981). With this advance, the technique will be a major tool in future structural studies.

The other technical development that has been proposed is the identification of carbohydrate units with a library of monoclonal antibodies. An example of an IgM monoclonal antibody, specific to the oligosaccharide portion of a surface glycoprotein of human neuroblastoma cells, has been described (Momoi *et al.*, 1980). The use of a monoclonal antibody (Kennett *et al.*, 1980) has the advantage over NMR in that a small quantity of an oligosaccharide can be identified. It has the disadvantage of the technical problems of selecting hybridomas producing monoclonal antibodies that will react with carbohydrates and then identifying the reacting oligosaccharides. However, the latter problem is inherent in all structural studies.

Extensive studies in many laboratories have shown that the membrane glycopeptides of virus-transformed cells have altered oligosaccharides (Glick *et al.*, 1980; Warren *et al.*, 1978). As predicted on the basis of overall carbohydrate composition (Glick, 1974), this alteration was the increase in more highly branched oligosaccharides accompanying transformation. The structure of the predominant glycopeptide from virus-transformed hamster fibroblasts (Fig. 2A) was found to be a triantennary oligosaccharide (Santer and Glick, 1979) and verified the prediction. This structure was elucidated with the use of exoglycosidases and quantitative recovery of the released monosaccharides by gas–liquid chromatography. Using hamster fibroblasts transformed by another virus, Ogata *et al.* (1976) reached a similar conclusion as the major oligosaccharides were proposed to be tetraantennary (Fig. 2E). The rationale for this proposed structure (Fig. 2E) was questioned, however, in view of size and charge separations (Glick, 1979). In fact, in a subsequent separation of the oligosaccharides after hydrazinolysis, bi- to hexa-antennary structures were reported for the transformed cells by sizing and enzyme sequencing (Takasaki *et al.*, 1980). Subsequently, others have reported 40 glycopeptides based on size, charge, and Con A affinity separations (Blithe *et al.*, 1980) in contrast to 20 classes previously separated by size and charge (Glick, 1979). The structural requirements of Con A affinity chromatography (Narasimhan *et al.*, 1978) were not considered and may be responsible for an artifactual separation of the glycopeptides.

Jeanloz and Codington (1974) using a TA3 tumor system defined a large glycoprotein, epiglycan, which masked cell surface antigens (Codington *et al.*, 1979). The structure of some of the carbohydrate residues of this glycoprotein has been elucidated (Fig. 2F). In another example of structures unique to tumor cells, an α-L-fucosidase that releases L-fucose residues linked $\alpha 1 \rightarrow 3$ or $\alpha 1 \rightarrow 4$ to GlcNAc was used to demonstrate this linkage

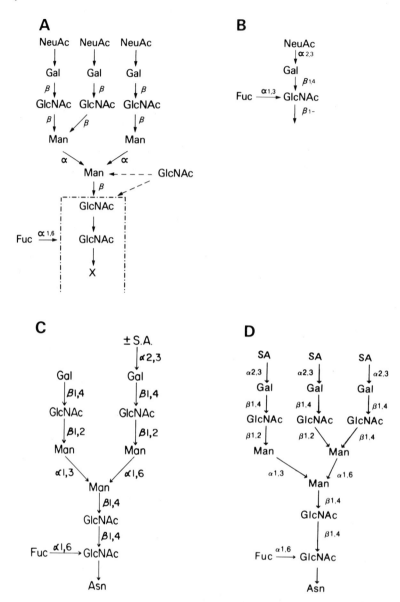

Figure 2. Partial structures proposed for membrane glycopeptides from cells associated with malignancy. (A) A predominant membrane glycopeptide from virus-transformed hamster kidney fibroblasts (Santer and Glick, 1979); (B) an oligosaccharide unit isolated from rat brain showing Fucα1→3GlcNAc on an antennary branch (Krusius and Finne, 1978) as proposed for human neuroblastoma cells (Santer and Glick, 1980); (C) the major glycopeptide of fibronectin (Fukuda and Hakomori, 1979); (D) a glycopeptide from vesicular stomatitis virus glycoprotein

in human neuroblastoma cells (Santer and Glick, 1980). Further sequencing by enzyme degradation of the oligosaccharides containing the α1→3 or α1→4 linked fucosyl residues showed that fucose was on the antennary branch in a configuration similar to that shown in Fig. 2B and reported by Krusius and Finne (1978) as a major oligosaccharide residue of rat brain glycoproteins.

Fibronectin is a large glycoprotein associated with the surface of fibroblasts and at

E

F

$$\text{NANA} \rightarrow \text{Gal} \rightarrow (\text{Gal, GNAc}) \rightarrow \text{Gal} \rightarrow \text{GalNAc} \rightarrow \overset{|}{\text{Ser}} \rightarrow (\text{Thr})$$

$$\text{Gal} \rightarrow \text{GalNAc} \rightarrow \overset{\vdots}{\underset{|}{\text{Ser}}} \rightarrow (\text{Thr})$$

G

grown in baby hamster kidney fibroblasts (Reading *et al.*, 1978) as shown in Kornfeld and Kornfeld (1980); (E) a model of the majority of the large glycopeptides from transformed cells (Ogato *et al.*, 1976); (F) a glycopeptide from TA3 mammary adenosarcoma cell surface (Jeanloz and Codington, 1974); (G) a glycopeptide from a glycoprotein of paramyxovirus grown in bovine kidney cells (Prehm *et al.*, 1979).

least in some cases is found predominantly in the culture medium after transformation. The structure of the major oligosaccharide residue (Fig. 2C) has been elucidated by Fukuda and Hakomori (1979). This fibronectin structure differs in monosaccharide linkage and substitution from those isolated from plasma (Kobata, 1979).

A detailed structure (Fig. 2D) has been assigned to a glycopeptide isolated from vesicular stomatitis virus grown in hamster cells (Reading *et al.*, 1978). The method of glycoprotein isolation takes advantage of the fact that the host membrane glycoproteins are incorporated into the virus coat (Sefton, 1976). Because large amounts of viruses can be readily produced, the oligosaccharide residues are obtained in quantities sufficient to study by methylation, periodate oxidation, mass spectroscopy, and NMR so that the complete structure could be elucidated. It is interesting that the structure (Fig. 2D) was similar to that isolated as a major glycopeptide from these hamster cells that were virus transformed (Fig. 2A). A glycopeptide with similar characteristics was seen only in small quantities in BHK_{21}/C_{13} cells that were not virus transformed (Glick, 1979). This suggests that virus infection may amplify certain oligosaccharide residues. The influence of cell transformation on the glycosylation of viral envelope glycoproteins has been described (Hunt *et al.*, 1981). A unique structure (Fig. 2G) has been presented for the glycoproteins of a paramyxovirus grown in bovine kidney cells, and it has been suggested that a similar glycopeptide is present in membranes of uninfected cells (Prehm *et al.*, 1979).

Thus, only a few structures have been elucidated for the oligosaccharide residues of glycoproteins or glycopeptides that have been isolated directly from the cell surface (Figs. 1 and 2). Nevertheless, the prospects for this decade are exceedingly bright for the structural analysis of the oligosaccharides of membrane glycoproteins. There is no doubt that much heterogeneity will be found and that the elucidation of these structures will help to define their putative biological roles.

ACKNOWLEDGMENT. Supported by grants from the United States Public Health Service.

REFERENCES

Atkinson, P. H., and Hakimi, J. (1980). In *The Biochemistry of Glycoproteins and Proteoglycans* (W. J. Lennarz, ed.), pp. 191–239, Plenum Press, New York.
Bhavanandan, V. P., Kemper, J. G., and Bystryn, J. C. (1980). *J. Biol. Chem.* **255**, 5145–5153.
Blithe, D. L., Buck, C. A., and Warren, L. (1980). *Biochemistry* **19**, 3386–3395.
Carson, D. D., and Lennarz, W. J. (1979). *Proc. Natl. Acad. Sci. USA* **76**, 5709–5713.
Carter, W. G., Rauvala, H., and Hakomori, S.-I. (1981). *J. Cell Biol.* **88**, 138–148.
Codington, J. F., Klein, G., Silber, C., Linsley, K. B., and Jeanloz, R. W. (1979). *Biochemistry* **18**, 2145–2149.
Culp, L. A., Murray, B. A., and Rollins, B. J. (1979). *J. Supramol. Struct.* **11**, 401–427.
Debray, H., Fournet, B., Montreuil, J., Dorland, L., and Vliegenthart, J. F. C. (1981). *Eur. J. Biochem.* **115**, 559–563.
Flowers, H. M., and Sharon, N. (1979). *Adv. Enzymol.* **48**, 29–95.
Frazier, W., and Glaser, L. (1979). *Annu. Rev. Biochem.* **48**, 491–523.
Fukuda, M., and Hakomori, S.-I. (1979). *J. Biol. Chem.* **254**, 5451–5457.
Gibson, R., Kornfeld, S., and Schlesinger, S. (1981). *J. Biol. Chem.* **256**, 456–462.
Giovanni, M. Y., Kessel, D., and Glick, M. C. (1981). *Proc. Natl. Acad. Sci. USA* **78**, 1250–1254.
Glick, M. C. (1974). In *Biology and Chemistry of Eucaryotic Cell Surfaces* (E. Y. C. Lee and E. E. Smith, eds.), pp. 213–240, Academic Press, New York.
Glick, M. C. (1979). *Biochemistry* **18**, 2525–2532.
Glick, M. C., and Flowers, H. (1978). In *The Glycoconjugates* (W. Pigman and M. I. Horowitz, eds.), Vol. 2, pp. 337–384, Academic Press, New York.
Glick, M. C., Momoi, M., and Santer, U. V. (1980). In *Metastatic Tumor Growth* (E. Grundmann, ed.), Vol. 4, pp. 11–19, Gustav Fischer Verlag, Stuttgart.
Hunt, L. A., Lamph, W., and Wright, S. E. (1981). *J. Virol.* **37**, 207–215.
Jeanloz, R. W., and Codington, J. F. (1974). In *Biology and Chemistry of Eucaryotic Cell Surfaces* (E. Y. C. Lee and E. E. Smith, eds.), pp. 241–258, Academic Press, New York.
Kawasaki, T., and Ashwell, G. (1976). *J. Biol. Chem.* **251**, 5292–5299.

Kennett, R. H., McKearn, T. J., and Bechtol, K. B. (eds.) (1980). *Monoclonal Antibodies,* Plenum Press, New York.

Kobata, A. (1979). *Cell Struct. Funct.* **4,** 169–181.

Kornfeld, R. (1978). *Biochemistry* **17,** 1415–1423.

Kornfeld, R., and Kornfeld, S. (1980). In *The Biochemistry of Glycoproteins and Proteoglycans* (W. J. Lennarz, ed.), pp. 1–34, Plenum Press, New York.

Krusius, T., and Finne, J. (1978). *Eur. J. Biochem.* **84,** 395–403.

Li, E., Gibson, R., and Kornfeld, S. (1980). *Arch. Biochem. Biophys.* **199,** 393–399.

Littauer, U. Z., Giovanni, M. Y., and Glick, M. C. (1980). *J. Biol. Chem.* **255,** 5448–5453.

Lotan, R., and Nicolson, G. L. (1979). *Biochim. Biophys. Acta* **559,** 329–376.

Momoi, M., Kennett, R. H., and Glick, M. C. (1980). *J. Biol. Chem.* **255,** 11914–11921.

Montreuil, J. (1980). *Adv. Carbohydr. Chem. Biochem.* **37,** 157–223.

Narasimhan, S., Wilson, J. R., Martin, E., and Schachter, H. (1978). *Can. J. Biochem.* **57,** 83–96.

Neufeld, E., and Ashwell, G. (1980). In *The Biochemistry of Glycoproteins and Proteoglycans* (W. J. Lennarz, ed.), pp. 241–266, Plenum Press, New York.

Ogata, S.-I., Muramatsu, T., and Kobata, A. (1976). *Nature (London)* **259,** 580–582.

Pearlstein, E., Salk, P. L., Yogeeswaran, G., and Karpatkin, S. (1980). *Proc. Natl. Acad. Sci. USA* **77,** 4336–4339.

Plummer, T. H., and Tarentino, A. L. (1981). *J. Biol. Chem.* **256,** 10243–10246.

Prehm, P., Scheid, A., and Choppin, P. W. (1979). *J. Biol. Chem.* **254,** 9669–9677.

Prujansky-Jakobovits, A., Gachelin, G., Muramatsu, T., Sharon, N., and Jacob, F. (1979). *Biochem. Biophys. Res. Commun.* **89,** 448–455.

Rauvala, H., and Hakomori, S.-I. (1981). *J. Cell Biol.* **88,** 149–159.

Rauvala, H., Carter, W. G., and Hakomori, S.-I. (1981). *J. Cell Biol.* **88,** 127–137.

Reading, C. L., Penhoet, E. E., and Ballou, C. E. (1978). *J. Biol. Chem.* **253,** 5600–5612.

Santer, U. V., and Glick, M. C. (1979). *Biochemistry* **18,** 2533–2540.

Santer, U. V., and Glick, M. C. (1980). *Biochem. Biophys. Res. Commun.* **96,** 219–226.

Sefton, B. M. (1976). *J. Virol.* **17,** 85–93.

Tabas, I., and Kornfeld, S. (1980). *J. Biol. Chem.* **255,** 6633–6639.

Takasaki, S., Ikehira, H., and Kobata, A. (1980). *Biochem. Biophys. Res. Commun.* **92,** 735–742.

Umemoto, J., Bhavanandan, V. P., and Davidson, E. A. (1977). *J. Biol. Chem.* **252,** 8609–8614.

Vaheri, A., and Mosher, D. F. (1978). *Biochim. Biophys. Acta* **516,** 1–25.

Vliegenthart, J. F. G., van Halbeek, H., and Dorland, L. (1981). *Pure Appl. Chem.* **53,** 45–77.

Warren, L., Buck, C. A., and Tuszynski, G. P. (1978). *Biochim. Biophys. Acta* **516,** 97–127.

Yoshima, H., Takasaki, S., and Kobata, A. (1980). *J. Biol. Chem.* **255,** 10793–10804.

169

Liposome-Mediated Delivery of Macromolecules into Eukaryotic Cells

Robert Fraley and Demetrios Papahadjopoulos

1. INTRODUCTION

In order to study the function, regulation, and interaction of macromolecules within the intracellular environment, a variety of techniques have been developed for introducing proteins and nucleic acids into living cells. These include the direct microinjection of macromolecules into the cytoplasm or nucleus with glass micropipets (Graessman *et al.*, 1979), the fusion of red blood cell ghosts containing entrapped molecules with recipient cells in the presence of sendai virus or other fusogens (Schlegel and Rechsteiner, 1978), and the uptake of nucleic acids by cells as coprecipitates with calcium phosphate or polycations (Graham and Van der Eb, 1973). Specific ligands (Cawley *et al.*, 1980) and antibody molecules (Thorpe *et al.*, 1978) have been used as targeted carriers for introducing molecules into cells, taking advantage of normal pathways for receptor-mediated endocytosis. While such techniques have been successful, each suffers disadvantages that limit its general applicability (requirement for special equipment or synthesis, limitation to working with certain types of macromolecules or small numbers or cells, low efficiency of delivery, etc.).

Liposomes have been used to introduce low-molecular-weight compounds such as drugs, nucleotides, etc. into cells (for recent review see Papahadjopoulos, 1979); however, recent developments in methodology for preparing large unilamellar liposomes have permitted the design of a flexible and generally applicable carrier system for introducing macromolecules into a variety of cell types. Liposomes as carriers encompass many of the advantages of the above techniques (simple preparation, high efficiency of encapsulation, targetable to specific cell types), while also overcoming several of their limitations (applicable to most cell types, increased efficiency of delivery, low toxicity). In addition, lipo-

Robert Fraley and Demetrios Papahadjopoulos • Cancer Research Institute and Department of Pharmacology, University of California School of Medicine, San Francisco, California 94143. *Present address of R.F.:* Monsanto Company, St. Louis, Missouri 63167.

somes can be used for unique applications, such as the introduction of integral membrane proteins into cells (Volsky *et al.*, 1979; Eytan and Eytan, 1980; Gad *et al.*, 1979).

In the following sections, we briefly review the methodology for encapsulating macromolecules in liposomes and describe certain *in vitro* applications as well as recent improvements in their use as carriers for introducing macromolecules into cells. The application of liposomes as drug carriers (Gregoriadis, 1977) or as a model system for membrane fusion studies (Papahadjopoulos *et al.*, 1979) has been discussed elsewhere.

2. ENCAPSULATION OF MACROMOLECULES IN LIPOSOMES

Large unilamellar vesicles (LUV) capable of encapsulating macromolecules can be prepared by several procedures, including Ca^{2+}-EDTA chelation, detergent dialysis, ether injection, and reverse-phase evaporation (for general review of these methods see Szoka and Papahadjopoulos, 1980). Because of their large size (diameter 0.15–0.6 μm) and high capture volume (5–20 liters/mole lipid) and efficiency (15–60% of the aqueous sample), LUV are superior to liposome preparations made by mechanical shaking or sonication.

The most versatile method for preparing LUV is reverse-phase evaporation, because liposomes can be prepared from a variety of phospholipids, and sample volumes may range from 0.05 to 5 ml. Further, this method has been used to encapsulate nucleic acids without degradation or loss of biological activity (Fraley *et al.*, 1980, 1982); several proteins have been encapsulated (Szoka and Papahadjopoulos, 1980) and others can be incorporated into the liposome membrane (Darszon *et al.*, 1979) without appreciable loss of activity.

Following vesicle preparation, LUV can be sized (and sterilized) by extrusion through polycarbonate membrane filters (Szoka *et al.*, 1980). Several methods have been used to remove encapsulated material from vesicle preparations, including dialysis, differential centrifugation (Wilson *et al.*, 1979), molecular sieve chromatography (Ostro *et al.*, 1978), and flotation on polymer gradients (Fraley *et al.*, 1980). The latter method allows liposome-macromolecule separations difficult to achieve by other procedures (Heath *et al.*, 1980). In addition, polymer flotation is rapid (15–30 min), adaptable for multiple samples, can be performed under sterile conditions, and results in a concentrated liposome preparation. Liposomes can be stored under argon or nitrogen at 4°C for several months without degradation or loss of activity of the encapsulated macromolecules (Fraley *et al.*, 1980; Keifer *et al.*, 1982; Gardas and MacPherson, 1979).

3. DELIVERY OF MACROMOLECULES TO CELLS

A major motivation for the introduction of proteins into cells has been to alter phenotype, for example, to correct certain metabolic deficiencies. Early success in enzyme-replacement therapy, using liposomes in *in vitro* model systems, has been dampened by more recent difficulties (for review see Finkelstein and Weissmann, 1978). However, the initial studies provided the first indication that liposomes might be useful carriers for delivering macromolecules into cells. They also suggested that intracellular delivery is a rather rare event; at most 10^4–10^5 liposome-encapsulated enzyme molecules became cell-associated and it is likely that only a small fraction of these actually entered the cell, escaped lysosomal degradation, and were active. Recent studies have indeed shown that only a few percent of cell-associated vesicles actually deliver their contents intracellularly (Wilson *et al.*, 1979; Szoka *et al.*, 1979, 1980).

The use of more potent, biologically active molecules, such as toxins and nucleic acids, compensates for the relatively low efficiency of liposomal delivery. The encapsulation of ricin in liposomes (Nicolson and Poste, 1978; Gardas and MacPherson, 1979; Dimitriadis and Butters, 1979) has been shown to increase its toxicity in cell lines that are normally ricin-resistant, indicating that liposome-mediated delivery bypasses the normal receptor pathway for uptake. In addition to their use in studying the mechanism of toxin action and resistance (Nicolson and Poste, 1978), liposomes containing these powerful cytotoxins might be useful as antineoplastic agents providing they can be targeted against tumor cells.

The introduction of membrane proteins and antigens into the plasma membrane of recipient cells represents a unique application of liposomes as macromolecular carriers. Newly incorporated membrane proteins have been shown to function normally (Eytan and Eytan, 1980; Gad *et al.*, 1979; Volsky *et al.*, 1979; Doyle *et al.*, 1979), providing new opportunities for studying the dynamics of plasma membrane components. In addition, the incorporation of specific antigens into cells has permitted investigation of the molecular requirements for activation of cytotoxic T lymphocytes and complement-mediated cell killing (Hale *et al.*, 1980).

A promising application of liposomes is in the introduction of nucleic acids into cells (Wilson *et al.*, 1977, 1979; Fraley *et al.*, 1979, 1980, 1981; Ostro *et al.*, 1978; Dimitriadis, 1978). Liposome entrapment of polio RNA (Wilson *et al.*, 1979) has resulted in higher values for infectivity than obtained using standard transfection techniques. Because uptake of liposomes (by fusion, endocytosis, etc.) bypasses the normal receptor-mediated pathway for uptake of polio virions by primate cells, encapsulated polio RNA is equally infectious when incubated with nonprimate cells. The infectivity obtained with liposome-encapsulated cowpea mosaic virus RNA (Keifer *et al.*, 1982) and tobacco mosaic virus RNA (Fraley *et al.*, 1982) is also higher than achieved with other methods for infecting plant protoplasts.

Liposome encapsulation has also been shown to facilitate the uptake of a variety of DNA molecules by both prokaryotic and eukaryotic cells (Fraley *et al.*, 1979, 1980, 1981, 1982; Wong *et al.*, 1980). The plasmid pBR322, which codes for tetracycline resistance, has been encapsulated and used to transform *E. coli* cells (Fraley *et al.*, 1979). The entrapment of metaphase chromosomes (Mukherjee *et al.*, 1978) and more recently of whole nuclei (Kondorosi and Duda, 1980) in liposomes increased the frequency of chromosome transfer and transformation in Chinese hamster ovary cells. Liposome-encapsulated SV40 DNA is 100-fold more infectious than free DNA when incubated with African green monkeys kidney (AGMK) cells (Fraley *et al.*, 1980). Recently, purified Ti DNA (isolated from *Agrobacterium tumefaciens*) encapsulated in LUV has been shown to transform tobacco protoplasts (Dellaporta *et al.*, 1982).

4. MAXIMIZING DELIVERY

Although it is clear from the studies described above that liposomes are useful for intracellular delivery of macromolecules in some systems, the relatively low efficiency of delivery has hindered their application to other systems in which the level of delivery is below the level of sensitivity of the assay employed. In order to extend the range of interesting macromolecules that can be studied, the efficiency of liposome delivery must be improved. There have been a number of recent reports describing different methods for increasing the intracellular delivery of liposome contents.

Because liposomes prepared from various phospholipids show large differences in their binding and leakage properties when incubated with cells (Fraley *et al.*, 1981), intracellular delivery can be increased most simply by proper choice of vesicle lipid composition. Liposomes prepared from acidic phospholipids, especially phosphatidylserine, have been shown to be the most efficient carriers for introducing nucleic acids into a variety of cell types (Fraley *et al.*, 1980, 1982; Martin *et al.*, 1981; Keifer *et al.*, 1982). Delivery is also dependent on vesicle stability (Fraley *et al.*, 1981), and including cholesterol in vesicle preparations increases the efficiency of delivery by severalfold.

A number of attempts have been made to increase specific binding of liposomes to cultured cells with lectins (Juliano and Stamp, 1976; Szoka *et al.*, 1981) and antibodies (Weinstein *et al.*, 1978; Leserman *et al.*, 1979, 1980; Heath *et al.*, 1980; Martin *et al.*, 1981). Although enhanced binding has often been the result of such efforts, in only a few instances has the enhancement been translated into markedly increased delivery (Leserman *et al.*, 1980; Szoka *et al.*, 1981). It is likely, though, that the recent development of new methodology for coupling large numbers of immunoglobulins (200/vesicle) to liposomes (Heath *et al.*, 1980; Martin *et al.*, 1981) and for stimulating their cellular uptake by manipulation of the incubation conditions (Fraley *et al.*, 1980; Szoka *et al.*, 1981) will increase the chances for success with this approach. An advantage of antibody-targeted liposomes over drug- or toxin-conjugates is the possibility for cooperative binding with multiliganded liposomes—a factor that may be important in some cases where the antigen density on cell surfaces is low or the binding is weak.

Because the majority of liposomes that interact with cells simply remain adsorbed to the cell surface (Szoka *et al.*, 1979, 1980), treatments that increase their internalization would be expected to enhance delivery greatly. Following incubation of liposomes with AGMK cells, SV40 DNA delivery can be increased (>200-fold) by exposing the cells to a glycerol solution (Fraley *et al.*, 1980). The glycerol treatment apparently enhances liposome uptake by stimulating an endocytosis-like process (Fraley *et al.*, 1981). Postincubation treatments with other polyalcohols such as polyethylene glycol (PEG) have also been used to increase (up to 10^7 molecules/cell) the intracellular delivery of proteins (Szoka *et al.*, 1981). An alternative approach for increasing vesicle uptake by cells is the incorporation into liposomes of viral proteins known to promote virus–cell fusion (Uchida *et al.*, 1979; Volsky *et al.*, 1979). Such reconstituted vesicles containing entrapped diphtheria toxin A fragment (Uchida *et al.*, 1979) have been shown to be more toxic (~100-fold) than toxin-containing liposomes not having the viral membrane protein.

Because a large fraction of the macromolecules delivered to cells by liposomes appear to be processed by lysosomes (Hoekstra *et al.*, 1978), the effective delivery of liposomal contents might be increased by preventing their intracellular degradation. It has been shown recently (Fraley *et al.*, 1981) that treatment of AGMK cells with chloroquine, a drug that interferes with lysosomal function, increases liposome-mediated SV40 DNA infectivity 5- to 10-fold. Such an observation illustrates the possibilities for altering cellular metabolism in order to engineer a more effective delivery of liposome contents.

It appears likely that further increases in liposome-mediated delivery will result from a combination of approaches described above, and that liposomes will emerge as a more efficient and generally applicable carrier for introducing macromolecules into cells.

ACKNOWLEDGMENTS. This work was supported by Grant CA 25526 from the National Cancer Institute and a postdoctoral fellowship (R.F.) from the Jane Coffin Child's Memorial Fund for Medical Research.

REFERENCES

Cawley, D., Herschman, H., Gilliland, D., and Collier, R. (1980). *Cell* **22**, 563–570.

Darszon, A., Vandenberg, C., Ellisman, M., and Montal, M. (1979). *J. Cell Biol.* **81**, 446–452.

Dellaporta, S., Fraley, R., Giles, K., Papahadjopoulos, D., Powell, A., Thomashow, M., Nester, E., and Gordon, M. (1981). Unpublished observations.

Dimitriadis, G. (1978). *Nature (London)* **274**, 923–924.

Dimitriadis, G., and Butters, T. (1979). *FEBS Lett.* **255**, 4992–4995.

Doyle, D., Hou, E., and Warren, R. (1979). *J. Biol. Chem.* **254**, 6853–6856.

Eytan, G., and Eytan, E. (1980). *J. Biol. Chem.* **255**, 4992–4995.

Finkelstein, M., and Weissmann, G. (1978). *J. Lipid Res.* **19**, 289–303.

Fraley, R., Fornari, C., and Kaplan, S. (1979). *Proc. Natl. Acad. Sci. USA* **76**, 3348–3352.

Fraley, R., Subramani, S., Berg, P., and Papahadjopoulos, D. (1980). *J. Biol. Chem.* **255**, 10431–10435.

Fraley, R., Dellaparta, S., and Papahadjopoulos, D. (1982). *Proc. Natl. Acad. Sci. USA*, in press.

Fraley, R., Straubinger, R., Rule, G., Springer, L., and Papahadjopoulos, D. (1981). *Biochem.* **20**, 6978–6987.

Gad, A., Broza, R., and Eytan, G. (1979). *FEBS Lett.* 102, 230–234.

Gardas, A., and MacPherson, I. (1979). *Biochim. Biophys. Acta* **584**, 538–541.

Graessman, A., Graessman, M., and Muller, C. (1979). In *Current Topics in Microbiology and Immunology* (W. Arber, S. Falkon, W. Henle, P. Hofschneider, J. Humphrey, J. Klein, P. Koldovsky, H. Koprowski, O. Maaloe, F. Melchers, R. Rott, H. Schweiger, L. Syrucek, and P. Vogt, eds.), pp. 1–21, Springer-Verlag, Berlin.

Graham, F., and Van der Eb, A. (1973). *Virology* **54**, 536–539.

Gregoriadis, G. (1977). *Nature* **265**, 407–410.

Hale, A., Ruebush, M., Lyles, D., and Harris, D. (1980). *Proc. Natl. Acad. Sci. USA* **77**, 6105–6108.

Heath, T., Fraley, R., and Papahadjopoulos, D. (1980). *Science* **210**, 539–540.

Hoekstra, D., Tomasini, R., and Scherphof, G. (1978). *Biochim. Biophys. Acta* **542**, 456–469.

Juliano, R., and Stamp, D. (1976). *Nature (London)* **261**, 235–237.

Keifer, M., Fraley, R., Papahadjopoulos, D., and Breuning, G. (1982). In preparation.

Kondorosi, E., and Duda, E. (1980). *FEBS Lett.* **120**, 37–40.

Leserman, L. Weinstein, J., Blumenthal, R., Sharron, S., and Terry, W. (1979). *J. Immunol.* **122**, 585–591.

Leserman, L., Barbet, J., Kourilsky, F., and Weinstein, J. (1980). *Nature (London)* **288**, 602–604.

Martin, F., Hubbell, W., and Papahadjopoulos, D. (1981). *Biochemistry* **20**, 4229–4238.

Mukherjee, A., Orloff, S., Butler, J., Tiche, T., Lalley, P., and Schulman, J. (1978). *Proc. Natl. Acad. Sci. USA* **75**, 1361–1365.

Nicolson, G., and Poste, G. (1978). *J. Supramol. Struct.* **8**, 235–245.

Ostro, M., Giacomoni, D., Lavelle, D., Paxton, W., and Dray, S. (1978). *Nature (London)* **274**, 921–923.

Papahadjopoulos, D. (1979). *Annu. Rep. Med. Chem.* **14**, 250–260.

Papahadjopoulos, D., Poste, G., and Vail, W. J. (1979). In *Methods in Membrane Biology* (E. D. Korn, ed.), Vol. 10, pp. 1–121, Plenum Press, New York.

Schlegel, R., and Rechsteiner, M. (1978). *Methods Cell Biol.* **20**, 341–354.

Szoka, F., and Papahadjopoulos, D. (1980). *Annu. Rev. Biophys. Bioeng.* **9**, 467–508.

Szoka, F., Jacobson, K., and Papahadjopoulos, D. (1979). *Biochim. Biophys. Acta* **551**, 295–303.

Szoka, F., Jacobson, K., Derzko, Z., and Papahadjopoulos, D. (1980). *Biochim. Biophys. Acta* **600**, 1–18.

Szoka, F., Magnusson, K., Wojcieszyn, J., Hou, Y., Derzko, Z., and Jacobson, K. (1981). *Proc. Natl. Acad. Sci. USA* **78**, in press.

Thorpe, P., Ross, W., Cumber, A., Hinson, C., Edwards, D., and Davies, A. (1978). *Nature (London)* **271**, 752–754.

Uchida, T., Kim, J., Yamaizumi, M., Miyake, Y., and Okada, Y. (1979). *J. Cell Biol.* **80**, 10–20.

Volsky, D., Cabantchik, Z., Beigel, M., and Loyter, A. (1979). *Proc. Natl. Acad. Sci. USA* **76**, 5440–5444.

Weinstein, J., Blumenthal, R., Sharrow, S., and Henkart, P. (1978). *Biochim. Biophys. Acta* **509**, 289–299.

Wilson, T., Papahadjopoulos, D., and Taber, R. (1977). *Proc. Natl. Acad. Sci. USA* **74**, 3471–3475.

Wilson, T., Papahadjopoulos, D., and Taber, R. (1979). *Cell* **17**, 77–84.

Wong, T.-K., Nicolau, C., and Hofschneider, P. (1980). *Gene* **10**, 87–94.

170

Enzyme Replacement Therapy

Specific Targeting of Exogenous Enzymes to Storage Cells

Roscoe O. Brady and F. Scott Furbish

1. INTRODUCTION

Several observations indicate that the infusion of an appropriately purified exogenous enzyme is beneficial to patients with lipid storage disorders. These indications include (1) decreased quantity of accumulating lipid in the circulation (Brady *et al.*, 1973, 1974; Johnson *et al.*, 1973; Desnick *et al.*, 1979), (2) diminished amount of lipid stored in various organs (Brady *et al.*, 1974, 1980), and (3) a suggestion of improvement in the clinical manifestations of the disorder (Brady *et al.*, 1981). The latter effect has been observed in young patients with Gaucher's disease who were treated on a prospective, long-term enzyme replacement regimen. These individuals showed improvement of their general health and vigor, arrest of the progressive organomegaly that is characteristic of this disorder, and restoration of diminished platelet levels to the normal range. Recipients tolerate these enzyme infusions satisfactorily and they do not become sensitized to the exogenous protein (Britton *et al.*, 1978). While these observations are highly encouraging concerning enzyme replacement for the treatment of heritable metabolic disorders, these trials are not yet conclusive. Much is still to be learned regarding the administration of exogenous enzyme. In this regard, it has become clear from animal studies that the majority of intravenously injected enzyme prepared by the method of Furbish *et al.* (1977) was incorporated by hepatocytes (Furbish *et al.*, 1978; Morrone *et al.*, 1981). This location is in contrast to the well-documented site of glucocerebroside storage in reticuloendothelial cells such as the Kupffer cells in the liver (Brady, 1978). Because we believed that the efficacy of the

Roscoe O. Brady and F. Scott Furbish • Developmental and Metabolic Neurology Branch, National Institute of Neurological and Communicative Disorders and Stroke, National Institutes of Health, Bethesda, Maryland 20205.

exogenous enzyme could be greatly improved, we have undertaken a series of investigations to increase the delivery of the enzyme to the specific cells in which the accumulating lipid is stored.

2. MODIFICATION OF THE TERMINAL GLYCOSYL RESIDUES OF ENZYMES BY REACTION WITH GLYCOSIDASES

Most, if not all, of the lysosomal enzymes that catalyze the catabolism of lipids, glycosaminoglycans, polysaccharides, and some proteins are glycoproteins. It has been convincingly established through the elegant work of Ashwell and others that glycoproteins are selectively removed from the circulation by interaction with carbohydrate-recognizing receptors on the surfaces of various cells. This association is then followed by endocytosis of the bound macromolecule. Thus, hepatocytes contain a receptor on their plasma (and internal) cell membranes that selectively promotes the uptake of galactose-terminated glycoproteins (Ashwell and Morell, 1974). Reticuloendothelial cells have a receptor that reacts with N-acetylglucosamine and mannose (Schlesinger *et al.*, 1978). Results obtained in our laboratory in studies with inhibitors indicated that the oligosaccharide portion of human placental glucocerebrosidase was comprised of the carbohydrate sequence or set of sequences consisting of N-acetylneuraminyl-galactosyl-(fucosyl)-N-acetylglucosaminyl-mannosyl-X, where X is the link to the polypeptide portion of the molecule (Fig. 1). It is not presently known whether any single sequence in the molecule has this structure. However, from the point of view of its recognition by cell surface receptors, the molecule demonstrates this order of saccharides. Removal of the terminal molecule of N-acetylneuraminic acid from glucocerebrosidase by immobilized neuraminidase caused an increase in the quantity of enzyme taken up by hepatocytes in experimental animals (Furbish *et al.*, 1978). Subsequent removal of galactose by bound galactosidase augmented the amount of enzyme taken up by sinusoidal cell preparations consisting endothelial lining cells and Kupffer cells (Steer *et al.*, 1978). This effect is assumed to be mediated by a receptor for terminal N-acetylglucosamine. This receptor has been demonstrated on nonparenchymal cells in the liver (Schlesinger *et al.*, 1978; Steer and Clarenburg, 1979) and pulmonary macrophages (Stahl *et al.*, 1978). The receptor has been isolated and purified from liver by Kawasaki *et al.* (1978). The receptor has specificity for both N-acetylglucosamine and mannose as ligands. Treating glucocerebrosidase from which sialic acid and galactose had been cleaved (N-acetylglucosamine terminated) with hexosaminidase further enhanced the delivery of enzyme to Kupffer cells (Furbish *et al.*, 1981). This finding indicated that uptake of the enzyme that was mediated via this N-acetylglucosamine/mannose receptor was more effective with mannose as the terminal hexose. This conclusion is supported by the observation that mannose-terminated orosomucoid was a more potent inhibitor of clearance of glucocerebrosidase from the circulation than was N-acetylglucosamine-terminated orosomucoid. The sequential cleavage of the oligosaccharide side chain to the mannose residues (Fig. 2) resulted in a fivefold augmentation in the delivery of placental glucocerebrosidase to the reticuloendothelial cells over that obtained with the unmodified enzyme (Table I).

Treatment of native, asialo, or agalactoglucocerebrosidase with α-fucosidase results

$$\text{NeuNAc} - \text{Gal} \xrightarrow{\beta} \text{GlcNAc} \xrightarrow{\beta} \text{Man} - \text{X}$$
$$\underset{\alpha|}{}$$
$$\text{Fuc}$$

Figure 1. Oligosaccharide sequence of the glycopeptide portion of human placental glucocerebrosidase. NeuNAc, N-acetylneuraminic acid; Gal, galactose; GlcNAc, N-acetylglucosamine; Fuc, fucose; Man, mannose.

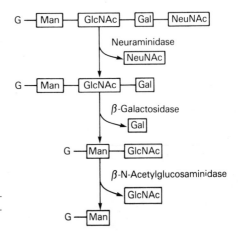

Figure 2. Sequential enzymatic cleavage of glycosyl residues from human placental glucocerebrosidase. Abbreviations as in Fig. 1.

in decreased incorporation of the enzyme by hepatocytes in all cases but has little effect on uptake by nonparenchymal cells (Furbish *et al.*, 1980). This observation implies that fucose residues play a role in adsorptive pinocytosis of this enzyme by hepatocytes. The observation of a fucose receptor on hepatocytes was originally made by Prieels *et al.* (1978).

Additional studies are under way that are designed to further reduce the delivery of exogenous enzyme to hepatocytes and presumably increase the incorporation by Kupffer cells. As hepatocytes possess a receptor for mannose-6-phosphate-containing glycoproteins (Ullrich *et al.*, 1978), dephosphorylation of the enzyme may result in decreased parenchymal cell uptake and increased Kupffer cell incorporation via the *N*-acetylglucosamine/mannose receptor. In the long run, it is our expectation that enzyme modified in an appropriate fashion will be more efficiently incorporated into the target cells and provide greater therapeutic effectiveness. Studies with the native and variously modified enzyme preparations with regard to interaction of the incorporated enzyme with stored substrate and survival of the enzyme *in situ* will be carried out to complement these efforts.

Table I. Uptake of Native and Glycosidase-Treated Human Placental Glucocerebrosidase by Rat Liver Cells [a]

Experiment	Treatment	Amount injected (units)	Cellular distribution (units/10^6 cells)	
			Nonparenchymal	Hepatocytes
1	Control (no enzyme)	0	6.6	104
2	None	8.8×10^5	48	256
3	Neuraminidase	7.0×10^5	16	337
4	Neuraminidase and galactosidase	6.5×10^5	170	380
5	Neuraminidase, galactosidase, and *N*-acetylglucosaminidase	6.0×10^5	245	290

[a] The glucocerebrosidase preparations were injected intravenously into adult male rats of the Osborne–Mendel strain. The animals were sacrificed 30 min following infusion and the hepatic cells were separated into the respective types according to Steer and Clarenburg (1979).

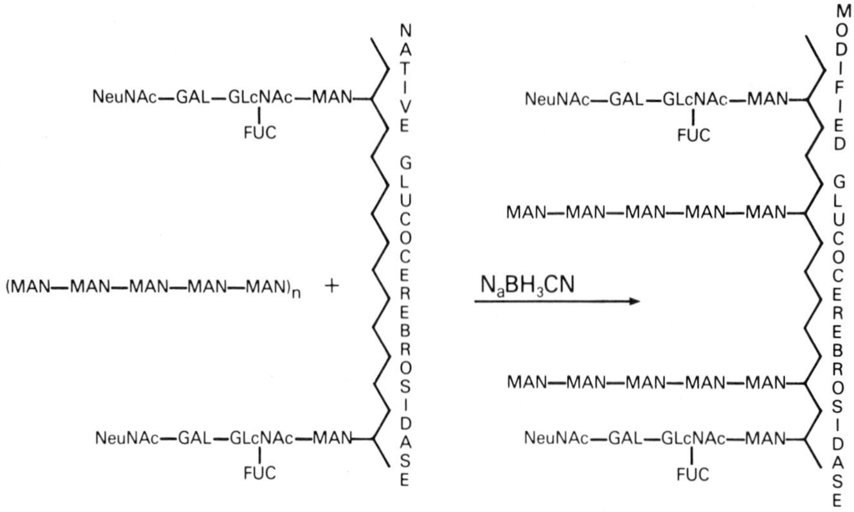

Figure 3. Chemical modification of human placental glucocerebrosidase. Abbreviations as in Fig. 1.

3. GLYCOCONJUGATES OF PLACENTAL GLUCOCEREBROSIDASE

Another approach that is under active investigation involves the covalent linking of additional molecules of mannose to the native enzyme (Brady *et al.*, 1980). The possibility of achieving this goal is strengthened by the observation of Murray and Neville (1980) that the attachment of mannose-6-phosphate to low-density lipoprotein (LDL) specifically promoted the uptake of LDL by the mannose-6-phosphate receptor that is present on cultured skin fibroblasts. There was negligible uptake of the protein by this pathway without the addition of the phosphorylated sugar which is specifically recognized by fibroblast receptors (Kaplan *et al.*, 1977). We propose to try to add mannose or mannose oligosaccharides to glucocerebrosidase to increase uptake of the enzyme by Kupffer and other reticuloendothelial cells (Fig. 3). We are aware that this approach may prove difficult as chemical modification of enzymes can lead to loss of catalytic activity. Glucocerebrosidase seems quite fastidious in this respect, for attempts to malelylate the enzyme, which might also have been expected to increase its delivery to reticuloendothelial cells (Goldstein *et al.*, 1979), resulted in major losses in catalytic activity (Brady, 1980, unpublished observations). This loss of enzymatic activity occurred even when the modification was carried out in the presence of the substrate glucocerebroside or simple sugars that were expected to protect the active site of the enzyme during the procedure.

4. CONCLUDING REMARKS

The modification of oligosaccharide side chains of glycoproteins so that they are specifically terminated with mannose, or conversely, selective enrichment of a protein with this covalently bound sugar should provide for significantly improved delivery to lipid-storing reticuloendothelial cells (Steer *et al.*, 1979). The effect of mannose termination is also seen in another important situation relating to the uptake of β-hexosaminidase A by neuronal cells. The ganglioside G_{M2} accumulates specifically in neurons in patients with

Tay–Sachs disease because of the deficiency of hexosaminidase activity (Kolodny *et al.*, 1969; Okada and O'Brien, 1969). Binding of labeled hexosaminidase A to brain synaptosomes and synaptic plasma membranes has been shown to be a specific and saturable process that is inhibited by mannose (Kusiak *et al.*, 1979). Furthermore, we have investigated the delivery of horseradish peroxidase, a histochemically demonstrable, mannose-rich glycoprotein, to the central nervous system (Barranger *et al.*, 1979). After temporary alteration of the blood–brain barrier to permit ingress of the enzyme from the bloodstream, we found that only the neuronal cells in the brain accumulated this protein. This selective delivery of enzyme based on its oligosaccharide structure provides considerable encouragement for enzyme replacement trials in Tay–Sachs disease and other heritable metabolic disorders, particularly where specific cell delivery of the required enzyme can be accomplished through a naturally occurring receptor (Kusiak *et al.*, 1980), or alternatively, when its targeting can be improved by specific modification procedures such as those outlined above.

Finally, one probably should think in terms of recombinant DNA techniques that have shown considerable potential and may have to be employed should natural sources of enzyme become insufficient or prohibitively expensive. Initially, one would expect that polypeptides coded by a recombined exon would not contain an oligosaccharide moiety (see Wade, 1980). With regard to the delivery of glucocerebrosidase, this situation may not be deleterious as it has long been known that intravenously injected nonglycosylated enzymes are preferentially taken up by reticuloendothelial cells (Wakim and Fleisher, 1963). This effect may be precisely what is desired in Gaucher's disease. Should the addition of hexoses be deemed helpful, the studies under way regarding chemical modification of the enzyme will provide useful information concerning the feasibility of such an approach. Finally, one could conceive of providing additional genetic information that would lead to the required glycosylation of the nascent polypeptide *in situ* or in the course of its extracellular translocation.

REFERENCES

Ashwell, G., and Morell, A. (1974). *Adv. Enzymol.* **41,** 99–128.

Barranger, J. A., Rapoport, S. I., Fredericks, W. R., Pentchev, P. G., MacDermot, K. D., Steusing, J. K., and Brady, R. O. (1979). *Proc. Natl. Acad. Sci. USA* **76,** 481–485.

Brady, R. O. (1978). In *The Metabolic Basis of Inherited Disease* (J. B. Stanburg, J. B. Wyngaarden, and D. S. Fredrickson, eds.), pp. 731–746, McGraw–Hill, New York.

Brady, R. O., Tallman, J. F., Johnson, W. G., Gal, A. E., Leahy, W. E., Quirk, J. M., and Dekaban, A. S. (1973). *N. Engl. J. Med.* **289,** 9–14.

Brady, R. O., Pentchev, P. G., Gal, A. E., Hibbert, S. R., and Dekaban, A. S. (1974). *N. Engl. J. Med.* **291,** 989–993.

Brady, R. O., Barranger, J. A., Gal, A. E., Pentchev, P. G., and Furbish, F. S. (1980). In *Enzyme Therapy in Genetic Diseases* (R. J. Desnick, ed.), Vol. 2, pp. 361–368, Liss, New York.

Brady, R. O., Barranger, J. A., Pentchev, P. G., Furbish, F. S., and Gal, A. E. (1981). In *Inborn Errors of Metabolism in Humans* (H. Aebi and N. N. Herschkowitz, eds.), MTP Press, London, in press.

Britton, D. E., Leinikki, P. O., Barranger, J. A., and Brady, R. O. (1978). *Life Sci.* **23,** 2517–2520.

Desnick, R. J., Dean, K. J., Grabowski, G. A., Bishop, D. F., and Sweeley, C. C. (1979). *Proc. Natl. Acad. Sci. USA* **76,** 5326–5330.

Furbish, F. S., Blair, H. E., Shiloach, J., Pentchev, P. G., and Brady, R. O. (1977). *Proc. Natl. Acad. Sci. USA* **74,** 3560–3563.

Furbish, F. S., Steer, C. J., Barranger, J. A., Jones, E. A., and Brady, R. O. (1978). *Biochem. Biophys. Res. Commun.* **81,** 1047–1053.

Furbish, F. S., Krett, N. L., Barranger, J. A., and Brady, R. O. (1980). *Biochem. Biophys. Res. Commun.* **95,** 1768–1774.

Furbish, F. S., Steer, C. J., Krett, N. L., and Barranger, J. A. (1981). *Biochim. Biophys. Acta* **673,** 425–434.

Goldstein, J. L., Ho, Y. K., Basu, S. K., and Brown, M. S. (1979). *Proc. Natl. Acad. Sci. USA.* **76,** 333–337.

Johnson, W. G., Desnick, R. J., Long, D. M., Sharp, H. L., Krivit, W., Brady, B., and Brady, R. O. (1973). In *Enzyme Therapy in Genetic Diseases* (R. J. Desnick, R. W. Bernlohr, and W. Krivit, eds.), pp. 120–124, Williams & Wilkins, Baltimore.

Kaplan, A., Achord, D. T., and Sly, W. S. (1977). *Proc. Natl. Acad. Sci. USA* **74,** 2026–2030.

Kawasaki, T., Etoh, R., and Yamashima, I. (1978). *Biochem. Biophys. Res. Commun* **81,** 1018–1024.

Kolodny, E. H., Brady, R. O., and Volk, B. W. (1969). *Biochem. Biophys. Res. Commun.* **37,** 526–531.

Kusiak, J. W., Toney, J. H., Quirk, J. M., and Brady, R. O. (1979). *Proc. Natl. Acad. Sci. USA* **76,** 982–985.

Kusiak, J. W., Quirk, J. M., and Brady, R. O. (1980). In *Enzyme Therapy in Genetic Diseases* (R. J. Desnick, ed.), Vol. 2, pp. 93–102, Liss, New York.

Morrone, S., Pentchev, P. G., Thorpe, S., and Baynes, J. (1981). *Biochem. J.* **194,** 733–742.

Murray, G. J., and Neville, D. M., Jr. (1980). *J. Biol. Chem.* **255,** 11942–11948.

Okada, S., and O'Brien, J. S. (1969). *Science* **165,** 698–700.

Prieels, J. P., Pizzo, S. V., Glascow, L. R., Paulson, J. C., and Hill, R. L. (1978). *Proc. Natl. Acad. Sci. USA* **75,** 2215–2219.

Schlesinger, P. H., Doebber, T. W., Mandell, B. F., White, R., deSchryver, C., Rodman, J. S., Miller, M. J., and Stahl, P. (1978). *Biochem. J.* **176,** 103–109.

Stahl, P. D., Rodman, J. S., Miller, M. J., and Schlesinger, P. H. (1978). *Proc. Natl. Acad. Sci. USA* **75,** 1399–1403.

Steer, C. J., and Clarenburg, R. (1979). *J. Biol. Chem.* **254,** 4457–4461.

Steer, C. J., Furbish, F. S., Barranger, J. A., Brady, R. O., and Jones, E. A. (1978). *FEBS Lett.* **91,** 202–205.

Steer, C. J., Kusiak, J. W., Brady, R. O., and Jones, E. A. (1979). *Proc. Natl. Acad. Sci. USA* **76,** 2774–2778.

Ullrich, K., Mersmann, G., Fleischer, M., and von Figura, K. (1978). *Hoppe-Seyler's Z. Physiol. Chem.* **359,** 1591–1598.

Wade, N. (1980). *Science* **209,** 1492–1494.

Wakim, K. G., and Fleisher, G. A. (1963). *J. Lab. Clin. Med.* **61,** 107–119.

171

Exocytosis under Conditions of Intracellular Solute Control

P. F. Baker and D. E. Knight

Exocytosis and endocytosis—the processes whereby vesicles fuse with or are retrieved from the plasma membrane—are examples of a much wider range of reactions that facilitate the ordered transfer of membrane and its trapped contents between intracellular compartments. In the case of exocytosis and endocytosis, the process of membrane fusion or retrieval permits the secretion or absorption of substances across the plasma membrane as well as the incorporation into or removal from the cell surface of specialized membrane. Unless the surface area of a cell is increasing or decreasing, exocytosis and endocytosis must always be in balance. It follows that while secretion of nervous transmitter substances and hormones by exocytosis must be followed by endocytosis (Nagasawa et al., 1971; Heuser and Reese, 1973; Nordmann et al., 1974), just as certainly receptor internalization by endocytosis (see Schlessinger et al., 1978) must also be matched by exocytosis. Indeed, it has been estimated that the whole cell surface of macrophages is turned over once very 30 min (Mellman et al., 1980). The extent to which exocytosis and endocytosis are strictly coupled in the sense of a push–pull arrangement is still unknown although there are a number of clues that both processes require calcium ions and both are sensitive to calmodulin antagonists such as trifluoperazine (Salisbury et al., 1980).

A major problem in the investigation of these processes is that they take place at the inner face of the plasma membrane, which is relatively inaccessible to experimental manipulation and control. In order to circumvent this problem we have recently introduced a new technique that permits the plasma membrane barrier to be bypassed without impairing the ability of the remaining membrane to undergo exocytosis and endocytosis (Baker and Knight, 1978, 1980). This selective and localized breakdown of the plasma membrane is achieved by subjecting suspensions of cells to intense electric fields of very brief duration (Zimmerman et al., 1975). By suitable choice of voltages the plasma membrane can be caused to undergo dielectric breakdown while the smaller membrane-bound intracellular

P. F. Baker and D. E. Knight • Department of Physiology, King's College, Strand, London WC2R 2LS, England.

organelles are unaffected. Working with cells isolated from the bovine adrenal medulla by enzyme digestion, we have found that exposure to fields of 2 kV/cm ($\tau = 200$ μsec) renders the cells freely permeable to molecules of low molecular weight (up to 1000) and the kinetics of uptake and loss of radioactive markers are quite consistent with the formation of two holes per discharge each about 4 nm in diameter. These holes remain patent for up to 1 hr in cells kept at 37°C. This technique permits, for the first time, control over the ionic environment of the intracellular sites at which exocytosis and endocytosis take place. We have established the following requirements for secretion in bovine adrenal medullary cells (see Baker and Knight, 1981):

1. Exocytosis is activated by buffered ionized Ca concentrations in the micromolar range (apparent $K_m = 1$ μM).
2. Exocytosis has a rather specific requirement for MgATP (apparent $K_m = 1$ mM). In the presence of MgATP, Ca-dependent exocytosis is not obviously affected by cAMP, cGMP, or S-adenosylmethionine.
3. Exocytosis and its activation by Ca seem unaffected by replacing K with Na as the major cation in the medium. Raising the ionized Mg reduces both the apparent affinity for Ca and also the maximum secretion attainable.
4. Exocytosis is sensitive to the nature of the major anion in the solution that has access to the cell interior. Ca-dependent exocytosis is greatest in media in which the major anion is glutamate or acetate and is largely inhibited when these anions are replaced by Cl, Br, I, or SCN. Ca-dependent exocytosis persists in isotonic sucrose.
5. Exocytosis is little affected even by very high concentrations of cytochalasin B, colchicine, vinblastine, TLCK, SITS, DIDS, D600, and procaine; it is partially inhibited by high concentrations of FCCP and completely inhibited by trifluoperazine, although this last substance tends also to promote spontaneous release of catecholamine, thus limiting its usefulness.

It follows that in order to obtain a viable preparation the choice of medium is very important. We normally subject isolated cells to brief intense electric fields in a medium of the following general composition (mM): K glutamate (or K acetate), 140; PIPES (pH 6.6 to 7.2), 20; MgATP, 5; free Mg, 2; and EGTA to give an ionized Ca concentration of 10^{-7} M, i.e., close to physiological (see Baker, 1972; Baker and Knight, 1981). Isolated adrenal medullary cells immersed in this medium and subjected to a series of brief intense electric fields release less than 1% of their total catecholamine. Raising the ionized Ca concentration into the micromolar range releases up to 30% of the total cellular catecholamine. Secretion seems to reflect exocytosis because catecholamine is released along with dopamine-β-hydroxylase whereas cytosolic proteins such as lactate dehydrogenase are not released in a Ca-dependent fashion. Exocytosis is followed by endocytosis just as it is in intact cells. These "leaky" cells have greatly facilitated studies of the mechanism of exocytosis and should also provide a convenient system for studying endocytosis. In addition to adrenal medullary cells, we have applied this technique successfully to the study of exocytosis in sea urchin eggs (Baker *et al.*, 1980) and human platelets (Knight and Scrutton, 1980), and Pace *et al.* (1980) have applied it to pancreatic β cells. In all instances, exocytosis seems to be activated by Ca in the micromolar concentration range, to have a specific requirement of MgATP, and to be sensitive to the nature of the major anion.

Experiments with inhibitors have so far proved rather unhelpful except in the negative sense that they provide no direct support for a number of widely discussed hypotheses such as the involvement in exocytosis of microtubules or microfilaments, the possible role of proteases or alterations in anion or cation permeability. Inhibition by trifluoperazine provides the only real clue but is perhaps not very surprising in view of this substance's widespread ability to interfere with Ca-dependent reactions in which the Ca receptor is a calmodulin (Weiss and Levin, 1978). Exocytosis is undoubtedly Ca dependent, and inhibition by trifluoperazine increases the likelihood that the Ca receptor in exocytosis is also a calmodulin. A similar conclusion has recently been reached for cortical granule exocytosis in the sea urchin (Baker and Whitaker, 1979) and insulin release in the pancreatic β cell (Gagliardino *et al.*, 1980).

One quite new and particularly interesting aspect of our data is the quantitative dependence of catecholamine release on Ca and ATP. In the presence of saturating concentrations on MgATP, raising the ionized Ca ultimately releases 30% of the total cellular catecholamine. Intermediate concentrations of Ca (say 1 μM) only release intermediate amounts of catecholamine even after prolonged periods of incubation. Similarly, in the presence of saturating concentrations of ionized Ca, intermediate concentrations of MgATP (say 1 mM) only cause part of the releasable catecholamine to be secreted. This failure of intermediate concentrations of Ca or MgATP to release all the available catecholamine does not result from some form of slow desensitization of the release mechanisms because a subsequent challenge with a higher concentration of the limiting reactant—Ca or MgATP—can cause the remaining releasable catecholamine to be secreted. If a ramp of Ca is applied—say from 10^{-7} M to 10^{-5} M—the percentage catecholamine secreted is the same irrespective of whether the time taken to reach the highest Ca concentrations is 1 min or 40 min.

At the level of each individual secretory vesicle, exocytosis is an all-or-none event, and one possible explanation for the above facts might be that individual vesicles have different thresholds for Ca or MgATP. If this were the case, the secretion elicited by 1 μM Ca in the presence of a saturating concentration of MgATP should have selectively discharged all the most Ca-sensitive vesicles, i.e., those with Ca thresholds of μM or lower, and the remaining ones should require a higher Ca concentration to cause them to react. This possibility can be tested by exposing "leaky" cells first to 1 μM Ca, then returning them to 0.01 μM Ca and subsequently challenging them a second time with 1 μM Ca. This second challenge always elicits secretion, implying that all the vesicles that can respond to 1 μ Ca could not have been released during the first exposure despite the fact that secretion had ceased.

One explanation consistent with these findings is that the chances of a vesicle undergoing exocytosis may only be increased while the level of ionized Ca (or some reactant dependent on Ca) is changing and during this time the probability of exocytosis may be proportional to the rate of change of Ca. Taking into account the dependence of exocytosis on MgATP, it may be that Ca promotes the phosphorylation of some key site in which case exocytosis may only take place when the steady-state level of phosphorylation is changing and the probability of an individual vesicle undergoing exocytosis may be linked to the rate of change of phosphorylation. In addition to our demonstration of a specific requirement of exocytosis for MgATP, there are a number of other observations drawing attention to the occurrence of membrane phosphorylation at the time of exocytosis (de Lorenzo and Freedman, 1978; de Lorenzo *et al.*, 1979; Michaelson *et al.*, 1980), and it seems important to establish whether phosphorylation always accompanies exocytosis and if so whether the kinetics of phosphorylation are at all consistent with the above hypothesis.

ACKNOWLEDGMENT. This work was supported by grants from the Medical Research Council of Great Britain.

REFERENCES

Baker, P. F. (1972). *Prog. Biophys. Mol. Biol.* **24,** 177–233.

Baker, P. F., and Knight, D. E. (1978). *Nature (London)* **276,** 620–622.

Baker, P. F., and Knight, D. E. (1980). *J. Physiol. (Paris)* **76,** 497–504.

Baker, P. F., and Knight, D. E. (1981). *Proc. R. Soc. London Ser. B* **296,** 83–103.

Baker, P. F., and Whitaker, M. J. (1979). *J. Physiol.* **298,** 55P.

Baker, P. F., Knight, D. E., and Whitaker, M. J. (1980). *Proc. R. Soc. London Ser. B* **207,** 149–161.

de Lorenzo, R. J., and Freedman, S. D. (1978). *Biochem. Biophys. Res. Commun.* **80,** 183–192.

de Lorenzo, R. J., Freedman, S. D., Yohe, W. B., and Maurer, S. C. (1979). *Proc. Natl. Acad. Sci. USA* **76,** 1838–1842.

Gagliardino, J. J., Harrison, D. E., Christie, M. R., Gagliardino, E. E., and Ashcroft, S. J. H. (1980). *Biochem. J.* **192,** 919–927.

Heuser, J. E., and Reese, T. S. (1973). *J. Cell Biol.* **57,** 315–344.

Knight, D. E., and Scrutton, M. (1980). *Thromb. Res.* **20,** 437–446.

Mellman, I. S., Steinman, R. M., Unkeless, J. C., and Cohn, Z. A. (1980). *J. Cell Biol.* **86,** 712–722.

Michaelson, D. M., Avissar, S., Ophir, I., Pinchasi, I., Angel, I., Kloog, Y., and Sokolovsky, M. (1980). *J. Physiol. (Paris)* **76,** 505–514.

Nagasawa, J., Douglas, W. W., and Schulz, R. (1971). *Nature (London)* **232,** 341–342.

Nordmann, J. J., Dreifuss, J. J., Baker, P. F., Ravazzola, M., Malaisse-Lagae, F., and Orci, L. (1974). *Nature (London)* **250,** 155–157.

Pace, C. S., Tarvin, J. T., Neighbors, A. S., Pirkle, J. A., and Greider, M. H. (1980). *Diabetes* **29,** 911–918.

Salisbury, J. L., Cordeelis, J. S., and Satir, P. (1980). *J. Cell Biol.* **87,** 132–141.

Schlessinger, J., Schechter, Y., Willingham, M. C., and Pastan, I. (1978). *Proc. Natl. Acad. Sci. USA* **75,** 2659–2663.

Weiss, B., and Levin, R. M. (1978). *Adv. Cycl. Nucl. Res.* **9,** 285–303.

Zimmerman, U., Riemann, F., and Pilwat, G. (1975). *Biochim. Biophys. Acta* **375,** 209–219.

XV

Membrane-Linked Metabolic and Transport Processes in Plants

172

The Proton-Linked ATPase of Chloroplasts

Richard E. McCarty

The H^+-ATPase of chloroplast thylakoids couples the synthesis of ATP to light-dependent electron flow, using an electrochemical proton gradient as the common intermediate. In this short review, I will outline the salient features of this enzyme. For more detailed treatments, the reader may consult recent reviews by Shavit (1980), Baird and Hammes (1979), and McCarty (1979).

H^+-ATPases, which are present in all coupling membranes, may be divided into two parts, a hydrophilic, oligomeric protein and a collection of hydrophobic polypeptides. The hydrophilic protein contains the active sites for ATP synthesis and hydrolysis and is called coupling factor 1 (CF_1). The hydrophobic components of the complex, which are integral membrane proteins, are collectively denoted as F_0. Thus, the chloroplast H^+-ATPase is often written as CF_1–F_0.

CF_1, an extrinsic thylakoid membrane protein, is readily extracted from thylakoids and is conveniently purified to homogeneity on a large scale (Binder *et al.*, 1978). It is a colorless, water-soluble protein that contains tightly bound ADP and, depending on storage conditions, ATP (Harris and Slater, 1975). The molecular weight of CF_1 is 325,000 (Farron, 1970; Paradies *et al.*, 1978). A much higher value was obtained by Yoshida *et al.* (1979), possibly because of the presence of 10% methanol during sedimentation.

The enzyme contains five different subunits. The stoichiometry of these subunits, labeled α–ϵ in order of decreasing molecular weight, appears to be $\alpha_2\beta_2\gamma\delta\epsilon_2$. Several independent lines of evidence support this stoichiometry, although there is more uncertainty about the δ and ϵ content because these subunits can dissociate from the complex. Nelson (1976) grew peas in $^{14}CO_2$ and determined the radioactivity in the subunits to be consistent with an $\alpha_2\beta_2\gamma$ composition. The cysteine composition of the subunits (α-2, β-2, γ-3–4) is in line with an $\alpha_2\beta_2\gamma$ stoichiometry (Binder *at al.*, 1978). Moreover, nucleotide binding (Cantley and Hammes, 1975) and chemical cross-linking studies (Baird and Hammes, 1976) point to an $\alpha_2\beta_2\gamma$ stoichiometry. As there is good evidence for an $\alpha_3\beta_3\gamma$

Richard E. McCarty • Section of Biochemistry, Molecular and Cell Biology, Division of Biological Sciences, Cornell University, Ithaca, New York 14853.

stoichiometry for the F_1 from PS-3, a thermophilic bacterium (Yoshida *et al.*, 1979), there is an intriguing possibility that the subunit structure of F_1 molecules may vary.

The H^+-ATPase has been extracted from thylakoids and purified in attempts to identify components of F_0. These preparations contain about three to four polypeptides in addition to the five CF_1 subunits (Pick and Racker, 1979; Nelson and Hauska, 1979) and are capable of P_i–ATP exchange and of net ATP synthesis when reconstituted into liposomes. To date, only the smallest polypeptide (molecular weight 8000) in the preparations has been shown to be a part of the H^+-ATPase. This polypeptide is soluble in organic solvents and binds the inhibitor N,N'-dicyclohexylcarbodiimide, which blocks ATP synthesis by preventing proton transport through the membrane sector of the H^+-ATPase. This polypeptide appears to be responsible for proton transport through F_0, as its incorporation into lipid vesicles promotes a dicyclohexylcarbodiimide-sensitive proton leak (Nelson *et al.*, 1977; Sigrist-Nelson and Azzi, 1980). There are about six of these polypeptides per H^+-ATPase.

The identification and function of other F_0 polypeptides remain to be established. At least one component is likely to function in the binding of CF_1, as CF_1 does not bind to vesicles containing the 8000-molecular-weight polypeptide. A polypeptide could also be required for assembly of the complex. With improvements in methodology, other components of F_0 will be elucidated. Some surprises may be in store. The turnover of CF_1 in the reconstituted systems for photophosphorylation is about 0.1% of the full potential of the enzyme in illuminated thylakoids. Although the poor efficiency of the reconstituted vesicles may in part be ascribed to their proton leakiness, it may also stem from a deficiency in a component of the H^+-ATPase lost during purification.

The ATPase activity of thylakoids, the extracted H^+-ATPase, and soluble CF_1 is latent and is induced by a variety of treatments, including heat, trypsin (Vambutas and Racker, 1965), and dithiothreitol. Even with soluble CF_1, the mechanism(s) of ATPase activation is unknown. These treatments may interfere with the interactions between the ϵ subunit and a catalytic subunit. The ϵ subunit can inhibit the ATPase activity of an ϵ-depleted enzyme (Nelson *et al.*, 1972). The activation by dithiothreitol points to a possible involvement of disulfides. This possibility is supported by the finding that a disulfide in γ exchanges with a dithiol in α during heat activation (Ravizzini *et al.*, 1980). An involvement of disulfide exchange in the activation process may also explain why the reaction of a group in the γ subunit with N-ethylmaleimide (McCarty and Fagan, 1973) partially inhibits ATP hydrolysis by isolated CF_1 (Moroney and McCarty, 1980).

The H^+-ATPase in thylakoids in the dark is inactive and is converted to an active form upon illumination. The mechanism of this activation is unclear, but may involve disulfide exchange for preillumination in the presence of dithiothreitol causes a more rapid activation (Harris and Crofts, 1978). Unless the thylakoids are illuminated in the presence of dithiothreitol, the rate of ATP hydrolysis decays very rapidly after illumination. An energy-dependent reduction of a disulfide bond in CF_1 probably converts the enzyme to a more stable form capable of ATP hydrolysis and P_i–ATP exchange in the dark. In intact-chloroplasts, the enzyme is apparently reduced by a thioredoxin system (Mills *et al.*, 1980). How wasteful ATP hydrolysis by the H^+-ATPase is prevented in intact chloroplasts in the dark is not known. However, the activated state is very sensitive to ADP, and the rebinding of ADP to CF_1 in the dark inactivates ATP hydrolysis (Shoshan and Selman, 1979).

ATP synthesis decreases the extent of the ΔpH across thylakoid membranes, and ATP hydrolysis in the dark causes inward proton translocation. An uncoupler-sensitive P_i–ATP exchange is associated with ATP hydrolysis, suggesting that ATP hydrolysis is the reverse of ATP synthesis. This conclusion has been questioned (Shavit, 1980). Photophorylation responds to a variety of treatments in a manner that is fully consistent with a chemiosmotic model (McCarty and Portis, 1976). Detailed analysis of both photophosphorylation and of ATP hydrolysis and P_i–ATP exchange provided evidence that 3 protons are translocated per ATP formed or hydrolyzed (H^+/P ratio). Thermodynamic measurements (Avron, 1978) also support an H^+/P ratio of 3.

Three distinct nucleotide binding sites have been detected in soluble CF_1. The properties of nucelotide binding to these sites are not consistent with those of a catalytic site. For example, the enzyme contains a site to which MgATP is remarkably firmly bound. The enzyme may be heat-activated in the presence of 40 mM ATP without loss of the bound ATP (Bruist and Hammes, 1981). ADP is firmly bound to native CF_1 and undergoes a very slow exchange with either ADP or ATP. After activation, the exchange is much faster (Carlier and Hammes, 1979), but is still too slow to be an obligate part of the catalytic mechanism. This exchange and the failure to recognize whether ATP is bound have confounded studies on nucleotide binding to isolated CF_1.

Nucleotide exchange also takes place in CF_1 in thylakoids (Harris and Slater, 1975). Energization of thylakoid membranes converts ADP from a tightly bound form to a loosely bound form:

$$CF_1 \sim ADP \underset{\text{energy}}{\rightleftharpoons} CF_1—ADP \rightleftharpoons CF_1 + ADP$$

where $CF_1 \sim ADP$ stands for the tightly bound form and CF_1—ADP for the loosely bound form (Strotmann et al., 1979). The rate of binding of ADP to the ADP-depleted enzyme is much slower than phosphorylation. There are several similarities between the site that undergoes energy-dependent exchange and the exchangeable site in CF_1. In addition, rapidly labeled, transitorily bound ATP has been detected in illuminated thylakoids (Rosen *et al.*, 1979).

The functions of these nucleotide sites are unknown. At least one site could be involved in the assembly of CF_1. Very likely, energy-dependent changes in the binding of nucleotides are involved in the regulation of the expression of ATPase activity.

The functions of the subunits of the chloroplast H^+-ATPase are being further defined (Table I). As mentioned previously, the smallest subunit of F_0 mediates transmembrane proton translocation and is thus likely to deliver protons to CF_1. The ϵ subunit of CF_1 may function in the regulation of the expression of ATPase activity, but in *E. coli* F_1, ϵ is required for binding of the enzyme to the membrane (Sternweis, 1978). CF_1 lacking the δ subunit does not bind to CF_1-deficient membranes and binding is restored by addition of

Table I. Properties of the Subunits of the Chloroplast H^+-ATPase

Component	Approximate molecular weight (10^{-3})	Possible functions
CF_1-α	59	Regulatory, activation (?)
CF_1-β	55	Contains catalytic sites
CF_1-γ	34	Proton transport; activation (?)
CF_1-δ	22	Binding to F_0
CF_1-ϵ	16	Regulatory; binding (?)
F_0-DCCD-binding protein[a]	8	Proton conductance to CF_1

[a] DCCD, *N,N'*-dicyclohexylcarbodiimide. There are likely to be other components of F_0 whose functions have not yet been investigated.

the δ subunit (Nelson and Karny, 1976). In view of the sensitivity of photophosphorylation to an antiserum against the γ subunit (Nelson *et al.*, 1973) and to a reaction of a group in γ with maleimides (McCarty and Fagan, 1973), the γ subunit plays an important role in photophosphorylation. The cross-linking of two groups within the γ subunit renders thylakoid membranes leaky to protons (Moroney and McCarty, 1979). The γ subunit may thus have a role in proton translocation by the ATPase and may also be involved in the conversion of the enzyme to an active form (Moroney and McCarty, 1980).

The active sites of the ATPase are likely to be on the β subunits. This conclusion is supported by inhibitor studies and by the fact that prolonged tryspin treatment of CF_1 digests the smaller subunits and partially clips the α subunit while leaving the β subunit apparently intact (Deters *et al.*, 1975). This preparation is nearly fully active as an ATPase. To date, the ATPase activity has not been reconstituted from isolated subunits. This has been accomplished with other F_1 molecules (see for example Futai, 1977). In these cases, the α, β, and γ subunits are required for reconstitution. Moreover, the β subunit of the H^+-ATPase of *R. rubrum*, a photosynthetic bacterium, is inactive in ATP hydrolysis, but is required for ATP hydrolysis (Philosoph *et al.*, 1977). Therefore, if the β subunit carries the active site, its association with other subunits is apparently required for its activity to be expressed. The possibility that the α subunit has the active site has not yet been excluded.

The several proposals for the mechanism of H^+-ATPases that have been put forth differ remarkably in the way in which protons are used in the reaction. A direct protonation of P_i bound to the active center of F_1 (Mitchell, 1977) was proposed. In contrast to this direct involvement of protons in the reaction, the proton gradient has been suggested to drive conformational changes that promote substrate binding and product release (Boyer, 1979; Slater, 1977). In this mechanism, called the binding change mechanism, energy is thought to be used not in the formation of ATP, but in the release of ATP and binding of P_i. Because soluble F_1 catalyzes both a slow intermediate P_i–HOH exchange (Choate *et al.*, 1979) and probably a P_i–ATP exchange (Bossard and Schuster, 1981), at least some covalent bond formation can apparently take place without the input of energy from the proton gradient.

Phosphorylated intermediates have not been detected in H^+-ATPases, although they are involved in the mechanism of other cation-translocating ATPases. Although marked advances in our understanding of the composition, structure, and nucleotide binding properties of the chloroplast H^+-ATPase have been made in the past 10 years, its detailed molecular mechanism remains elusive.

REFERENCES

Avron, M. (1978). *FEBS Lett.* **96**, 225–232.

Baird, B. A., and Hammes, G. G. (1976). *J. Biol. Chem.* **251**, 6953–6962.

Baird, B. A., and Hammes, G. G. (1979). *Biochim. Biophys. Acta* **549**, 31–54.

Binder, A., Jagendorf, A. T., and Ngo, E. (1978). *J. Biol. Chem.* **253**, 3094–3100.

Bossard, M. J., and Schuster, S. M. (1981). *J. Biol. Chem.* **256**, 1518–1521.

Boyer, P. D. (1979). In *Membrane Bioenergetics* (C. P. Lee, G. Schatz, and G. Ernster, eds.), pp. 461–479, Addison–Wesley, Reading, Mass.

Bruist, M. F., and Hammes, G. G. (1981). *Biochemistry* **20**, 6298–6305.

Cantley, L. C., and Hammes, G. G. (1975). *Biochemistry* **14**, 2968–2975.

Carlier, M. F., and Hammes, G. G. (1979). *Biochemistry* **18**, 3446–3451.

Choate, G. L., Hutton, R. L., and Boyer, P. D. (1979). *J. Biol. Chem.* **254**, 286–290.

Deters, D. W., Racker, E., Nelson, N., and Nelson, H. (1975). *J. Biol. Chem.* **250**, 1041–1047.

Farron, F. (1970). *Biochemistry* **9**, 3823–3829.

Futai, M. (1977). *Biochem. Biophys. Res. Commun.* **79**, 1231–1237.

Harris, D. A., and Crofts, A. R. (1978). *Biochim. Biophys. Acta* **502**, 87–102.

Harris, D. A., and Slater, E. C. (1975). *Biochim. Biophys. Acta* **387**, 335–348.

McCarty, R. E. (1979). *Annu. Rev. Plant Physiol.* **30**, 79–104.

McCarty, R. E., and Fagan, J. (1973). *Biochemistry* **12**, 1503–1507.

McCarty, R. E., and Portis, A. R., Jr. (1976). *Biochemistry* **15**, 5110–5114.

Mills, J. D., Mitchell, P., and Schürmann, P. (1980). *FEBS Lett.* **112**, 173–177.

Mitchell, P. (1977). *Annu. Rev. Biochem.* **46**, 996–1005.

Moroney, J. V., and McCarty, R. E. (1979). *J. Biol. Chem.* **254**, 8951–8955.

Moroney, J. V., and McCarty, R. E. (1980). *Proceedings, Fifth International Congress on Photosynthesis*, in press.

Nelson, N. (1976). *Biochim. Biophys. Acta* **456**, 314–338.

Nelson, N., and Hauska, G. A. (1979). In *Membrane Bioenergetics* (C. P. Lee, G. Schatz, and L. Ernster, eds.), pp. 189–202, Addison–Wesley, Reading, Mass.

Nelson, N., and Karny, O. (1976). FEBS Lett. **70**, 249–253.

Nelson, N., Nelson, H., and Racker, E. (1972). *J. Biol. Chem.* **247**, 6506–6510.

Nelson, N., Deters, D. W., Nelson, H., and Racker, E. (1973). *J. Biol. Chem.* **248**, 2049–2055.

Nelson, N., Eytan, E., Notsani, B.-E., Sigrist, H., Sigrist-Nelson, K., and Gitler, C. (1977). *Proc. Natl. Acad. Sci. USA* **74**, 2375–2378.

Paradies, H. H., Zimmerman, J., and Schmidt, U. (1978). *J. Biol. Chem.* **253**, 8972–8979.

Philosoph, S., Binder, A., and Gromet-Elhanan, Z. (1977). *J. Biol. Chem.* **252**, 8747–8752.

Pick, U., and Racker, E. (1979). *J. Biol. Chem.* **254**, 2793–2799.

Ravizzini, R. A., Andreo, C. S., and Vallejos, R. H. (1980). *Biochim. Biophys. Acta* **591**, 135–141.

Rosen, G., Gresser, M., Vinkler, C., and Boyer, P. D. (1979). *J. Biol. Chem.* **254**, 10654–10661.

Shavit, N. (1980). *Annu. Rev. Biochem.* **49**, 111–138.

Shoshan, V., and Selman, B. R. (1979). *J. Biol. Chem.* **254**, 8801–8807.

Sigrist-Nelson, K., and Azzi, A. (1980). *J. Biol. Chem.* **255**, 10638–10643.

Slater, E. C. (1977). *Annu. Rev. Biochem.* **46**, 1015–1026.

Sternweis, P. C. (1978). *J. Biol. Chem.* **253**, 3123–3128.

Strotmann, H., Bickel-Sandkötter, S., and Shoshan, V. (1979). *FEBS Lett.* **101**, 316–320.

Vambutas, V. K., and Racker, E. (1965). *J. Biol. Chem.* **240**, 2660–2667.

Yoshida, M., Sone, N., Hirata, H., Kagawa, Y., and Ui, N. (1979). *J. Biol. Chem.* **254**, 9525–9533.

173

Protein Complexes in Chloroplast Membranes

Nathan Nelson

1. INTRODUCTION

Energy conduction by biological membranes is based on a functional asymmetry of the system (Mitchell, 1966). The asymmetry of the chloroplast membranes is determined by the uneven distribution of membrane proteins between the two surfaces. This functional asymmetry is maintained by tight protein complexes that have specific biochemical function in the overall activity of energy transduction. The chloroplast inner membrane contains four well-defined and biochemically active protein complexes. Two of them, the photosystem II and photosystem I reaction centers, are associated with chlorophyll, and the other two, cytochrome b_6-f complex and the H^+-ATPase complex, are free of chlorophyll. Each one of these four complexes is composed of several polypeptides. The structure of the complexes and the function of individual subunits are the subject matter of this review.

2. PHOTOSYSTEM II REACTION CENTER

Photosystem II reaction center can be defined as the minimal structure that is active in photooxidation of water, when the electron acceptor is plastoquinone. The minimal unit that can perform the primary event of charge separation associated with photosystem II can be called P_{680} reaction center. Neither photosystem II reaction center nor P_{680} reaction center has been purified in an active state. Satoh (1979, 1980) reported a highly purified protein complex containing three polypeptides with apparent molecular weights of 6500, 27,000, and 43,000. This preparation possesses the capacity of photoxidize 1,5-diphenyl-carbazide, a reaction specific for photosystem II. The larger of the three peptides (43K) contains the chlorophyll a molecules associated with photosystem II, and the low-molecular-weight (6.5K) polypeptide is believed to be the cytochrome b_{559}. The possible function

Nathan Nelson • Department of Biology, Technion-Israel Institute of Technology, Haifa, Israel.

of the 27K polypeptide is not known. For the time being, Satoh's preparation can be considered as P_{680} reaction center, but it is quite likely that in the future a purified 43K polypeptide will meet the expectations of such a reaction center. The organization of the three polypeptides in the membrane is not known. One can predict that the 43K polypeptide is situated in the membrane at least as a dimer and that the P_{680} chlorophyll is located in the internal side of the membrane. Cytochrome b_{559} should be located within the membrane but part of it might protrude outside (antibody against cytochrome b_{559} interacts with it from outside).

Recently, Spector and Winget (1980) have purified a manganese-containing protein that functions in photosynthetic oxygen evolution. The purified protein has an apparent molecular weight of 65,000 and contains two atoms of manganese bound to each monomer. If this protein, reconstituted together with Satoh's reaction center, causes light-dependent oxygen evolution, the structure of photosystem II reaction center will turn out to be surprisingly simple.

An additional candidate for taking part in photosystem II reaction center is the "photogene" product. This rapidly metabolized 32K protein was recently found to be functionally involved in oxygen evolution (Renger, 1976; Mattoo *et al.*, 1981). However, the fact that this protein does not stain in SDS gels makes it difficult to evaluate its functional assembly in photosystem II reaction center.

A schematic representation of the organization of protein complexes in the chloroplast membranes is shown in Fig. 1.

3. CYTOCHROME b_6-f COMPLEX

Plastoquinone is the electron donor of the cytochrome b_6-f complex and plastocyanine serves as its natural electron acceptor. Initial attempts to purify this protein complex yielded inactive preparations (Nelson and Neumann, 1972), but we learned that evolution sympathized with it and maintained its general structure and function in various organisms. Recently, Hauska *et al.* (1980b) have purified an active cytochrome b_6-f complex with plastoquinol-plastocyanine oxidoreductase activity. The purified complex has five polypeptide subunits with apparent molecular weights of 34, 33, 23, 20, and 17.5K. The largest polypeptide (34K) is cytochrome f, which can be visualized as a purple band on SDS gels (Nelson and Racker, 1972). Two molecules of cytochrome b_6 are present in each complex. The 23K polypeptide might be the apoprotein of this cytochrome. The purified complex contains about two nonheme iron proteins (Rieske type, see Malkin and Posner, 1978) per cytochrome f. Good candidates for their apoproteins are the 20K polypeptides.

Using an antibody against cytochrome f has revealed that this polypeptide is located in the internal side of the thylakoid membrane. Assuming a high degree of similarity in the structure and function of the mitochondrial cytochrome b-c_1 complex and the chloroplast cytochrome b_6-f complex, Fig. 1 depicts a possible organization of the latter in the membrane. Some properties of individual subunits are given in Table I.

4. PHOTOSYSTEM I REACTION CENTER

Photosystem I reaction center can be defined as the minimal structure that is active in NADP photoreduction when the electron donor to the system is reduced plastocyanine

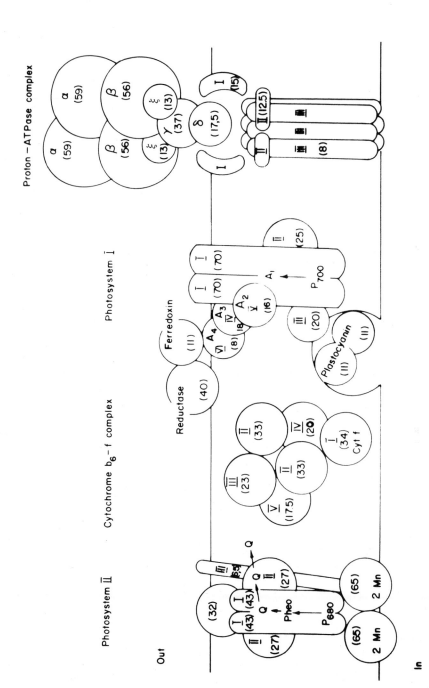

Figure 1. Schematic representation of the organization of protein complexes in the chloroplast membranes.

Table I. Size and Function of Individual Polypeptides in the Chloroplast Membrane

	Subunit	Molecular weight	Pigments and functional groups	Function
Photosystem II				
Photosystem II reaction center	I	43,000	Chlorophyll a P_{680}? Pheophytin? Primary quinone acceptor?	Primary event of light harvesting and charge separation in photosystem II
	II	27,000	—	?
	III	6,500	Cytochrome b_{559}	Water oxidation?
Manganese-containing protein		65,000?	^{M}n	Water oxidation?
Photogene product		32,000	—	Regulation of oxygen evolution? Conferral of DCMU sensitivity?
Electron transport chain				
Cytochrome $b_6 f$ complex	I	34,000	Cytochrome f	Reduction of plastocyanin
	II	33,000	—	?
	III	23,000	Cytochrome b	Q cycle?
	IV	20,000	Nonheme iron	Q cycle?
	V	17,500	—	?
Plastocyanin		11,000	Cu	Electron donor to photosystem I
Photosystem I				
Photosystem I reaction center	I	70,000	Chlorophyll a, β-carotene P_{700} Primary acceptor -chlorophyll a?	Primary events of light harvesting and charge separation in photosystem I Primary acceptor A_1
	II	25,000	—	?
	III	20,000	—	Required for the reduction of P_{700}^+ by plastocyanin
	IV	18,000	Nonheme iron?	Electron acceptor A_3?
	V	16,000	Nonheme iron?	Electron acceptor A_2?
	VI	8,000	Nonheme iron?	Electron acceptor A_4?
	VII?	11,500	?	?
Ferredoxin		11,000	Nonheme iron	Electron transport from photosystem I reaction center to ferredoxin-NADP reductase. Cyclic electron transport
Ferredoxin-NADP-reductase		40,000	FAD	Electron transport from ferredoxin to NADP
H$^+$-ATPase complex				
CF$_1$	α	59,000	—	High-affinity regulatory binding sites
	β	56,000	ADP?	Active site
	γ	37,000	—	Energy transduction?
	δ	17,500	—	Binding of CF$_1$ to the membrane
	ϵ	13,000	—	ATPase inhibitor
CF$_0$	I	15,000	—	Binding of CF$_1$?
	II	12,500	—	Tightening hexamers of subunit III?
	III	8,000	—	Proton channel

(Nelson and Bengis, 1975). Such a reaction center was isolated from chloroplasts of several plant species (Bengis and Nelson, 1975, 1977; Nelson and Notsani, 1977; Okamura *et al.*, 1981). The isolated protein complex is active in NADP photoreduction by ascorbate when the system is supplemented with purified plastocyanine, ferredoxin, and ferredoxin-NADP-reductase. Upon reconstitution of the purified photosystem I reaction center together with the chloroplast H^+-ATPase complex, photophosphorylation was obtained (Hauska *et al.*, 1980a). These experiments show that purified photosystem I reaction center preserved all of the photobiochemical activities of photosystem I.

The complex is composed of six to seven different polypeptides that were designated as subunits I to VII in the order of decreasing molecular weights of 70,000 to 8,000 (Bengis and Nelson, 1975, 1977; Nelson and Notsani, 1977). Subunit I (70K) of photosystem I reaction center was isolated in pure form (Bengis and Nelson, 1975). The preparation is active in light-induced P_{700} oxidation and also contains the primary electron acceptor A_1 (Nelson and Notsani, 1977). This preparation has been denoted as P_{700} reaction center because it is the minimal unit that can perform the primary event of light-induced charge separation. There are two copies of subunit I, and together they contain one P_{700} pigment along with about 40 chlorophyll *a* molecules and about one β-carotene molecule. These pigments serve as the primary light-harvesting antenna of photosystem I (Bengis and Nelson, 1977). The mutual orientation of pigments in photosystem I reaction center has been studied by Junge and his colleagues (Junge *et al.*, 1977; Junge and Schaffernicht, 1979). It was demonstrated that most of the primary antenna chlorophyll *a* molecules are highly oriented toward P_{700}. The P_{700} pigment is probably situated in the internal side of the chloroplast membrane. Studies with specific antibody against subunit I indicated that this subunit protrudes out of the membrane and therefore is a transmembrane polypeptide (Nelson and Bengis, 1975; Bengis and Nelson, 1975, 1977).

Treatment of photosystem I reaction center with a high concentration of Triton X-100 specifically liberated subunit III (20K) from the preparation. Parallel loss of subunit III, cytochrome 552 photooxidation activity, and plastocyanine dependent NADP photoreduction activity was observed (Bengis and Nelson, 1977). It was concluded that subunit III functions in the oxidizing side of photosystem I reaction center by providing the binding site for plastocyanine or by induction of the proper conformation to subunit I (Bengis and Nelson, 1977; Haehnel *et al.*, 1980).

The function of the individual subunits of the remaining polypeptides in photosystem I reaction center is not clear. It was suggested that subunits IV (18K), V (16K), and VI (8K) are the secondary electron acceptors $A_2(X)$, $A_3(B)$, and $A_4(A)$ (Bengis and Nelson, 1977; Okamura *et al.*, 1982). This suggestion awaits further experimental evidence. Figure 1 and Table I depict the suggested structure and function of the individual subunits of photosystem I reaction center.

5. THE CHLOROPLAST H^+-ATPase COMPLEX

One of the functions of the above-mentioned three protein complexes is to create membrane potential and proton gradient across the chloroplast membrane. The function of the H^+-ATPase complex is to convert the energy stored in the protonmotive force into ATP. The structure and function of H^+-ATPase complexes were strictly preserved during long periods of evolution and complexes similar to that of the chloroplast can be found in almost every living creature. The main functional difference that can be found among the H^+-ATPase complexes, from various sources, lies in their ATPase activity. The ATPase activity of the chloroplast complex should be strictly controlled and this activity should be totally blocked especially at nighttime. On the other hand, many bacterial species and the mitochondria make use of the ATPase activity of the complex for amino acid uptake and reverse electron transport.

The chloroplast ATPase is composed of two distinct structures, a catalytic sector that is hydrophilic in nature and a membrane sector that is hydrophobic in nature (Nelson, 1976, 1981). The function of the catalytic sector (CF_1) is to catalyze the formation of ATP at the expense of energy expressed in a flux of protons. The function of the membrane sector (CF_0) is to provide the catalytic sector with a directed flux of protons across the membrane. CF_1 is composed of five subunits with molecular weights of about 59K (α), 56K (β), 37K (γ), 17.5K (δ), and 13K (ϵ) (Nelson, 1976). From the subunit composition of the purified chloroplast H^+-ATPase, it is apparent that CF_0 is comprised of three different polypeptides designated subunit I, II, and III in the order of decreasing molecular weight (15, 12.5, and 8K, respectively) (Pick and Racker, 1979; Nelson et al., 1980). A specific function for each individual subunit has been proposed (see Table I and Nelson 1976, 1981). The proposed functions for some of the individual subunits have been confirmed and extended for several H^+-ATPase complexes from various sources.

REFERENCES

Bengis, C., and Nelson, N. (1975). *J. Biol. Chem.* **250**, 2783–2788.
Bengis, C., and Nelson, N. (1977). *J. Biol. Chem.* **252**, 4564–4569.
Haehnel, W., Hesse, V., and Propper, A. (1980). *FEBS Lett.* **111**, 79–82.
Hauska, G., Samoray, D., Orlich, G., and Nelson, N. (1980a). *Eur. J. Biochem.* **111**, 535–543.
Hauska, G., Orlich, G., Samoray, D., Hurt, E., and Sane, P. V. (1980b). *Proceedings of the 5th International Photosynthesis Congress*, in press.
Junge, W., and Schaffernicht, H. (1979). In *Chlorophyll Organization and Energy Transfer in Photosynthesis*, pp. 127–146, Excerpta Medica, Amsterdam.
Junge, W., Schaffernicht, H., and Nelson, N. (1977). *Biochim. Biophys. Acta* **462**, 73–85.
Malkin, R., and Posner, H. B. (1978). *Biochim. Biophys. Acta* **501**, 552–554.
Mattoo, A. K., Pick, U., Hoffman-Falk, H., and Edelman, M. (1981). *Proc. Natl. Acad. Sci. USA* **78**, 1572–1576.
Mitchell, P. (1966). *Biol. Rev.* **41**, 445–502.
Nelson, N. (1976). *Biochim. Biophys. Acta* **456**, 314–338.
Nelson, N. (1981). *Curr. Top. Bioenerg.* **11**, 1–33.
Nelson, N., and Bengis, C. (1975). In *Proceedings of the Third International Congress on Photosynthesis* (M. Avron, ed.), pp. 609–620, Elsevier, Amsterdam.
Nelson, N., and Neumann, J. (1972). *J. Biol. Chem.* **247**, 1817–1824.
Nelson, N., and Notsani, B. (1977). In *Bioenergetics of Membranes* (L. Packer et al., eds.), pp. 233–244, Elsevier, Amsterdam.
Nelson, N., and Racker, E. (1972). *J. Biol. Chem.* **247**, 3848–3853.
Nelson, N., Nelson, H., and Schatz, G. (1980). *Proc. Natl. Acad. Sci. USA* **77**, 1361–1364.
Okamura, M. Y., Feher, G., and Nelson, N. (1982). In *Integrated Approach to Plant and Bacterial Photosynthesis* (Gavendjee, ed.), Academic Press, New York, in press.
Pick, U., and Racker, E. (1979). *J. Biol. Chem.* **254**, 2793–2799.
Renger, G. (1976). *Biochim. Biophys. Acta* **440**, 287–300.
Satoh, K. (1979). *Biochim. Biophys. Acta* **546**, 84–92.
Satoh, K. (1980). In *Proceedings of the 5th International Photosynthesis Congress*, in press.
Spector, M., and Winget, G. D. (1980). *Proc. Natl. Acad. Sci. USA* **77**, 957–959.

174

Export of Light-Generated Reducing Equivalents from Illuminated Chloroplasts by Shuttle Mechanisms

J. W. Anderson

The transformation of light energy into chemical energy by the photosynthetic apparatus of plants is associated with the phosphorylation of ADP and reduction of the oxidized forms of Fd* and NADP. Fd_{red} and NADPH can be regarded as the source of the reducing equivalents (and hence the main energy sources) for the reductive assimilation of inorganic carbon, nitrogen, and sulfur (e.g., CO_2, NO_2^-, and SO_4^{2-}) into sugars, amino acids, and other essential biological molecules. In plants the formation of reduced forms of NADP and Fd in light-dependent processes and their expenditure in the reduction of CO_2 and inorganic nitrogen and sulfur occurs in the chloroplasts of photosynthetic cells. The association of most of the enzymes of CO_2, NO_2^-, and SO_4^{2-} assimilation with chloroplasts is consistent with this view (Anderson, 1981).

Another factor relevant to the compartmentalization of the assimilatory processes of chloroplasts is the impermeability of the chloroplast envelope to $Fd_{ox/red}$ and NADP/NADPH. This is instanced by the very slow rate of reduction of exogenous NADP and Fd by illuminated intact chloroplasts but rapid rates of reduction when the membrane is disrupted by osmotic shock (Heber, 1974). Similarly, the membrane is impermeable to endogenous NADPH; the NADPH/NADP ratio of the chloroplasts of darkened leaf cells is enhanced by illumination but a similar change in the cytoplasm is not observed (Heber and Santarius,

---

*Abbreviations: DCMU, 3-(3,4-dichlorophenyl)-1,1-dimethylurea; DHAP, dihydroxyacetone phosphate; Fd_{red} and Fd_{ox}, reduced and oxidized forms of ferredoxin; GAP, glyceraldehyde 3-phosphate; GAPDH, glyceraldehyde 3-phosphate dehydrogenase; GDH, glutamate dehydrogenase; GSH and GSSG, reduced and oxidized forms of glutathione; LDH, lactate dehydrogenase; MDH, malate dehydrogenase; OAA, oxaloacetate; PGA, 3-phosphoglycerate; diPGA, 1,3-diphosphoglycerate. Enzyme abbreviations preceded by NAD or NADP refer to the specificity of the enzyme with respect to the nucleotide (e.g., NADP-MDH refers to NADP-specific MDH). Glutathione refers nonspecifically to GSH and/or GSSG.

J. W. Anderson • Botany Department, La Trobe University, Bundoora, Victoria 3083, Australia.

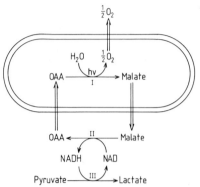

Figure 1. Experimental model for examining the C_4 dicarboxylic acid shuttle. Isolated chloroplasts are incubated in the light with catalytic concentrations of OAA or malate, catalytic concentrations of NAD, substrate amounts of pyruvate and exogenous NAD-MDH and LDH. Reaction I involves chloroplast NADP-MDH using light-generated NADPH as reductant with concomitant evolution of O_2. The malate so produced is oxidized externally to the chloroplast by the reaction sequence involving NAD-MDH and LDH (reactions II and III, respectively). The OAA so formed can recycle via reaction I. Entry and exit of OAA and malate are facilitated by the dicarboxylic acid translocator. Malate can replace OAA as the source of dicarboxylic acid but must first be oxidized by the extrachloroplast system involving reactions II and III. Reaction III can be replaced by other exergonic reactions involving the oxidation of NADH (e.g., reduction of nitrate to nitrite in the presence of nitrate reductase). [From Anderson and House, 1979b.]

1965). The impermeability of the chloroplast envelope to the Fd_{red} and NADPH produced within the chloroplast through the action of light implies that these substrates are not available for reductive reactions in the cytoplasm. Illuminated ruptured chloroplasts reduce GSSG to GSH with the concomitant evolution of O_2 (Jablonski and Anderson, 1978), suggesting that GSH could serve as a carrier of light-generated reducing equivalents. However, because intact chloroplasts do not support GSSG-dependent O_2 evolution, this implies that the chloroplast membrane is impermeable to GSSG and that GSH/GSSG does not act as a redox pair for the transfer of reducing equivalents between chloroplasts and other metabolic compartments of plant cells.

The chloroplast envelope is freely permeable to OAA and PGA (Heldt, 1976). Addition of OAA and PGA to intact chloroplasts results in the immediate evolution of O_2 at rapid rates (Fig. 1). The PGA-dependent reaction is attributed to the reduction of PGA to GAP, DHAP, etc. via the reactions of the CO_2 assimilation pathway; the evolution of O_2 is associated with the production of NADPH by the light reactions and its oxidation by diPGA in a reaction catalyzed by NADP-GAPDH. Similarly, exogenous OAA effects the oxidation of the endogenous NADH/NADPH pool of illuminated chloroplasts in a light-dependent reaction (Heber, 1974). It is now generally agreed that the light-dependent reaction with OAA involves its reduction to malate by light-coupled NADP-MDH of chloroplasts (Fig. 1); in particular, 20 to 100 μM NADPH (but not NADH) supports OAA-dependent O_2 evolution by osmotically shocked chloroplasts (Anderson and House, 1979a).

In C_3 plants, the reduction of OAA to malate does not form part of a major assimilatory process as described for the reduction of PGA, and thus other functions for the highly active light-coupled NADP-MDH of chloroplasts have been sought. In the presence of exogenous NAD-MDH and substrate amounts of NAD, addition of OAA to chloroplasts initiates the reduction of NAD outside the chloroplast in the light (Heber and Krause, 1970). This implies that the NADP(H) and NAD(H) pools inside and outside the chloroplast respectively are in equilibrium and that malate/OAA serves as the redox pair in the transfer of reducing equivalents between the two pools. This hypothesis, which is known as the C_4 dicarboxylic acid shuttle, predicts that light-generated reducing equivalents are directed from NADPH into OAA via chloroplast NADP-MDH and transferred (as malate) across the chloroplast envelope to other metabolic compartments (e.g., cytoplasm) where they effect the reduction of NAD through the action of an NAD-MDH associated with the other metabolic compartment. It follows from this hypothesis that trace amounts of OAA/malate will support the reduction of a suitable substrate (in the presence of an appropriate enzyme) outside the chloroplast in a light-dependent reaction. The reduction of the eventual electron acceptor in the second metabolic compartment would be stoichiometrically related to the evolution of O_2 by the chloroplasts but neither parameter need bear a fixed relation to the amount of C_4 dicarboxylic acid because the latter can recycle indefinitely. In the presence of NAD-MDH, LDH, and micromolar concentrations of NAD and OAA/malate, isolated chloroplasts support the light-dependent reduction of pyruvate outside the chloroplast in a reaction with characteristics consistent with those described above (Anderson and House, 1979b). Similarly, isolated pea chloroplasts support light-dependent reduction of nitrate in the presence of exogenous nitrate reductase, NAD-MDH, and micromolar concentrations of NAD and OAA/malate though under these conditions the nitrite so formed is further reduced to NH_3 within the chloroplast itself (House and Anderson, 1980).

In leaves, the reduction of nitrate (not to be confused with nitrate reductase activity) is enhanced by light (Beevers and Hageman, 1969) and by infiltrating leaves with malate and intermediates of the CO_2 assimilation pathway (Kleeper et al., 1971; Neyra and Hageman, 1976; Nicholas et al., 1976). Because nitrate reductase is a cytoplasmic enzyme (Dalling et al., 1972; Wallsgrove et al., 1979), it has been proposed that the C_4 and other shuttles of chloroplasts could participate in the transfer of light-generated reducing equivalents to the cytoplasm and the reduction of nitrate. The experiments demonstrating the role of C_4 acids in the reduction of nitrate by isolated chloroplasts (House and Anderson, 1980) are consistent with this proposal.

Another shuttle mechanism for the export of reducing equivalents from chloroplasts involves the redox pair PGA/DHAP as carrier (Heber, 1974; Krause and Heber, 1976). Light-dependent reduction of PGA entails the processes of the CO_2 assimilation pathway; phosphorylation of PGA by ATP, reduction of diPGA via NADP-GAPDH, and isomerization of GAP to DHAP. DHAP is oxidized in an analogous series of reactions in another metabolic compartment and the PGA so formed is recycled to the chloroplast. An important feature of the PGA/DHAP shuttle is that it serves as a mechanism for the transport of high-energy phosphate in addition to reducing equivalents; these are recovered as ATP and NADH by the oxidative mechanism outside the chloroplast. Thus, isolated chloroplasts in the presence of phosphoglyceryl kinase, NAD-GAPDH, and triose phosphate isomerase incorporate P_i into ATP in a light-dependent reaction outside the chloroplast; the rate of phosphorylation is enhanced by trapping the exported reducing equivalents with pyruvate in the presence of LDH (Heber and Santarius, 1970).

Like malate, DHAP and GAP promote nitrate reduction in leaves and protoplasts (Klepper et al., 1971; Rathnam, 1978), suggesting that the PGA/DHAP shuttle could proceed in vivo. However, in this event the PGA/DHAP shuttle would cause a net loss of reducing equivalents and phosphorylation potential from the CO_2 assimilation pathway whereas the C_4 shuttle would not. A variation of the PGA/DHAP shuttle could entail oxidation of GAP by the irreversible (nonphosphorylating) NAD-GAPDH (Kelly and Gibbs, 1973) outside the chloroplast. In this event, the PGA/DHAP shuttle would also be irreversible and would not serve as a carrier of high-energy phosphate. Conversely, it has been proposed that the (reversible) PGA/DHAP shuttle can act in concert with the C_4 shuttle to effect the net export of high-energy phosphate potential only (Heber and Kirk, 1975). According to this hypothesis, the reducing equivalents that are exported via the PGA/DHAP shuttle are returned to the chloroplast via the C_4 shuttle operating in the reverse direction. This model would afford an explanation for the elevated ATP/ADP ratios in the cytoplasm when leaf cells are transferred from dark to light but why a similar increase in the NAD(P)/NAD(P)H ratio is not observed (Heber and Santarius, 1965, 1970).

The C_5 dicarboxylic acid shuttle reported recently (Dawson and Anderson, 1980) involves glutamate as the redox carrier for the export of reducing equivalents from illuminated chloroplasts (Fig. 2). α-Ketoglutarate is reduced to glutamate via light-coupled glutamate synthase (Anderson and Done, 1977a) and exported to another cellular organelle/compartment where it is oxidized via GDH. The NH_3 produced in the latter reaction is reincorporated in the chloroplast via glutamine synthetase (Anderson and Done, 1977b), which forms an integral part of the shuttle (Fig. 2). This hypothesis was examined with isolated chloroplasts in the presence of GDH, trace amounts of glutamate (or α-ketoglutarate and NH_3), and an electron trap outside the chloroplast (NAD, LDH, and pyruvate). Under these conditions, chloroplasts directed light-generated reducing equivalents into the electron trap (Dawson and Anderson, 1980). Further, the reaction was sensitive to inhibitors of glutamine synthetase and glutamate synthase as predicted for the C_5 shuttle. However, in view of the relatively low activities of glutamate synthase and glutamine synthetase in chloroplasts relative to NADP-GAPDH and NADP-MDH, it is unlikely that the C_5 shuttle is quantitatively important in the transfer of reducing equivalents; reassimilation of photorespired NH_3 (Keys *et al.*, 1978) would seem a more likely function.

Light is essential for the generation of reducing equivalents, which are exported from chloroplasts by shuttle mechanisms. However, the functions of many of the light-coupled processes of chloroplasts appear to be controlled by light by additional more specific mechanisms. The activities of several chloroplast enzymes including NADP-MDH and NADP-

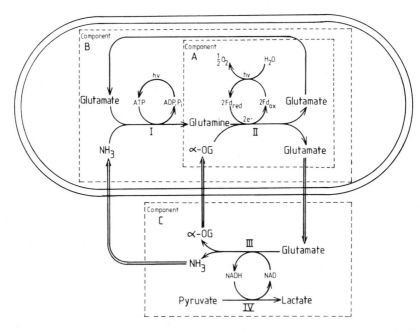

Figure 2. Model for the export of reducing equivalents from illuminated chloroplasts involving a C_5 dicarboxylic acid shuttle. The area enclosed within the double lines represents a chloroplast. Reactions I and II involve light-coupled glutamine synthetase and glutamate synthase, respectively. Glutamate, synthesized from NH_3 and α-ketoglutarate by reactions I and II, is exported from the chloroplast with the countertransport of α-ketoglutarate by the dicarboxylic translocator and oxidized to α-ketoglutarate and NH_3 by component C, which involves NAD-GDH (reaction III), LDH (reaction IV), and catalytic amounts of NAD with pyruvate as eventual electron acceptor. Other models for component C involving a range of electron acceptors are also possible. [From Dawson and Anderson, 1980.]

GAPDH are subject to light modulation (Buchanan, 1980). For example, chloroplast NADP-MDH is active in the light but essentially inactive in the dark (Johnson and Hatch, 1970). In the absence of some form of light regulation of chloroplast NADP-MDH, the C_4 dicarboxylic acid shuttle could, in theory, proceed in the reverse direction in the dark, thereby directing cytoplasmic sources of energy into the chloroplast in the dark. Similar remarks would apply to the PGA/DHAP shuttle if NADP-GAPDH was not light-activated although the latter enzyme is only partially inactivated in the dark. Thus, depending on the characteristics of the light-modulated enzyme involved, the C_4 and PGA/DHAP shuttles could be subject to varying degrees of light regulation. The C_5 dicarboxylic acid shuttle could be controlled by light regulation of glutamine synthetase activity (Tischner and Hutterman, 1980).

Experimental evidence concerning the proposal that the transfer of reducing equivalents between chloroplasts and other metabolic compartments via the C_4 and DHAP/PGA shuttles is light-regulated is inconclusive. For example, DCMU inhibits light activation of NADP-MDH and NADP-GAPDH (Anderson and Avron, 1976) in addition to light-dependent reduction of NADP, but chloroplasts treated with DCMU assimilate CO_2 in the presence of malate or DHAP (Rathnam, 1978; Rathnam and Zilinskas, 1977). On the other hand, certain quantitative aspects of these experiments are consistent with the greater degree of light activation of NADP-MDH relative to NADP-GAPDH (Anderson, 1981). Here it seems that attention should be directed toward devising conditions for specifically activating and inactivating each of the light-modulated enzymes to determine the effect of malate and DHAP as exogenous sources of reducing equivalents for assimilatory processes within the chloroplast. It would also be necessary to control the photoreduction of NADP in these experiments. In this respect the activation of NADP-MDH by dithiothreitol (Johnson and Hatch, 1970; Anderson *et al.*, 1978) and inhibition of light activation of NADP-MDH (but not photosystem II) by disalicylidenepropanediamine (Anderson and Avron, 1976) could be useful.

REFERENCES

Anderson, J. W. (1981). In *The Biochemistry of Plants* Vol. 8 (M. D. Hatch and N. K. Boardman, eds.), pp. 473–500, Academic Press, New York.

Anderson, J. W., and Done, J. (1977a). *Plant Physiol.* **60**, 354–359.

Anderson, J. W., and Done, J. (1977b). *Plant Physiol.* **60**, 504–508.

Anderson, J. W., and House, C. M. (1979a). *Plant Physiol.* **64**, 1058–1063.

Anderson, J. W., and House, C. M. (1979b). *Plant Physiol.* **64**, 1064–1069.

Anderson, L. E., and Avron, M. (1976). *Plant Physiol.* **57**, 209–213.

Anderson, L. E., Nehrlich, S. C., and Champigny, M.-L. (1978). *Plant Physiol.* **61**, 601–605.

Beevers, L., and Hageman, R. H. (1969). *Annu. Rev. Plant Physiol.* **20**, 495–522.

Buchanan, B. B. (1980). *Annu. Rev. Plant Physiol.* **31**, 341–374.

Dalling, M. J., Tolbert, N. E., and Hageman, R. H. (1972). *Biochim. Biophys. Acta* **283**, 505–512.

Dawson, J. C., and Anderson, J. W. (1980). *Phytochemistry* **19**, 2255–2261.

Heber, U. (1974). *Annu. Rev. Plant Physiol.* **25**, 393–421.

Heber, U., and Kirk, M. R. (1975). *Biochim. Biophys. Acta* **376**, 136–150.

Heber, U., and Krause, G. H. (1970). In *Photosynthesis and Photorespiration* (M. D. Hatch, C. B. Osmond, and R. O. Slatyer, eds.), pp. 218–225, Wiley–Interscience, New York.

Heber, U., and Santarius, K. A. (1965). *Biochim. Biophys. Acta* **109**, 390–408.

Heber, U., and Santarius, K. A. (1970). *Z. Naturforsch.* **25b**, 718–728.

Heldt, H. W. (1976). In *The Intact Chloroplast* (J. Barber, ed.), pp. 215–234, Elsevier/North-Holland, Amsterdam.

House, C. M., and Anderson, J. W. (1980). *Phytochemistry* **19**, 1925–1930.

Jablonski, P. P., and Anderson, J. W. (1978). *Plant Physiol.* **61,** 221–225.

Johnson, H. S., and Hatch, M. D. (1970). *Biochem. J.* **119,** 273–280.

Kelly, G. J., and Gibbs, M. (1973). *Plant Physiol.* **52,** 111–118.

Keys, A. J., Bird, I. F., Cornelius, M. J., Lea, P. J., Wallsgrove, R. M., and Miflin, B. J. (1978). *Nature (London)* **275,** 741–743.

Klepper, L., Flesher, D., and Hageman, R. H. (1971). *Plant Physiol.* **48,** 580–590.

Krause, G. H., and Heber, U. (1976). In *The Intact Chloroplast* (J. Barber, ed.), pp. 171–214, Elsevier/North-Holland, Amsterdam.

Neyra, C. A., and Hageman, R. H. (1976). *Plant Physiol.* **58,** 726–730.

Nicholas, J. C., Harper, J. E., and Hageman, R. H. (1976). *Plant Physiol.* **58,** 736–739.

Rathnam, C. K. M. (1978). *Plant Physiol.* **62,** 220–223.

Rathnam, C. K. M., and Zilinskas, B. A. (1977). *Plant Physiol.* **60,** 51–53.

Tischner, R., and Hutterman, A. (1980). *Plant Physiol.* **66,** 805–808.

Wallsgrove, R. M., Lea, P. J., and Miflin, B. J. (1979). *Plant Physiol.* **63,** 232–236.

175

Membrane Protein–Proton Interactions in Chloroplast Bioenergetics

Localized pH Effects Deduced from the Use of Chemical Modifiers

Richard A. Dilley

1. INTRODUCTION

Chloroplast thylakoid (inner) membranes produce a protonmotive force (inside positive, inside low pH) driven either by light-induced electron transport or by ATPase function (reviews: Trebst, 1974; McCarty, 1979). The chemiosmotic hypothesis provides a conceptual scheme that accounts for much of the structure–function interactions of the chloroplast membrane system (Mitchell, 1966). Figure 1 shows a representation of photosynthetic electron and proton flow in accord with the original concept of Mitchell. Proton release in the photosystem II (PS II) water oxidation mechanism and in the oxidation of plastohydroquinone (PQH_2) were proposed to be directly into the inner aqueous phase, from where a transmembrane protonmotive force drops across the CF_0–CF_1 coupling complex. The flux of protons through the CF_0–CF_1 complex is believed to drive ATP formation. The reverse reaction of ATP hydrolysis causes proton accumulation (McCarty, 1979).

Problems in bioenergetics not yet understood include the molecular interactions of protons with the thylakoid membrane proteins and lipids. Recent research indicates that protons involved in the protonmotive force may exert their effects via localized pH gradients rather than obligatorily through a transmembrane gradient. This report will summarize experiments done in the author's laboratory supporting the concept of localized pH effects in chloroplast membranes. Space will not permit detailed discussion of related work from other laboratories.

Richard A. Dilley • Department of Biological Sciences, Purdue University, West Lafayette, Indiana 47907.

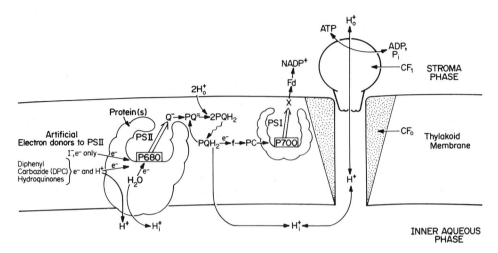

Figure 1. A diagrammatic model of a chloroplast thylakoid membrane, showing direct proton insertion into the inner aqueous phase. The proteins associated with the photosystems are shown as the wavy structures. Electron donors to PS II that can replace water are shown; I⁻, an electron-only donor or diphenylcarbazide, an electron- and a proton-releasing donor (Izawa and Ort, 1974). [After Mitchell, 1966.]

2. MEMBRANE PROTEIN CONFORMATIONAL CHANGES LINKED TO PHOTOSYSTEM II PROTON ACCUMULATION

Giaquinta *et al.* (1975) showed that protons released by PS II caused changes in labeling of membrane proteins with diazobenzene [^{35}S]sulfonate (DABS), which could not be caused by PS I-linked proton fluxes. The labeling change was interpreted as due to conformational changes in membrane proteins caused by protonation of acid–base ionizable groups (Giaquinta *et al.*, 1974). Accompanying the increased derivatization of membrane proteins was an inhibition of water oxidation activity, a point consistent with the notion that sensitive groups associated with PS II function became exposed to the water-soluble DABS reagent via a conformational change driven by proton release in the water oxidation mechanism.

Those conclusions were supported by similar experiments using two unrelated chemical modification reagents, iodoacetic acid (primarily SH directed) and acetic anhydride (Ac$_2$O, primarily NH$_2$ group directed). The most interesting results have been obtained using Ac$_2$O and the discussion will therefore center on this reagent. A summary list of the main points and experimental results may be helpful for the subsequent discussion:

A. Labeling of chloroplast membranes by [^{3}H]-Ac$_2$O in the dark is increased by treatments causing increased H$^+$ permeability (Baker *et al.*, 1981). Typical labeling data for 3.5 mM Ac$_2$O treatment for 30–45 sec are:

1. Dark − uncoupler: 60 nmoles acetate/mg protein. Dark + 0.05 μM gramicidin: 90 nmoles acetate/mg protein. Such an uncoupler addition was observed to cause approximately 20–40 nmoles H$^+$ to be effluxed from the membranes, measured by pH electrodes in a chloroplast suspension. Bearing in mind that the Ac$_2$O reaction conditions are at pH 8.6 and the membranes were isolated and stored at pH 8, it is likely that a pool of protons at pH ≈ 8 may have been present within the lipid barrier of the membrane.

2. Similar results were found with a dark temperature transition, 20°C→30°C, for 1 min→20°C followed by Ac₂O treatment. Similar treatments have been shown by other workers to cause H⁺ release from chloroplasts (Takahama *et al.*, 1976).

3. Inducing H⁺ leakiness by removal of the coupling factor, CF₁, with EDTA treatment also results in the higher (90 nmoles/mg Chl) labeling level in the dark (Bhatnagar *et al.*, in preparation).

B. *Light-induced electron flow given before and during the Ac₂O treatment restores the lower level of labeling* (Prochaska and Dilley, 1978a; Baker *et al.*, 1981). If the uncoupler and EDTA treatments (above) were not too harsh, the labeling level in the light returns to that characteristic of the lower level found in the dark, uncoupler-absent case.

Using the 0.05 μM gramicidin conditions, typical data are:

Dark − uncoupler	59 nmoles acetate/mg protein
Dark + uncoupler	83 nmoles acetate/mg protein
Light − uncoupler	59 nmoles acetate/mg protein
Light + uncoupler	62 nmoles acetate/mg protein

C. *The uncoupler-mediated increase in Ac₂O labeling in the dark is accompanied by inhibition of water oxidation (PS II) activity* (Baker *et al.*, 1981).

1. Figure 2 shows the correlation between inhibition of water oxidation and the uncoupler-induced extra derivatization of the membrane by Ac₂O.

2. The inhibition is protected against by light (electron transport) as is the increased labeling (see B above), suggesting that the inhibition is due to the derivatization of functional groups of the protein(s) associated with PS II water oxidation.

3. Only PS II electron transport (H⁺ release) protects against the inhibition (Tables III and IV of Baker *et al.*, 1981). PS I redox function giving equivalent H⁺ accumulation as PS II activity is not effective either in maintaining the lower level of derivatization or in protecting PS II against inhibition.

4. The requirement for protons, released by PS II oxidation, to drive the conformational change is shown by the fact that an electron-only PS II donor, I⁻ (in place of water, the normal donor), does not potentiate the light–dark derivatization changes normally found with DABS, iodoacetate, or Ac₂O (Giaquinta *et al.*, 1975; Prochaska and Dilley, 1978a). However, using diphenylcarbazide, a proton and an electron donor to PS II, does give the iodoacetate and Ac₂O derivatization change (the DABS changes were not tested for with this system). The I⁻ donor system results in H⁺ transport at the plastoquinone loop (Izawa and Ort, 1974), but not in

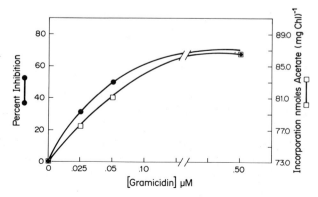

Figure 2. Uncoupler dependence of acetic anhydride inhibition of water oxidation and incorporation of [³H]acetate from acetic anhydride. The inhibition refers to the reaction H₂O→methyl viologen, assayed after exposure of the chloroplasts to 3.5 mM acetic anhydride for 45 sec in the dark with the indicated concentrations of gramicidin present. [Taken from Baker *et al.*, 1981.]

PS II. In the presence of DCMU to block PS II function, several PS I proton-transporting systems were tested and one of them potentiated the effects (Prochaska and Dilley, 1978a).

3. INTERPRETATION

The model we propose to explain the data posits that PS II proton release in a sequestered domain causes protonation of acid–base ionizable groups, with subsequent conformational rearrangement of the protein chains to accommodate the different distribution of charged and neutralized amino acid groups. The DABS labeling increase, occurring at or close to the aqueous-phase membrane surface, could be due to exposure of –SH, tyrosine, –NH$_2$, or histidine side groups by the conformational change (Giaquinta *et al.*, 1974). The iodoacetate (Prochaska and Dilley, 1978a) and nitrothiophenol (Kobayashi *et al.*, 1976) light-dependent labeling increases similarly could be due to exposure of otherwise buried reactive groups. Other evidence that significant conformational rearrangements of PS II occur in the light comes from studies by Harth *et al.* (1974).

For the membrane-permeable reagent acetic anhydride, the above data can be explained by postulating that the Ac$_2$O inhibition of PS II water oxidation and the increased derivatization are the result of anhydride attack on the unprotonated form of protein amine groups associated with PS II function. The Ac$_2$O attack requires the lone pair of electrons being free to act as a nucleophilic center (Means and Feeney, 1971). The requirement that the membrane be either slightly leaky to protons or be heat-treated to release protons can be understood if the –NH$_3^+$ groups are normally in the charged, unreactive form, but the proton leak to the pH 8.6 outer medium allows these groups to convert to the reactive – N̈H$_2$ form. This implies that the region containing those amine groups is buried behind the lipid barrier of the membrane, in a metastable state. The fact that PS I-linked H$^+$ accumulation cannot restore the protected state, while PS II activity does, implies that the sensitive groups are sequestered in a region not identical with, nor in equilibrium with the inner aqueous space (which is reached by protons released by either photosystem, Kraayenhof *et al.*, 1972). This leads to the postulate that the H$^+$ release site associated with water oxidation is buried in a domain separated from both the inner and the outer aqueous phases. This domain could be an intramembrane domain, or possibly an interfacial surface region (Kell, 1979), although the latter possibility is more difficult to defend, given that the time course for the treatments is of the order of tens of seconds, seemingly time enough for equilibration between a bulk phase and surface layers. It seems clear that H$^+$ ions released by either PS II or PS I equilibrate rather rapidly with the inner aqueous phase (Junge *et al.*, 1978), a point that seems to be counter to Kell's electrodic view, which posits a kinetically slow exchange of protons between the interfacial layer and the bulk phase. However, if the protons rapidly move to the CF$_0$ channel via a site-specific pathway, and then equilibrate with the inner aqueous phase (as shown in Fig. 3), the data of Junge and co-workers could be accommodated.

4. MEMBRANE PROTEINS INVOLVED

If the movement of protons is associated with localized domains, it is of interest to try to identify the membrane proteins involved in the phenomenon. A search for which membrane proteins show the change in derivatization under light–dark transition revealed

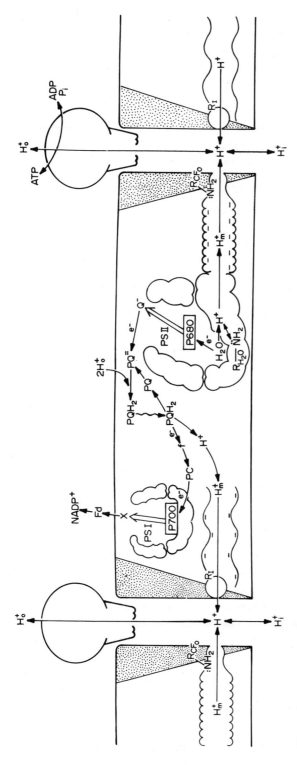

Figure 3. A model depicting an intramembrane domain connecting the PS II apparatus and the 8KD CF_0 complex. The proton release site of the water oxidation mechanism is hypothesized to be into this "domain," wherein the protons can traverse to the CF_0. An amine group associated with water oxidation, shown as $R_{\overline{H_2O}}NH_2$, and the Lys-48 amine groups, $R_{\overline{CF_0}}NH_2$, can react with acetic anhydride, and the reactivity of those amines is influenced by the local pH, as discussed in the text. For symmetry, PS I is shown associated with a separate proton-processing domain, but no data are available that directly deal with that issue.

Table I. Acetic Anhydride Labeling of Thylakoid Membranes and the 8KD CF$_0$ Protein [a]

Treatment	Whole membrane nmoles acetyl per mg protein	CF$_0$ nmoles acetyl per	
		mg CF$_0$	nmole CF$_0$
3.5 mM acetic anhydride			
Light − uncoupler	59	89	0.71
Dark − uncoupler	59	88	0.70
Light + uncoupler	63	89	0.71
Dark + uncoupler	83	120	0.96

[a] The [^{3}H] acetic anhydride labeling was carried out as described in Fig. 2 of Baker *et al.* (1981). The 8KD CF$_0$ protein was isolated according to Nelson *et al.* (1977) and purified on SDS-PAGE following Prochaska and Dilley (1978b). The uncoupler was 0.05 μM gramicidin.

the startling finding that the major protein is the 8000-molecular-weight hydrophobic component of the CF$_0$ complex (Prochaska and Dilley, 1978b,c). The 8KD CF$_0$ protein can be easily isolated by the butanol extraction, ether precipitation, and SDS-PAGE purification of Nelson *et al.* (1977). Following that procedure we find that the labeling of the 8KD CF$_0$ protein is strongly affected by proton release in PS II- but not PS I-linked redox activity (Prochaska and Dilley, 1978b). Moreover, the labeling of the 8KD CF$_0$ protein responds to uncoupler addition in the dark and the reversal by light similarly as the whole membrane labeling and inhibition of water oxidation activity (Table I).

These data imply that the 8KD CF$_0$ protein shares a restricted perhaps intramembrane, domain with the water oxidation proton release site. A diagrammatic model depicting this is shown in Fig. 3. The model shows what the data require; namely, that protons released by PS II water oxidation interact with an amine group of the 8KD CF$_0$ in a way that cannot be duplicated by protons released in the PS I system.

There is a single primary amine group (Lys-48) in the 8KD CF$_0$ protein (Sebald and Wachter, 1980). We are currently doing experiments to more critically test whether (1) the protein we have isolated showing the Ac$_2$O labelling change is the 8KD CF$_0$ protein, and (2) whether the anhydride labeling is restricted to the lysine amine group. Preliminary data suggest that this is the case (Dilley *et al.*, 1980).

The model shown in Fig. 3, while consistent with the data, raises certain problems. It is not clear how protons might move from PS II to the CF$_0$–CF$_1$ complex. Protein surfaces within the membrane may provide structured water as a pathway; or the hypothesis of Nagle and Morowitz (1978) concerning a proton charge relay mechanism along appropriate amino acid side chain groups may be involved.

In any event, the implication of the data so far leads to considering that PS I may have its own, but different, localized domain pathway along which PS I-generated protons move into the CF$_0$–CF$_1$ complex, as shown in Fig. 3. However, there are no data from the experiments discussed above that lead to such a conclusion.

Other studies suggest that chloroplasts and other energy-transducing membrane systems utilize proton gradients that are not primarily bulk-phase to bulk-phase. Ort and Dilley (1976), Ort *et al.* (1976), Kell (1979), and Melandri *et al.* (1980) have presented data along this line, but space does not permit discussion of those works. Suffice it to say that considerable evidence is mounting that requires serious consideration of localized pH effects as being involved in linking proton gradients to the membrane energy-coupling apparatus.

ACKNOWLEDGMENTS. The author is grateful for the collaborative experimental work done by Mr. Gary Baker, Dr. Deepak Bhatnagar, and Mr. Norman Tandy on this project.

This work was supported in part by Grant PCM 76-01640 from the National Science Foundation and Grant GM 19595-06 from the U.S. Public Health Service.

REFERENCES

Baker, G. M., Bhatnagar, D., and Dilley, R. A. (1981). *Biochemistry* **20**, 2307–2315.

Dilley, R. A., Baker, G. M., Bhatnagar, D., Tandy, N., and Hermodson, M. (1980). In *5th International Congress on Photosynthesis*, Greece, in press.

Giaquinta, R. T., Dilley, R. A., and Anderson, B. (1974). *Arch. Biochem. Biophys.* **162**, 200–209.

Giaquinta, R. T., Ort, D. R., and Dilley R. A. (1975). *Biochemistry* **14**, 4392–4396.

Harth, E., Reimer, S., and Trebst A. (1974). *FEBS Lett.* **42**, 165–168.

Izawa, S., and Ort, D. R. (1974). *Biochim. Biophys. Acta* **357**, 127–143.

Junge, W., McGeer, A., and Ausländer, W. (1978). In *Frontiers of Biological Energetics* (L. Dutton, J. Leigh, and A. Scarpa, eds.), Vol. 1, pp. 275–283, Academic Press, New York.

Kell, D. (1979). *Biochim. Biophys. Acta* **549**, 55–99.

Kobayashi, Y., Inoue, Y., and Shibata, K. (1976). *Biochim. Biophys. Acta* **423**, 80–90.

Kraayenhof, R., Izawa, S., and Chance, B. (1972). *Plant Physiol.* **50**, 713–718.

McCarty, R. E. (1979). *Annu. Rev. Plant Physiol.* **30**, 79–104.

Means, G. E., and Feeney, R. E. (1971). *Chemical Modification of Proteins,* p. 69, Holden-Day, San Francisco.

Melandri, B. A., Venturi, G., De Santis, A., and Baccarini-Melandri, A. (1980). *Biochim. Biophys. Acta* **592**, 38–52.

Mitchell, P. (1966). *Biol. Rev. Cambridge Philos. Soc.* **41**, 445–540.

Nagle, J. F., and Morowitz, H. J. (1978). *Proc. Natl. Acad. Sci. USA* **75**, 298–302.

Nelson, N., Eytan, E., El-Notsano, B., Sigrist, H., Sigrist-Nelson, K., and Gitler, C. (1977). *Proc. Natl. Acad. Sci. USA* **74**, 2375–2378.

Ort, D. R., and Dilley, R. A. (1976). *Biochim. Biophys. Acta* **449**, 95–107.

Ort, D. R., Dilley R. A., and Good, N. E. (1976). *Biochim. Biophys. Acta* **449**, 108–124.

Prochaska, L. J., and Dilley, R. A. (1978a). *Arch. Biochem. Biophys.* **187**, 61–71.

Prochaska, L. J., and Dilley, R. A. (1978b). *Biochem. Biophys. Res. Commun.* **83**, 664–672.

Prochaska, L. J., and Dilley, R. A. (1978c). In *Frontiers of Biological Energetics* (L. Dutton, J. Leigh, and A. Scarpa, eds.), Vol. 1, pp. 265–274, Academic Press, New York.

Sebald, W., and Wachter, E. (1980). *FEBS Lett.* **122**, 307–311.

Takahama, U., Shimizu, M., and Nishimura, M. (1976). *Biochim. Biophys. Acta* **440**, 261–265.

Trebst, A. (1974). *Annu. Rev. Plant Physiol.* **25**, 423–458.

176

Organization and Function of the Chloroplast Thylakoid Membrane Obtained from Studies on Inside-Out Vesicles

Bertil Andersson, Hans-Erik Åkerlund, and Per-Åke Albertsson

1. INTRODUCTION

Asymmetric arrangement of components across the lipid bilayer has been shown to be a general feature of membrane organization and function. Experimentally this can be best studied for membranes from which sealed inside-out vesicles can be isolated. Such vesicles could not be obtained for the chloroplast thylakoid membrane until the aqueous polymer two-phase partition technique was introduced for its subfractionation. This review deals with the isolation of everted thylakoids and their use for obtaining new information concerning membrane organization and reactions taking place at the inner thylakoid surface.

2. PHASE PARTITION

There is a great need for efficient separation methods for the isolation of membrane vesicles. The different methods should complement each other so that their use in sequence produces an efficient purification.

Phase partition is a method that separates particles according to their surface properties and it therefore complements centrifugation. Phase partition involves the selective distribution of particles between two immiscible aqueous liquid phases. These are obtained by mixing two different water-soluble polymers in water. To the phase systems can be added low-molecular-weight compounds such as sugars to give the desired tonicity and to pre-

Bertil Andersson, Hans-Erik Åkerlund, and Per-Åke Albertsson • Department of Biochemistry, University of Lund, S-220 07 Lund, Sweden.

serve biological activity. The polymer combination that has been mainly used is dextran–polyethylene glycol. The polymer phase systems are mild toward biological particles and seem to preserve the structure and function of cell organelles. There are several ways to steer the partition of particles to get maximal separation. Phase partition can also be used to characterize the surface properties of membranes. For example the isoelectric point of membranes can be determined by cross partition (Albertsson, 1970). For a general treatment of the method, see Albertsson (1971, 1978).

When cells are homogenized, many membranes are broken into fragments. These often form small vesicles by fusion of "open ends" of the membrane fragments. In this manner, some inside-out vesicles may be obtained depending on the mechanism of breakage and fusion. Separation by centrifugation will yield vesicles having the same size and density. These may, however, be a mixture of inside-out and right-side-out vesicles. By having different orientation the two vesicles expose different surfaces and can therefore be separated by phase partition. Many membrane vesicle preparations used so far are probably mixtures of inside-out and right-side-out vesicles. Such preparations should therefore be tested for homogeneity by a surface-dependent method such as phase partition, before a meaningful study of the vectorial properties of the membrane can be made.

3. ISOLATION AND EVIDENCE FOR INSIDE-OUT THYLAKOID VESICLES

A mixture of inside-out and right-side-out thylakoid vesicles is obtained after mechanical disintegration of chloroplast thylakoids stacked by cations (Andersson *et al.*, 1977; Andersson and Åkerlund, 1978). The vesicles of opposite sidedness are separated from each other by partition in an aqueous polymer two-phase system.

Functional and structural studies have shown that the vesicles that prefer the lower phase (designated B vesicles) are turned inside out while the material preferring the upper phase (designated T vesicles) is of normal sidedness. The sidedness of B and T vesicles has been established by: (1) direction of the light-induced proton gradient (Andersson *et al.*, 1977; Andersson and Åkerlund, 1978). When supplied with an electron acceptor, phenyl-*p*-benzoquinon, T vesicles show light-induced reversible proton uptake as is observed for normal thylakoids. Under identical conditions, light-induced reversible extrusion is observed for the B vesicles. The proton uptake by the T fraction and proton extrusion by the B fraction can also be demonstrated by the absorbance change of bromcresol purple after a single turnover flash. Also the internal pH changes have been measured using neutral red (cf. Ausländer and Junge, 1975). The T vesicles show acidification of their internal compartments in the same way as normal thylakoids, in contrast to the B vesicles, which show an internal pH rise.

(2) The direction of the light-induced primary charge separation, of the B vesicles measured by the macroscopic electrode method (Witt and Zickler, 1974), is reversed to that of thylakoids of normal orientation (Gräber *et al.*, 1978).

(3) Freeze-fracture electron microscopy revealed that the B fraction contains vesicles with a predominance of concave EF and convex PF fracture faces (Fig. 1), as is expected from an everted thylakoid vesicle (Andersson *et al.*, 1978). The sidedness of the T and B vesicles has also been examined by freeze-etching, in which the latter vesicles show convex membrane surfaces exposing the tetrameric particles typical of the inner thylakoid surface (ES).

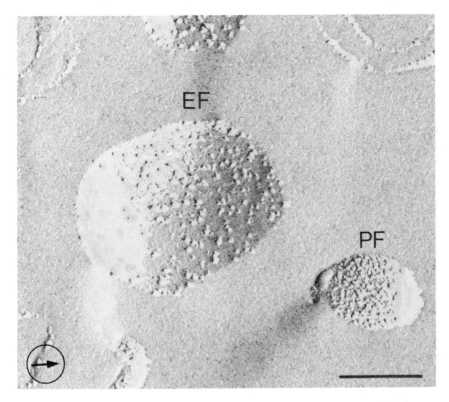

Figure 1. Freeze-fracture appearance of thylakoid vesicles found in the lower phase after phase partition. The concave EF and convex PF faces show that these vesicles are turned inside-out. Bar = 0.25 μm. Courtesy of D. J. Simpson.

4. MECHANISM OF FORMATION OF INSIDE-OUT THYLAKOIDS

Composition and structure of the inside-out vesicles indicate that these originate from the grana stacks (Åkerlund *et al.*, 1976; Andersson *et al.*, 1978). The following mechanism for their formation was therefore proposed (Fig. 2). Due to the shearing forces, some grana stacks rupture at their edges but remain appressed. Resealing of such membranes with their nearby appressed neighbors would form inside-out vesicles. Such resealing is favored by the two membranes being closely held together by the attractive stacking forces. Recently, this mechanism was experimentally supported (Andersson *et al.*, 1980) by studying the yield and properties of inside-out vesicles under various conditions affecting the stacking of the thylakoid membranes. If chloroplasts, stacked by cations, are mechanically fragmented and fractionated by phase partition, inside-out vesicles highly enriched in photosystem II are obtained. If chloroplasts are experimentally destacked and randomized (Staehelin, 1976) prior to fragmentation, no inside-out vesicles are produced and all material is recovered in the upper phase during the subsequent phase partition (Andersson *et al.*, 1980). Restacking of chloroplasts by addition of Mg^{2+} before fragmentation again allows the photosystem II-enriched vesicles to be isolated. These observations demonstrate the necessity of grana partitions for the formation of inside-out thylakoid vesicles and support the mechanism outlined in Fig. 2.

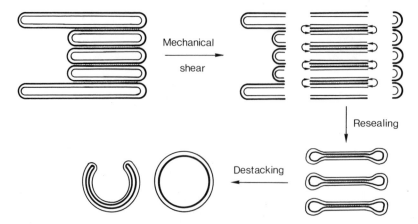

Figure 2. Suggested mechanism for the formation of inside-out thylakoid vesicles from stacked chloroplast thylakoids.

An alternative way to isolate inside-out thylakoids is to subject destacked and random‐ ized chloroplasts to pH 4.7 prior to fragmentation and phase partition (Andersson *et al.*, 1980). The low pH induces membrane appressions due to diminished electrostatic repul‐ sions. In contrast to the inside-out vesicles obtained from normal, stacked thylakoids, these everted vesicles contained the photosystems in about equal proportions compared to the starting material. The modified method therefore gives two vesicle populations differing in sidedness only, making them ideal for studies on the transbilayer organization. Thus, even artificially induced membrane paring is sufficient to allow formation of everted vesicles after disruption. The approach of inducing appressed membranes before fragmentation may be a way to obtain everted vesicles also from other biological membranes.

5. STUDIES ON THYLAKOID TRANSBILAYER ORGANIZATION

Conclusions regarding the transbilayer organization for several components of the thy‐ lakoid membrane have been based on the effects of nonpermeable agents on the outer thylakoid surface (Sane, 1977). Such an approach has certain limitations that may explain many conflicting results presented in the literature. For example, modification of a com‐ ponent at the outer surface may cause rearrangements in the membrane that lead to inacti‐ vation of internal catalytic sites. On the other hand, the absence of effects does not tell whether a component is internally located, shielded behind some bulky outer surface com‐ ponents, or simply unable to react with the added effector. Thus, a direct comparison between the effect of nonpermeating agents on the outer and inner membrane surfaces is advantageous. The location of several thylakoid components has therefore been reinvesti‐ gated using the inside-out thylakoid vesicles.

One main controversy concerns the location of the water-splitting complex (Trebst, 1980). By trypsin treatment of thylakoid membranes of opposite sidedness, Jansson *et al.* (1979) obtained strong evidence that the water-splitting site is located at the inner thylakoid surface. This was done by comparing the effect of proteolysis on the rate of 2,6-dichloro‐ phenol indophenol (DCIP) reduction with water as sole electron donor and that obtained after addition of the electron donor diphenyl carbazide, known to bypass the water-splitting site. Trypsin inhibited the electron transport from water to DCIP in both right-sided and everted material. Upon addition of diphenyl carbazide, the DCIP reduction was completely restored in the inside-out vesicles in contrast to the right-side-out material.

The inside location of the water-splitting system is also supported by the observation that a 34,000-molecular-weight polypeptide can be degraded by trypsin or eliminated by Tris washing in inside-out but not in right-side-out thylakoids (Åkerlund and Jansson, 1981). A 34,000-molecular-weight polypeptide has been ascribed to the water-splitting system by a recent study on mutants of *Scenedesmus* (Metz *et al.*, 1980).

The location of plastocyanin, the primary donor of photosystem I, has been another area of controversy (Hauska, 1978). Kinetic and immunological studies with vesicles of opposite sidedness were therefore performed (Haehnel *et al.*, 1981). The accessibility of the photosystem I donor site to externally added plastocyanin was compared for vesicles of opposite sidedness (prepared from randomized thylakoids, Andersson *et al.*, 1980). In the right-side-out fraction, only 15% of the total P_{700} (reaction-center chlorophyll of photosystem I) was rapidly reduced, while as much as 70% was rapidly reduced in the inside-out fraction. When the plastocyanin content of the vesicles was determined by rocket electroimmunodiffusion, it was shown that the right-sided material retained around 25% of the specific plastocyanin content whereas the everted material retained only 5%. The selective release of plastocyanin from, and high accessibility to the everted vesicles provide strong evidence for a location at the inner thylakoid surface.

By combining proteolysis and antibody agglutination of vesicles of opposite sidedness, two chlorophyll–protein complexes—the light-harvesting complex and the presumed reaction-center complex of photosystem II—were shown to span the membrane (Andersson *et al.*, 1982).

A charge asymmetry across the thylakoid membrane was indicated from the different isoelectric points of the outer (pH 4.5) and inner (pH 4.1) surfaces obtained by cross partition (Åkerlund *et al.*, 1979).

6. STUDIES ON THYLAKOID LATERAL HETEROGENEITY

The structural heterogeneity in the lateral plane of stacked chloroplasts is apparent. In addition, more and more evidence points to a functional heterogeneity (Sane, 1977). In this respect, the old nomenclature of grana stacks and stroma lamellae has become insufficient as pointed out by Berzborn *et al.* (1981), who suggest the use of exposed (nonappressed) and nonexposed (appressed) thylakoid regions. The nonappressed regions are the stroma thylakoids, end granal membranes, and grana margins, while the appressed regions are the grana partitions. Subfractionation of chloroplasts by conventional procedures like differential centrifugation separates stroma lamella vesicles from grana stacks (Park and Sane, 1971). A preparation containing only grana partitions was not obtained, for the grana stacks still contained the nonappressed margins and end membranes. As indicated above, there is convincing evidence that the inside-out vesicles isolated from normal stacked chloroplasts originate from the partitions. Thus, these vesicles may be used not only for studies on thylakoid transverse asymmetry but also on lateral heterogeneity.

In a recent study (Andersson and Anderson, 1980) SDS-polyacrylamide gel electrophoresis was used to estimate the content of chlorophyll–protein complexes in subchloroplast fractions derived from different thylakoid regions. It was demonstrated that the photosystem I reaction-center complex was excluded to a great extent from the grana partitions and concentrated in the nonappressed regions. The opposite was found for the photosystem II reaction-center complex and the light-harvesting complex. The light-harvesting complex was found to be associated mainly with the photosystem II complex rather than the photosystem I complex. These observations raise new questions concerning the distribution of radiation energy and transport of electrons between photosystems being segregated into different regions.

Immunological studies (Berzborn *et al.*, 1981) on the vesicles showed that the coupling factors (CF_1) are mainly excluded from the grana partitions, confirming previous morphological studies (Miller and Staehelin, 1976).

7. STUDIES ON OXYGEN EVOLUTION

Despite great efforts the components and the mechanism of the photosynthetic oxygen-evolving complex are still largely unknown. One main reason is the fragility of this system, which makes functional studies difficult. As shown by Jansson *et al.* (1979), the inside-out vesicles expose the oxygen-evolving complex. This offers unique possibilities for investigations on this reaction, as the water-splitting components can be reached and modified or removed by mild treatments. Therefore, it is not necessary to apply drastic methods like detergents or sonication, which lead to irreversible inactivation and changes in the gross structure of the membrane.

Åkerlund (1981) could inhibit water oxidation specifically in inside-out thylakoids by removing proteins by washing with 50–500 mM NaCl. By readdition of the protein extract to the treated inside-out thylakoids, it was possible to obtain a partial reconstitution of oxygen evolution. The extract had no effect on either right-side-out thylakoids or untreated inside-out thylakoids. The reconstituting factor could be inactivated by proteolytic digestion with subtilisin, demonstrating its protein nature. That the stimulating factor can be released from the membrane by increasing the ionic strength suggests that it is bound primarily by electrostatic forces and is therefore not an intrinsic membrane protein.

The photosynthetic water oxidation leads to the release of four protons to the thylakoid lumen per O_2 produced. The nature of the proton release is therefore of fundamental importance for understanding the water oxidation mechanism. In the inside-out thylakoids, the protons are released to the surrounding medium (Andersson *et al.*, 1977; Andersson and Åkerlund, 1978), and the kinetics and extent of proton release can therefore be determined more accurately as assumptions about for example internal volume are not necessary. Using inside-out thylakoids, Tieman *et al.* (1981) were able to measure the kinetics of the rapid proton release with an external pH indicating dye, bromcresol purple. The rapid proton release (half time 0.6 msec) showed a lag phase (250 μsec), indicating that it was not directly caused by the oxidation of the primary donor of reaction center II but possibly by a secondary donor like a water-oxidizing component.

REFERENCES

Åkerlund, H.-E. (1981). In *Proceedings, 5th International Congress on Photosynthesis* (G. A. Akoyunoglou, ed.), Vol. II, pp. 465–472, Balaban International Science Services, Philadelphia.

Åkerlund, H.-E., and Jansson, C. (1981). *FEBS Lett.* **124**, 229–232.

Åkerlund, H.-E., Andersson, B., and Albertsson, P.-Å. (1976). *Biochim. Biophys. Acta* **449**, 525–535.

Åkerlund, H.-E., Andersson, B., Persson, A., and Albertsson, P.-Å. (1979). *Biochim. Biophys. Acta* **552**, 238–246.

Albertsson, P.-Å. (1970). In *Advances in Protein Chemistry* (C. B. Anfinsen, Jr., A. T. Edsall, and F. M. Richards, eds.), Vol. 24, pp. 309–341, Academic Press, New York.

Albertsson, P.-Å. (1971). *Partition of Cell Particles and Macromolecules* (2nd ed.), Wiley, New York.

Albertsson, P.-Å. (1978). *J. Chromatogr.* **159**, 111–122.

Andersson, B., and Åkerlund, H.-E. (1978). *Biochim. Biophys. Acta* **503**, 462–472.

Andersson, B., and Anderson, J. M. (1980). *Biochim. Biophys. Acta* **593**, 427–440.

Andersson, B., Åkerlund, H.-E., and Albertsson, P.-Å. (1977). *FEBS Lett.* **77**, 141–145.

Andersson, B., Simpson, D. J., and Høyer-Hansen, G. (1978). *Carlsberg Res. Commun.* **43**, 77–89.

Andersson, B., Sundby, C., and Albertsson, P.-Å. (1980). *Biochim. Biophys. Acta* **599**, 391–402.

Andersson, B., Anderson, J. M., and Ryrie, I. J. (1982). *Eur. J. Biochem.*, in press.

Ausländer, W., and Junge, W. (1975). *FEBS Lett.* **59**, 310–315.

Berzborn, R. J., Müller, D., Roos, P., and Andersson, B. (1981). In *Proceedings, 5th International Congress on Photosynthesis* (G. A. Akoyunoglou, ed.), Vol. III, pp. 107–120, Balaban International Science Services, Philadelphia.

Gräber, P., Zickler, A., and Åkerlund, H.-E. (1978). *FEBS Lett.* **96**, 233–237.

Haehnel, W., Berzborn, R. J., and Andersson, B. (1981). *Biochim. Biophys. Acta* **637**, 389–399.

Hauska, G. (1978). In *Proceedings, 4th International Congress on Photosynthesis* (D. O. Hall, J. Coombs, and T. W. Goodwin, eds.), pp. 185–196, The Biochemical Society, London.

Jansson, C., Andersson, B., and Åkerlund, H.-E. (1979). *FEBS Lett.* **105**, 177–180.

Metz, J. G., Wong, J., and Bishop, N. J. (1980). *FEBS Lett.* **114**, 61–66.

Miller, K. R., and Staehelin, L. A. (1976). *J. Cell Biol.* **68**, 30–47.

Park, R. B., and Sane, P. V. (1971). *Annu. Rev. Plant Physiol.* **22**, 395–430.

Sane, P. V. (1977). In *Encyclopedia of Plant Physiology, Photosynthesis I* (A. Trebst and M. Avron, eds.), Vol. 5, pp. 522–542, Springer-Verlag, Berlin.

Staehelin, L. A. (1976). *J. Cell Biol.* **71**, 136–158.

Tieman, R., Renger, G., and Gräber, P. (1981). In *Proceedings, 5th International Congress on Photosynthesis* (G. A. Akoyunoglou, ed.), Vol. III, pp. 85–95, Balaban International Science Services, Philadelphia.

Trebst, A. (1980). *Methods Enzymol.* **69C**, 675–715.

Witt, H. T., and Zickler, A. (1974). *FEBS Lett.* **39**, 205–208.

177

The Plasma Membrane ATPase of Plant Cells: Cation or Proton Pump?

Robert T. Leonard

1. INTRODUCTION

The plasma membrane of plant cells has two major types of ion transport functions that are fundamental to cellular physiology. One, this membrane controls the transport of substances needed to provide the proper ionic environment for cellular biochemistry. This includes not only the transport of nutrient ions such as K^+, $H_2PO_4^-$, and NO_3^- (Lüttge and Higinbotham, 1979), but also the fluxes of H^+ (and/or OH^-) for the regulation of cytoplasmic pH (Smith and Raven, 1979) and the generation of the membrane potential difference (Poole, 1978). Two, the accumulation of K^+ in the cell provides the osmotic pressure needed for turgor-driven cell growth and for other turgor-generated plant responses such as stomatal and leaf movements (MacRobbie, 1977). Both types of function require metabolic energy, and the energy transduction appears to be performed by a plasma-membrane—associated ATPase (Leonard and Hodges, 1980). A major unresolved question is whether the plasma membrane has an ATPase that is a primary cation pump analogous to cation-ATPases of animal cells, or only a primary proton pump similar to the electrogenic H^+-ATPase proposed for fungal cells. Here, the information presently available to support the various ideas on the function of the plasma membrane ATPase in plant cells is summarized.

2. K^+-ATPase AND CATION TRANSPORT

Plant cells contain several ATP-hydrolyzing activities that are not related to transport processes at the plasma membrane. For example, ATPases in mitochondria and chloroplasts and nonspecific phosphatase in the central vacuole readily hydrolyze phosphate from

Robert T. Leonard • Department of Botany and Plant Sciences, University of California, Riverside, California 92521.

ATP and confound attempts to study "transport ATPase" in membrane preparations from homogenates of cells. The first convincing evidence for the existence of a membrane-bound ATPase that might function in cation transport was provided by Hodges and colleagues in 1970 (Hodges, 1976, for review). They reported an excellent correlation between membrane-associated, K^+-stimulated ATPase (K^+-ATPase) activity and K^+ transport rates into roots of several plant species. The K^+-ATPase activity was later shown (Hodges *et al.*, 1972) to copurify with plasma membrane vesicles from oat roots. Since then, many workers have confirmed the observation that K^+-ATPase activity is characteristically associated with plasma-membrane-rich fractions from various tissues and species of the higher plants (Leonard and Hodges, 1980; Perlin and Spanswick, 1980; Travis and Berkowitz, 1980; Quail, 1979, for references).

The K^+-ATPase on the plasma membrane from oat and corn roots has been extensively characterized (Leonard and Hodges, 1973; Leonard and Hotchkiss, 1976; DuPont *et al.*, 1981, for references). The enzyme requires Mg^{2+} in concentrations equal to ATP (presumably, MgATP is the substrate, see Balke and Hodges, 1975), and is stimulated by monovalent cations with a sequence ($K^+ > NH_4^+ > Rb^+ >> Na^+ > Cs^+ > Li^+$) similar to that observed for the specificity of monovalent cation transport into roots (Sze and Hodges, 1977). There is no indication of a synergistic stimulation by K^+ and Na^+, but none is expected as there is no evidence for the coupled transport of K^+ and Na^+ in tissues of higher plants. Enzyme activity is maximal at about pH 6.5 and 38–40°C. Activity is not inhibited by oligomycin, azide, or ouabain but is sensitive to Ca^{2+}, N,N'-dicyclohexylcarbodiimide, diethylstilbestrol, octylguanidine, orthovanadate, and various sulfhydryl inhibitors (Balke and Hodges, 1979; DuPont *et al.*, 1981, for references). K^+-stimulated ATPase activity is much more sensitive to these inhibitors than ATPase activity in the absence of K^+. This is consistent with the view that K^+ stimulation of the enzyme is a more specific measure of transport function.

With the exception of orthovanadate, none of the ATPase inhibitors has been specific enough to be useful for studies seeking to correlate reduction in transport rates with effects on K^+-ATPase activity. Vanadate appears to be a relatively specific inhibitor of the ATPase, but because it is a recent discovery, it has not as yet been widely utilized in studies on ATPase function (see DuPont *et al.*, 1981, for references). Vanadate should prove to be a useful tool for studies on the plant K^+-ATPase. At the very least, one can now use vanadate-sensitive, oligomycin-insensitive activity at pH 6.5 to distinguish plasma membrane ATPase from oligomycin-sensitive, vanadate-insensitive ATPase in mitochondria.

The K^+-ATPase shows simple Michaelis–Menten enzyme kinetics with respect to MgATP concentration. The apparent K_m for MgATP is about 1 mM. The major effect of K^+ in stimulating ATPase activity is to increase V_{max} (Leonard and Hodges, 1973; Leonard and Hotchkiss, 1976).

The kinetic data for K^+ stimulation of ATPase activity are complex, but very similar to kinetic data for K^+ transport into plant cells (Leonard and Hodges, 1973; Leonard and Hotchkiss, 1976). The interpretation of the kinetic data for both K^+ transport and K^+ stimulation of the ATPase is debatable (see Leonard and Hotchkiss, 1976, for references), but the similarity between the two supports the view that the enzyme functions, in a direct way, in cation transport.

Recently, Sze (1980) reported that K^+-ATPase activity associated with sealed vesicles (possibly of plasma membrane origin) from tobacco callus microsomes is stimulated by nigericin or by valinomycin in combination with a protonophore. This is consistent with the idea that the K^+-ATPase mediates an electrogenic or electroneutral K^+–H^+ exchange.

That is, in the sealed vesicles, ATPase activity may be limited by a gradient of K^+, H^+, or both, which develops across the membrane as a result of ATPase action. Addition of a K^+/H^+ ionophore like nigericin collapses the gradient and allows the enzyme to turn over at a higher rate. The K^+–H^+ exchange is electrogenic, for ATP-stimulated uptake of a permeant anion, SCN^-, can be measured in these vesicles (Sze and Churchill, 1981). While these results may have other interpretations, the important point is that a sealed-vesicle transport system is now available for studies on the function of the plasma membrane ATPase in plants. The lack of an isolated vesicle system has inhibited progress in studying ion transport in plants.

Another approach to the determination of the function of the K^+-ATPase is to solubilize, purify, and reconstitute the enzyme into tight lipid vesicles. After reconstitution, a cation-ATPase should show ATP-dependent K^+ transport in the presence of a protonophore. The K^+-ATPase of corn root plasma membranes can be solubilized with the detergent octyl-β-D-glucopyranoside, but only partial purification has been achieved (DuPont and Leonard, 1980). It appears that the K^+-ATPase is a relatively small percentage of the protein associated with the plant plasma membrane and that it will be difficult to purify the enzyme. However, it is clear from the results available that the K^+-ATPase is an intrinsic membrane protein that is intimately associated with membrane lipids, and that the enzyme retains activity only when associated with lipids.

To summarize this section, there are several reasons to believe that K^+-stimulated ATPase activity of the plant plasma membrane represents a pump involved, perhaps directly, in K^+ transport. The levels of K^+-ATPase in root tissues correlate with K^+ transport rates into those tissues. The sequence with which various monovalent cations stimulate the ATPase is similar to the specificity observed for monovalent cation transport. The nature of the kinetic data for K^+ stimulation of ATPase and K^+ transport is similar. K^+-stimulated ATPase activity is more sensitive to inhibitors of ion transport than ATPase activity measured without K^+. Also, developing research utilizing sealed vesicles suggests that ATPase activity is stimulated when K^+–H^+ exchange is facilitated. However, none of these observations is inconsistent with the view that K^+ stimulation is simply a salt effect unrelated to transport, and that the enzyme functions only as an electrogenic proton pump.

3. ATPase AND THE PROTONMOTIVE FORCE

There is compelling evidence for ATP-dependent, electrogenic H^+ extrusion through the plasma membrane of plant cells (Poole, 1978; Smith and Raven, 1979; Mercier and Poole, 1980, for references). The electrogenic H^+ extrusion is presumed to establish a charge and pH gradient (protonmotive force) that represent energy conserved for the transport of inorganic ions, including K^+, and metabolites such as amino acids and sugars. The transport of these substances is assumed to be facilitated by, as yet unidentified, "Mitchell-type porters," which conduct uni-, co-, and antitransport. It is quite possible that the K^+-ATPase described above and thought by some to participate directly in K^+ transport is, in fact, simply an electrogenic H^+ pump.

Research results on the plasma membrane ATPase of the fungus *Neurospora* have been very influential in the development of the proposal that the plant K^+-ATPase is a primary proton pump. ATPase activities of plasma membrane preparation from *Neurospora* and plants have remarkably similar characteristics (Bowman *et al.*, 1980). There is also evidence that the plasma membrane of *Neurospora* contains a major ATPase activity that represents the action of an electrogenic proton pump (Scarborough, 1980). Similar evidence for plants is lacking, but the sealed-vesicle system described by Sze (Sze, 1980; Sze and Churchill, 1981) may soon provide the needed information. However, the questions remain, even for *Neurospora*; is K^+ the counter ion moved in near-stoichiometric amounts to H^+, and is the H^+ pump also a K^+ pump?

4. ATPase-MEDIATED ELECTROGENIC H⁺–K⁺ EXCHANGE

In general, treatments (e.g., inhibitors, activators, etc.) that affect the electrogenic H^+ pump have a corresponding effect on K^+ influx suggesting that there may be a tight coupling of H^+ efflux and K^+ influx in cells of higher plants (Poole, 1978; Gronewald *et al.*, 1979; Marré, 1979). While the correlation between H^+ pump activity and K^+ influx is not perfect (e.g., Cheeseman *et al.*, 1980; Mercier and Poole, 1980), the data are consistent with the view expressed by Hodges (1973) that the electrogenic H^+ pump is also a K^+ influx pump, and that the two pump activities are mediated by the K^+-ATPase of the plant plasma membrane. Hence, it is conceivable that the plant ATPase may be functionally (and perhaps structurally) similar to the Na^+/K^+-ATPase of animal cells, except in plants efflux of H^+ rather than Na^+ is, in most circumstances, coupled to K^+ influx.

5. SUMMARY

Historically, the K^+-stimulated ATPase associated with the plasma membrane has been thought of as a primary cation pump that functions in the transport of K^+ and other monovalent cations in plant cells. Recent research clearly indicates that electrogenic H^+ extrusion is also a prominent and vital function of the plasma membrane. This information, along with the demostration of an electrogenic H^+-pumping ATPase in fungi, has led to the view that the K^+-ATPase in plants may, in fact, only be a primary proton pump that establishes a protonmotive force to energize a number of secondary transport systems, including one for K^+. However, there are growing indications that H^+ efflux is tightly coupled to K^+ influx. It appears likely that the K^+-ATPase of the plant plasma membrane mediates an electrogenic H^+–K^+ exchange. Final resolution of the question will depend on success with a newly developed "tight" vesicle transport system and/or with attempts to reconstitute the purified K^+-ATPase.

REFERENCES

Balke, N. E., and Hodges, T. K. (1975). *Plant Physiol.* **55**, 83–86.
Balke, N. E., and Hodges, T. K. (1979). *Plant Physiol.* **63**, 53–56.
Bowman, B. J., Blasco, F., Allen, K. E., and Slayman, C. W. (1980). In *Plant Membrane Transport: Current Conceptual Issues* (R. M. Spanswick, W. J. Lucas, and J. Dainty, eds.), pp. 195–205, Elsevier/North-Holland, Amsterdam.
Cheeseman, J. M., LaFayette, P. R., Gronewald, J. W., and Hanson, J. B. (1980). *Plant Physiol.* **65**, 1139–1145.
DuPont, F. M., and Leonard, R. T. (1980). *Plant Physiol.* **65**, 931–938.
DuPont, F. M., Burke, L. L., and Spanswick, R. M. (1981). *Plant Physiol.* **67**, 59–63.
Gronewald, J. W., Cheeseman, J. M., and Hanson, J. B. (1979). *Plant Physiol.* 63, 255–259.
Hodges, T. K. (1973). *Adv. Agron.* 25, 163–207.
Hodges, T. K. (1976). In *Transport in Plants II, Part A Cells* (U. Lüttge and M. G. Pitman, eds.), Vol. II, pp. 260–283, Springer-Verlag, Berlin.

Hodges, T. K., Leonard, R. T., Bracker, C. E., and Keenan, T. W. (1972). *Proc. Natl. Acad. Sci. USA* **69,** 3307–3311.

Leonard, R. T., and Hodges, T. K. (1973). *Plant Physiol.* **52,** 6–12.

Leonard, R. T., and Hodges, T. K. (1980). In *The Biochemistry of Plants: A Comprehensive Treatise* (P. K. Stumpf and E. E. Conn, eds.), Vol. 1, pp. 163–182, Academic Press, New York.

Leonard, R. T., and Hotchkiss, C. W. (1976). *Plant Physiol.* **58,** 331–335.

Lüttge, U., and Higinbotham, N. (1979). *Transport in Plants,* Springer-Verlag, Berlin.

MacRobbie, E. A. C. (1977). In *International Review of Plant Biochemistry II* (D. H. Northcote, ed.), Vol. 13, pp. 211–247, University Park Press, Baltimore.

Marré, E. (1979). *Annu. Rev. Plant Physiol.* **30,** 273–288.

Mercier, A. J., and Poole, R. J. (1980). *J. Membr. Biol.* **55,** 165–174.

Perlin, D. S., and Spanswick, R. M. (1980). *Plant Physiol.* **65,** 1053–1057.

Poole, R. J. (1978). *Annu. Rev. Plant Physiol.* **29,** 437–460.

Quail, P. H. (1979). *Annu. Rev. Plant Physiol.* **30,** 425–484.

Scarborough, G. A. (1980). *Biochem.* **19,** 2925–2931.

Smith, E. A., and Raven, J. A. (1979). *Annu. Rev. Plant Physiol.* **30,** 289–311.

Sze, H. (1980). *Proc. Natl. Acad. Sci. USA* **77,** 5904–5908.

Sze, H., and Churchill, K. A. (1981). *Proc. Natl. Acad. Sci. USA* **78,** 5578–5582.

Sze, H., and Hodges, T. K. (1977). *Plant Physiol.* **59,** 641–646.

Travis, R. L., and Berkowitz, R. L. (1980). *Plant Physiol.* **65,** 871–879.

178

Kinetic Analysis of Regulation of Membrane Transport

Ulf-Peter Hansen

1. INTRODUCTION

Kinetic analysis seeks to interpret measured time-courses or frequency responses on the basis of reaction kinetic schemes, and constitutes a useful tool for membrane physiologists.

Because of the inherent correlation of temporal analysis with signal processing, kinetic analysis has been widely applied in the field of neurophysiology, particularly with respect to nervous excitation (Goldman and Hahin, 1978; Conti *et al.*, 1980), synaptic transmission (Rosenberry, 1979), and transduction in sensory organs (Brodie *et al.*, 1978). Other examples of its application have been the investigation of the initial membrane-bound reactions of photosynthesis (Witt, 1979), and the establishment of reaction schemes for carrier-mediated transport (Hladky, 1979).

Besides communication theory, it was mainly control theory that created the demand for the development of the system theoretical tools for the analysis of temporal behavior. Therefore, it is expected that kinetic analysis should possess special strengths with respect to the investigation of regulation of membrane transport. However, in spite of the well-documented biological importance of feedback control of transport (Cram, 1976; Macknight and Leaf, 1977; Slayman, 1980), and of the need of recognizing and avoiding interference from feedback loops during the study of transport mechanisms, the full power of a rigorous temporal analysis has not been extensively utilized.

Nowadays, the conditions for its application are improving. Inexpensive computers ease the mathematical burden, and fast experimental procedures, such as microelectrode techniques, allow the analysis of time courses of internal pH (Boron, 1980) or of membrane potential even in cells as small as *E. coli* (Felle *et al.*, 1978).

This article intends to promote the use of temporal analysis by giving an introduction to its theoretical background and by discussing some of its applications.

Ulf-Peter Hansen • Institut für Angewandte Physik, Neue Universität, 23 Kiel, West Germany.

2. THEORETICAL BACKGROUND—REACTION KINETICS

As long ago as 1920, Einstein postulated that chemical reactions should modulate the temporal behavior of physical parameters (in a sound wave). However, it was not until the 1950s that Eigen and co-workers (Eigen, 1968) provided the theoretical and experimental tools that gave rise to the widespread application of "relaxation analysis" in chemistry.

The theory starts from the law of mass action and the rate equations derived from it. This is illustrated for the case of the reaction.

$$A + B \underset{k_{-1}}{\overset{k_1}{\rightleftharpoons}} C \tag{1}$$

Equation (1) leads to the rate equation

$$\frac{d[C]}{dt} = k_1[A][B] - k_{-1}[C] \tag{2}$$

If a "small" perturbation (e.g., a change in temperature or in light intensity) pushes the steady-state concentrations from $[A]_0$, $[B]_0$, and $[C]_0$ to $[A]_{00}$, $[B]_{00}$, and $[C]_{00}$, respectively, the time course of $\Delta[C]$, the change in [C], is given by

$$\Delta[C] = \Delta[C]_0(1 - \exp(-t/\tau)) \tag{3}$$

with

$$\Delta[C]_0 = [C]_{00} - [C]_0 \text{ and } \tau^{-1} = k_1([A]_0 + [B]_0) + k_{-1} \tag{4a,b}$$

Equation (4b) gives the relationship between the observed quantity τ (time-constant) and the parameters of the chemical reaction. Studying the influence of different $[A]_0$ and $[B]_0$ on τ yields the exact values of k_1 and k_{-1}. This method is called extended kinetics. An example of its application to the formation of the gramicidin channel is given by Kolb *et al.* (1975).

In more complicated systems, several reactions are involved in shaping the time course of the observed parameter. In that case, equation (2) is replaced by a system of differential equations. For small perturbations they result in a time course $h_{(t)}$, which is a superposition of exponentials with different time constants τ_i and can be described as

$$h_{(t)} = a_0 + \sum_{i=1}^{n} a_i \exp(-t/\tau_i) \tag{5}$$

Relationships corresponding to equation (4b) are given for many reaction schemes in textbooks (e.g., Capellos and Bielski, 1972; Bernasconi, 1976; Eigen, 1968).

In order for linear equations of the form of equations (3) or (5) to apply, the experimental design must be such that either the system responds linearly to a stimulus or that linear and nonlinear effects can be separated by mathematical methods.

Most commonly, experimental linearization of the responses is achieved by the application of small perturbations (Eigen, 1968; Bernasconi, 1976). Linearity is tested by the independence of curve shape and of "amplification" from the amplitude of the signal (Martens *et al.*, 1979).

Mathematical separation of linear effects is possible if white noise is used as input signal (Marmarelis and Marmarelis, 1978). This allows the analysis of nonlinear properties, too, such as the sequence of linear and nonlinear reactions in sensory transmission (French and Wong, 1976). Noise can be applied either from external sources (French and Wong, 1976), or from internal sources, as in the field of synaptic transmission (Neher and Stevens, 1977) or of transport mechanisms in artificial membranes (Kolb *et al.*, 1975).

3. CHOICE OF INPUT SIGNAL—TIME AND FREQUENCY DOMAIN

Stimulation by pulses or stepwise changes is very often preferred because of the simplicity of its application. The analysis of the evoked responses on the basis of equation (5) is called "treatment in the time domain." However, other signals can provide other benefits, as mentioned above for the case of white noise as input signal.

The high resolution required for the analysis of multicomponent systems can be provided solely by sine waves (Milsum, 1966). The sinusoidal modulation of the input signal causes sinusoidal changes of the observed quantity (output), which is delayed by the phase shift $(-\varphi)$. The dependence of φ and H_0, the ratio of output amplitude to input amplitude, on the frequency f is given by the "transfer function" or "frequency response" (Milsum, 1966; Capellos and Bielski, 1972) $(p = 2\pi\sqrt{-1}f)$:

$$H_{0(p)}\exp(\sqrt{-1}\ \varphi_{(p)}) = H_{(p)} = H_{00}\frac{(1+pn_1)\ (1+pn_2)\ \cdots}{(1+p\tau_1)\ (1+p\tau_2)\ \cdots} \tag{6}$$

with H_{00} being the "dc amplification" (for steady states), n_i the so-called inverse zeros, which are related to the a_i in equation (5), and τ_i the time constants, which are identical to those in equation (5). Equations (5) and (6) are interconvertible by use of the Laplace transform, which is similar to the Fourier transform (Milsum, 1966). Thus, experimental results can be presented in the "time domain," using equation (5), or in the "frequency domain," using equation (6).

4. THE INVERSE PROBLEM

Finding a reaction scheme that generates the proper τ_i and a_i in equation (5) [or n_i in equation (6)] is called a solution of the "inverse problem." As a guideline, inverse zeros, n_i, arise from parallel pathways or from backflow from subsequent reaction chains (see appendices of Hansen, 1978; Martens *et al.*, 1979).

5. APPLICATION OF TEMPORAL ANALYSIS IN THE FIELD OF REGULATION OF TRANSPORT

It has been known for many years that illumination causes changes in membrane potential in green plant cells (Haake, 1892). It is now well established that the membrane potential in plants is dominated by an outwardly directed H^+ pump (Slayman, 1980; Walker, 1980). Thus, the observation that light induces parallel changes in membrane potential and in cytoplasmic pH (Davis, 1974), coupled with the a priori consideration that transmembrane transport is required for pH regulation (Smith and Raven, 1979), has led to the proposal that light-induced changes in membrane potential are related to the regulation of cytoplasmic pH (Hansen, 1978, 1980).

The time course of the light-induced changes in membrane potential is complicated, indicating the involvement of a multicomponent reaction scheme. For the freshwater alga *Nitella*, it has been shown that the kinetic analysis of the linearized light responses can be used to extract time constants and to propose a reaction scheme of the processes involved in the putative pH_i regulating system (Hansen, 1980).

Frequency responses [equation (6)] were measured by recording the potential changes caused by the modulation of light intensity by sine waves (Hansen, 1978; Martens *et al.*, 1979), noise (Korff, 1979), or step functions (Hansen and Wittmaack, 1980). Curve-fitting on the basis of equation (6) resulted in a frequency response with seven time constants (τ_1 to τ_7) and five inverse zeroes (n_1 to n_5).

From these investigations, the following model has been extracted (Hansen, 1980). Light activates a process A ($\tau_1 \approx 10$ min, probably biosynthesis by photosynthesis) and blocks a process B ($\tau_4 \approx 30$ sec, probably respiration). When operating, A and B release protons into the cytoplasmic pool. Rapid perturbations are smoothed out by a buffer ($\tau_7 \approx 30$ msec). Long-term net adjustment is achieved by transmembrane transport. Both A and B send feed-forward signals to membrane transport ($\tau_6 \approx 0.3$ sec and $\tau_5 \approx 3$ sec, respectively). A feedback loop sensing pH_i [$\tau_{2,3} \approx 3 \pm 10\sqrt{-1}$ min (complex numbers)] achieves the final adjustment of transport. The indication of the existence of the feedback loop by the complex time constants τ_2 and τ_3 is supported by the observation of spontaneous oscillations (Hansen, 1978). The number of major processes involved is seven as given by the number of measured time constants. It is a minimum number as the linearization eliminates the influence of side chains. They can be investigated by extended kinetics [see above, equation (4b)]. The existence of parallel pathways is a major message of the model. The adopted solution is highly reliable as five measured inverse zeros determine only four model parameters (Martens *et al.*, 1979). Normally, this would lead to contradictions, if the model were wrong. In contrast, this model is "scatter reducing": the model parameters display much less scatter than the measured ones from which they are calculated (Hansen, 1980).

Kinetic analysis per se cannot tell whether it is pH that is regulated by transport. However, this interpretation is supported by the fact that current injection which probably exchanges K^+ for H^+ stimulates the same feedback system (Hansen, 1978).

The model of multiple pathways of light action can easily explain the findings of other authors. Kawamura *et al.* (1980) found large simply shaped hyperpolarizations with light in internally perfused cells of *Chara* (a closely related genus to *Nitella*), in contrast to small responses of complicated shape in intact cells. In terms of the above model, it can be suggested that perfusion washes out the messenger of the feedback system (and probably of pathway B), thus leaving the large hyperpolarizations of pathway A uncompensated. The same proposal accounts for the findings of Fujii *et al.* (1979) that large light-induced hyperpolarizations persist in perfused *Chara*, but membrane potential does not respond to changes in internal pH. Washing out of a soluble feedback messenger is not an unlikely assumption, as it has been found in the case of calmodulin (Sarkadi, 1980). Furthermore, the transition from small, complicated responses to light to large simple ones is also observed when changing external pH from pH 5 to pH 9 (Denesh *et al.*, 1980). On the basis of the model, it appears that metabolic OH^- production due to the assimilation of bicarbonate at external pH 9 (Bisson and Walker, 1980) compensates the light-induced proton load on the cytoplasm and prevents the stimulation of the feedback loop.

The search for corresponding components in light-induced changes in membrane potential and in membrane resistance (Keunecke, 1974; Hansen and Keunecke, 1977) revealed no common time constants at low light intensities (1 W/m²). This suggests that changes in potential are brought about by carrier-mediated transport operating in the saturation region of their IV curves (high resistance) (Gradmann *et al.*, 1981), whereas changes in resistance are related to an ion being close to electrochemical equilibrium (K^+?).

This independence of light-induced changes in potential and in resistance was found by Kawamura *et al.* (1980) in ATP-depleted *Chara*. Strange hyperpolarizations toward E_K in illuminated ATP-depleted *Chara* filled with neutral red (Kawamura and Tazawa, 1980) can also be explained by a separate light action on K^+ permeability.

Testing predictions of the model can be done by the method of extended kinetics [see equation (4b)], which is based on the study of the dependence of the measured time constants on experimental conditions. The above model predicts that all time constants may change with light intensity, because of the nonlinearity of the system, except τ_2 and τ_3, which "see" the regulated pH, but not light. Hansen and Wittmaack (1980) showed that this prediction was verified by the experimental findings. The progress of this kind of research depends on the availability of poisons that modify some of the measured time constants selectively. Inhibitors of photosynthesis act like darker light (see Fig. 1 of Hansen and Keunecke, 1977), as expected. The use of fusicoccin (Marrè, 1979) may be more promising as it is supposed to act on the control mechanism of the H^+ pump directly.

ACKNOWLEDGMENTS. I am grateful to Dr. D. Sanders (New Haven) for contributing many ideas, and to Drs. C. L. Slayman (New Haven) and D. Gradmann (Munich) for helpful discussions.

REFERENCES

Bernasconi, C. F. (1976). *Relaxation Kinetics*, Academic Press, New York.

Bisson, M. A., and Walker, N. A. (1980). *J. Membr. Biol.* **56**, 1–7.

Boron, W. F. (1980). *Curr. Top. Membr. Transp.* **13**, 1–22.

Brodie, E., Knight, B. W., and Ratliff, F. (1978). *J. Gen. Physiol.* **72**, 167–202.

Capellos, C., and Bielski, B. H. J. (1972). *Kinetic Systems*, Wiley–Interscience, New York.

Conti, F., Neumcke, B., Nonner, W., and Stämpfli, R. (1980). *J. Physiol.* **308**, 217–239.

Cram, W. J. (1976). In *Encyclopedia of Plant Physiology*, New Series (U. Lüttge and M. G. Pitman, eds.), Vol. II, Part A, pp. 284–316, Springer-Verlag, Berlin.

Davis, R. F. (1974). In *Membrane Transport in Plants* (U. Zimmermann and J. Dainty, eds.), pp. 197–201, Springer-Verlag, Berlin.

Denesh, M., Andrianow, V. K., Bulychev, A. A., and Kurella, G. A. (1980). *Biophysics* **24**, 677–683; *J. Exp. Bot.* **29**, 1185–1195.

Eigen, M. (1968). *Q. Rev. Biophys.* **11**, 3–33.

Einstein, A. (1920). *Sitzungsber. Preuss. Akad. Wiss. Phys. Math. Kl.* **1920**, 380–385.

Felle, H., Stetson, D. L., Long, S. W., and Slayman, C. L. (1978). In *Frontiers of Biological Energetics* (L. P. Dutton, L. S. Leigh, and A. Scarpa, eds.), Vol. 2, pp. 1399–1408, Academic Press, New York.

French, A. S., and Wong, R. K. S. (1976). *Biol. Cybernetics* **22**, 33–38.

Fujii, S., Shimmen, T., and Tazawa, M. (1979). *Plant Cell Physiol.* **20**, 1315–1328.

Goldman, L., and Hahin, R. (1978). *J. Gen. Physiol.* **72**, 879–898.

Gradmann, D., Hansen, U.-P., and Slayman, C. L. (1981). In *Electrogenic Pumps* (C. L. Slayman, ed.), Academic Press, New York.

Haake, O. (1892). *Flora (Jena)* **75**, 455–487.

Hansen, U.-P. (1978). *J. Membr. Biol.* **41**, 197–224.

Hansen, U.-P. (1980). In *Plant Membrane Transport* (R. M. Spanswick, W. J. Lucas, and J. Dainty, eds.), pp. 587–588, Elsevier, Amsterdam.

Hansen, U.-P., and Keunecke, P. (1977). In *Transmembrane Ionic Exchanges in Plants* (M. Thellier, A. Monnier, M. Demarty, and J. Dainty, eds.), pp. 333–340, CNRS, Paris.

Hansen, U.-P., and Wittmaack, W. (1980). *Biophys. Struct. Mech.* **6**(Suppl.), 75.

Hladky, S. B. (1979). *Curr. Top. Membr. Transp.* **12**, 53–164.

Kawamura, G., and Tazawa, M. (1980). *Plant Cell Physiol.* **21**, 547–559.

Kawamura, G., Shimmen, T., and Tazawa, M. (1980). *Planta* **149**, 213–218.

Keunecke, P. (1974). *Ber. Dtsch. Bot. Ges.* **87**, 529–536.

Kolb, H.-A., Läuger, P., and Bamberg, E. (1975). *J. Membr. Biol.* **20**, 133–154.

Korff, H.-M. (1979). Ph.D. thesis, Kiel.

Macknight, A. O. C., and Leaf, A. (1977). *Physiol. Rev.* **57**, 510–573.

Marmarelis, P. Z., and Marmarelis, V. Z. (1978). *Analysis of Physiological Systems: The White Noise Approach,* Plenum Press, New York.

Marrè, E. (1979). *Annu. Rev. Plant Physiol.* **30**, 273–288.

Martens, J., Hansen, U.-P., and Warncke, J. (1979). *J. Membr. Biol.* **48**, 115–139.

Milsum, J. H. (1966). *Biological Control Systems Analysis,* McGraw–Hill, New York.

Neher, E., and Stevens, C. F. (1977). *Annu. Rev. Biophys. Bioeng.* **6**, 345–381.

Rosenberry, T. L. (1979). *Biophys. J.* **26**, 263–290.

Sarkadi, B. (1980). *Biochim. Biophys. Acta* **604**, 159–190.

Slayman, C. L. (1980). In *Plant Membrane Transport* (R. M. Spanswick, W. J. Lucas, and J. Dainty, eds.), pp. 179–194, Elsevier, Amsterdam.

Smith, F. A., and Raven, J. A. (1979). *Annu. Rev. Plant Physiol.* **30**, 289–311.

Walker, N. A. (1980). In *Plant Membrane Transport* (R. M. Spanswick, W. J. Lucas, and J. Dainty, eds.), pp. 287–300, Elsevier, Amsterdam.

Witt, H. T. (1979). *Biochim. Biophys. Acta* **505**, 355–427.

179

Membrane Transport in Charophyte Plants: Chemiosmotic but Electrically Versatile

N. A. Walker

The electric potential difference across the plasmalemma (ψ_{co}) in the large internodal cells of charophyte plants is relatively easy to measure, as it is nearly equal to that between vacuole and medium (ψ_{vo}). It has often been studied in an attempt to understand the processes transporting charge across the plasmalemma. This understanding is far from complete, but it has reached an interesting stage. The dependence of ψ_{co} on external factors (e.g., pH_o) is seen now to be important in characterizing different modes of behavior of the membrane (Walker, 1980a). In some cases the whole membrane behaves differently at different times, while in others different parts of the cell membrane behave differently at the same time.

1. TEMPORALLY SEPARATED MODES FOR THE WHOLE CELL

1.1. Passive K⁺ Mode (Depolarized State)

The early physiologists (see Osterhout, 1931) discovered the role of external potassium concentration ($[K^+]_o$) in determining the value of ψ_{co}; this was put on the present quantitative basis by Hope and Walker (1961), who showed that, under some conditions, ψ_{co} depended on the passive uniport of K^+ and Na^+ with a permeability ratio of about 20. That is, ψ_{co} is (sometimes) given by

$$\Psi_{co} \doteq \frac{RT}{F} \ln \frac{[K^+]_o + \alpha[Na^+]_o}{[K^+]_c} \tag{1}$$

N. A. Walker • Biophysics Laboratory, School of Biological Sciences, University of Sydney, Sydney, New South Wales 2006, Australia.

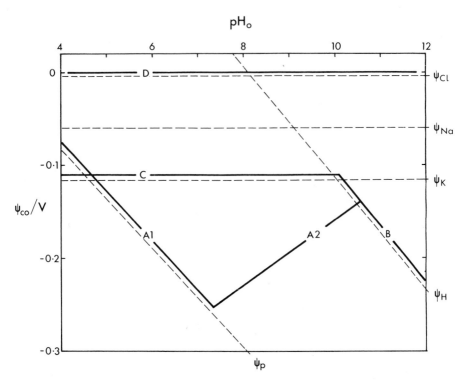

Figure 1. The dependence of the membrane potential difference (ψ_{co}) in charophyte cells on the external pH (pH_0), showing schematically: *solid lines*, the four modes of behavior discussed in the text: A1 and A2, active H^+; B, passive H^+; C, passive K^+; and D, passive Cl^-; *broken lines*, equilibrium potentials: ψ_p, $2H^+$-ATPase; ψ_H, H^+ passive uniport; ψ_K, K^+ passive uniport; ψ_{Na}, Na^+ passive uniport; and ψ_{Cl}, Cl^- passive uniport. The equilibrium potential for passive uniport of Ca^{2+} is at about $+0.1$ V.

where R, T, and F have their usual meanings and α is about $1/20$. The membrane conductance g_{co} also depends on $[K^+]_0$ in a way not yet well quantified, and is usually in the range 0.1–0.5 S/m^2; it also seems to depend strongly on ψ_{co} (Smith and Walker, 1981). As $[K^+]_c$ is normally near 100 moles/m^3 (100 mM), ψ_{co} is often about -0.15 V in the dilute media used in experiments.

For *Chara*, this behavior occurs for cells in media without added Ca^{2+} (Hope and Walker, 1961) or for some time after an action potential (Oda, 1962); for *Nitella*, it is normal for cells in the dark, and for cells in the light with high $[CO_2]$ (Spanswick, 1972); for *Hydrodictyon*, a chlorophyte alga, it normally occurs in the dark (G. P. Findlay, unpublished data). In other cases, ψ_{co} is often more negative than the equilibrium potential for K^+ (ψ_K), which is the most negative ion equilibrium potential (Fig. 1).

1.2. Active H⁺ Mode (Hyperpolarized State)

Under normal conditions for *Chara*, light and low $[CO_2]$ for *Nitella*, and in light for *Hydrodictyon*, ψ_{co} does not vary significantly with $[K^+]_0$—instead it depends directly on pH_0 (Fig. 1, A1 and A2); g_{co} also depends on pH_0—it is usually in the range 0.3–1.0 S/m^2. Kitasato (1968) attributed the hyperpolarization to active H^+ efflux: Spanswick (1972, 1974) and Richards and Hope (1974) have shown that this "normal" state of the membrane is dominated by an active efflux of H^+ that is probably potential-dependent. This pump is almost certainly an ATPase (Shimmen and Tazawa, 1977; Spanswick, 1980; Smith and Walker, 1981): the latter two groups have shown that in the perfused plasmalemma, the addition of MgATP to the perfusion medium produces a pH-dependent outward movement of positive charge. For intact cells, in the range 4 to 7.5, pH_o determines ψ_{co} in a way that fits the equation (Walker and Smith, 1975)

$$\psi_{co} \doteqdot (1/F)\{\Delta G/2 - 2.303RT(pH_c - pH_o)\} \qquad (2)$$

where ΔG is the free energy of hydrolysis of ATP, taken to be 50–55 kJ/mole, and where the steady-state value of pH_c is given by the correlation (Smith and Walker, 1976; Smith, 1980):

$$pH_c \doteqdot 6.28 + 0.22 pH_o \qquad (3)$$

Equation (2) implies that ψ_{co} is near the equilibrium (stalling) potential for a $2H^+$-ATPase. A stoichiometric ratio of 2 is probable on other grounds (Smith and Walker, 1981), but is not certain; it may in any case vary (cf. *Neurospora*, Warncke and Slayman, 1980). The implication that ψ_{co} depends directly on $[ATP]_c$ has been investigated by Reid (1980), who found most inhibitors to reduce ψ_{co} before they reduced $[ATP]_c$: Keifer and Spanswick (1979) found that under certain conditions a correlation did exist between $[ATP]_c$ and ψ_{co}. Between ph 7 or 8 and pH 10, ψ_{co} depolarizes as pH_o is raised (Fig. 1, A2); this regime is not yet understood.

If chemiosmotic, primary H^+ transport (Mitchell, 1970; Smith, 1970) is a feature of the "normal" mode of behavior of the charophyte membrane, one would expect to find transport systems for other substances that use as driving force ψ_{co}, ΔpH_{co}, or a combination of these.

The most important osmoticum for vacuolate charophyte cells is KCl (and NaCl); it appears that Cl^- is taken up by a transport system carrying net positive charge inwards, as judged by depolarization (Sanders, 1980) and by the direction of clamp current in a rapid-flow setup (Beilby and Walker, 1981). Proton symport is probable, as there is no direct requirement for external K^+ or Na^+, and the likely ratio is $2 H^+ : 1 Cl^-$. The K_m for Cl^- influx is of the order of 25 μM; the saturated velocity increases with "Cl starvation" to about 100 nmoles/m² per sec—this seems to be an effect of change in pH_c or in $[Cl^-]_c$, or both.

Ammonium, an important source of reduced N, is taken up by passive uniport of the ion NH_4^+ as well as by diffusion of the free base (Walker *et al.*, 1979b; Smith and Walker, 1980). The ion uniport has a K_m of the order of 3 μM for NH_4^+, and 250 μM for $CH_3NH_3^+$, with a saturated velocity as high as 1–3 μmoles/m² per sec (Walker *et al.*, 1979a). Both K_m and V_{max} are weakly dependent on ψ_{co}, as if the ammonium binding site were below the membrane surface. The current–voltage curves for this transport system show clearly the effect of unstirred layers of solution next to the membrane.

Saturable transport of other N-containing substances is known (Wilson, 1980), but the low values of V_{max} make the question of H^+-coupling difficult to decide. Arguments exist (Walker, 1980a) for H^+-coupled antiport of both K^+ and Na^+, but evidence is lacking. There are no reports of H^+-coupled sugar transport.

1.3. The Transition between Modes

The transition from the active H^+ mode to the passive K^+ mode involves a reduction in pump current and conductance, leading to depolarization, and an increase in K^+ conductance (g_K). There is evidence that g_K depends on ψ_{co}, increasing as ψ_{co} becomes more positive and approaches ψ_K (Smith and Walker, 1981; Bisson and Walker, unpublished). Thus, a positive change in ψ_{co} towards ψ_K can lead to an increase in g_K that will enable K^+ passive uniport to dominate ψ_{co}. However, other signals are likely to be involved as well, for it is possible to depolarize *Chara* (by low pH_o) without an increase of g_K "grabbing" ψ_{co} as it passes ψ_K (Bisson and Walker, unpublished). An action potential, on the other hand, which appears to inhibit the pump (M. J. Beilby and J. R. Smith, unpublished), does allow ψ_{co} to be "grabbed" near ψ_K by an increase in g_K at the end of the recovery.

2. SPATIALLY SEPARATED MODES: STATIC

An anion of importance is bicarbonate, often used as the source of carbon for photosynthesis under natural conditions. In the light, during assimilation of HCO_3^-, relatively large steady electric currents—to about 0.1 A/m² or 1.0 μmole/m² per sec—circulate in the medium (Walker and Smith, 1977) between successive annular segments of the cell surface. Where positive charge leaves the cell, the medium is often more acid (e.g., pH 6 to 7) than the bulk, and where positive charge enters the cell, the medium is more alkaline (e.g., pH 10 to 11) than the bulk pH (Spear *et al.*, 1969; Lucas and Smith, 1973). The effective uptake of HCO_3^- seems to occur in the zone of membrane under an acid band; the zone under an alkaline band exports equivalent electric charge as OH^- or imports it as H^+.

The speed of the reactions between different C species in the $CO_2/H_2CO_3/HCO_3^-/CO_3^-$ system, and the existence of unstirred layers of solution, have so far prevented any clear decision as to the C species transported across the membrane. While primary active transport of HCO_3^- has been accepted (cf. Lucas, 1980), it is a simpler postulate that CO_2 uptake by diffusion follows extracellular conversion of HCO_3^- to H_2CO_3 by H^+ extruded by the pump (Walker *et al.*, 1980). Other possibilities include H_2CO_3 uptake by diffusion or uniport and symport of HCO_3^- with H^+. In each case about the same amount of charge must be transported in the acid zone, either as HCO_3^- inwards or as H^+ outwards. The postulate is here made that in the acid zone the membrane is in active H^+ (chemiosmotic) mose. However, in the alkaline zone another mode of behavior appears.

Passive H^+ Mode. The membrane of the charophyte internode becomes reversibly permeable to H^+ (or, indistinguishably, to OH^-) when pH_o is raised to a high value, i.e., above some critical value that lies between 9.5 and 10.5 for different cells (Bisson and Walker, 1980). This mode of behavior of the whole cell, under experimental manipulation, seems to be the mode in the alkaline zone of the banded cell. The value of ψ_{co} is, in this mode, close to ψ_H (Fig. 1, B), while g_{co} is high (about 1.5 S/m²) and g_K is low (Bisson and Walker, unpublished). The H^+ uniport system can be thought of as "grabbing" ψ_{co} as it approaches ψ_H from more negative values.

The result of this and of the passive K^+ uniport is that ψ_{co} is confined to values close to or more negative than ψ_K and ψ_H, except during an action potential (cf. Fig. 1).

3. SPATIALLY SEPARATED MODES: TRANSIENT

Charophyte internodal cells normally produce a slow action potential when depolarized beyond a threshold value near -0.09 V. The excitation is electrically propagated; currents flow that under voltage clamp are of the order of 1A/m², so that the excited mode can grab ψ_{co} whatever other transport system is in action.

Passive Cl⁻ (Excited) Mode. The peak of the action potential is at about 0.0 V, and the excited mode is associated with transient increases in permeability to Cl^- (Gaffey and Mullins, 1958; Findlay and Hope, 1964) and to a second ion that may be Ca^{2+} (Beilby and Coster, 1979), although its reversal potential is difficult to understand. The Cl^- channels are not very selective (Shimmen and Tazawa, 1980), and it may be that the Ca^{2+} channels are also indiscriminate. Both flows (Cl^- out and Ca^{2+} in) depolarize the membrane, which is repolarized by a transition to passive K^+ mode.

4. FUNCTIONS OF THE MODES

In active H^+ mode, the plasmalemma of the charophyte plant is apparently a normal chemiosmotic membrane, resembling that of fungi, yeasts, and prokaryotes. It seems to resemble the bacteria (V. Skulachev, personal communication) in possessing a mode of operation in which ψ_{co} is supported by K^+ uniport; the use of a permeant ion for the purpose of increasing the apparent electric capacitance of a membrane has been discussed by Mitchell (1968). The passive K^+ mode is thus to be seen as a backup system that can maintain some transport systems in operation without the need for rapid H^+ transport.

The charophyte plants share with other aquatic macrophytes the problem of photosynthesizing in media of low $[CO_2]$ but higher $[HCO_3^-]$. The solution adopted seems to be that HCO_3^- diffuses in through the unstirred layer of solutions outside the membrane, to be converted into the permeant species CO_2 by H^+ exported from the cell—neither H^+ nor CO_2 occurring in the medium at high enough concentration to reach the membrane through the unstirred layer at the rates involved. The passive H^+ mode then plays the role of returning the exported H^+ to the cell; in *Chara* as in the freshwater angiosperms, different parts of the cell or leaf are simultaneously in active H^+ and in passive H^+ mode.

The action potential has almost always the result that cytoplasmic streaming stops—this may be the result of Ca^{2+} influx into the cytoplasm. It has been suggested (Walker, 1980b) that the function of the action potential is to stop streaming and consequently intercellular transport (Bostrom and Walker, 1976). Thus, injury to the membrane of one cell, which normally causes an action potential, will not cause consequent changes in the cytoplasm of its immediate neighbors.

REFERENCES

Beilby, M. J., and Coster, H. G. L. (1979). *Aust. J. Plant Physiol.* **6**, 323–335.
Beilby, M. J., and Walker, N. A. (1981). *J. Exp. Bot.* **32**, 43–54.
Bisson, M. A., and Walker, N. A. (1980). *J. Membr. Biol.* **56**, 1–7.
Bostrom. T. E., and Walker, N. A. (1976). *J. Exp. Bot.* **27**, 347–357.
Findlay, G. P., and Hope, A. B. (1964). *Aust. J. Biol. Sci.* **17**, 400–411.
Gaffey, C. T., and Mullins, L. J. (1958). *J. Physiol.* **144**, 505–524.
Hope, A. B., and Walker, N. A. (1961). *Aust. J. Biol. Sci.* **14**, 26–44.
Keifer, D. W., and Spanswick, R. M. (1979). *Plant Physiol.* **64**, 165–168.
Kitasato, H. (1968). *J. Gen. Physiol.* **52**, 60–87.
Lucas, W. J. (1980). In *Plant Membrane Transport: Current Conceptual Issues* (R. M. Spanswick, W. J. Lucas, and J. Dainty, eds.), pp. 317–327, Elsevier/North-Holland, Amsterdam.
Lucas, W. J., and Smith, F. A. (1973). *J. Exp. Bot.* **24**, 1–14.
Mitchell, P. (1968). *Chemiosmotic Coupling and Energy Transduction*, Glynn Research, Bodmin.
Mitchell, P. (1970). *Symp. Soc. Gen. Microbiol.* **20**, 121–166.
Oda, K. (1962). *Sci. Rep. Tohoku Univ. Ser. 4* **28**, 1–16.
Osterhoust, W. J. V. (1931). *Biol. Rev.* **6**, 369–411.

Reid, R. J. (1980). Ph.D. thesis, University of Sydney.

Richards, J. L., and Hope, A. B. (1974). *J. Membr. Biol.* **16,** 121–144.

Sanders, D. C. (1980). *J. Exp. Bot.* **31,** 105–118.

Shimmen, T., and Tazawa, M. (1977). *J. Membr. Biol.* **37,** 167–192.

Shimmen, T., and Tazawa, M. (1980). *J. Membr. Biol.* **55,** 223–232.

Smith, F. A. (1970). *New Phytol.* **69,** 903–917.

Smith, F. A. (1980). *J. Exp. Bot.* **31,** 597–606.

Smith, F. A., and Walker, N. A. (1976). *J. Exp. Bot.* **27,** 451–459.

Smith, F. A., and Walker, N. A. (1980). *J. Exp. Bot.* **31,** 119–133.

Smith, P. T., and Walker, N. A. (1981). *J. Membr. Biol.*, 223–236.

Spanswick, R. M. (1972). *Biochim. Biophys. Acta* **288,** 73–89.

Spanswick, R. M. (1974). *Biochim. Biophys. Acta* **332,** 387–398.

Spanswick, R. M. (1980). In *Plant Membrane Transport: Current Conceptual Issues* (R. M. Spanswick, W. J. Lucas, and J. Dainty, eds.), pp. 305–313, Elsevier/North-Holland, Amsterdam.

Spear, D. C., Barr, J. K., and Barr, C. E. (1969). *J. Gen. Physiol.* **54,** 397–414.

Walker, N. A. (1980a). In *Plant Membrane Transport: Current Conceptual Issues* (R. M. Spanswick, W. J. Lucas, and J. Dainty, eds.), pp. 287–300, Elsevier/North-Holland, Amsterdam.

Walker, N. A. (1980b). In *Plant Membrane Transport: Current Conceptual Issues* (R. M. Spanswick, W. J. Lucas, and J. Dainty, eds.), p. 346, Elsevier/North-Holland, Amsterdam.

Walker, N. A., and Smith, F. A. (1975). *Plant Sci. Lett.* **4,** 125–132.

Walker, N. A., and Smith, F. A. (1977). *J. Exp. Bot.* **28,** 1190–1206.

Walker, N. A., Beilby, M. J., and Smith, F. A. (1979a). *J. Membr. Biol.* **49,** 21–55.

Walker, N. A., Smith, F. A., and Beilby, M. J. (1979b). *J. Membr. Biol.* **49,** 283–296.

Walker, N. A., Smith, F. A., and Cathers, I. R. (1980). *J. Membr. Biol.* **57,** 51–58.

Warncke, J., and Slayman, C. L. (1980). *Biochim. Biophys. Acta* **591,** 224–233.

Wilson, M. R. (1980). B.Sc. (Hons) thesis, University of Sydney.

180

Electrogenic Transport at the Plasma Membrane of Plant Cells

Ronald J. Poole

This article reviews specific transport processes at the plant cell membrane that can be detected by their electrical properties. Superimposed on a background diffusion potential, changes as large as 80 mV are produced by the switching on of additional transport systems. These electrogenic effects play a major role in the energy coupling of membrane transport.

1. PROTON TRANSPORT

It is now clear that a primary electrogenic proton efflux driven by ATP hydrolysis is present in the plasma membrane of most plant cells (Poole, 1978; Bentrup, 1980). Although this pump has not yet been isolated and characterized from green plants, its inhibitor sensitivity (e.g., Jacobs and Taiz, 1980) resembles that of the proton pump in the plasma membrane of fungi. These eukaryotic proton pumps may functionally resemble those of prokaryotes and organelles, but structurally they are very different, and more closely resemble the Na^+/K^+-ATPase of animal cells in molecular weight and catalytic mechanism (Dame and Scarborough, 1980).

1.1. Proton Transport in Characeae

The freshwater algae *Chara* and *Nitella* have giant coenocytic cells, 1-mm diameter × several centimeters long, which are much favored for electrical studies. The potential is highly sensitive to external pH (more than 50 mV/pH unit) and the conductance of the pump itself is large compared with the conductance of other membrane channels (Spanswick, 1972). This pump may be switched on and off by changes in light intensity (Spanswick, 1972) or inhibited by removal of external Ca^{2+}, increase of external K^+, or appli-

Ronald J. Poole • Biology Department, McGill University, Montreal H3A 1B1, Canada.

cation of inhibitors such as N,N'-dicyclohexylcarbodiimide (DCCD), diethylstilbestrol, or 2,4-dinitrophenol (Keifer and Spanswick, 1978). The dramatic effect of light on pump activity, conductance, and potential (Spanswick, 1972) indicates an important regulatory mechanism that is as yet poorly understood, but is not attributable to changes in ATP level (Kikuyama *et al.*, 1979) or cytoplasmic pH (Fujii *et al.*, 1979). Perfusion of these giant cells has demonstrated the dependence of H^+ efflux, as well as membrane potential and conductance, on internal ATP (Shimmen and Tazawa, 1980; Kawamura *et al.*, 1980). Although the cytoplasmic level of ATP is of the order of 1 mM, the ATP concentration required to activate the pump is only about 30 μM (Kikuyama *et al.*, 1979), unlike the situation in the fungus *Neurospora*, where the K_m of the H^+ pump is about 2 mM (Scarborough, 1980).

1.2. Effects of Bicarbonate in Characeae

The rapid consumption of CO_2 in photosynthesis by these giant cells has some interesting repercussions on membrane transport, especially in solutions at neutral or high pH, where the available carbon is mainly present as HCO_3^-. Uptake of HCO_3^- does not occur uniformly over the cell surface, but in bands that become readily visible to the naked eye if a pH indicator dye is added to the external solution. The regions of HCO_3^- uptake acidify the surrounding solution, and because charge balance is maintained by H^+ influx (or OH^- efflux) in the intervening areas of cell membrane, these areas become alkaline. Because transport in both the acid and the alkaline bands is electrogenic, large electric currents circulate between them. The magnitude of these currents corresponds to peak surface ionic fluxes of several hundred pmoles per square centimeter per second (Lucas and Nuccitelli, 1980), or about 100 times higher than typical ion flux rates measured in higher plant cells.

It has proven difficult to distinguish between influx of HCO_3^- as such, and influx of CO_2 associated with efflux of H^+ (which would decompose the HCO_3^- at the membrane surface). It now seems likely that the latter is the case, especially because carbon uptake is inhibited by the same conditions (low Ca^{2+} or high K^+) that inhibit H^+ efflux (Walker *et al.*, 1980). CO_2 influx may therefore be driven by circulating proton currents. H^+ efflux in the acid bands is clearly active, while the pH conditions of the alkaline bands have been shown to create a high passive permeability to H^+ in these regions of the membrane (Bisson and Walker, 1980).

1.3. Proton Transport in Higher Plants

The proton pump of higher plants can now be recognized with some certainty as a result of the discovery of a specific activator, fusicoccin (Marrè, 1979), and some moderately specific inhibitors: DCCD, diethylstilbestrol, and vanadate (e.g., Cheeseman *et al.*, 1980; Jacobs and Taiz, 1980). Regulation of pump activity by these agents or by respiratory inhibitors, anoxia (Cheeseman and Hanson, 1979), or low temperature (Bentrup, 1980) shows that in a wide range of experimental material the proton pump hyperpolarizes the membrane by up to 80 mV, and, in the presence of external K^+, drives net H^+ efflux and K^+ influx.

In favorable experimental material, the current carried by the pump can be measured over a range of membrane potentials. This has been done in rhizoids of *Riccia fluitans* by Felle (Bentrup, 1980). The pump current was obtained by subtraction of the current–voltage curve at 5°C from that at 20°C. The result indicates that the pump resembles that of the fungus *Neurospora* more than that of the Characeae, in that the reversal potential (equilibrium with ATP hydrolysis) occurs at about −360 mV, suggesting less than 2

H^+/ATP, and at normal membrane potentials the pump acts like a high-resistance current source.

The pattern of ATP dependence in beet cells, as well as limited data on current–voltage relationships, also suggest a high-resistance pump far from its equilibrium potential (Mercier and Poole, 1980). In this material, the potential produced by the pump depends sharply on ATP level at a critical concentration of ATP estimated at about 0.4 mM in the cytoplasm, although the potassium transport attributed to the same pump depends linearly on ATP up to the highest levels obtained (see the following discussion).

1.4. Proton–Potassium Exchange and Electrogenesis

It has been difficult to determine whether the proton pump of higher plants transports H^+ alone, or to what extent it performs a linked exchange of H^+ and K^+. Influx of K^+ against its electrochemical gradient has been frequently reported (Poole, 1978; Cheeseman and Hanson, 1979), and this active influx is sensitive to fusicoccin and to many other treatments that affect the H^+ pump. In addition, maximal activity of the ATPase in membrane vesicles is not obtained by addition of valinomycin or carbonyl cyanide m-chlorophenylhydrazone alone, but requires both these ionophores or nigericin, suggesting a linked H^+/K^+ transport (Sze, 1980).

On the other hand, a number of observations indicate the lack of a fixed ratio of H^+ to K^+ transported. For example, it has been shown that lipophilic cations can partially substitute for external K^+ in evoking net H^+ efflux by the proton pump (Marrè, 1979). In red beet, the potential generated by a given rate of pump activity varies with pH (Poole, 1974), with ATP level (Mercier and Poole, 1980), and with K^+ concentration (unpublished data).

A detailed study of corn roots by Cheeseman and colleagues gives insight into the relationship between the electrogenic pump and the influx of K^+ (Cheeseman and Hanson, 1979; Cheeseman *et al.*, 1980). The effects of anoxia and of ATPase inhibitors on the potential and on K^+ influx were determined over a wide range of external K^+ concentrations. It was found that in the range of K^+ concentrations from 10 μM to about 1 mM, K^+ influx is against its electrochemical gradient, and is inhibited by the H^+-ATPase inhibitor DCCD. This active K^+ flux increases with concentration, reaching saturation at about 1 mM. The potential, on the other hand, showed a DCCD-inhibitable component, attributed to the proton pump, which *decreases* from a maximum of 40 mV at 10 μM K^+ or less, down to zero at about 1 mM. The degree of depolarization of this pump potential corresponds closely with the degree of saturation of K^+ influx. This suggests that the H^+ efflux pump carries K^+ inwards to a variable extent, and that its contribution to the potential decreases as its K^+ saturation increases.

In the K^+ concentration range from 1 mM to about 30 mM, a component of potential inhibited by anoxia but not by DCCD was found to increase with increasing K^+ concentration, and to provide the electrical driving force for an increasing passive influx of K^+. The origin of this potential is unknown (Cheeseman *et al.*, 1980).

2. CHLORIDE TRANSPORT

The plant materials discussed above have all been freshwater or land plants, in which proton transport is the dominant electrogenic mechanism. Two giant-celled marine algae that have been studied in detail show a different pattern of transport. The membrane electrical properties of *Acetabularia* (Gradmann, 1975) and *Halicystis* (Graves and Gutknecht,

1977) are dominated by chloride pumps, which carry large electric currents and contribute about -80 mV to the membrane potential. The membrane current–voltage relationships have been investigated, and related to isotope fluxes—a comparison that is impossible for the proton pump. In the case of *Halicystis*, there is good agreement between the short-circuit current and the flux of Cl^-. The chloride pumps of these photosynthetic organisms, like the proton pumps of the Characeae, are sensitive to light as well as to metabolic inhibitors and low temperature.

Chloride transport in freshwater and land plants is not well understood but is not conspicuously electrogenic. An apparent exception is the Cl^- pump in the salt glands of *Limonium* (Hill and Hill, 1973). Although the electrical properties of this material were measured across the whole tissue rather than a single membrane, the results were difficult to explain except in terms of electrogenic Cl^- transport.

3. INTERACTION OF SECONDARY TRANSPORT SYSTEMS WITH THE MEMBRANE POTENTIAL

In principle, any transport system that carries a net charge across the membrane will both affect the potential and be affected by the potential. One way to quantify these interactions is to construct equivalent circuits (Poole, 1978) that express the equilibrium potential and the conductance of each pathway. Here I will summarize in a qualitative way the types of effects that have been observed.

3.1. Direct Interactions with the Potential

Passive ion fluxes create a diffusion potential, usually described by the Goldman equation, and often dominated by K^+. This diffusion potential is identified as the component of the membrane potential that is insensitive to inhibitors. Cheeseman *et al.* (1980) give a justification for considering the normal membrane potential to be the algebraic sum of the contributions of the active and passive pathways.

Some ions, notably NH_4^+, an important ion in plant nutrition, other amines, and some basic amino acids are transported by a uniport mechanism, i.e., a saturable carrier with no energy coupling other than the membrane potential (Walker *et al.*, 1979; Bentrup, 1980; Kinraide and Etherton, 1980). The current carried by NH_4^+ is large, but is saturated at an external concentration of 2 or 3 μM. Both V_{max} and K_m are affected by the potential, the effect on K_m being attributed to the location of the binding site within the electrical field across the membrane. These transport systems have a marked depolarizing effect on the potential.

Proton cotransport of sugars (Komor and Tanner, 1976) and neutral amino acids (Kinraide and Etherton, 1980) will also cause an initial depolarization of the potential. In the case of sugar transport by *Chlorella*, the membrane potential clearly contributes to the free energy of transport (Komor and Tanner, 1976). However, the potential does not seem to affect the transport V_{max}, but changes the apparent K_m for proton binding, perhaps again because of the location of the binding site (Schwab and Komor, 1978).

3.2. Indirect Effects on the Potential

Some other effects of secondary transport on the potential may be due to changes in cytoplasmic pH, which alter the rate of operation of the electrogenic proton pump. This is the usual interpretation of a partial repolarization observed after the initial depolarization by sugars and neutral amino acids (Komor and Tanner, 1976; Kinraide and Etherton, 1980). In support of this idea, Kinraide and Etherton (1980) showed that the repolarization, but not the initial depolarization, can be eliminated by treatment with cyanide.

Changes in cytoplasmic pH may perhaps also be invoked to explain some interesting observations by Bowling *et al.* (1978) and Bowling and Dunlop (1978a,b). These workers observed that in roots of both sunflower and clover the activity of the electrogenic pump was closely linked to the influx of phosphate, a phosphate concentration of only 1 μM being sufficient to change the potential by -50 mV. The failure of other groups to observe this effect is ascribed to the fact that Bowling and colleagues worked with intact plants, whereas most electrical studies by other workers have been made on excised tissue. It seems probable that phosphate is taken up by neutral cotransport with protons, and that the resulting acidification of the cytoplasm activates the electrogenic proton pump.

REFERENCES

Bentrup, F. W. (1980). *Biophys. Struct. Mech.* **6,** 175–189.

Bisson, M. A., and Walker, N. A. (1980). *J. Membr. Biol.* **56,** 1–7.

Bowling, D. J. F., and Dunlop, J. (1978a). *J. Exp. Bot.* **29,** 1139–1146.

Bowling, D. J. F., and Dunlop, J. (1978b). *J. Exp. Bot.* **29,** 1147–1153.

Bowling, D. J. F., Graham, R. D., and Dunlop, J. (1978). *J. Exp. Bot.* **29,** 135–140.

Cheeseman, J. M., and Hanson, J. B. (1979). *Plant Physiol.* **64,** 842–845.

Cheeseman, J. M., LaFayette, P. R., Gronewald, J. W., and Hanson, J. B. (1980). *Plant Physiol.* **65,** 1139–1145.

Dame, J. B., and Scarborough, G. A. (1980). *Biochemistry* **19,** 2931–2937.

Fujii, S., Shimmen, T., and Tazawa, M. (1979). *Plant Cell Physiol.* **20,** 1315–1328.

Gradmann, D. (1975). *J. Membr. Biol.* **25,** 183–208.

Graves, J. S., and Gutknecht, J. (1977). *J. Membr. Biol.* **36,** 65–81.

Hill, A. E., and Hill, B. S. (1973). *J. Membr. Biol.* **12,** 129–144.

Jacobs, M., and Taiz, L. (1980). *Proc. Natl. Acad. Sci. USA* **77,** 7242–7246.

Kawamura, G., Shimmen, T., and Tazawa, M. (1980). *Planta* **149,** 213–218.

Keifer, D. W., and Spanswick, R. M. (1978). *Plant Physiol.* **62,** 653–661.

Kikuyama, M., Hayama, T., Fujii, S., and Tazawa, M. (1979). *Plant Cell Physiol.* **20,** 993–1002.

Kinraide, T. B., and Etherton, B. (1980). *Plant Physiol.* **65,** 1085–1089.

Komor, E., and Tanner, W. (1976). *Eur. J. Biochem.* **70,** 197–204.

Lucas, W. J., and Nuccitelli, R. (1980). *Planta* **150,** 120–131.

Marrè, E. (1979). *Annu. Rev. Plant Physiol.* **30,** 273–288.

Mercier, A. J., and Poole, R. J. (1980). *J. Membr. Biol.* **55,** 165–174.

Poole, R. J. (1974). *Can. J. Bot.* **52,** 1023–1028.

Poole, R. J. (1978). *Annu. Rev. Plant Physiol.* **29,** 437–460.

Scarborough, G. A. (1980). *Biochemistry* **19,** 2925–2931.

Schwab, W. G. W., and Komor, E. (1978). *FEBS lett.* **87,** 157–160.

Shimmen, T., and Tazawa, M. (1980). *Plant Cell Physiol.* **21,** 1007–1013.

Spanswick, R. M. (1972). *Biochim. Biophys. Acta* **288,** 73–89.

Sze, H. (1980). *Proc. Natl. Acad. Sci. USA* **77,** 5904–5908.

Walker, N. A., Beilby, M. J., and Smith, F. A. (1979). *J. Membr. Biol.* **49,** 21–55.

Walker, N. A., Smith, F. A., and Cathers, I. R. (1980). *J. Membr. Biol.* **57,** 51–58.

181

Ion Transport across the Root

M. G. Pitman

1. INTRODUCTION

Plant roots are remarkably efficient at absorbing ions from the soil or from solutions. Current views on uptake and on mechanisms of transport are reviewed elsewhere in this volume. But the root is more than a scavenger for ions and nutrients from the soil. It also supplies nutrients to the rest of the plant. Lüttge (1974) aptly compared the root with a gland, as both the excised root and a gland can exude salt-loaded solutions and both have analogous structures, including barriers against diffusion of ions back from the secretory areas into the bulk of tissue. This comparison illustrates the role of the root in export of ions and the need to study many aspects of the overall process in order to understand transport across the root.

Features of recent studies of transport have been the location of transport processes in the root, their relation to structure, and control by the rest of the plant.

2. ORGANIZATION OF THE ROOT

In recent years the role of the epidermis has been emphasized as the physiological interface between soil and root. It has become evident that at the low concentrations normal in the soil, uptake must take place largely at the epidermal cells rather than at the surfaces of inner cortical cells after diffusion into the root (e.g., Bange, 1973; see Nye and Tinker, 1977, for consideration of diffusion up to the root). The special role of the epidermis is also seen by development of transfer cell structures under certain conditions. This occurred in saltbush under salt stress (Kramer *et al.*, 1978) and in iron-deficient sunflower roots (Kramer *et al.*, 1980), and is likely to be found in other roots when they are investigated. It has also shown that there can be a cuticle on the surface of the root, which has been

M. G. Pitman • School of Biological Sciences, University of Sydney, Sydney, New South Wales 2006, Australia.

isolated and studied, though it does not appear to be an effective barrier to movement of water and solutes (Clarkson et al., 1978).

The boundary between the stele and cortex at the endodermis has been studied using tracers and heavy-metal precipitates (La^{2+}). Such experiments have shown the reality of the barriers to movement of water and solutes at the casparian strip and at the suberin lining of the endodermal cells, a function suspected for years but only recently unambiguously established (see review by Läuchli, 1976).

Many cell-to-cell connections occur between cells of the cortex, the endodermis, and cells of the stele. The interconnected cytoplasms of the cells are considered to form a symplast (Spanswick, 1976; Tyree, 1970).

Within the stele, the symplasmic pathways terminate in the xylem parenchyma cells, which are interfaces between the water-filled cavities of the xylem and the cytoplasmic continuum in the root. One of the major problems in current studies of transport across the root is the role of the xylem parenchyma cells. Are they sites of controlled solute release to the xylem? Are they responsible for removal of solutes from the xylem? Are these cells responsible for the glandlike secretion from the root? Structural features of these cells that are of possible importance to transport across the root are the large number of plasmodesmata between cells in the stele and endodermis; the number of mitochondria in the xylem parenchyma and development of endoplasmic reticulum, as evidence of cytoplasmic activity; and to the presence of transfer cell structures (see discussion by Läuchli and Pflüger, 1978).

3. PATHWAYS OF MOVEMENT ACROSS THE ROOT

Two routes are available for passage of solutes across the root. One (the apoplast) is in the cell walls or intercellular spaces, outside of the cell membranes. The other (the symplast) is through the interconnected sets of cytoplasms, via the plasmodesmata.

The idea of transport across the root in the symplast has been considered by physiologists for many years, but it has only been possible to support it in recent years, due to development of improved techniques for structural and biophysical studies. For example, demonstration of the abundance and dimensions of plasmodesmata in roots, and study of the effectiveness of transport through the plasmodesmata have shown that the symplast can be a pathway for a major proportion of transroot transport of water and solutes (see reviews by Tyree, 1970; Spanswick, 1976; Robards and Clarkson, 1976; Pitman, 1977).

Particularly useful evidence about the symplast has come from study of uptakes of K^+, $H_2PO_4^-$, and Ca^{2+} over different parts of the root surface and their relation to structural changes in the endodermis.

The effectiveness of the barrier at the endodermis changes from the apex, where cells are maturing, to the older parts of the root. The endodermis continues to develop and differentiate well after other cells have reached maturity (e.g., Clarkson et al., 1971). In the older parts of the root, transport of ions and water across the root is limited to the symplast, for deposition of suberin in the cell walls of the endodermis means the cells can only be traversed through the plasmodesmata and hence in the symplast. Uptake of K^+ and $H_2PO_4^-$ occurred in this part of the root.

In younger parts of the root, the whole surface of the endodermal cell is accessible to water and solutes, and it is here that most of the Ca^{2+} and water uptake to the root takes place. The realization of the physiological importance of this variation in structure along the root owes much to the work of Clarkson, Robards, and colleagues.

Figure 1. Model system illustrating processes involved in transport across the root. Ions enter the cytoplasm of the cortical cells by influx ($\emptyset_{oc}$) or can leave by efflux ($\emptyset_{co}$). Within the cytoplasm, ions can exchange with the vacuoles ($\emptyset_{cv}$) or pass in the symplasm across the endodermis to the stele. The cell wall free space is shown blocked at the endodermis. Within the stele, ions may enter the vacuoles of xylem parenchyma or else be transported to xylem vessels ($\emptyset_{cx}$). [From Schaefer *et al.*, 1975.]

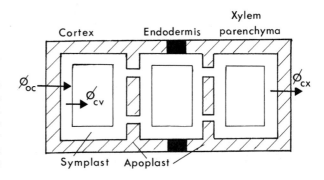

A model commonly used to relate transport across the root and the processes involved is shown in Fig. 1. Advantages of considering the symplast as a major pathway for transport of certain ions are that it explains the ability of the root to transport ions selectively to the xylem; it allows for *outward* movement of sugars or organic acids; it explains why uptake of certain ions (K^+, phosphate, NO_3^-) may be largely independent of water flow; and it explains how there can be a ready exchange between solutes in the vacuoles of cortical cells and solutes transported across the root.

4. TRANSPORT AND WATER FLOW

In a normal, intact plant, the transpiration stream removes solutes from the xylem vessels in the roots and may increase transport of solutes across the root. Plants show a whole range of responses from independence to complete dependence of uptake of particular solutes on water flow. Interactions between solute transport and water flow may be due to concentration changes near the surface of the root (see Nye and Tinker, 1977), or it may be due to greater transport within the root. Water-dependent uptake seems to be particularly important for uptake of NaCl by halophytes, but at present there are no clear ways of explaining how these interactions arise in the root. This topic should become particularly relevant in studies of plant tolerance and adaptation to salinity.

Transroot transport of solutes also has an effect on water flow, producing the spectacular exudation from excised roots, also referred to as "root pressure," and seen as guttation in whole plants at early morning or when humidity is high. Exudation is best described as a standing osmotic flow (Anderson *et al.*, 1970) in which continued transport of solutes (mainly ions) into the xylem vessels draws water across the root due to the osmotic concentration in the xylem.

Exudation depends on both the net rate of release of solutes into the xylem and the permeability of the root to water. It can be stopped either by inhibiting ion release to the xylem or by decreasing the permeability of the root to water. As export of ions from excised roots depends on there being some water flow out of the xylem, it is important to know whether inhibitors of transport are having an effect due to action on ion transport or on permeability of the root to water. Inhibitors of respiration can severely reduce the hydraulic conductivity of the root (e.g., Ginsburg and Ginzburg, 1970). Interpretation of effects of inhibition of transport from roots also needs some caution, for the hydraulic conductivity of young seedlings appears to be proportional to the rate of flow in the root (e.g., Aston and Lawler, 1979).

Attempts to induce flow by application of differences of hydrostatic pressure to excised root systems have not entirely escaped the criticism that the results obtained do not correspond with the behavior of intact plants.

Exudation from excised roots has been shown to respond qualitatively the same as guttation from whole plants (Diefenbach *et al.*, 1980) and has certainly posed some useful questions about the mechanism of transroot transport that are still not resolved to everyone's satisfaction.

5. MECHANISMS OF TRANSROOT TRANSPORT

Exudation from a root could be achieved in the simple model of Fig. 1 if there were uptake of solutes from the solution into the cytoplasm (ϕ_{oc}) to a much higher concentration, movement of the solutes in the symplast across the cortex and stele, and then diffusion out of the symplast into the xylem vessels. Concentrations of K^+ of about 60 to 150mM in the cytoplasm would readily account for observed concentrations of K^+ in the xylem (30 to 50 mM). This "one-pump" model is simple and has been supported by measurements of electrochemical potential of K^+ in individual cells across the root, which showed no evidence of any "uphill" step other than at entry to the outer cortical cells (Dunlop and Bowling, 1971; see also Bowling, 1976).

Though this simple model is adequate to account for the overall energetics of transport, it does not account for certain specific observations nor for the degree of control that the root seems to have over solute release to the xylem in the root. The alternative is to locate active transport both at entry to the symplast and at the site of release into the stele, i.e., possibly from the xylem parenchyma to the xylem vessels.

Davis and Higinbotham (1976) produced evidence that transport of K^+ into the xylem was against its electrochemical gradient, though transport of Cl^- was not.

Transport of K^+ and Cl^- out of the root can be inhibited by the uncouplers carbonyl cyanide *m*-chlorophenylhydrazone or 2,4-dinitrophenol, but they also inhibit influx to the root and reduce the hydraulic conductivity. This complex set of interactions requires careful analysis to demonstrate inhibition of ϕ_{cx} separate from the other effects of the inhibitors (Pitman *et al.*, 1981).

Better evidence for control of release from the stele comes from certain compounds that act on protein synthesis. Cycloheximide and the amino acid analogs *p*-fluorophenyl alanine and azetidine 2-carboxylic acid can inhibit transport across the root *without* inhibition of uptake *or* reduction in hydraulic conductivity. Streaming of the cytoplasm has also been shown to be unaffected and it is considered that inhibition takes place in the stele. The plant hormone abscisic acid also can inhibit transport in the root independently of reduction in uptake or hydraulic conductivity (Schaefer *et al.*, 1975; Pitman, 1980; Pitman and Wellfare, 1978).

It has often been considered that streaming in the cytoplasm would be necessary for transport across the root, and inhibition of streaming has been claimed to be a reason why various inhibitors blocked the transport. However, streaming could be observed when fluorophenyl alanine was inhibiting release to the xylem (Schaefer *et al.*, 1975) and inhibition of streaming with cytochalasin does not stop transport across the root (Glass and Perley, 1979).

Hanson (1978) has considered application of chemiosmotic theory to transport across the root based on H^+-coupled ion uptake at entry to and release from the cytoplasm. He points out that the symplast could be a means of coupling the gradients of proton electrochemical activity at the two membranes. This is an interesting idea that brings together the mechanistic models with pumps at different locations. It is of interest too for its relation to control of H^+ in the plant as a whole, as H^+ or OH^- release to the xylem could require subsequent adjustment during passage through the stem or in the leaves.

The root can exhibit selectivity in transport between K^+ and Na^+, between univalent and divalent cations, or between Cl^- and NO_3^-. This selectivity also appears to be a property of symplasmic transport. It has been suggested that K^+ influx/Na^+ efflux processes at the entry to the symplast and at the exit from the symplast to the xylem are involved in this selectivity (e.g., Jeschke, 1980), but there is also some degree of nonselective transport of Na^+ and K^+ that seems to be related to water flow (Pitman, 1977). The role of these various processes in selectivity shown by intact plants needs further investigation.

6. LOCALIZATION STUDIES

The application of X-ray microanalysis to frozen fractures of roots has allowed estimation of ionic content of vacuoles of single cells. Such studies confirm that there can be large differences in ratios of ions between cells of the endodermis and adjacent cells of the cortex. More surprising is the observation that some cells in the cortex may have a high concentration of K^+ while adjacent cells have a relatively higher concentration of Na^+. Differences have also been found in phosphate and Cl^- content of cortical cell vacuoles (e.g., Van Steveninck *et al.*, 1980). The relevance of these observations to transport across the root is that it shows the symplast may not be a *complete* continuum of cytoplasms but that cells, or blocks of cells, may be isolated from others, perhaps by blockages of the plasmodesmata, or even by differences in operative genome of the cells. X-Ray microanalysis also shows that there can be gradients in the ratio of K^+/Na^+ across the root in the cortex, again implying that the simple view of the symplast (Fig. 1) as a transroot continuum needs modification to allow for changes in selectivity and gradual separation of K^+ from Na^+ (e.g., Stelzer and Läuchli, 1978).

This technique has also been useful for demonstrating the role of xylem parenchyma in removing Na^+ from the xylem vessels (see Läuchli and Pflüger, 1978). Ideally, it should be possible to determine concentrations in individual xylem vessels, allowing proper tests of mechanistic models for transport across the root.

7. CONTROL AND REGULATION OF UPTAKE

Transport across the root supplies much of the inorganic solutes to the stem and leaves of the plant, and in recent years there has been emphasis on the extent to which release of solutes into the xylem is controlled or regulated by the plant. There is ample evidence to show that the release of solutes can be affected by energy supply via the phloem to the root. It has also been suggested that ion transport into the root could be affected by plant hormones such as abscisic acid or benzyl adenine. (See reviews by Lüttge, 1974; Pitman and Cram, 1977; Läuchli and Pflüger, 1978; Clarkson and Hanson, 1980).

Although it has become widely accepted that the concentration of solutes in the shoot appears to be regulated in response to such factors as water potential of the soil, or the demand of the plant for nutrients, there is no clear model of how this is achieved or how transport across the root is functionally integrated with growth.

8. PROSPECTS

Current work on transport across the root puts particular emphasis on the structure of the system in which transport takes place, the role of symplast and apoplast in transport, and interactions between solute and water flow. X-Ray microanalysis promises to be a useful tool for investigation and chemiosmotic theory a challenge to the model-makers. Of particular interest to me is the increasing concern with the way the root operates as part of the whole plant.

REFERENCES

Anderson, W. P., Aikman, D. P., and Meiri, A. (1970). *Proc. R. Soc. London Ser. B* **174**, 445–458.
Aston, M. J., and Lawler, O. W. (1979). *J. Exp. Bot.* **30**, 169–181.
Bange, G. G. J. (1973). *Acta Bot. Neerl.* **22**, 529–542.
Bowling, D. J. F. (1976). *Uptake of Ions by Plant Roots*, Chapman & Hall, London.
Clarkson, D. T., and Hanson, J. B. (1980). *Annu. Rev. Plant Physiol.* **31**, 239–298.
Clarkson, D. T., Robards, A. W., and Sanderson, J. (1971). *Planta* **96**, 292–305.
Clarkson, D. T., Robards, A. W., Sanderson, J., and Peterson, C. A. (1978). *Can. J. Bot.* **56**, 1526–1532.
Davis, R. F., and Higinbotham, N. (1976). *Plant Physiol.* **57**, 129–136.
Diefenbach, H., Lüttge, U., and Pitman, M. G. (1980). *Ann. Bot.* **45**, 703–712.
Dunlop, J., and Bowling, D. J. F. (1971). *J. Exp. Bot.* **22**, 434–444.
Glass, A. D. M., and Perley, J. G. (1979). *Planta* **145**, 399–401.
Ginsburg, H., and Ginzburg, B. Z. (1970). *J. Exp. Bot.* **21**, 580–592.
Hanson, J. B. (1978). *Plant Physiol.* **62**, 402–405.
Jeschke, W. D. (1980). In *Plant Membrane Transport: Current Conceptual Issues* (R. M. Spanswick, W. J. Lucas, and J. Dainty, eds.), pp. 17–28, Elsevier/North-Holland, Amsterdam.
Kramer, D., Anderson, W. P., and Preston, J. (1978). *Aust. J. Plant Physiol.* **5**, 739–747.
Kramer, D., Römheld, V., Landsberg, E., and Marschner, H. (1980). *Planta* **147**, 335–339.
Läuchli, A. (1976). In *Encyclopedia of Plant Physiology* (U. Lüttge and M. G. Pitman, eds.), Vol. IIB, pp. 3–34, Springer-Verlag, Berlin.
Läuchli, A., and Pflüger, R. (1978). In *Potassium Research—Review and Trends*, pp. 111–163, International Potash Institute, Bern.
Lüttge, U. (1974). In *Membrane Transport in Plants* (U. Zimmermann and J. Dainty, eds.), pp. 353–362, Springer-Verlag, Berlin.
Nye, P. H., and Tinker, P. B. (1977). *Solute Movement in the Soil–Root System*, Blackwell, Oxford.
Pitman, M. G. (1977). *Annu. Rev. Plant Physiol.* **28**, 71–88.
Pitman, M. G. (1980). *Plant Cell Environ.* **3**, 59–61.
Pitman, M. G., and Cram, W. J. (1977). In *Integration of Activity in the Higher Plant*, pp. 391–424, Symposium of Society for Experimental Biology XXXI, Cambridge University Press, London.
Pitman, M. G., and Wellfare, D. (1978). *J. Exp. Bot.* 29, 1125–1138.
Pitman, M. G., Wellfare, D., and Carter, C. (1981). *Plant Physiol.* **67**, 802–808.
Robards, A. W., and Clarkson, D. T. (1976). In *Intercellular Communication in Plants: Studies on Plasmodesmata* (B. E. S. Gunning and A. W. Robards, eds.), pp. 181–202, Springer-Verlag, Berlin.
Schaefer, N., Wildes, R. A., and Pitman, M. G. (1975). *Aust. J. Plant Physiol.* **2**, 61–74.
Spanswick, R. M. (1976). In *Encyclopedia of Plant Physiology* (U. Lüttge and M. G. Pitman, eds.), Vol. IIB, pp. 35–56, Springer-Verlag, Berlin.
Stelzer, R., and Läuchli, A. (1978). *Z. Pflanzenphysiol.* **88**, 437–448.
Tyree, M. T. (1970). *J. Theor. Biol.* **26**, 181–214.
Van Steveninck, R. F. M., Van Steveninck, M. E., Stelzer, R., and Läuchli, A. (1980). In *Plant Membrane Transport: Current Conceptual Issues* (R. M. Spanswick, W. J. Lucas and J. Dainty, eds.), pp. 489–490, Elsevier/North-Holland, Amsterdam.

182

The Synthesis of Fatty Acids in Plant Systems

Paul K. Stumpf

Much is now known about the molecular structure of fatty acid synthetases in bacteria, yeast, and animal systems (Lynen, 1980). Most bacterial systems consist of nonassociated polypeptides and include the acyl carrier protein (ACP), β-ketoacyl ACP synthase, β-ketoacyl ACP reductase, enoyl ACP reductase, β-hydroxy acyl ACP dehydratase, acetyl CoA : ACP transacylase, malonyl CoA : ACP transacylase. All of these polypeptides have been purified, crystallized, and their enzymatic characteristics carefully studied (Vagelos, 1974). In contrast, the yeast system has been identified as a heterodimer consisting of an α polypeptide (212,000 molecular weight) and a β polypeptide (203,000 molecular weight) (Stoops and Wakil, 1980). As each polypeptide is present six times, the heterodimer is actually a massive molecular structure of 2,400,000 molecular weight (consisting of $\alpha_6\beta_6$ in its native form). The α subunit contains three enzymatic domains, namely, the acyl carrier site, the β-ketoacyl synthase site, and the β-ketoacyl reductase site. The β subunit has the remaining domains, that is, the enoyl reductase activity, the β-hydroxy acyl dehydratase activity, the acetyl CoA : transacylase activity, the malonyl CoA : transacylase site, and the palmitoyl transferase activity. In contrast, a typical animal system such as pigeon liver appears to be a homodimer (450,000 molecular weight) consisting of two similar chains each with a molecular weight of 220,000 and each containing all the catalytic domains for fatty acid synthesis (Stoops et al., 1979). The yeast system releases its final product, palmitic acid, as a CoA thioester whereas the animal system synthesizes free palmitic acid as the final reaction product. The chemistry of acyl synthesis is identical in all systems.

In terms of molecular structure, plant systems have not been examined in detail. Whereas ACP is in a free form in a bacterial system such as E. coli, it is covalently linked in both yeast and animal synthetases. In plant systems, ACP occurs in the free form. It has a molecular weight slightly higher than its bacterial counterpart (Simoni et al., 1967).

Paul K. Stumpf • Department of Biochemistry and Biophysics, University of California, Davis, California 95616.

ACP obtained from *E. coli* can be used as a replacement for plant ACPs in studying plant fatty acid synthesis. Interestingly, however, cross-reactivity of antibodies of *E. coli* ACP and plant ACPs is poor. Thus, whereas the biochemical reactivity of the 4'-phosphopantotheine in both bacterial and plant ACP is the same, the immunoproperties differ markedly (Ohlrogge *et al.*, 1979).

A feature unique to all plant fatty acid synthetases is the localization of these enzymes in specialized organelles. In the leaf cell, the chloroplast was known for many years to have a high capacity for the synthesis of oleic acid from acetate (Stumpf, 1980), the product being free oleic acid. Because ACP is an essential component of all fatty acid synthase systems, it is thus a specific marker for these systems, and the localization of ACP defines the specific presence of all the enzymes that require ACP, i.e., the fatty acid synthase system. Recent studies revealed that indeed in the leaf cell, the only site for ACP was the chloroplast (Ohlrogge *et al.*, 1979). Thus, in the leaf cell, the sole site for fatty acid synthesis must be the chloroplast. We can now conclude that the chloroplast is a multifunction organelle that carries out a number of very important biochemical syntheses. These include: (1) photooxidation of water to release molecular O_2; (2) photophosphorylation to generate ATP; (3) photoreduction to form $NADPH_2$; (4) generation of triose phosphates via the reductive pentose phosphate cycle to provide the precursors for sucrose synthesis that occurs in the cytosolic compartment of the leaf cell; (5) the sole site for the synthesis of palmitic and oleic acids in the leaf cell; (6) the sole site for sulfate reduction; (7) the sole site for the reduction of nitrite to ammonia and its incorporation via glutamine synthetase and glutamate synthase to glutamic acid; (8) the site of synthesis of a number of essential amino acids including cysteine, valine, leucine, isoleucine, and the aromatic amino acids; and finally (9) the synthesis of carotenoids and in part the prenyl quinones. Thus, the chloroplast is the principal, if not the only, site for biosynthesis of very important biomolecules in the plant and could even be considered as the most important organelle in nature.

We can now summarize the events of lipid biosynthesis in the chloroplast. A free acetate anion is probably generated in leaf mitochondria via oxidative decarboxylation of pyruvate to acetyl CoA. Acetyl CoA is then hydrolyzed to free acetate that now can freely diffuse out of the mitochondria to the chloroplast. The chloroplast is the only site for acetyl CoA synthetase, which converts the unreactive acetate anion to a fully reactive acetyl CoA (Kuhn *et al.*, 1981). All the enzymes necessary for the conversion of acetyl CoA to free oleic acid are solely localized in the chloroplast (Stumpf, 1980). These appear to be nonassociated and quite analogous to those found in eubacterial systems. However, the formation of oleic acid in plants differs in a number of respects to those found in bacterial and animal systems. In a number of bacteria, an anaerobic pathway introduces the *cis* double bond at the C_{10} or C_{12} level. Elongation of these acids to the C_{16} and C_{18} level then produces the monoenoic acid characteristic of that bacterial system (Vagelos, 1974). The formation of oleic acid in animal systems is via the microsomal enzyme stearoyl CoA desaturase, which involves an NADH : cytochrome b_5 reductase, cytochrome b_5, and stearoyl CoA desaturase required for the desaturation of stearoyl CoA to oleoyl CoA in the presence of molecular O_2. In plants, the system is quite different. All the enzymes involved are completely soluble: NADPH : ferredoxin reductase, ferredoxin, and stearoyl ACP desaturase are the soluble components and stearoyl ACP is the highly specific substrate (McKeon and Stumpf, 1981). Molecular oxygen is employed as the final electron acceptor. Of further interest is the observation that oleoyl ACP formed by the desaturase system is rapidly hydrolyzed by a specific oleoyl ACP hydrolase that is also localized in the chloroplast (Ohlrogge *et al.*, 1978a). Once free oleic acid is formed, it readily passes through the chloroplast into the cytosol where further desaturation reactions may occur.

A similar picture unfolds in developing seeds where, instead of chloroplasts, the organelle for fatty acid synthesis is the proplastid (Simcox *et al.*, 1977; Nakamura and Yamada, 1974; Weaire and Kekwick, 1975). This organelle contains all the enzymes necessary for the synthesis of the fatty acid, oleic acid. Indeed, oleic acid can be considered as a central substrate in many seed systems. For example, proplastids in developing castor bean endosperm tissue convert acetate to oleic acid. This acid then diffuses to the cytosolic compartment where it is readily hydroxylated by a microsomal hydroxylase to ricinoleic acid and this product is inserted into triacyl glycerols as a storage lipid. In developing rapeseeds, there is good evidence that these seeds synthesize oleic acid in their proplastids and then convert this acid to oleoyl CoA in the cytosolic compartment. Elongation with malonyl CoA now occurs to erucic acid. In the developing cotyledons of jojoba seeds, its proplastid synthesizes oleic acid; this substrate again enters the cytosolic compartment there to be elongated via CoA thioester derivatives, reduced to corresponding alcohols, and then both the alcohols and acyl thioesters are condensed to form the storage wax esters characteristic of this seed (Ohlrogge *et al.*, 1978b; Pollard *et al.*, 1979).

In sharp contrast to animal and bacterial systems, in plants specialized organelles generate oleic acid as the prime precursor. This compartment is termed the synthesizing compartment. Then the product is transferred to another compartment where it undergoes modification. In the castor bean, the modifying reaction is hydroxylation; in the rapeseed, the modifying reaction is elongation; in the jojoba seed, the modifying reactions are elongation, reduction, condensations; and in the leaf cell, the modifying reactions are further desaturations of oleic to linoleic and linolenic acids. These ideas are summarized in Fig. 1.

In summary, most bacterial systems have nonassociated enzymes localized in the cytosolic compartment responsible for the synthesis of saturated fatty acids by conventional

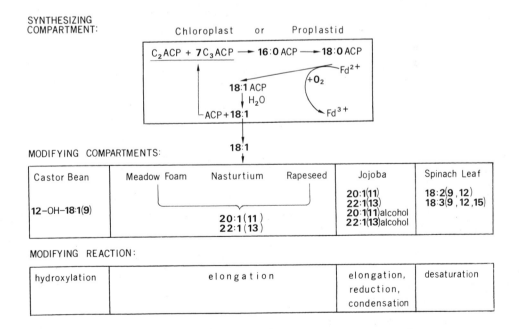

Figure 1. The relationship between the chloroplast compartment in leaf tissue and the proplastid compartment in seed tissues to their respective modifying compartments in the cell cytosol.

reactions; in *E. coli*, the synthesis of monoenoic fatty acids is via an anaerobic pathway. In animals, a homodimer is responsible for the synthesis of free palmitic acid presumably in the cytosol, whereas in yeasts and fungi heterodimers, massive structures in the cytosol, are involved in the synthesis of palmitoyl CoA. In plants, these systems are localized in special organelles that contain all the necessary enzymes to synthesize both palmitic and oleic acids. Once these are formed, they are transferred to other compartments where the modifications typical of a given plant will occur. Future investigations will reveal the molecular nature of the plant systems. The comparative aspects of all the known systems in terms of evaluation should be an interesting story.

REFERENCES

Kuhn, D. N., Knauf, M., and Stumpf, P. K. (1981). *Arch. Biochem. Biophys.* **209,** 441–450.
Lynen, F. (1980). *Eur. J. Biochem.* **112,** 431– 442.
McKeon, T., and Stumpf, P. K. (1982). *Arch. Biochem. Biophys.*, submitted.
Nakamura, Y., and Yamada, M. (1974). *Plant Cell Physiol.* **15,** 37– 48.
Ohlrogge, J. B., Shine, W. E., and Stumpf, P. K. (1978a). *Arch. Biochem. Biophys.* **189,** 382–391.
Ohlrogge, J. B., Pollard, M. R., and Stumpf, P. K. (1978b). *Lipids* **13,** 203–210.
Ohlrogge, J. B., Kuhn, D. N., and Stumpf, P. K. (1979). *Proc. Natl. Acad. Sci. USA* **76,** 1194–1198.
Pollard, M. R., McKeon, T., Gupta, L., and Stumpf, P. K. (1979). *Lipids* **14,** 651– 662.
Simcox, P. D., Reid, E. E., Canvin, D. T., and Dennis, D. T. (1977). *Plant Physiol.* **59,** 1128–1132.
Simoni, R. D., Criddle, R. S., and Stumpf, P. K. (1967). *J. Biol. Chem.* **242,** 573–581.
Stoops, J. K., and Wakil, S. J. (1980). *Proc. Natl. Acad. Sci. USA* **77,** 4544– 4548.
Stoops, J. K., Ross, P., Arslanian, M. J., Aune, K. C., Wakil, S. J., and Oliver, R. M. (1979). *J. Biol. Chem.* **254,** 7418–7426.
Stumpf, P. K. (ed.) (1980). In *The Biochemistry of Plants,* Vol. IV, pp. 177–204, Academic Press, New York.
Vagelos, P. R. (1974). In *MTP International Review of Science, Biochemistry of Lipids* (T. W. Goodwin, ed.), Vol. 4, pp. 100–140, Butterworths, London.
Weaire, P. J., and Kekwick, R. G. O. (1975). *Biochem. J.* **146,** 425– 437, 439– 445.

Index